# Edexcel (A)

# advanced Geography

LITTLE HEATH SCHOOL
LITTLE HEATH ROAD
TILEHURST, READING
BERKSHIRE, RG31 5TY

*Edexcel (A) Advanced Geography*

To our families, who have tolerated our many hours in front of the PC

Philip Allan Updates
Market Place
Deddington
Oxfordshire
OX15 0SE

Tel: 01869 338652
Fax: 01869 337590
e-mail: sales@philipallan.co.uk
www.philipallan.co.uk

© Philip Allan Updates 2005

ISBN-13: 978-1-84489-205-1
ISBN-10: 1-84489-205-0

All rights reserved; no part of this publication may be reproduced, stored in a retrieval system, or transmitted, in any form or by any means, electronic, mechanical, photocopying, recording or otherwise without either the prior written permission of Philip Allan Updates or a licence permitting restricted copying in the United Kingdom issued by the Copyright Licensing Agency Ltd, 90 Tottenham Court Road, London W1P 9HE.

This textbook has been written specifically to support students studying Edexcel (A) Advanced geography. The content has been neither approved nor endorsed by Edexcel and remains the sole responsibility of the authors.

All efforts have been made to trace copyright on items used.

Front cover photographs reproduced by permission of Science Photo Library, Corel Corporation and PhotoDisc.

Printed by Scotprint, Haddington, East Lothian

The paper on which this title is printed is sourced from managed, sustainable forests.

# Contents

**Introduction** .................................................................................................................. **vi**

## AS Unit 1  Physical environments

### Chapter 1  Earth systems .................................................................................. 2
The dynamic Earth: continental drift and plate tectonics ................................. 2
Igneous activity ........................................................................................................ 15
Weathering ............................................................................................................... 20
Examination question ............................................................................................ 30
Synoptic link: relationships between people and tectonic environments ........ 31

### Chapter 2  Fluvial environments ..................................................................... 44
The hydrological cycle ........................................................................................... 44
Rivers and their response to changing conditions ............................................. 49
River processes and their impacts ........................................................................ 58
Examination question ............................................................................................ 72
Synoptic link: the use and management of the hydrological cycle ................. 73

### Chapter 3  Coastal environments .................................................................... 80
The coast .................................................................................................................. 80
Coastal landforms ................................................................................................... 85
Changes in sea level ............................................................................................... 93
Coastal ecosystems ................................................................................................. 100
Examination question ............................................................................................ 106
Synoptic link: the management of coastal processes ........................................ 106

## AS Unit 2  Human environments

### Chapter 4  Population characteristics ........................................................... 112
Variations in populations ...................................................................................... 112
The components of population change .............................................................. 133
The demographic transition model ..................................................................... 141
The characteristics of ageing and youthful populations ................................... 144
The population debate .......................................................................................... 147
Overpopulation, underpopulation and optimum population ......................... 149
Examination question ............................................................................................ 152
Synoptic link: the influence of governments on population change .............. 153

### Chapter 5  Settlement patterns ....................................................................... 158
Rural settlements .................................................................................................... 158
Urban settlements .................................................................................................. 165
Settlement hierarchies: central place, range and threshold ............................. 169
Spatial variations in land-use patterns in urban settlements ........................... 176
The development of cities in MEDCs .................................................................. 178
Urban land-use models and theories ................................................................... 183
Urban development in LEDCs ............................................................................. 189
The changing city ................................................................................................... 190

*Edexcel (A)* *Advanced Geography*

Examination question .................................................. 200
Synoptic link: influence of government policies on settlement
　　characteristics and patterns .................................. 200

### Chapter 6  Population movements ........................ 206
Migration .................................................................... 206
Theories of migration ................................................. 212
The impact of migration on the physical environment ... 219
Different types of migration: American case studies ..... 221
Rates and volumes of migration: contrasts in modern Europe ..... 232
The problem of generalisation ..................................... 233
Consequences of migration ........................................ 237
Examination question ................................................ 238
Synoptic link: government policies influence migration patterns ..... 238

## AS Unit 3  Personal enquiry/Applied geographical skills

### Chapter 7  Personal enquiry ................................. 248

### Chapter 8  Applied geographical skills: Section A ..... 254
Examination question ................................................ 276

### Chapter 9  Applied geographical skills: Section B ..... 279
Examination question ................................................ 284

## A2 Unit 4  Physical systems, processes and patterns

### Chapter 10  Atmospheric systems ........................ 286
The dynamic atmosphere ........................................... 286
Moisture in the atmosphere ....................................... 294
Local effects of human and physical factors on the atmosphere ..... 307
Examination questions .............................................. 315
Synoptic link: weather and human activity are interdependent ..... 316

### Chapter 11  Glacial systems ................................ 323
Glaciers are dynamic systems ..................................... 323
Glacial processes and landforms ................................. 329
Fluvioglaciation ......................................................... 343
Examination questions .............................................. 356
Synoptic link: Opportunities and challenges associated with glacial and
　　periglacial areas ................................................. 357

### Chapter 12  Ecosystems ..................................... 364
Ecosystems are dynamic systems ............................... 364
Grassland and forest biomes ..................................... 369
Soil characteristics .................................................... 381
Examination questions .............................................. 389
Synoptic link: Opportunities and challenges associated with ecosystems ..... 390

## A2 Unit 5  Human systems, processes and patterns

### Chapter 13  Economic systems ........................... 398
The classification of economic activity ......................... 398
Classical location theory ............................................ 406

# Contents

The rise and decline of consumer industries in MEDCs .................. 419
The rise of manufacturing in NICs .................. 431
The emergence of large trading blocs .................. 441
The emergence of a new international division of labour .................. 445
Transnational corporations and globalisation .................. 448
Examination questions .................. 455
Synoptic link: industrialisation and de-industrialisation have an impact
 on the physical and human environment .................. 455

## Chapter 14  Rural–urban interrelationships .................. 461
The world's major urban areas .................. 461
Urbanisation today .................. 469
The management of cities .................. 474
Agriculture and rural areas .................. 478
The globalisation of food production .................. 493
The interdependence between urban and rural environments .................. 501
The rural/urban fringe .................. 509
The influence of urban economies on the socioeconomic characteristics
 of rural areas .................. 516
Examination questions .................. 518
Synoptic link: rapid change has created pressure on rural–urban
 interdependence .................. 518

## Chapter 15  Development processes .................. 522
Global variations in development .................. 522
Measurements of development .................. 527
Development theory .................. 539
Variations in economic growth and development within countries .................. 552
Trade and aid .................. 566
The consequences of development .................. 574
The debt crisis .................. 577
Examination questions .................. 581
Synoptic link: the world is increasingly interdependent .................. 582

# A2 Unit 6  Synoptic: People and their environments

## Chapter 16  Section A .................. 588
Synoptic links .................. 588
The synoptic assessment .................. 590
Examination question .................. 594

## Chapter 17  Section B .................. 599
Synoptic themes for essays .................. 599
Developed ideas for synoptic essay answers .................. 604
Sample questions .................. 613

## Index .................. 614

*Edexcel (A) Advanced Geography*

# Introduction

This book has been written for the Edexcel (A) Geography specification, and covers both AS and A2 halves of the course. It is organised to reflect the exact structure of the specification — each chapter covers one sub-section of a unit. These chapters include abundant **case studies** and **full explanations of the main processes** of both physical and human geography, providing an invaluable resource for students and teachers.

Within a chapter, each specification concept is addressed and full coverage of the content requirement is provided. Where case studies are prescribed by the specification, they are included. These typically include description and explanation as well as an appropriate amount of located detail. Definitions of **key words** are provided to assist students in developing their grasp of geographical terminology.

Each chapter includes at least one **examination-style question**, written to reflect the demands placed on students by the unit tests.

Coverage of the **synoptic links** is provided at the end of each relevant chapter, although it is recognised that, at AS, these may not be taught at the same time as the unit content. To assist in the application of these links to the examination questions, and to help students synthesise information, **examiner's tips** are provided.

As with all textbooks, material provided here should be supplemented with the use of resources gathered from a wide variety of media, including periodicals, newspapers and the internet.

# The specification

The Edexcel (A) specification provides students with an opportunity to study geographical themes and contemporary issues within the framework of a recognisable structure. Although the division between physical and human fields of study are reflected in the unit titles, the detailed content reflects the impact of people on the physical world, and vice versa.

The synoptic content is delivered through the teaching of 'synoptic links' attached to each of Units 1, 2, 4 and 5, although these links will not necessarily be taught alongside the rest of the content of their respective units. Given that the synoptic links can only be examined in Unit 6, coverage of many of the links in the AS units can be left until Year 13.

Units 1 and 2 are divided into three compulsory sections. The remaining AS unit, Unit 3, can be delivered either through a piece of coursework (Personal enquiry) or through a written exam (Applied geographical skills).

At A2, Units 4 and 5 are also divided into three sections each, but given that only two essays need to be written in each examination, one section can be omitted by students.

The synoptic examination (Unit 6) involves a compulsory resource-based question and an essay drawn from four possible titles.

# Introduction

## Assessment objectives

In common with all other geography specifications, there are four assessment objectives. Candidates should:
- show knowledge of the specified content
- show critical understanding of the specified content
- apply knowledge and critical understanding to unfamiliar contexts
- select and use a variety of skills and techniques, including communication skills, appropriate to geographical studies

These objectives are weighted differently across the units, while within the units there are questions that address different aspects of these objectives.

## Scheme of assessment

Each of AS and A2 consists of three units and makes up 50% of the total award. The structure is outlined below:

|  | Unit | Duration and length | Mode | Unit code | Advanced GCE (AS) weighting |
|---|---|---|---|---|---|
| AS | 1 Physical environments | 1 hour 15 minutes | Written examination | 6461 | 15% (30%) |
|  | 2 Human environments | 1 hour 15 minutes | Written examination | 6462 | 15% (30%) |
|  | 3a Personal enquiry | Coursework (2500 words) | Internal assessment and external moderation | 6463/01 | 20% (40%) |
|  | or |  |  |  |  |
|  | 3b Applied geographical skills | 1 hour 30 minutes | Written examination | 6463/02 |  |
| A2 | 4 Physical systems, processes and patterns | 1 hour 30 minutes | Written examination | 6464 | 15% |
|  | 5 Human systems, processes and patterns | 1 hour 30 minutes | Written examination | 6465 | 15% |
|  | 6 Synoptic | 2 hours | Written examination | 6466 | 20% |

## Units 1 and 2

These are divided into three sections as follows:

*Unit 1 Physical environments*
1.1  Earth systems
1.2  Fluvial environments
1.3  Coastal environments

*Unit 2 Human environments*
2.1  Population characteristics
2.2  Settlement patterns
2.3  Population movements

In each of these papers, there are two questions set on each section. Candidates are required to answer **one from each section**. Each question comprises a resource and a series of question parts.

## Edexcel (A) Advanced Geography

Questions are frequently structured as follows:
- An (a) part in which the student is asked to respond to the resource and/or provide a definition. This may be divided into as many as four sub-sections, at least one of which will involve a descriptive skill.
- A (b) part, frequently divided into two or three sub-sections, that generally involves more understanding of process or application of knowledge and process.
- A (c) part that allows some extended writing in which candidates can show deeper knowledge and understanding.

## Unit 3

This has two options available:

### Unit 3a  Personal enquiry
This is a piece of coursework based on an approved title and limited to 2500 words.

### Unit 3b  Applied geographical skills
This paper is divided into two parts, both of which are compulsory:
- Question 1 is based on a resource booklet that replicates the fieldwork experience with questions about the design, execution and conclusions of that work. Statistical and presentational skills are frequently tested.
- Question 2 requires candidates to give an account of some part of their own fieldwork experience, asking them to describe and explain key aspects of that work, including the design, execution and conclusions drawn.

## Units 4 and 5

Each of these is subdivided into three sections as follows:

| Unit 4 Physical systems, processes and patterns | Unit 5 Human systems, processes and patterns |
|---|---|
| 4.1  Atmospheric processes | 5.1  Economic systems |
| 4.2  Glacial systems | 5.2  Rural–urban interrelationships |
| 4.3  Ecosystems | 5.3  Development processes |

In each of these papers there are two questions set on each section. Candidates are required to answer **two questions drawn from different sections**. Each question comprises a resource (intended as a stimulus) and is divided into two parts. The (a) part, which is marked out of 5, is frequently definitional in character. The (b) part is an essay title and is marked out of 20.

## Unit 6  Synoptic: People and their environments

The synoptic links detailed in each unit are subdivided into four main themes:
6.1   Physical environments influence human activity
6.2   Human activities modify physical environments
6.3   Physical and human resources may be exploited, managed and protected
6.4   Communities and their governance influence geographical interrelationships at a range of scales.

# Introduction

These links provide the topics for the four essay titles from which candidates need to select one, worth 25 marks out of the 75 available on the paper. The other 50 marks come from Question 1, a compulsory question usually divided into four parts. This question requires candidates to interpret and synthesise the material offered to them in the resource booklet and does not assume prior knowledge of either the specific location or the specific issues and problems identified.

## Exam technique at AS

The short-answer style questions that are dominant in Units 1, 2 and 3 require a number of specific techniques. Although candidates will not be short of time, it remains important not to waste precious minutes. Frequent problems are dealt with below.

### Failure to follow the command words

Command words instruct you as to what to do. At AS these include:
- **compare** — set items side by side and show their similarities and differences. A balanced, objective answer is expected.
- **consider** — describe and give your thoughts on the subject.
- **contrast** — point out only the differences between two items.
- **define** — explain the precise meaning of a concept. This should include, at this level, a recognition of any difficulties in defining the term.
- **describe** — say what something is like, how it works and so on.
- **discuss** — explain an item or concept, and then gives details about it, using supportive information, examples, points for and against, and explanations for the facts put forward. It is important to give both sides of an argument and come to a conclusion.
- **examine** — investigate in detail, offering evidence for and against a point of view or a judgement
- **explain** — offer a detailed and exact explanation of an idea or principle, or a set of reasons for a situation or attitude.
- **identify** ('what…', 'name…', 'state…') — express the relevant points briefly and clearly without lengthy discussion or minor details.
- **illustrate** — provide examples to demonstrate or prove the subject of the question. This command is often added to another instruction.
- **justify** — give only the reasons for a position or argument, and the main objections likely to be made to them.
- **summarise/outline** — provide a summary of all the available information about a subject. Questions of this type often require short answers.

The commonest errors include confusing description with explanation, and explaining when not asked to do so. No marks are lost for doing this, but time is wasted.

*Edexcel (A) Advanced Geography*

### Not spotting a key word

Key words often provide the focus of the question. The topic might be 'internal migration', but the focus might be the 'changing **rate** of (internal migration)'. Students often ignore key words such as rate. Others include:

- changes in
- distribution
- variation
- structure
- characteristics
- pattern
- relationship
- volume

### Lack of locational knowledge

Case studies are needed to reinforce and illustrate the understanding of processes. Thus in the extended writing sections of the papers, locational evidence is almost always required for higher-band responses.

### Lack of precision in descriptions or explanations

In your answers, you should avoid words such as 'good' and 'bad', or 'easy' and 'hard'. It is, for example, better to outline in what ways mountainous areas might be difficult for human habitation rather than just saying that they are and then leaving the examiner to deduce why.

### Repeating the question in the answer

No marks are ever awarded for rewriting questions. It is simply a waste of time.

# Exam technique at A2

## Essay writing

At A2, essay writing skills are paramount. Five essays are required in the exams — two each for Units 4 and 5 and one for Unit 6. Students often do worse than they should in examination essays, not because their writing skills are weak or because their knowledge of the subject matter is insufficient, but because they have not fully understood what they have been asked to do. To score high marks in an examination or an essay, it is important to understand what a question means and how it should be answered. In order to understand the question it is useful to deconstruct it and to search for certain key components, using the following steps:

1. Identify the *topic*. What, in broad terms, is this about.
2. Identify the *focus*. What is the precise target area of the question? What aspect of the topic should be explored?
3. If the *topic* has a *restriction*, identify it. Is it in any way necessary to limit the discussion to certain areas, processes or phenomena? Watch out for an absence of restrictions too. Do not restrict yourself if the title does not tell you to do so.
4. What is the *command word* or phrase? What treatment of the topic is requested?
5. Is there a *viewpoint?* This refers to the requirement, in the question, to write from a particular point of view. This will be incorporated in the type of command words used, as in 'Criticise the view that…'

## Command words at A2

In addition to those used at AS, a number of extra terms are employed at A2:
- **analyse** — take apart an idea, concept or statement in order to consider all the factors of which it consists. Answers of this type should be very methodical.
- **consider** — describe and give your thoughts on the subject.
- **criticise (the view)** — point out mistakes or weaknesses, but *also* indicate any favourable aspects of the subject of the question. However, the widespread use of *criticise* as only pointing out weaknesses means that it will not prejudice you if you take this single-sided approach.
- **evaluate/assess** — decide and explain how great, valuable or important something is. The judgement should be backed by a discussion of the evidence.
- **explore (the view)** — examine the subject thoroughly and consider it from a variety of viewpoints.

## The structure of essays

### Essays have a beginning: the introduction

The introduction should be a context statement, which says something meaningful. Avoid statements such as 'In this essay, I am going to…' An introduction is not a contents page. It is often possible to clarify the terms used in the question through extending definitions or to point out how some of the terms are difficult or problematic. If you can, try to set the tone of the essay by stating how this topic interrelates with others.

### The 'middle': the analytical part of the essay

The core of the essay should be organised by paragraphs, each of which takes on an aspect of the question, with examples. Structure it logically by tackling one point in each paragraph. Do not drift off into themes or evidence that are well known to you but not what the question requires.

Use evidence constructively and make it work for you. Remember that you can never lose marks. A bold guess is better than nothing. You may have only a hazy recall of the precise rate at which the South Korean economy grew in the 1980s, but better a hazy idea than none at all. If in doubt put it in!

Each paragraph should end with a comment that reflects on how the evidence offered relates to the title set. Below are three sample paragraphs in response to the question: 'Explain why rates of urban growth vary from place to place and from time to time'.

*Sample paragraph 1*
Cities grow at different rates. Some cities are growing very rapidly, e.g. in Africa.

*e* Better than nothing, but has no detail or qualification.

*Sample paragraph 2*
Some cities are growing very rapidly, e.g. many African cities, but especially those in countries that have rapid rates of population growth, e.g. Lagos in Nigeria or Addis Ababa in Ethiopia. This has been rapid in recent years as internal migration has accelerated.

*e* Much stronger, indicating recognition of internal variety on the African continent, although not very specific.

*Edexcel (A)* Advanced Geography

*Sample paragraph 3*

Some cities are growing very rapidly, e.g many African cities, but especially those in sub-Saharan countries that have very rapid population growth rates (sometimes 3% per annum). This has occurred through natural increase, but also high levels of rural–urban migration. The rates are frequently hard to measure, but might involve 1000 migrants arriving each day, as in Lagos (Nigeria) or Addis Ababa in Ethiopia. Internal migration, fuelled by the growth of commercial agriculture, has been the primary cause and this process has affected countries at different times.

**e** *Shows a high level of understanding and although the figures are not, in fact, very accurate, the distinction between natural increase and migration is clear and the recognition of the problem of accurate measurement is thoughtful.*

In your analysis of an essay title, try to bear in mind some of the following factors — not all will be relevant to all titles, but watch out for those that are:

- **contrasts in scale.** Does your answer tackle the fact that some explanations work at some scales but not at others? For example, climate might be the key variable in controlling agricultural land use at a global scale but not at a local scale.
- **contrasts in space.** Does your answer allow for the fact that what is true for one part of the world may not apply to another? A common contrast here would be between MEDCs and LEDCs.
- **contrasts in time.** Does your answer allow for the fact that explanations will vary from one period to another? What was true in 1945 may not be true today. For example, a theory such as Weber's may be applicable in the circumstances of one period but not another.

**The end: the conclusion**

This should *not* be a précis of the middle. Do not start conclusions by writing 'As I have said…' Try to refer back to the original question and its key aspects or proposition — for example, 'Therefore it can be seen that…' or 'Thus it is clear that…'

Then, try to leave the examiner with the distinct impression that you would have had much more to say if only you had been allowed the time — for example, 'However, it should be remembered that…' or 'Nonetheless, some have argued that…' All conclusions in human geography are liable to be partial, tentative and incomplete, so you are perfectly entitled to say so.

# Using this book

The structure of this book means that you should always know how the topics you are studying relate to the specification content and how you will be assessed in the examinations. It should provide you with essential knowledge, help you to develop a secure understanding and enable you to have high-quality skills at your disposal.

# AS Unit 1
## Physical environments

# Earth systems

## The dynamic Earth: continental drift and plate tectonics

### Evidence for continental drift

**Continental drift** is the theory that the continental landmasses have changed position over time. Although Francis Bacon had observed the apparent jigsaw fit of the continents back in the seventeenth century, the theory is largely attributed to Alfred Wegener, a German meteorologist. In 1912 he wrote a paper offering a range of evidence in support of the idea. The evidence fell into three categories:
- geological
- biological
- climatological

### Geological
- The way in which continents such as South America and Africa appear to 'fit' together, especially when the edge of the continental shelf is used rather than the present coastline, suggest that they may once have been joined together.
- Rock types and geological structures are similar on the two sides of the Atlantic. For example, the Appalachian Mountains of North America and the Caledonian Mountains of Scotland both have the same sequence of igneous and sedimentary rocks, suggesting that they were formed at the same time and in the same place.

### Biological
- Some fossils, for example that of *Mesosaurus*, a small freshwater reptile, are found in both southwest Africa and Brazil. This suggests that the landmasses were once joined and such creatures lived in the whole combined area.
- Fossils of a small fern, *Glossopteris*, are found widely across all the southern continents, suggesting that these continents were once joined.
- Fossils of Triassic period reptiles have been found in areas that are now far apart. Examples include *Cynagnathus* in South America and Africa, and *Lystrosaurus* in India, Africa and Antarctica.

### Climatological
- The coal and oil reserves found in Antarctica suggest that this area was once in a different climatic zone.

- Glacial striations in Brazil and west Africa and glacial deposits in India, South America and the Vaal Valley in South Africa indicate that these areas once had very similar climatic conditions, even though they are now far apart and in very different climatic zones.

## Developing Wegener's ideas

Based on this evidence, Wegener suggested that all the continents were joined together 300 million years ago as a single supercontinent. Wegener called this supercontinent Pangaea (Figure 1.1). About 200 million years ago, the continents drifted apart forming a northern hemisphere assemblage known as Laurasia, comprising Eurasia and North America, and a southern hemisphere group called Gondwanaland, consisting of Africa, Australasia, India, South America and Antarctica. The present day continental landmasses were formed as these two supercontinents continued to separate into their present-day arrangement.

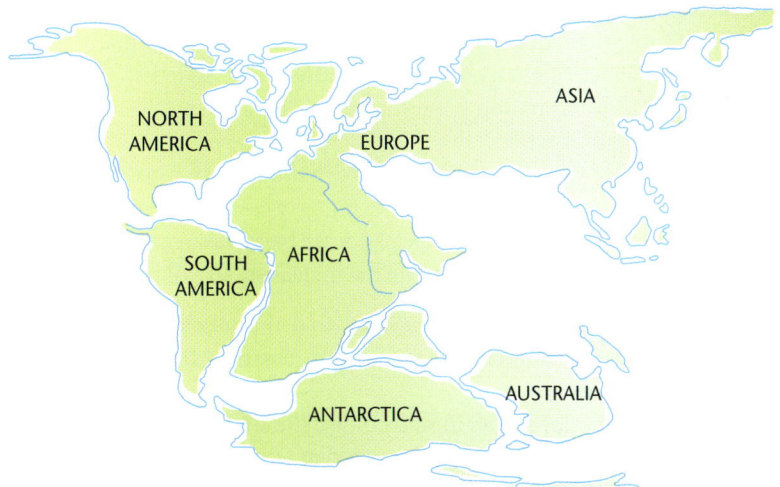

*Figure 1.1 Pangaea: note the fit between South America and the west coast of Africa*

Wegener suggested that the mechanism for the movements was a combination of centrifugal force and the gravitational pull of the moon. Physicists quickly proved that these forces were insufficient to move continents and, partly for this reason, the theory was not widely accepted. Geologists were critical of Wegener's ideas, not least because he was a meteorologist! At the time little was known about the nature of the sea floor and this made it difficult to explain the movement. In addition, it could not be explained why continents appeared to be strong enough to plough through ocean basins even though it was the continents themselves that were deformed by folding and faulting.

Later research, particularly research on the Atlantic sea floor by Harry Hess and others, led to the formulation of **plate tectonic** theory. This theory was based on the idea that the Earth's crust is divided into several major and a number of minor slabs (plates) of relatively rigid crustal material. These slabs move in relation to each other, riding on the weak, soft and partially molten rocks beneath. The major evidence for this theory came from investigations into sea-floor spreading, which was seen to be taking place in the Atlantic Ocean.

# Chapter 1  Earth systems

Figure 1.2 The pattern of parallel palaeomagnetism and the age of the rocks on the Atlantic Ocean floor

- Near the mid-Atlantic ridge the basaltic rocks of the sea floor are much younger than those found towards the continental shelves at the edge of the ocean (Figure 1.2). This can be taken as evidence that the rocks were formed at the ridge and have subsequently moved apart in both easterly and westerly directions, being replaced by newer rocks which form at the ridge.
- Palaeomagnetism of the basalts also suggests that rock has been formed at the ridge and then moved away in both directions. There is a parallel pattern of magnetism in the rocks (Figure 1.2). This develops because, as magma cools and solidifies, the metallic elements within it are magnetised in the direction of the Earth's magnetic field. It is known that there have been many reversals of the Earth's magnetic field over time. The pattern of magnetism found in the rocks relates to this alternating pattern of northerly and southerly magnetism of the Earth's magnetic field.
- Sea-floor spreading implies that the Earth must be increasing in size. However, satellite imagery confirms that this is not the case. Instead, tectonic plates are being destroyed in subduction zones to accommodate the increasing size of the oceanic crust.

Research has also suggested that the mechanism underlying these movements is convection currents beneath the Earth's surface.

Developments in seismology, the study of how seismic waves pass through the Earth, mean that we now know more about the internal structure of the planet. A number of distinct layers have been identified: the crust, the mantle, the outer core and the inner core (Figure 1.3). These layers display different physical and chemical properties.

The **crust** itself is of two types — continental and oceanic. Their characteristics are summarised in Table 1.1.

Figure 1.3 The internal structure of the Earth

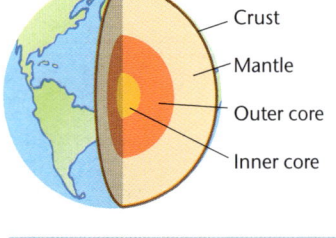

|  | Continental | Oceanic |
|---|---|---|
| Thickness | 30–70 km | 5–9 km |
| Density | 2.7 g cm$^{-3}$ | 3.0 g cm$^{-3}$ |
| Main rock type | Granite | Basalt |

Table 1.1 Differences between continental and oceanic crust

## Key terms

**Collision margin** The boundary between two plates of continental crust that are moving towards each other. No subduction occurs, and so no crust is destroyed and no new crust is created.

**Conservative margin** The boundary between two plates that are sliding past each other, either in opposite directions or in the same direction but at different speeds. Crust is conserved; no new crust is created and no existing crust is destroyed.

**Constructive margin** The boundary between two crustal plates that are moving apart. Rising magma creates new crustal rock.

**Crustal plate** A rigid slab of crustal rock 'floating' on the semi-molten material of the asthenosphere.

**Destructive margin** The boundary between two crustal plates that are moving towards each other. One of the plates subducts beneath the other and is destroyed.

## Tectonic plates

**Plate tectonic theory**, a term coined by geologists MacKenzie and Palm[er] widely accepted. The Earth's crust is believed to be divided into a numbe[r] of different sizes, many of which consist of a combination of oc[eanic and] continental crust. These plates move slowly and irregularly in relati[on to each] other, typically at rates of a few centimetres a year:

- in some locations plates move away from each other (a constructive [margin])
- in some locations plates move towards each other (a destructive ma[rgin])
- in a few places plates move past each other, either in opposite directions or in the same direction but at different speeds (a conservative margin).

Figure 1.4 shows the pattern of plates, the directions they are moving in and, for the major plates, their rates of movement. The types of plate margins are discussed in more detail below.

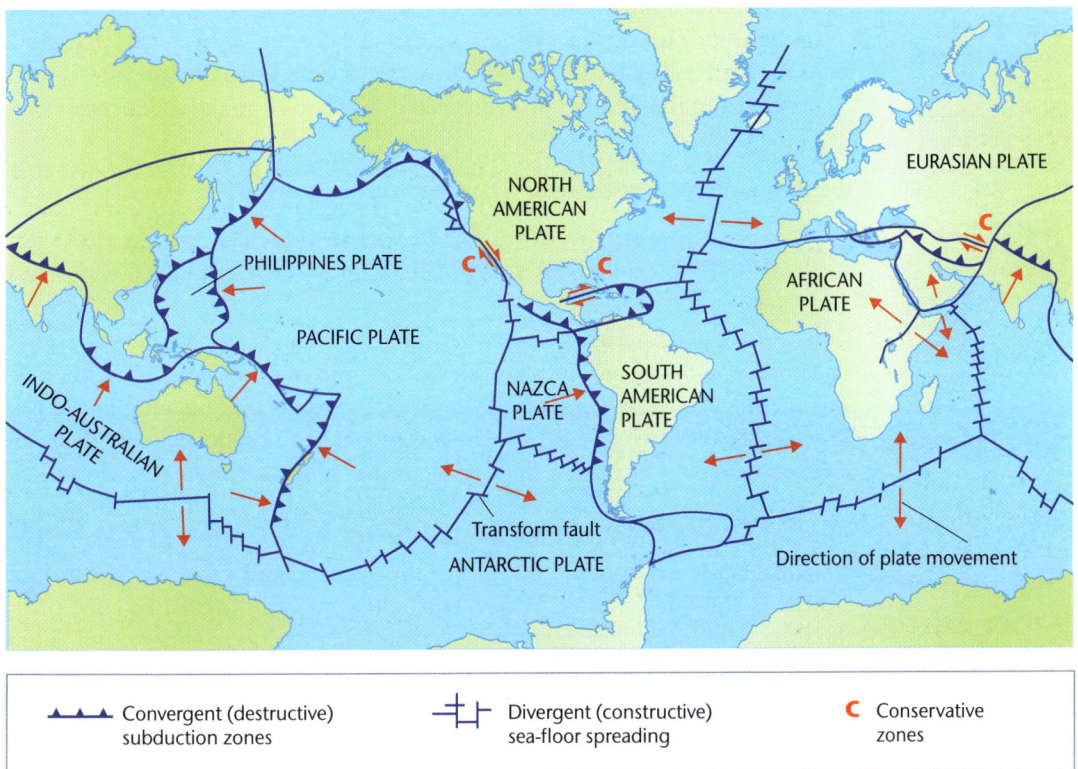

Figure 1.4
The global pattern of tectonic plates

## Convection currents

It is now known that the outer layers of the Earth cannot simply be divided into crust and mantle. Rather, the outer part of the mantle consists of rigid material that is attached to the crust. Together, these two elements form the **lithosphere**. The lower part of the mantle, which is molten or semi-molten, is known as the **asthenosphere**.

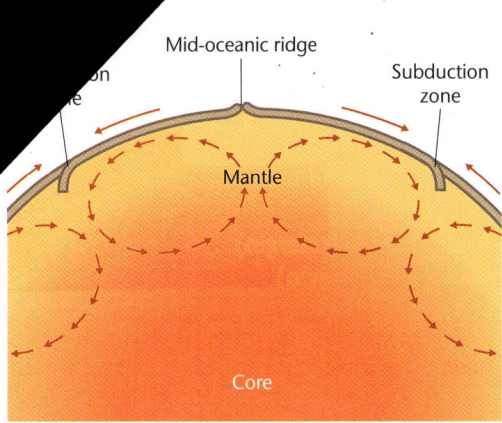

*Figure 1.5 Convection currents and plate movement*

The most likely cause of plate movement is the existence of convection cell currents in the asthenosphere, caused by heat from the core. This heat comes from a combination of radioactive decay in the core and residual primary heat. The convection currents cause the magma to circulate and this moves the lithosphere. There is uncertainty about the forces involved, but the movement is thought to be due to either the pushing apart of plates at places where two rising limbs of convection cells diverge below the surface or the pulling downwards (known as drag) of the edges of plates at places where descending limbs exist (Figure 1.5).

- Plates are at their hottest nearest the mid-oceanic ridges and they cool down as they move away from the mid-oceanic ridges. This means that material will rise near the mid-oceanic ridges.
- As the material cools its density increases, so it sinks lower into the molten rock beneath and is relatively easily dragged downwards into the subduction zone.

Plate movement may well be caused by a combination of both these forces, but the downward drag does seem to have the greater strength.

## Plate margins

### Constructive plate margins

**Constructive plate margins** occur where plates are moving away from each other. This may be a divergence of two plates of oceanic crust, such as along the mid-Atlantic ridge, or of two plates of continental crust, as is happening in the East African rift valley (see Figure 1.4). The movement apart of the plates is due to the divergence of two convection cells, which brings magma from the asthenosphere towards the surface. The pressure from the rising magma leads to a doming up of the Earth's surface and the formation of a ridge. As the plates are moved apart, tensional faults are produced into which the rising magma can enter. The magma cools and solidifies, producing new crust either within the existing crust or on the surface following volcanic eruptions. Volcanoes can occur either at individual vents or along the plate margin as fissure eruptions. The subsidence of sections of crust between the fault lines creates rift valleys within the ridge.

Earthquake activity occurs at all plate margins. An earthquake is a movement or shaking of the ground caused by the release of pressure that builds up as the plates try to move in relation to each other but are hindered by friction between them. The nature of earthquake activity varies with the type of plate margin. At constructive margins most earthquake activity is shallow focus, low magnitude and high frequency. This is because the movements are at or near the surface and the pressure is easily released as the plates diverge.

*Physical environments* **AS Unit 1**

### Case study 1.1  Constructive margin: oceanic

The divergence of the Eurasian plate and the North American plate in the mid-Atlantic is a good example of a constructive margin (Figure 1.6). A ridge and rift system extends along the mid-Atlantic for about 10 000 km. This was created about 60 million years ago as Greenland (on the North American plate) and northwest Scotland (on the Eurasian plate) separated to form the Atlantic Ocean. A series of underwater volcanoes exists along the margin, occasionally appearing above sea level as volcanic islands. An example of such islands is Iceland, much of which is a lava plateau up to 200 m above sea level, with none of its rocks more than 3 million years old. The rift valley is clearly visible at Pingvellir, and active volcanoes such as Hekla and Grimsvötn have erupted within the last 30 years. A significant earthquake, measuring 6.5 on the Richter scale, occurred on the south coast on 17 June 2000.

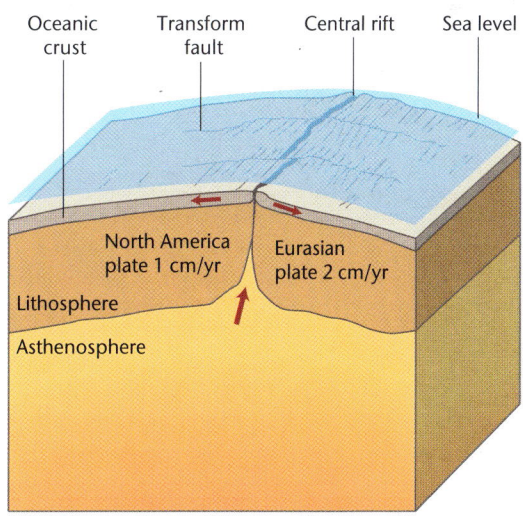

*Figure 1.6 Cross-section of the mid-Atlantic constructive margin*

### Case study 1.2  Constructive margin: continental

The East African rift valley is an example of a constructive margin in an area of continental crust. Eastern Africa is moving in a northeasterly direction, diverging from the main African plate, which is heading north. The valley, which actually consists of two broadly parallel rifts, extends for 4000 km from Mozambique to the Red Sea. Inward-facing scarp slopes reach heights of 600 m above the valley floor (Figure 1.7). Rifts form when sections of crust between parallel fault lines subside.

The area experiences volcanic activity, suggesting that the crust has been weakened and thinned by tension, resulting in rising magma escaping onto the surface at volcanoes such as Mt Kilimanjaro.

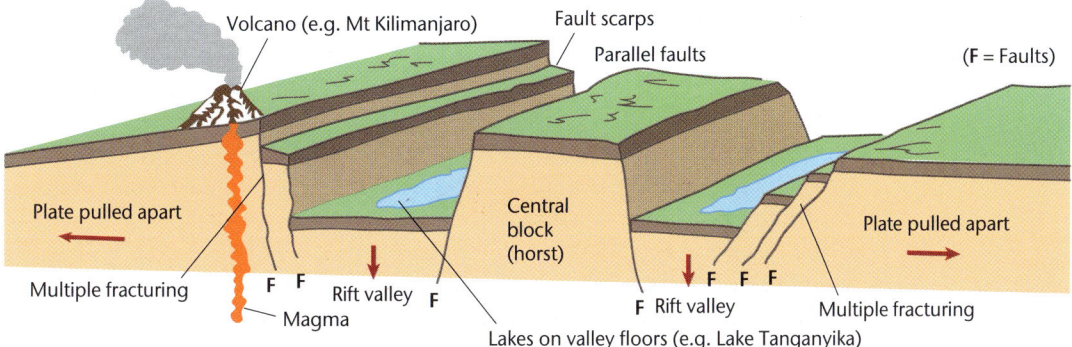

*Figure 1.7 Idealised cross-section of the East African rift valley*

7

# Chapter 1  Earth systems

### Destructive plate margins

**Destructive plate margins** occur where two plates converge due to the existence of descending limbs of convection cell currents in the asthenosphere, beneath the lithosphere. This can happen where oceanic crust meets continental crust, such as the convergence of the Nazca plate and the South American plate, or where two plates of oceanic crust come together, as happens where the Pacific plate and Philippines plate converge (see Figure 1.4).

Oceanic crust is more dense than continental crust (see Table 1.1). When the two types of crust converge, the more dense crust is subducted down into the asthenosphere beneath the less dense crust. An ocean trench is formed on the sea floor at the point of subduction. The continental crust, being more buoyant, is not subducted but is uplifted, buckled and folded, and so forms a range of fold mountains. The subducted plate is heated and eventually melts under pressure at about 100 km below the surface. This melted material is less dense than the surrounding rocks and so rises through any lines of weakness towards the surface. It may cool and solidify beneath the surface, forming intrusive igneous rocks such as granite, or it may eventually reach the surface under great pressure, giving rise to violent, infrequent volcanic eruptions.

## Case study 1.3  Destructive margin: oceanic/continental

The oceanic Nazca plate is moving east at approximately 12 cm yr$^{-1}$ and converges with and subducts beneath the continental South American plate, which is moving west at 1 cm yr$^{-1}$, to the west of South America. The Andes, a chain of fold mountains rising to nearly 7000 m above sea level, has been formed as the continental crust has been buckled and uplifted (Figure 1.8). Volcanoes, such as Cotopaxi, occur along the chain of mountains. The Peru–Chile trench, which reaches depths of 8000 m, occurs at the point of subduction. Earthquakes, such as that in northern Peru in 1970 which killed 67 000 people, are common and are often of high magnitude.

*Figure 1.8 Cross-section of an oceanic/continental destructive margin*

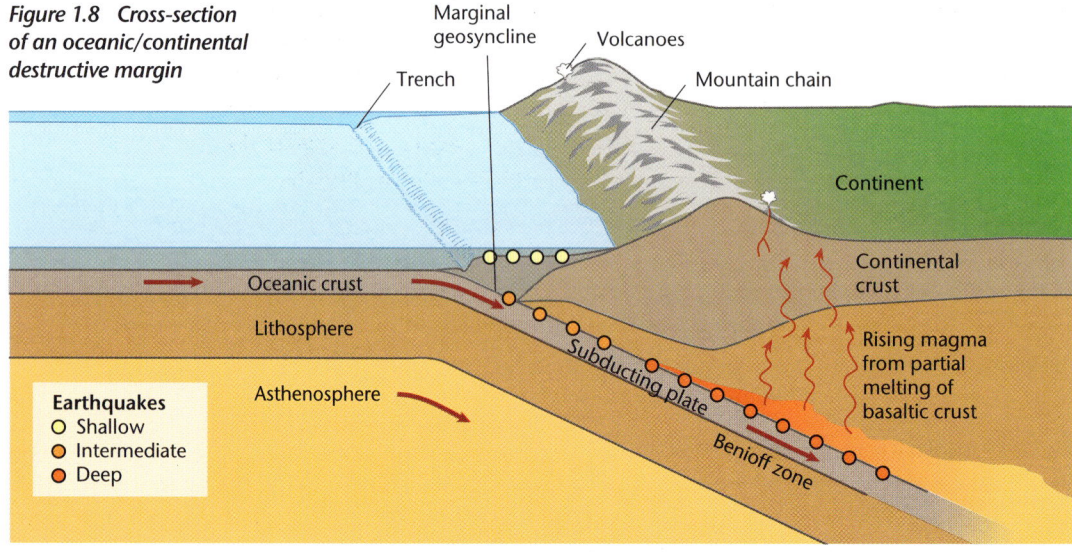

Earthquakes are a common feature of destructive margins. They occur at a range of depths along the Benioff zone, the boundary between the subducting plate and the overlying crustal rocks, from shallow focus events at the ocean trench down to a depth of 700 km. Beyond this depth the mantle is molten and so the subducting plate is no longer subject to friction.

Where oceanic crust converges with oceanic crust, subduction still occurs, as one plate is likely to be slightly older and slightly denser than the other. The landforms and features that result are very similar to those at oceanic/continental margins, except that volcanic activity leads to the formation of a chain of volcanic islands above the subduction zone, known as an **island arc**. The Mariana Islands have been formed in this way as a result of the convergence of the Pacific plate and the Philippines plate, with the Pacific plate subducting below the Philippines plate (Figure 1.9).

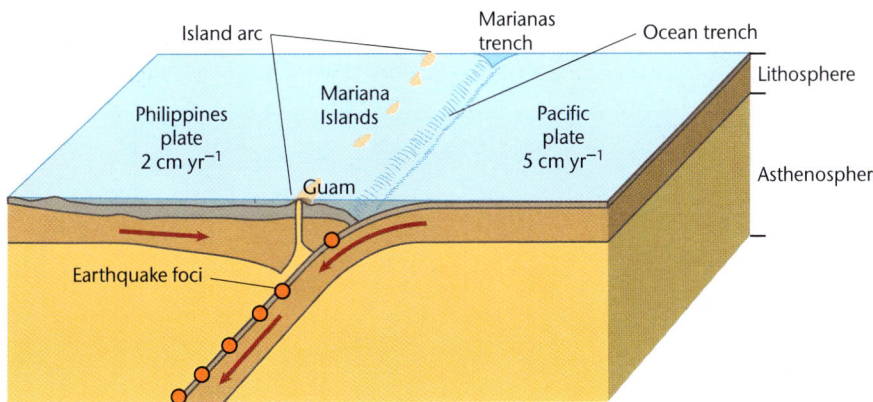

*Figure 1.9 Cross-section of an oceanic/oceanic destructive margin*

### Collision margins

The convergence of two plates of continental crust is known as a **collision margin**. No subduction occurs, as both plates are buoyant and of low density. However, intervening oceanic sediments trapped between the two converging plates are heaved upwards, resulting in the formation of major fold mountain ranges. No volcanic activity is found at this type of margin as no crust is being destroyed by subduction and no new crust is being created by rising magma. Earthquakes do occur, although they are often deep focus and have limited surface impact.

### Conservative plate margins

A **conservative margin** occurs when two plates move laterally past each other. This is also known as a **transform margin**. As with collision margins, no volcanic activity is found here because no crust is being destroyed by subduction and no new crust is being created by rising magma. Shallow focus earthquakes of varying frequency and magnitude do occur. Low magnitude, high frequency events occur when pressure along the margin is relatively easily released. Occasional major events take place after a significant build-up of pressure, typically when high levels of friction restrict movement of the crust along fault lines.

Most plates have a constructive margin at one edge and a destructive margin at another, with conservative margins making up the other two sides.

# Chapter 1 Earth systems

## Case study 1.4 Collision margin

The Indo-Australian plate is moving northwards at a rate of about 5 cm per year. It collides with the Eurasian plate, which is moving southeastwards at a slightly slower rate. Prior to their collision, the two continental land masses were separated by the remnants of the Tethys Sea, which originated at the time of the break up of Pangaea 300 million years ago. As the two plates collided the Himalayas, a range of fold mountains, were thrust upwards to a height of 9000 m (Figure 1.10). Rocks found high up in the Himalayas contain fossils of small sea creatures, confirming the existence of the previously intervening ocean.

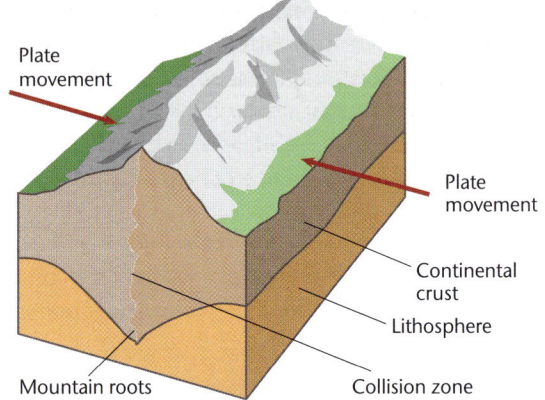

*Figure 1.10 Cross-section of a collision margin*

Photograph 1.1 The Himalayas are a range of fold mountains

### Hot spots

**Hot spots** are places where plumes of magma are rising from the asthenosphere, even though they are not necessarily near a plate margin. If the crust is particularly thin or weak, then the magma may escape onto the surface as a volcanic eruption. Lava may build up over time until it is above the present-day sea level, giving rise to a volcanic island.

## Tectonic landforms

Each type of plate margin considered above displays its own distinctive combination of processes and resultant landforms. The global pattern of these landforms is, therefore, closely linked to the pattern of plate margins.

Physical environments  AS Unit 1

## Case study 1.5 Conservative margin

By far the best-known example of a conservative margin is the one between the Pacific plate and the North American plate along the coast of California. The Pacific plate is moving northwest at approximately 6 cm yr$^{-1}$ while the North American plate, although moving in the same general direction, is only moving at about 1cm yr$^{-1}$ (Figure 1.11).

Movements of the crust occur along fault lines, such as the San Andreas and Hayward faults. Low magnitude earthquakes occur very frequently, 20 per day being not uncommon, with major events such as those of 1906 and 1989 happening very occasionally.

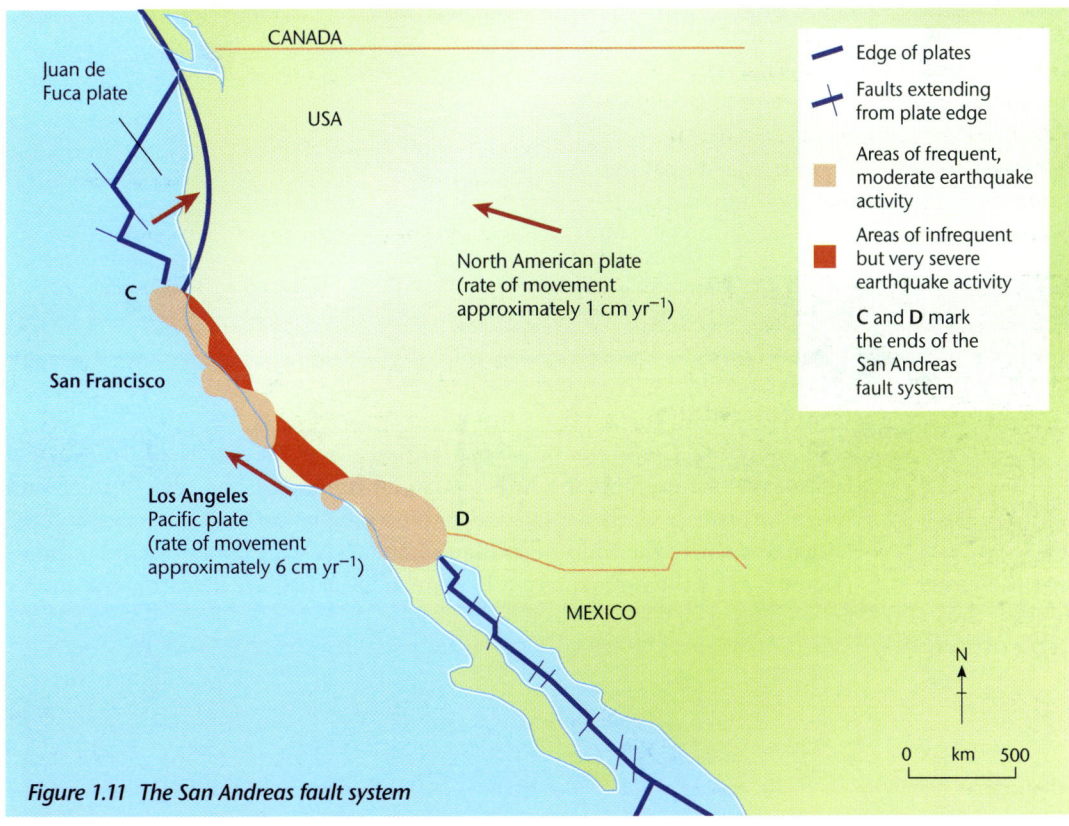

Figure 1.11 The San Andreas fault system

## Case study 1.6 Hot spot

The Hawaiian islands are a chain of volcanic islands lying over a stable hot spot. The Pacific plate has been moving over the hot spot for about 70 million years and a succession of volcanic islands and underwater volcanoes has formed over that time. As the plate has moved, so the volcanoes have been carried away from the hot spot in a north-westerly direction, forming a chain of extinct underwater volcanoes, called seamounts, often eroded into flat-topped remnants called guyots, extending all the way to the Aleutian islands (Figure 1.12). Currently a new volcano is erupting 35 km southeast of Hawaii. Loihi is only 3000 m tall at the moment and rises to 2000 m below sea level. It is predicted that Loihi will reach the sea surface in 10 000–100 000 years time.

# Chapter 1 Earth systems

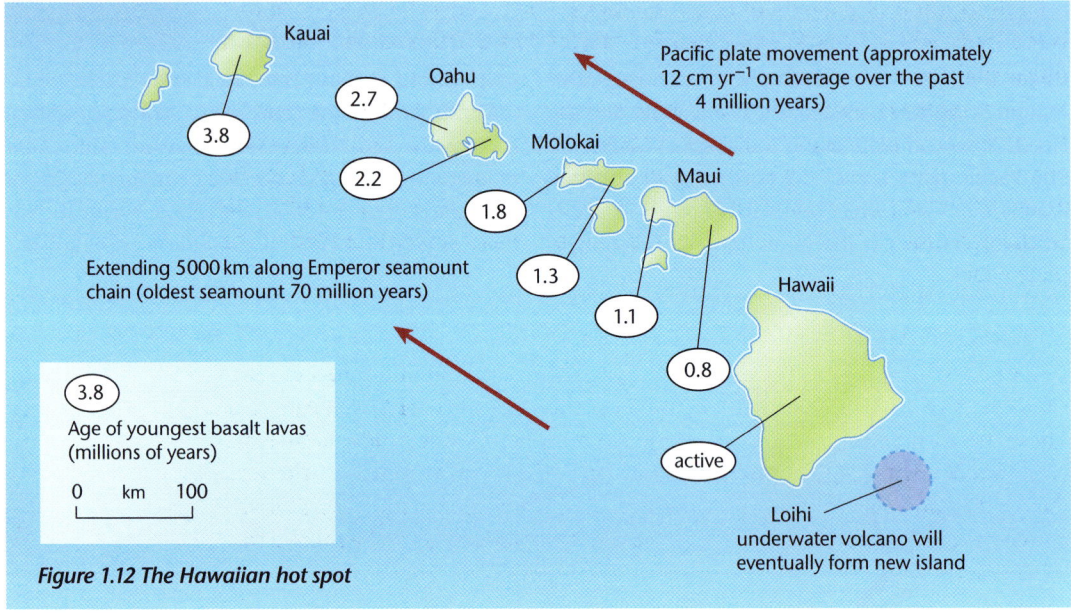

Figure 1.12 The Hawaiian hot spot

## Fold mountains

Fold mountains tend to be found either along the edges of continental landmasses, for example the Andes of South America at the edge of the South American plate, or where continental landmasses abut each other, for example the Himalayas where the Indo-Australian plate meets with the Eurasian landmass (Figure 1.13). Fold mountains tend to form a linear pattern, hence the commonly

Figure 1.13 The global distribution of fold mountains

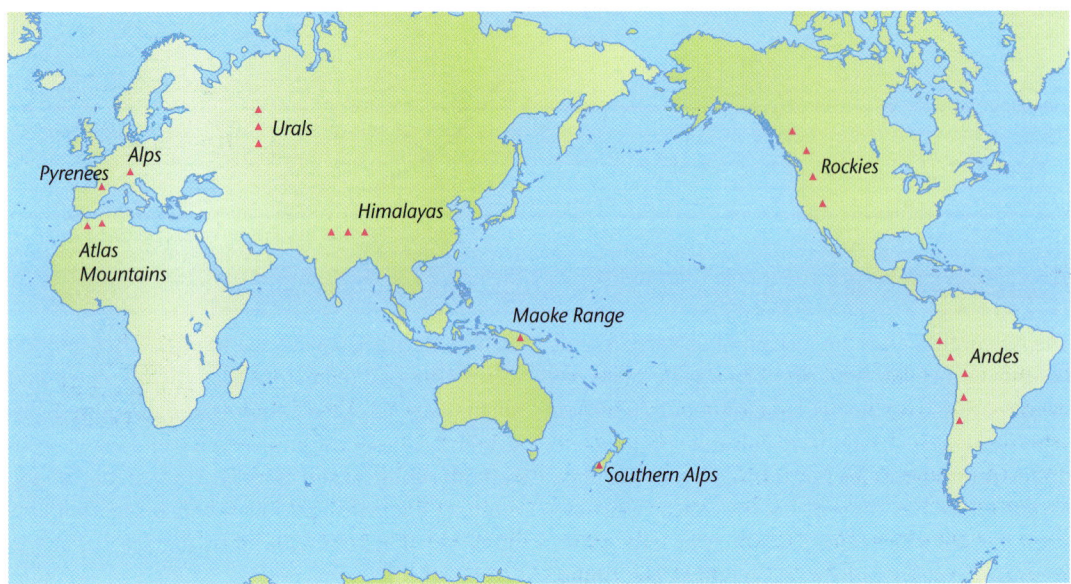

*Physical environments* **AS Unit 1**

used description of a 'chain' of fold mountains. The anomaly to the general rule is the Urals, which lie away from the edge of a landmass, although they are still a linear chain.

The reason for their pattern in shape is that fold mountains are formed by compression resulting from the convergence of two crustal plates. This occurs either at a collision margin (e.g. the Himalayas) or a destructive margin (e.g. the Andes). The anomaly of the Urals is difficult to explain, but they are likely to be the outcome of crustal crumpling due to compressional forces exerted at the plate edges.

## Ocean trenches

Ocean trenches are also found in a linear pattern and most are located just offshore, such as the Peru–Chile trench off the west coast of South America. These are formed at subduction zones where oceanic crust subducts beneath continental crust. Some ocean trenches such as the Mariana trench in the west Pacific Ocean (Figure 1.14) are further offshore and are the result of the convergence of two areas of oceanic crust and the subduction of the more dense of the two.

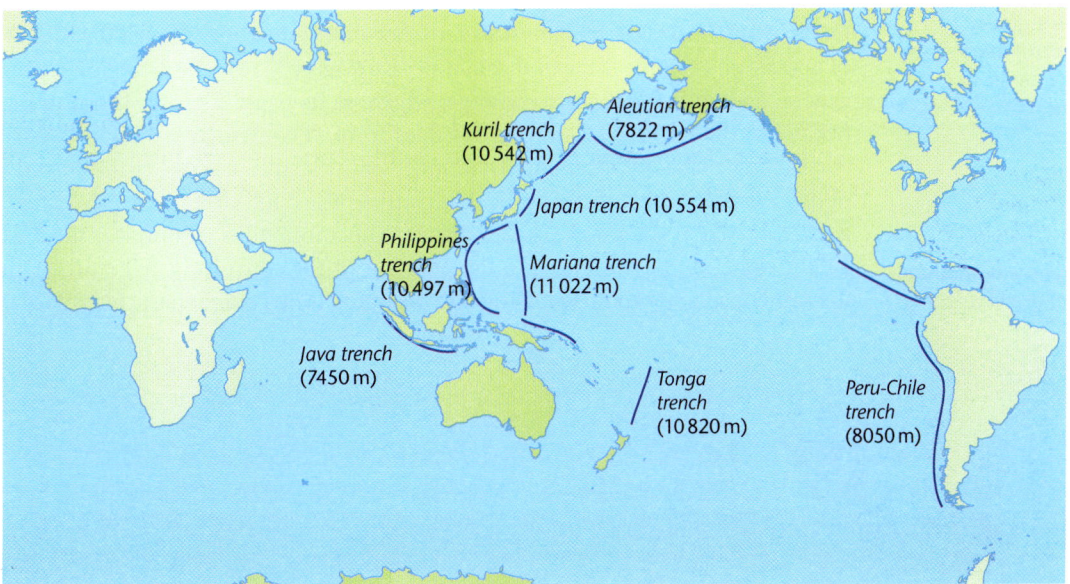

*Figure 1.14 The global distribution of ocean trenches*

## Island arcs

Island arcs are again found in a linear pattern, although they tend to be curved or arc-shaped. They are usually located offshore, and an example is the Mariana Islands (Figure 1.15). Island arcs are formed at destructive margins where two areas of oceanic crust converge. The subducted crust melts and the low-density molten material rises, aided by compressional forces from the convergence, to form a chain of volcanic islands. Where the edge of the subducting plate is curved in shape, as is true of the Pacific plate, the islands form an arc.

# Chapter 1  Earth systems

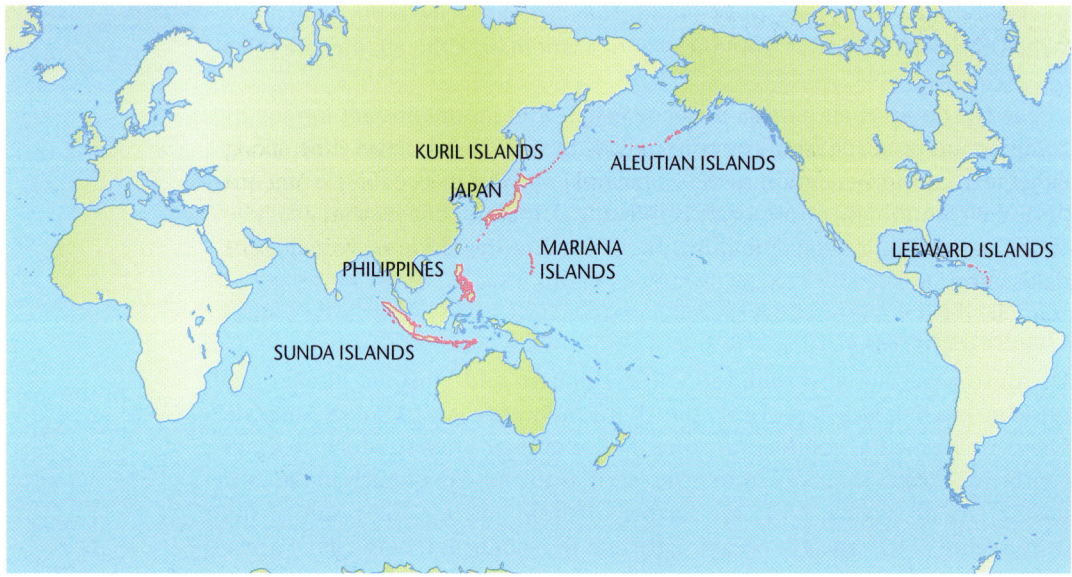

*Figure 1.15 The global distribution of island arcs*

*Figure 1.16 The global distribution of ocean ridges*

## Ocean ridges

Ocean ridges also occur in a linear pattern, normally extending through the middle of oceans; an example is the mid-Atlantic ridge (Figure 1.16). The exception to this general pattern is the East Pacific ridge, which, as its name suggests, is not in the middle of the Pacific Ocean. Ocean ridges form at constructive margins where two oceanic plates are diverging. Their mid-oceanic position is due to the creation of the oceans in which they now lie. The continents, which were previously joined, first separated at the break up of Pangaea. The anomalous position of the East Pacific ridge can be attributed to the position of the convection cell currents in the asthenosphere producing rising and diverging limbs at that particular place.

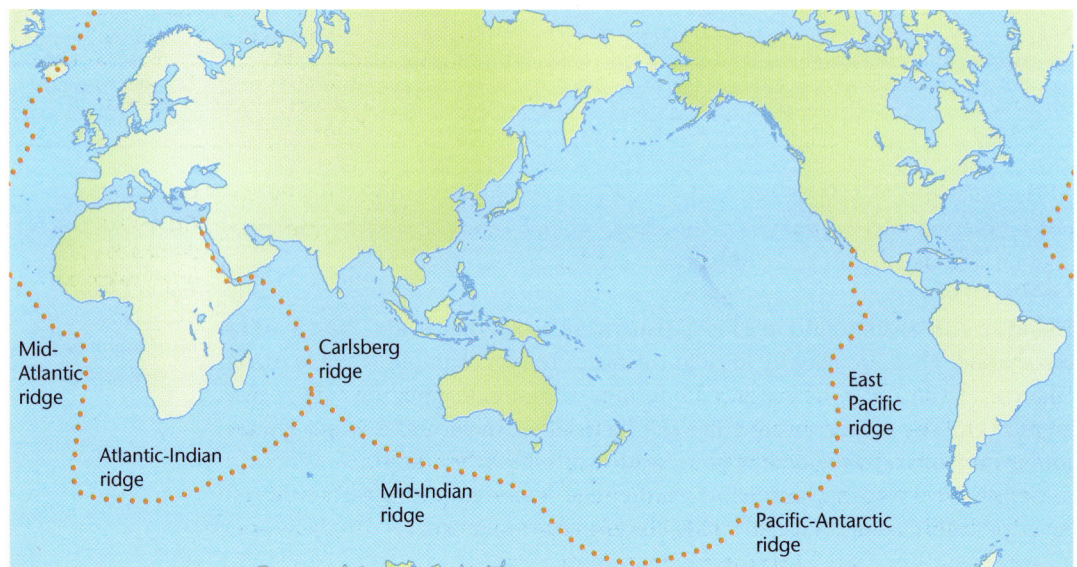

14

Physical environments — AS Unit 1

# Igneous activity

## Distribution of volcanic activity

Volcanoes occur in a series of broad bands. These bands tend to be either along the edge of continental landmasses, for example the west coast of South America (including the volcanoes Cotopaxi and Nevado del Ruiz), or through the middle of oceans, such as along the mid-Atlantic (including the Icelandic volcanoes of Hekla and Haeimaey). There are exceptions to this general pattern, however. For example, the volcanoes of the Hawaiian islands (such as Mauna Loa) are rather more isolated (Figure 1.17).

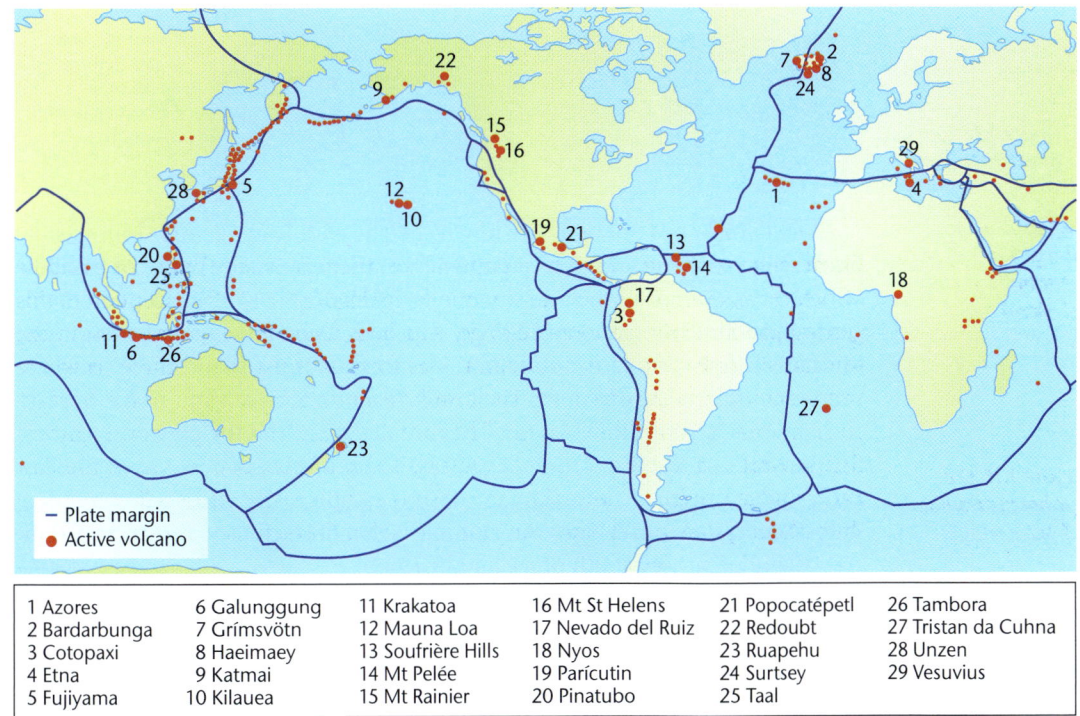

| 1 Azores | 6 Galunggung | 11 Krakatoa | 16 Mt St Helens | 21 Popocatépetl | 26 Tambora |
| --- | --- | --- | --- | --- | --- |
| 2 Bardarbunga | 7 Grímsvötn | 12 Mauna Loa | 17 Nevado del Ruiz | 22 Redoubt | 27 Tristan da Cuhna |
| 3 Cotopaxi | 8 Haeimaey | 13 Soufrière Hills | 18 Nyos | 23 Ruapehu | 28 Unzen |
| 4 Etna | 9 Katmai | 14 Mt Pelée | 19 Parícutin | 24 Surtsey | 29 Vesuvius |
| 5 Fujiyama | 10 Kilauea | 15 Mt Rainier | 20 Pinatubo | 25 Taal | |

*Figure 1.17 The global distribution of active volcanoes*

The pattern in the distribution of volcanic activity is largely explained by the position of the various types of plate margin, as volcanoes occur at constructive margins and destructive margins, as well as at hot spots.

## Extrusive landforms

### Volcanoes

The type of volcanic cone produced by eruptions depends upon the type of lava erupted and the nature of the eruptions themselves. In turn, the eruptions are greatly influenced by the type of plate margin on which they occur.

Constructive plate margins tend to give rise to fissure eruptions and shield (or basic) volcanic cones (Figure 1.18). **Fissure eruptions** occur along fault and fracture lines, while **shield volcanoes** erupt from a vent.

15

# Chapter 1    Earth systems

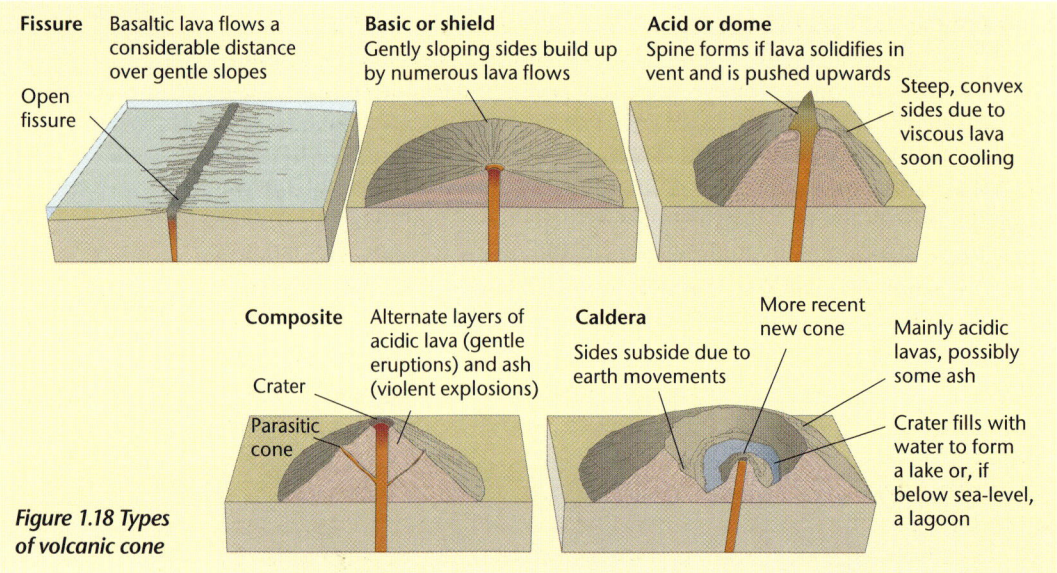

Figure 1.18 Types of volcanic cone

Shield volcanoes are typically low in height with long, gently sloping sides and a wide base. The lava that is erupted from them is usually basic (or basaltic), which means it has a low viscosity due to its low (<50%) silica content. Being quite fluid and hot (about 1200°C), this lava flows quickly and covers long distances before it cools and solidifies, which explains the shape of the cone. The eruptions are generally frequent but low in magnitude as magma is able to reach the surface relatively easily — the plates are diverging and the crust is fracturing. This means that there is seldom a great build-up of pressure. However, pressure build-up can occur if the magma does not escape easily and so more violent eruptions do sometimes occur. An example is that of Haeimaey, which created the new island of Surtsey south of Iceland in 1963.

Shield volcanoes also occur at 'hot spots'. These are locations often well away from plate margins at which plumes of hot magma are rising towards the surface having been differentially heated by the core. Volcanoes may occur where the crust is thin and/or weak, such as those of the Hawaiian islands, including Mauna Loa. Mauna Loa is one of the most active volcanoes on Earth, and has built up a cone that rises 9000 m from the sea floor. It has erupted continuously since 1983 and has a diameter of 120 km at its base, sloping gently at about 6° to the top.

Volcanoes at destructive margins are rather different in character and in cone shape. Eruptions tend to be less frequent and much more explosive. Rising magma here often has a much greater thickness of crust through which to pass while fractures, which would provide easy routes for the rising magma, tend to be less common. The lava itself is typically acidic, with more than 50% silica content. It has a lower temperature of about 800°C, and is much thicker and more viscous. This means that it flows

## Key terms

**Extrusive igneous landforms** Features formed on the surface from magma flowing as lava before cooling and solidifying.

**Hot spot** A location where a plume of rising magma reaches the surface.

**Intrusive igneous landforms** Features formed by rising magma cooling and solidifying beneath the surface, which can be exposed at a later date.

**Volcanic activity** Processes and features associated with rising magma nearing or reaching the surface.

slowly and soon solidifies, giving rise to cones with steeply sloping sides, a narrow base and a greater height. This type of volcano is known as an **acid cone** volcano.

Sometimes acid cone volcanoes have secondary or parasitic cones on their sides. They form when the passage of rising magma through the main vent is blocked, probably due to magma from earlier eruptions having solidified in the vent before it was able to escape. This leads to a build-up of pressure and the magma forces its way through the sides of the vent, often creating cracks.

Acid cone volcanoes are often composite in their structure, with alternating layers of ash and lava. The ash is produced by a highly explosive eruption, often after the vent has been blocked, which fragments parts of the cone or the plug of solidified magma. Some of the world's largest volcanoes are of this type, including Etna, Vesuvius and Popocatépetl. Etna has slopes of 50° near its base but slopes of only 30° at the summit. It has numerous secondary cones on its flanks from earlier eruptions.

The various types of volcanic cone, influenced by the plate margin on which they occur, are summarised in Figure 1.18. There is no volcanic activity, however, at conservative margins or collision margins: no new crust is being created by rising magma and no existing crust is being destroyed by subduction.

### Case study 1.7 Lava plateau

The Deccan plateau of India is an excellent example of this extrusive landform. It is a huge expanse of lava covering 700 000 km². It consists of 29 lava flows, which have recently been dated and found to have all occurred within a period of less than 2 million years.

It is suggested that this may have been the result of a series of major eruptions from a mantle plume of rising magma. The plateau lies at about 700–900 m above sea level and has a number of major river valleys eroded into it, such as those of the Wardha and Manira Rivers.

Photograph 1.2 The Deccan plateau, India

#### Lava plateaux

The other major extrusive landform produced by volcanic activity is a lava plateau. Lava plateaux are extensive areas of basaltic lava, often with a layered structure. They are formed by major eruptions from vents or, more usually, from a fissure. The layered structure is caused by the accumulation of lava from a series of

# Chapter 1 Earth systems

eruptions over a period of time. The plateau itself tends to be flat and featureless, with limited soil and vegetation cover. Eruptions from oceanic ridges produce huge abyssal plains on the sea floor.

## Intrusive landforms

Intrusive landforms are formed by magma rising towards the surface but cooling and solidifying before being extruded. This is likely to be the case if the magma is rising slowly, if there is a great thickness of crust to pass through and if there are few weaknesses in the crust through which it can flow out. The magma cools slowly because it is not exposed to the air and so mineral crystals, for example quartz crystals in granite, grow to a large size.

**Batholiths** are large masses of intrusive rock that may cause a general doming up of the surface as they are forming. They are only exposed after the gradual weathering and erosion of the less resistant overlying 'country rock'; this weathering is facilitated by the fractures and cracks that develop due to the tensional forces the surface experiences as it is stretched during uplift. Similar, but smaller, features are known as **bosses**. An example is the boss at Shap in Cumbria, which crops out over an area of just a few kilometres.

The heat and/or pressure exerted on the country rock causes metamorphic rock to be produced around the intruding magma. Examples of this include sandstone being metamorphosed into schists and limestone into marble.

**Sills** are intrusions that are formed parallel to bedding planes in the country rock, often, but not always, lying horizontally. The bedding planes provide a line of weakness along which the magma will flow before cooling and solidifying. The magma contracts as it cools, producing cracks in the resultant rock. When the overlying rock is weathered and eroded the sill is exposed, sometimes forming steep coastal cliffs or rock outcrops, including cap rocks on waterfalls.

**Dykes** cut across the bedding planes of the country rock, often vertically. Magma flows through cracks and weaknesses but again cools and solidifies before reaching the surface. Contraction joints develop parallel to the surface as the magma solidifies. Once exposed, the dykes can appear as linear outcrops of resistant rock. Figure 1.19 summarises these features.

*Figure 1.19 Batholiths, sills and dykes*

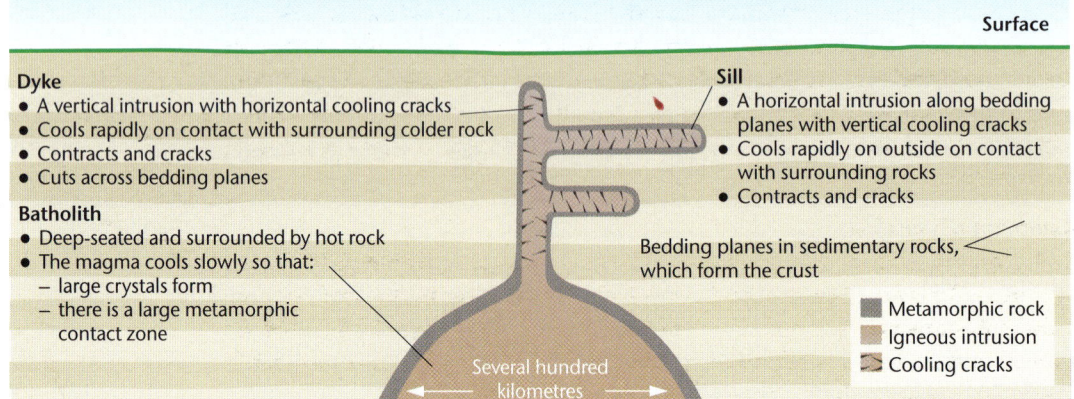

18

## Case study 1.8 Intrusive igneous landforms

The Isle of Arran, off the west coast of Scotland, was created by the intrusion of a large granite batholith (Figure 1.20). This batholith domed the sandstone surface about 60 million years ago, as Greenland separated from Scotland during the creation of the Atlantic Ocean. The fractured overlying sandstone and metamorphic schists have subsequently been weathered and eroded to expose the resistant granite, peaking at a height of 874 m on Goat Fell. A series of dykes, 2–3 m wide, are exposed across the beach at Kildonan, looking rather like natural rock groynes.

At Drumadoon a sill has been exposed on the coast, forming a 50 m high cliff.

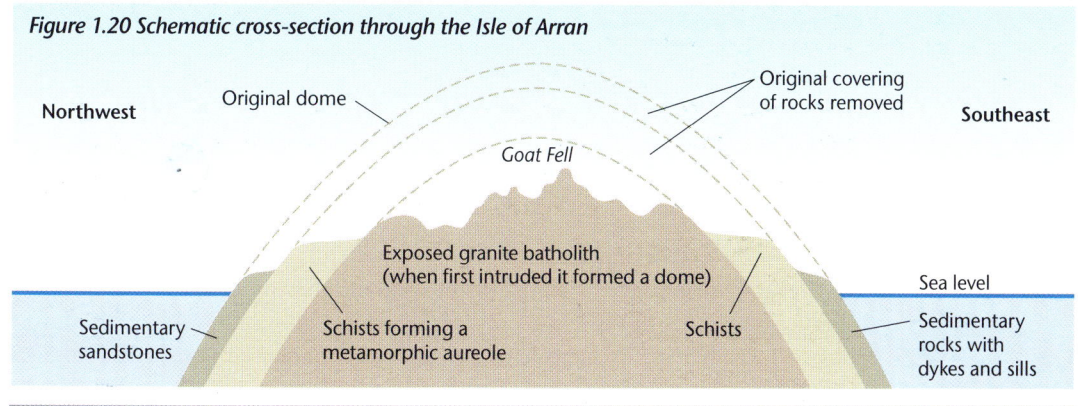

Figure 1.20 Schematic cross-section through the Isle of Arran

## Economic benefits of igneous activity

Igneous activity, both intrusive and extrusive, provides a range of resources that can be economically exploited.

### Building materials

Granite is a resistant igneous rock that is widely used for buildings, kerbstones and gravestones due to its ability to survive attack from weathering and erosion. Aberdeen is known as 'granite city' as so many buildings, including the city hall and many houses, are constructed from this rock.

### Minerals

Slow cooling of magma permits the growth of mineral crystals in the rock, leading to a significant development of potentially valuable raw materials. Tin and copper ores occur widely in the southwest of England and used to be extracted from mines in the batholiths of Dartmoor and Bodmin Moor. One such mine, Wheal Jane, operated from the 1500s until its closure in 1913. The copper ore was sent to Redruth for export.

### Geothermal energy

In Iceland, most of Reykjavik's domestic heating is supplied by hot water from a geothermal plant at Nesjavellir. Due to Iceland's position on the mid-Atlantic ridge, sub-surface temperatures are very high, typically 150°C within 100 m of the ground surface. Hot water from the geothermal fields is too corrosive to use, but it heats cold water, which is then piped 28 km to Reykjavik.

# Chapter 1  Earth systems

*Included in Coasts*

### Tourism
On the Isle of Arran the scenic nature of the dykes and sills attracts visitors, who create a demand for tertiary employment in hotels, guesthouses and field centres. This can increase spending in the local economy and lead to a multiplier effect.

# Weathering

## Processes of weathering

**Weathering** can be defined as 'the breakdown and decay of rocks by the elements of the weather (except wind) acting in situ'. Weathering should not be confused with *erosion*, which involves the wearing away of rock by moving forces such as running water, breaking waves, advancing glaciers and the wind. Weathering can be divided into three types:

- physical weathering
- chemical weathering
- biological weathering

'Breakdown' is largely achieved by **physical weathering** processes that produce smaller fragments of the same rock. No chemical alteration takes place during physical weathering. By increasing the exposed surface area of the rock, physical weathering allows further weathering to take place.

'Decay' is the result of **chemical weathering**, which involves chemical reactions between the elements of the weather and some minerals within the rock. It may reduce the rock to its chemical constituents or alter the chemical and mineral composition of the rock. Chemical weathering processes produce weak residues of material that are easily removed by erosion or transportation processes.

**Biological weathering** may consist of physical actions, such as the growth of plant roots, or chemical processes, such as chelation by organic acids. Although this arguably does not fit with the precise definition of weathering, biological processes are usually classed as a type of weathering. The effects are very similar to some of the physical and chemical processes, even if it may be difficult to relate them directly to 'the elements of the weather' as stated in the definition.

### Key terms
**Biological weathering** The actions of plants and animals in the breakdown and decay of rock.
**Chemical weathering** The in situ decay of rock due to chemical processes.
**Physical weathering** The in situ breakdown of rock into smaller fragments of the same material by mechanical processes.
**Weathering** Breakdown and decay of material in situ.

### Physical weathering

*Freeze–thaw*
Water that enters cracks and joints in rock expands by nearly 10% when it freezes. In confined spaces this exerts pressure on the rock, causing it to split or pieces to break off, even in very resistant rocks.

*Pressure release*
When overlying rocks are removed by weathering and erosion, the underlying rock expands and fractures parallel to the surface. This is significant in the exposure of granite batholiths and is also known as dilatation. The parallel fractures are sometimes called pseudo-bedding planes.

*Included in Coasts*

*Thermal expansion*

Rocks expand when heated and contract when cooled. If they are subjected to frequent cycles of temperature change, then the outer layers may crack and flake off. This is also known as insolation weathering, although experiments have cast doubts on its effectiveness unless water is present.

*Salt crystallisation*

Solutions of salts can seep into the pore spaces in porous rocks. Here the salts precipitate, forming crystals. The growth of these crystals creates stress in the rock, causing it to disintegrate. Sodium sulphate and sodium carbonate are particularly effective, expanding by about 300% in areas where the temperature fluctuates around 26–28°C.

## Chemical weathering

*Oxidation*

Some minerals in rocks, especially iron-based minerals, react with oxygen in the air or in water. The rock becomes soluble under extremely acidic conditions and the original structure is destroyed. Oxygen often attacks the iron-rich cements that bind sand grains together in some sandstones.

*Carbonation*

Carbon dioxide from the atmosphere dissolves in rainwater to produce a weak acid, called carbonic acid. Carbonic acid reacts with calcium carbonate in rocks such as limestone to produce calcium bicarbonate, which is soluble. This process is reversible; the precipitation of calcite occurs during the evaporation of calcium-rich water in caves, forming stalactites and stalagmites.

*Solution*

Some salts are soluble in water. Other minerals, such as iron, are only soluble in very acidic water, with a pH of about 3. Any process by which a mineral dissolves in water is known as solution, although mineral-specific processes such as carbonation can be identified.

*Hydrolysis*

This is a chemical reaction between rock minerals and water. Silicates combine with water, producing secondary minerals such as clays. Feldspar in granite reacts with hydrogen in water to produce kaolin (china clay).

*Hydration*

Water molecules added to rock minerals create new minerals of a larger volume. This happens to anhydrite, forming gypsum. Hydration causes surface flaking in many rocks, partly because some minerals also expand by about 0.5% during the chemical change because they absorb water.

## Biological weathering

*Tree roots*

Tree roots grow into cracks or joints in rocks and exert outward pressure. This operates in a similar way to freeze–thaw, and with similar effects. When trees topple, their roots can also exert leverage on rock and soil, bringing them to the surface and exposing them to further weathering. Burrowing animals may have a similar effect.

### Organic acids

Organic acids produced during the decomposition of litter cause soil water to become more acidic and react with some minerals in a process called chelation. Blue-green algae can have a weathering effect, producing a shiny film of iron and manganese oxides on rocks — this is known as 'desert varnish'.

## Combining weathering processes

In many cases, fracturing of the rock by one weathering process will expose a greater surface area of the rock to the elements of the weather, thereby increasing the incidence of weathering by other processes.

## Factors influencing the rate of weathering

A range of factors controls the rate at which different weathering processes operate. These factors may be physical, human or temporal and they often interact with each other and operate in tandem in influencing the rate of weathering. The most important factors are climate and geology, although relief, soil and vegetation cover, and human activity can also be significant.

### Climate

The types and rates of weathering vary with climate, in particular moisture availability and temperature. This is summarised in Peltier's diagram (Figure 1.21).

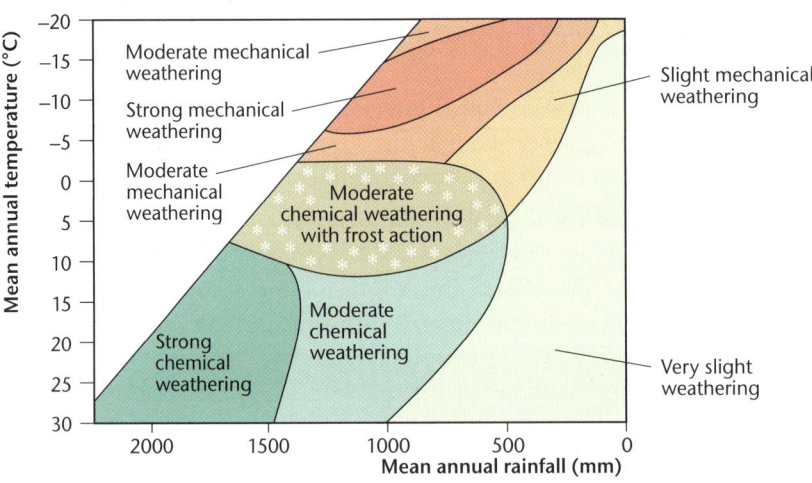

Figure 1.21 Peltier's diagram, showing variations in weathering with climate

Moisture is important, as water is the medium for most chemical reactions. These include carbonation, in which carbon dioxide dissolves in water. Water also provides the hydrogen for hydrolysis. Therefore, the greater the availability of moisture, the greater the rate of most chemical reactions. Water is also present in most physical processes, including freeze–thaw and salt crystallisation.

Temperature is important, as most chemical reactions take place faster at higher temperatures. van't Hoff's law states that a 10°C increase in temperature leads to a 2.5-times increase in the rate of chemical reaction (up to 60°C). Warm, moist tropical environments therefore experience the fastest rates of chemical weathering and cold, dry locations experience the slowest. However, it is worth noting that carbonation can be more effective at low temperatures as carbon dioxide is more readily absorbed in cold water than in warm water.

Temperature fluctuations are also significant. As a general rule, the greater the number of freeze–thaw cycles that occur, the faster the rate of freeze–thaw weathering. So in areas where the temperature fluctuates around 0°C, more freeze–thaw weathering occurs. This is why locations such as northern Scandinavia have higher rates of physical weathering than more extreme locations, such as Antarctica where the temperatures remain below zero and few, if any, freeze–thaw cycles occur.

On a rather different scale, the diurnal (daily) range of temperature affects the rate of thermal expansion. This process is particularly effective in hot desert environments where day-time temperatures may be 40°C or more but, due to the lack of cloud cover to trap out-going radiation, night-time temperatures can fall well below freezing. If these cycles of temperature change occur frequently, the rocks expand and contract repeatedly, causing significant amounts of breakdown.

However, experiments carried out by Professor David Griggs in 1936, during which he simulated the typical diurnal temperature fluctuations of a hot desert environment over the equivalent of 100 years, revealed that rocks only weathered to any great degree when water was present. This suggests that expansion and contraction alone may not be effective without the additional contribution of freeze–thaw. Water is available in hot deserts, particularly at night, as condensation takes place at the very low temperatures.

### Geology

Rates of weathering are also strongly influenced by the physical and chemical characteristics of the rock type being weathered. Some rock types are susceptible to specific processes because of their mineral composition. For example, limestone contains a high percentage of calcium carbonate and so is prone to carbonation; granite suffers from hydrolysis due to the presence of feldspar. These specific processes will weather these particular rock types much more rapidly than others with different chemical compositions. The presence of resistant minerals, such as quartz, slows weathering.

The existence of joints, cracks, bedding planes and pores all permit water to enter a rock and increase the surface area being weathered by water-based processes (both physical and chemical). The greater the concentration of such lines of weakness in a rock, the faster it will be weathered by these processes. This is known as the **fissility** of the rock.

Rocks also vary in the strength of the bonding between individual particles. If particles are weakly bonded or are bonded by soluble cements, rates of weathering will be more rapid than they are in rocks with strong bonds.

Grain size is a factor too. Coarse-grained rocks have more void spaces and a higher permeability than fine-grained rocks, and so can be weathered more quickly, especially by water-based processes. However, fine-grained rocks do have

# Chapter 1: Earth systems

a greater surface area exposed to weathering and so more of the rock is subjected to temperature fluctuations.

### Relief
Relief does not have much impact on weathering, but it may affect microclimates. An example is that north-facing slopes in the northern hemisphere that are shaded from direct sunlight may have increased numbers of freeze–thaw cycles. Relief can also cause water to collect at the base of slopes, providing increased amounts of moisture for chemical weathering processes compared to the top of a slope. There may even be down-slope movement of weathered debris, exposing the fresh rock beneath to weathering processes.

Photograph 1.3 Plant roots are growing into the joints of this limestone in the Yorkshire Dales

### Soil/vegetation cover
The presence of a layer of soil or significant vegetation cover tends to increase rates of chemical weathering. This is because the acidity of rainwater draining through the soil is increased as it dissolves organic acids released by decayed plant and animal matter, leading to higher rates of weathering by processes such as carbonation on susceptible rocks beneath. Significant tree cover provides more roots, which can lead to greater breakdown of underlying rocks by biological weathering. However, soil, and vegetation cover can decrease rates of some types of physical weathering by protecting the rock from extremes of temperature. Where rock is covered by soil, thermal expansion is less likely to occur and temperatures may fall below zero less often, meaning fewer freeze–thaw cycles.

Experiments investigating rates of limestone weathering in the Burren in western Ireland produced typical figures of 0.015 cm per year on exposed rock and 0.045 cm per year under soil and grass cover, because of the increased acidity of rainwater.

### Human activity
Human impacts on weathering have become increasingly important in recent years. Greater atmospheric pollution, due to emissions of gases such as carbon dioxide and sulphur dioxide from power stations and vehicles, can increase the acidity of rainwater and so speed up the rate of some chemical weathering processes. Vegetation cover may be removed, for example by deforestation in tropical environments, which can reduce rates of chemical and biological weathering. The reverse is also true if reforestation takes place.

Global warming, due to the influence of humans in enhancing the greenhouse effect, may increase rates of chemical weathering as chemical reactions are speeded up at higher temperatures. However, rates of freeze–thaw weathering may be reduced in areas where global warming keeps temperatures above zero and reduces the number of freeze–thaw cycles experienced.

## Physical environments — AS Unit 1

### Time

Rates of weathering may vary over time as the influence of other factors changes. However, time is not really a factor in itself but is the medium through which the other factors operate. As a general rule, rates of weathering are very slow.

In any particular location a range of different weathering processes may be acting and a combination of different factors will control their rates.

## The impact of weathering on the landscape

### Regolith

**Regolith** is the collective term for all weathered material that accumulates at or near the surface. It may be made up of fragments of rock from physical processes or chemical residues from chemical processes and is often the basis of soil development.

### Scree

**Scree** consists of angular fragments of weathered rock, often accumulating at the base of a slope. In Britain scree commonly forms below vertical cliffs in upland areas, although many screes are relict landforms and are no longer actively forming. Evidence that screes are no longer forming comes from the amount of vegetation cover they possess and the lack of unweathered faces on the rock particles. Scree fragments form when freeze–thaw weathering, possibly during earlier, colder climatic conditions, breaks down the rocks of the exposed cliff face and the weathered fragments fall under gravity to accumulate at the bottom of the slope. The gradient of scree slopes is typically 30–40° and they often display an element of sorting, as larger particles move more easily towards the bottom of the slope.

*Photograph 1.4 A scree slope near Austwick, Yorkshire Dales*

# Chapter 1  Earth systems

### Soil
**Soil** is a mixture of mineral and organic matter. The organic material is derived from the decay of plant and animal remains, but the mineral matter comes mainly from weathering. As bare rock is slowly weathered by chemical and physical processes, a soil starts to form if the resultant weathered debris accumulates on the surface. Further mineral inputs are added as the underlying rock continues to be weathered. Once the rock is covered with a layer of soil, it is more likely to weather chemically than physically. The rate at which soils develop depends largely on the climatic conditions and the geology, but it typically takes about 10 000 years for a 1 m deep mature soil to develop in the mid-latitudes.

### Granite landscapes
**Granite landscapes** are usually upland areas, partly due to the rock's general resistance to weathering and erosion and partly because it causes the surface rocks to dome upwards when it is intruded as a batholith.

One of the most distinctive features of granite landscapes is the **tor**. There are differing theories about how tors are formed, but a possible sequence of events is as follows.
- The intrusion of a granite batholith beneath the surface.
- Deep chemical weathering by hydrolysis, most rapidly where a closely spaced pattern of cooling joints exists.
- Removal of overlying rocks by surface weathering, erosion and possibly solifluction, a mass movement process common in periglacial climates.

## Case study 1.9  Granite landscape

One of the most distinctive granite landscapes in Britain is Dartmoor in Devon. Approximately 60 million years ago an igneous intrusion took place here. The surface sedimentary rocks, mainly sandstones and shales, were domed up and a granite batholith formed beneath them. The overlying rocks, already relatively weak compared to the resistant granite, were further weakened by the tensional forces exerted during the upward doming, leading to the development of fractures and faults. These overlying rocks have been subsequently removed by weathering and erosion, exposing the resistant granite and producing a distinctive landscape. The batholith forms an irregular surface, reaching a maximum height of 621 m. It is cut with deep valleys formed by rivers such as the Dart and the Plym. Numerous tors have been formed, Hay Tor and Yes Tor being two of the best known. Both of these are surrounded by blockfields and clitter. Significant accumulations of kaolinite are found in many of the valleys, including that of a tributary of the River Teign near Bovey Tracey.

*Photograph 1.5 Hay Tor*

- Present-day weathering by processes such as freeze–thaw further enlarging pressure release joints.
- Chemically weathered debris, for example kaolinite, is removed to the surrounding valley bottoms where it accumulates.
- Physically weathered debris accumulates around the tor. If this consists of large blocks of granite, it is known as a **blockfield**. Accumulations of smaller fragments are called **clitter**.
- The tor remains exposed as an isolated mass of bare rock, often up to 20 m high, consisting of a number of individual boulders resting on top of the core stones beneath.

### Limestone landscapes

Limestone landscapes are distinctive, largely on account of their permeability and susceptibility to carbonation weathering. Very little surface water is found in these areas, as streams originating on neighbouring areas of impermeable rock go underground at sink holes on reaching the limestone. A striking feature of limestone landscapes is limestone pavement, an area of bare, exposed limestone formed by the following sequence of events.

- Glacial action removes soil and vegetation cover, exposing the highly jointed limestone.
- Water enters the joints and enlarges them by carbonation to form **grykes**.
- The blocks between the joints remain upstanding as **clints**.
- Water may sit on the surface of the clints and form small depressions called **solution hollows**.
- As this water runs off the clints and flows into the grykes, it erodes small grooves called **karren**.
- Vegetation may colonise the grykes, increasing weathering rates by producing organic acids and possibly by the action of tree roots.

Freeze–thaw weathering may also exploit the joints on cliff faces, leading to the formation of scree slopes at their base.

Other distinctive features of limestone landscapes are shake holes and dolines (Figure 1.22). These are surface depressions caused by the collapse of underlying limestone that has been severely affected by carbonation, perhaps due to very close joint spacing. Overlying soil and any superficial deposits subside, leaving circular depressions on the surface. Dolines are larger than shake holes and may be formed by the collapse of an underground cave or by significant weathering of one master joint in the limestone, in which case the depression is more of a funnel shape.

*Figure 1.22 Formation of shake holes and dolines*

Glacial deposits holding water against rock surface

Limestone

Water infiltrates down to the limestone and weathers the joints by carbonation

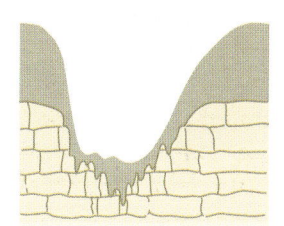

Where joints are closely spaced, the limestone disintegrates and the glacial deposits collapse, forming a shake hole or doline

# Chapter 1 Earth systems

## The impact of weathering on human activity

Weathering has both positive and negative impacts on human activity. The negative impacts include damage to buildings and hazards for transport, while the positive impacts include the creation of landscapes that attract tourists and the extraction of raw materials. In both cases there is likely to be a direct and immediate impact, with indirect or knock-on effects that follow afterwards.

### Case study 1.10 Limestone landscape

The area around Malham in the Yorkshire Dales displays many limestone landforms and is a popular destination for field studies as well as for tourists. There is a good example of limestone pavement above Malham Cove. Soil and vegetation cover was stripped off by the glaciers that covered the area approximately 10 000 years ago. The exposed carboniferous limestone has since been weathered by carbonation and now forms a pavement with grykes typically 0.5 m deep, occasionally colonised by plants and even the odd tree.

On the hillside below Ingleborough, a limestone hill capped with millstone grit, there are a large number of shake holes and dolines of varying sizes, but generally 3–5 m in diameter and 2–3 m deep.

In the valley of Goredale, exposed cliffs of limestone on the valley side have been physically weathered, mainly by freeze–thaw, leading to the development of significant scree slopes at their base.

*Photograph 1.6 Limestone pavement in the Yorkshire Dales*

*Physical environments* **AS Unit 1**

### Building damage
St Paul's Cathedral in London suffered rapid weathering in the last century, particularly on the south side which faces the Bankside Power Station. Measured rates of weathering were 50% higher on the south side than on the north during the 1970s. Expensive repairs to the stonework were needed. The problem has now receded as Bankside Power Station has been closed and redeveloped as the Tate Modern gallery.

### Transport
Netting has had to be applied to bare limestone cuttings alongside the M5 in the Gordano Valley near Bristol to stop scree falling on the motorway. Fences have also been erected at the base of the cuttings to trap particles dislodged by weathering processes.

*Photograph 1.7 Weathering of ornamental stonework on a Bristol church*

### Scenic value
The Yorkshire Dales attracts visitors in large numbers, with 8.3 million visitor days in 2001. This creates a demand for tertiary employment and it is estimated that over 1000 full-time jobs depend upon tourism and £89m per year is spent in the local economy. However, the large number of visitors has caused severe footpath erosion around Malham Cove and there have been major traffic problems in the area. A new car park, with capacity for 168 cars and eight coaches, has been built on the edge of Malham village.

### Raw materials: china clay extraction
The hydrolysis of granite on Bodmin Moor, leading to the accumulation of kaolinite, used in paint, ceramics and paper, has meant that jobs and trade have been created at Lee Moor on the southern edge of Dartmoor. The quarry has been open since 1830 and is currently operated by Watts Blake and Bearne. This has been an important benefit to the economy of a remote, rural area, but there is much opposition to the tipping of unsightly waste in large mounds after the kaolinite has been extracted.

*Photograph 1.8 China clay workings in Cornwall*

# Chapter 1  Earth systems

## Examination question

Study Figure 1.23, which shows the margin between the Philippines plate and the Eurasian plate.

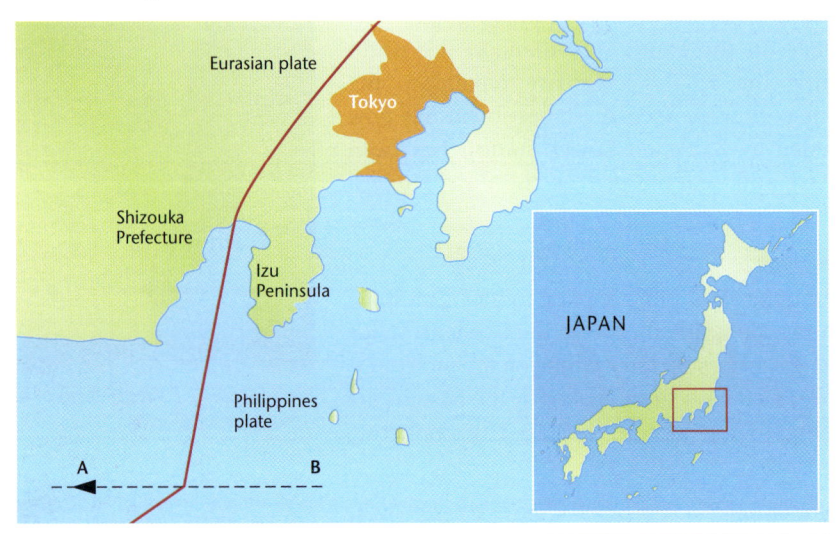

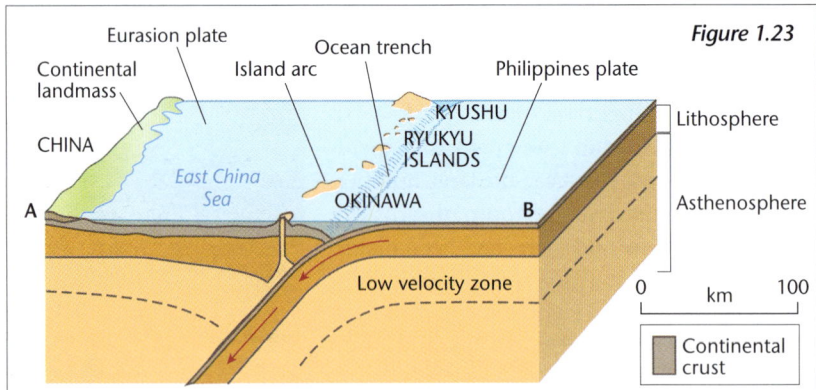

*Figure 1.23*

**a** With reference to the diagram:
  (i) Describe the movement of the Philippines plate. (2)
  (ii) Explain this movement. (4)
  (iii) Explain the formation of the island arc. (3)

**b** (i) What is a sill? (1)
  (ii) Explain how it is formed. (2)

**c** (i) Define the term 'weathering'. (2)
  (ii) With reference to a located example, describe and explain the impact of weathering on the landscape of a limestone area. (6)
  *You may use a diagram to help your answer.*

Total = 20 marks

*Physical environments* **AS Unit 1**

# Synoptic link
## Relationships between people and tectonic environments

- Positive and negative impacts of earthquakes and volcanoes on human activity, both short-term and long-term
- The importance of risk assessment, prediction and monitoring, and their limitations

## The impacts of earthquakes

### Ground displacement

Ground displacement is, in itself, not life threatening, but the impact of ground movements on buildings and other structures, such as bridges and roads, most certainly is. The worst earthquake catastrophe in an MEDC for many years occurred on western Honshu Island, Japan in 1995. More than 5000 people perished in southern Hyogo prefecture, most of them in the city of Kobe, Japan's most important port. The loss of so many lives, in a country where so much effort had been made to prepare for earthquakes, shocked observers worldwide. In the Kobe earthquake, surface rupture of the fault was observed only in a rural area of Awaji Island, with displacements of up to 3 m. Few structures were near enough the fault to be damaged by the displacement itself, although underground utilities, fences and ditches were cut. Many earthquake faults never break the surface, ruling out direct effects.

*Photograph 1.9 The impact of the Kobe earthquake*

### Landslides

Landslides are movements of a mass of rock, earth or debris down a slope. Such slope failure can be triggered by a number of events, including earthquakes. One of the best known and most documented was the landslide on Mt St Helens, USA in 1980. On May 18, at 8:32 a.m., a magnitude 5.0 earthquake triggered a rapid series of events. The entire northern slope of the mountain above the bulge failed and the north flank of the volcano began to slide downward from almost the exact site of the east–west fracture at the summit. This huge landslide released a tremendous mass from above the hydrothermal system that had driven the earlier steam eruptions. The sudden loss of confining pressure above the super-heated

# Chapter 1    Earth systems

groundwater caused a massive hydrothermal blast that was directed sideways through the landslide scar.

## Liquefaction

Liquefaction of loose, water-saturated sands and other granular soils due to earthquake shaking is a major cause of damage to and destruction of the built environment. One of the major impacts of liquefaction-induced types of ground failure is the sideways spreading of sloping ground.

The 1985 earthquake centred on Mexico City occurred when the Cocos plate beneath the Pacific Ocean ruptured from the North American plate. The earthquake measured 8.1 on the Richter scale and occurred as two events 26 seconds apart. When the shaking stopped, the 18 million people in Mexico City (one of the most populated urban areas on Earth) began assessing the damage. Thousands lay buried. In the Benito Jaurez Hospital alone, over 1000 people died. A shift change of doctors, nurses and other staff had been taking place just as the hospital collapsed.

*Photograph 1.10 The lateral blast of the 1980 eruption of Mt St Helens flattened trees up to 30 miles away*

Downtown Mexico City is built on a foundation of soft sedimentary material, the remnants of an ancient lake bed. An unstable subsoil of sand and mud on top of clay and gravel are all held in a shallow geological saucer of bedrock. This mixture is so soft that buildings in Mexico City can sink a few centimetres every year even in normal conditions. During an earthquake the sediments amplify seismic waves, causing the ground to liquefy. In the 1985 eathquake, buildings swayed for over 3 minutes as the ground liquefied. Mexico City lies over the remains of another city that once flourished. The Aztecs built their capital, Tenochtitlan, on the same location. The excellent construction techniques used by the Aztecs were demonstrated during this earthquake. Both the Metropolitan Cathedral, built by the Aztecs in 1525 AD, and the National Palace, built around 1700 AD, remained standing. Neither suffered severe damage from this major earthquake, even though modern buildings on the same block collapsed completely.

*Photograph 1.11 The effect of liquefaction in the Mexico City earthquake*

*Physical environments* **AS Unit 1**

*Photograph 1.12 The impact of the 2004 tsunami on a Sumatran village*

## Tsunamis

Tsunamis are waves, most often created by sea-floor earthquakes, volcanic activity and landslides. These waves travel rapidly across the ocean and have extremely long wavelengths. They are often, mistakenly, called tidal waves, but they are not related to tides and so should be called seismic sea waves. On Boxing Day 2004, a magnitude 9.3 earthquake occurred off the northwest coast of Sumatra, Indonesia at 07:58 local time. The earthquake had a relatively shallow focus — 18 km below the surface — and produced a megathrust along the subduction zone between the Indo-Australian plate and the Burma microplate. The Burma plate was uplifted by around 5 m, displacing huge amounts of water, which created the tsunami. The wave progressed at speeds of up to 1000 km hr$^{-1}$, principally to the east (towards Indonesia and Thailand) and the west (towards Sri Lanka, India and beyond) perpendicular to the plate margin fault line. The wave heights are estimated to have been only 1–1.15 m high initially, but they increased to 10–15 m as they neared the coast of Sumatra 15 minutes later, forced upwards by the frictional effects of the increasingly shallow seabed. To the west the wave heights were much lower with increased distance, reaching an average of 1.25 m on the east coast of Sri Lanka and only 0.75 m in the Maldives, partly due to the protection offered by the fringing coral reefs.

# Chapter 1

*Earth systems*

The tsunami caused over 180 000 deaths with, at the time of writing, over 125 000 people still missing. The vast majority of these, over 127 000, were in Indonesia, with Sri Lanka, India and Thailand also suffering greatly. In addition to the loss of life and the destruction of homes, 316 km of major roads, over 500 bridges and around 20 ports were either damaged or destroyed in northern Sumatra. The World Bank estimates that US$4.5 billion worth of damage and economic loss will have occurred in Indonesia, with up to 1 million Indonesians sinking into poverty because of the disaster.

The total financial costs are estimated at over US$14 billion, with only about US$3.5 billion covered by insurance.

## Fires

Large sections of San Francisco remained more or less intact after the earthquake of 1906. Yet when it was all over, much of the city had suffered some damage, and many areas were almost completely devastated. As in the same city in 1989, in Northridge in 1994, and in Kobe in 1996, fire was responsible for more loss of property than any other factor. In 1906 it was unavoidable. Fire protection was rudimentary and many buildings, particularly in the Marina district, were constructed of wood.

In more recent earthquakes, the largest fires have started primarily as a result of ruptured gas pipes and fallen power lines. Fire protection is expensive and the costs of providing full coverage to a city are prohibitive given that the number of machines and personnel likely to be necessary to cover a major earthquake far exceeds 'normal' daily requirements. This raises significant issues about the costs and benefits of such life-saving systems.

*Photograph 1.13 Fires in San Francisco, 1906*

*Physical environments* **AS Unit 1**

> **Examiner's tip**
>
> ■ None of these physical events can be regarded as *positive*. It is possible that the destruction of a human environment regarded as deeply unattractive might provide an opportunity for urban renewal, but any loss of life would outweigh this benefit.
>
> ■ The destruction caused by the Kobe earthquake is a useful reminder that MEDCs are not immune from heavy loss of life in such events, despite the well-known contrast between human and economic costs of hazards in MEDCs and LEDCs. The example of Kobe could be used in a 'However, it should be remembered that…' type sentence.
>
> ■ The Mexico City earthquake effectively illustrates the significance of time of day in determining fatalities and the importance of ground conditions. It also reinforces a point that could be made about Kobe, that despite our assumptions that modern technology is superior, we are still vulnerable. Given the survival of ancient buildings, we may be in fact more vulnerable than ever.
>
> ■ Mt St Helens is such a well-known case study that it might be beneficial to add a new dimension to it. One way would be to use it to illustrate how hazards are multifaceted; in the case of the Mt St Helens eruption this is well illustrated by the sequence of events: landslide, steam (phreatic) eruption, followed by volcanic eruption. It is also worth pointing out how floods, mudslides and gas clouds were responsible for almost all the loss of life in the eruption.

# The impacts of volcanoes

Volcanic hazards are complicated by the fact that volcanic areas are often attractive for settlement, thus increasing the population of the area likely to be affected. Unlike earthquakes, therefore, volcanoes do have positive impacts.

## Minerals and natural resources

Volcanoes bring many important and valuable resources, such as diamonds, copper and gold, to the surface of the Earth. For example, ancient deposits found in association with volcanoes are some of the most important sources of gold in Canada. Obsidian (volcanic glass) was the hardest substance known to Palaeolithic man and was much prized. Ancient sea-floor volcanoes contributed to massive accumulations of base metals such as lead, zinc and copper. In a longer time frame, volcanoes are also perhaps responsible for most of the water that exists on the surface of the Earth, having erupted huge amounts of water vapour into the atmosphere during the period immediately after the initial formation of the Earth, 3.5 billion years ago. Hence, they might be considered to have been absolutely necessary for life itself.

## Fertile soils

Volcanoes provide nutrients to the surrounding soil. Volcanic ash often contains minerals that are beneficial to plants. If the ash is very fine, it is broken down quickly and gets mixed into the soil. Areas that obviously benefit from volcanic soils include Indonesia, the Philippines, Japan and parts of Italy. The most notable example is probably the island of Java, with its high rural population density

# Chapter 1

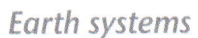

*Earth systems*

based on wet-rice agriculture, which has thrived for millennia in the volcanic soils of the island.

### Geothermal energy

Iceland is located on the north mid-Atlantic ridge. It is crossed by an active volcanic zone, from southwest to northeast, which stores an enormous potential source of thermal energy. Water running through the Earth's crust brings this geothermal energy to the surface and emerges as hot springs and fumaroles. Large numbers of hot springs are spread widely all over Iceland; these are classified as high or low temperature areas, depending on whether the temperature 100 m from the surface is above or below 150°C. This geothermal energy is used extensively for space heating, for electricity generation and for industrial applications. The exploitation of geothermal power for heating is largely responsible for the growth of Iceland's capital Reykjavik in the last 100 years. It is a good example of turning a potential hazard into something useful.

### Tourism

In modern Western culture volcanoes are seen as beautiful and threatening in about equal measure. In many MEDCs volcanic regions, both active and extinct, generate considerable interest from visitors, and this is exploited by local authorities wishing to generate employment in what are frequently quite remote regions. One obvious example is Iceland, which has capitalised on its tectonic activity in recent years in an attempt to diversify its economy away from fishing. Over 300 000 people visit the island each year and nearly 5% of the country's employment is now in hotels and restaurants. Organised trips to the many volcanoes, such as Hekla, and to witness the thermal activity at Geysir, are critical to the economy of the country, creating a demand for tertiary employment and increasing spending in the local economy.

> **Examiner's tip**
>
> - Unlike earthquakes, volcanoes can be said to have a positive impact. You can make the point that this positive impact can be treated as both long term (e.g. the development of a tourism industry) and medium term (e.g. creation of fertile soils), whereas the costs (losses) are almost all short term.

## Volcanoes as hazards

Volcanic hazards vary with the nature of the magma. Acidic magma (>66% silica content) produces explosive volcanic action, where the main risks are from pyroclastic flows, ash and lahars (volcanic mud flows). Basic magma (45–52% silica) produces effusive volcanic action, where the main risks are from ash and lava flows. There is a magnitude/frequency relationship for volcanic eruptions; unsurprisingly, larger eruptions occur less frequently. Sometimes the plumbing and the chemistry of a volcano are more complicated and so the volcano does not fit expectations. For example, Vesuvius, which is on a destructive boundary, has acidic explosive magma as one would expect. However, nearby Stromboli has relatively basic magma coming from a deeper source nearer the mantle.

A volcanic *hazard* refers to any potentially dangerous volcanic process. On the other hand, a volcanic *risk* is any potential loss or damage as a result of the volcanic hazard that might be incurred by people or property or which negatively impacts the productive capacity or sustainability of a population. Risk not only includes the potential monetary and human losses, but also includes a population's vulnerability.

Volcanoes may be associated with a range of related hazards. These include:
- pyroclastic flows (Mt Pelée)
- lahars (Nevado del Ruiz)
- ashfall (Hekla)
- lava flows (Haeimaey)
- torrential rain and floods (Pinatubo)
- earthquakes and tsunami (Krakatoa)
- gas asphyxiation (Lake Nyos, Cameroon)
- Jökulhlaup (Vatnajökull, Iceland)

In 2000 it was estimated that 500 million people were at risk from volcanic hazards. In the past 500 years, more than 200 000 people have lost their lives due to volcanic eruptions. The number of deaths in recent years runs at about 1000 per year, which is far greater than the number of deaths for previous centuries. The reason behind this increase is not increased volcanism, but an increase in the number of people populating the flanks of active volcanoes and the valley areas near those volcanoes.

## Human costs of tectonic hazards

The human impact of both earthquakes and volcanic activity is dependent upon:
- the severity of the event and its duration
- the population density of the affected region
- the preparedness of the population
   In common with all other hazards, this involves:
- human costs
- economic costs

These costs are related in that human loss of life has a profound economic impact, as well as social and psychological impacts.

The human costs can be sub-divided into primary, secondary and tertiary casualties.

### Primary casualties

Primary casualties are those people who are killed or injured directly as a result of the earthquake itself — whether by fire, by structural collapse or by other physical hazards related to the actual shaking. Primary casualties tend to be much higher in LEDCs than in MEDCs because of poor construction methods.

# Chapter 1  Earth systems

Table 1.2 The death tolls of some volcanic eruptions

| Date | Location | Death toll |
|---|---|---|
| 1815 | Tambora | 92 000 |
| 79 AD | Vesuvius | 16 000 |
| 1985 | Nevado del Ruiz, Colombia | 23 000 |
| 1980 | Mt St Helens | 57 |

Table 1.3 The death tolls of some earthquakes

| Date | Location | Death toll |
|---|---|---|
| 1556 | Shen-Shu, China | 830 000 |
| 1783 | Calabria, Italy | 50 000 |
| 1906 | San Francisco | 700 |
| 1989 | Loma Prieta, California | 40 |
| 1999 | Turkey | 15 000 |
| 2004 | Indian Ocean tsunami | 180 000* |

* with 125 000 missing

## Secondary casualties

Secondary casualties are people who survive the initial incident, but are injured during it, and who die because of insufficient resources and the inability to access effective and immediate emergency medical care. Once again, secondary casualties tend to be higher in LEDCs than MEDCs because of limited resources.

## Tertiary casualties

Tertiary casualties are those people who suffer from pre-existing medical conditions that are aggravated by the earthquake and those who become ill as a result of the post-disaster environment, largely through infectious diseases, and who may or may not die as a result. In LEDCs, the tertiary casualties are often the largest group.

# Economic costs of tectonic hazards

Economic costs can be evaluated as direct costs and indirect costs.

## Direct costs

Direct costs include the immediate costs of repairing damage caused by the event itself. For example, the estimated repair costs for the Kobe earthquake have been said to range between $95 billion and $147 billion.

## Indirect costs

Indirect costs include the loss of earnings brought about by the disaster. If the disruption is prolonged and the infrastructure complex, then they can become very substantial, running into billions of dollars. Increasingly, major natural hazards are causing secondary technological and industrial accidents and emergencies. Following the 1999 earthquake in Turkey, a fire in a major oil refinery greatly reduced refinery capacity for the entire country and threatened the safety and health of earthquake victims. These costs may run on for many years and, although real enough, they are more or less impossible to evaluate with accuracy.

> **Examiner's tip**
>
> - What makes tectonic activity hazardous is an important issue. The impact of volcanic eruptions is rising largely because there are more of us and we are, by and large, better insured. The frequency of the events is probably no greater now than it was a century ago. Hazards only become disasters when we 'get in the way'.
>
> - The 'costs' of loss of life and bereavement are calculated for insurance purposes through a macabre algebra that, amongst other things, multiplies earning potential by life expectancy. A point could be made about the 'costs' of hazards to illustrate the inherent problems of such calculations.

## Earthquake risk assessment

Earthquake records suggest that, over time, there have been more and more earthquakes with greater levels of destructiveness. The explanation for this is not that earthquakes have increased in frequency but that human populations have expanded into earthquake risk zones and human structures are increasingly expensive and vulnerable. Since earthquakes have a frequency/magnitude relationship, some areas of infrequent but very violent earthquakes have been occupied with little awareness of the dangers. The eastern seaboard of the USA is an example.

Where written records of earthquakes do not exist it is possible to use geological maps and soil maps to identify clues to past earthquake activity. Once the geological record is understood, areas of special risk can be mapped. Within earthquake zones, areas of high risk include steep slopes, sensitive soils and low-lying coastal areas.

Risk assessment also has to take the nature of the settlement and infrastructure into account. For example, medium-height buildings are more vulnerable than either tall or small buildings. Masonry buildings are more vulnerable than wooden or steel framed buildings, yet wooden buildings are more prone to fire risk. The location of mains services (electricity, water or gas) may have an impact on potential damage. The size and design of structures such as roads and bridges will have considerable impact on evacuation, emergency access and potential loss of life.

## Volcano risk assessment

Volcanic risk assessment includes (a) monitoring current levels of activity and (b) mapping evidence for destruction in previous eruptions. It is feasible to modify lava flows by damming, freezing (with water) and bombing, but the only realistic approach to living with volcanoes is avoiding high risk sites (e.g. mudflow tracks) and evacuating as necessary. It is obvious from the recent history of human settlement on the slopes of Vesuvius that avoidance of high-risk areas is a strategy that is widely ignored.

## Earthquake prediction

There are many potential indicators for earthquakes, but none has proved reliable. Key areas of research and monitoring for earthquake prediction include the following.
- P wave/S wave ratio — this ratio drops prior to a large earthquake.
- Warning activity — there is an increase in the number of small earth tremors before a major shock. These are known as **foreshocks**, and can be of different types.
- Water levels in wells — these rise or fall as the rocks are squeezed by the increasing strain before an earthquake.
- Radon gas levels in wells — this radioactive gas is squeezed out from rock pores by the build-up of strain.

# Chapter 1  Earth systems

- Recent research by Swedish geologists in northern Iceland found that levels of manganese, zinc and copper in basaltic rocks at a depth of 1000 m increased by over 1000% before an earthquake and fell rapidly afterwards.
- Changes in the electrical properties of rock — increasing strain causes the crystals in rocks to rearrange their structures. This may result in changes in the rock's electrical properties. In extreme cases this produces light displays (the so-called 'earthquake lights' phenomenon).
- Ground deformation occurs as rocks are strained. This can be surveyed on the ground by using laser ranging techniques or can be measured by accurate radar imagery from satellites. This method is still being developed and global positioning systems (GPS) are increasingly being used.
- Animals are often reported as having unusual behaviour before earthquakes. We can either monitor animal behaviour or identify the sensory cues they use (which may be subsonic vibrations, changes in magnetism or electrical changes) and monitor these cues directly.

Some earthquakes have been predicted with spectacular accuracy. Others have come with no obvious warning. Hence there is a certain public disbelief about the reliability of any predictions.

## Monitoring volcanoes

Methods used to predict volcanic eruptions include the following:
- inflation monitoring to check the bulging of the volcano as magma approaches the surface
- gas sampling, since changes in gas composition reflect activity levels in the magma underground
- geothermal monitoring from space to register changes in heat flow as magma approaches the surface
- seismic monitoring to 'listen' to the rising blobs of magma as they force their way upwards

The main difficulties in predicting volcanic activity include:
- magma may be generated from different depths and arrive at the surface by different routes
- the nature and chemistry of the magma changes over time and may be different from one phase of a volcano's life to another

An eruption takes place when the pressure of upwelling magma is no longer balanced by the weight and rigidity of the overlying rocks. Since these rocks are riddled with fractures and inconsistencies it becomes extremely difficult to predict:
- whether an eruption will take place
- when an eruption will take place
- where an eruption will take place

*Physical environments*  AS Unit 1

In fact, for much of the time we are making educated guesses. However, recent research by Bernard Chuoet into the existence of so-called long-period seismic events as a precursor to an eruption enabled the eruption of Popocatepetl in Mexico during 2000 to be successfully predicted and the population evacuated.

## Planning for earthquakes

The most logical assumption one could make when planning for an earthquake is that everything is destroyed. Imagine, for instance, that a magnitude 8 or greater earthquake strikes San Francisco and flattens everything. It is better to plan to have nothing available rather than to depend on things that might not be functioning. Even a 'moderate' earthquake of 6.0 magnitude is enough to severely damage many structures and disrupt life.

The impact of a projected 8.3 magnitude earthquake occurring near a major population centre has been predicted as follows.

- Severe damage will be widespread, perhaps ranging for hundreds of kilometres. Moderate damage could occur over 300 km from the epicentre.
- Surface waves produced by subduction zone earthquakes can last for several minutes, accentuating the effects of soil liquefaction, and so may cause extensive structural collapse, particularly of tall buildings.
- Infrastructure (communications, utilities, transportation systems) may be disrupted for at least 24 hours, perhaps longer.
- Landslides will occur in areas with significant slope hazard.
- Tsunamis will be generated, resulting in major damage both locally and at distant sites, frequently without any warning. The Pacific Ocean has a tsunami warning system, but there is currently no such system in the Indian Ocean. Lack of warning was a major contributor to the high death toll in the 2004 event.
- Injuries and death will come as a result of structural collapse, and some of the collapses will be buildings of critical importance: schools, hospitals, emergency operations and communications centres (e.g. police and fire stations). While many buildings are strengthened against earthquakes, few are capable of withstanding a magnitude 8 earthquake. Many homes will be destroyed, although newer structures (wood-frame, rectangular footprint) may survive with only moderate damage.

Regardless of the severity of the earthquake, the following assumptions can be made about all damaging earthquakes:

- they occur without warning, and pre-event response activity will not be possible
- the probability of the event occurring during non-working hours is greater than 3:1
- damage to sensitive communications systems will interfere with response management
- aftershocks are likely, and will cause additional damage, interfere with response efforts and cause unease among the population

# Chapter 1

## Earth systems

The key strategies that need to be adopted in order to reduce the impact of earthquakes lie in the hands of governments.

- Land-use zoning — the zones most at risk of damage from earthquakes should be used for activities that will suffer limited damage impact, such as open spaces, recreational areas, low density/low rise buildings.
- Building regulations — buildings can be constructed with features such as flexible steel frames, rubberised foundations, or counterbalance weights in order that they move with the ground rather than resisting the shaking (Figure 1.24).

*Figure 1.24 Building strategies for earthquake zones*

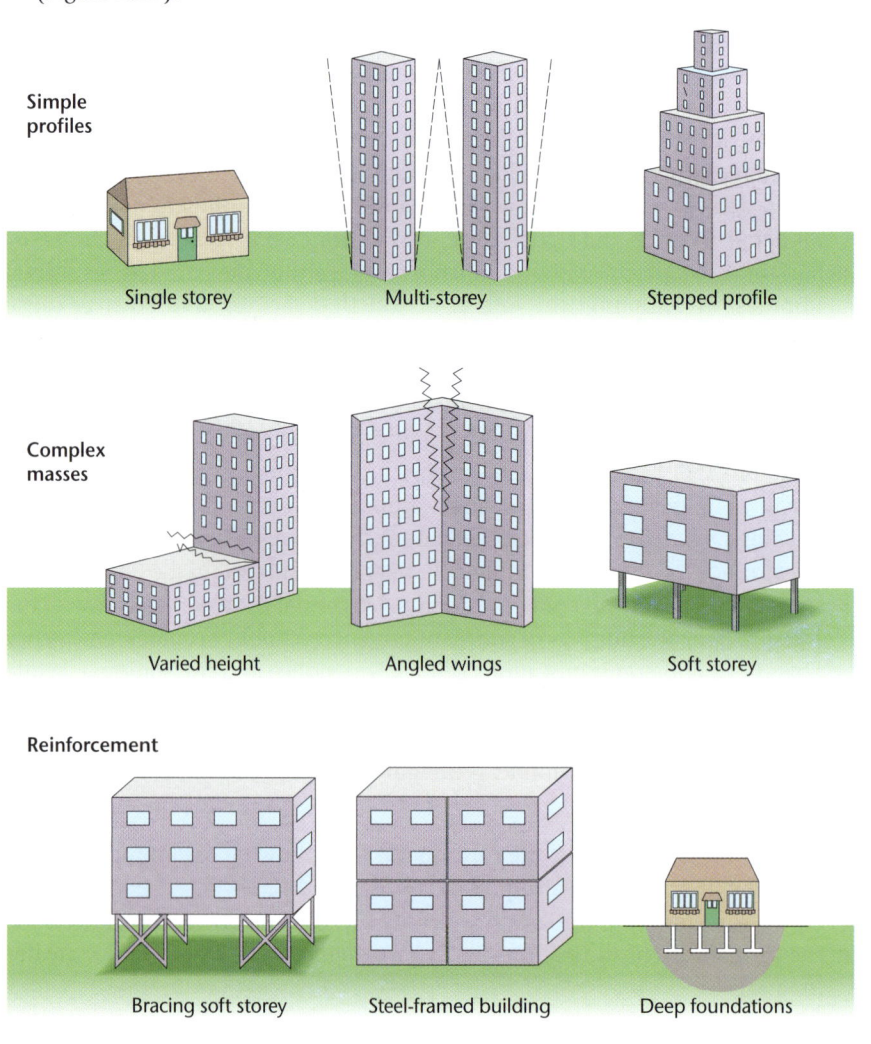

**Simple profiles**
Single storey | Multi-storey | Stepped profile

**Complex masses**
Varied height | Angled wings | Soft storey

**Reinforcement**
Bracing soft storey | Steel-framed building | Deep foundations

-- Extent of possible movement    ⌇ Line of possible separation during movement

42

*Physical environments* **AS Unit 1**

- Evacuation drills — those in areas at risk, especially schools, need to be trained to respond correctly in an earthquake. This can include knowing safe routes and rendezvous locations, having emergency supplies available and being aware of how to protect oneself, by hiding under a desk for example.
- Emergency service provision — sufficient medical and fire services should be provided, fully trained in how to deal with the hazards.

> **Examiner's tip**
>
> - Try to appreciate that the most obvious problem with planning for any disaster, whether tectonic or otherwise, is that disasters are, by their very nature, not something one can anticipate. The Japanese Earthquake Research Council has stated that there is a 90% chance a major earthquake will strike Japan in the next 50 years, but this number is essentially meaningless because it is based on averages rather than a measurement of the causal factors.
>
> - Recognise that because earthquakes are unpredictable and will occur with no warning, the risk is somewhat imprecise. As a general rule, people are not easily excited by non-specific risks that have no precise date attached to them. People are also reluctant to think about the things that may kill or injure them or their loved ones.
>
> - Be aware of the fact that emergency planning is expensive, and, with virtually no return on investment, it is difficult to encourage spending on preparedness and equipment. Although the question is one of 'pay now, or pay big later', it is often not seen in that light. Politicians and citizens, more concerned with short-term problems, do not see the potential threat as a major concern. There are few votes to be had in raising taxes to pay for improvements in a fire service, for example, that may not be needed in one's lifetime.

# Chapter 2

# Fluvial environments

# The hydrological cycle

## The global hydrological cycle

The **global hydrological cycle** is a closed system — it does not have external inputs and outputs. There is a fixed amount of water in the Earth–atmosphere system. This water can exist in different states: vapour, liquid and solid, although the proportions in each state can change over time.

Water can exist in **stores** and move between these stores in a series of **transfers**. Some of these transfers involve a change in state, while others are merely a movement of water from one store to another.

The stores are the atmosphere, the land and the oceans.

- **The atmosphere:** water can exist in the atmosphere either as water vapour (gas) or minute water droplets in clouds (liquid). However, this is a tiny fraction of the total water in the whole system (about 0.03%).
- **The land:** water may be stored in rivers and streams, lakes and reservoirs, intercepted by and/or stored in vegetation and stored beneath the surface in the soil or the bedrock (groundwater). Water generally exists in a liquid state in these stores. It can also exist as a solid in snow and ice, for example in glaciers and ice sheets. Stores on the land account for about 80% of the world's fresh water.

### Key terms

**Condensation** The change from a gas to a liquid as water vapour changes into water droplets.

**Evaporation** The change in state of water from a liquid to a gas.

**Evapotranspiration** The combination of evaporation and transpiration as water is transferred from vegetation and the ground surface.

**Groundwater flow** Water moving underground through pores, joints and underground streams and rivers.

**Infiltration** The movement of water from the ground surface into the soil.

**Percolation** The transfer of water from the surface or from the soil into the bedrock beneath.

**Precipitation** The movement of water in any form from the atmosphere to the ground.

**Surface runoff** The movement of water that is unconfined by a channel across the surface of the ground.

**Transpiration** The diffusion of water from vegetation into the atmosphere involving a change from a liquid to a gas.

- **The oceans:** the vast majority of water in the system, about 98%, is stored in the oceans, either in liquid form or as a solid in icebergs.

In total there is about 1385 million km$^3$ of water in the system. This amount cannot change, although the proportion held in each store can vary considerably. During the last glacial period, for instance, more water was held on the land in solid form as snow and ice and so sea levels were as much as 140 m lower than they are today.

The transfers that exist between stores are the following.

- **Evaporation** is the change in state of water from a liquid to a gas, usually from standing surfaces of water. Evaporation occurs due to the availability of heat energy. It is particularly important in the transfer of water from the ocean store into the atmosphere. Several meteorological factors influence rates of evaporation, including temperature, humidity and wind speed. Of these, temperature is the most important. Other less important factors include water depth, vegetation surface, surface colour and water quality.
- **Transpiration** is the diffusion of water from vegetation into the atmosphere involving a change from liquid to gas.
- **Evapotranspiration** is the combination of evaporation and transpiration. Evapotranspiration can account for 100% of precipitation disposal in arid areas and 75% in humid areas.
- **Condensation** is the change from a gas to a liquid as water vapour changes into water droplets. Condensation occurs when air is cooled to its dew point or when it becomes saturated by further evaporation into it. Condensation takes place around nuclei such as dust particles.
- **Freezing** is the change from liquid to solid as temperatures fall below 0°C.
- **Melting** is the change from solid to liquid as temperatures rise above 0°C.
- **Sublimation**, confusingly, is the term for changes from solid directly to gas and from gas directly to solid. Sublimation can be significant on ice sheets and glaciers.

All the above involve a change of state. The transfers below are movements without a change of state.

- **Precipitation** is the movement of water in any form from the atmosphere to the ground. Precipitation includes rain, snow, sleet, hail, frost and dew.
- **Surface runoff** (or overland flow) is the movement of water across the ground surface unconfined by a channel. Surface runoff occurs either when the rate of rainfall exceeds the rate of infiltration or when the soil is saturated.
- **Groundwater flow** (or base flow) is water moving underground through pores, joints and underground streams and rivers. The upper layer of the saturated zone beneath the surface is called the **water table**. The height of the water table varies seasonally in response to varying precipitation inputs. Rates of movement vary enormously, depending on the permeability of the geology, but it can take tens or even hundreds of years for water to be recycled in this way.
- **Throughflow** is the movement of water laterally through the soil through pore spaces, fissures and pipes created by roots and animal burrows. These movements are very slow compared to surface runoff. Typically throughflow

# Chapter 2
## Fluvial environments

occurs at speeds of 0.005–0.3 m h⁻¹, but in temperate latitudes it accounts for most of the transfer to rivers.
- **Infiltration** is the movement of water from the ground surface into the soil. Infiltration rates normally decrease with time during a period of rainfall until a more or less constant value, the **infiltration capacity**, is reached. Infiltration is affected by the duration of the rainfall, the existing soil moisture conditions, soil porosity, vegetation cover, raindrop size and slope angle.
- **Percolation** is the transfer of water from the surface or from the soil into the bedrock beneath.

Figure 2.1 shows the global hydrological cycle and gives figures for the transfers between the three stores.

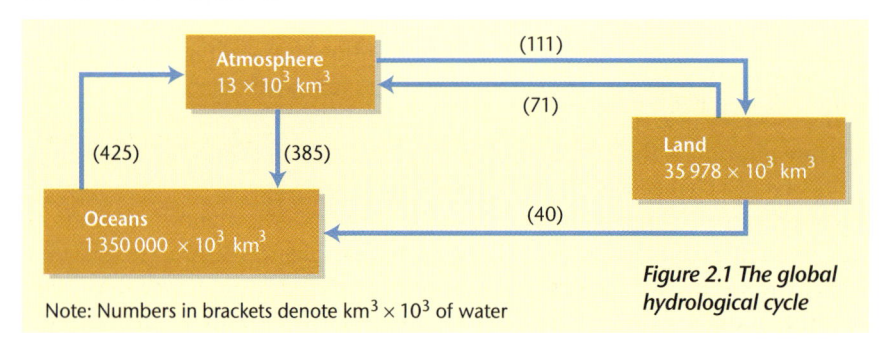

**Figure 2.1 The global hydrological cycle**

Note: Numbers in brackets denote km³ × 10³ of water

## The drainage basin cycle

On a smaller scale, the drainage basin cycle is an open system in that it has external inputs and outputs and the amount of water in the basin varies over time (Figure 2.2).

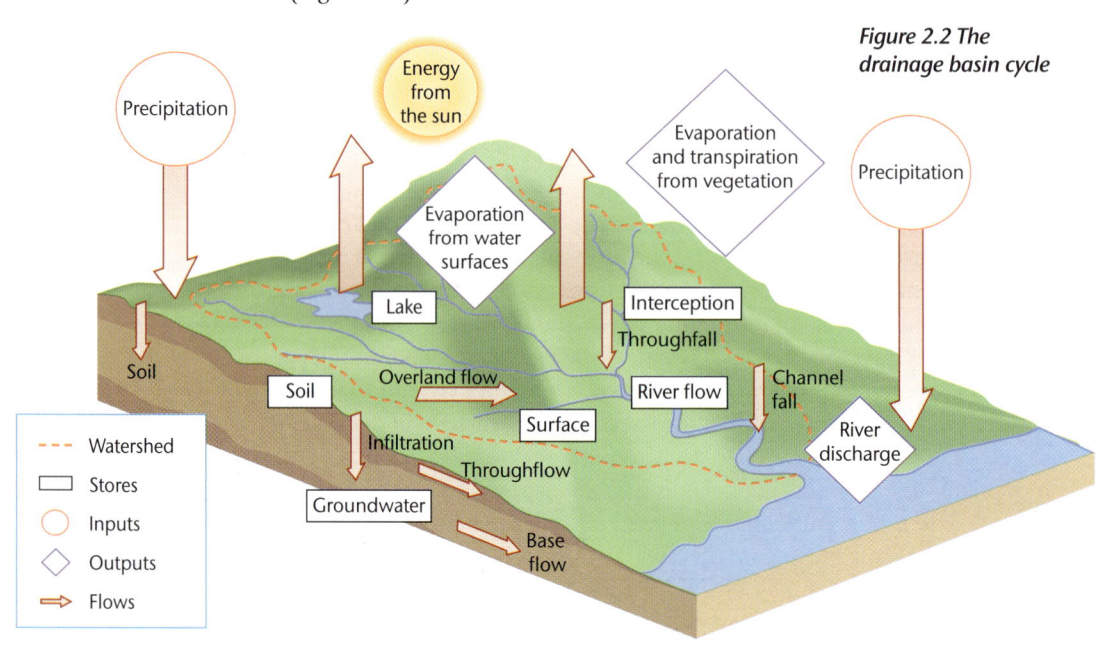

**Figure 2.2 The drainage basin cycle**

46

*Physical environments* **AS Unit 1**

A **drainage basin** can be defined as the area of land drained by a river and its tributaries; its boundary is known as a **watershed**. Drainage basins can be of any size, from a small area drained by a single stream to that of a major river system such as the Amazon or the Mississippi. Determining the size of a drainage basin depends only on identifying a measuring point somewhere on the stream or river. This may or may not be the river mouth or a confluence point where two rivers meet. The basin is the area of land that is drained through that particular point.

Interception occurs when vegetation prevents precipitation from reaching the ground surface directly. It has three components:

- **Interception loss:** once leaf surfaces trap falling precipitation, the water may be evaporated away or absorbed by the plant. This water never reaches the ground surface.
- **Throughfall:** water which either falls through gaps in the vegetation or drips from leaves, twigs or stems.
- **Stemflow:** water which trickles along branches and stems before running down the main trunk.

Interception loss is normally greatest at the start of a precipitation event as the capacity of the leaves to intercept is greatest when they are dry. As they become wetter the weight of water reduces surface tension and the leaves may bend downwards under the weight. Interception losses also vary with different types of vegetation. Losses are less from grasses than from deciduous woodland, while the greatest losses come from coniferous forests. This is because pine needles allow for large amounts of individual accumulation and the freer air circulation permits more evapotranspiration.

## The causes of rainfall

The key to understanding the causes of rainfall is to be able to explain cloud formation.

- The main reason that clouds form is that as air rises, it cools.
- As cool air can hold less water vapour than warm air, the air will eventually become saturated as it rises.
- Once the air is saturated, any further cooling leads to condensation as the air is unable to hold any more water vapour.
- Tiny water droplets form as condensation occurs around condensation nuclei.

This sequence of cloud formation occurs under three different trigger mechanisms that cause the air to rise initially. These are known as the three types of rainfall (Figure 2.3).

- **Orographic** occurs when air is forced to rise over a relief barrier such as a hill or a mountain range (also known as **relief rainfall**).
- **Frontal:** when bodies of air of different temperatures meet, the warmer air rises over the colder air because the warm air is less dense than the cold air (also known as **cyclonic rainfall**).

## Chapter 2 Fluvial environments

- **Convectional:** when the ground surface is heated by solar radiation, heat is transferred to the air near the ground, which is warmed, becomes less dense and so rises.

*Figure 2.3 The three types of rainfall*

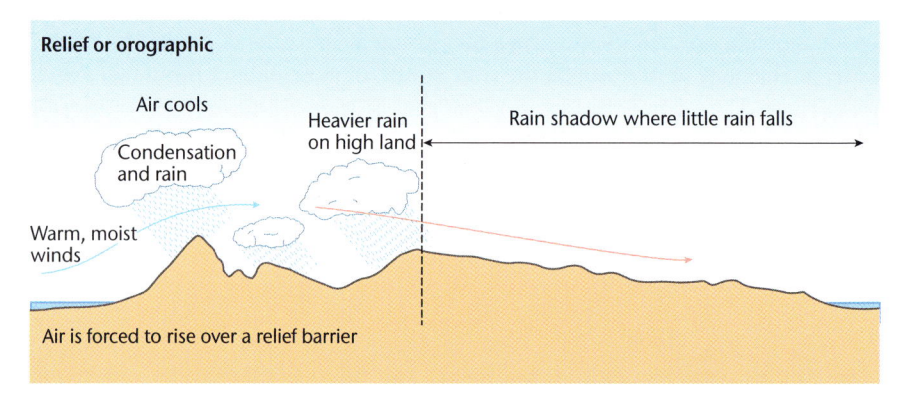

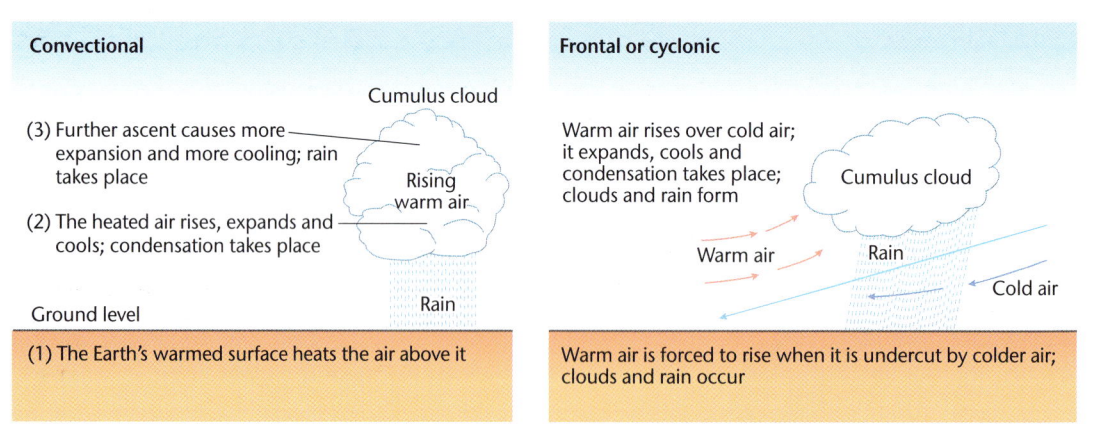

**Relief or orographic**

Air cools — Condensation and rain — Heavier rain on high land — Rain shadow where little rain falls — Warm, moist winds — Air is forced to rise over a relief barrier

**Convectional**

Cumulus cloud
(3) Further ascent causes more expansion and more cooling; rain takes place
Rising warm air
(2) The heated air rises, expands and cools; condensation takes place
Rain
Ground level
(1) The Earth's warmed surface heats the air above it

**Frontal or cyclonic**

Warm air rises over cold air; it expands, cools and condensation takes place; clouds and rain form
Cumulus cloud
Warm air  Rain  Cold air
Warm air is forced to rise when it is undercut by colder air; clouds and rain occur

Although the above mechanisms are usually known as the three types of rainfall, they only explain why air rises and clouds form. Cloud formation does not always lead to rainfall.

For rain to fall, the tiny water droplets in the clouds need to increase in size until they are heavy enough to fall under gravity through the rising air beneath. This is thought to happen when the droplets collide with each other and coalesce; turbulent currents in the air cause movement and mixing within the clouds.

> **Examiner's tip**
>
> - The key to explaining each type of rainfall is to know why the air rises initially, and then to describe the sequence of cooling, saturation and condensation that follows. The use of correct terminology helps.
>
> - You do not need to know what causes the other forms of precipitation, such as snow, hail and dew.

*Physical environments* **AS Unit 1**

# Rivers and their response to changing conditions

## River regimes

The **regime** of a river is its pattern of discharge, usually over a period of a year. The **discharge** is the volume of water passing a point on a river over a period of time. It is calculated by the following formula:

discharge = cross-sectional area × velocity

Discharge is usually expressed in cumecs ($m^3 s^{-1}$). The cross-sectional area must be measured in $m^2$ and the velocity in $m\ s^{-1}$.

### Examiner's tip

- You need to have studied two rivers with contrasting regimes. For each you should be able to *describe* the regime (peaks, troughs, seasonal variations, etc.) and *explain* the physical and human factors that influence the shape of the regime.

### Case study 2.1   The Colorado River, USA

The Colorado has its source in the Rocky Mountains and it flows over 1000 km to its mouth in the Gulf of California.

The annual regime is uniform for most of the year with a constant discharge of about 500 cumecs below the Glen Canyon dam. However, in April there is a sudden and rapid rise to a peak of 2000 cumecs in May, followed by an almost equally rapid return to the usual discharge in June (Figure 2.4).

The reason for the constant discharge for most of the year is that the Colorado is heavily managed and its discharge is regulated by a series of 11 major dams, including the Glen Canyon and Hoover dams. These are part of a multi-purpose management scheme that controls flooding, provides irrigation water, generates hydroelectric power and offers opportunities for leisure and recreation.

However, in April there is significant snow melt as temperatures rise above zero in the Rockies, and the dams are unable to cope with the sudden increase in water moving through the system. Over 2400 mm of snowfall can fall in the winter and when it melts, water moves rapidly down the steep relief from heights of over 4000 m near the source of the river. Once this water has been allowed to pass downstream, the dams are again used to regulate the flow for the remainder of the year.

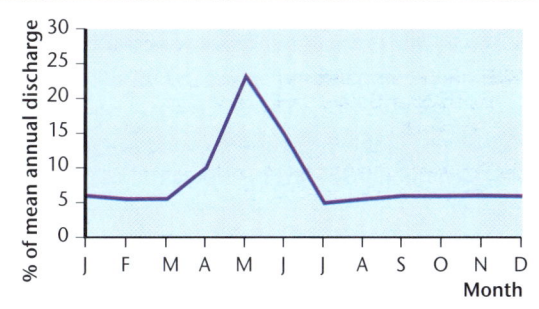

*Figure 2.4 The annual regime of the Colorado River*

*Photograph 2.1 The Hoover Dam*

# Chapter 2  Fluvial environments

> ### Key terms
> **River regime**  Variations in the discharge of a river over the course of a year.
>
> **Storm hydrograph**  A graph showing variations in the discharge of a river in response to a single rainfall event.

## Case study 2.2  The River Thames, UK

The Thames rises in the Cotswold Hills in Gloucestershire and flows about 300 km to its mouth in London.

The regime is fairly 'flat', with little variation through the year. At Reading (about halfway along the river), there is a peak of about 150 cumecs in February and a trough of about 10 cumecs in September. The discharge changes slowly and gradually between these two levels (Figure 2.5).

The main reason for the seasonal differences in discharge is the change in rainfall and temperature. In January, the Cotswold Hills receive about 100 mm of rainfall compared to about 60 mm in June. However, these differences become less significant further east in the basin until in London rainfall is very even through the year. Temperature differences influence the amount of water lost to evapotranspiration. The typical January temperatures of 8°C are much lower than the 18°C of June and so much of the summer rainfall does not reach the mouth to be recorded as discharge.

Human factors also play a part. There is more abstraction of water from the basin in the summer due to a higher demand from agriculture, for irrigation, and from domestic users. Approximately 60% of the water consumed in the Thames region comes from surface sources.

The slow rates of change in discharge are influenced by the physical characteristics of the basin. The Cotswold Hills are limestone hills, which means the rock is permeable. Much of the rainfall enters the ground and moves slowly towards the river. The same is true of the other upland parts of the drainage basin, such as the chalk hills of the Chilterns. Overall, two-thirds of the drainage basin is permeable rock. The basin also has a fairly low relief, with the highest parts only 300 m above sea level. This does not encourage rapid surface runoff and so, again, much of the water moves slowly towards the river either above or below ground.

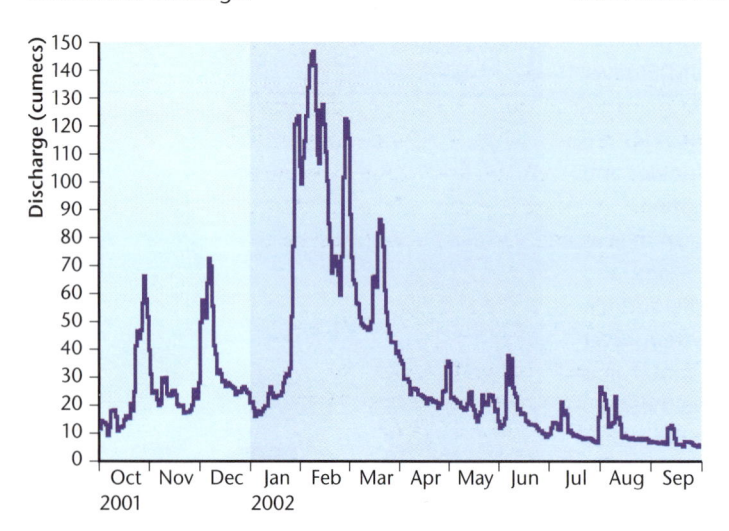

*Figure 2.5 The annual regime of the Thames at Reading, 2001–02*

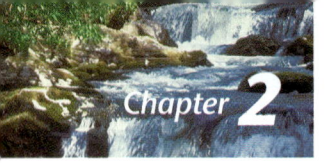

*Photograph 2.2 The Thames near Reading*

*Physical environments*  **AS Unit 1**

## Storm hydrographs

A **storm hydrograph** is a graph which records the changing discharge of a river in response to a specific input of precipitation. The main features of a typical storm hydrograph are shown in Figure 2.6.

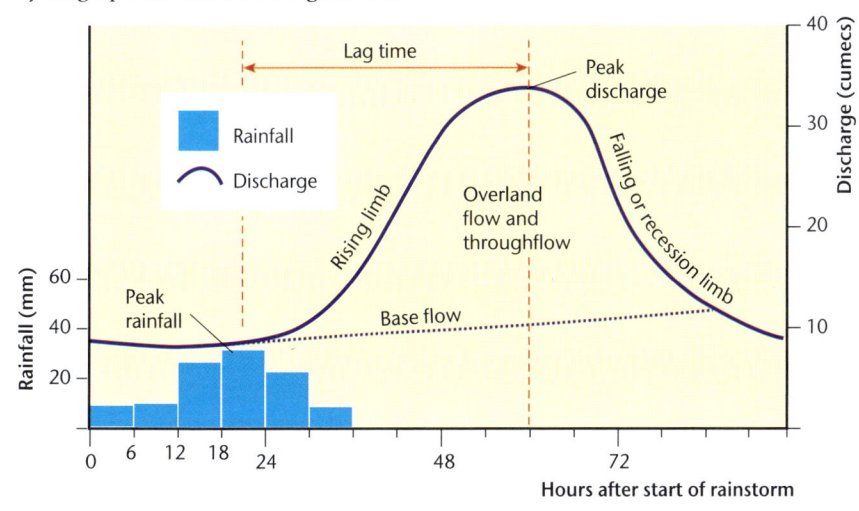

*Figure 2.6 Features of a typical storm hydrograph*

Before the input of a particular precipitation event, a river contains water that it has received from earlier inputs, but which has moved slowly through the basin, largely as underground, base flow movement. Once a new input is received, the discharge starts to rise — this is shown by the **rising limb** on the storm hydrograph. A peak discharge is eventually reached; this occurs some time after the peak of the input because the water takes time to move through the system to the measuring point of the basin. This time interval is known as the **lag time**. Once the input has ceased, the amount of water in the river starts to decrease; this is shown by the **falling or recession limb** of the storm hydrograph. Eventually, the discharge returns to its 'normal' level, fed by base flow.

The shape of a storm hydrograph may vary from event to event on the same river (temporal variation) and also from one river to another (spatial variation). The factors that influence the shape of a storm hydrograph are the same as those that influence the shape of a regime and are detailed in Table 2.1. It should be appreciated that these factors vary in importance and that they also interact with each other.

### Examiner's tip

- The key to explaining the shape of a hydrograph is to be able to explain the speed of transfer of water through the basin. Surface runoff movements are much more rapid than base flow movements and so produce a much 'flashier' shape.

- You do not need to have studied the hydrograph of a particular river/storm but you need to be able to apply your knowledge of the factors that influence hydrographs to an unfamiliar example.

- Try applying what you know to the storm hydrograph in Figure 2.7.

51

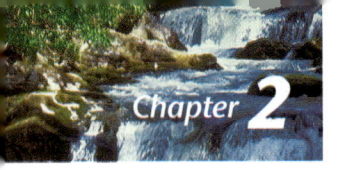

# Chapter 2 Fluvial environments

*Figure 2.7 Storm hydrograph of River Wyre, 19 December 1993*

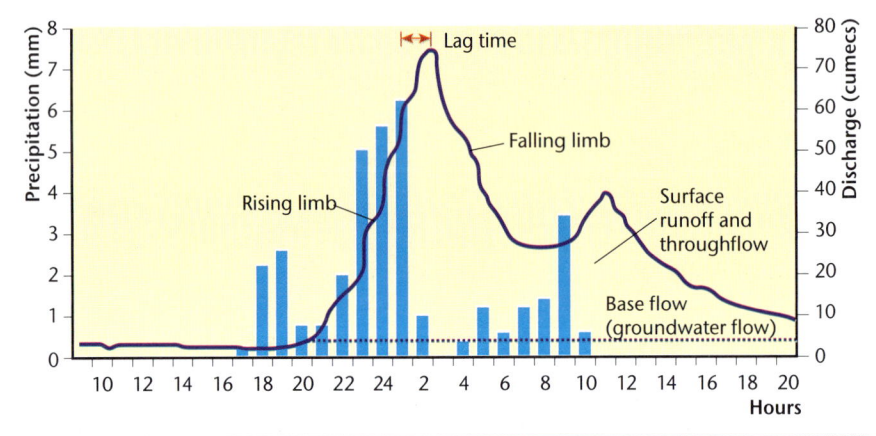

*Table 2.1 Factors that influence storm hydrographs*

| Factor | 'Flashy' river | 'Flat' river |
|---|---|---|
| (Description of hydrograph) | (Short lag time, high peak, steep rising limb) | (Long lag time, low peak, gently sloping rising limb) |
| Weather/climate | Intense storm which exceeds the infiltration capacity of the soil | Steady rainfall which is less than the infiltration capacity of the soil |
| | Rapid snow melt as temperatures suddenly rise above zero | Slow snow melt as temperatures gradually rise above zero |
| | Low evaporation rates due to low temperatures | High evaporation rates due to high temperatures |
| Rock type | Impermeable rocks such as granite, which restrict percolation and encourage rapid surface runoff | Permeable rocks such as limestone, which allow percolation and so rapid surface runoff is limited |
| Soil | Low infiltration rate, such as clayey soils (0–4 mm h$^{-1}$) | High infiltration rate, such as sandy soils (3–12 mm h$^{-1}$) |
| Relief | High, steep slopes which promote surface runoff | Low, gentle slopes which allow infiltration and percolation |
| Drainage density | High drainage density means many streams and rivers per unit area, so water will move quickly to the measuring point | Low drainage density means few streams and rivers and so water is more likely to enter the ground and move slowly through the basin |
| Vegetation | Bare/low density, deciduous in winter means low levels of interception and more rapid movement through the system | Dense, deciduous in summer means high levels of interception and a slower passage through the system; more water lost to evaporation from vegetation surfaces |
| Pre-existing (antecedent) conditions | Basin already wet from previous rain, water table high, soils saturated and so low infiltration/percolation | Basin dry, low water table, unsaturated soils and so high infiltration/percolation |
| Human activity | Urbanisation producing impermeable concrete and tarmac surfaces | Low population density, few artificial, impermeable surfaces |
| | Deforestation reducing interception | Reforestation increasing interception |

*Physical environments*  AS Unit 1

## The causes of flooding

Rivers flood when they exceed the capacity of their banks and water leaves the river channel to flow on to the surrounding area. The causes of flooding are varied and floods usually result from the combined effect of several factors, both human and physical. These factors are likely to include those listed in Table 2.1 as leading to a 'flashy' hydrograph. A 'flashy' catchment often experiences flooding, because a precipitation input moves rapidly through the basin and into the river. This causes water levels to rise rapidly and significantly.

> **Examiner's tip**
>
> ■ You need to have a good knowledge of the *causes* of a particular flood event. A good example is the event at Boscastle, Cornwall, in August 2004 (see Case study 2.3). You should be able to provide facts, figures and local detail, but you do not need to have knowledge of the *impact* of the flood on people, lives and property.

### Case study 2.3  Flooding in Boscastle, Cornwall

Boscastle is a small coastal village with a harbour on the north coast of Cornwall, near Bude (Figure 2.8). On Monday 16 August 2004 it experienced a dramatic flash event in the middle of the afternoon. Over 100 cars were washed out to sea, trees were uprooted and several buildings, such as Clovelly Clothing and Harbour Lights, were destroyed. Local residents and tourists had to be rescued from rooftops by helicopter.

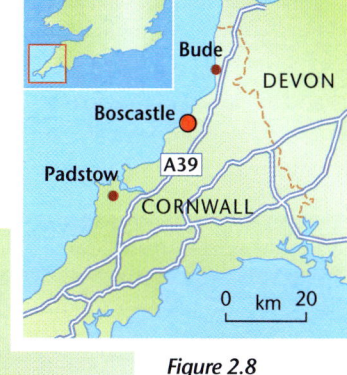

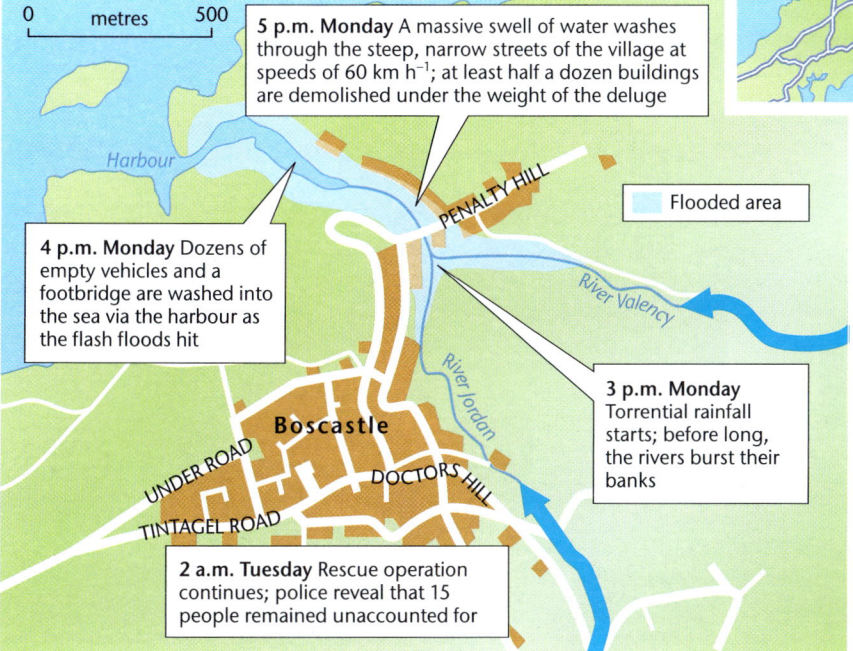

*Figure 2.8 The location of Boscastle*

**5 p.m. Monday** A massive swell of water washes through the steep, narrow streets of the village at speeds of 60 km h$^{-1}$; at least half a dozen buildings are demolished under the weight of the deluge

**4 p.m. Monday** Dozens of empty vehicles and a footbridge are washed into the sea via the harbour as the flash floods hit

**3 p.m. Monday** Torrential rainfall starts; before long, the rivers burst their banks

**2 a.m. Tuesday** Rescue operation continues; police reveal that 15 people remained unaccounted for

*Figure 2.9 Map of Boscastle, showing the position of the two rivers*

53

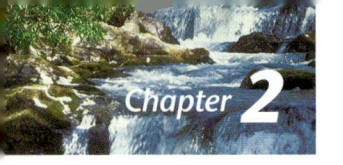

## Chapter 2  Fluvial environments

The flooding resulted from a combination of factors.
- The village sits at the output point of a small drainage basin, only 21.5 km² in area. Small basins tend to react rapidly to inputs, often giving rise to 'flashy' storm hydrographs.
- Two rivers converge in the village. The Jordan, fed by its tributary the Paradise River, and the Valency feed their combined flow into the centre of the village (Figure 2.9).
- The village sits in the base of a very steep valley, with heights dropping over 300 m in about 6 km. As a result, surface runoff is rapid under gravitational forces.
- The valley floor is narrow and there is no floodplain onto which excess water can spread.
- Higher than average rainfall during July and early August meant that the water table was high and the soil almost saturated. When the storm hit, the additional input of water was unable to infiltrate the ground and so rapid surface runoff occurred. River levels rose quickly, rising over 2 m in 1 hour, and the rivers overflowed.
- 130 mm of rain fell between noon and 6 p.m., with over 50 mm falling between 4 p.m. and 6 p.m. In comparison, the average rainfall in the area for the whole of August is 70–90 mm.
- An estimated 1 million tonnes of water rushed through the village.
- The intense rainfall was caused by two warm, moist airstreams converging and rising rapidly, one from the northwest and one from the southeast. Clouds formed up to a height of 40 000 feet.
- A high tide prevented water from easily leaving the basin at the river's mouth.
- A small landslide in the upper basin temporarily blocked the basin, leading to a build up of water that suddenly burst free.
- A tarmac-surfaced car park at the top of the village caused rapid surface runoff due to its impermeable surface.

An Environment Agency report early in 2005 concluded that the floods in Boscastle were a 1 in 400 year event. The agency further concluded that all the houses and buildings in Boscastle should be restored and that it is not too dangerous to live or work in the area.

Photograph 2.3 Damage caused by the flood in Boscastle

# Physical environments — AS Unit 1

## Downstream changes in river channel variables

The key river channel variables are:
- velocity
- discharge
- efficiency (hydraulic radius)
- channel shape

Theoretically, the variables listed above should change with distance downstream, as shown in Bradshaw's model (Figure 2.10).

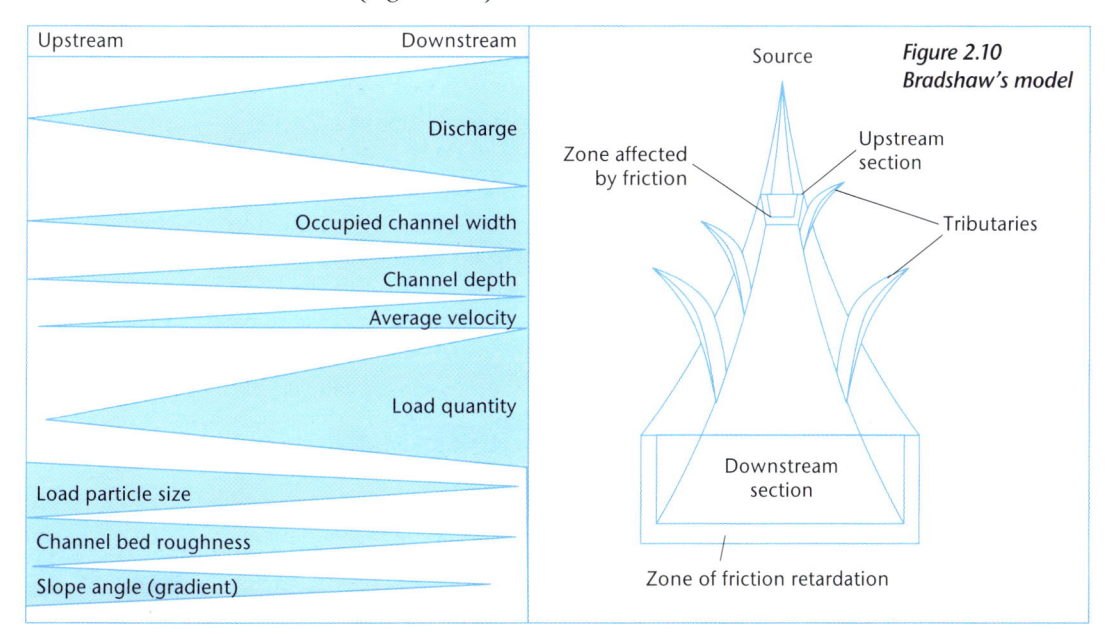

*Figure 2.10 Bradshaw's model*

## Velocity

**Velocity** is the speed of flow of water in the channel. It is usually expressed in metres per second ($m\,s^{-1}$). Velocity is influenced by gradient, channel roughness and efficiency. It increases downstream despite the fact that gradient normally decreases downstream. This suggests that the other factors, which are both related to friction, are more important. The channel shape typically becomes more efficient and so there is proportionally less friction from the banks and bed of the channel. Erosion also reduces the size of bedload material and decreases the roughness of the channel.

## Discharge

**Discharge** is the rate of transfer of water through the channel. It increases downstream as more water is added to the channel by tributaries and groundwater flow and because the shape of the channel usually becomes more efficient with distance downstream. As velocity is also a component of discharge, the higher velocities downstream also affect it.

### Key terms

**Discharge** The rate at which water moves through a channel.

**Efficiency** A measure of the ability of a channel to move water.

**Velocity** The speed of flow of water.

55

# Chapter 2 Fluvial environments

## Hydraulic radius

**Hydraulic radius** is calculated using this equation:

$$\text{hydraulic radius} = \frac{\text{cross-sectional area}}{\text{wetted perimeter}}$$

Hydraulic radius is a way of quantifying the efficiency of a channel containing moving water. The higher the hydraulic radius, the more efficient is the channel shape. The most efficient shape a channel can have is semi-circular. Although this is unlikely to happen in the natural world (except in streams flowing on the surface of clean glaciers), it does explain why pipelines are circular; they produce the maximum area for a given circumference, hence minimising friction. Hydraulic radius increases downstream as river channels are eroded and the width to depth ratio approaches the optimum of 2:1.

## Channel shape

**Channel shape** tends to become more efficient (indicated by a higher hydraulic radius value) with distance downstream. This means that proportionally there is less water in contact with the frictional effects of the bed and banks. Size itself is not the key, as large channels are not necessarily efficient, particularly if they are very wide compared to their depth (a high width to depth ratio).

Figure 2.11 contains some calculations showing the effect of channel shape on hydraulic radius.

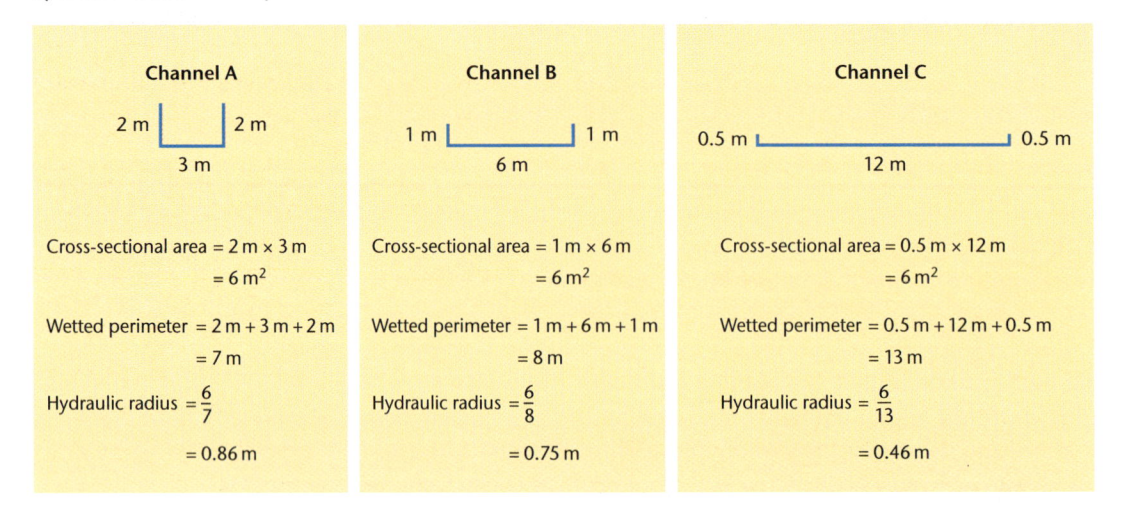

*Figure 2.11 Effect of channel shape on hydraulic radius*

| Channel A | Channel B | Channel C |
|---|---|---|
| 2 m ⌐ ⌐ 2 m<br>3 m | 1 m ⌐ ⌐ 1 m<br>6 m | 0.5 m ⌐ ⌐ 0.5 m<br>12 m |
| Cross-sectional area = 2 m × 3 m<br>= 6 m² | Cross-sectional area = 1 m × 6 m<br>= 6 m² | Cross-sectional area = 0.5 m × 12 m<br>= 6 m² |
| Wetted perimeter = 2 m + 3 m + 2 m<br>= 7 m | Wetted perimeter = 1 m + 6 m + 1 m<br>= 8 m | Wetted perimeter = 0.5 m + 12 m + 0.5 m<br>= 13 m |
| Hydraulic radius = $\frac{6}{7}$<br>= 0.86 m | Hydraulic radius = $\frac{6}{8}$<br>= 0.75 m | Hydraulic radius = $\frac{6}{13}$<br>= 0.46 m |

### Examiner's tip

- You need to have investigated the downstream changes in these variables on a particular river. This allows you to compare the actual changes with those expected and to explain why they differ. This may well be done by undertaking fieldwork on a local river. If not, Case study 2.4 below is suitable.

*Physical environments*  **AS Unit 1**

## Case study 2.4 The River Hooke, Dorset

The Hooke is a small tributary of the River Frome. It rises above the village of Hooke and has its confluence with the Frome at Maiden Newton, about 10 km downstream (Figure 2.12).

The changing channel variables were recorded by a group of students undertaking fieldwork at 10 sites, sampled systematically.

The variables show many similarities to Bradshaw's model (Figure 2.10), but there are some differences. By comparing the actual changes to those predicted by the model, these differences can be highlighted and explanations for them can be sought by looking at the distinctive characteristics of this particular river and its basin.

The Hooke rises at a spring where chalk meets clay at Toller Whelme, above the village of Hooke. The river is very small initially and flows through the village in a narrow, confined channel alongside the road.

Just above site 3 (Table 2.2), there is a significant input of water from a fish farm that obtains its water from another spring. This greatly increases the discharge.

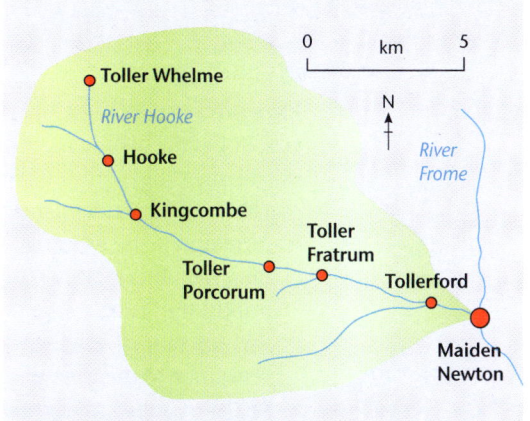

Below this the river becomes wider and shallower in the weak clay deposits of the valley floor. This

*Figure 2.12 Sketch map of the Hooke basin*

makes the channel shape less efficient and the river slows a little here.

Around sites 4 and 5, near Toller Porcorum, the river meanders across an increasingly wide floodplain, which also keeps mean velocity relatively low.

The final major change comes at Maiden Newton, site 10, where the Hooke joins the Frome. Below the confluence the river is much wider and deeper and with a much higher discharge.

*Photograph 2.4 Students collecting data on the Hooke*

| Site | Cross-sectional area (m$^2$) | Hydraulic radius (m) | Velocity (m s$^{-1}$) | Discharge (cumecs) |
|---|---|---|---|---|
| 1 | 0.01 | 0.01 | 0.22 | 0.002 |
| 2 | 0.02 | 0.08 | 0.25 | 0.005 |
| 3 | 0.13 | 0.19 | 0.64 | 0.083 |
| 4 | 0.39 | 0.14 | 0.43 | 0.168 |
| 5 | 0.24 | 0.28 | 0.44 | 0.106 |
| 6 | 0.56 | 0.35 | 0.78 | 0.437 |
| 7 | 0.75 | 0.30 | 0.95 | 0.713 |
| 8 | 1.52 | 0.23 | 0.68 | 1.034 |
| 9 | 2.70 | 0.20 | 0.82 | 2.214 |
| 10 | 6.50 | 0.24 | 1.20 | 7.800 |

*Table 2.2*

# Chapter 2

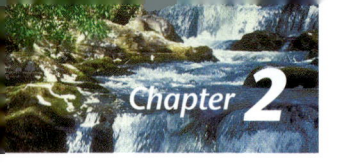

*Fluvial environments*

# River processes and their impacts

## The channel processes

### Erosion

**Erosion** is the wearing away and/or removal of material by moving forces. In the case of rivers, **fluvial erosion** results from the force exerted by the water as it flows downstream through the river channel.

The specific mechanisms of fluvial erosion include the following.

- **Abrasion** is the wearing away of the river bed and banks by the load carried in the river. It involves a mechanical impact and in most rivers it is the most important process. The rate of erosion depends upon the velocity of the flow, the amount of load, the shape, size and hardness of the particles, and the resistance of the river bed/bank material.
- **Attrition** occurs as the particles of the load collide with the river bed and banks, as well as with each other, and sharp edges are knocked off. This makes the particles smoother and smaller. The further the load is carried downstream the smoother and smaller the load becomes.
- **Solution** occurs when rivers flow over calcareous rocks, such as chalk and other limestones. Soluble minerals such as calcium carbonate are dissolved and carried away downstream in solution. For this to be an effective process, not only do the rocks need to contain soluble minerals but the pH of the water must be acidic.
- **Hydraulic action** involves air and/or water being compressed into cracks and crevices in the river bank. The cracks and crevices are widened by the force, causing the bank to weaken and collapse. This is only really effective in turbulent and high velocity flow.
- **Cavitation** occurs when bubbles form in the water as velocity increases and pressure falls. If the bubbles implode, tiny jets of water are evicted which can hit the bed and banks at speeds of up to 130 m s$^{-1}$. This process is significant at the base of waterfalls and in areas of rapids.

Fluvial erosion may act vertically, deepening the river bed and valley floor. This is particularly true in areas of high relief where fast-flowing rivers carry large angular bedload. Lateral erosion is common in winding and meandering rivers, where the highest velocity occurs on the outside of the bends. Erosion causes the bank to be undercut and to collapse, widening the channel and the valley floor.

### Key terms

**Deposition** The dropping and laying down of material.

**Erosion** The wearing away and/or removal of material by a moving force.

**Fluvial deposition** The dropping and laying down of material by water flowing in a channel.

**Fluvial erosion** The wearing away and/or movement of material by water flowing in a channel.

**Fluvial transportation** The movement of material by water flowing in a channel.

**Transportation** The movement of material by a moving force.

## Transportation

**Transportation** is the movement of material, in this case by the energy of flowing water. The river obtains its **load** from two main sources: material that has been washed or has fallen into the river from the valley sides and material eroded by the river itself from its bed and banks.

There are four main mechanisms by which rivers transport their load.

- **Solution:** minerals that have been dissolved in the mass of moving water. This type of load is invisible and the minerals will remain in solution until water is evaporated and they precipitate out of solution.
- **Suspension:** small particles of silt and clay can be carried along within the flow of the moving water. In most rivers this is the major element of the load and it accounts for the brown or muddy appearance of many rivers. Larger particles can also be carried in this way if the velocity is high enough to give the river sufficient energy. This is particularly common during times of flood.
- **Saltation:** a series of irregular movements of slightly larger material that is too heavy to be carried continuously in suspension. Turbulent flow may enable a particle to be picked up (entrained) and carried for a short while, only to be dropped back down. There may be a slight rebound on impact that enables water to get under the particle and lift it back up again.
- **Traction:** the largest particles in the load may be pushed along the bed by the force of the flow. Large boulders may undertake a partial rotation before coming to rest again. The flow may not be powerful enough for the movement to continue until some time later, particularly if the flow is turbulent.

## Competence and capacity

The load carried by a river can also be described in terms of its competence and capacity. Both of these are influenced by the velocity and, hence, the discharge of the flow.

The **capacity** of a river is a measure of the total volume of the material that it can carry. Research suggests that the capacity is proportional to the third power of the velocity. This means that if the velocity doubles, the capacity increases by eight times ($2^3$).

The **competence** of a river is the diameter of the largest particle that the river can carry at any given velocity. As fast-flowing rivers generally have greater turbulence and a greater ability to entrain material, there seems to be a relationship between competence and the sixth power of the velocity. For example, if the velocity doubles, the competence increases by 64 times ($2^6$).

## Deposition

**Deposition** is the laying down of material carried by the river. This occurs when there is a decrease in the energy of the river, caused by a reduction in velocity and/or volume. A number of situations exist in which deposition is likely to occur:

- where a river enters a lake or the sea and loses velocity as the moving water enters the static body of water
- where a decrease in gradient occurs, for instance at the base of a valley side where it meets the floor, causing a reduction in velocity

## Chapter 2 Fluvial environments

- when the volume of water decreases during a drought or other period of low precipitation
- when a river floods and spreads out on its floodplain; this increases friction and reduces velocity
- on the inside of a meander bend where water becomes shallower and slower moving
- when there is an increase in the load to be carried, perhaps following a landslide or other input of extra sediment
- where the river crosses an area of permeable rock which increases percolation and leads to a reduction in volume
- after a flood event when the volume and velocity both decrease

### The Hjulström curve

The Hjulström curve (Figure 2.13) shows the relationship between the three river processes (erosion, transportation and deposition) and velocity for particles of different sizes. Logarithmic scales are used on both axes, to allow a wide range of values to be plotted without losing accuracy and detail at the lower end of the range.

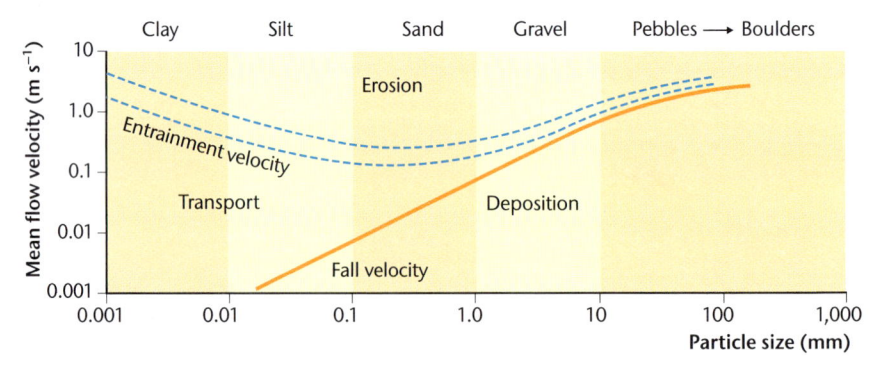

*Figure 2.13 The Hjulström curve*

The graph reflects the general principles that:
- river energy increases as velocity increases
- erosion requires more velocity than transportation for particles of the same size
- larger particles have a greater mass and so higher velocities are required for them to be eroded and transported
- deposition occurs sequentially as velocity decreases

Note, however, that the graph shows that the relationships are not linear. The entrainment and fall velocity curves flatten out at higher velocities. This is because of the sixth power law mentioned earlier.

The other anomaly is the relatively high velocity needed to entrain the smallest particles. This is because these very small clay and silt particles are cohesive and stick together, making them harder to pick up. They may also be electrically bonded and they have a small surface area on which the force of the water is operating.

It is also worth noting that once entrained, particles can be carried at lower velocities than those needed initially for entrainment to occur. However, this difference is very small at high velocities and so the particles may be deposited soon after they have been picked up.

Very small particles can be carried almost indefinitely until the river enters the sea. At this point the mixing of fresh and salt water causes particles to flocculate (join together) and settle. Material carried in solution is not dependent on velocity and so is carried out to sea.

There are weaknesses in the application of the Hjulström curve to describe these relationships:
- the flow velocity stated is the mean figure for the whole channel, whereas most entrainment takes place from the river bed where velocity is slower due to friction
- the graph relates to smooth channels, whereas most channels are irregular
- it can be argued that velocity itself is not the key factor in entrainment, but the drag exerted on the particles — this is dependent on other variables, such as water depth
- the density, shape and material of which the particles are composed are variables that the graph does not take into account

However, despite these weaknesses, the Hjulström curve provides a useful, simplified summary of these important relationships.

## Fluvial landforms

Fluvial processes shape the landscape by producing distinctive surface features or landforms.

### River valleys

River valleys have both long profiles and cross profiles.

The **long profile** shows the changes in the altitude of a river along its course from source to mouth (Figure 2.14). Long profiles generally have a smooth concave shape, with the gradient steeper in the upper part of the course and becoming progressively less steep towards the mouth. However, there will often be irregularities in this profile, such as waterfalls and lakes.

Where there are irregularities, these can be explained in terms of:
- **varying rock types**, particularly where resistant rocks produce waterfalls (see below)
- **natural lakes** or artificially created reservoirs, which cause a flattening of the profile
- **rejuvenation**, where a relative fall in sea level gives rivers increased energy and a greater ability to erode vertically, producing steps in the long profile, such as waterfalls or knickpoints

*Figure 2.14 The long profile of a river*

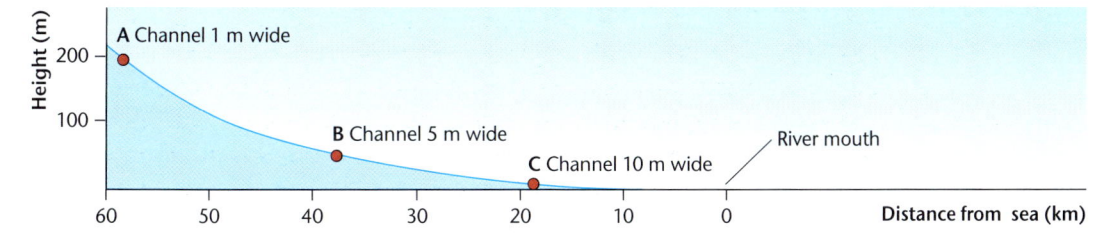

## Chapter 2 Fluvial environments

The **cross profile** is the view of the valley from one side to the other. Generally the cross profile changes with increasing distance downstream from the upper course, near the source, through the middle course to the lower course, near the mouth (Figure 2.15).

*Figure 2.15 Valley cross profile characteristics*

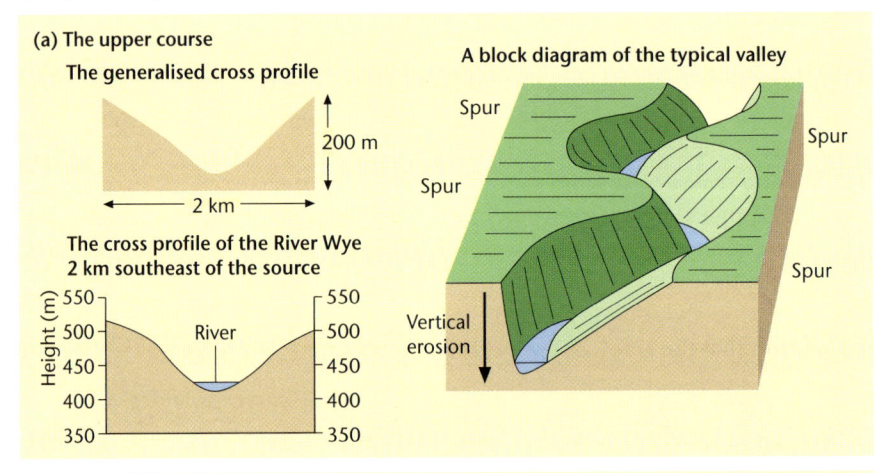

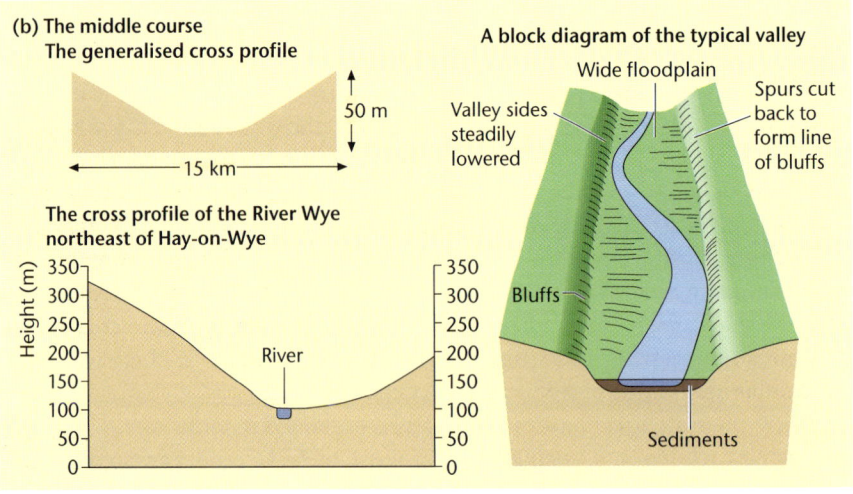

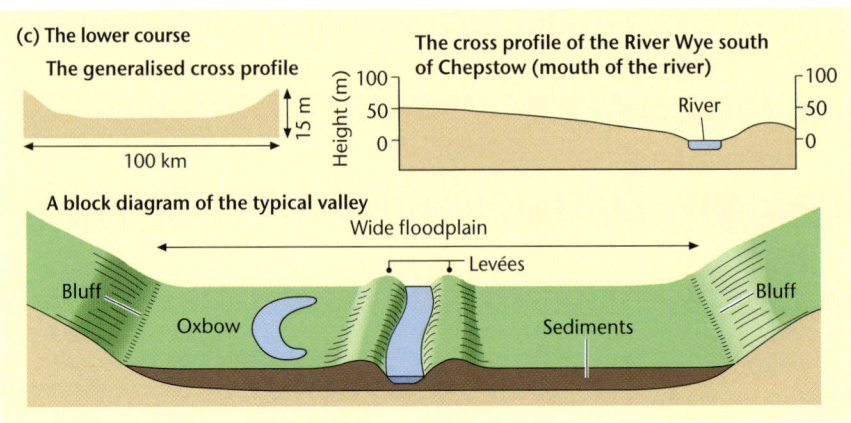

**(a) The upper course**
The generalised cross profile

A block diagram of the typical valley

The cross profile of the River Wye 2 km southeast of the source

**(b) The middle course**
The generalised cross profile

A block diagram of the typical valley

The cross profile of the River Wye northeast of Hay-on-Wye

**(c) The lower course**
The generalised cross profile

The cross profile of the River Wye south of Chepstow (mouth of the river)

A block diagram of the typical valley

62

Typically the characteristics of the profile in each part of the course are distinctive and can be explained.

- **Upper course:** a V-shaped valley which has steep sides and a narrow floor, often completely occupied by the river. This is the result of the dominance of vertical erosion. At this high altitude above sea level, the point at which the water is trying to reach the river has plenty of potential energy.
- **Middle course:** a much more open V-shape, with a wider valley floor and less steep sides. The valley floor may consist of a small floodplain with bluffs forming at the edge. This is the result of the dominance of lateral erosion as the river swings from side to side and is now at a lower altitude.
- **Lower course:** a wide, flat-floored valley consisting of a large floodplain. The valley sides may be difficult to locate. The dominant process here is deposition.

## Waterfalls

**Waterfalls** are steep steps in the profile of a river's course. They may be formed by:
- a band of resistant rock occurring on the river bed, leading to differential rates of erosion
- the river reaching the edge of a plateau
- the rejuvenation of the area, giving the river increased erosional power as sea level falls

Where rocks of differing resistance occur together, the weaker rock is eroded more rapidly and a step develops in the bed. The increased velocity gained by the water as it falls over the step further increases the rate of erosion of the weaker, downstream band of rock. Abrasion and hydraulic action in the increasingly turbulent flow cause undercutting and the formation of a plunge pool at the base of the fall. Eventually the overhanging, more resistant rock collapses due to a lack of support and the position of the waterfall retreats in an upstream direction. If this process is repeated over time, a gorge will be formed downstream of the waterfall (Figure 2.16).

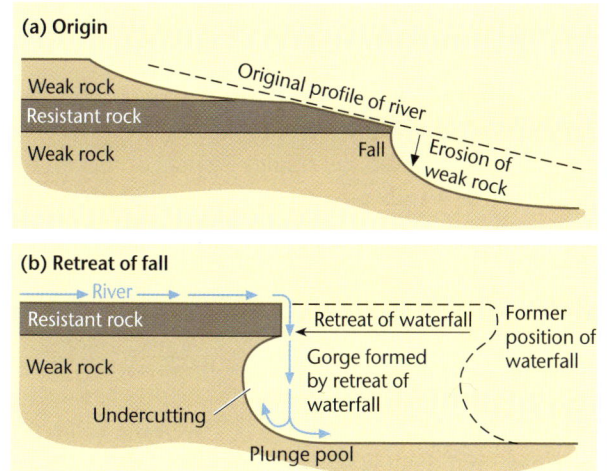

*Figure 2.16 The formation of a waterfall and gorge*

# Chapter 2 Fluvial environments

> **Examiner's tip**
>
> ■ Diagrams are particularly useful in examinations and you may be specifically invited to draw one to help your answer. They do not need to be works of art, but should be clear, neat and labelled or annotated appropriately in response to the demands of the question. A *label* simply identifies a characteristic of the landform, e.g. the layers of different rock, while an *annotation* is used to explain the characteristic, e.g. the weaker rock being eroded more rapidly, leading to undercutting.

### High Force waterfall, River Tees

The River Tees rises in the Pennines and flows eastwards to its mouth at Middlesbrough. In upper Teesdale an outcrop of resistant, igneous rock called the Whin Sill overlies weaker sandstone, shale and limestone. Undercutting, collapse and retreat have led to the formation of a 22 m-high waterfall with a gorge stretching 500 m downstream.

*Photograph 2.5 High Force waterfall*

## Rapids

Rapids are small steps in the long profile, usually found in the upper course. They are, in effect, miniature waterfalls and are also the result of bands of rock in the river bed of differing resistance. The bands are often at an angle rather than horizontal or vertical and the differences in resistance between different bands may be quite small, so the rates of erosion are not that dissimilar. They are sometimes the first stage in the formation of a waterfall. Most upper course rivers have rapids, even if they are only very minor features.

## Meanders

Meanders are sinuous bends in a river. Their formation has been the subject of much debate among geographers, and there is no single, simple explanation for it. Meanders occur on a variety of bed materials and in different parts of a river's course, and it should be noted that meandering is the normal behaviour of fluids and gases that are in motion. However, meanders seem to develop where channel slope, discharge and load combine to make meandering the only way that a river can use up its available energy throughout the section of the channel.

The formation of meanders does seem to be linked to the existence of alternating pools (deep parts of the channel eroded by turbulent currents) and riffles (shallow parts resulting from deposition of coarse sediment). This causes the flow of the river to swing from side to side, directing the line of maximum velocity (**thalweg**) towards one of the banks. Erosion then occurs by undercutting on that side and an outer, concave bank is created, called a **river cliff**. Deposition takes place in the slower-moving water on the other side of the river, leading to the formation of a convex bank, known as a **slip-off slope** or **point bar**. The cross-section of a meander is, therefore, asymmetrical (Figure 2.17).

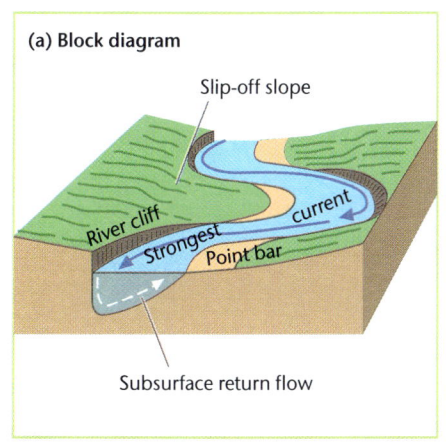

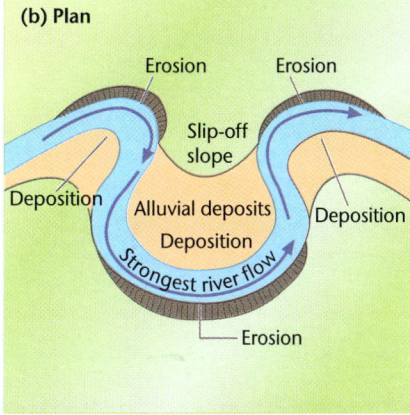

Figure 2.17
A meander

The process continues and so, although the width of the river does not increase, its sinuosity does. Sinuosity can be expressed as the ratio between the actual river length and the straight-line distance downstream. Meandering channels are usually classified as having a sinuosity ratio of greater than 1.5.

Once they have been formed, meanders are perpetuated by the surface flow of water across to the outer bank and the compensatory subsurface return flow to the inner bank. This corkscrew-like motion is known as **helical flow**. As a result of this, sediment eroded from the outer bank is transported downstream and deposited on the next inner bank.

A number of relationships have been observed by those studying meanders. Not all of these relationships can be easily explained.
- Meander wavelengths are usually 6–10 times the channel width and the square root of the discharge.
- Meander wavelengths are generally 5 times the radius of curvature.
- Meander amplitude is usually 14–20 times the channel width.
- Riffles occur at about 6 times the channel width.
- Meandering is more likely on gentle gradients.
- Meandering is more pronounced when bedload varies.

### Examiner's tip
- When you are offering possible explanations for meander formation, you should phrase your answer in tentative terms. Expressions such as 'it is possible that...' or 'it is thought likely that...' are useful in this context.

## Chapter 2 — Fluvial environments

Once meanders are formed, a number of possible developments can take place.
- The meander may migrate downstream as the point of maximum erosion is just downstream of the mid-point on the outside of the bend.
- The meander may migrate laterally as erosion is greater on the outside of the bend than on the inside. This widens the valley floor, helping to increase the size of the floodplain and forming bluffs on the valley sides.
- The meander may become so exaggerated that it is cut off, forming an **oxbow lake**. This is especially likely to happen during a time of high discharge, such as a flood, which causes the meander neck to be breached. The former channel is eventually sealed off by deposition. The stagnant water left in the oxbow will deposit its load and if the water evaporates or percolates, the oxbow lake will dry up and leave a meander scar (Figure 2.18).

*Figure 2.18 The development of an oxbow lake*

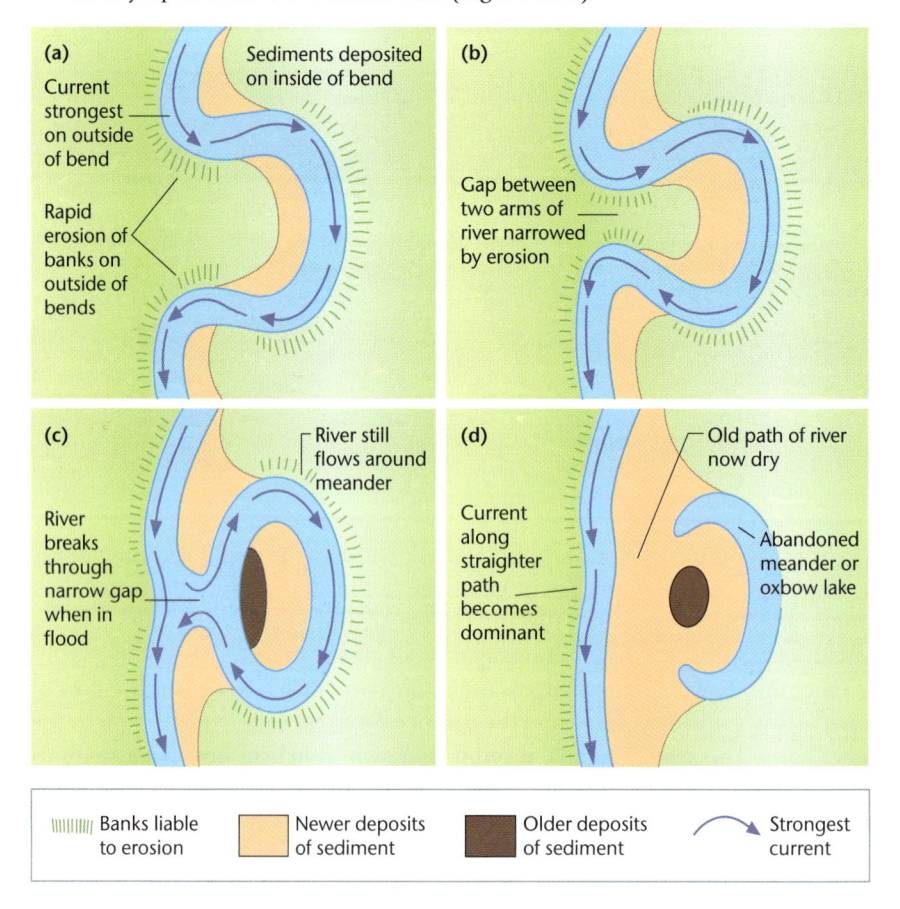

### The River Avon, Hampshire

The Hampshire Avon rises on the Marlborough Downs and flows south to its mouth at Christchurch.

In the lower course, south of Ringwood, there is a series of meanders and oxbow lakes as well as visible meander scars. The meanders have a typical wavelength of 50–100 m and the river itself is 5–10 m wide.

*Physical environments*  **AS Unit 1**

*Photograph 2.6 Meanders and meander scars on the River Avon, Hampshire*

## Braids

**Braids** are channels which are divided by small islands or bars (Figure 2.19). Braiding usually occurs as a result of a combination of different factors operating together:
- a moderately steep channel gradient (steeper than meandering channels)
- a large load of predominantly coarse sediment
- a highly variable discharge

Braiding begins with a mid-channel bar which grows downstream. Discharge decreases after a flood or a period of snow melt, causing the coarsest particles in the load to be deposited first. As discharge continues to decrease, finer material is added to the bar, increasing its size. When exposed at times of low discharge the bars can become stabilised by vegetation and then exist as more permanent features. The river divides around these islands and then rejoins. Unvegetated bars lack stability and often move, form and re-form with successive floods or high discharge events.

These landforms are particularly common in both periglacial and semi-arid areas.

# Chapter 2 Fluvial environments

*Figure 2.19 Braided channels*

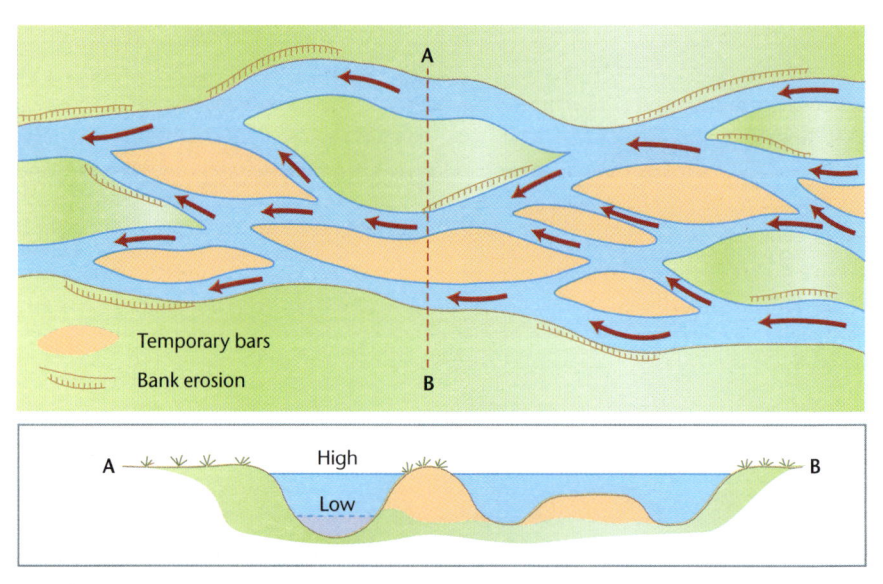

*Photograph 2.7 Braided stream in southern Iceland*

### Braided channels in southern Iceland

The coastal plain of southern Iceland is crossed by a series of braided streams and rivers. Their high sediment load is provided by both glacial erosion from the glaciers inland, such as Myrdalsjökull, and active weathering processes such as freeze–thaw. The discharge is very variable due to seasonal changes in temperature, giving rise to significant snow melt in the spring and summer as well as the occasional and dramatic jökulhlaup (glacier burst due to volcanic activity). Photograph 2.7 shows a heavily braided meltwater stream.

## Levées

**Levées** are low ridges or embankments that run parallel to river channels. They form during periods of high discharge when the capacity of the river is exceeded. As the water leaves the confines of the channel, it loses energy and deposits sediment on the bank. Sequential deposition occurs, meaning that the largest material is deposited first and the finer material is carried away onto the floodplain. Levées often have a layered structure as sediment builds up during successive flood events. On the Yangtze River in China, the levées are as much as 20 m above the level of the floodplain.

Levées can be artificially strengthened or increased in height in order for them to act as a defence against flooding. On some rivers they are completely artificial, traditionally made of earth and stone but nowadays from concrete. The levées alongside the Mississippi at St Louis, USA are 15 m high. (See also Case study 2.5, p. 71.)

## Floodplains

**Floodplains** are common features of river valley floors. In the lower course of a river the floodplain can be particularly extensive. They are formed by both erosion and deposition.
- Erosion on the outside of meander bends causes the meander to migrate and to widen the floodplain by eroding the edge of the valley.
- Deposition of river alluvium can build up the level of the floodplain by vertical accretion. Sediment carried by flood water spreading out onto the floodplain is deposited because it loses energy due to the high friction in the shallow water.
- Lateral accretion also occurs. This takes place on the inside of the meander bends as point bars which become abandoned as the meander migrates.

The floodplain of the River Ganges in northern India is over 300 km wide in places.

## Deltas

**Deltas** are relatively flat, low areas of sediment at the mouth of some rivers (Figure 2.20). They also occur in some lakes, where they are known as **lacustrine deltas**. An example is where the Volga River enters the Caspian Sea. Deltas develop when the supply of fluvial sediment is greater than the rate at which marine processes are able to remove it from the mouth of the river. When a river enters the sea, it suffers a significant loss of energy as its velocity decreases, and so deposition occurs. Deposition occurs sequentially and so the largest material is found on the landward side of the delta and the finest material on the seaward side. The mixing of fresh water and salt water causes the flocculation of clay particles, which are then heavy enough to be deposited.

Deltas often show a distinctive vertical structure:
- **bottom-set beds:** the lower part of the delta built outwards along the sea floor by turbid currents laden with fine sediment
- **fore-set beds:** inclined layers of coarse material, carried by traction and saltation and then deposited on top of the bottom-set
- **top-set beds:** a continuation of the floodplain and made of fine material that was carried in suspension

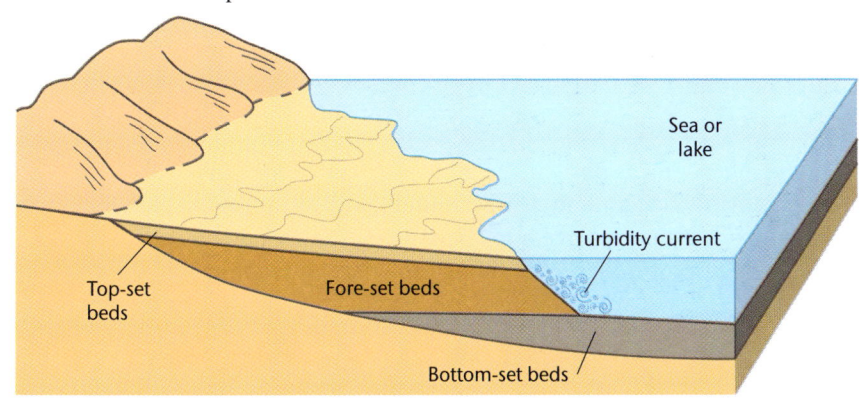

*Figure 2.20 The structure of a delta*

# Chapter 2

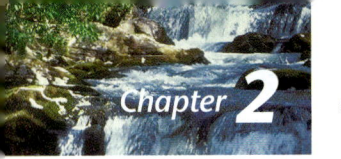

## Fluvial environments

Colonisation by vegetation can assist the process, as the roots of the plants further reduce the movement of water and increase rates of deposition. This is known as **bioconstruction**.

### The Mississippi delta

Deltas vary in form, from the curving shoreline of arcuate deltas, such as the Nile, to the projecting bird's foot delta of the Mississippi (Figure 2.21). The Mississippi delta covers an area of 26 000 km² and the river carries 450 million tonnes of sediment annually. At the delta the river has divided into a series of distributaries, often formed by crevasse splaying as levées are breached. Each of these distributaries deposits at its own mouth, causing a seaward extension that has increased over recent times. This growth can be traced by studying historical maps, which show that in West Bay there was 16 km of growth between 1839 and 1875 (an average growth of over 440 m per year). However, rising sea levels are causing higher rates of marine erosion and the delta is now retreating at a rate of 25 m per year.

*Figure 2.21 Types of delta*

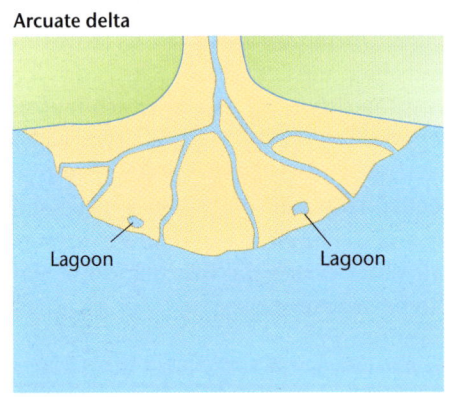

Arcuate delta — Lagoon, Lagoon

Bird's foot delta

### Examiner's tip

- You should be able to name and locate an example of each of the landforms listed here. You do not need to be able to give a detailed case study, but a small amount of specific information about the location is useful. This information could be a specific rock type, some indication of the dimensions of the landform or perhaps the name of a tributary or a settlement.

## Managing river processes

Rivers are often a focus for settlement because they offer many opportunities for human activity, such as transport, water supply and flat land for building. This often means that the river needs to be managed so that humans can continue to utilise these opportunities, even though the river is dynamic and the situation changes.

The need for the management of river processes might include the prevention of bank erosion to protect riverbank settlements, or the trapping of river deposits to add fertility to the soil on the floodplain.

The methods used include hard engineering solutions such as concrete revetments, and soft engineering methods such as planting appropriate vegetation (an example is willow) to help stabilise eroding banks.

> **Examiner's tip**
>
> ■ Do not confuse this topic with managing flooding! Some of the methods may be the same and schemes on some rivers may address both issues, but the emphasis here is on processes such as erosion and deposition.

## Case study 2.5 Managing river processes on the Mississippi

The Mississippi River in the USA is a major river system that drains almost one-third of the country. It has a number of major tributaries, such as the Missouri and the Tennessee.

The Mississippi provided an important natural route for cotton growers upstream to export their cotton via New Orleans at the river mouth. To ensure that the required 9-foot depth of water existed for the paddle steamers used in the 1930s, the river was frequently dredged to remove deposited sediment from the centre of the channel. In the late 1960s, as much as 63 000 m³ of sediment was dredged per kilometre each year. This was time consuming and costly and so an alternative method was sought. Wing dykes were constructed, especially along the Greenville reach of the lower Mississippi, from one side of the river bank extending out into the channel. These trapped sediment behind them, forced the water in the channel to the other side and made it flow at a higher velocity. The increased volume and velocity led to increased rates of erosion and the natural deepening of the channel on the other side.

To shorten the journey time for boats and barges, many of the Mississippi's meanders were straightened, reducing the distance by over 300 km. However, the river kept trying to meander and erosion was occurring on one side of the channel. To prevent this, a whole series of concrete mattress revetments were laid on the bank, at a cost of over $1 billion.

Photograph 2.8 The Mississippi River, showing levées and revetments

# Chapter 2
## *Fluvial environments*

## Examination question

Study Figure 2.22, which is a flow diagram of the passage of water through a drainage basin.

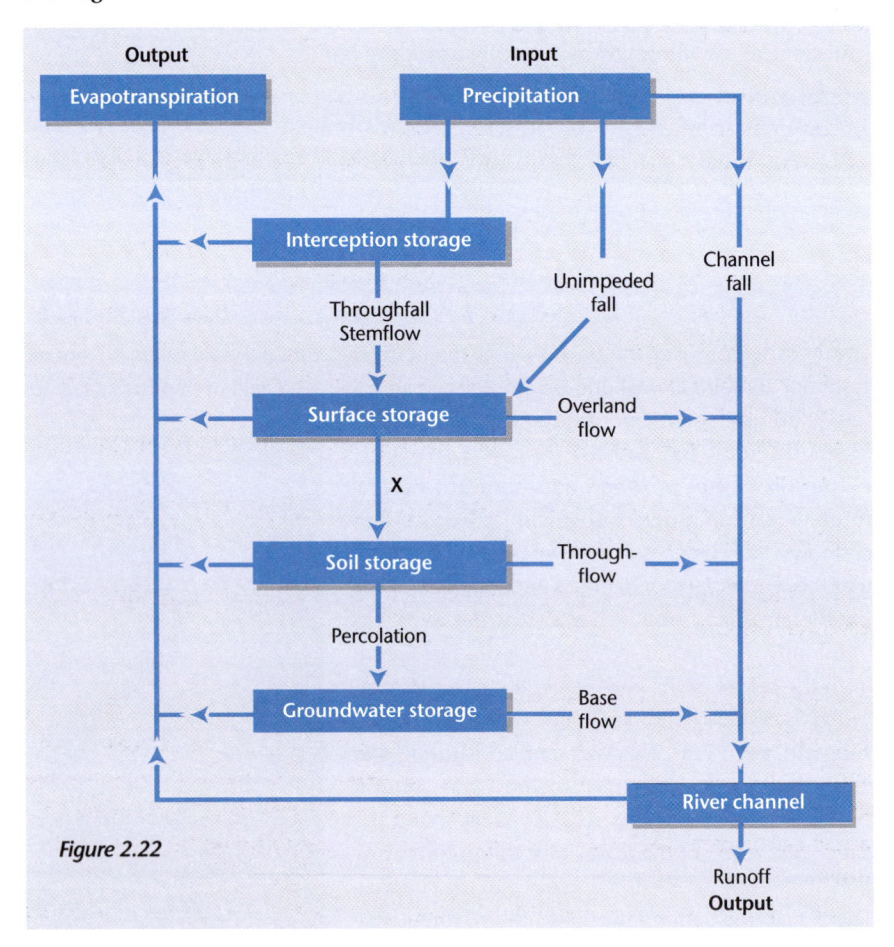

**Figure 2.22**

*[Flow diagram showing: Inputs (Precipitation) flowing via Unimpeded fall and Channel fall; through Interception storage (with Throughfall/Stemflow), Surface storage (with Overland flow), Soil storage (with Throughflow), Groundwater storage (with Base flow via Percolation), to River channel; Output: Evapotranspiration and Runoff. Flow X between Surface storage and Soil storage.]*

**a** With reference to the diagram:
  (i) Define the term 'drainage basin'. (2)
  (ii) Identify flow X. (1)
  (iii) Describe the process of throughflow. (2)
  (iv) Why is the drainage basin cycle referred to as an open system? (2)
**b** Explain how rainfall can be produced by frontal processes. (4)
**c** Name and outline *three* processes by which rivers erode. (3)
**d** With reference to a named and located example, describe the appearance of and explain the formation and development of a waterfall. (6)

*Total = 20 marks*

*Physical environments*  **AS Unit 1**

# Synoptic link
## The use and management of the hydrological cycle

- The reasons for and methods of groundwater and river management in countries at different states of development.
- Decision-making issues related to management of the hydrological cycle.

## Reasons for groundwater management

The reasons for the management of water resources are self-evident.
- Water is essential for human existence: for drinking, the irrigation of crops and the provision of power.
- There is frequent competition between consumers of water, both within nations and between nations.
- Growing populations in many LEDCs pose special problems for the future demand on water resources.
- Population movements in some MEDCs (e.g. the USA) have altered the pattern of demand within those countries.

Two issues dominate water management: the quantity and the quality of this essential resource.

## Groundwater management

Groundwater is water that is found underground in gaps and spaces in soil, sand and rocks. The area where water fills these spaces is called the saturated zone. The top of this zone is called the water table. The water table may be only a few centimetres below the ground's surface or it may be hundreds of metres down.

Groundwater can be found almost everywhere, even in arid regions. The water table may be deep or shallow; and may rise or fall both seasonally and in the longer term. Heavy rain or melting snow may cause the water table to rise, or an extended period of dry weather may cause the water table to fall. Groundwater moves slowly through layers of soil, sand and rocks known as aquifers. The speed of groundwater flow depends on the size of the spaces in the soil or rock and how well the spaces are connected. Aquifers typically consist of gravel, sand or sandstone, or fractured rock, like limestone. Water in aquifers is brought to the surface naturally through springs, or can be discharged into lakes and streams. Water can also be extracted through a well drilled into the aquifer. A well is a pipe in the ground that fills with groundwater. This water can be brought to the surface using a pump. Shallow wells may go dry if the water table falls below the bottom of the well. Artesian wells rely on natural pressures that force the water up and out of the well.

## Chapter 2 Fluvial environments

Groundwater supplies are replenished, or recharged, by rain and snow melt. In some areas of the world, people face serious water shortages because groundwater is being used faster than it is naturally replenished. In areas where material above the aquifer is permeable, pollutants can sink into the groundwater. Groundwater can be polluted by landfills, septic tanks and leaky underground gas tanks, and from the overuse of fertilisers and pesticides. If groundwater becomes polluted, it is no longer safe to drink.

## Groundwater management in MEDCs

In many MEDCs the issue of water contamination is at least as important as that of water supply. The following is a list of contaminants found in groundwater in the state of Michigan, USA.

*Organic chemicals*
- industrial solvents (e.g. trichlorethylene (TCE), vinyl chloride)
- petroleum products (including benzene and gasoline)
- pesticides

*Inorganic substances*
- nitrates
- fluorides
- arsenic
- brine/salt
- sulphates
- chlorides

*Other*
- metals
- radioactive materials (some occurring naturally)
- bacteria

### Examiner's tip

- It is important to appreciate that in arid regions of the world, both in MEDCs and LEDCs, the issue of groundwater management has a growing significance. Neglect of this resource was almost universal until 20 years ago, and only now is the rate of groundwater depletion being recognised. Groundwater has been the most neglected of all resources.

## Water supply and quality in an MEDC

The southwest of the USA has experienced rapid growth in population in the last 50 years, largely through in-migration. Much of this region is arid, with precipitation levels at less than 300 mm per year. One of the greatest and most serious issues facing the region is the availability, use and quality of groundwater. In large sections of the west, groundwater is the only dependable source of water for agriculture and domestic consumption. Yet many of the aquifers are being depleted at a rate that will suck them dry within a century. The Ogallala Aquifer underlies approximately 360 000 km$^2$ in the Great Plains region, particularly in the High Plains of Texas, New Mexico, Oklahoma, Kansas, Colorado and Nebraska (Figure 2.23). The depth of the aquifer from the surface of the land varies from region to region.

Use of the aquifer began around 1900, and since the Second World War reliance on groundwater from the aquifer has steadily increased as the population of the area has grown. Abstraction has now greatly surpassed the aquifer's rate of natural recharge. Some places overlying the aquifer have already exhausted their underground supply as a source of irrigation. Other parts have more favourable saturated thickness and recharge rates, and so are less vulnerable.

What is certain is that dependence upon groundwater in many areas will only increase in the future. This dependence is already having serious consequences for small towns on the Great Plains. Faced with growing costs associated with deeper wells and the need for ever more advanced technology for extracting water, these towns find they lack the resources to maintain current agricultural practices.

*Figure 2.23 The Ogallala Aquifer, USA*

## Water supply and quality in an LEDC

Groundwater is a vital source of domestic water supply in most LEDCs. In India, roughly 80% of rural water supply for domestic uses is met from groundwater. Wells in villages and towns free women, who do most of the agricultural work, from long daily walks to fetch water from springs or rivers for livestock and domestic uses. This frees time and labour for other, more productive, activities. Furthermore, since water no longer has to be carried over long distances, more is used. This can have major health benefits. In addition, because of the filtering nature of the soil and frequent long residence time underground, groundwater is commonly much cleaner than surface water sources.

Access to groundwater is a key resource for economic development. In areas dependent on irrigated agriculture, the reliability of groundwater sources and the high crop yields generally achieved as a result often enable farmers with small landholdings to increase their income. In India small and marginal farmers (those with less than 2 hectares) own 29% of the agricultural area. Their share in net area irrigated by wells is, however, 38.1% and they account for 35.3% of the tube wells fitted with electric pump sets.

### Examiner's tip

- Groundwater is a useful example of a resource that has been neglected in both MEDCs and LEDCs.

## River management

### Irrigation

Rivers have provided water for irrigation for at least 6000 years. Even today, in an era of piped water, the alluvial floodplains of the world are the main centres of human population and activity. The complex irrigation systems of both Chinese and Indian agriculture gave rise to well-organised states and empires with highly sophisticated management systems before Europeans had even evolved nation states. China's irrigated areas increased from 27 million ha in 1957 to about

# Chapter 2

*Fluvial environments*

51 million ha today, or 54% of the total cultivated area. Water availability is a factor limiting the policy goal of bringing an additional 20 million ha under irrigation, half in new areas and half by increasing efficiency in existing areas. Overexploitation of groundwater sources and growing competition for surface water, particularly in the North China Plain, are major challenges. About 20% of China's irrigated areas suffer the effects of water-logging or soil salinity to varying degrees.

*Photograph 2.9 Irrigated farmland in China*

## Flood management

Historically, floods have been viewed as a major obstacle to economic development. Attempts to control rivers have a long and somewhat chequered history. If the destruction of wetlands and forests can unintentionally increase river flooding, what about intentional building in rivers and on floodplains? In the USA, as in many other countries, floodplains have been a focus of development. People need to live near rivers, for agriculture, water, transportation and waste disposal.

The US Army Corps of Engineers has built a range of navigation and flood-control structures in many major American rivers:

- dams create lakes for irrigation, navigation and flood control
- levées (raised river banks built parallel to the flow) constrain the river, helping form deep navigation channels and offering some protection to land on the floodplain (see Photograph 2.8)
- wing dykes built at right-angles to the direction of flow on one side of a channel trap sediment and force water towards the other side of the channel, where it promotes bed scour, making navigation easier (see Case study 2.5).

Throughout the twentieth century, the corps built and maintained an extensive system of locks, dams and levées on the Mississippi and Missouri Rivers, rivers which drain most of the USA east of the Rockies.

To farmers, transport companies and property owners, levées are critical for taming rivers. But to environmentalists, they are artificial restraints that encourage development and farming in the floodplain. The natural role of a floodplain is to carry excess water during periods of heavy runoff, but when the floodplain is walled off behind levées, the artificially narrow river must rise higher to compensate for its lack of ability to spread out laterally. While levées do protect farms and towns during minor floods, the Mississippi flood of 1993 demonstrated that giant floods would still reclaim the floodplain, no matter how high the levées.

The logical move, say the opponents of river management, is to get off the floodplain, to move away from rivers that cannot be controlled. After the 1993 flood, several towns were relocated above the Mississippi and Missouri River floodplains. Despite this, further development is occurring on the floodplain. Malls and homes are being built behind restored levées. Eventually, these levées are likely to break again when a large flood event comes along, as it inevitably will one day.

## Hydroelectric power

Hydroelectric schemes have long been a central part of regional development projects. The Tennessee Valley Authority (TVA) of 1933 was, in many respects, the 'father' of all subsequent schemes, many of which have been the keystone of development policy. The reasoning for including hydroelectric schemes runs as follows:

- they create a more dependable water resource for agriculture, and irrigation improves yields
- this increases farm income, permitting greater investment and mechanisation
- this releases labour from the land
- the dam also produces cheap electricity which attracts industries
- these industries soak up the labour released from the land
- industries using large amounts of power, such as aluminium smelting and chemical manufacturing, attract other industries that use these materials
- regional take-off is achieved, accelerated by the multiplier effect

Hydroelectric power plants convert the kinetic energy contained in falling water into electricity. The energy in flowing water is ultimately derived from the sun and is therefore constantly being renewed. Energy contained in sunlight evaporates water from the oceans and deposits it on land in the form of rain. Differences in land elevation results in rainfall runoff, and allow some of the original solar energy to be captured as hydroelectric power.

Hydroelectric power is currently the world's largest renewable source of electricity, accounting for 6% of worldwide energy supply or about 15% of the world's electricity. In Canada, for example, hydroelectric power is abundant and supplies 60% of all electrical needs. It has traditionally been thought of as a cheap and clean source of electricity, but most large hydroelectric schemes being planned today are coming up against a great deal of opposition from environmental groups and native peoples.

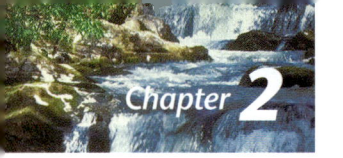

## Chapter 2 Fluvial environments

There is much unused hydropower in LEDCs. In China, for example, only 12% of the estimated hydropower potential has been developed. Hydropower supplies about 19% of power needs; coal supplies 80%. Plans are to increase hydropower to 40% of supply by 2020. Most large hydroelectric potential is in the upper Yangtze and western rivers, where the controversial Three Gorges dam scheme is being developed.

*Photograph 2.10 Three Gorges dam construction on the Yangtze, China*

### Examiner's tip

- Although hydroelectric schemes are less polluting and possibly more sustainable than power generation from fossil fuels, they do displace large numbers of people. These people are frequently poor and often politically weak, but they have managed to organise powerful campaigns in countries as politically diverse as China, Brazil and India.

## River management issues in an LEDC: the Mekong

The Mekong River is the longest river in southeast Asia. From its source in China near the border with Tibet, the Mekong flows generally southeast to the South China Sea, a distance of 4200 km. The Mekong crosses Yunnan Province in China, forming the border between Myanmar (formerly Burma) and Laos and most of the border between Laos and Thailand. It then flows through Cambodia and southern Vietnam into a rich delta before emptying into the South China Sea. In the upper course are steep descents and rapids, but the river is navigable south of Louangphrabang in Laos. Because the Mekong flows through several countries, the issues of river management are especially significant. Decisions made in one country clearly have an impact in others.

The management issues, values and attitudes vary from country to country and the level of development and populations vary significantly. In northeast Thailand, with over 20 million people, the water resources are well developed and, partly as a consequence of this, problems are emerging associated with salinisation of arable lands as a result of over-clearing of native vegetation and poor irrigation. In Laos (5 million people), which is a much poorer country, the water resources are, as yet, undeveloped. Cambodia (10 million people) is recovering from 40 years of war. In the Mekong delta, some 20 million Vietnamese live on some of the most fertile agricultural land in the world.

## River management issues in an MEDC: the Colorado

The Colorado River drains nearly one-twelfth of the continental USA. Initially the Colorado was developed to promote irrigated agriculture, but the river now provides a variety of consumptive and non-consumptive resources to a diverse set of groups and individuals. The Colorado is the principal river in an arid region, and has always been vital to life. But, while its consumptive values for agriculture and drinking water remain high, over the past several decades the river's waters have come to mean more than simply economic prosperity.

- The Colorado increasingly provides recreation and aesthetic values to a rapidly urbanising region.
- For more than 800 km, the river flows through National Parks.
- Drinking water needs of rapidly growing regional populations and Native American water rights claims increase demands on the river.
- Threatened or endangered native species and their habitats exist in many sections of the river.

The rising importance of non-consumptive uses, such as recreation and aesthetics, and the recognition of the need to protect cultural and natural resources, complicate management. Fifty years ago, the challenge was harnessing the river to deliver its water efficiently. Today the challenges are even greater and, consequently, management is undergoing great changes. There are complicated and sometimes controversial issues about who should have the greatest claims on the water. There have been fundamental changes in the economics and demographics of the region. While the basin itself remains largely rural and agricultural, fast-growing urban areas dominate the surrounding regions. For example, a city such as Las Vegas has very high water consumption, often for extravagances such as watering golf courses in the desert.

### Examiner's tip

- The Colorado and the Mekong provide a useful comparison of the demands made by rapidly growing urban populations on water resources in an arid climate. It is a useful reminder that the consequences of rapid urban growth are not confined to LEDCs, and that the arguments over scarce resources are almost as fierce between states within a nation as they are between nation states themselves.

# Chapter 3

# Coastal environments

## The coast

The coast is the dynamic zone that forms the interface between the land and the sea. Changes can occur rapidly, suggesting that some coasts have yet to achieve equilibrium. Given that the present day coastline is only about 6000 years old (the time since the sea-level rise at the end of the last glacial period), this is not really surprising. The coastal system is an open system, in that it receives inputs of energy from waves, winds and tides.

### Coastal processes

#### Marine erosion

**Marine erosion** is the wearing away and/or removal of material by the action of sea water. The basic principle of marine erosion is similar to that of rivers, given that they both involve moving water. However, whereas fluvial processes are due to the continual flow of water, marine processes are largely the result of wave action, and different forces are involved. The specific mechanisms differ as a result.

- **Abrasion** (or corrasion) is when the load of the sea is thrown against the rocks of the coast by breaking waves. This wears the rocks away.
- **Attrition** happens when particles of load become worn away by colliding with each other and as they hit the coastal rocks. The particles become progressively smoother and more rounded as well as becoming smaller.
- **Hydraulic action** occurs when waves break against a cliff face: air and water in cracks and crevices become compressed. As the wave recedes the pressure is released, the air suddenly expands and the crack is widened.

### Key terms

**Backwash** The movement of water down a beach under gravity from the top of the swash.

**Constructive wave** A wave that adds more material to a coastline than it removes.

**Destructive wave** A wave that removes more material from a coastline than it adds.

**Groyne** A wooden, concrete or rock structure constructed perpendicular to the coast to trap sediment being moved by longshore drift.

**Longshore drift** The movement of material along a coastline by wave action.

**Swash** The movement of water up a beach by the energy of a breaking wave.

- **Pounding** occurs when a breaking wave exerts pressure on the rock, causing it to weaken, even without any material to wear it away. Pressures of as much as 30 000 kg m$^{-2}$ can be exerted by high-energy waves.
- **Solution** (or corrosion) involves the taking into solution of soluble minerals in coastal rock. As the pH of sea water is invariably about 7, this process is usually of limited significance unless the water is heavily polluted and acidic. Even then, only rocks containing significant amounts of soluble minerals, such as calcareous rocks (which contain calcium carbonate), are likely to be affected by solution.

## Transportation

**Transportation** is the movement of material, in this case by the sea. It occurs by the same processes that rivers transport material, namely traction, saltation, suspension and solution. However, there is also an important movement of material on beaches by waves, which is called **longshore drift**. Waves may approach the coast at an angle, due to the direction of the dominant wind. After such a wave has broken, the swash carries particles diagonally up the beach. Under gravity the backwash moves them perpendicularly back down the beach. If this movement is repeated, the net result is a movement of material along the beach (Figure 3.1). This process also leads to the attrition of sediment on the beach and so particles tend to become smaller and more rounded with increasing distance along the beach. Obstacles to this movement, such as groynes and piers, cause sediment to accumulate on their windward side, leading to entrapment of beach material. Material may also accumulate in sheltered locations, such as at the head of a bay.

*Examiner's tip*
- If you are explaining the process of longshore drift, you are expected to refer to the gravitational force acting on the backwash as the reason why backwash occurs perpendicularly to the coastline.

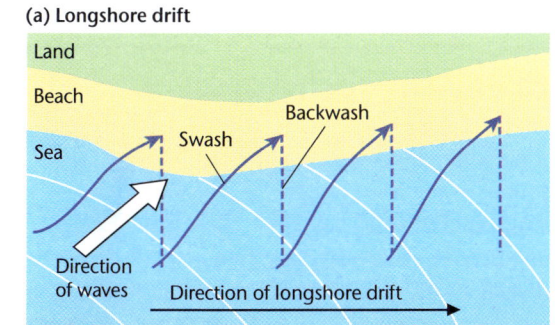

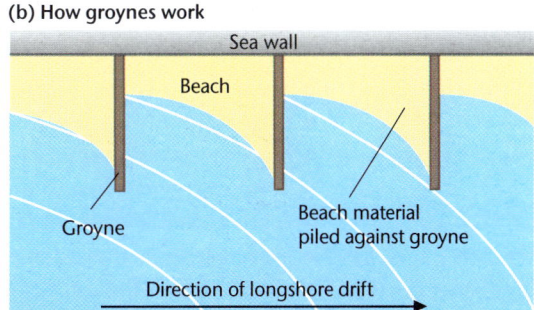

Figure 3.1 Longshore drift

## Deposition

**Marine deposition** is the laying down of material on the coast by the sea. Again, the basic principles are the same as those in fluvial deposition; material is deposited when there is a loss of energy caused by a decrease in velocity and/or volume of water.

Deposition tends to take place:
- when the rate of sediment accumulation exceeds the rate of removal
- when waves slow down immediately after breaking
- at the top of the swash where the water stops moving for a brief moment
- during the backwash when water percolates into the beach material

# Chapter 3
## Coastal environments

### The factors influencing marine processes

The rate (how fast) and the location (where) of marine processes depends upon the interaction of a range of factors. These factors vary spatially (from place to place) and temporally (over time) in their influence and their relative importance. Figure 3.2 provides a useful summary.

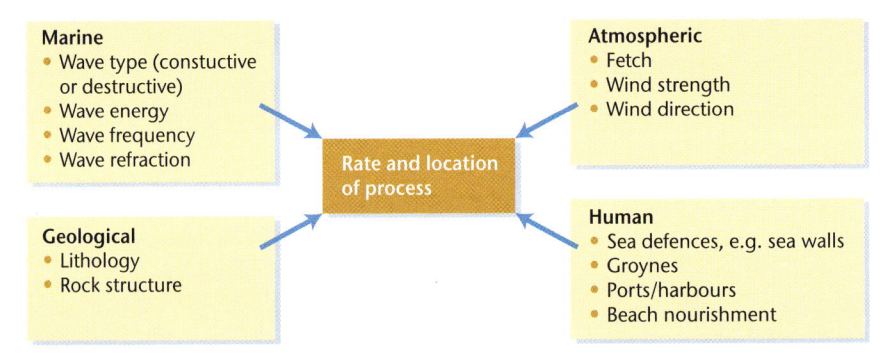

*Figure 3.2 Factors influencing rate and location of coastal processes*

**Marine**
- Wave type (constuctive or destructive)
- Wave energy
- Wave frequency
- Wave refraction

**Atmospheric**
- Fetch
- Wind strength
- Wind direction

**Geological**
- Lithology
- Rock structure

**Human**
- Sea defences, e.g. sea walls
- Groynes
- Ports/harbours
- Beach nourishment

→ Rate and location of process

### Waves

The transfer of energy from the wind as it blows over the sea surface produces waves. The energy acquired by a wave depends upon the strength of the wind, the length of time it has been blowing and the distance over which it blows (**fetch**). The wave itself is only a form on the surface of the water; until the wave breaks there is no net forward movement of water. When you stand on a beach and see a wave coming towards you, you are seeing the movement of a form, not a movement of the water itself. Within the wave there is an orbital movement of water.

Waves can be described in terms of a number of different characteristics and dimensions (see Figure 3.3):

- **wave length** is the average distance between successive wave crests
- **wave height** is the vertical distance between a wave trough and a wave crest
- **wave steepness** is the ratio of wave height to wave length
- **wave energy** is equal to the square of the wave height
- **wave power** is wave energy multiplied by wave velocity
- **wave period** is the average time between successive waves
- **swash time** is the interval between a wave breaking and the swash reaching its highest position on a beach

As waves approach the coast, they enter shallower water. Friction with the sea bed increases and the orbital movement of the water in the wave becomes increasingly elliptical. This has the effect of reducing the wave length and increasing the wave height; hence the wave steepens. When the height to length ratio equals 1:7 the wave will break. The breaking of the wave produces a net forward movement of water, called the **swash**, which runs up the beach. When the water has no energy left, it runs back down the beach under gravity as **backwash**.

*Figure 3.3 Wave terminology*

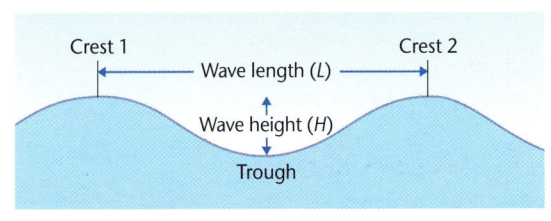

Physical environments  AS Unit 1

Table 3.1 Wave characteristics

| Characteristic | Constructive | Destructive |
| --- | --- | --- |
| Height | Low; often less than 1 m | High; over 1 m |
| Length | Long; often up to 100 m | Short; typically 20 m |
| Steepness | Gentle | Steep |
| Period | Long | Short |
| Fetch | Long | Short |
| Frequency | 6–8 per min | 10–14 per min |
| Water movement | Surging | Spilling or plunging |
| Strength of swash and backwash | Swash > backwash | Backwash > swash |

Waves can be classified as either **constructive** or **destructive**. Each type of wave displays a series of typical characteristics (Table 3.1 and Figure 3.4).

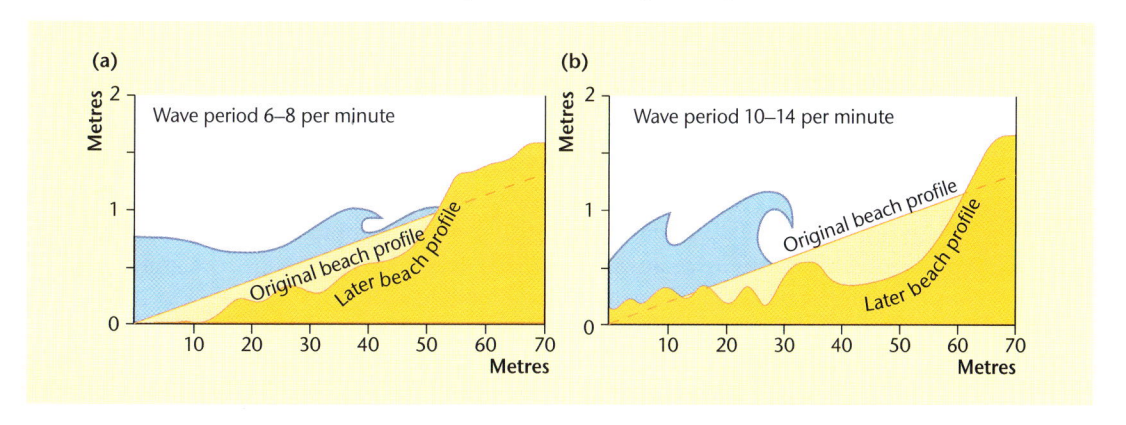

Figure 3.4 (a) Constructive and (b) destructive waves

Constructive waves move material up the beach, as the swash is stronger than the backwash. This gradually increases the gradient of the beach. Destructive waves comb material back down the beach and gradually reduce the gradient of the lower beach. Material may collect as a breakpoint or longshore bar. For waves of a particular size and steepness, the swash and backwash adjust the beach profile until it is in equilibrium.

When waves approach an irregularly shaped coastline, **wave refraction** takes place and they become increasingly parallel to the coastline. This is particularly true on coastlines with bays and headlands. As each wave nears the coastline, it is slowed by friction in the shallower water off the headland. The part of the wave crest in the deeper water approaching the bay moves faster as it is not being slowed by friction.

- Waves bend, or refract, round a headland and the orthogonals (imaginary lines of equal energy at right angles to the wave crest) converge. The sides of the headland therefore receive a concentration of wave energy and erosion takes place.
- In the bays, the orthogonals diverge and energy is dissipated, leading to deposition.

# Chapter 3
## Coastal environments

As the waves break on the sides of the headland at an angle there is a longshore movement of eroded material into the bays, adding to the build up of a beach (Figure 3.5).

*Figure 3.5 Wave refraction, leading to erosion of a headland*

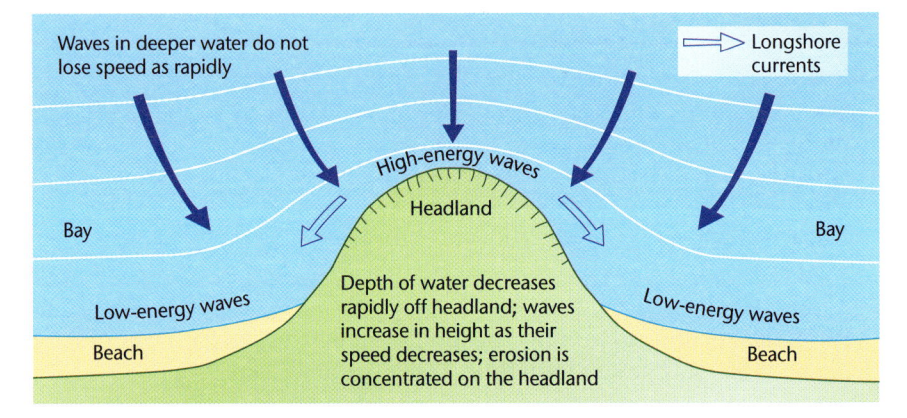

## Geology

**Lithology** refers to the characteristics of rocks, particularly their resistance to erosion and their permeability. Rocks that are tightly bonded, such as granite, have a physical strength that resists the action of the erosion processes. Some minerals in rocks, such as calcium carbonate in limestone, may be taken into solution if sea water is weakly acidic. Rocks with joints, bedding planes and other lines of weakness can be quite rapidly eroded as water can enter the rock and come into contact with a greater surface area of the material.

The **structure** of the rocks is also a factor. Rocks that lie parallel to the coastline produce a concordant coastline. If the most seaward rock is resistant, it protects any weaker rocks behind from erosion. The resultant coastline is quite straight and even, occasionally punctuated by inlets eroded at points of weakness. A good example of this can be found along the south-facing coast of the Isle of Purbeck in Dorset (Figure 3.6). The resistant limestone protects the weaker clays behind, except where erosion has broken through to form inlets such as Lulworth Cove and the larger Worbarrow Bay. However, where the rocks lie perpendicular to the coastline, a series of bays and headlands is likely to be formed as rocks of differing resistance are eroded at different rates. This forms a discordant coastline, such as the east-facing part of the Purbeck coast.

*Figure 3.6 The Purbeck coast*

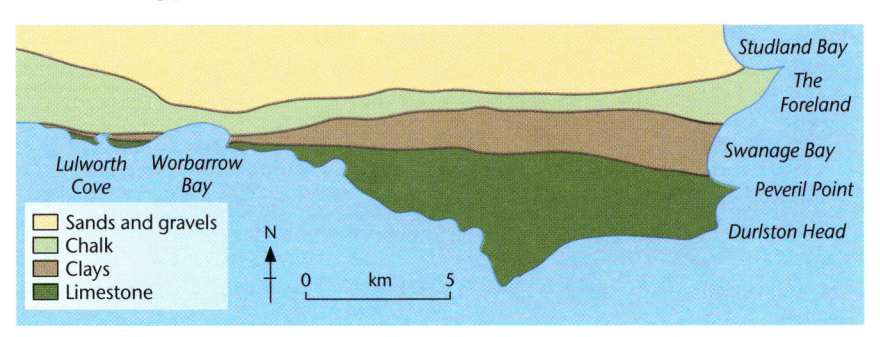

*Physical environments*  **AS Unit 1**

## Human activity

**Human activity** can have significant effects on marine processes. The coastline is increasingly used and managed both because of its potential for tourism and trade and to protect threatened coastal ecosystems. In areas where destructive waves cause high rates of erosion, defences are often employed to protect coastal settlements and transport links.

- **Sea walls**, particularly if they are recurved, can reflect wave energy, leading to its dissipation in the water.
- **Groynes**, wooden or concrete barriers at right angles to the coast, can be effective at reducing rates of longshore drift. This ensures that a beach remains, which can slow down approaching waves and absorb their energy as they break.
- **Beach replenishment** or nourishment involves bringing sand or shingle to beaches to replace sediment lost to erosion, again ensuring that a beach exists on which the waves then break.

Although these methods are all designed to reduce rates of erosion, poor management techniques can have effects elsewhere on the coastline. For example, the dredging of sediment from offshore can lead to increased wave energy and higher rates of erosion as the sea tries to regain the lost sediment.

Rates of erosion, therefore, tend to be highest in areas with high energy, destructive waves and weak geology that are unprotected by human activity (Table 3.2).

| Location | Geology | Erosion rate (m per 100 years) |
|---|---|---|
| Holderness | Glacial deposits | 120 |
| Folkestone | Clay | 28 |
| Beachy Head | Chalk | 106 |
| Isle of Portland | Limestone | 0.2 |
| Land's End | Granite | 0.1 |

*Table 3.2 Typical erosion rates*

# Coastal landforms

## Landforms of erosion

### Bays and headlands

**Bays** and **headlands** typically form adjacent to each other, usually due to the presence of bands of rock of differing resistance to erosion. The weaker rocks are eroded more rapidly, forming bays, while the more resistant rocks remain between bays as headlands. This can be seen on the east-facing Purbeck coast (see Figure 3.6).

- Bays have been eroded in the weaker rocks, such as Swanage Bay in the clay.
- Headlands remain in the more resistant rock, such as The Foreland, which is chalk.
- There is also a small bay, Durlston Bay, in the resistant limestone due to the presence of a fault line, which has been exploited as a line of weakness by wave action.

### Key terms

**Orthogonal** An imaginary line of equal wave energy perpendicular to the wave crest.

**Wave refraction** The bending of wave crests around headlands on irregular coastlines, causing an uneven distribution of wave energy.

# Chapter 3

## Coastal environments

**Photograph 3.1** An aerial view of the Swanage Bay area

Once bays have been formed, rates of erosion may reduce as the headlands either side protect and shelter the bay from some wind/wave directions. The headlands themselves then become the focus of erosion, leading to the formation of a sequence of distinctive landforms.

### Caves, arches, stacks and stumps

**Caves, arches, stacks** and **stumps** form on the sides of headlands as wave refraction occurs. Any points of weakness, such as faults or joints, are exploited by erosion processes and a small cave may develop on one, or even both, sides of the headland.

- Wave attack is concentrated between high and low tide levels and it is here that caves will form.
- If the cave enlarges to such an extent that it extends backwards through to the other side of the headland, possibly meeting another cave, an **arch** is formed.
- Continued erosion widens the arch and weakens its support. Weathering processes may also contribute and the roof may collapse, leaving an isolated **stack** separated from the headland.
- Further erosion at the base of the stack may eventually cause it to collapse, leaving a small, flat portion of the original stack as a **stump**. This may only be visible at low tide.

An excellent example of these landforms is Old Harry Rocks at the end of the headland known as The Foreland, near Swanage on the Isle of Purbeck (Figures 3.6 and 3.7). Wave refraction around the headland has led to concentrated erosion at lines of weakness in the otherwise resistant chalk rock.

*Physical environments*  **AS Unit 1**

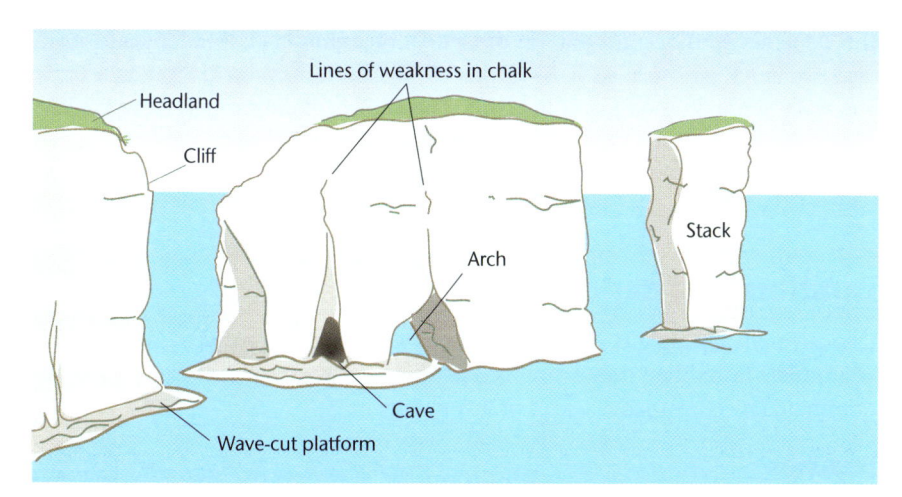

*Figure 3.7
A labelled field sketch of the Old Harry Rocks*

### Examiner's tip

- Simple line diagrams or field sketches can really help enhance an answer when you are writing about landforms. Sometimes they are explicitly asked for, sometimes you are invited to draw them. If they are annotated effectively, explanation as well as description can be included.

## Cliffs and wave-cut platforms

**Cliffs** and **wave-cut platforms** also tend to be produced together. When destructive waves break repeatedly on relatively steeply sloping coastlines, undercutting can occur between the high and low tide levels. This forms a wave-cut notch. Continued undercutting weakens the support for the material above and it eventually collapses, producing a steeper profile and leaving a cliff. The sequence of undercutting, collapse and retreat continues and so the cliff becomes higher. At its base, a gently sloping wave-cut platform is cut into the rock. Although these platforms appear to be flat and even, they are often deeply cut into by abrasion from the large amount of rock debris that is dragged across the surface.

Some rock debris from cliff collapses may accumulate on the platform if it is too large to be removed by wave action. Eventually, the platform will reach a size where it produces shallow water and small waves, even at high tide, and slows down approaching waves sufficiently for them to break on the platform rather than at the base of the cliff. At this point no further undercutting takes place. Research suggests that wave-cut platforms reach a maximum width of about 500 m before this happens.

Most coastlines that are subjected to destructive waves develop cliffs and wave-cut platforms. Flamborough Head in Yorkshire provides a particularly good example. The profile of any

*Photograph 3.2
Flamborough Head, Yorkshire*

cliff depends upon a combination of factors, including rock lithology and structure, sub-aerial processes such as weathering and mass movement, human activity, and wave action at the base of the cliffs. The cliffs at Flamborough Head are very steep (almost vertical in places) and over 25 m high. This is partly due to the fact that the chalk is horizontally bedded so that undercutting leads to parallel retreat and significant rock fall from the cliff face.

## Landforms of deposition

Depositional landforms are produced on coastlines where sand and shingle accumulate faster than they are removed. This usually occurs in areas of coastline dominated by constructive waves of low energy.

### Beaches

**Beaches** are the most common landform of deposition and represent the accumulation of material deposited between the positions of the lowest tides and the highest storm waves. Beach material, in the form of sand, shingle, pebbles and cobbles, comes from three sources:
- **cliff erosion:** from the land behind the beach; typically only about 5%
- **offshore:** combed from the seabed, often during periods of rising sea levels; again about 5%
- **rivers:** the remaining 90% is carried into the coastal system as suspended and bedload through river mouths

| Material | Diameter (mm) | Beach angle (°) |
|---|---|---|
| Cobbles | 32 | 24 |
| Pebbles | 4 | 17 |
| Coarse sand | 2 | 7 |
| Medium sand | 0.2 | 5 |
| Fine sand | 0.02 | 3 |
| Very fine sand | 0.002 | 1 |

*Table 3.3 Beach profiles and particle size*

Sand produces beaches with a gentle gradient; usually less than 5° because its small particle size means that it becomes compact when wet, allowing little percolation during backwash. As little energy is lost to friction, material is carried back down the beach, leading to the development of **ridges** and **runnels** parallel to the shore. These are occasionally breached by **channels** draining the water off the beach.

Shingle produces steeper beaches; often 10–20°. Shingle may make up the upper part of the beach, where rapid percolation due to larger air spaces means that little backwash occurs and so material is left at the top (Table 3.3).
- At the back of the beach, very strong swash during storm conditions may deposit larger material, forming a **storm beach** or storm ridge.
- **Berms** are smaller ridges that develop at the position of the mean high tide mark, again resulting from deposition at the top of the swash. Here, though, the waves have less energy than storm waves and so the material is smaller.
- **Cusps** are semi-circular depressions; they are smaller and more temporary features formed by a collection of waves reaching the same point when swash and backwash have similar strength. The sides of the cusp channel the incoming swash into the centre of the depression and this produces a strong backwash, which drags material down the beach from the centre of the cusp.

- Further down the beach, **ripples** may develop in the sand due to the orbital movement of water in waves.

Figure 3.8 summarises these beach features.

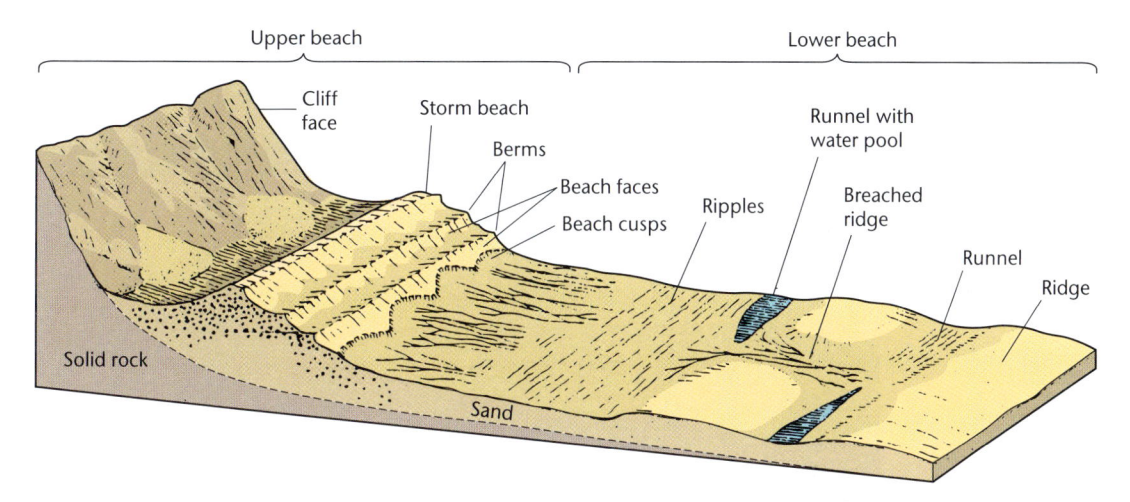

*Figure 3.8 Beach profile features*

Beaches are dynamic and their profiles change over time, especially seasonally as the wind strength changes and wave energy changes. Beaches are systems that seek an equilibrium form:
- steep beaches which help produce destructive waves are flattened by the movement of material down the beach in the strong backwash
- gently sloping beaches are steepened by deposition at the top of the swash by constructive waves

Beaches also vary in their plan form.
- **Swash aligned** beaches are usually straight and lack longshore drift movements as waves approach perpendicularly to the coastline or are fully refracted.
- **Drift aligned beaches** are dominated by waves approaching at an oblique angle, possibly due to partial refraction, and by the movement of sand by longshore drift.

## Spits

**Spits** are long, narrow beaches of sand or shingle that are attached to the land at one end and extend across a bay, estuary or indentation in a coastline at the other end.
- Spits are generally formed by longshore drift occurring in one dominant direction, which carries beach material to the end of the beach and then beyond into the open water.
- Storms build up more and larger material, helping to make the feature more substantial and permanent.
- The end of the spit often becomes curved as a result of wave refraction around the end of the spit and, possibly, the presence of a secondary wind/wave direction.

# Chapter 3  Coastal environments

**Photograph 3.3**
*An aerial view of Orford Ness*

**Figure 3.9 Growth of the spit at Orford Ness**

- Over time spits may continue to grow and a number of curved or hooked ends may develop.
- If the spit is developing across an estuary, its length may be limited by the actions of the river current.
- In the sheltered area behind the spit, deposition occurs as wave energy is reduced.
- The silt and mud deposited builds up and eventually salt-tolerant vegetation species may colonise, leading to the formation of a salt marsh.

A good example of a spit can be seen at Orford Ness in East Anglia. Here the coastline is east-facing and so is largely unaffected by Britain's prevailing south-westerly winds. Instead, northeasterly winds and waves are locally dominant, which has resulted in longshore drift from north to south. A spit has formed across the estuary of the River Ore, but the river current has prevented it reaching the land on the other side. Instead, the spit has continued to grow parallel to the coastline, diverting the river mouth some 12 km further south (Figure 3.9). Historical maps provide evidence of this growth.

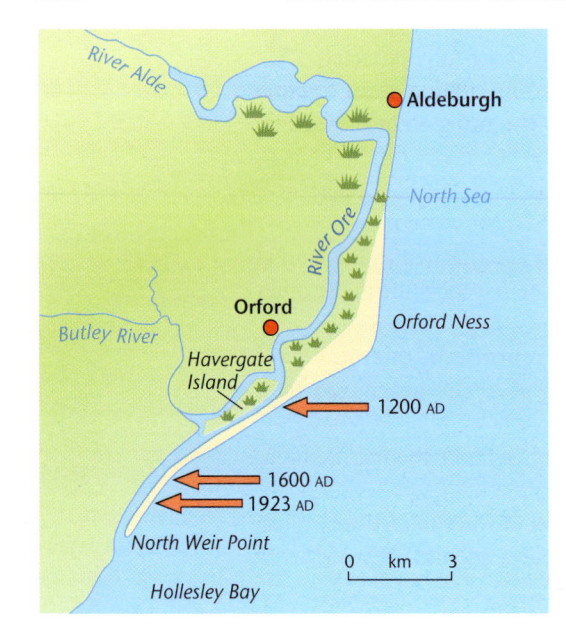

## Onshore bars

**Onshore bars** can develop if a spit continues to grow across an indentation until it joins onto the land at the other end. This forms a lagoon of water behind the bar. The bar at Slapton Sands in Devon may have been formed in this way. However, there does not appear to be a significantly dominant direction of longshore drift on this east-facing coastline and so it is thought that this feature may, at least in part, have been formed by the onshore movement of offshore sediment, perhaps during the post-glacial sea level rise that occurred about 10 000 years ago.

*Photograph 3.4 Slapton Ley and Slapton Sands, Devon*

## Tombolos

**Tombolos** are spits that have continued to grow seawards until they reach and join an island. The 30 km long Chesil Beach near Weymouth, Dorset, was once thought to have been formed from a tombolo (Figure 3.10). However, the movement onshore of offshore sediments is now thought more likely to have been the cause, with Chesil Beach reaching its present position some 6000 years ago.

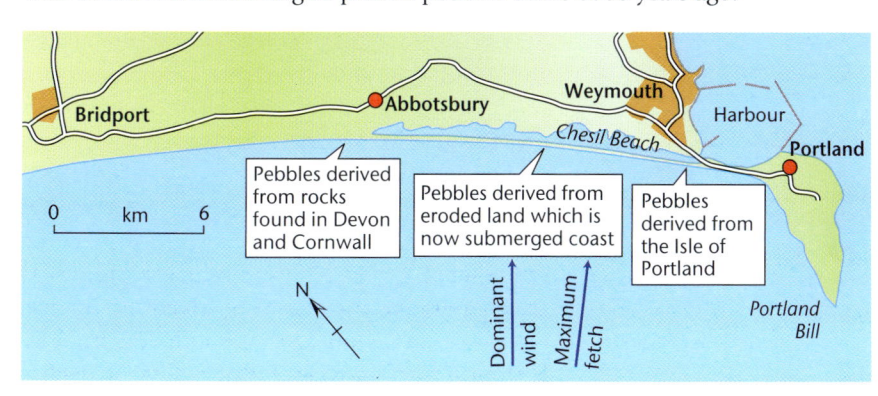

*Figure 3.10 Chesil Beach*

# Chapter 3 Coastal environments

## Offshore bars

**Offshore bars** or barrier beaches are elongated ridges of sand and/or shingle that lie parallel to the coastline and unattached to it. They are found on gently sloping, low-energy coastlines and are thought to have been formed in one of two ways.

- Some offshore bars may be the result of rising sea levels, causing beaches to become separated from the mainland.
- Offshore bars may be formed by deposition at the wave breakpoint in shallow water. As the waves lose energy immediately after they have broken, sediment is deposited. Once an accumulation of sediment begins, further deposition is encouraged, as more and more waves break on reaching the shallow water created by the bar. As most of the deposition takes place after the wave has broken, material is added mainly to the onshore side of the bar. This can lead to the bar 'rolling' landwards, again, perhaps, aided by rising sea levels.

Good examples of this landform can be seen off the east coast of North America between Miami and Texas, as well as in the Friesian Islands of the Netherlands and Germany.

## Cuspate forelands

**Cuspate forelands** are triangular-shaped accretions of sand and shingle that extend seawards (Figure 3.11). There is much debate about how they are formed, and different explanations may be appropriate for different examples. It is quite likely that many of them develop as a result of longshore drift occurring from two different directions. This may be because of the existence of a significant secondary wind direction in addition to a dominant wind, which may or may

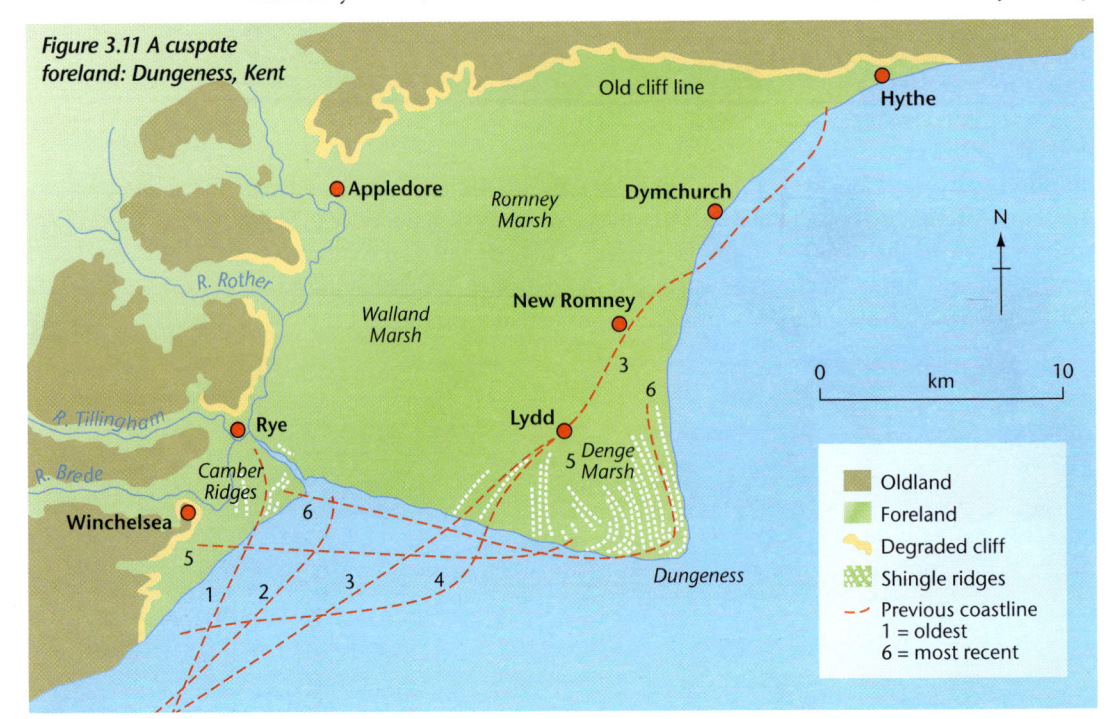

*Figure 3.11 A cuspate foreland: Dungeness, Kent*

92

*Physical environments*  **AS Unit 1**

*Photograph 3.5
Dungeness*

not be the prevailing wind. At a change of direction in the coastline, sediment is brought from both directions, causing the foreland to develop. This may even involve the formation of two spits which join together, trapping more sediment in the sheltered water behind. Sedimentation here may be further enhanced by rivers bringing sediment to their mouths.

An alternative explanation, which may be applied to the best example in the British Isles, Dungeness in Kent, is that of the migration of shingle ridges as wind directions have varied over time (Figure 3.11).

# Changes in sea level

## Causes of long-term sea-level change

Tides are responsible for the daily changes in the level of the sea, which are usually small. However, there are also long-term changes in the average position of sea level relative to the land. For example, sea level was 100–120 m lower than it is today during the last glacial period, and large areas of the continental shelf around the British Isles were dry land.

There are two main causes of changes in relative sea level:
- eustatic changes
- isostatic changes

### Key terms

**Emergence** The impact of a relative fall in sea level.

**Eustatic** Global-scale sea level change caused by a change in the volume of water in the ocean store.

**Isostatic** Local-scale sea level change caused by a change in the level of the land relative to the level of the sea.

**Submergence** The impact of a rise in relative sea level.

# Chapter 3

## Coastal environments

### Eustatic changes

**Eustatic changes** are:
- global in scale
- caused by changes in the volume of water in the world's oceans
- caused by physical factors, such as global climate change

A decrease in global temperature leads to more precipitation occurring in the form of snow. Eventually this snow turns to ice and so water is stored on the land rather than being returned to the ocean store. As a consequence there is a global fall in sea level.

If global temperatures subsequently rise, the water stored on the land is returned to the ocean store as glaciers retreat and ice caps thaw. Thus there is a rise in sea level.

There are a number of physical factors that can affect changes in global temperature, including:
- variations in the Earth's orbit around the sun, typically every 400 000 years
- variations in the amount of energy produced by the sun, every 11 years or so
- changes in the composition of the atmosphere due to major volcanic eruptions, the debris from which blocks out sunlight
- variations in the tilt of the Earth, occurring every 41 000 years

These factors may be added to by human factors, such as possible changes in the greenhouse effect due to the burning of fossil fuels in recent years.

### Isostatic changes

**Isostatic changes** are:
- local in scale
- caused by changes in the height of the land leading to a relative sea level change
- caused by physical factors such as uplift due to plate collision or sinking/rising of the crust as the weight of ice sheets is added/removed during a glacial period

During a glacial period, the great mass of ice in ice sheets and glaciers adds significant weight to the Earth's crust at particular points, causing the crust to sink lower into the mantle rock beneath. This results in an apparent fall in the level of the sea adjacent to the area being glaciated. In fact, it is a relative rather than an actual change, but it can be very significant. At the end of the glacial period, the ice masses melt and the weight is lost from the crust, causing it to slowly rise, readjusting its position. A relative fall in sea level will therefore occur in the areas affected. In the British Isles, the ice was thickest in Scotland during glaciation and so it is here that isostatic readjustment has been greatest over the last 10 000 years or so. It is still happening, with some places on the east coast of Scotland rising by as much as 7 mm per year in recent times.

> **Examiner's tip**
> - You should appreciate that changes in relative sea level may be the result of a combination of isostatic and eustatic causes. You should also be aware that not only has sea level changed in the past, but that it may well change again in the future.

## Landforms of submergence

Landforms of submergence are caused by a relative rise in sea level; the sea spreads over the land (marine transgression) and the coastline retreats.

### Rias

**Rias** are submerged river valleys. The lowest part of the river's course and the floodplains alongside the river may be completely drowned, but the higher land forming the tops of the valley sides and the middle and upper part of the river's course remain exposed. In cross section the ria has relatively shallow water, becoming increasingly deep towards the centre. The exposed valley sides are quite gently sloping. In long section rias are quite even with a smooth profile and water of uniform depth. In plan view they tend to be winding, reflecting the original route of the river and its valley (Figure 3.12).

*Figure 3.12 Plan, cross section and long section of a ria*

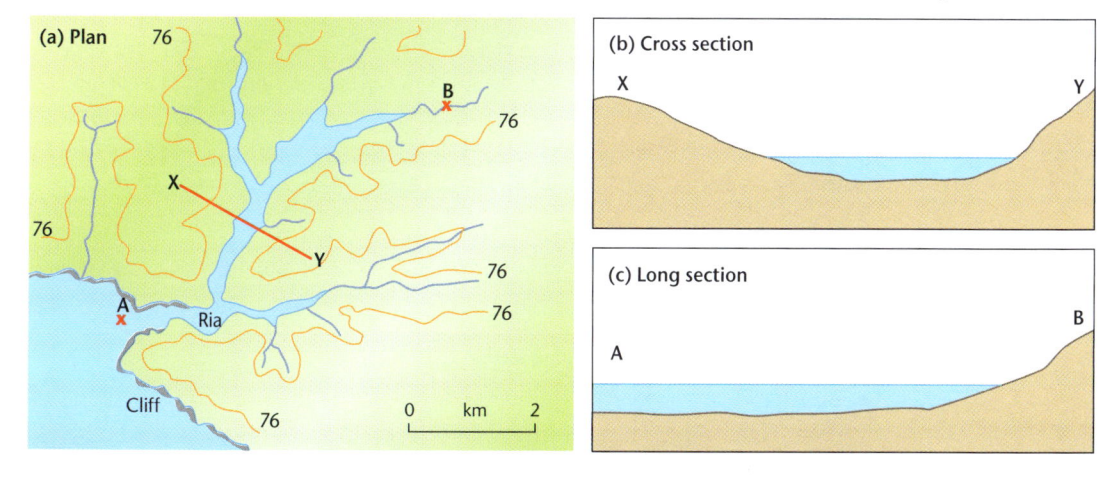

A number of rias can be found on the south coasts of Devon and Cornwall, including those at Kingsbridge in Devon and Fowey and Carrick Roads in Cornwall.

*Photograph 3.6 The Kingsbridge estuary, south Devon*

## Chapter 3 Coastal environments

### Fjords

**Fjords** are submerged glacial valleys. They have steep, almost cliff-like, valley sides and the water is uniformly deep, often reaching over 1000 m in depth. The U-shaped cross section reflects the original shape of the glacial valley. Unlike rias, fjords are not deepest at their mouths, but generally consist of a glacial rock basin with a shallower section at the seaward end, known as the **threshold** (Figure 3.13). Fjords also tend to have much straighter routes, as the glacier would have truncated any interlocking spurs present. The Sogne Fjord in Norway is nearly 200 km long, although similar features in Scotland are less well developed — the ice was not as thick here during the glacial period.

Figure 3.13 (a) Plan, (b) cross section and (c) long section of a fjord

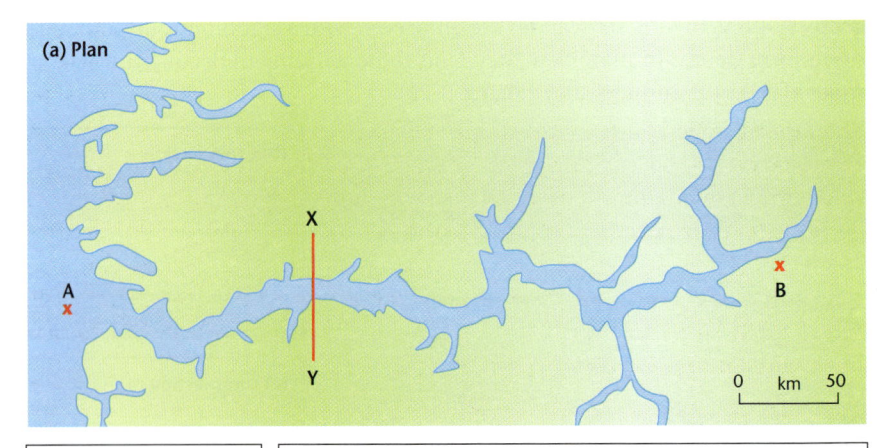

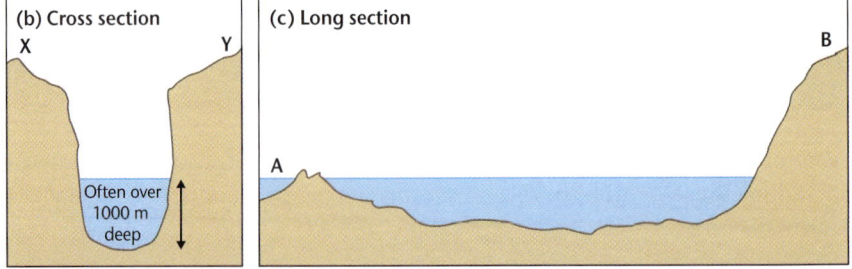

Photograph 3.7 The entrance to Milford Sound fjord, New Zealand

### Other landforms

Other, less common, landforms include **fjards**, which are submerged glacial lowlands, and **Dalmatian coasts**, which consist of submerged valleys running parallel to the coastline.

## Landforms of emergence

Landforms of emergence are formed as a result of a relative fall in sea level during a period of marine regression in which the coastline advances.

### Raised beaches

**Raised beaches** are areas of former wave-cut platforms and their beaches that are left at a higher level than the present sea level. They are also found a distance inland from the present coastline. Behind the beach it is not uncommon to find **abandoned (relict) cliffs** with wave-cut notches, caves and even arches and stacks that were formed by marine erosion during periods of higher relative sea level.

On the southern tip of the Isle of Portland, near Weymouth in Dorset, there is a distinct raised beach at a height of about 15 m above the present day sea level. This is thought to have been formed about 125 000 years ago during an interglacial period when sea levels were much higher. On top of the abandoned cliff is a 1–1.5 m layer of frost-shattered limestone debris, deposited when the area experienced periglacial conditions during the last glacial period.

*Photograph 3.8 Raised beach at Little Gruinard near Ullapool in the far northwest of Scotland*

# Chapter 3  *Coastal environments*

## Case study 3.1  *The effect of sea-level rise: Chichester and Selsey, Sussex*

The OS map of the area under discussion in this case study is shown in Figure 3.14.

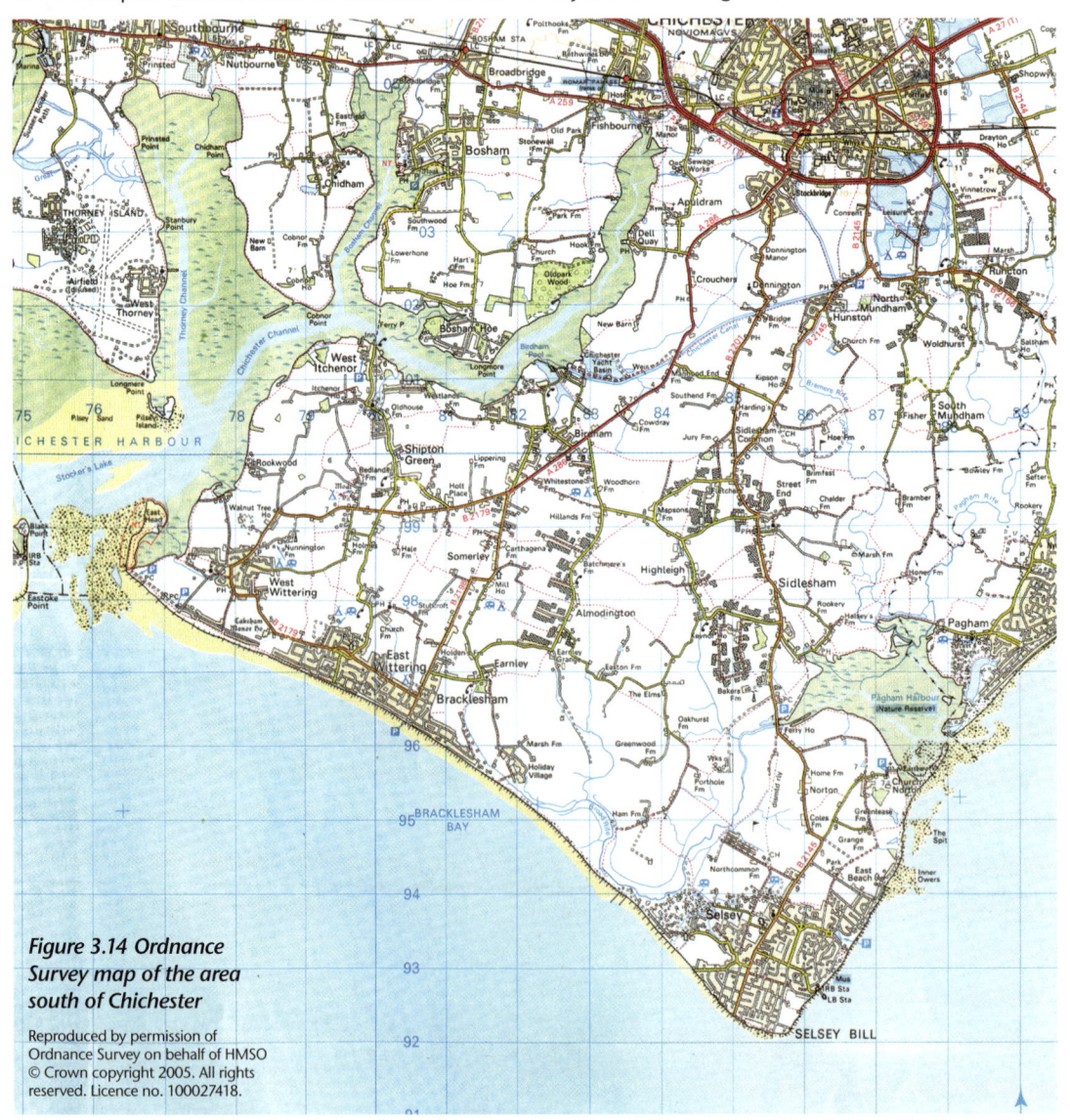

Figure 3.14 Ordnance Survey map of the area south of Chichester

Reproduced by permission of Ordnance Survey on behalf of HMSO © Crown copyright 2005. All rights reserved. Licence no. 100027418.

South of the A259/A27(T), much of this area is less than 5.0 m above sea level. It is estimated that mean sea level might rise by as much as 1.0 m over the next 100 years or so, although spring tides would be even higher than that.

### Agriculture

Much of the farmland is close to sea level, for example in the valley of Broad Rife (8495), and would become flooded. Even areas slightly further above sea level would be affected by salinisation of

98

groundwater and increasingly saline soils, which would reduce crop yields. Much market gardening takes places and there are extensive glasshouses, for example at Batchmere's Farm (8298). These would have to be abandoned.

## Settlement

Chichester, the largest settlement, lies at about 16 m above sea level and so would be safe. However, many villages, such as West Itchenor, would become uninhabitable. Pagham is also at risk, being only 2 m above sea level.

## Transport

Many of the roads across the lowlands are on slightly higher ground, but those on the valley floor would be submerged, including the B2145 and the B2166. The road to Selsey would be under water, so even though Selsey is safe at 9 m above sea level, it would effectively be cut off.

## Industry

The chief industry, apart from agriculture, is tourism, particularly water sports, beach holidays and bird watching. A sea-level rise would deepen the Chichester Channel and so improve yachting facilities in the short term. Salt marshes, such as Pagham Harbour, home to many wading birds, would be lost. Chichester, however, might benefit in becoming a coastal resort, with the potential for a tourist-based economy. Increased coastal erosion, though, would be a possible problem.

*Photograph 3.9 An aerial photograph of Selsey, Sussex*

## Chapter 3 Coastal environments

### The impact of rising sea level on human uses of the coastline

There is growing concern over the prospects of future sea-level rises. This concern has partly arisen because of our increased understanding of how the greenhouse effect can be enhanced by human activity such as burning fossil fuels. Predictions vary considerably, but at present rates of atmospheric pollution, the sea level may rise by 1 m or so in the next 100 years. Such rises will have a significant impact upon a whole range of human activities, including agriculture, settlement, transport and the tourist industry, particularly in low-lying coastal areas such as large parts of Essex and north Kent around the Thames estuary.

Major decisions need to be made about how to cope with this threat. The range of options includes the following.

- **Hold the line:** maintain and strengthen existing defences. These include measures such as sea walls.
- **Advance the line:** construct new defences in front of the coastline. The Thames Barrage is a relatively recent response to the threat of rising sea levels to London.
- **Managed retreat:** no defences are constructed but people and businesses at risk are moved away from the coastline. This can be more cost effective than trying to fight a (possibly) losing battle against rising sea levels, especially if the land is of low value and is little used.

Until the mid-1990s, approximately £60 million per year was spent on coastal protection in the UK.

# Coastal ecosystems

## Plant succession in a sand dune ecosystem (psammosere)

**Plant succession** is the long-term change in a plant community as initial, colonising species are gradually replaced by others. Eventually, a **climax community** is created, which is in equilibrium with the environmental conditions. When this happens in a sand dune ecosystem, it is known as a **psammosere**.

Coastal sand dunes are accumulations of sand shaped into mounds by the wind. They develop best when there is a wide beach and a prevailing onshore wind.

- When the beach dries out at low tide, some sand is blown to the back of the beach by the wind.
- The sand can accumulate, often initially around an object such as driftwood or seaweed. This produces an embryo dune.
- Over time the dune grows and becomes more stable.
- Another embryo dune may then develop on the seaward side of the original dune. The original dune is now relatively further inland and away from the influence of the sea.
- If this sequence continues, a series of dunes will eventually develop; the one furthest inland is the oldest and the one nearest the sea is the youngest.

Conditions on an embryo dune are harsh and few plants can survive. They have to cope with:
- salinity
- submergence in high tides
- a lack of moisture, as the sand drains freely
- wind
- mobility of the sand

Initial colonising species include sea twitch and sea couch grass. Sea couch grass is able to store water in its succulent leaves and it grows close to the surface to prevent possible wind damage.

The next species to develop is often marram grass, which is **xerophytic** (can withstand a lack of water). Marram grass has a number of adaptations to help it survive the challenging conditions:
- deep tap roots to find water
- extensive roots to trap and stabilise the sand
- sunken stomata (pores) on its leaves to reduce transpiration losses
- leaves that curl inwards to increase humidity around the stomata, reducing the drying effect of the wind

Marram grass not only traps sand and stabilises the dune, but it also provides shelter which allows other, less well-adapted species, such as plantains, to develop. With increased vegetation, more humus is produced as dead organic matter decays. This increases the fertility and reduces the pH of the 'soil', again permitting other species to grow. The typical yellow colour of the young dunes is gradually replaced by the noticeably grey colour of the more humus-rich older dunes. This is known as **bioconstruction**. In time other plants such as red fescue, sea buckthorn, ragwort and then heather and gorse replace marram. Eventually, slow-growing trees can flourish in the thicker soil that develops, and a woodland community may be able to establish itself.

Over time, the plant community typically becomes taller, denser, more diverse and woodier as the environmental conditions become less severe.

Dune **slacks** are depressions between dunes where the water table is near the surface. They are sheltered and protected by the ridges either side of them. They tend to have more favourable environmental conditions and so the plant community is much better developed here. Water-loving species such as bog myrtle, reeds and willow are often found.

## Plant succession in a salt marsh ecosystem (halosere)

The principles of succession in a salt marsh ecosystem are the same as those in a sand dune ecosystem. The plant community becomes taller, denser, more diverse and woodier over time as the environmental conditions improve.

Salt marsh environments develop in sheltered, low-energy environments such as behind spits. Rivers and waves deposit sediment in the area between the mean high tide level and mean low tide level, gradually building up an area of mudflats.

# Chapter 3 Coastal environments

## Case study 3.2 Sand dunes: Studland, Dorset

Historical maps show that the series of dune ridges developed at Studland over a period of about 400 years. As each new ridge developed on the seaward side, changing conditions on the older dunes allowed the plant community to develop further.

Fieldwork undertaken by a group of students in 2000 resulted in a set of data that can be used to summarise the changes in the plant communities with increasing distance inland, and hence increasing age (Table 3.4).

|  | Zero ridge | First ridge | Second ridge | Third ridge | Fourth ridge |
| --- | --- | --- | --- | --- | --- |
| Key species | Sea couch | Marram | Marram, red fescue | Heather, gorse | Pine |
| % cover | 5 | 20 | 70 | 85 | 100 |
| Height (cm) | 2 | 65 | 95 | 120 | 500+ |
| pH | 8 | 7.5 | 7 | 6 | 5 |

Table 3.4 Data from Studland

Photograph 3.10 Sand dunes at Studland, Dorset

### Examiner's tip

- It is *vital* that you avoid confusing salt marshes and sand dunes: these two types of ecosystem are different! If a question refers to a salt marsh and you write about a sand dune, you are likely to score no marks at all.

## Case study 3.3 Salt marsh: Keyhaven Marshes, Hampshire

A salt marsh has developed behind the spit near Hurst Castle in Hampshire (Figure 3.15). Plant succession has taken place in a series of identifiable stages over time.

### Stage 1

Initial colonisers such as algae, *Salicornia* (glasswort) and eelgrass develop on the bare mud flats. They have to tolerate saline conditions and submergence for as much as 11 hours in every 12-hour tidal cycle.

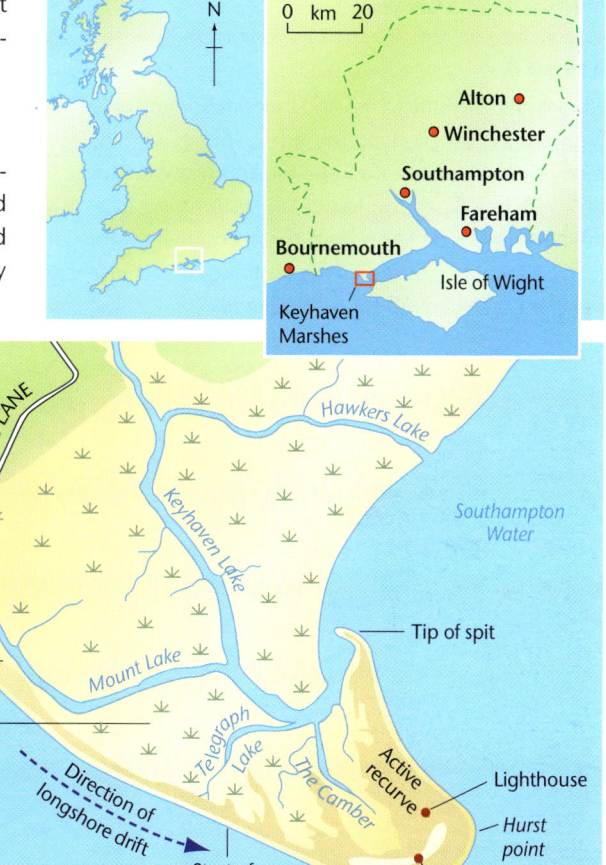

Figure 3.15 The location of Keyhaven Marshes

The presence of vegetation slows down the movement of water and waves lose energy and deposit their sediment. This increases the height and extent of the marsh.

### Stage 2

As the height of the mud flats increases, submergence by the sea is shorter, perhaps 6–8 hours in each tidal cycle. Species such as *Spartina* (cord grass) can now grow. This has two root systems — a fine surface mat which binds mud and a deeper, thicker root that secures it. This helps build up the marsh by as much as 5 cm per year. *Spartina* is taller than *Salicornia* but not as halophytic (salt tolerant), and so can only survive a shorter period of submergence.

### Stage 3

A discontinuous mat of surface vegetation about 15 cm in height develops, with species such as sea lavender and sea aster. These are less salt-tolerant and so occur once the marsh has built up enough that submergence only occurs at high tide. Salt pans are also found here. These are small depressions that trap salt water and prevent vegetation growth as

# Chapter 3 Coastal environments

conditions are too saline. Dead organic material added to the mud increases the height further, as well as improving fertility. The marsh grows by as much as 25 mm per year.

## Stage 4

Mud levels continue to rise and inundation now only happens in very high, spring tides. Species such as reeds and rushes can now survive and grow to 1 m or so in height. Complex creek systems such as Keyhaven Lake develop, and these channel water across the marsh. As the marsh continues to increase in height the channels become deeper.

## Stage 5

The marsh is now high enough, dry enough and fertile enough for trees such as ash and alder, which are non-halophytic, to dominate in the area around Saltgrass Lane. This area is seldom, if ever, submerged by tidal water. Now that trees dominate, lesser species may be killed off by competition, reducing the diversity a little. These stages can be seen in a transect across the marsh over a distance of about 2 km (Figure 3.16). The parts of the marsh furthest inland are the oldest and so have the most well-developed plant community, furthest from the restrictive conditions of the inter-tidal zone.

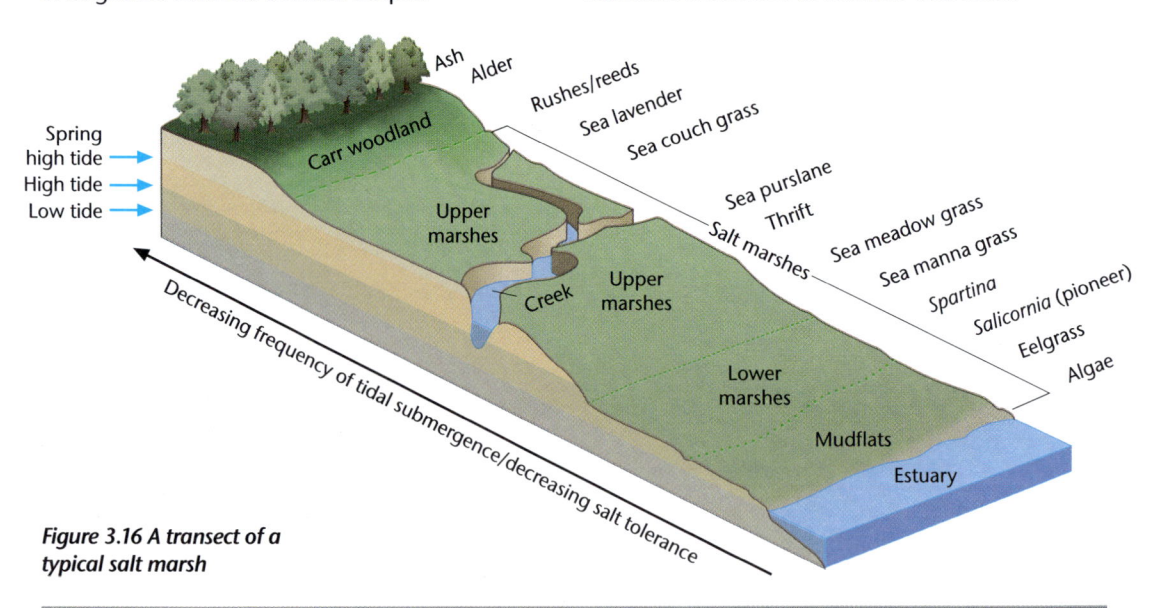

Figure 3.16 A transect of a typical salt marsh

## Coastal ecosystems and human activity

Modification of coastal ecosystems may be:
- short term or long term
- deliberate or accidental
- positive as well as negative

In areas that are popular tourist destinations, it may well be that the visitors themselves cause damage. The landowners then have to adopt strategies for managing the ecosystem in order to protect it.

> **Examiner's tip**
>
> - You must focus on the impact on the *ecosystem*, not the environment in general. References to specific plant and animal species can really help.

*Physical environments* **AS Unit 1**

### Key terms

**Halosere** Plant succession in a salt marsh environment.

**Plant succession** The long-term evolution of a plant community from the initial colonisers to the climax plant species.

**Psammosere** Plant succession in a sand dune environment.

*Photograph 3.11 A fenced boardwalk path across the dunes*

| Human activity | Impact on the ecosystem | Solution |
| --- | --- | --- |
| Trampling of vegetation such as marram grass by the 1 million visitors per year | Dunes become bare and subject to greater rates of erosion. Loss of habitat for rare species such as smooth snakes and the Dartford warbler | National Trust has installed boardwalk paths to stop this problem |
| Fire caused by campfires and barbecues lit by visitors in the dunes | Vegetation such as marram is burnt, again exposing the sand to erosion and causing a loss of habitat | National Trust has a designated barbecue area on the edge of the dunes with beaters and extinguishers available to prevent fires spreading |
| Litter dropped by visitors | Birds and animals can injure themselves on broken glass and cans. Birds also get their heads stuck in the plastic bands that hold 4-packs together | Litter bins have been installed and patrols collect 20 tonnes of litter a week in the summer |
| Visitors stray off the paths between car parks and the beach | Vegetation is damaged and dunes become unstable | National Trust has fenced off areas of young marram and some replanting has taken place. This protects the vegetation from trampling and encourages growth and succession |
| National Trust has cut back gorse and felled invading pine trees | This maintains species diversity, which the National Trust regards as an attractive feature of the nature reserve on the dunes, by preventing plant succession | — |
| Non-native species, such as rhododendrons, have been introduced | The natural plant community is changed and different habitats are created | — |

*Table 3.5 The impact of human activity on the sand dune ecosystem at Studland, Dorset*

# Chapter 3
*Coastal environments*

## Examination question

Study Photograph 3.12 of coastal landforms in southern Iceland.

*Photograph 3.12*

**a** With reference to the photograph:
   (i) Name landform X. (1)
   (ii) Mark and label a stack. (1)
   (iii) Describe what is likely to happen to landform X over time. (2)
   (iv) Explain why this is likely to happen. (4)

**b** (i) What is a berm? (2)
   (ii) Explain how berms are formed. (4)

**c** With reference to a named example, describe and explain the sequence of succession in a sand dune environment (psammosere). (6)

*Total = 20 marks*

# Synoptic link
# The management of coastal processes

- The need for coastal management schemes.
- Issues of management, including methods and strategies used, and their possible impact.

## The need for coastal management

In the last 50 years, the need for coastal management has become increasingly obvious, for a number of reasons:
- The coastal zone has the most complex environmental, resource and physical systems in comparison to the Earth's other environments.
- A nation's coastal zone is frequently its most valued and contentious area.
- Almost half the world's population lives within 150 km of a coastline.

## Physical environments — AS Unit 1

- Important tropical coastal resources and environments continue their downward spirals of degradation, non-sustainable levels of exploitation, elimination and extinction.
- The growing reality of global sea-level rise poses both short-term and long-term problems for almost all coastal regions.

Sea level is rising rapidly worldwide. It is estimated that along the Gulf and Atlantic coasts of the USA, a 30 cm rise in sea level is likely by 2050. In the next 100 years a 1 m rise is possible and sea level will probably continue to rise for several centuries, even if global temperatures were to stop rising a few decades from now.

Rising sea level inundates wetlands and other low-lying lands, erodes beaches, intensifies flooding and increases the salinity of rivers, bays and groundwater tables. Other effects of changing climate may add to these problems. Measures that people take to protect private property from rising sea level may have negative effects on the environment and on public uses of beaches and waterways. State and local governments are already starting to take measures to prepare for the consequences of rising sea levels.

## Coastal management in an MEDC: New England, USA

Almost all the coastal communities of New England have restated their intent to 'hold the line' on their residential land and on all developed areas; very few advocate drawing back in a managed retreat.

*Photograph 3.13 The Atlantic coast of New England*

It is widely believed that sea levels will continue to rise, and probably at an increasing rate. Various sectors of the coastal zone, tourism, fishing, the salt marshes which nurture many young fish, water supply — and the threat of rising

# Chapter 3 Coastal environments

insurance rates — are all important to the region's economy. This coast has suffered a 40 cm sea level rise over the past 100 years and might have to cope with as much as a 2 m rise in the next 100 years. Since 1980, ten northeasterly storms have hit the heavily populated coastline, plus several hurricanes. The coast is continually changing, and flooding is increasing in area and damage.

The years from 1870 to 1920 were comparatively calm, no storms struck, and summerhouses were built up and down the coast. Ferocious storms followed the 50-year lull: a fierce northeasterly storm in 1920 was followed by a severe hurricane in 1944, which destroyed all the timber jetties, and yet another large storm struck in March 1962. All of this pales into insignificance compared with the storms of the past 20 years.

Storms that once would have been rated as 100-year storms now occur every 30 years. There is general inundation at the water's edge: deeper water offshore causes higher wave energies at the beach, and this mobilises more sand. The areas with no 'buffer' will experience changes first and most dramatically.

In common with many coastal regions of MEDCs, one response to rising sea level and the increasing cost of protecting the shoreline in New England has been to advocate managed retreat. This avoids throwing good money after bad by trying to prevent the 'inevitable'. But after years of watching their beaches wash away, and following many expensive and often pointless efforts to halt the damage, the coastal communities of this coastline still do not agree on solutions.

Scientists, engineers and many local planners see coastal erosion and flooding as a serious issue, but previously viewed the concept of sea-level rise as a codeword for 'retreat', giving up installing protective measures for the barrier island communities. There has been polarisation in the past between:

- those who favour public expenditure to hold back the sea
- those who are sceptical about erosion-fighting programmes
- those who believe people should not be allowed to rebuild in hazardous areas

### Examiner's tip

- Be aware that this sort of calculation is complex because the prediction of future change is so central to making decisions about present policy. Such predictions are notoriously unreliable and it is easy to find sceptics who disbelieve in human-enhanced global warming, and thus take a different view about the type of coastal management needed.

## Coastal management in an LEDC: Bangladesh

Bangladesh has one of the most densely populated, low-lying, coastal zones in the world, with 20–25 million people living within a 1-m elevation from the high-tide level. In general, the topography of the country is extremely flat, except for marginal hills along the northeast and southeast region. The average land elevation is about 7.5 m above mean sea level and so many of the waterways are under tidal influence. The coastline in Bangladesh totals about 735 km, of which 125 km are covered by the Sunderbans — the natural mangrove forest.

The sectors of Bangladesh coastal resources identified as most vulnerable to climate change and sea-level change are agriculture, aquaculture, fisheries, forestry and tourism. The possible effects of sea-level change are inundation, salt-water intrusion, flash floods, droughts and storm surges. The coastal zone is also vulnerable to the impact of tropical cyclones, such as those of 1997, which caused tremendous flooding and severe damage to coastal embankments.

Over recent decades, a large number of projects, by both government and NGOs, have tried to reduce the threat of coastal erosion and flooding. An example is the Coastal Embankment Rehabilitation Project, which ran from 1989 to 2002 and was implemented by the Bangladesh Water Development Board and Forest Department with a budget of US$87 800 000.

In this project, 930 km of coastal embankments were raised and strengthened. They were also repaired after the severe damage caused by the 1997 cyclones. An economic rate of return of 15.2 % was calculated, excluding the value of saving human lives. The Bangladesh government approved a Coastal Zone Policy in January 2005. This is designed to harmonise the work of all government departments and NGOs with the aim of reducing poverty and developing sustainable livelihoods in the 17 designated coastal districts.

*Photograph 3.14 Bangladeshi coastline*

### Examiner's tip
- Don't forget that the ability of a country to manage resources effectively depends upon its level of development. In the case of Bangladesh, a country with very few resources, the coastline needs to be exploited to encourage further economic development.

## Management issues

The management of marine, coastal and estuarine environments is an important component of government in all countries with a coastline, most especially in MEDCs, which have the resources to invest in such strategies. Opportunities for these environments are explored in terms of management strategies to promote principles of ecologically sustainable development and the protection and development of specific coastal resources. Management issues include:
- the protection of sensitive areas
- tourism development
- coastal and marina development
- coastal residential development
- re-vegetation and weed eradication programmes
- education
- local government involvement and community programmes
- climate change and sea-level rise, and its effect on coastal processes
- water quality
- pollution and waste management
- wildlife management

# Chapter 3 Coastal environments

> **Examiner's tip**
> 
> ■ Try and acknowledge that resource management and conservation mean minimising the impact of port and residential development, oil transportation and runoff pollution. To deal with coastal problems, many coastal management policies attempt to oversee almost all activities on the coast, from adding a new porch to a private home to building a new refinery.

Many coastal areas are facing levels of marine erosion that are threatening coastal environments, especially tourist resorts, ports and harbours and other industrial sites. A number of possible methods can be used to try and reduce rates of erosion, each of which has potential positive and negative impacts (Table 3.6).

Table 3.6 Methods for tackling coastal erosion

| Method | Cost | Positive impacts | Negative impacts |
| --- | --- | --- | --- |
| Recurved sea wall | £10 000 per metre | Effectively reflects wave energy<br>Long lasting | Unattractive<br>May limit beach access |
| Groynes | £5000 each | Prevent longshore drift<br>Cheap to maintain | Unattractive<br>Restrict walking along the beach |
| Beach nourishment | £5 per m$^3$ | Natural appearance<br>Absorbs wave energy | Needs regular replacement<br>Not very effective on high-energy coasts |
| Gabions | £100 per metre | Cheap<br>Easy to repair | Need steps to allow beach access<br>Need regular maintenance as wires break |

> **Examiner's tip**
> 
> ■ Ensure that the distinction is clear between hard engineering methods that use constructions to fight the forces of the sea, and soft engineering methods, which try to work with the natural environment and to modify the operation of the system in a more environmentally friendly way.

# AS Unit 2
## Human environments

# Chapter 4

# Population characteristics

## Variations in populations

### Population distribution

**Population distribution** refers to the way in which people are spread out across the Earth's surface. It is, to put it simply, a description of where people live. **Population density** can be used to quantify the differences between areas. There are several population density measures; generally these indicate the number of people per unit area. In effect, they show the number of people who would be living in each square kilometre (km$^2$) if the population was evenly distributed over the land (Figure 4.1).

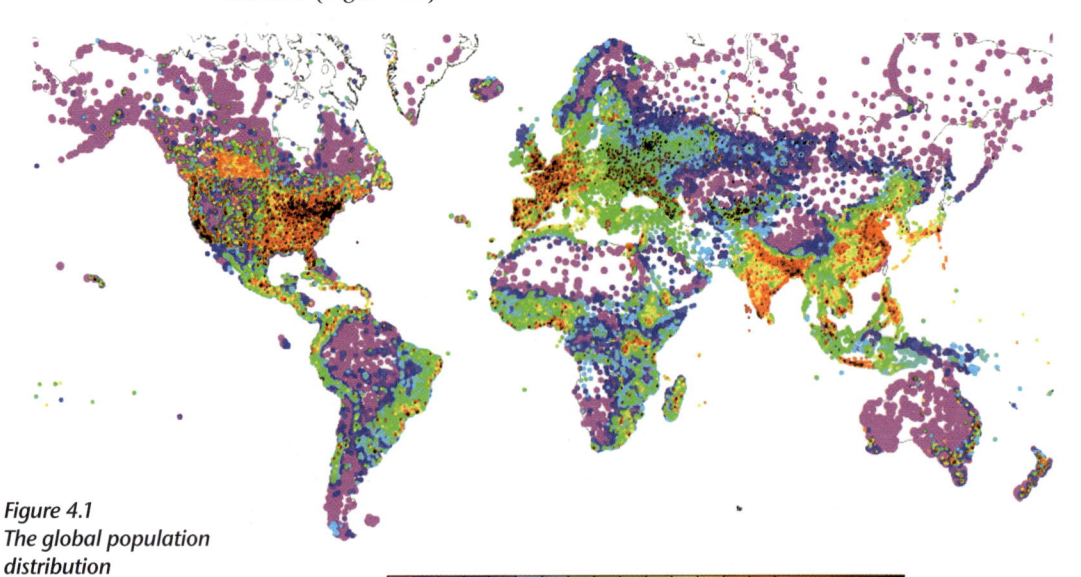

Figure 4.1
The global population distribution

0  20  40  60  80  100 120 140 160 180 200 220 240 260 280 300         400      People km$^{-2}$

### Key terms

**Population density** The number of people per unit area of land.

**Population distribution** Where people live.

But, of course, people are not evenly spread, whatever unit area is used. There are, for example, uninhabited areas in densely populated cities — Central Park in New York or Hyde Park in London are obvious examples. In the UK it is easy enough to find a square kilometre on almost any 1:50 000 map that has no permanent residents. It is much harder to find an area of 10 km$^2$

without any houses (10 grid boxes) and you need to go to the Scottish Highlands and other remote upland areas to find a 100 km² area without human habitation. The variables that control population distribution can be very broadly classified as either physical or human.

## Physical factors

**Climate** is the dominant explanatory factor for differences in population distribution at a global scale and it is also important at a national level. The impact of climate is not directly upon humans but on plants and therefore agriculture. The density and variety of plant life depends on inputs of solar insolation and moisture. The lower the inputs, the less plant growth there is and the less food available either for humans to consume directly or for them to feed animals.

The most important elements of climate in relation to population are temperature, rainfall and water availability. Other physical factors include relief, aspect, altitude, distance from the coast and type of soil.

### *Temperature*

Temperature, which limits the length of the growing season and thus controls agricultural production, was and is critical in determining the global distribution of population. It explains the major variations reasonably well. The length of the winter rather than the length of the summer is the key, for this controls the 'thermal growing season', the length of time in which crops can grow. Most cereal crops require at least 90 days of soil temperatures above 6°C to develop and produce food in the form of grain. If these conditions are not met, only pastoral agriculture or hunter-gathering are possible. Our ancestors were hunter-gatherers and this form of human activity is still practised by a small number of indigenous people, such as the Inuits in northern Canada. Their numbers are limited in the same way as the number of top predators, lions for example, are limited by the food supply of the grazing animals on which they feed.

### *Rainfall and water availability*

Although it is obvious that we require water to live, that water does not have to be delivered by rain.

Arid regions, where rainfall is less than 250 mm per year, are generally very lightly populated because of the limited growth of plant matter. However, there are some obvious exceptions, of which the Nile valley is the best known. The high population densities in the Nile valley are supported by river water used to irrigate agriculture.

It is more complicated at the upper end of the spectrum of world rainfall figures. Some of the least populated areas of the planet are very wet. Amazonia is an obvious example, with more than 2000 mm of rain per year in many areas. However, other areas that are equally wet due to monsoon rainfall support high populations with, once again, irrigated agriculture. Assam in India is a good example.

### *Relief*

**Relief**, the angle at which the land slopes, operates on a local scale. Steep slopes can be limiting both because of their influence on soil depth and soil fertility

# Chapter 4 Population characteristics

and because there are practical problems working on them. Relief is frequently confused with altitude, but steep slopes can be found in lowland areas, for example the escarpments of the South Downs in Kent and Sussex.

### Aspect
**Aspect** is closely related to relief and describes the direction in which a slope faces. Thus south-facing slopes of 30° receive some 30% more solar insolation than north-facing slopes in the British Isles. On a local scale, aspect can be a key influence on the distribution of settlements and explains local variations in population density and distribution. In the Pennine valleys of Swaledale and Wensleydale, the majority of villages are on the northern side of the valley and so have a southerly aspect. The same is true in the Norwegian fjord areas.

### Altitude
Population density decreases with increasing **altitude** in most parts of the world (Figure 4.2). This has little to do with altitude in itself. Although there is a reduction in available oxygen with height above sea level, this has limited impact below an altitude of 2000 m. The real significance of altitude is the impact it has on temperature. As a rough rule of thumb there is a reduction of 0.6°C in average temperatures for every 100 m of altitude. This leads to a shortening in the thermal growing season. There are also significant increases in rainfall amounts as orographic (relief) rainfall increases. Heavy rainfall is problematic for agriculture when combined with colder temperatures. There are some significant anomalies to this 'rule' close to the equator (e.g. in Ecuador and Kenya), where the coastal climate is tropical and higher, cooler altitudes can be better for human habitation.

*Figure 4.2 Population and altitude*

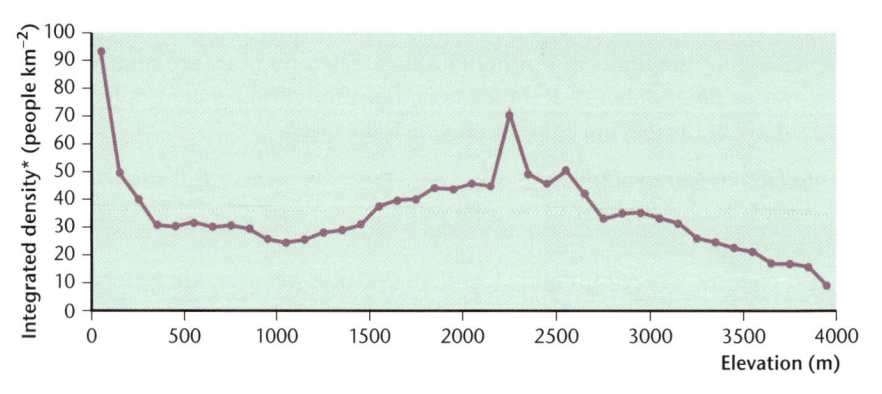

*Based on total land area available at each altitude

### Distance from the coast
Population density generally declines with **distance from the coast**. Three factors play a part here:
- the most important, and related, factor is the impact of altitude and the fertility of floodplains which, of course, get wider the closer they get to the sea
- on large land masses, temperatures (both high and low) are moderated by proximity to the sea and become more extreme inland
- sea trade and ports have been of economic significance from a very early period

*Soil*

**Soils** become more significant at the national level and, especially, the local level. Soils of low fertility such as podzols tend to be closely related to particular climate types and cannot support large populations, unless the population is supported from outside the area by trade. Similarly, desert soils are thin and lack nutrients as a consequence of limited organic matter and low rates of chemical weathering. The relationship between soil and population is quite complex in a vast region like Amazonia, where the soil becomes infertile if the natural vegetation is removed for farming. It is not that the soil is inherently infertile, but its inability to support intensive agriculture limits population.

## Human factors

*Economic development*

The key idea of economic development is that the growth of manufacturing industries frees people from a dependence on agriculture. This means that large concentrations of population can be supported on a small area of land by producing goods of sufficient value to exchange for the food that they are unable to grow themselves. This involves trade both within national boundaries and internationally. Industrialisation meant areas no longer had to be self-sufficient and specialisation could develop, first by region and then internationally.

Economic development also increases the productivity of the land because more 'scientific' farming methods and machinery can be used. As a consequence, yields per hectare rise and more people can be supported by smaller and smaller areas of land. This releases people from the need to produce food. Thus the Industrial Revolution in Britain was associated with a sharp increase in population. Small farms now produced more food than previously by applying rotation techniques, improved seed varieties and, little by little, the use of machinery to replace human labour. People moved away from the farms; meanwhile factories developed using local power sources and soaked up this 'surplus' labour released from food production. The sale of the manufactured goods supported a growing urban population.

*Historical factors*

Some parts of the world are relatively lightly populated because humans have not been in occupation as long as in other areas. The last continent settled permanently by humans was South America, although the last currently occupied island reached was Iceland. This provides some explanation for the relatively low overall population density of some regions of the world. The very high population densities in parts of Asia, such as the Ganges or Hwang Ho valleys, are a result of the development of irrigation systems and the wet-rice agriculture that these systems have supported.

In Java, for example, wet-rice agriculture supports a population of more than 1000 per km$^2$ in many areas. This is a particularly interesting example because Java has more or less the same climate as Amazonia. It therefore appears to undermine the often quoted 'too hot, too humid and too many diseases' explanation for the Amazon basin's very low population density. However, Java has immensely rich

# Chapter 4  Population characteristics

volcanic soils and wet-rice agriculture can, and has been, maintained for many millennia because the irrigation process naturally fertilises the soil, bringing in silts from the river's load. Rice is not a native species of South America, but is becoming increasingly important there.

### Government policies

Governments can move people around by direct action or indirectly. An example of direct action is the millions of people forcibly moved by Stalin in the former USSR in the 1930s. Planning and economic management can alter population distribution indirectly. Decisions are made about where people can and cannot live and there are very many examples of governments forcing people to move, as you will find out later in this unit (Chapter 6).

> **Examiner's tip**
>
> ■ The factors that affect population distribution vary according to the scale of the area under consideration. Thus climate dominates on a global scale, but locally soil, human history and government are at least as important.

## Case study 4.1  High population density: India

- India's population recently exceeded 1 billion people (1 085 000 000 was the estimate for July 2005).
- It is a distant second to China in terms of total population (China has a population of about 1.3 billion), but India is only about one-third the size of China. India is also about one-third the size of the USA.
- An average of about 320 people live on every km$^2$ in India, more or less the same as in the UK. For the USA, the average is around 30 people for each km$^2$.
- Approximately one of every six people in the world lives in India, yet it comprises only 2.2% of the Earth's surface.
- The population distribution within India is highly uneven.
- The Ganges River drains about 1 700 000 km$^2$ of land, in which about 300 million people live, nearly 5% of the world's population.

By the middle of this century India will be the most populated country in the world, with the greatest increase in population being found in the lower class. The population has increased by about 21% over the last decade, but this represents a slowing rate of 1.93% per annum, the lowest over the last five decades. India still has a largely rural population, with about one-third of the population living in 85 large-sized districts and only 28% of the population in urban areas. Both the life expectancy and the literacy rate have increased over the last decade. The literacy rate has gone up to 65.4%.

The most productive agricultural regions are the great river basins of the Ganges, Brahmaputra and Indus Rivers, each more than 2400 km in length and originating in the Himalayan mountains/Tibetan plateau. These rivers carry rich, alluvial soil to the plains below, which are then well watered by monsoon rains. The Ganges plain, a belt of flat and fertile lowlands as much as 300 km wide bordering the Ganges River, includes some of the most productive agricultural land in India.

India has industrialised rapidly in recent years and, far from being the 'hopeless cause' predicted by many a few years ago, it is now seen as a genuine contender, with China, as a global industrial power. In the 60 years since the withdrawal of the British, India has experienced rapid population growth. It has also been a period in which death from famine and disease have been at a historic low and growth of GDP per capita has been rapid.

Human environments  AS Unit 2

## Changing global distribution

As the global population has doubled over the past 40 years, the shifts in geographical distribution of that population have been remarkable (Table 4.1). In 1960, 2.1 billion of the world's 3 billion people lived in LEDCs (70% of the global population). By late 1999, the population of the LEDCs had grown to 4.8 billion (80% of the total). By 2025, 98% of the projected growth of world population will occur in these regions.

| Year | Europe | Asia | Africa | North America | Latin America | Total |
|---|---|---|---|---|---|---|
| 1650 | 103 | 327 | 100 | 12 | 1 | 545 |
| 1750 | 144 | 475 | 100 | 11 | 1 | 728 |
| 1800 | 192 | 597 | 100 | 19 | 10 | 906 |
| 1850 | 274 | 741 | 100 | 33 | 26 | 1171 |
| 1900 | 423 | 915 | 141 | 63 | 81 | 1608 |
| 1920 | 485 | 997 | 151 | 92 | 120 | 1834 |
| 1930 | 530 | 1069 | 164 | 110 | 134 | 2008 |
| 1940 | 579 | 1173 | 183 | 132 | 145 | 2216 |
| 1950 | 594 | 1272 | 219 | 162 | 166 | 2406 |
| 1960 | 641 | 1665 | 273 | 208 | 199 | 2972 |
| 1970 | 702 | 2027 | 352 | 283 | 226 | 3580 |
| 1980 | 751 | 2558 | 469 | 359 | 250 | 4416 |
| 1990 | 776 | 3003 | 645 | 412 | 280 | 5355 |
| 2000 | 806 | 3675 | 799 | 497 | 315 | 6092 |

Table 4.1 Global population (millions), 1650–2000

- Africa, with an average fertility rate over the entire period exceeding five children per woman, has grown the fastest among these regions. There are almost three times as many Africans alive today (799 million) as there were in 1960.
- Asia, by far the most populous region, has more than doubled in population since 1960.
- By contrast, the population of Europe has increased by only 20% and is now roughly stable.

Figure 4.3 shows the population growth rates for the period 1950–2050.

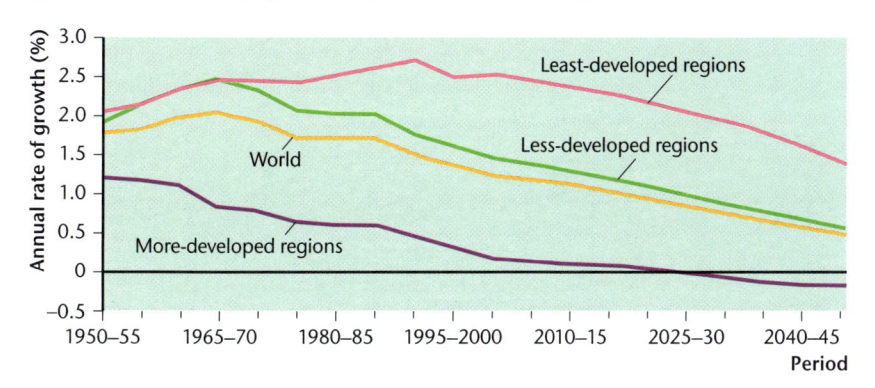

Figure 4.3 Population growth rates (historic and projected)

# Chapter 4  Population characteristics

Africa's share of the global population is projected to rise to 20% in 2050 (from only 9% in 1960), although many of these projections are being rapidly readjusted downwards in the light of Africa's profound AIDS crisis. On the other hand, Europe's share is projected to decline from 20% to 7% over the same period. In 1960, Africa had less than half the population of Europe; in 2050 it may have as much as three times as many people.

The altered balance of population distribution among global regions does not pose a problem, as long as development progresses everywhere and population growth is balanced by the development of social and economic capacity. The challenge is to create conditions that enable countries in all regions to adopt policies and strategies that foster equitable development.

## UK population distribution

The population density in the UK is 246 people per km², which is relatively high when compared with many other countries (Table 4.2).

*Table 4.2 Comparing population density*

| Country | People per km² | Country | People per km² |
| --- | --- | --- | --- |
| Singapore | 6751 | Pakistan | 188 |
| Maldives | 1004.9 | China | 134.1 |
| Bangladesh | 926 | France | 110 |
| Taiwan | 627 | Spain | 79 |
| Mauritius | 588 | USA | 31 |
| South Korea | 491 | Brazil | 20.7 |
| Japan | 335 | Argentina | 14 |
| India | 319.3 | Russia | 8.5 |
| UK | 246 | Australia | 3 |

*Figure 4.4 The distribution of the UK population*

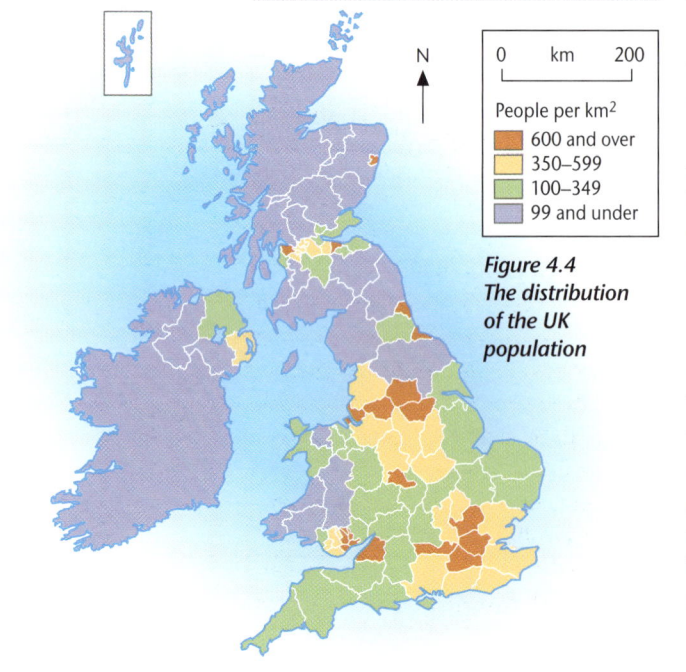

People per km²: 600 and over; 350–599; 100–349; 99 and under

There is also a great deal of internal variation within the UK and the average figure gives a poor reflection of the overall distribution (Figure 4.4). This variation depends on the scale adopted, but some areas are very lightly populated, for example much of the Highlands of Scotland has densities below 10 people per km².

Most of the areas of low population density have physical limitations that would have made settlement difficult in the pre-industrial era. These areas did not have sufficient attractions to lead to growth once people were released from the need to grow food locally. Thus the broad distribution still reflects the historical physical constraints.

*Human environments* **AS Unit 2**

## Key terms

**Centrifugal forces** Forces that lead to the population spreading out.

**Centripetal forces** Forces that lead to the population coming together.

**Counterurbanisation** The movement of people from cities to adjacent rural areas.

**Inertia** The tendency for a settlement to remain on its original site long after the reasons for the development of that site have disappeared.

**Thermal growing season** The time in which crops can grow in a particular area. It is usually calculated as the number of days when soil temperatures are above 6°C.

## Physical constraints

- The fact that the **thermal growing season** declines with altitude, combined with the windiness of exposed mountain areas, makes farming marginal in many areas of Wales, Scotland and northern England. These upland areas of Britain are not unlike tundra environments. The reduction in temperature with height is frequently about 0.6°C for each 100 m gained, making it one the fastest rates of decline in the world. This corresponds to a loss of about 13 days off the growing season per 100 metres.
- **Rainfall** increases with altitude. On average the coasts of eastern Scotland experience about 500 hours of rain a year while the northwest Highlands have over 2000 hours (out of a total 8760 hours in a year). The intensity of rainfall also increases with altitude, so in total there is a nine-fold difference in volume of rainfall between the two areas. The impact of these higher rainfall amounts is substantial, reducing soil fertility through leaching and making most arable crops uneconomic or impossible.
- On a local scale **soils** are a significant factor. The very nutrient-poor sandy soils of Breckland in East Anglia supported little pre-modern settlement and the impact of this has persisted into recent times. The lowland farmers who arrived from the European mainland from the fifth century AD did not favour the chalk soils of southern England, which once supported relatively high prehistoric populations. Thus the Vale of Pewsey in Wiltshire has a higher population density than the adjacent chalk uplands of Salisbury Plain.
- The location of **raw materials** has also played a key part in determining the distribution and density of the population. The coal reserves discovered in south Wales led to the rapid growth of population in the late eighteenth and nineteenth centuries, as this natural resource was exploited for a growing industrial and domestic market. Since the 1970s the decline in supply and fall in demand for this resource has led to a serious drop in employment and there has been considerable out-migration of people from the region.
- **Accessibility** remains a significant factor that limits the economic development, and thus the population, of the remoter corners of Britain. Even in a world of internet communication, videophones and virtual conferences, and even when there are no manufactured goods to be transported, personal contact remains

## Chapter 4 Population characteristics

significant. Thus, the southeast corner of the UK is much more attractive to inward investment than all other parts of the country. London's dominance has become more marked rather than less marked in the last two decades.
- Coastal areas have also been attractive to settlers, especially estuaries that provided shelter for shipping. As they are low-lying, coastal areas are often fertile, providing floodplains with rich soils.

These physical factors provide the background upon which economic forces played a major part in shaping the pattern of settlement, especially over the past 200 years. At the end of the seventeenth century, population density reflected soil fertility and climate very closely. Since the transformation of agriculture into commercial farming and the growth of manufacturing industry in the Industrial Revolution, the close relationship between physical conditions, especially climate, and population distribution has been weakened.

**Human**, largely **economic**, factors have had a growing impact on population distribution over the last 200 years. In modern industrial economies the growth of large cities has been a critical factor in redistributing the population as well as stimulating its growth. London and the southeast are highly populated, with an average population density of over 1500 km$^{-2}$. At a local scale, this region is even more polarised. Salisbury Plain has a population density of 1–5 km$^{-2}$ while some wards in the London Borough of Kensington and Chelsea have 20 000 km$^{-2}$. The population distribution broadly reflects the industrial pattern, with a dense area in the southeast and the Midlands stretching up to Lancashire/Yorkshire. Outside that area, high population density is found in the old industrial conurbations, such as Newcastle and Glasgow. Industrialisation is a powerful **centripetal** force, bringing people together at a place of work. Highly labour intensive industries, such as coal mining, textiles, steel and shipbuilding, were a characteristic feature of the Industrial Revolution.

Since the 1930s, the growth of consumer industries and the breakthroughs made with production lines and automation have reduced the demand for labour, and the location of industry has became more footloose. These **centrifugal** forces lead to people dispersing as a response to better transport facilities, cheaper travel and industry relocating to greenfield sites. Such forces have led to a decline in population in the UK's major conurbations since the 1950s, especially in the 1970s and 1980s, with the exception of London which arrested its decline in the late 1980s and has subsequently grown (Figure 4.5).

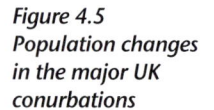

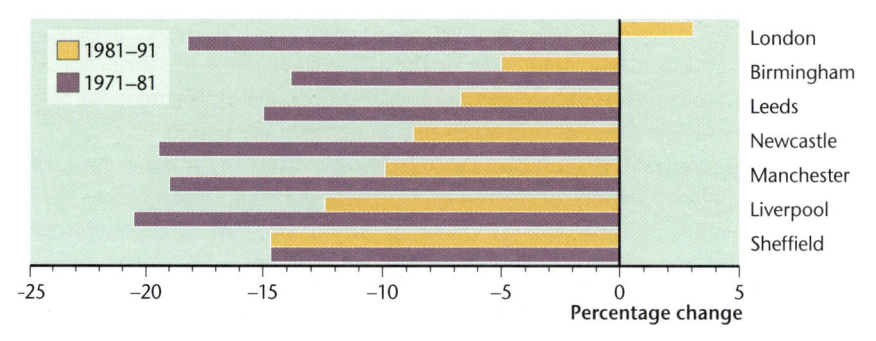

*Figure 4.5 Population changes in the major UK conurbations*

*Human environments* **AS Unit 2**

**Inertia** is the tendency for the pattern of distribution to reflect old forces and processes. People migrate and the population distribution changes, but this happens slowly, as in south Wales. The population declines, but not as fast as one would expect given the rapid rate of industrial loss.

'New' areas of rising population density include the coastal areas and attractive rural areas which are popular for retirement, and rural areas around large cities. Both these changes reflect **counterurbanisation** trends. South Devon, Dorset and the North Yorkshire dales are examples of retirement areas; Buckinghamshire is an example of commuter belt. It is important to realise that these areas are not densely populated compared with urban areas, but they have experienced growth — thus illustrating the point about changing distribution.

Far more important is the drift south in the last 30 years as the economic 'centre of gravity' has shifted from northern manufacturing industry to southern services and 'hi-tech' industries. The fastest growing settlements in the UK are medium-sized towns and cities of between 50 000 and 150 000 people, especially those in the southeast and central southern England: Swindon is a well-known example.

Some of this growth is encouraged by **government**, which has had a role in population distribution for many years. The Town and Country Planning Act of 1946 gave government powerful tools of social engineering; the 'New Towns' are an obvious product of this. Governments use planning to influence the growth and development of settlements, and thus alter the distribution of population. This also works on a local scale, with many counties pursuing a 'key settlement' policy that encourages the growth of some rural settlements but more or less prevents it in others.

## Reasons for variations in population density in rural and urban areas

### Urban or rural?

*Photograph 4.1 Can you define what it is that makes this area rural?*

The definition of a **rural area** is not as straightforward as it might seem. There is no generally agreed rule allowing us to distinguish between urban areas and rural

# Chapter 4 Population characteristics

areas. In some parts of the world, especially in the USA, urban and rural can visibly merge into one another as one travels for miles through a sort of suburbanised countryside.

The rural–urban distinction can be based on **form**, that is to say what an area looks like in terms of population density, or on **function**, which is, broadly speaking, what happens in an area. In medieval times this distinction was more straightforward.

- The 'countryside' was devoted, by and large, to the production of food and basic necessities. Each village had a range of services, from carpenters through to blacksmiths, but was dominated by farms devoted to food production.
- Towns and cities were **central places** that provided a market for the exchange of goods, administration, some higher order manufacturing such as goldsmiths and armourers, and, on many occasions, a place of refuge in times of war and unrest.

In MEDCs today, very few villages have more than 15% of their population dependent upon agriculture. In some parts of southern England, agricultural workers now live in local towns, commuting to the countryside to work because they cannot afford rural house prices. These have been raised by commuters who have moved to rural areas and generally have much higher salaries.

Thus, the economic distinction between urban and rural areas has broken down and we are left with size of settlement, which is an arbitrary and rather unsatisfactory measure.

## Urban areas

In many MEDCs, the older, inner-city areas have higher population densities than the twentieth-century suburban areas. The inner-city areas were often built during the nineteenth century as industries grew alongside rivers, docks or railway termini. The houses were small, tended to be terraced and were often built on poorer land that had been ignored in previous growth periods. The 'new' industries were frequently co-located with the housing. The sanitary conditions left a great deal to be desired: sewers were uncommon until the 1850s and water supplies were primitive.

The later development of suburban industry led to the loss of population from these central areas. This trend was later exacerbated in many cities by the spread of central service functions into surrounding areas. Meanwhile, the suburbs grew at explosive rates, fuelled by mass transport, standardised housing, newly available

### Key terms

**CBD** Central business district.

**Central places** Places that develop to serve the countryside as market and administrative centres.

**Form** The shape of a settlement.

**Function** What a settlement does.

**Rural areas** Areas which are lightly populated and have a relatively larger dependence on agriculture.

**Urban areas** Areas that are relatively more densely populated and have a range of economic activities but usually limited amounts of agriculture.

mortgages and rapidly growing employment. Some twentieth-century developments involved large-scale municipal housing schemes built on cheaper land on the city margins. This was especially the case during postwar reconstruction from 1945 onwards and during inner-city 'slum' clearance. These municipal housing schemes often had higher population densities than the privately built middle-class estates.

Many CBDs have a low permanent population but high daytime populations. This is not always the case; in central Paris, some local districts have high permanent populations. Inner-city redevelopment schemes, such as Cardiff Bay and London Docklands, can increase urban population densities in inner-city areas. Some areas in cities have remained free from any development and have very low population densities (e.g. Hyde Park in London and Central Park in New York). These areas have been protected first by private owners and then by planning regulations. There are still a few privately owned protected areas: Buckingham Palace is an obvious example.

## Case study 4.2 London

London is growing at an increasing pace. Its population has increased by about 700 000 since 1989 and its role as a global city is likely to provide the momentum for further growth in the next few decades.

London's population density is quite high compared to other European cities (Table 4.3). The figures in Table 4.3 reflect not so much the real built-up area but the administrative area of the city, which doesn't necessarily correspond to the 'real' city boundary. However, there are clear differences between the cities. Paris has retained a much higher population in the CBD, with far fewer office developments in the centre and a limit on building height that has inhibited the dominance of commercial usage.

It is also important to recall that there is much internal variety within a particular city. Inner London has a population density of 8613 km$^{-2}$ while outer London has a population density of only 3519 km$^{-2}$. The variation in London's population density is shown in Figure 4.6. The lightly populated CBD is concentrated in the City of London itself, which has a low night-time population and a density of only

*Table 4.3 Population densities of some European capital cities*

| City | Density of population (km$^{-2}$) |
|---|---|
| Paris | 20 135 |
| Brussels | 5 898 |
| **London** | **4 554** |
| Berlin | 3 854 |
| Lisbon | 1 738 |
| Amsterdam | 1 262 |
| Rome | 709 |
| Madrid | 628 |
| Stockholm | 270 |

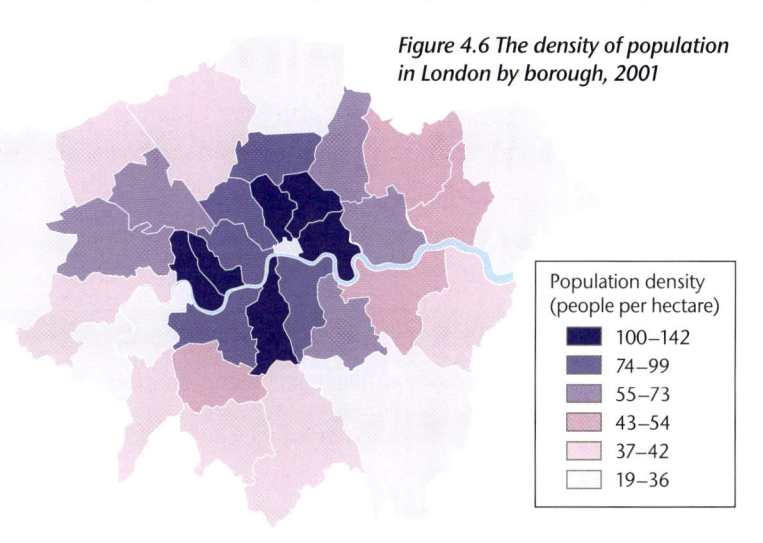

*Figure 4.6 The density of population in London by borough, 2001*

Population density (people per hectare)
- 100–142
- 74–99
- 55–73
- 43–54
- 37–42
- 19–36

# Chapter 4   Population characteristics

1900 km$^{-2}$, and the adjacent areas of the West End, which are also dominated by commercial uses. This old core is surrounded by an area of much higher density with the most densely populated borough being Kensington and Chelsea at 14 200 km$^{-2}$. Despite having some of the most expensive and exclusive properties in the UK, much of it at quite high densities, Kensington and Chelsea is also an area with considerable pockets of poorer housing at even higher densities. Outer boroughs, developed later, especially in the inter-war years of the 1920s and 1930s, and dominated by semi-detached housing, have markedly lower densities (Table 4.4).

There are many anomalies. The greater densities to the north of London reflect the industrial growth along the Lea Valley, and some of the higher density pockets on the outer fringes are a result of postwar municipal housing developments on the fringes of the city. Within the denser areas there are areas of undeveloped space, such as the great parks and Heathrow airport, which show up as lower densities at a more local scale.

| Area | Area (km$^2$) | Population (in 1998) | Density of population (km$^{-2}$) |
|---|---|---|---|
| Inner London | 322 | 2 760 800 | 8 574 |
| Outer London | 1 258 | 4 426 600 | 3 519 |
| Total Greater London | 1 580 | 7 187 400 | 4 554 |
| Outer suburban | 6 019 | 5 076 500 | 843 |
| London urbanised area | 7 599 | 12 263 900 | 1 614 |

*Table 4.4 London's population densities*

## Population structure

The make-up of a population in terms of age and gender is often displayed in the form of population pyramids. Populations can also be divided according to religion, ethnic group, language and educational attainment.

### Population pyramids

Population pyramids show a cross section of a population at a particular moment, dividing the population by gender and by age, usually into 5-year groups or cohorts. Pyramids use either per cent of population or absolute numbers on the *x*-axis. These graphs alone explain nothing about what has happened or what is going to happen; they simply describe populations as they were at the time of the last census, which in some cases may be considerably out-of-date. It is dangerous to make predictions from pyramids because there is no law that allows us to say that the population structure of a country will evolve in any particular way. If one cohort is smaller than the one below it, this could be explained by:

- a decline in birth rate
- an increase in death rate
- migration
- any combination of the above

### Key terms

**Ethnicity** A measure of the variety of a population, according to the racial, religious and geographical origin of its inhabitants.

**Population pyramid** A method of dividing a population into groups according to age and gender.

Human environments    AS Unit 2

The main purpose of drawing population pyramids is to show contrasts between populations, either over time or from place to place. For instance, LEDCs tend to have many more people in the youthful pyramid base. This contrasts markedly with MEDCs, which tend to have more people in the top half of the pyramid and lower numbers of younger people, giving a rather top-heavy appearance.

Once patterns have been established, it is possible to use the information for planning purposes. For instance, the requirements for school places can be assessed and cohorts of pensioners can be provided for. Mobility can be assessed, and the effects of individual 'events' can be chronicled: wars and famines are obvious examples. Population pyramids record and summarise much information about the demography of a country.

## Pyramids change over time

The pyramids for France shown in Figure 4.7 are typical of those of an MEDC, both in shape and in the fact that the populations are relatively stable and exhibit few changes over time other than the gradual ageing of the population.

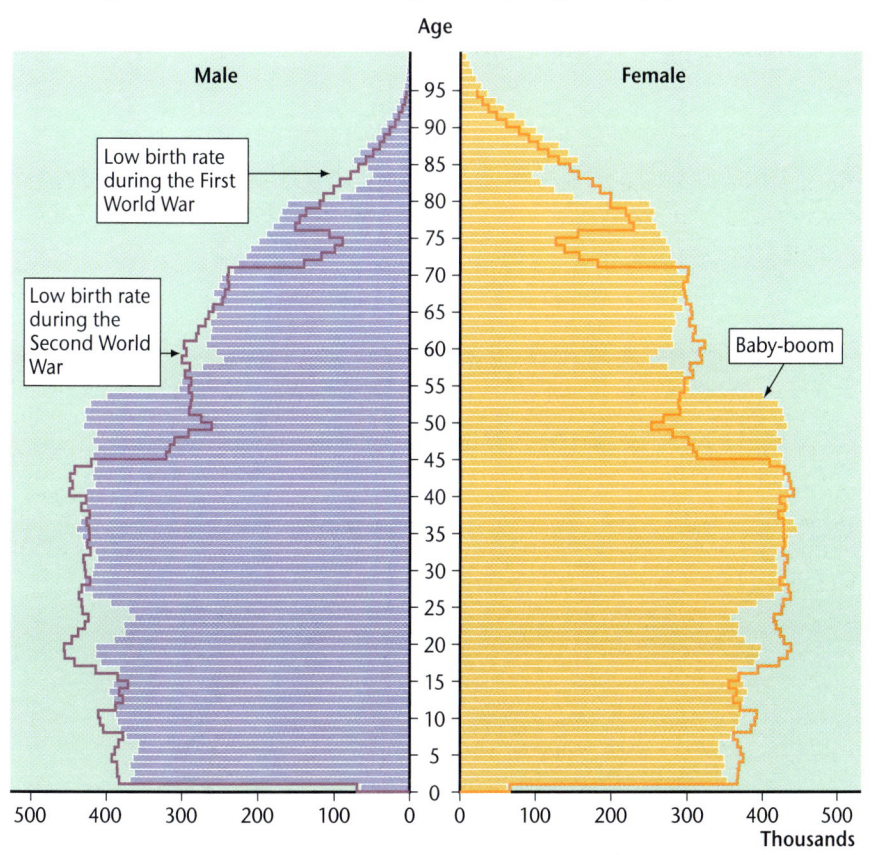

*Figure 4.7 Population pyramid for France in 2000 with the outline of that for 1990 shown for comparison*

Figure 4.7 shows total numbers rather than per cent of population and also shows each year group rather than 5- or 10-year cohorts. For both genders the pyramid is fairly straight-sided, suggesting a stationary or perhaps falling

# Chapter 4
## Population characteristics

population. There are clear signs of falling fertility in that the pyramid is undercut at the base. Thus, as is common in many MEDCs where birth rate has fallen, the largest generation is not the most recent.

The changes between 1990 and 2000 are largely a consequence of the population ageing, so the various bumps and bulges in the pyramid appear further up in 2000 than in 1990 simply because people aged 10 years between the two censuses. Thus the baby-boomers who were aged between 45 and 55 in 1990 were between 55 and 65 by the year 2000. The same can be seen in the indents caused by wartime reductions in birth rates from both world wars. In both pyramids there are substantially more older women than older men because of greater female life expectancy.

The Botswana pyramids are very different from those for France. The 2000 pyramid is comparable with many others for sub-Saharan Africa, showing expansion but at a slowing rate, so the 'steps' between each cohort become smaller as we go down the pyramid (Figure 4.8a). The largest generation remains the most recent, but by a far smaller margin than previously. The size of this step is driven

*Figure 4.8 Population pyramids for Botswana (a) in 2000 and (b) predicted for 2025*

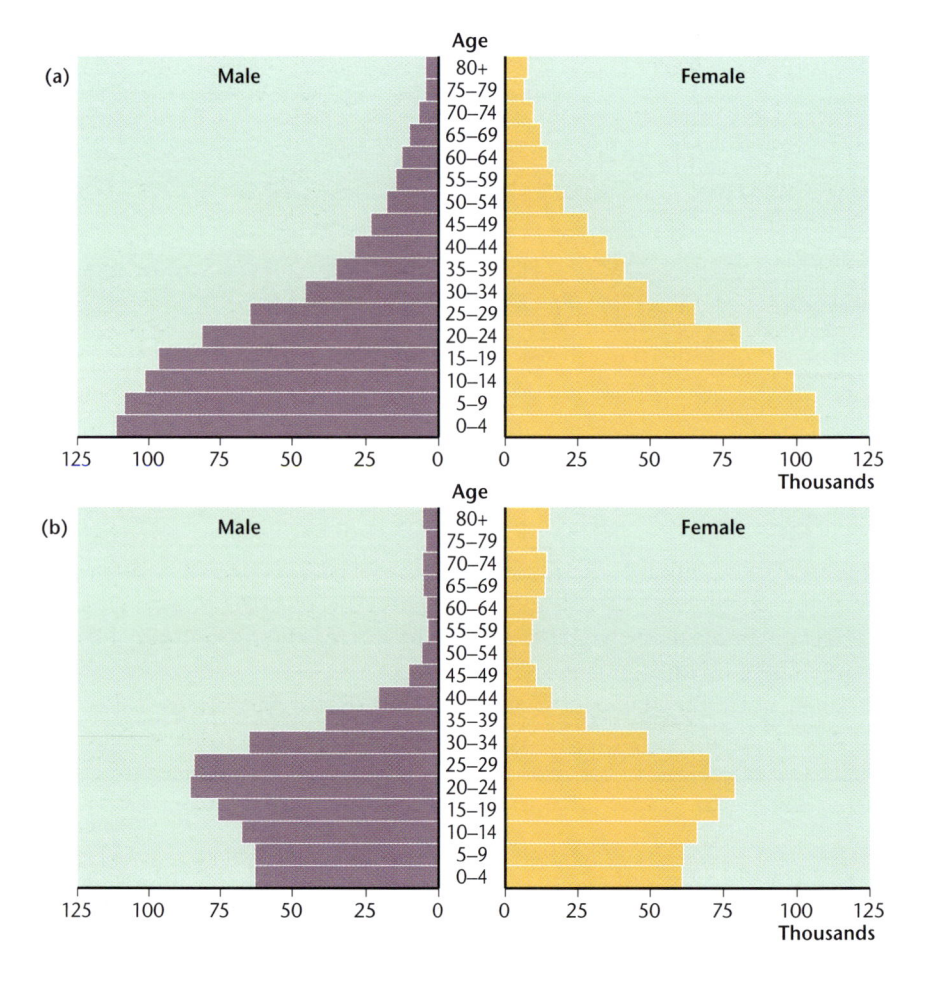

by both changes in birth rate, which has been falling for some years, and changes in infant mortality, which has also been falling. The impact of both of these has been to increase the 'steepness' of the sides of the pyramid, suggesting a slowdown in the rate of population increase.

The predictions in Figure 4.8(b) for 2025 are grim, showing a fall in absolute numbers of population. The generation born between 1995 and 2000 forms the bottom cohort in Figure 4.8(a) and numbers about 220 000 people. It is predicted that only 150 000 will be left alive in 2025. Even more startling is the decline in the 20–24 year old generation of 2000, which is predicted to fall from about 160 000 to about 20 000 45–50 year olds by 2025. There is also a notable gender imbalance, with fewer women than men in the cohorts between 15 and 44. This haemorrhaging of population could be explained by out-migration, or perhaps by war, but in this example it shows the expected results of the HIV/AIDS epidemic (Table 4.5).

| Region | People with HIV/AIDS | Newly infected | Adult prevalence rate (%) |
|---|---|---|---|
| Sub-Saharan Africa | 28 100 000 | 3 400 000 | 8.4 |
| North Africa and Middle East | 440 000 | 80 000 | 0.2 |
| South and southeast Asia | 6 100 000 | 800 000 | 0.6 |
| East Asia and Pacific | 1 000 000 | 270 000 | 0.1 |
| Latin America | 1 400 000 | 130 000 | 0.5 |
| Caribbean | 420 000 | 60 000 | 2.2 |
| Eastern Europe and Central Asia | 1 000 000 | 250 000 | 0.5 |
| Western Europe | 560 000 | 30 000 | 0.3 |
| North America | 940 000 | 45 000 | 0.6 |
| Australia and New Zealand | 15 000 | 500 | 0.1 |
| Total | 40 000 000 | 5 000 000 | 1.2 |

Source: *New Internationalist*, June 2002

*Table 4.5 Global HIV/AIDS data*

## Pyramids vary from place to place within a country

There may be significant differences in the population structures of different parts of the same country. These differences are usually generated not so much by differences in birth rate or death rate but by internal migration, which has an impact on both fertility and mortality.

Alaska has a younger population than Florida (Figure 4.9). Alaska is also characterised by a larger number of children and fewer elderly. It has experienced in-migration of people employed in the oil industry and their families. Given its severe climatic conditions, it is not a state that attracts people for retirement. On the contrary, it loses people who retire and move southwards.

At a local level, differences can be even more profound. Figure 4.10 compares three suburbs of Los Angeles. The data show the sharp differences within some urban areas.

# Chapter 4 Population characteristics

*Figure 4.9 Population pyramids for (a) Alaska, 1990 and (b) Florida, 1990*

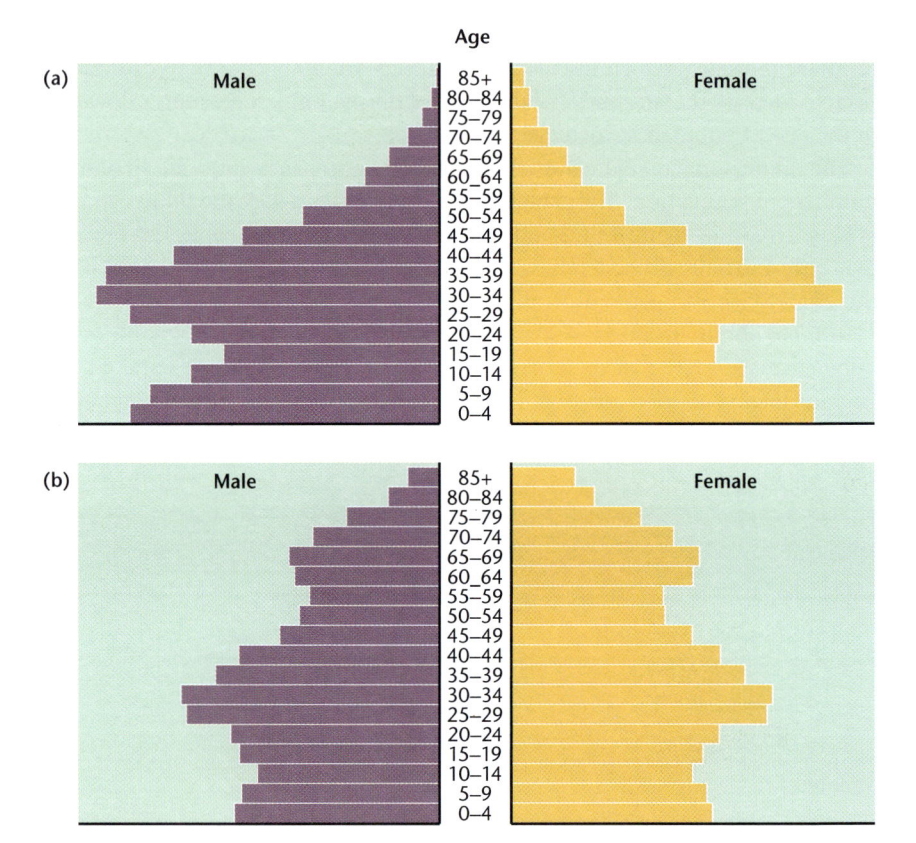

- Marina del Rey is a fashionable area on the ocean, a typical 'Friends' landscape of bars and cafés. The population tends to be young and mobile, living in shared apartments but moving out to fringe suburbs when they start families.
- By contrast, East Los Angeles is a poor inner-city area dominated by Hispanic in-migrants (mostly Mexicans) who are young, with more males than females, and who have higher fertility rates than most groups in the city. This is shown by the wide base to the pyramid.
- Beverly Hills is well known as an area associated with Hollywood and its wealth. It has a far higher number of middle-aged people, considerably fewer children, and has more women than men.

The contrasts between the three areas are largely a consequence of different migration patterns into the city and within it.

Urban areas in the UK have equally contrasting pyramids. Army towns such as Aldershot have a structure that looks very different from that for Worthing, which has a high percentage of elderly retirees. University towns and older industrial areas, such as some of the old coal-mining centres in south Wales, show similar contrast.

In MEDCs, rural areas in many remote regions have lost their younger population through out-migration, giving them an older median age. At the same time, some rural areas have experienced in-migration of the elderly (retirement).

*Human environments* **AS Unit 2**

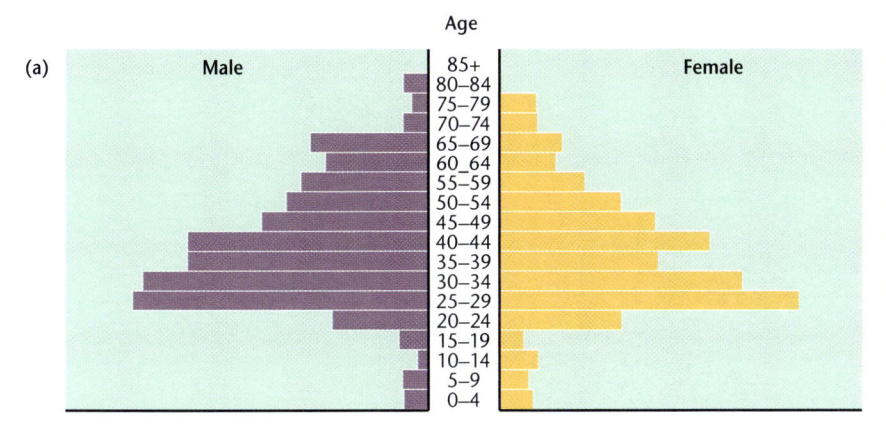

*Figure 4.10 Population pyramids for three areas of Los Angeles: (a) Marina del Rey, 1990, (b) East Los Angeles, 1990 and (c) Beverly Hills, 1990*

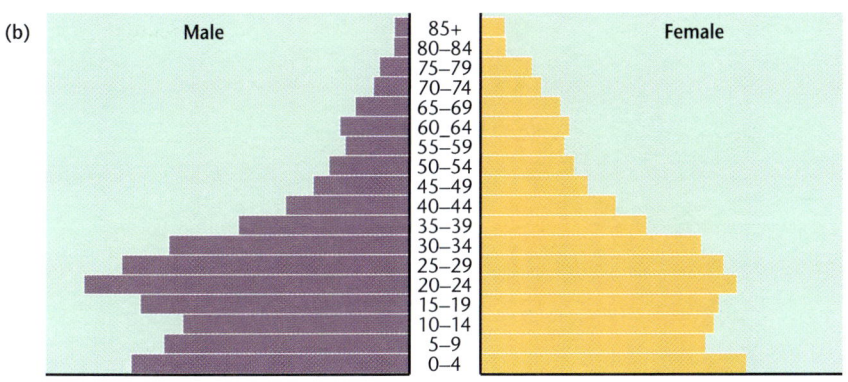

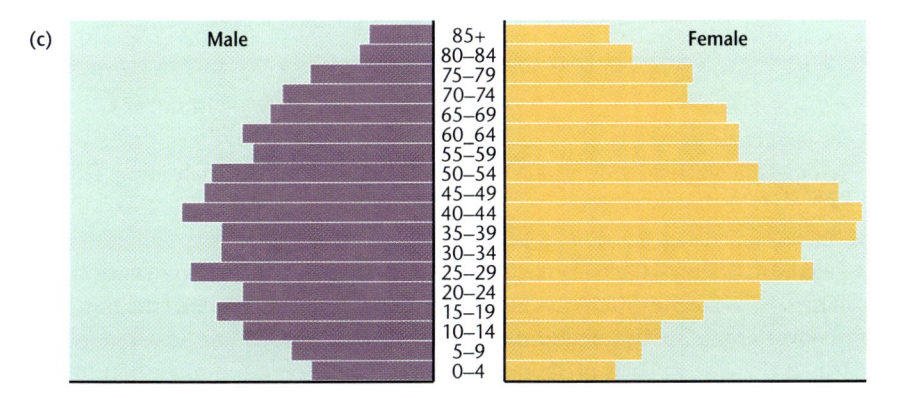

### Examiner's tip

- The median is the middle value in a sequence. It gives a better representation of population than the mean if the distribution is skewed by an extreme value. Just think about the average age of the people in a nursery school classroom when the teacher is there and contrast it with the average age when the teacher is not there. The median age gives a better description when the teacher is there (see Chapter 8).

129

# Chapter 4  Population characteristics

Figure 4.11
The percentage of the British Isles population aged over 65, by county

Legend:
>22.0%
20.0–21.9%
18.0–19.9%
16.0–17.9%
14.0–15.9%
0–13.9%

Figure 4.11 shows the percentage of the population aged over 65 in each county of the UK. Note the differences between the southwest and parts of the northeast. In some areas, the in-migration of other younger groups has occurred as well. This is the case in parts of Texas, which have experienced in-migration of both retirees and younger people from Mexico, moving to work in local industries.

In LEDCs there is a comparable degree of variety, but the forces and factors involved are rather different. In many of the poorer LEDCs, especially in sub-Saharan Africa, the dominant theme is rural–urban migration. This movement is selective in terms of age and gender, creating very different pyramids for urban and rural areas. Figure 4.12 shows three population pyramids for Namibia. These show changes in the urban population between 1970 and 1991 and the contrasts between the urban and rural populations in 1991. In Namibia, apart from rural–urban migration, employment in neighbouring South Africa and

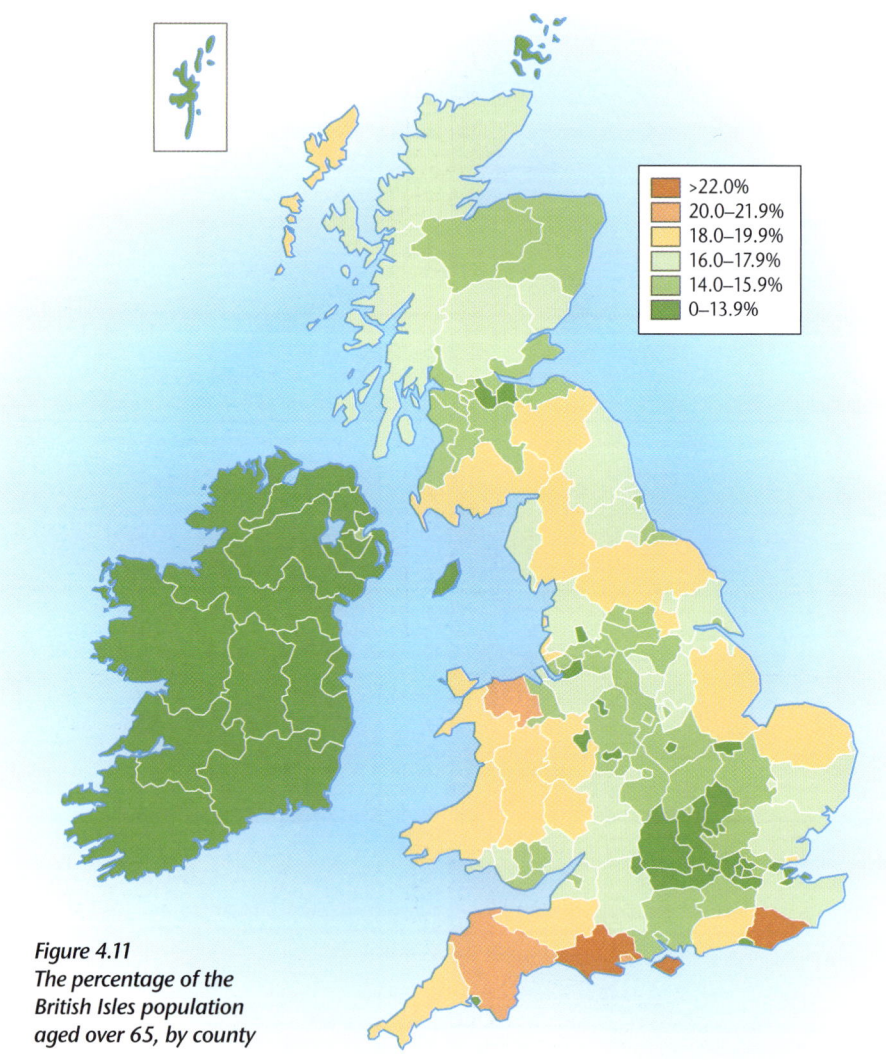

*Examiner's tip*

- Do not assume that rural populations are declining in LEDCs. In many cases, out-migration is more than compensated for by a higher than average birth rate.

*Human environments* **AS Unit 2**

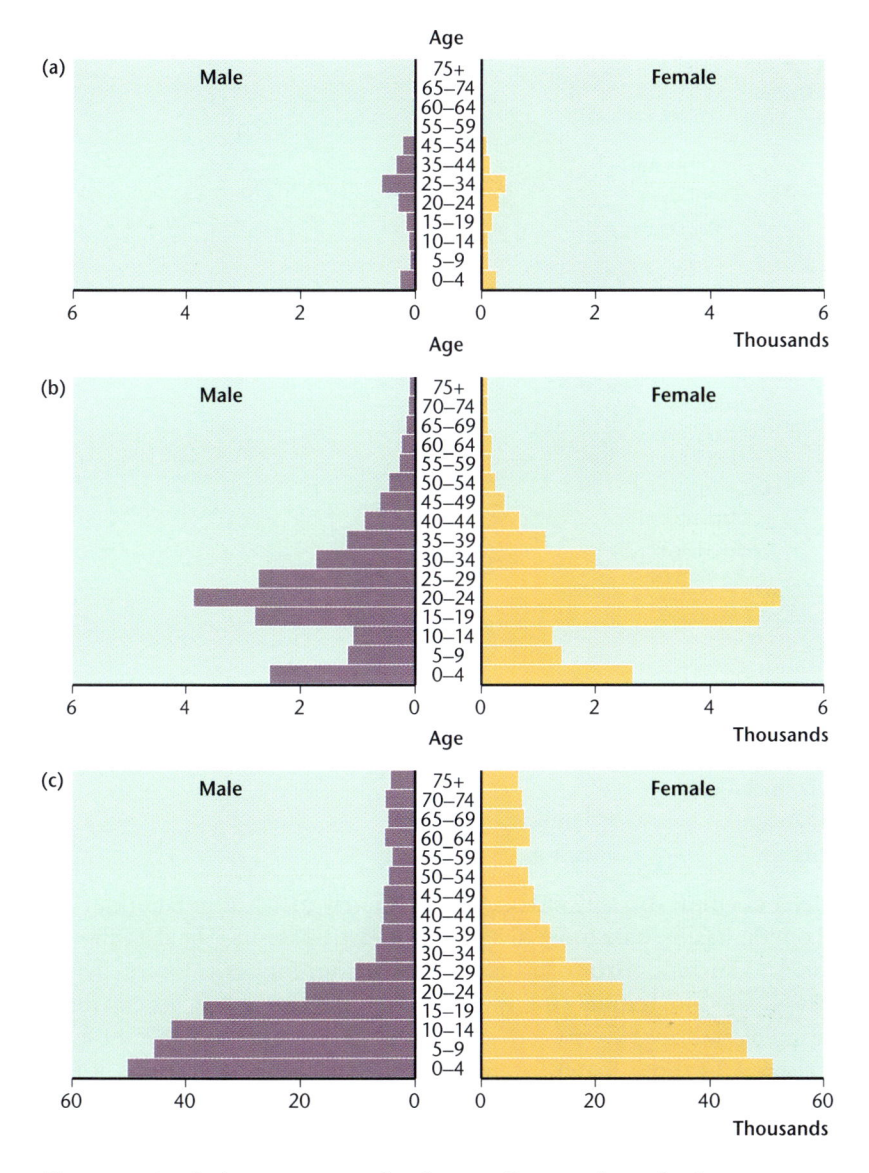

*Figure 4.12 Population pyramids for Namibia: (a) urban, 1970, (b) urban, 1991, and (c) rural, 1991*

military service help to account for the smaller number of urban young men compared to young women. This region is now deeply affected by the HIV/AIDS crisis. The population of urban areas in LEDCs is characteristically younger, reflecting age-selective migration and the impact of this on natural increase. This is also true in far wealthier LEDCs such as Mexico, which has experienced out-migration to the USA, especially in northern regions of the country.

## Ethnicity

In general terms, cities show much more ethnic variety than rural areas. This is largely because migration into cities is more common than migration into rural areas. The most ethnically diverse country is the USA, with its long history of

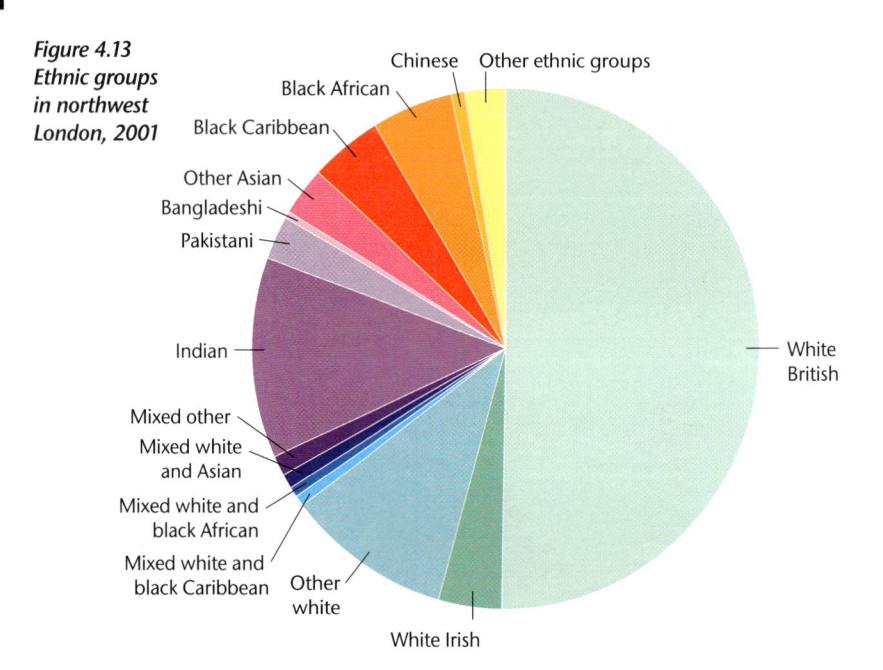

# Chapter 4  Population characteristics

**Figure 4.13 Ethnic groups in northwest London, 2001**

[Pie chart showing ethnic groups: White British, White Irish, Other white, Mixed white and black Caribbean, Mixed white and black African, Mixed white and Asian, Mixed other, Indian, Pakistani, Bangladeshi, Other Asian, Black Caribbean, Black African, Chinese, Other ethnic groups]

in-migration, and the cities of the southwest are among the most diverse of all. In Los Angeles there is a markedly segregated structure, because people wish to live among others of the same ethnic background. This wish is driven by a number of factors, including:

- language
- religion
- family ties
- culture

The level of ethnic diversity in Los Angeles is rarely matched in European cities, but London has become more diverse in the last 100 years. The data shown in Figure 4.13 show a mixture of people, many from former colonies, living in northwest London. Fertility rates vary from group to group. As a general rule the more recent the arrival the lower the wages they receive and the higher the fertility rate. Figure 4.14 shows that there are significantly more younger people (20–44) in the city than in the country as a whole.

**Figure 4.14 Age profile: northwest London compared with England and Wales, 2001**

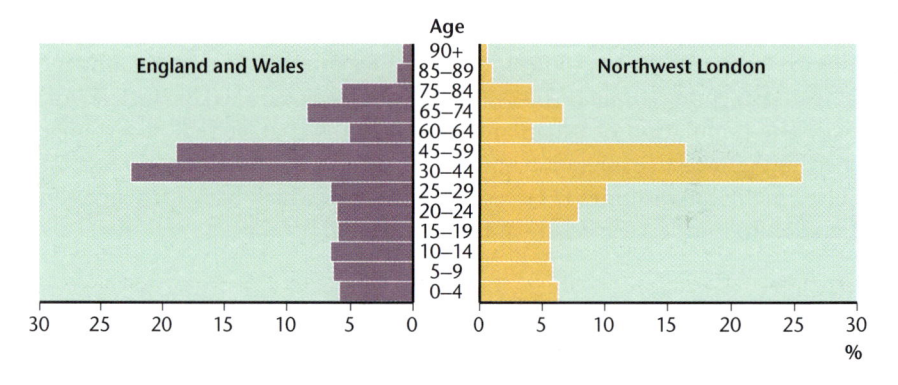

[Population pyramid comparing England and Wales with Northwest London by age groups: 0–4, 5–9, 10–14, 15–19, 20–24, 25–29, 30–44, 45–59, 60–64, 65–74, 75–84, 85–89, 90+]

132

# The components of population change

**Population change** is the change in population over a given period, expressed in absolute figures or proportional figures (rate of change).

**Birth rate** (BR) is the number of live births in a given time (usually 1 year), expressed as a rate (per 1000 population). Birth rate is calculated for the whole population.

**Death rate** (or mortality rate) (DR) is the number of deaths in a given time (usually 1 year), expressed as a rate (per 1000 population).

> *Examiner's tip*
>
> - Don't forget that birth rate and death rate are *rates*. If the BR is 10 per 1000, then expect 10 births in a 'typical' village of 1000 people, 100 in a town of 10 000, and so on. Larger places will have larger numbers of births (and deaths) but the rate may be exactly the same.

**Infant mortality** is usually expressed as the chance of dying in one's first year of life. In most societies this is the riskiest year of existence until old age. In the UK the figure is 7 in 1000, meaning that seven children die before the age of one. In Sierra Leone it is 180 per 1000, meaning that nearly one child in five dies before its first birthday. Infant mortality is wrapped up with death rate in published figures, so the death rate includes infant mortality.

**Life expectancy** is the average age of people at death. If 1000 people die, 500 at the age of 40 and the other 500 at the age of 60, the life expectancy for that group is 50. Life expectancies vary globally, from 35 in Sierra Leone to 82 in Japan and Iceland. The most critical factor controlling life expectancy is the rate of infant mortality. Where infant mortality is high, life expectancy is low, as in Sierra Leone. Odd as it might seem, life expectancy for an individual in Sierra Leone goes up in the first couple of years of life if they survive those dangerous years.

> *Examiner's tip*
>
> - Life expectancy is an average. A life expectancy of 35 at birth in a country does not mean that very few people survive to old age; it means that very many people die when they are still young.

**Fertility rate** is the average number of children born to a woman in her reproductive life. This figure is very illuminating. Should a woman produce two children, she has, in effect, produced replacements for herself and her partner. (Obviously, not all women who have two children will have one girl and one boy, but it does even out over the whole population.) If the fertility rate is greater than 2, the population is expanding; if the fertility rate is below 2, the population is contracting. The fertility rate in the UK is currently 1.7.

# Chapter 4 Population characteristics

|  | Desired fertility rate | Actual fertility rate |
|---|---|---|
| Austria | 2.0 | 1.69 |
| Belgium | 2.1 | 1.84 |
| Finland | 2.2 | 1.95 |
| France | 2.3 | 2.10 |
| Germany | 2.0 | 1.65 |
| Hungary | 2.1 | 2.02 |
| Italy | 2.1 | 1.65 |
| Netherlands | 2.1 | 1.85 |
| Norway | 2.2 | 2.09 |
| Poland | 2.3 | 2.18 |
| Portugal | 2.1 | 1.90 |
| Spain | 2.2 | 1.75 |
| Sweden | 2.5 | 2.04 |
| Switzerland | 2.2 | 1.77 |
| USA | 2.3 | 2.02 |

*Table 4.6 Comparison of actual fertility rates and desired fertility rates*

It is also worth distinguishing between the **actual fertility rate** (what happens) and the **desired fertility rate**, which is evaluated by asking women how many children they would like (Table 4.6).

**Natural increase** is the difference between births and deaths, expressed numerically or as a rate, usually as a percentage. Thus if BR = 20 per 1000 and DR = 10 per 1000, then the rate of increase is 10 per 1000 or 1%. When births exceed deaths, the change is a natural increase; when deaths exceed births, the change is a natural decrease, although some textbooks still use the term natural increase and have a negative figure.

The total change of population (on any scale other than global) incorporates migration. People move frequently in modern societies such as the UK or the USA. Each year one person in seven moves in the USA, meaning that nearly 43 million people move house each year out of a total population of about 300 million.

**Net migration** equals the number of **immigrants** less the number of **emigrants**.

- **Immigrants** are those who arrive in an area, sometimes referred to as **in-migrants**; these people arrive in **receiving areas**.
- **Emigrants** are those who leave **donor areas** to go elsewhere.

Rates of population increase can be deceptive in that seemingly unimpressive annual rates of increase can become substantial over time, rather like compound interest with a bank account. The world's population, currently just over 6 billion, is growing at a rate of about 1.3% per annum (per year). This can be translated as a **doubling time** of 54 years if the current rates are maintained, so the world's population might become 12 billion by 2058. The world's growth rate peaked in the 1960s at 2%, a doubling time of 35 years. In reality, the rate is still falling year on year and the doubling time may very well be a good deal longer than 54 years.

At a national level, doubling time can yield unlikely results. Afghanistan has a birth rate of 48 per 1000 and a death rate of 21 per 1000, yielding a rate of natural increase of 2.7%. This represents a doubling time of about 26 years.

The short-cut route to calculating doubling time is to divide 70 by the natural increase. Thus a country with a 1% rate of natural increase will double its population every 70 years; with a 2% rate of natural increase, it will double its population in 35 years. This is a rough and ready method, but, given all the other problems of such predictions, it will do. Table 4.7 gives a range of data for a selection of countries.

There are a number of technical issues to address with all such numerical data:
- census data, even if accurate for that year, age rapidly
- the forecasts are based on best guesses about future trends, which may prove to be quite some way from reality
- the quality of the original data is highly variable

**Examiner's tip**

- Don't forget that migration is a balance between people arriving and people leaving.

134

*Human environments* **AS Unit 2**

| Country | Population (in 2004) | Birth rate (per 1000) | Death rate (per 1000) | Rate of natural increase (%) | Projected population 2025 (millions) | Fertility rate | Life expectancy at birth | Population density (km$^{-2}$) |
|---|---|---|---|---|---|---|---|---|
| Afghanistan | 28.5 | 48 | 21 | 2.7 | 50.3 | 6.8 | 43 | 44 |
| Botswana | 1.7 | 27 | 26 | 0.1 | 1.1 | 3.5 | 36 | 2.7 |
| France | 60.0 | 13 | 9 | 0.4 | 63.4 | 1.9 | 79 | 109 |
| Indonesia | 218.7 | 22 | 6 | 1.6 | 275.5 | 2.6 | 68 | 115 |
| Mexico | 106.2 | 25 | 5 | 2.0 | 131.7 | 2.8 | 75 | 54 |
| Romania | 21.7 | 10 | 12 | –0.2 | 18.1 | 1.2 | 71 | 91 |
| UK | 59.7 | 12 | 10 | 0.2 | 64.0 | 1.7 | 78 | 244 |
| USA | 293.6 | 14 | 8 | 0.6 | 349.4 | 2.0 | 77 | 31 |

However, the figures are reliable enough for us to make a number of points about these contrasts:

*Table 4.7 Some population data*

- some countries, of which Romania is one, are now experiencing falling populations
- low fertility rates in a number of other countries suggest that populations will soon be falling there, e.g. the UK and France, unless in-migration exceeds the negative rate of natural increase
- birth rates appear to be more variable than death rates
- fertility rates and life expectancies both show large variations, as do population densities
- there is no relationship between population density and population size

## Reasons for variations in fertility and mortality

Fertility rates and death rates vary:
- from time to time
- from place to place

When considering the so-called 'crude' birth rates and 'crude' death rates, the most important factor explaining variation in both dimensions, space and time, is the age structure of the population. The highest death rates occur where the population is oldest and the lowest where the population is youngest. Thus in Afghanistan 45% of the population is under 15 and only 2% over 65. Even though the country is war-torn and suffers great poverty, this age structure keeps death rate well below birth rate. In some African states death rates are equivalent to those in western Europe and other MEDCs. In Ghana, for example, the death rate is 10 per 1000, because 40% of the population is under 15 and the risk of dying at that age, even in a poor LEDC, is quite small. Of course the HIV/AIDS crisis has radically altered these figures in some countries, as the Botswana data suggest (see Figure 4.8).

Demographers tend to use age-adjusted data to explore variations in both birth and death rates. It is not surprising to find that places with large numbers of elderly people have higher death rates and lower birth rates

*Examiner's tip*

- A high death rate may not indicate unfavourable living conditions. On the contrary, the higher death rate in Cornwall or Florida reflects the fact that elderly people wish to retire there because of the perceived attractions of the areas. By doing so they push up the death rate and push down the birth rate.

## Chapter 4 Population characteristics

than neighbouring towns with younger profiles. Thus the birth rate in Beverly Hills is lower than that in East Los Angeles, while the death rate is higher (see Figure 4.10 for the population pyramids for these areas).

### Factors explaining variations in death rate

Broadly speaking, death rate is determined by factors we cannot control. Unlike birth rate, it is not determined by social and cultural processes but by factors that are **exogenous** or outside the social system. Death rates have varied historically — they have been as high as 100–200 per 1000 during the great pandemics of bubonic plague and are as low as 2 per 1000 in some modern societies that are also youthful, such as Kuwait.

*Diet*

There are two relevant factors here: malnourishment and undernourishment.
- **Malnourishment** is when the diet is not properly balanced. It makes the body less able to resist disease and thus makes death more likely. It might also provoke certain diseases such as cardiac conditions and cancers. The 'burger and chips' diet is as threatening to a middle-aged businessman in an MEDC as a rice diet is to a middle-aged peasant farmer in Bangladesh.
- **Undernourishment** is shortage of calorific intake which, if allowed to continue, will ultimately result in death. Children die more rapidly of undernourishment as they have lower resistance to many diseases.

*Fresh water*

The availability of fresh water is widely seen to be the most important factor explaining variations in death rate from place to place and time to time, other than the age structure of the population. The absence of clean fresh water is especially dangerous for children and contributes to the fact that the world's major killing diseases are almost all avoidable. Examples are diarrhoea and dysentery. One form of diarrhoea alone, rotavirus, kills over 500 000 children a year. A lack of fresh water is believed to cause the deaths of more than 2000 children under 1 year every day, almost all of them in the poorest countries (Figure 4.15).

*Figure 4.15 Percentage of deaths under 5 years due to diarrhoea for countries at different income levels*

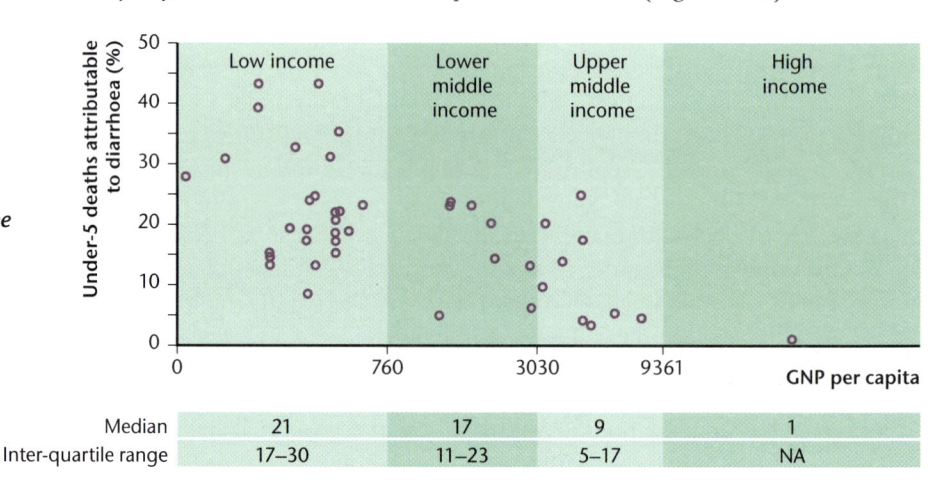

| | Low income | Lower middle income | Upper middle income | High income |
|---|---|---|---|---|
| Median | 21 | 17 | 9 | 1 |
| Inter-quartile range | 17–30 | 11–23 | 5–17 | NA |

136

*Access to medicines*

The quality of medical provision can certainly prolong life. Immunisation programmes for children have led to marked reductions in the incidence of smallpox (which has been eradicated) and other epidemic diseases. Malaria remains a big problem in many countries, as does HIV/AIDS, but in both cases progress is being made. Variations in death rate may very well reflect the affordability of treatment rather than the existence of the treatment itself.

*War and civil disorder*

Wars can have a profound impact on death rate. The 20 million Russians who died between 1940 and 1945 in the Second World War made a heavy indent in the age/gender population pyramid. Historically, war used to be gender selective — almost all those killed in the First World War were men — but the era of 'total war' has tended to lessen this distinction. Unlike the other causes of death, periods of war tend to be episodic in nature and show up as sharp peaks in the timelines of death rate as recorded on demographic transition graphs.

*Endogenous factors*

In most MEDCs there are relatively small differences in the above variables from place to place within a country. Some places may have better healthcare than others, but the other variables are unlikely to show marked variation. However, death rate does vary spatially, and when the age structure is disregarded the highest death rates are found in poor inner-city areas. The suggested reasons for this are problems such as alcoholism, smoking, drug abuse, poor diet, suicide and murder. In the UK, for example, central Glasgow has a higher age-adjusted death rate than Surrey, and Liverpool has a higher age-adjusted rate than the Fylde coast. At an international level, life expectancy for men has been falling in many eastern European countries as unemployment has risen during the post-Soviet era.

## Factors explaining variations in fertility rates

Variations in birth rate tend to be less extreme than variations in death rate because, for the most part, birth rates are controlled by economic, social, cultural and even political factors. Generally, these are **endogenous** (internal) factors. Nonetheless, birth rate does vary widely at a global level and in some countries is in the 40s and even low 50s per 1000. The west African state of Niger is the current hot spot, with a birth rate of 55 per 1000 and a fertility rate of 8. Low birth rates of 8 or 9 per 1000 are found in Japan, Hong Kong and, for very different reasons, many eastern European countries.

Once again the main factor that explains these differences is the age structure of the population, and, at a local scale, the gender balance. Given that 50% of the population is biologically incapable of having children, a strongly masculine population will have a lower birth rate than a more balanced one. What might be called the feasible upper limit for birth rate is much lower than that for death rate. Obviously the death rate can reach 1000 per 1000 if there is a cataclysmic disaster, such as that which overwhelmed the islands neighbouring Krakatoa when it erupted in 1883. This cannot happen with birth rate.

# Chapter 4
## Population characteristics

- Males cannot have children and therefore the maximum rate can only be 500 per 1000.
- Only females of menstruating age can have children; in the UK today this is 60% of females, reducing the maximum rate to 300 per 1000.
- Perhaps as many as 10% of females of child-bearing age are infertile, reducing the rate to 270 per 1000.

For the rest, it becomes a question of decision making, usually in a complex social context. If everyone who could have a baby had one *every year*, the birth rate would be 270 per 1000.

### The cost and benefits of children

The most important factor controlling variations in birth rate is whether or not people need children. In some societies, more children might mean more wealth. In areas relying on subsistence agriculture, children from the age of about 5 often produce more with their labour than they consume. This is achieved largely by carrying out small chores around the vegetable plots, and fetching and carrying firewood or water, thus releasing elder siblings or parents for more productive work. This dynamic is especially true if land is in good supply. Even in the early stages of industrial societies, when child labour was common, a bigger family meant a richer family.

In most MEDCs today, having children satisfies a deep-seated genetic need and a number of social norms, but is certainly not 'cost effective'. Thus the decision *to have* children will be made for reasons other than economic. On the other hand, in these countries the decision *not to have* children, or to delay having children, may very well be economic.

### Other factors

- **Welfare and pensions provision.** If there is a limited social welfare system, or if that system is being dismantled, one might expect the birth rate to be higher. This is because people create their own social welfare system by having children who can look after them in their declining years.
- **More tickets in the lottery.** The larger the family size then the greater are the chances that one child will become successful and promote the whole family out of poverty.
- **Attitudes to women.** In many societies, the birth of a girl was or still is no great cause for celebration. In such societies girls eventually move out of the family and live with their husbands, taking with them a valuable dowry (a gift) in the form of land or a few head of livestock. Male children, on the other hand, marry and bring dowries into the family. The impact of this attitude to children is to increase family size as couples seek to have boys to 'compensate' for expensive girls.
- **Male dominance.** There are a number of societies in which the social status of a man is still measured in terms of the number of children (especially boys) he can produce, and his wife (or wives) does not have much choice in the matter. A number of surveys on contraceptive use in both South America and Africa have revealed significant differences between the sexes in response to the

## Case study 4.3 Falling fertility in the UK

Birth rates (12 per 1000) and fertility rates (1.7) are high in the UK compared with many neighbouring countries. However, note that a fertility rate of 1.7 is well below replacement level and thus both rates are likely to go on falling in the future. There are a number of significant social trends behind these figures in the UK.

- Rates of divorce are at 40%.
- In 1960, 90% of the population of 60 or over either was married or had been married. Projections suggest that this figure will fall to 65% by 2010.
- The mean age at which women have their first child rose from 24.6 in 1976 to 28.9 in 1998 and is believed to have topped 30 in 2004.
- In 1976 two-thirds of all births were to women in their twenties; this fell to less than half by 1998.
- In contrast, the proportion of births to women in their thirties rose from 20% to 40% of all births in the same period.
- Childlessness was unusual in women born between 1945 and 1955. Only 1 in 10 of these women had no children. Current projections suggest that 1 in 5 women born between 1975 and 1985 will remain childless.
- The Institute of Policy Studies in London found that the three dominant reasons offered for this decline were:
  - an inability to find a suitable partner
  - being with a partner who didn't want children or had them already in a former relationship
  - not feeling economically secure enough to start a family
- Life-style decisions may mean that material objects are regarded as more rewarding than babies, or women wish to pursue careers rather than motherhood.

The falling fertility in the UK seems to arise partly from social difficulties related to marriage and partnership, but mainly from economic factors, particularly for those who have not enjoyed rapidly rising living standards.

In reality, the actual fertility rate seems to be below the desired rate; almost all research suggests that, on average, women would like to have 2.1 children. For the many who now require two incomes to satisfy their material ambitions, child rearing is delayed as it inevitably takes one partner out of employment.

---

question: 'Do you and your partner practise any form of birth control?' Unsurprisingly, not all women appear to be telling their machismo-driven partners the whole truth.
- **Polygamy.** In some societies men take several wives and father many children.
- **High rates of infant mortality** are also significant. If you expect some of your children to die before adulthood, you adjust your reproductive habits accordingly. This factor tends to have a generational delay built into it. Your expectations of child survival are likely to be driven by your own experiences in childhood. Thus if two in five of your siblings died, the number of children you have is likely to be that much higher than if none had died.
- **Education** works on several levels:
  - if children go to school, they become a cost rather than a benefit to their parents, at least in the short run
  - if girls are educated to the same level as boys, they may seek to gain employment, and this empowering process leads to a reduction in birth rate as the age of marriage rises and women delay child rearing
  - education at the village level in remote rural communities in LEDCs may well bring information about contraception that will lead to a fall in the number of accidental or unplanned births

# Chapter 4  Population characteristics

## Case study 4.4  Kerala: an object lesson

Kerala is one of India's most southerly states. It has a tropical climate, a high population density of 820 people km$^{-2}$ and an overall population of 31.8 million, according to the 2001 Indian national census. It is the second most densely populated region of India.

Kerala is best known for the so-called 'Kerala model', whereby population growth, life expectancy and literacy levels have been brought up to levels that compare favourably with parts of the developed world. Its GDP per capita is low compared to MEDCs, but much higher than the average for India.

This demonstrates that Kerala has all the social and developmental hallmarks of advanced MEDCs, despite an income level about a tenth of that in the UK.

The reasons for this include the following.

- A progressive education policy, which included females, was instituted in the 1920s and continued in the 1960s by the first elected communist regime in an Indian region. More Keralan girls go to university than Keralan boys — a rare situation.
- Land reform programmes led by the 'land to the tiller' reform in the 1960s broke up the old estates and redistributed them to landless poor, often Muslims. No family was allowed to own more than 8 hectares. This allowed the poor to grow their own food rather than rely on their large and growing families to produce higher incomes.
- The 'right to literacy' became a popular mass movement. In the 1970s the Kerala People's Science Movement (the KSSP) set up study classes, medical camps and literacy classes in villages. By the early 1980s there were nearly 5000 village libraries which were free to the users.
- Family planning was facilitated, along with cheaper health provision and a widespread education programme.
- Inspired political leadership: Kerala's socialist and communist elected politicians pursued a non-violent democratic path to achieve the change they sought. Influenced by the great Italian writer Gramsci, they sought the reform of institutions. They were also quick to see the importance of women's rights and environmental issues in their peaceful drive to reform Kerala.

Figure 4.16 The location of Kerala in southern India

| Indicator | Kerala | India as a whole | Low-income countries | UK |
| --- | --- | --- | --- | --- |
| GDP ($ per capita at PPP*) | 2950 | 460 | 432 | 27 000 |
| Adult literacy rate (%) | 91 | 58 | 39 | 98 |
| Life expectancy (years) | 71 | 64 | 59 | 79 |
| Infant mortality (per 1000) | 12 | 65 | 80 | 7 |
| Fertility rate (per 1000) | 1.8 | 3.1 | 3.5 | 1.7 |

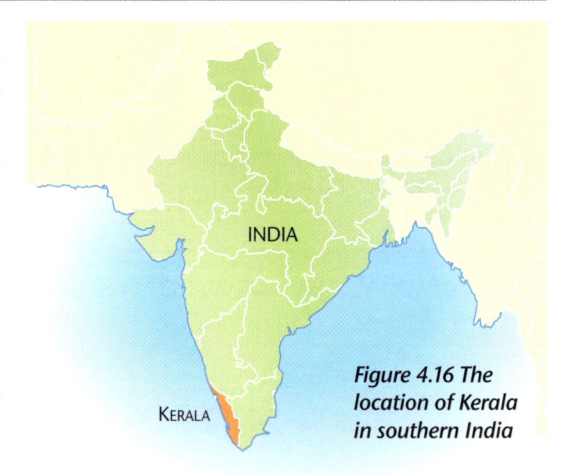

Table 4.8 Development data for Kerala

* Purchasing power parity

- **Religious beliefs** are important, although not perhaps as significant as some would think. In Northern Ireland the Catholic community has a significantly higher fertility rate than the Protestant. On the other hand, two countries with

Human environments · AS Unit 2

among the lowest fertility rates in the world are Italy and Spain, both of which are Catholic. Religion is obviously not a controlling factor.
- **Contraception.** There is no doubt that the availability of contraception, especially the oral contraceptive pill for women, has modified social and sexual behaviour and facilitated the control of family size. Most of the evidence suggests that where contraception is freely available and couples are able to use it, the gap between desired fertility and actual fertility is quite small.

In many parts of the world declining fertility is a goal of government policy rather than an unlooked-for consequence of social trends. One of the most successful schemes to reduce fertility rates has been in the province of Kerala in southern India.

# The demographic transition model

The **demographic transition model** is a simplified, visual representation of how birth and death rates in a country change as it develops. The model was based on the experience of the UK population as the country industrialised but, as with any model, it simplifies some of the changes.

Figure 4.17 shows the stages of the demographic transition model.

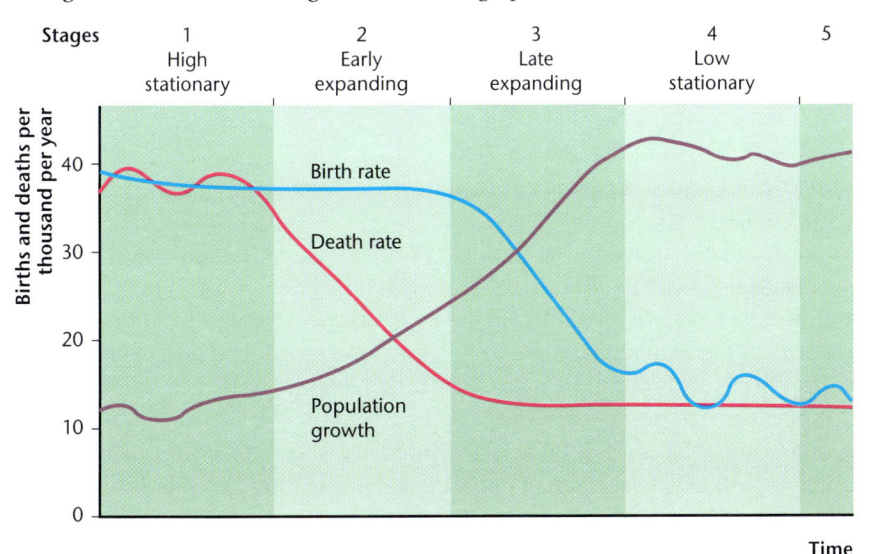

*Figure 4.17 The demographic transition model*

- In Stage 1, the **high stationary stage**, both the birth rate (BR) and the death rate (DR) are high and fluctuate. Natural increase is low because the huge number of deaths cancels out the high BR. These societies have high rates of infant mortality, low life expectancies and high birth rates driven by economic need. Death rates tend to fluctuate according to the waves of pandemic diseases that affect such societies. There are no modern examples of countries or regions in Stage 1.

# Chapter 4 Population characteristics

- In Stage 2, the **early expanding stage**, the DR falls but the BR remains high. The rate of natural increase goes up and size of the population begins to rise. The fall in DR is caused by improvements in diet and a reduction in infant mortality. A number of countries are in this stage, examples being Afghanistan and Sierra Leone.
- In Stage 3, the **late expanding stage**, the BR begins to fall. The DR continues to fall, albeit at a slower rate. The rate of natural increase, having peaked at the end of Stage 2, begins to fall so population growth slows down. Very many LEDCs are in this stage, with China and India being notable examples.
- In Stage 4, the **low stationary stage**, the BR and DR are both low and fluctuating, albeit within fairly narrow limits. Thus the rate of natural increase is low. The USA is the best modern example.
- In some societies it may be possible to add a Stage 5, the **slow declining stage**, when the birth rate has fallen and, because of the increasing average age of the population, the death rate has risen. At this stage the population begins to decline. Many MEDCs, especially in Europe, now appear to be in this stage.

## What the model does

- The model describes population changes over time in a generalised way that can be applied to a number of MEDCs.
- It allows a few simple statements to be made about the population characteristics of societies at various stages of development.
- It is possible to link population structure and migration to the demographic transition model.

## What the model does not do

- It does not forecast changes or provide explanations as to why one stage might lead to the next.
- It does not claim any universality. Just because it works in some places does not mean it will work in all.
- It does not explain anything. Because a country is in Stage 3 of the model, it does not mean that it will inevitably pass into Stage 4. It is not pre-ordained.
- The model only looks at natural increase. For example, it is important to remember that during the time period covered for the UK, there was a major migration overseas involving many millions who did not find work in the cities. This, of course, had an effect on the population as a whole and on its age structure.

## The demographic model and the UK

The demographic model is a considerable simplification of what actually took place in the UK, ironing out some significant bumps and jumps in the changes. Figure 4.18 shows actual data for the UK, with the stages of the demographic model added in.

Even though the model is based on the UK, the UK data differ from the model in a number of ways.

*Human environments*  **AS Unit 2**

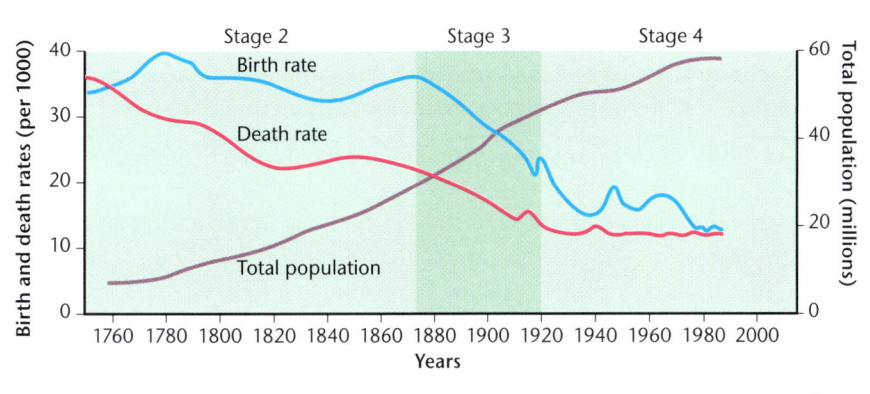

*Figure 4.18 Demographic transition data for the UK, 1750–1985*

- During the long fall in death rate in Stage 2 there were two 20-year periods in which it rose:
  - between 1780 and 1800 a series of poor harvests led to a rise in death rate, especially in rural areas
  - between 1820 and 1840 early industrialisation led to a rise in ill-health, with cholera and other vector-borne diseases common in the unsanitary cities
- The death rate rose for a short period in Stage 3, when the combined impact of the First World War and the postwar influenza epidemic led to more than 1 million deaths in the space of 5 years, in addition to the 'normal' deaths.
- Birth rate dipped sharply around 1800 and then rose again in the mid-nineteenth century. Levels of child labour declined with the onset of industrialisation, and so the cost of bringing up children rose and the BR fell. This was followed by a rise in BR that may have been a belated reaction to the rise in death rates a generation earlier but was also a consequence of improved incomes, younger marriage and better diet.
- The well-known 'baby booms' after both major twentieth-century wars created a surge in birth rates. Family formation had been delayed during wartime and the depressions that had preceded both wars. Postwar optimism made itself felt after both wars.
- Many city areas did not fit the pattern described by the model. For instance, there were differences in BR between skilled and unskilled workers during the period of the Industrial Revolution and there are variations today in BR and DR between different social groups.

There are three key questions to answer regarding this model.

### Question 1 Why did death rate start to decline at the end of Stage 1?
*Answer*
- Diets improved as a consequence of improved agricultural output, which began in the mid-eighteenth century. The gradual commercialisation of agriculture moved people off the land, but the new industries that were beginning to emerge at the end of the century were able to absorb many of them.
- During the nineteenth century, local businesses started to reform the poor housing and sanitary conditions in the major cities. They required a fitter and more skilled workforce, and so they built sewers and improved the water

# Chapter 4

## Population characteristics

supply. These improvements became more commonplace, especially after the mid-century Public Health Acts passed the costs on to local government.
- Immunisation and reduced infant mortality reduced the death rate.

*Question 2 Why did birth rate stay high until the 1870s (end of Stage 2)?*
*Answer*
- Large families remained a logical decision for most households. The move to labour-intensive industrialisation changed little. Child labour, high infant mortality and low incomes combined to ensure that birth rate was still driven by poverty.

*Question 3 Why did birth rate start to fall in the 1870s (start of Stage 3)?*
*Answer*
- The demand for labour was changing. As machines replaced labour, the requirement was for a smaller but more skilled workforce.
- The various Factory Acts of 1833, 1844 and 1867 had made child labour less widespread, although it was still quite common.
- In response, the Elementary School Act of 1870 had made school attendance to age 12 compulsory. This rapidly fed through to changed social and reproductive behaviour, reinforcing the earlier shift from children being a benefit to becoming a cost.

> **Examiner's tip**
> - Don't forget that all models simplify reality, though this alone is not a very illuminating criticism. You need to be able to develop your criticism beyond this simple statement.

# The characteristics of ageing and youthful populations

## Dependency

The concept of dependency is simple enough. It refers to those people in a population who are unable to provide for themselves. They might be too young, too old, too ill or too preoccupied with child rearing to expend the effort necessary to find their own provisions. Of course in a modern economy few of us do this in any event, but the idea is firmly embedded. The usual way of estimating dependency within a population is to calculate a dependency ratio.

## Dependency ratio

As a shorthand measure of dependency, those between 15 and 65 are known as the active/working population, and those under 15 and over 65 are known as the dependent population, respectively young and old dependants. The dependency ratio is worked out using this formula:

$$\text{dependency ratio} = \frac{(\% \text{ under } 15) + (\% \text{ over } 65)}{(\% \text{ between } 15 \text{ and } 65)} \times 100$$

A couple of worked examples should make this clearer.

Syria, which is an LEDC, has 41% of its population younger than 15, and 4% over 65. This leaves 55% between the ages of 15 and 64 (100 − (41 + 4)).

$$\text{dependency ratio} = \frac{41 + 4}{55} \times 100$$
$$= \frac{45}{55} \times 100$$
$$= 81.8$$

Iceland, a developed country, has 23% of its population younger than 15, and 12% over 65. This leaves 65% between 15 and 64.

$$\text{dependency ratio} = \frac{23 + 12}{65} \times 100$$
$$= \frac{35}{65} \times 100$$
$$= 53.8$$

Generally, the dependency ratio for MEDCs is between 50 and 70, whereas for some LEDCs it is close to 100.

However, only 62% of the officially 'active' population of the EU is actually in work. Thus, the dependency ratio is far from being a satisfactory measurement, because it uses age as the only criterion for assessing dependency.

### Problems of using the dependency ratio in MEDCs
- Many young people stay in education well beyond 15. In fact the compulsory school leaving age in many countries is 16.
- There are many who stay in work beyond 65, whereas others retire before that age.
- Many people who are 'active' in terms of age choose not to work; the largest single category is parents at home with young children.
- The ratio does not differentiate between the young dependants and the elderly dependants, and this is an important distinction for government planners.

### Problems of using the dependency ratio in LEDCs
- Many children in the poorest countries contribute economically at a young age, especially in subsistence agriculture or the informal sector in cities.
- Retirement is a luxury that many elderly cannot afford in the poorest countries, and they continue to work into old age.
- Unemployment or underemployment may mean that, despite being 'active', people are not able to support 'dependants'.
- Again, the ratio does not differentiate between the young dependants and the elderly dependants.

## The ageing world

With widespread falling fertility rates and, outside sub-Saharan Africa, significant increases in life expectancy, the statistics tell an obvious story. The median age is rising and it will affect the economies of all countries. Today's younger generation is outnumbered by its parents' generation. Given that fertility rates are at unprecedented lows in so many countries, the 'greying' of the population is bound to continue for quite a while, whatever the reproductive behaviour of

# Chapter 4  Population characteristics

|  | 1975 | 2000 | 2025 |
|---|---|---|---|
| Africa | 18.5 | 17.5 | 22.0 |
| Latin America | 20.0 | 24.0 | 30.0 |
| North America | 30.0 | 36.0 | 38.0 |
| East Asia | 23.5 | 31.0 | 38.5 |
| South Asia | 20.5 | 24.0 | 31.5 |
| Europe | 30.5 | 38.0 | 41.0 |
| World | 23.5 | 26.5 | 32.0 |

*Table 4.9 Median ages*

your generation. In the UK the median age is set to rise to 46 by 2060, with 26% of the population over 65. To maintain current levels of production it is estimated that the present generation of 15–19-year-olds will need to work until they are 74. Table 4.9 shows the median ages of the populations of continental areas.

The median is a better measure than the mean in distributions that are skewed because it takes the middle value in a series, not the average. The average is distorted if there are large numbers of either young or old (see Chapter 8).

## Case study 4.5  Europe's ageing population

The trend shown in Table 4.10 is a result of:
- a falling fertility rate and
- a gradual edging-up of life expectancy

There are variations both between countries and within them. In the UK overall, 7.5% of the population is older than 75. However, in some areas, Fylde in Lancashire for example, the figure reaches 12%. Fylde is a coastal town, popular for retirement. Unsurprisingly, Fylde also has a death rate above the national figure and a birth rate well below the national figure.

Many issues are raised by the ageing population of Europe.
- The EU countries spend an average of 12% of their GDP on pensions (only 4% is spent in the USA), so the elderly are relatively well provided for, but at the cost of a huge strain on national budgets.
- There are increasing numbers of very old people (aged 80+). For example, by 2025 it is estimated that 7.1% of Italians will be over 80. The very old are far more expensive in terms of healthcare and social support.

*Table 4.10 The population of western Europe, by age and sex, 1996–2025 (predicted)*

|  | 1996 | 2003 | 2010 | 2025 |
|---|---|---|---|---|
| 0–14, all (%) | 17.3 | 16.4 | 15.4 | 14.2 |
| 15–64, all (%) | 67.0 | 66.7 | 66.4 | 63.1 |
| 65+, all (%) | 15.6 | 16.9 | 18.2 | 22.7 |
| 65+, male (%) | 12.8 | 14.2 | 15.6 | 20.1 |
| 65+, female (%) | 18.4 | 19.4 | 20.7 | 25.3 |

- Most European states have highly mobile populations; caring for elderly relatives places constraints upon labour mobility and family life.
- High and increasing rates of family break-up complicate the allocation of caring responsibilities as families become ever more fractured.
- There is a lack of skills in some areas and a lack of labour, especially in service jobs at the lower end of the labour market. Raising the age of retirement is one solution already being pursued.
- The influence of the older population in countries such as the UK is considerable. Brought up to 'save', the 55–64 year old cohort saves 85% more than the average UK household and represents 68% of those with £10 000 or more saved in stocks or in banks and building societies. They own their homes and have a large disposable income.
- Increasing numbers of older people in the electorate mean that politicians need to be aware of their potential power when allocating resources.

The greying of the European population also has an impact on specific locations.
- Retirement migration to coastal resorts and southern Europe brings a higher demand for specialist housing and services.
- A poorer elderly population is left in inner cities or declining rural areas as a consequence of out-migration.
- The influx of retirement migrants can regenerate declining rural areas, bringing wealth and spending power into otherwise remote regions, such as Cornwall or Brittany.

Despite the obvious fact that an ageing population is not necessarily a 'bad thing', it does require some policy changes on the part of European governments. Governments need to aim at:

- ensuring economic security in old age
- combatting social exclusion from age discrimination
- providing long-term care in the context of changing family and residence patterns

## Youthful populations

Many LEDCs in earlier stages of demographic transition still support extremely youthful populations. For example, Ethiopia and Indonesia have 50% of their populations under 15 years of age. In countries such as Mexico, the population is not expected to stabilise until about 2035, as those who are of reproductive age have to 'work through' the system. In other words, although fertility rates and birth rates are beginning to fall, the cohorts of each successive generation are larger, reflecting birth rates in the past. Thus, although the birth rate might be declining, the number of children being born is still increasing.

Countries with youthful structures have to organise their distribution of resources rather differently, with a much greater emphasis on education. In countries such as Tanzania, crippling debt burdens have inhibited this effort.

# The population debate

Many argue that the world contains sufficient resources to support a potential population of 10–12 billion. However, inequalities in distribution mean that many people (about 1.5 billion) are never far away from famine, and about 4000 children under the age of 1 year die every day from a lack of fresh water and basic sanitation.

The question of whether there are too many people in the world may seem simple enough, but in fact it is one of the most controversial of all issues. You don't need to take sides in the debate (although you can if you want to), but you do need to understand the different views.

## The Malthusian viewpoint

**Thomas Malthus** wrote his essay on population at the very end of the eighteenth century, which was a time of great change, both economically and politically. He wanted the Poor Law abolished (this was an early form of social welfare), because he believed that by artificially prolonging the life of the poor, the 'least useful' in society would multiply, whereas his own social class would be better able to exercise restraint.

Malthus predicted that the rapid (geometric) increase in population would not be matched by an equivalent growth in food supply. As a result there would be a gap between food demand and supply that would lead to famine or, perhaps, civil war. The French Revolution was the big political event of the time and Malthus, in common with many others, was fearful of the spread of radical ideas like democracy and liberation, ideas which were, of course, particularly popular with those who did not hold power.

*Photograph 4.2 Thomas Robert Malthus, 1766–1834*

## Chapter 4 — Population characteristics

Those with strictly Malthusian views today tend to be reacting on instinct to the seductive idea that poverty is a function of too many people in the world. They argue that the current shortages of food in LEDCs are the result of overpopulation; in Malthusian terms, 'the power of the population to increase is greater than that of the Earth to sustain it'. Malthusian views, although intellectually influential (Darwin was an admirer), were rather overtaken by the large increases in agricultural yields experienced in the nineteenth century. An ever-growing population in the cities was fed by both domestic and global food production. Hence, at the end of the nineteenth century there were more people in the world and they were also, by and large, better off.

If there was a relationship between people and resources, it was obviously not quite as Malthus had argued. The same historical, positive correlation between population and income levels was repeated in the twentieth century:
- population growth was more rapid than at any other time in human history (from about 1 billion to about 6 billion)
- living standards and quality of life globally rose more than in any other century

As a result, it is not surprising that purely Malthusian views have become untenable. In the past 40 years the so-called **neo-Malthusians** (represented by Professor Ehrlich and many others) have argued that the increased agricultural and industrial activity needed to feed and maintain the population will eventually lead to environmental disaster and intense financial problems. They have abandoned blaming the poor for their own plight but, in some cases, have shifted the focus to the rich 20% of the planet who consume 80% of the resources. Thus the argument is no longer a simple question of numbers, but more to do with the potential of the planet to sustain economic development.

To put it crudely, can the environment survive examples like the simple equation suggested by Table 4.11?

Table 4.11 What could happen if China became as developed as the USA

| Country | Population in 2000 (millions) | Projected population in 2025 (millions) | Number of cars in 2000 (millions) | Number of cars per person in 2000 | Number of cars in 2025, based on current US rates (millions) |
|---|---|---|---|---|---|
| USA | 294 | 349 | 205 | 0.69 | 240 |
| China | 1300 | 1476 | 10 | 0.007 | 1018 |

From Table 4.11 it can be calculated that if China, the fastest growing economy in the world, were to reach US levels of car ownership, there would be a further billion cars on the world's roads by 2025. With 'only' 600 million cars on the world's roads today, and ignoring the other rapidly growing economies such as India, this a frightening statistic for those concerned about the effects of global warming and atmospheric pollution. We are, so the neo-Malthusians believe, heading for catastrophe because we are using up the world's resources at a rate that threatens the sustainability of the planet.

In essence, it is a simple argument: the Earth might technically be able to support many more people, but only by destroying the resource base. It is not just a matter of global warming; there is also desertification, deforestation, salinisation and the destruction of species. The neo-Malthusians develop this argument by warning that some environmental problems might accelerate very rapidly. For example, climatic change might take place over decades, not centuries, and it might be your children rather than some unimaginably distant descendant who has to face the problems caused by our casual waste of planetary resources.

Unlike Malthus, who wanted to control the 'unbridled lust of the poor', the neo-Malthusians identify the rich who have an 'unbridled lust for consumption' as the culprits, and most of these live in MEDC countries (especially the USA, Europe and Japan). This does not make the neo-Malthusian approach universally popular.

### The Boserupian view

Ester Boserup took a wholly different view of the relationship. Writing from a background in economics, she suggested that population growth has a positive impact on people. As we approach a resource boundary (to put it simply, when something starts to run out), we 'invent' or innovate our way out of the problem. Boserup was part of the staff at the United Nations and she wrote the book *The Conditions of Agricultural Growth: The Economics of Agrarian Change under Population Pressure* out of her experience as a consultant in developing countries. Her views are usually labelled as 'anti-Malthusian' and can be encapsulated in a phrase such as 'population growth causes agricultural growth'. This is undoubtedly an implication of her model and is useful to those who do not believe that the (human) carrying capacity of a given area is fixed. By extension of her argument, we (humans) 'invented' farming because we were starting to run out of hunted food. Population growth thus becomes something positive and, what is more, quite central to our development as a species.

This rather reassuring and optimistic view has not gone unchallenged. The neo-Malthusians take no particular issue with Boserup in the context of explaining something of our prehistory and would, by and large, go along with her general idea. What they do dispute is the applicability of such an apparently relaxed attitude in the twenty-first century. For some, the more-people–more-wealth point, which seems to be the incontestable conclusion of the past 200 years, ignores the environmental impact of population growth.

# Overpopulation, underpopulation and optimum population

The concepts of **overpopulation**, **underpopulation** and **optimum population** refer to the balance between population, resources and development. Non-Malthusian theories of population (such as that of Boserup) place little value on such terminology. In fact they treat it as a distraction from the main issue of what causes poverty. At a basic level, these concepts vary depending on the level of economic development. Modern examples of overpopulation are almost

# Chapter 4 Population characteristics

## Case study 4.6 Easter Island

Easter Island is a small and remote island in the Pacific, best known for its statues called Moai. It appears that humans first reached it about 1200 years ago and that the early settlers discovered an environment of dense palm forest quite unlike the treeless and rather bleak landscape of today.

The population of Easter Island reached its peak at perhaps more than 10 000, far exceeding the capacity of the small island's ecosystem. Resources became increasingly scarce, and the once lush palm forests were destroyed. The land was cleared for agriculture and the timber was used for building, for fuel and as the rollers and levers necessary to move the massive Moai. It is thought that as the forests were destroyed, soil erosion took place and the thin topsoil was stripped off. Increasingly desperate, the islanders turned to cannibalism and civil war broke out.

For many, Easter Island has become a story with a clear message for the modern world which has distinct Malthusian overtones. A twist to the story is that the first contacts with Western 'civilisation' proved almost as bad, as disease and slavery reduced the population to a mere 100 or so individuals. It is also worth remarking that it is impossible to find recent examples of such a clear and decisive relationship between local resources and population. Trade and the movement of people has made things more complex in the modern world.

*Photograph 4.3 Easter Island as it is today*

---

impossible to find, given that in almost all societies there is no obvious connection between rising population and falling living standards. Unfortunately for the Malthusians, the obvious examples turn out to be not at all obvious.

**Overpopulation** exists when the population of a country or area is greater than that which its resources can support (sometimes known as its **carrying capacity**), at a given standard of living. Although a high density of population often accompanies overpopulation, it is not the same thing. The implication here is that a reduction in the population would lead to an increase in average wealth.

Frequently quoted examples drawn from sub-Saharan Africa offer an apparently straightforward example of 'overpopulation', but these are complicated when we consider the role of national politics (civil war perhaps), agribusiness involvement in export-crop development on irrigated land, and uneven land distribution on the 'carrying capacity' of a given area.

## Case study 4.7 Is the USA underpopulated?

The enormous supply of land in the USA is the crucial point in understanding American economic history and geography. With a land area of over 7 million km², America is the size of Europe. The amount of land available for settlement was always higher than the growth in population during the period from 1600 to 1900. The economic consequences were profound; labour was always in short supply, and thus was expensive, so Americans soon learnt the value of labour-saving devices, developing an obsession with gadgets that is still obvious.

As land was so abundant it was much easier to become wealthy in the USA than it was in Europe, and colonial America enjoyed a very high standard of living. The USA was a country rooted in its fertile and extensive land supply. Upward social mobility was possible for a large majority of the population, and a vast middle class developed out of the largely poor peasant immigrants.

So is the USA still underpopulated today? There is strong evidence in favour of the idea.

- It has a very low population density when compared to most European countries and Japan.
- It is very well resourced.
- It is less dependent on trade than any other MEDC.
- It is clearly so short of labour in key areas that it recruits skilled labour from around the Pacific for the hi-tech industries of California and less skilled labour from Mexico for other industries.
- It is a high wage economy: certainly at the top end of the labour market.
- It accepts migrants, albeit selectively.
- The increase in its population over recent years corresponds with an increase in living standards and income.

Opponents argue that this is simplistic:
- environmental damage is increasing
- poverty is increasing among a substantial sector of the population
- the growth is not sustainable

Few topics are as controversial as population and few have such an enormous diversity of opinion. Thus for some the USA is obviously underpopulated while for others it is obviously overpopulated.

## Case study 4.8 The UK: underpopulated or overpopulated?

For those persuaded by neo-Malthusian arguments, there is no doubt that the UK is overpopulated. They argue that oil is likely to run out by 2050 and carbon dioxide emissions pose a threat to global temperatures and thus to food supplies. Countries therefore need to calculate the population that can be supported in their own living space without intruding on other countries with an overlarge ecological footprint. According to the Optimum Population Trust, this figure for the UK is about 30 million. This makes the UK very much an overpopulated country as the current population is nearly 60 million. The Trust suggests that the target population of 30 million could be achieved, excepting high rates of immigration, without coercion by 2100 if fertility rates reduced to about 1.3, in line with many other European states.

On the other hand, critics point out that this logic would make highly successful states like Singapore impossibly overpopulated, even though the evidence suggests that as their population has increased so have their living standards. It might also be pointed out that, with a population of nearly 60 million, full employment and a high rate of economic growth, the UK is, in reality, underpopulated. The evidence for underpopulation is the large number of job vacancies and obvious skill shortages that have stimulated a relatively high rate of in-migration in recent years. If there were open borders, which is not the case in the UK, it would seem logical to suggest that an oversupply of labour (too many people) would lead to out-migration, whereas a high demand for labour would lead to in-migration.

# Chapter 4
## Population characteristics

The Optimum Population Trust would reply that one only has to drive to work, or look at house-price rises, to see that we are facing overpopulation, not underpopulation. Supporters of the idea that the UK is overpopulated need then to explain who is going to fill the jobs in London and the southeast currently being filled by international migrants. If the migrants are prevented from coming, which is recommended by many adherents of the 'overpopulation' view, will this lead to an increase in economic activity or a decrease in economic activity?

**Underpopulation** exists when the population of a country or area is insufficient to utilise fully the resources in that area or country. The potential is present to increase national wealth, but there are too few people to do so. This is not at all the same thing as a low density of population. The implication here is that an increase in the population would lead to an increase in average wealth.

## Examination question

Figure 4.19 shows the changes in the Russian population since 1991 and the prediction for 2015.

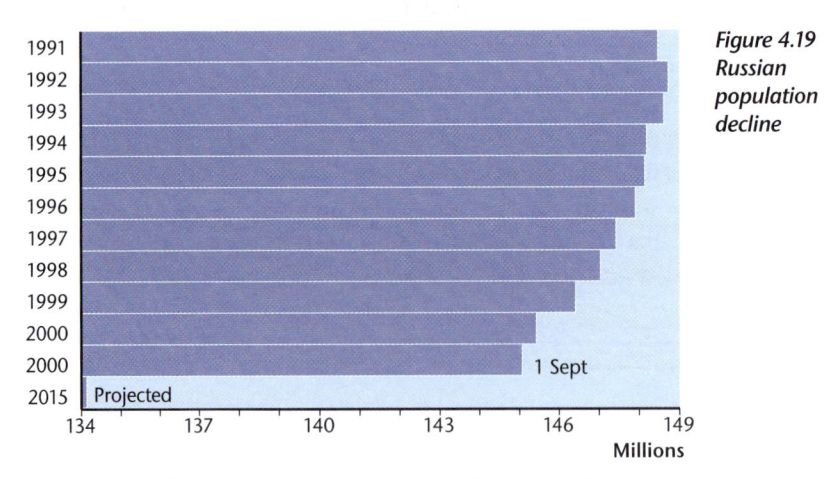

*Figure 4.19 Russian population decline*

You need to refer to Figure 4.19 to answer these questions.

**a** (i) Identify the year in which the population fell by the largest number. (1)
   (ii) Calculate the projected percentage fall in population between 2000 and 2015. (1)
   (iii) Describe the changes in population between 1991 and 2000. (3)

**b** (i) Define the median age of a population. (1)
   (ii) Describe the changes in birth rate and death rate that might lead to a rise in the median age. (4)
   (iii) Explain the social and economic impact of a rise in the median age. (4)

**c** With reference to examples, describe and explain the impact of a rapidly growing population on the physical environment. (6)

*Total = 20 marks*

*Human environments* **AS Unit 2**

# Synoptic link
## The influence of governments on population change

- Data collection (e.g. census) and its role in planning and providing services for a changing population.
- Government policies to increase and reduce birth rates: the reasons and effects.

## Government impact on population change

### Censuses

Information about a country's population, growth, characteristics, living conditions, distribution and physical resources is essential for the development of policies, and the planning and implementation of those policies. Thus the collection and analysis of population and development data constitute a fundamental part of government activity in every state. Over the past two decades, many LEDCs have made real progress in obtaining such data to match those gathered in MEDCs for almost a century.

Planners need population and development data and information:

- to assess demographic trends
- to assess the socioeconomic situation of women
- to design population policies, strategies and programmes
- to integrate population factors into development planning
- to monitor and evaluate the effectiveness of policies and programmes in light of national and international development goals
- to help promote population awareness among government decision makers and the population at large

All these tasks require a vast amount of information, which, in turn, requires a national statistical institutional framework as well as training and research in demographic and statistical methods. The data collected are not perfect. Under-enumeration (undercounting) is a well-known phenomenon in even the most advanced societies, especially in urban areas where the population is frequently on the move and not always keen to be 'counted'. For example, no census form was returned by about 3.0% of the population of Northern Ireland. Some households were also missed by enumerators, and some people, especially those who sleep 'rough', were not included in census returns. The Census Coverage Survey has estimated that this represents a further 1.8% of the population. It is thus estimated that about 95% of the population responded to the 2001 Census.

This so-called 'data barrier' is a particular obstacle for planners in many developing countries, where the quality of the data is often even less reliable. The reliability of census information varies greatly. In some countries self-completion is not possible because of high levels of illiteracy. It is obviously not possible to say quite how inaccurate such data are, but ±5% is a fair margin to allow.

# Chapter 4

## Population characteristics

### The UK census

The 2001 census of the UK took place on the 29 April, and counted almost 60 million people, living in approximately 24 million households. It marked the 200th anniversary of the first census taken in March 1801. The 2001 census asked 40 questions and produced over 2 billion items of information, which will be used to guide about £50 billion of public spending each year. Two centuries of census taking have produced a record of remarkable changes in British society. Here are some examples:

- In 1931, 16% of the population lived in urban areas, compared with 90% in 2001.
- The average size of households has fallen by half in 100 years from 4.6 persons in 1901 to approximately 2.4 persons in 2001.
- We are living much longer: in 1821 almost half of the population was under 20 years of age compared with just over a quarter under 20 years of age in 2001.

Britain was by no means the first country to take a census. The Babylonians and the Chinese collected statistics about their people for taxation purposes more than 3000 years ago, as did the Egyptians and the Romans.

In more modern times, many countries took censuses as the state developed. Quebec completed a census as early as 1666; Iceland did so in 1703. However, there were objections.

- Many people objected to census-taking on biblical grounds — the census had ominous overtones for Christians with unhappy reminders of the Roman census of the Holy Land.
- Others feared that such data would give away too much information to foreign enemies.
- Some saw it as an intrusion on civil liberties as the state grew in power. There are echoes of this in the objections to identity cards today.

The first UK census was approved by Parliament in an atmosphere of fear after the publication of Malthus's essay on population and the dreadful harvest of 1800, which many thought to be an early sign of famine.

Over the past 200 years census questions have developed to reflect changes in society. A religion question was added to the 2001 census, which reflects the fact that increasing numbers of people identify themselves in terms of their religion. Historical examples include the addition of a question about the number of rooms in each household in 1891, in response to fears of overcrowding in industrial cities. A question about 'journey to work' was introduced in 1921 as people moved out of cities to the suburbs and began to take more bus and train journeys.

During the past 200 years, census information has been vital for planning public policy. For example, in 1851 detailed data available in the census were used to classify people by occupation and age. A government statistician concluded that 'miners die in undue proportions,... tailors die in considerable numbers at younger ages'. Such results had an impact on the opinion of politicians and added fuel to movements for social reform. The census was thus a largely progressive phenomenon.

### Major gaps in data in LEDCs

Accurate population information and statistics are indispensable in the study of how demographic trends are affected by, and have an impact on, social, economic and environmental factors. In many countries major gaps in the data exist in the following areas:
- women's reproductive health, especially frequency of AIDS and HIV
- women's work
- migration, particularly at the regional and international levels
- data relevant to the effects of population growth and distribution on poverty, environment and natural resources

## Population policies

Governments can adopt one of three attitudes or sets of policies with respect to their population:
- a pro-natal view, by which an increase in fertility is promoted
- an anti-natal view, by which a decrease in fertility is promoted
- a neutral view, which is indifferent about population growth or decline

Governments can deliberately attempt to change population trends by applying any of the following policies:
- exhortation, by which women (and men) are encouraged or discouraged from having children
- fiscal policies (especially tax rates), which either promote or discourage large families
- social policies, as in the provision of daycare and relatively generous maternity leave
- coercive policies, which punish women (and men) for having too many children
- migration policies, which either encourage or discourage population loss or gain

## Pro-natal policies

The motives for increasing fertility are complex but include the following:
- compensation for wartime losses, as in France after 1918
- to preserve the labour force in an ageing population, as in modern Japan
- to promote a particular race in pursuit of racially motivated policies, as in Nazi Germany in the 1930s
- frontier development, as in modern Brazil with its large areas of under-exploited resources

## Anti-natal policies

### Voluntary policies

Early examples of voluntary anti-natal policies occurred in India and Pakistan in the 1950s. They included the spreading of information about family-planning clinics. In Indonesia the government has set up 2000 family-planning clinics with funding from the World Health Organization, World Bank and United Nations.

# Chapter 4    Population characteristics

The clinics have been promoted through the mass media and fieldworkers. Since 1976 the country has had a rapid drop in the fertility rate.

Land reform and the improvement in the legal status of woman have had similar effects in Kerala (in southern India) where a Marxist-dominated regional government has seen a reduction in poverty as the key to controlling high fertility. It has therefore implemented reforms to eradicate extremes of wealth (see Case study 4.4).

## Coercive policies

The one-child policy in China could be thought of as a good example of a coercive policy. However, look at Case study 4.9.

### Case study 4.9  China and the one-child myth

It is almost received wisdom that in China:
- there was and is a population problem
- its strong government responded with a coercive policy
- this policy forced Chinese families to have only one child
- as a result, China has almost solved its population problem

In fact almost none of this is true.

Annual population growth in China exceeded 2% for most years between 1949, when the communists took over, and 1974. Broadly speaking, the policy of the government was pro-natal. Beginning in the mid-1970s, however, China abruptly shifted gears and fertility declined dramatically. The annual population growth rate has remained around 1.5% since the mid-1970s. Fertility rates dropped sharply in the 1970s in response to rising agricultural output and more opportunities for the rural population. The birth rate had already dropped from 34 per 1000 in 1970 to 18 per 1000 in 1979.

The one-child policy was introduced in 1979, a long time after the significant fall in Chinese fertility rates. Its introduction saw few notable changes, although in fact China experienced a slight increase in birth rate in the 1980s, with the figure fluctuating around 21 births per 1000. There were marked differences between urban and rural areas, with fertility rates running at about 2.5 in rural areas but below replacement levels in cities, at 1.2.

By the mid-1980s, fewer than 20% of all eligible married couples had signed the one-child certificate, which is a contract granting a couple and their child economic and educational advantages in return for promising not to have more than one child. The one-child policy seems to have been strongly and successfully resisted by people, especially couples living in rural regions. Enforcing the one-child policy in the face of such an obstinate desire to have more than one child would have required more forceful measures than the Chinese government was willing to use.

Thus in summary:
- the one-child policy had little impact
- it was resisted by the people
- China has had fewer famines in the past 50 years than for many centuries

## Changing policies

Policies can change over time, and much depends on the attitude of governments, the performance of the economy and demographic trends. With falling fertility it might be predicted that many western European countries will shift towards more obviously pro-natal policies in the next few years. The other possibility is that attitudes towards immigration will change. A recent case of such changing attitudes is to be found in Singapore.

## Case study 4.10 Singapore

Population planning in Singapore is related to economic planning. On an island only 536 km² in size and with few natural resources, the physical carrying capacity of the country is very small. As a result, the government adopted economic and population planning policies. The 'stop at two' population policy of 1965–87 was designed to improve the quality of life of the people.

Restraint of population growth was sold to the people as a necessary move, serving the common good of the country as it struggled to build a nation in the first years of independence from the UK.

However in 1983, Singapore changed dramatically from being anti-natal to being pro-natal. A new attitude was taken, which saw the population as the only asset that the country had. Rapid advances in technology have allowed Singapore to do well in the global economy, concentrating on the high value-added service 'products', such as merchant banking and currency markets, but only because of its highly educated population. The concern now is falling fertility (currently down to 1.4) and the risk that Singapore will experience economic decline if its greatest (and only) asset, its population, begins to decline. The old policy of 'stop at two' has been replaced by the policy of 'have three children or more if you can afford it'. This has now been extended by the so-called 'Baby Bonus' scheme.

The Children Development Co-Savings Scheme, or 'Baby Bonus' for short, aims to make it easier to raise a family. The government is introducing the following measures.

- When a couple has a second or third child, the government will open a Children Development Account for the family. Money will be credited into the account annually until the child is 6 years old.
- For the second child, the government will contribute SG$500 annually into the account. In addition, the government will match dollar-for-dollar any contributions made by the parents into the account, up to SG$1000. The money in the account can be used to pay for the development and education of all the children in the family; it is not just restricted to the second and third child. The Ministry of Community Development and Sports is still ironing out the list of approved uses.
- Under previous policies, the Employment Act only provided paid maternity leave for the first two children. Many Singaporeans felt that the lack of paid maternity leave for the third child was a major obstacle to having more children. To alleviate this problem, mothers will be given 8 weeks paid maternity leave for the third child as well. However, instead of shifting the burden to the employers, the government will pay the wage cost of the maternity leave for the third child, up to a maximum of $20 000.
- The government will cap the Further Tax Rebate for working mothers, which is based on 15% of a woman's annual income, at $20 000 for the third child and $40 000 for the fourth child. These monetary incentives take effect for babies born from 1 April 2005. However, as the problem is a multi-dimensional one, other non-monetary measures will be implemented as well. Three days of paternal leave will be given to married men annually for the first three children and there will be an attempt to achieve a more family-friendly working environment in the civil service.

Singapore is a good example of how changing values and attitudes can lead to a complete reversal of policy; in this case from an almost Malthusian view to that argued by Ester Boserup. Alternatively, one could read this case study as support for the view that different periods need different policies.

# Chapter 5

# Settlement patterns

## Rural settlements

### Nucleated and dispersed settlements

**Nucleated** settlements take the form of large villages, often separated from other villages by expanses of open country with very limited settlement. In contrast, **dispersed** settlements take the form of scattered hamlets or isolated farms, generally with shorter distances between them.

Settlement patterns — nucleated or dispersed settlements, distributed evenly or unevenly in the landscape — reflect both human and physical factors.
- In the uplands of northern England, dispersed settlements are unevenly distributed, with a concentration in the river valleys and a scarcity of settlement at higher altitudes.
- By contrast, the nucleated villages of the Beauce area of the Paris basin are quite evenly distributed across this fertile but relatively featureless plain.

### Case study 5.1 Rural settlements in Ireland

In 1996, the total rural population of Ireland was 1.5 million or 42% of the total population.
- 250 000 people (8% of the Irish population) lived in small towns and nucleated villages with fewer than 1500 inhabitants but with a cluster of streets containing at least 50 inhabited houses.
- 1.25 million people (34% of the Irish population) lived in isolated rural houses or dispersed hamlets.

These dispersed hamlets frequently consist of about 200 houses, with an average of three occupants per house, and the houses are distributed unevenly in the parish area. This type of settlement is more common in the west and north of Ireland, where the soils are poorer and the climate wetter. However, dispersed settlements existed in greater numbers in the past before 'improving' landlords (mostly English) removed them, especially in the south and east of Ireland.

### Key terms

**Dispersed settlement** Pattern of settlement that is made up of small hamlets and many single farmhouses.

**Function** What a settlement does.

**Nucleated settlement** Pattern of settlement where population is concentrated in large villages with few intervening settlements.

**Primate city** The city that dominates its region or country in terms of both size and economic importance.

**Site** The physical characteristics of where a settlement is built.

**Situation** Where a settlement is in relation to other places.

*Human environments*  **AS Unit 2**

*Photograph 5.1
A typical dispersed rural settlement — Garsdale, Cumbria*

These two broad categories can co-exist within the same small area, as shown in Case study 5.2.

Settlement pattern can be affected by the following factors.

- **The physical environment**: for example, terrain and water supply controlled the early distribution of people, with dispersal frequently a function of more difficult physical conditions that limited farming to a mainly pastoral activity. The excellent quality of the soils in some areas of the UK led to the development of arable farming, with large villages distributed at regular intervals across a relatively uniform physical landscape. In some less-favoured areas of a region the wetter clay soils were used for dairying, which led to a dispersed pattern of farms and hamlets rather than large nucleated villages. Cows need to be near the farm to be milked twice a day and so a dairy farm is surrounded by its fields, leading to a dispersed form of settlement.
- **Historical influences**: land ownership and changes in ownership are key factors here. Many settlements on the west coast of Scotland were removed by landlords who cleared the land for sheep grazing in the so-called Highland Clearances.

# Chapter 5  Settlement patterns

## Case study 5.2  Historic Worcestershire

The county of Worcestershire in central England can be sub-divided into three zones according to its early settlement pattern. Much of this early pattern has survived, at least in outline, to the present day.

- The western area is largely made up of dispersed settlements. Nucleated settlements here are mainly market towns or small hamlets. The area was densely settled back in the eleventh century, according to the Domesday Book (a valuable historic record of the situation in 1086).
- The West Midlands area to the north-east is made up of low plateaus and escarpments. It was lightly settled at the time of the Domesday Book. This area has some nucleated settlements, but is largely made up of small hamlets and isolated, stand-alone farms. The nucleated settlements tend to be small. There is some surviving medieval woodland in the area. Much of this settlement pattern has been overwhelmed by rapid urban growth in the last 200 years.
- The Cotswold scarp and vale area to the south-east is dominated by large nucleated villages surrounded by large fields, although some dispersed settlement does exist.

*Photograph 5.2  Worcestershire's Malvern hills*

- **Current human influences**, for example patterns of employment and service and leisure needs. In the 1930s more stand-alone houses were built in the countryside because at this time planning restrictions were rare and mobility was increasing as car ownership grew.
- **Levels of technology**, for example telecommunications and transport networks. These release people from an immediate dependency on the local area for food and employment, and so facilitate the dispersion of a population. Technology might allow the population of remote areas to increase — for example, people on Scottish Islands such as Skye can use the internet to work from home.

- **Government policies** often control settlement distribution and growth. Governments can force people to move. One of the most spectacular cases occurred in Stalinist Russia in the 1930s, when rural areas were depopulated and large collective villages were created. Another example is the transmigration policies of Indonesia — in this case, creating new 'dispersed' settlements.

Once developed, settlement patterns show a high level of continuity because they generate further growth in a feedback mechanism. In simple terms, bigger settlements offer more employment opportunities than smaller ones. This attracts more people; the settlement becomes bigger, and the process goes on.

This process may very well be supported by government policy, as in the UK today. Many counties have a 'key-settlement' policy that promotes growth in some villages, where the services are concentrated, but not in other, usually smaller, villages, where planning is much more restrictive. Thus the big villages get bigger and the small villages either stay the same size or contract as occupation rates decline.

In some villages the presence of a powerful landlord who has retained ownership of most of the properties prevents growth. An example is Sandy Lane in Wiltshire, where the vast majority of properties are rented from the Bowood estate. This is a settlement that has not physically expanded and, as a consequence, it has lost population.

In the urban-based economies of most MEDCs, settlement is strongly affected by urban influences; towns have an impact on rural areas and dormitory settlements develop in nearby villages. These villages have clearly changed their function.

## Site, situation and function: some definitions

### Site

**Site** is the position of a settlement in terms of its immediate physical location — in other words, where it is. Factors affecting site include:
- water availability
- ground conditions and drainage
- altitude
- relief and aspect
- soil fertility

### Situation

**Situation** is the location of a place in its broader regional context — that is, in relation to its wider surroundings.

### Function

**Function** tells you what a place does, in terms of its industrial, residential, commercial and administrative activities. Maps may provide some evidence about a settlement's previous main function, but only by inference, for example the presence of mines or a port in the area. All settlements have multiple functions and these functions change over time.

# Chapter 5   Settlement patterns

## Case study 5.3   Site: three Wiltshire villages

The three villages of Tilshead, Little Cheverell and Worton are situated within a few kilometres of each other in Wiltshire, between the market town of Devizes to the north and the small city and regional centre of Salisbury some 25 km to the south (Figure 5.1). However, the three villages are sited very differently.

### Tilshead

This settlement has developed at a wet-point site on the chalk uplands of Salisbury plain around one of the rare places where surface water is found. The small village sits in a shallow saucer-shaped depression at the head of a valley on very gently sloping ground (less than 5°) at an altitude of 110 m. Growth of the village has been minimal in recent years as it is surrounded by land taken over by the army during the First World War and still used by the military. There has been some growth due to **infill**, with new houses being 'squeezed' between existing properties.

### Little Cheverell

This is a spring-line settlement sited on upper greensand close to the foot of the chalk escarpment of Salisbury Plain. Soils are fertile and the village occupies gently sloping ground on a north-facing slope. This village has lost most of its services, although the pub and the church remain. Property prices are high and new development rare and very much at the top end of the housing market.

### Worton

This village is on a dry-point site, elevated on a narrow outcrop of sandstone a few metres above the heavy and very water-retentive clays of the Vale of Pewsey. The site is on more or less flat ground at an altitude of 75 m. Worton has grown a good deal in recent years as properties have occupied the ground to the north of the village, which is less elevated. The village has a pub, a church and a village school but the shop has disappeared, probably because of the village's proximity to Devizes.

*Photograph 5.3  The village of Tilshead*

162

Human environments  AS Unit 2

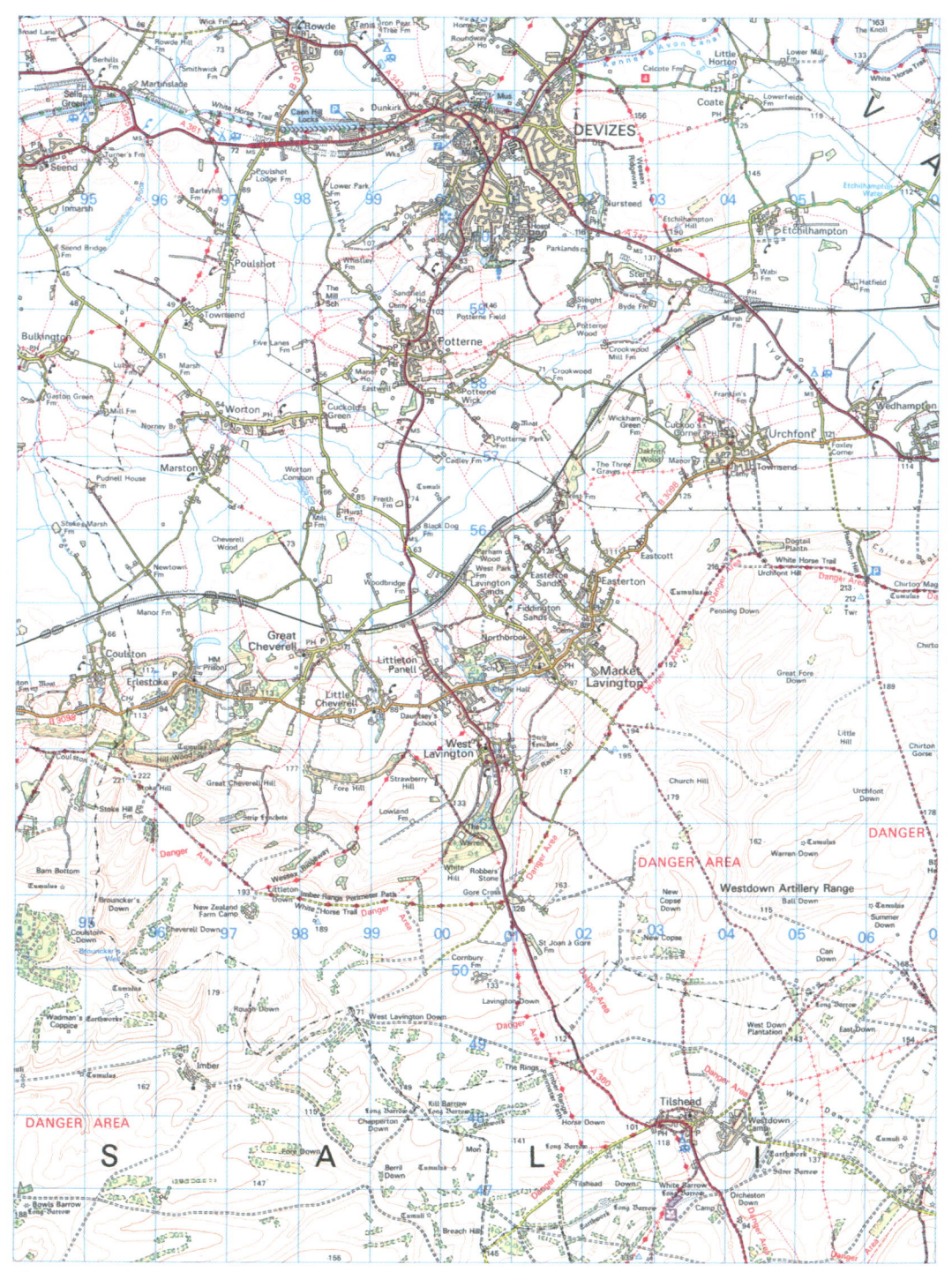

*Figure 5.1 Ordnance Survey map showing the villages of Tilshead, Little Cheverell and Worton*

Reproduced by permission of Ordnance Survey on behalf of HMSO. © Crown copyright 2005. All rights reserved. Licence no. 100027418.

# Chapter 5 Settlement patterns

## Information sources

A good deal of information about settlements can be gathered from map evidence (Table 5.1), although it is sometimes difficult to be definite.

*Table 5.1 Information about settlements available from a map*

| Map information | Source | Quantifiable? | Comment |
| --- | --- | --- | --- |
| Altitude | Contours and spot heights can be used | Yes | Avoid qualitative terms such as 'mountain' unless certain; use comparative comments |
| Relief | Contours and distances | Yes | Often confused with altitude; use terms such as flat, steeply sloping, undulating; comparative comments help |
| Aspect | Contours and compass point; direction a slope faces; scale might obscure detail | Yes | Remember that south-facing slopes are found on the northern sides of valleys in the northern hemisphere |
| Surface drainage | Rivers, lakes, drainage ditches | Yes, through drainage density | Not all rivers are marked on maps, but a large number of surface ponds and rivers might, for example, lead one to suspect that the rock is impermeable |
| Land use (partial) | Map symbols indicating marsh, moor, woodland etc. can only guess usage of 'white' areas | Yes, partially | Deduction possible, but conclusions will always be tentative — beware of inferring too much from place names |
| Geology and soils | Can only infer these | No | Deduction possible, but tentative; woodland is a key indicator here; lots of woodland suggests limitations on productive agriculture; the distribution of farms may indicate a pastoral rather than arable farming system; place names may help |
| Settlement size | In area terms only; no information on population because densities will vary from place to place | Yes | Have no idea about population density, because don't know about height (and therefore population capacity) of buildings |
| Settlement form | Shape from map | No | Use terms like 'linear' or 'nucleated' |
| Settlement distribution | From map | Yes, through nearest neighbour analysis and other techniques | Use distances for measurement; one grid square = 1 km$^2$ |
| Settlement function (partial only) | A few guesses possible using other information from the map such as mines, ports, tourist information | No | Use tourist information, location, routeways as clues |
| Transport systems | Roads, tracks, waterways, railways | Network analysis possible | No idea of usage of these |

## Continuity

There is a **continuity** of settlement in most countries in Europe; many settlements have occupied the same site for centuries. In lowland England waves of Saxon settlers probably established many of the villages between the sixth and ninth centuries. These settlements have changed greatly in both form and function, but the actual site is often unchanged. Of course some settlements have disappeared, either by gradual decline or, occasionally, due to a dramatic event such as the Black Death. New settlements have appeared and will continue to appear to satisfy a growing demand for housing in some parts of the country. The original site factors are no longer significant because water is now supplied by pipe and the fertility of the soil is now of more concern to gardeners than it is to the economic life of villages, which rarely have much dependence on agriculture.

With the exception of a few settlements founded more recently as holiday resorts or industrial sites, most built-up areas in Britain put down their roots about 1000 years ago. Other than mining towns and villages, rarely do we find a settlement owing its location to just one geographical factor. Generally, factors combine and enable us to describe any settlement in terms of site, situation and function.

# Urban settlements

## Variations in the size of urban settlements

Urban settlements have many functions, but these can be usefully divided into two groups.

- **City-serving**, or basic, functions or activities. These are the functions or employment generated by the population of the settlement; primary schools are an obvious example. If a town has 25 000 inhabitants of whom 10% (2500) are of primary school age, and assuming a pupil/teacher ratio of about 25 to 1, there will be 100 primary-school teachers in the town.
- **City-forming** or non-basic functions or activities. These are the functions or employment generated by people who live outside the settlement. Cambridge, for example, is widely known as a university and hi-tech city. The number of university teachers living in Cambridge is far beyond the needs of the population aged 18–21 who actually list Cambridge as their permanent address. Similarly, Great Yarmouth is a holiday resort and Aldershot is a garrison town; these are two more city-forming activities.

There is an obvious overlap between these functions. For instance, major shopping centres attract customers from outside the city, but also provide retail functions for residents. Similarly, some of the students who go to Cambridge University were born and brought up in the city. Figure 5.2 shows the factors that affected the initial settlement and subsequent development of Cambridge. To a considerable extent, the larger the number of city-forming activities found in a town or city, then the larger the city in terms of its population.

# Chapter 5

## Settlement patterns

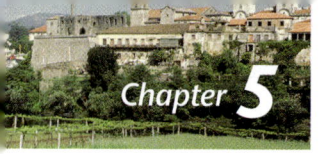

*Figure 5.2 The development of Cambridge*

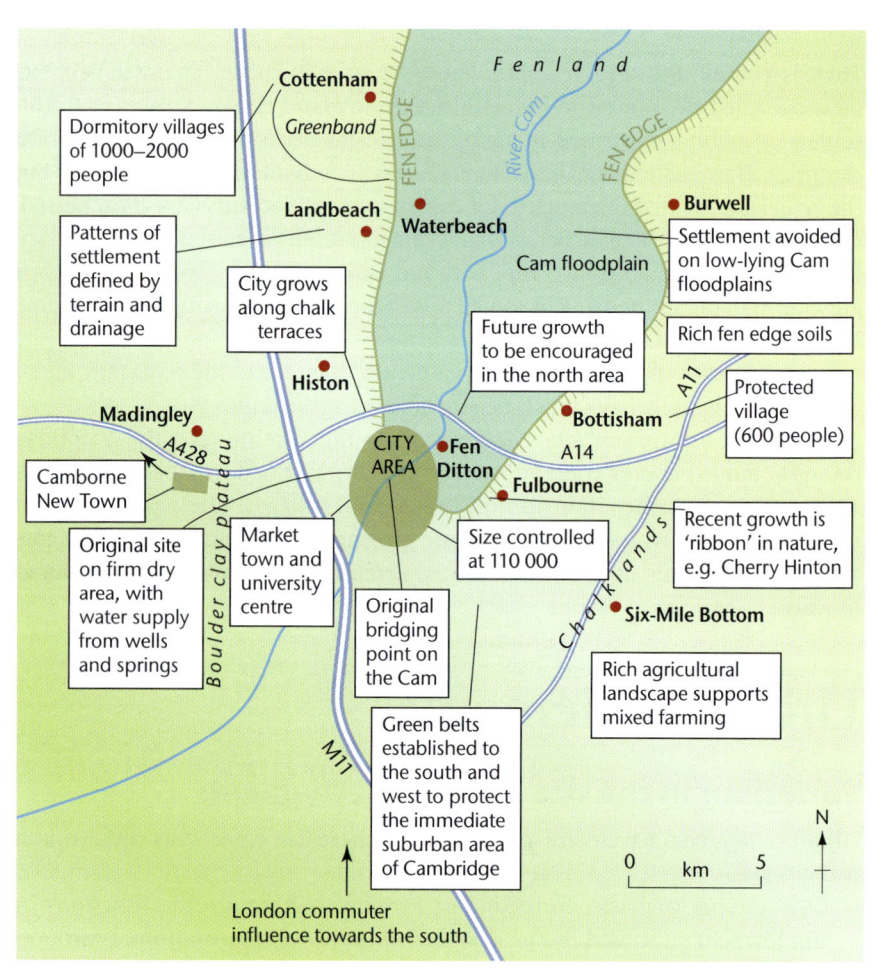

## The rank–size rule

Zipf developed Jefferson's 'law of the primate city' by recognising a pattern, which he then developed into the **rank–size rule**. The rule suggests that the second largest settlement in an area is half the size of the largest, the third largest is one-third the size of the largest, the fourth largest is one-fourth the size of the largest, and so on. Thus, Zipf commented, there is an observable inverse relationship between the size and rank of a given settlement. Table 5.2 shows the rank–size rule applied to the ten largest cities in the USA, using census data from 2000.

There has been a good deal of speculation about why the rank–size relationship exists, but no single explanation has ever made much impact. The US data give some clues as to why this is. Although the rule seems to work well for a few cities ranked between 3 and 6, it is less reliable outside that narrow band. It is also considerably compromised by the fact that deciding on the population of an urban settlement is something of a lottery. The listed population of a town or city corresponds to the population of that administrative district at the time of the census. However, the administrative district may include areas well beyond the city

| City | Population (2000 census) | Rank by size (2000) | 'Expected' population by rank size rule | Difference (real – predicted) | Difference (%) |
|---|---|---|---|---|---|
| New York | 8 008 278 | 1 | 8 008 278 | – | – |
| Los Angeles | 3 694 820 | 2 | 4 004 139 | –300 319 | –7.5 |
| Chicago | 2 896 016 | 3 | 2 669 426 | +226 590 | +8.5 |
| Houston | 1 953 631 | 4 | 2 002 069 | –48 438 | –2.4 |
| Philadelphia | 1 517 550 | 5 | 1 601 656 | –84 106 | –5.3 |
| Phoenix | 1 321 045 | 6 | 1 334 713 | –13 668 | –1.0 |
| San Diego | 1 223 400 | 7 | 1 144 040 | +79 360 | +6.9 |
| Dallas | 1 188 580 | 8 | 1 001 035 | +187 545 | +18.75 |
| San Antonio | 1 144 646 | 9 | 889 808 | +254 838 | +28.64 |
| Detroit | 951 270 | 10 | 800 828 | +150 442 | +18.87 |

*Table 5.2 Applying the rank–size rule in the USA*

limits or, conversely, the 'administrative' city might be only a small part of the built-up area. Quite apart from the fact that administrative districts change boundaries over time, it all depends on what is included in the 'city' and what is not. As an example, strictly speaking the City of Los Angeles comprises 1215 km$^2$ and had a population of about 3.7 million people at the 2000 census. However, the greater metropolitan area had a much larger population; Los Angeles County, which covers 10 518 km$^2$ and encompasses 88 'cities', had a population of about 9.5 million people at the 2000 census. If figures like these were to be 'mapped' into the rank–size rule table, they would give a very different set of results.

The most frequent statistical discrepancy from the rank–size rule occurs because of the tendency to **primacy** in many countries, especially countries that have experienced relatively recent urbanisation. Primacy occurs when the largest city is several times larger than the next largest settlement and clearly dominates the urban hierarchy in a particular way. The largest primate cities are many times bigger than other settlements further down the rank and dominate the country in terms of economic, political and social affairs. They are generally super-dominant in the number of higher-order professions and services on offer, well beyond their dominance in terms of population. The explanations of primacy include the following.

- A history of colonialism, with government and investment concentrated in the main city, which is often a port. The other parts of the country historically served as centres for agricultural production for export, and the development of local markets was not encouraged, so urban growth did not take place (e.g. Dublin in Ireland).
- A history of highly centralised government, with little or no independence for the regions. As a result, a whole range of city-forming activities were concentrated in the capital (e.g. Paris in France).
- A lack of industrial development, so a rural economy did not develop any urban centres to speak of, other than the capital city. The capital was also the administrative centre and was usually a port as well (e.g. Montevideo in Uruguay).

*Chapter* **5**　　*Settlement patterns*

## Case study 5.4　*A primate city: Dublin*

*Photograph 5.4*
*The centre of Dublin*

Dublin is a primate city within Ireland, and is four times the size of the next largest city, Cork. Even these figures underestimate its true dominance.

- The Dublin trade area effectively becomes the whole country for the highest-order goods and services.
- The greater Dublin area, defined as an area within an hour's journey of the capital, includes over 900 000 people, 39% of the total Irish population, and is six times the size of greater Cork.
- As is often the case with primate cities, Dublin's importance goes beyond just the numbers involved. The number of higher-paid professionals and company headquarters in Dublin and the flow of investment is well in excess even of Dublin's population dominance.

As in other countries, the super-dominance of one city is a function of Ireland's history. Much of Ireland was very late to urbanise (as was Iceland — see Case study 6.11, Chapter 6). However, Dublin, unlike Reykjavik, grew as the point of colonial control, the port of exit and entry, and was the military and administrative capital of one of England's oldest colonies. The country was organised as an agricultural economy, designed to grow crops for export to England, and industry was not encouraged, given that Ireland was a captive market for English goods in the same way that India was to become later.

In these circumstances, urban central place functions are unlikely to develop because there isn't a 'surplus' of local production to be sold 'locally', and this is the main reason for the growth of 'market' towns. Nor are they going to develop workshops when English-made goods are being brought into the country. Much the same applied to the economy of the 'deep south' of the USA during the plantation period, where towns were also a rarity. The reduction of the Irish population from 6 million at the time of the potato famine in 1847 to a little over 3 million today is another obvious reason for the lack of urban development.

- The **positive feedback** of large cities generating their own rapid growth at the expense of other cities as they established linkages with other large cities in neighbouring countries. This is true of many of the so-called 'global cities' that have emerged in the past 30 or 40 years (e.g. Taipei in Taiwan or Seoul in Korea).
- The size of a country, in particular small countries within larger trading areas. An example is Belgium, which supports the large primate city of Brussels. Brussels is a central place for many who live outside Belgium and has grown very fast. It is also a second-rank global city with many international headquarters, not the least of which is its role as a 'capital' of the European Union.

By contrast, lack of primacy, where the major urban centre is scarcely more significant than the next in the rank order, is often the result of the following factors.

- **Relatively recent national integration**. For example, both Italy and Germany only became nation states in the nineteenth century when they were formed out of a collection of much smaller states, some of which were virtually 'city states', such as Venice. Before that final amalgamation, each area consisted of many competing states, and each state had its own capital and administrative centre. Many of these had developed as major centres before integration. Milan, Turin, Genoa, Naples and Rome were all 'capitals' and primate within their own states before the formation of the modern Italian state. The same was true for Munich, Cologne, Berlin and Frankfurt in Germany.
- A **system of government**, possibly stemming from late national integration which allows more independence to the individual regions of a country. This 'federal' system is present in the USA, Australia and Brazil as well as in Germany. The federal system promotes the development of larger-than-average regional capitals where power and jobs, and thus population, are concentrated.
- **Large size** can also reduce the chance of primacy. The USA is a continental-sized country. It is hardly surprising that no one city completely dominates the urban system. The dominance of New York has declined as Chicago and then Los Angeles have grown to dominate their regions of this very large country. The same is true in Brazil.

# Settlement hierarchies: central place, range and threshold

Just over 70 years ago, Walter Christaller published his famous study of **central places** in southern Germany. 'Central places' are settlements that develop to provide a surrounding region with specific market functions. People come to them to purchase goods that are not available in their local villages. The relative importance of a central place is measured in terms of the numbers and varieties of goods and services provided for the people who live there and in the surrounding **catchment area**. The theory was disregarded in the years following

# Chapter 5
## Settlement patterns

the Second World War because it was used by the Nazis to plan their settlement of conquered eastern provinces. Since then, however, it has had a huge impact on geographical study. These days, the details of the theory are less important than the reasons why Christaller's predictions simply do not fit the modern settlement landscape, which has changed greatly since the 1930s.

The two fundamental concepts that underpinned Christaller's study were the threshold population and the range.

- The **threshold population** is the minimum number of people needed to support an outlet in a central place. In other words, if the threshold population is not met, then it is not profitable to offer the good or service for sale.
- The **range** of a good or service is the distance of the boundary (the sphere of influence or catchment area) needed to enclose the threshold population for the outlet that provides it, i.e. the distance measured in time and cost that a customer is prepared to travel to obtain a good or service.

Conventionally, goods and services are divided into low-order, middle-order and high-order goods.

- **Low-order goods** are cheap and frequently purchased items, sometimes slightly misleadingly known as **necessities** or **convenience goods**, such as a newspaper or a bottle of milk. Not much time or money will be expended travelling to obtain such an item. The range of these low-order goods is therefore small, but because they are bought frequently the threshold population is also quite small.
- **Middle-order goods** are not daily purchases, but may be bought weekly or even monthly. Middle-order goods include clothes, shoes and a whole range of other household items from CDs to DVDs that warrant a short journey of a few kilometres, but not much more.
- **High-order goods** are the more rarely bought items that represent significant purchases in most households. Like most middle-order goods, they can be categorised as **comparison goods** because choices have to be made about model, type, performance, colour and so forth. The most obvious example, although the most expensive, is a car.

Christaller made a number of assumptions to make the theory workable:
- an isotropic surface (uniform transportation costs in all directions); thus no physical obstacles to movement
- a uniform distribution per capita of demand and population (demand density); no richer areas and no poorer areas
- no trade with areas outside the study area
- profit-maximising producers
- consumers who made rational decisions
- an economy free of government or social classes
- ubiquitous (available at every place) production inputs at the same price
- no shopping externalities (no supermarkets)

Christaller identified a hierarchy of seven levels. The *Landstadt* (or regional centre) was at the top, with a population of 500 000 and a catchment population of

3.5 million. In southern Germany, where he did his research, these were centres such as Munich, Nürnberg, Stuttgart and Frankfurt. Below this came the *Provinzstadt*, with a population of 100 000 and a catchment population of 1 million, including cities like Augsburg, Ulm, Würzburg and Regensburg. The lowest category in the hierarchy was the tiny *Marktorte*, with a typical population of 1000 and a catchment area of 3000 people. Christaller did not include the national capital in his study. Table 5.3 summarises his hierarchy.

| Type | Market area radius (km) | Number of settlements | Population of town | Population of market area |
|---|---|---|---|---|
| M (*Marktorte*) | 4.0 | 729 | 1 000 | 3 500 |
| A (*Amtsort*) | 6.9 | 243 | 2 000 | 11 000 |
| K (*Kreisstadt*) | 12.0 | 81 | 4 000 | 35 000 |
| B (*Bezirkstadt*) | 20.7 | 27 | 10 000 | 100 000 |
| G (*Gaustadt*) | 36.0 | 9 | 30 000 | 350 000 |
| P (*Provinzstadt*) | 62.1 | 3 | 100 000 | 1 000 000 |
| L (*Landstadt*) | 108.0 | 1 | 500 000 | 3 500 000 |

Table 5.3 Christaller's hierarchy

A number of conclusions can be drawn from Christaller's work. Each of these conclusions is testable.
- There will be a given number of settlements at each level of the hierarchy, providing a stepped hierarchy. These settlements will be very similar in the goods and services offered. For example, Christaller showed that each market town in a region would offer the same number of goods and services and that if there were 48 market towns, there would be a predictable number of larger regional centres — 16 in the most commonly used system.
- At each level of the hierarchy, the goods and services offered will include all those offered by settlements at a lower rank on the hierarchy but also a number of goods and services from a higher order that are not offered at the lower-ranked settlements.
- On a flat featureless plain (an isotropic surface) there will be a given distance between each settlement.

There is an apparent contradiction here with the rank–size rule, which ranks the populations of towns and cities in a country. The rank–size rule rarely shows a hierarchy that would be represented by having 48 market towns all of roughly the same population (say 20 000) and 16 regional centres all of roughly the same, much larger, population (say 50 000). In other words, intermediate settlements should not exist, according to the rank–size rule.

### Examiner's tip

- In fact, however, there is no contradiction, as Christaller recognised that most towns and cities have more than one function. A town might be both an industrial city and a tourist centre as well as being a central place, and therefore it would be larger in population than an equivalent central place without those other functions.

# Chapter 5  Settlement patterns

Where such featureless plains do occur (e.g. southern Germany, East Anglia and the Great Plains of the USA), Christaller's theory works reasonably well, generating a pattern of market towns at regular intervals, determined by the distance that could be travelled (there and back) in a day using a horse and cart. Figure 5.3 shows how Christaller's ideas can be applied to the area around Norwich.

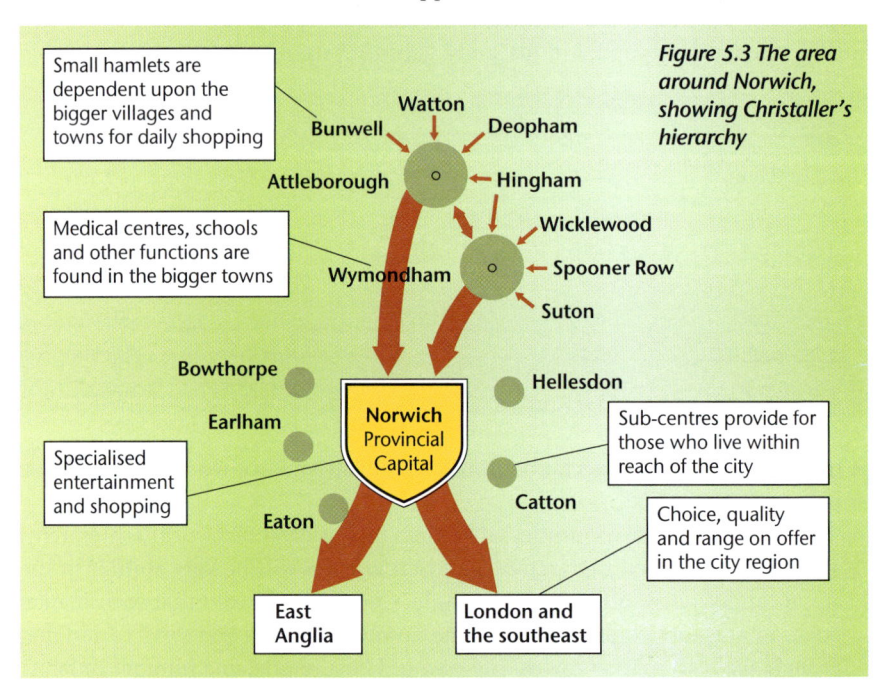

*Figure 5.3 The area around Norwich, showing Christaller's hierarchy*

- Small hamlets are dependent upon the bigger villages and towns for daily shopping
- Medical centres, schools and other functions are found in the bigger towns
- Specialised entertainment and shopping
- Sub-centres provide for those who live within reach of the city
- Choice, quality and range on offer in the city region

Settlements shown: Watton, Bunwell, Deopham, Attleborough, Hingham, Wicklewood, Wymondham, Spooner Row, Suton, Bowthorpe, Hellesdon, Earlham, Norwich (Provincial Capital), Eaton, Catton; arrows to East Anglia and London and the southeast.

Ironically, in one of the more appropriate study areas of the American West, many of the smaller settlements did not develop historically from an earlier distribution of villages. On the contrary, the smaller settlements grew after the railway towns were established. In other words, the 'countryside' was 'filled up' after the towns had been formed.

However, there are more obvious reasons why Christaller's ideas do not work so well in modern advanced capitalist societies.

- Car-ownership has risen and the mobility of individuals has transformed the retail business. There are close on 200 million privately owned cars in the USA, for a population of under 300 million.
- Out-of-town shopping centres have transformed the geography of retailing, as did supermarkets in the 1960s. People do travel long distances to buy a newspaper now because it is purchased at the same time as a shopping trolley filled with other 'low-order' goods that are bought in bulk.
- At the same time, antique shops and gift shops selling high- or middle-order goods have opened in small villages; these are often run by retired people or people with other income. These 'hobby retailers' do not profit-maximise and do not seek the best possible locations for their enterprises. It is not possible for central place theory to cope with such deviations from logical economic behaviour.

Human environments  **AS Unit 2**

## Case study 5.5 Applying Christaller's theory: Bristol

### Regional centre
*Bristol*

The major focus for regional services, shopping, cultural activities, education and tourism:
- good communications
- high quality and range of leisure and cultural facilities
- high-quality transport interconnections

### Sub-regional centres/principal centres
*Bath, Weston-super-Mare, Swindon, Gloucester and Cheltenham, Taunton, the Bournemouth–Poole conurbation*

Important economic, social, cultural and service centres:
- principal growth centres
- large number of shops, services and employment

### Major towns/district centres

#### Gloucestershire
*Stroud, Tewkesbury and Cirencester and district centres in Gloucester and Cheltenham*

#### Somerset
*Bridgwater, Burnham-on-Sea and Highbridge, Chard, Crewkerne, Frome, Glastonbury, Ilminster, Minehead, Shepton Mallet, Street, Wellington, Wells, Wincanton, Yeovil*

#### Former Avon (towns and neighbourhoods)
*Keynsham, Midsomer Norton, Bedminster, Clifton, Fishponds, Whiteladies Rd, Clevedon — the Triangle, Nailsea, Portishead, Thornbury, Yate*

Locations for employment and shopping, cultural, community and education services:
- secondary schools and further education establishments
- hospitals and specialist social service facilities
- professional and commercial services such as banks, insurance companies, building societies and solicitors

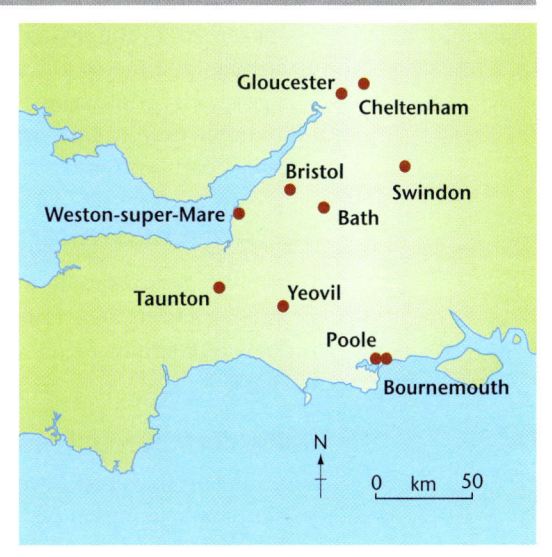

*Figure 5.4 The area around Bristol*

- entertainment and recreation facilities such as cinemas, theatres and sports centres
- a wide range of shopping facilities
- good accessibility to national and county routes and regular public transport services

### Smaller centres/market towns

#### Former Avon (towns and neighbourhoods)
*Moorland Road, Radstock, Gloucester Rd, Henleaze, Lodge Causeway, St George, Shirehampton, Wells Road, Westbury-on-Trym, Yatton, Clevedon Hill Rd, Portishead, Winscombe, Worle High Street, Chipping Sodbury, Downend, Hanham, Staple Hill*

Local employment and shopping, social and community activity:
- variety of small shops
- health, cultural, financial, administrative and education services

### Villages/local centres

Limited local services:
- few shops
- sub-post office
- primary school or community hall

# Chapter 5

*Settlement patterns*

- Local authorities have encouraged the centralisation of services such as schools and doctors' practices. Centralisation is advantageous for the services concerned, in the saving on overhead costs such as receptionists, caretakers, heating bills and so forth. However, it is less clear that centralisation is advantageous for people using those services, who have to travel further to reach them and may need to take more time off work, or pay a child-minder for longer, quite apart from the cost of travel.

## Testing Christaller's theory

Christaller demonstrated the influence a settlement had for a certain distance in its surrounding area. His theory gave rise to attempts at calculating spheres of influence or urban fields, with gravity models enabling such calculations to be made.

### Reilly's law of retail gravitation

Imagine you have two towns. Between these two towns there exists a breaking point: people living on one side of this point shop in one centre (town A) and people on the other side shop in the other town (town B). Reilly devised a formula, using population size, to demonstrate the position of this breaking point:

*BP* is the distance from town A to the breaking point.

$$BP = \frac{\text{distance between towns A and B}}{1 + \sqrt{\frac{\text{population of town B}}{\text{population of town A}}}}$$

By calculating the breaking points of several towns of the same order around a larger settlement, the spheres of influence can be drawn in.

Unfortunately, Reilly's model assumes away a number of problems that have become more rather than less significant.

- People don't always act in a logical way. They may prefer to shop in a smaller town because they have friends or family there or the parking is easier, which is sensible enough, but not part of the model.
- Different populations may have different incomes and thus quite different expenditure patterns.
- Distance is a poor criterion in itself. It would be better to use travel time as a measure.
- Some cities, for example mining towns, are large but lack city-forming central-place functions.
- The advent of the car has changed people's shopping habits completely.

A good deal of work has been done in establishing the existence of urban hierarchies. There is no doubt that they exist, although they differ a good deal from those predicted by Christaller.

*Human environments* **AS Unit 2**

## Case study 5.6 *Measuring the urban hierarchy in Belgium*

Two approaches have been used to establish the urban hierarchy of this small European country.

- The first involved questionnaires to establish the pattern of retail and service behaviour. The 1993/94 survey was extensive and was filled in by a randomly selected 2% sample in each municipality (an administrative division that might include a wide range of settlement types). For retail and basic services, a subdivision was made between everyday goods, periodical goods and exceptional goods. Questions were also asked about locations for sporting activities, visiting theatres and cinemas, doctors (specialist or polyclinic), hospitals, education, secondary schools and restaurants.
- The second approach consisted of a 'points scoring' system in which the facilities available in towns and cities were measured by giving a rating based on the quantitative level of each function (say the number of restaurants) or the qualitative level (university, provincial capital, post-sorting centre). This type of method has many variations but gives an objective measure based on relatively easily established 'facts'. There are, of course, problems. For example, the number of restaurants in a town or city is not the same thing as the number of tables available in restaurants, so one large and expensive restaurant may have a turnover many times that of ten smaller and less prestigious restaurants. Also, not all services are a function of demand so that, for example, a police headquarters may be located in a town or city simply because of government policy.

The two approaches gave rise to different results.

- Some municipalities appear to be urban centres on the basis of their facilities but not by user survey. These centres, which are seen as 'small towns' on the basis of their facilities, include municipalities which are either parts of conurbations or municipalities of some importance but do not have a significant sphere of influence. Often this is because of a high population density and the presence of similar adjacent municipalities that have each developed facilities for their own populations.
- Other places defined as urban centres on the basis of the user survey disappear when the facilities method is used. This situation tends to occur in sparsely populated rural areas: they have their own catchments despite the low level of facilities, precisely because of the geographical characteristics of their situation and the lightly populated surrounding area.

The levels devised were:
- large cities
- regional cities
- small towns
  - small towns a
  - small towns b
  - small towns c

Once the urban hierarchy has been defined, a sphere of influence can be calculated and geographically defined for each town or city on the basis of the user survey.

- Belgium has five large cities, each of which forms the centre of an urban region. The five large cities can be further divided:
  - two very large cities: Brussels, which is dominant, and Antwerp
  - two large cities: Ghent and Liège
  - Charleroi, which is clearly the smallest of the five, and is best described as a 'large city with fewer facilities'
- Below that level there are 17 regional cities, 13 of which form the core of an urban region, and the 4 that do not are dominated by the Brussels region.
- Finally, 81 small towns are identified, whose spheres of influence cover the country.

# Chapter 5

*Settlement patterns*

# Spatial variations in land-use patterns in urban settlements

There are various land uses in all cities:
- residence, segregated into different social classes
- administration
- manufacturing industry
- retail and service provision
- commercial space in the form of offices
- recreational space

In some ancient, pre-industrial, cities, it is probable that the land use was more mixed geographically because it was not possible to move far on foot during the working day. However, some segregation obviously took place: for example, the rich senatorial class in Rome did not live next door to the urban poor. Segregation was not complete by Victorian times as the limited transport availability meant that craftsmen and artisans still had to live in close proximity to the classes that they served. The early 'utopian' garden cities made a conscious effort to mix up land use in this way, with residential areas offering a range of housing type and quality. This is a long way from today's gated suburbs and, at the other end of the spectrum, 'no-go areas' in some cities.

A series of factors determine land use, and they can be divided into physical influences and social factors.

## Physical influences

### Rivers and estuaries

Heavy industry was often sited near rivers, sometimes on land liable to flooding and thus not suitable for residential use. The attraction of port locations, which were often the most significant **centripetal** force in city growth in the nineteenth century, made these areas 'break-of-bulk' points and a whole series of processing industries developed. Examples include paper-making, printing, furniture making and sugar refining. The ports were often adjacent to administrative areas, which had grown up around the core areas close to bridging points on the river. Thus there was an uneasy but economically productive proximity of the CBD and the industrial zones. London is an obvious example of such a city, although Paris and New York show the same pattern. In industrial cities, the lower-class housing grew alongside these facilities.

As sea transport has become less significant and the need for deepwater access and the use of container ships has made riverine and estuarine ports less significant, river frontages have become an attractive rather than a repulsive force for high-class housing. The transformation of the London Docklands area, Cardiff Bay, Salford Quays and dozens of other areas worldwide show this in action.

## Altitude

Broadly speaking, social class and housing costs increase with altitude, especially in industrial and post-industrial cities. The reasons are:
- views
- avoidance of noxious fumes
- avoidance of waterlogged ground

However, shifts in the location and type of industrial base in cities have made low-lying river frontages desirable too.

## Slope angle

Building on steep slopes is more expensive, if it conforms to building regulations. As a result the steepest land may well be left undeveloped. However, this depends on the level of demand for land. Steep slopes are used intensively in Hong Kong, for example, because land is so scarce.

## Coasts and lakesides

Coasts and lakesides are similar to rivers and estuaries in their effect on land use, for the same reasons. They also affect the shape of the city and give rise to accessibility issues, because the city takes on a form that reflects the shape of the coast. For example, Chicago's growth was largely semi-circular, with its CBD on the shores of Lake Michigan.

In tourist areas, the linear nature of the coastal development tends to lead to a strip of housing that is highly desirable, with housing costs reducing with distance from the sea, as it does in Brighton.

# Social factors

## Age and cultural value of housing

In many urban areas in Europe and North America, the age of the housing defines the social status of the inhabitants. Lower classes inhabit older housing in the centre of the urban area, and middle and higher classes live in newer housing towards the urban margin. The construction of municipal housing (council housing) in large estates on the urban margin distorts the ideal pattern of residential zones by introducing lower-class housing into areas of higher- and middle-class housing.

Within this crude pattern, changes appear as cultural values and norms change. A fashion for restoring Victorian and Georgian housing, known as **gentrification**, has meant that the inner cities have been invaded by a higher social class.

## Land ownership

In a free market, where all users of land try to maximise their earnings, it would be hard to explain parks and open spaces in cities. Indeed, where the land market is 'free', then few parks are evident. This can be seen in Las Vegas, where there is a great deal of 'private' open space which either one has to pay to use, such as golf courses and hotel gardens, or is simply inaccessible, such as private gardens.

## Chapter 5 Settlement patterns

The existence of large open spaces in urban areas is often a result of landowners preventing development of the land. These landowners include the Church of England, private landlords and local councils. The advent of planning in most cities in the past 50 years has protected these areas, in part at least, from encroachment, although they do tend to get 'nibbled' at the edges.

### Planning

Almost all countries have planning legislation. Its enforcement varies from place to place, as does the rigour of its rules. The Town and Country Planning Act, which has been applied in the UK since 1946, tends to reinforce the existing land-use pattern and prevents major changes. The main theme of UK planning in the past 60 years has been the **neighbourhood** principle, which tends to reinforce the likelihood that land uses will become segregated, with clear demarcation between, for example, residential areas and industrial areas. This principle was attacked by Alice Coleman and Jane Jacobs in the 1980s and has been substantially revised in the light of these criticisms. Planning has also come under a sustained attack as being both inefficient and bureaucratic by those who argue that the free market will do the job better.

# The development of cities in MEDCs

Between 1875 and 1920, many cities in Europe and North America grew dramatically in land area and in population (Figure 5.5). Some cities grew at rates that are just as startling as LEDC city growth in recent years.

Land use within the cities became differentiated by area. Increasing agricultural surpluses released more farmers from their fields. The demand for labour in the cities was high from the new industries that developed there. In-migration and rural–urban migration were high and most MEDCs were becoming urban nations. Changing technology influenced city form as well as growth. Most important among these new technologies were trams, electricity, assembly lines, gas and electric street lighting and residential lighting, steel-frame building construction and the elevator (the last two making the skyscraper possible). Vastly expanding industrial production was driving the new urbanisation. Many countries were moving from traditional agrarian societies to modern capitalist economies.

### Key terms

**Externalities** The impacts of an economic activity on people and the environment.

**Multiplier** Money is spent several times, so any increase in income generates a further increase.

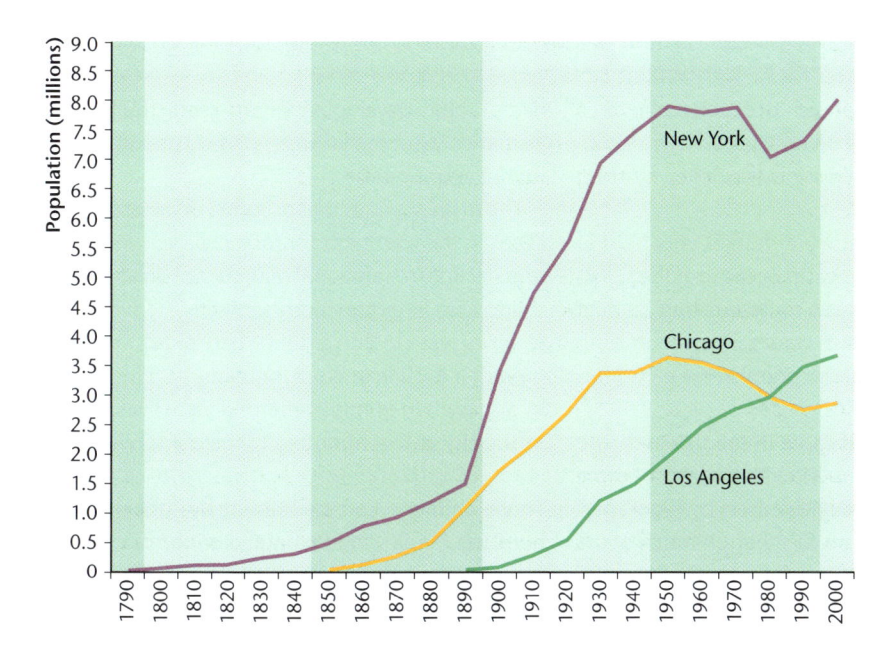

*Figure 5.5 The growth of Chicago, Los Angeles and New York, 1790–2000*

The development of factories proved to be a powerful centripetal force, separating production from consumption, which had previously taken place under the same roof. New technologies were generating new products as well as new ways to manufacture them. New products and expanding markets were creating greater employment and higher incomes, and leading to more consumption and more profits for the companies, which they could re-invest in new equipment. This was the **multiplier** in operation and was the first great boom of capitalism.

Specialisation of task became more common within the workforce, leading to production lines and large factories which enjoyed **economies of scale**. The skill levels required rose, which placed a higher stress on education. Schooling became compulsory in many MEDCs at the end of the nineteenth century. This, in turn, reduced birth rates as children no longer worked (and no longer earned a wage) and it became economically less viable to have large families. Part of the higher incomes and new wealth was taxed away to build infrastructure in order to support the cities. This included roads, bridges, tunnels, sewers, water lines, schools and street lighting. These made the whole system work more efficiently.

There was still a great deal of conflict between various users of land in the city due to the lack of planning or other land-use controls. One of the most important changes that occurred during the period was the specialisation of land use by type. Remember that in the early nineteenth century city, all land uses were mixed together. These had started to separate out in the period between 1840 and 1875, but most of this spatial separation occurred after the introduction of railways and omnibuses as the means of moving people into and through the city. The importance of the rail system in shaping the urban core was vital.

# Chapter 5  Settlement patterns

## Case study 5.7  City growth and development: Leeds

The earliest written record of Leeds shows a Domesday population of 35 families, mostly distributed around the Kirkgate as this was a key crossing point of the River Aire.

By the thirteenth century a market was held in Leeds. Presumably this grew up because of the significance of the small town as a route-meeting point at the bridge. There were also trades involved in cloth finishing, with the first mill recorded in the early fourteenth century.

The first 'boom' period came in the sixteenth and seventeenth centuries, when cloth making became important. The trade was organised by wool merchants who lived in the town, although much of the actual cloth making took place in households rather than factories, the so-called 'putting-out' system.

These same merchants petitioned the Crown to make Leeds a 'free' borough, giving them greater control over their own affairs. A century later they provided the funds to make the River Aire navigable so that cloth could be exported overseas via Hull.

By the end of the seventeenth century, the population was around 6000, by 1770 it was 16 000 and by 1800 it was 50 000. By now Leeds was the sixth largest settlement in England.

Diversification of industry was already occurring during the eighteenth century, with pottery making, linen manufacture, soap boiling, sugar refining and chemical manufacture all making an appearance in the town. At the same time a whole series of city-serving activities developed such as brick manufacture, wood working, dressmaking and tailoring, shoe making and printing.

Growth in the nineteenth century was built around the textile industry, mostly in finishing trades such as dyeing and then marketing the cloth. In-migrants came from local villages, although the birth rate also rose fast in the middle of the nineteenth century. Steam slowly replaced water as the dominant source of power, and factory development took place. At this period of its history Leeds became a successful **import-replacing** city, with a wide range of industries from engineering to machine tools and leather goods as well as the traditional textiles. It also had a rapidly developing tertiary sector, built around insurance services that were a direct spin-off from the manufacturing base.

The main changes that occurred in the economy of Leeds in the twentieth century include:

- a decline in traditional manufacturing and a rapid increase in tertiary employment
- an increase in the part-time and female workforce, with a decline in work for men
- a growth in employment of people living outside the city and commuting to the city to work

These changes can be shown by looking at the numbers employed in manufacturing:

- in 1951, at the peak of manufacturing employment in the UK, 50% of the workforce in Leeds was employed in manufacturing
- by 1971 the figure was 33%
- 64 000 (about 20%) were employed in manufacturing in 1991
- by 1999 manufacturing accounted for 13% of the working population; 51 600 people worked for 1600 firms

Engineering, clothing and textiles employed 45% of the population in 1911. In 1999 engineering employed 19 800 people (5%) and clothing and textiles 3500 people (less than 1%). At its height, textiles and clothing employed 50 000 people.

There are still some large engineering firms. The largest make turbine blades, components, alloys, valves, switchgear, printers' supplies, surgical and hospital equipment and axles.

The service sector now accounts for 80% of employment (no less than 312 900 people). Leeds has turned itself into a centre of financial and business services (FBS) for the northeast of England and is second only to London in this sector within the UK. The financial sector employs approximately 91 600 people (24%). This represents an astonishing 2400% rise from the 1920s. A further 6600 people are employed in legal services and another 6100 in insurance. Once again, Leeds has turned itself into a satellite city for London in this category of employment.

At the other end of the service sector, nearly 18 000 people are employed in call centres. Leeds is also the fourth largest retail centre in England.

Thus the **functions** of Leeds have been:
- a subsistence village
- a market town and central place for its surrounding area
- an important base for merchants who organised the wool trade
- a manufacturing centre, finishing and dyeing woollen cloth
- a manufacturing centre for a wide variety of manufacturing industries
- a centre for financial services and a major centre for administration, education and retailing

## Bid-rent theory

In the nineteenth century there were very few planning controls, so competing land users would find themselves in an unconstrained competition for space. Fortunately, these users did not all want exactly the same space, but they all preferred centrality. This is the background for what developed into 'bid-rent theory'. Since there was no planning to control where particular uses went in the city, those who were willing to pay the most got the best locations.

Willingness to pay was, and to an extent still is, based on the **yield**, or return, offered by a particular location. Land in itself has no value. The price that people are prepared to pay for land is controlled by the use to which they can put it and, more particularly, how much revenue they can generate from it. Thus, selling sweets generates a given revenue per square metre per year, while selling computers generates a different income.

Some types of land use are more intensive than others. One small office of only a few square metres might generate considerable income from the six or seven people who work in it exchanging foreign currencies or moving money about the world. Similarly, a busy high street store might generate a high income from a relatively small space. For most retailers, the more central their location the larger the number of potential customers and thus the larger their turnover. They are, in theory, prepared to pay more for central sites.

The bid-rent curves (see Figure 5.6) are what economists call **indifference curves**. Should a number of sites be available, the rent bid will reflect the expected retail turnover, which, in turn, is driven by accessibility. This is greatest in the city centre and declines with distance from the centre. In order to preserve the same profit level, the rent bid will decline with distance from the centre. The steeper the curve the more important centrality is to that land-using category. Even farmers might be prepared to pay slightly more for central city sites because it would reduce their transport costs getting food to the market, but they would be unable to use their land as intensively as other users as they would always be outbid for the space. A hierarchy of bid-rents occurs in this way.
- Office users, such as banks and insurance companies, are willing to pay the most for central locations because they use space very intensively and they need to have access to the maximum possible workforce to maintain their competitiveness. They also like to be co-located with other financial enterprises and want the prestige that a central address brings.

# Chapter 5  Settlement patterns

- Large retailers also require central sites in order to maximise their retail turnover. More customers lead to more sales per square metre of floor space. They are unlikely to be quite as intensive in their land usage as offices. Broadly speaking, retail use and offices (known collectively as 'commerce') will pay highly for central sites, whereas other users are more indifferent to the attractions of being central and so will bid relatively less.
- Manufacturers were at one time able to pay for centrality because of a relatively intense use of space compared with residential users, but they could not make as much money per square metre of space as either retailers or office users. They were more likely to want a site on the water, if this existed, or along railway lines, but not too far from the railway terminal because some of their labour force, especially managers, had to travel to work.
- Residents desire centrality to minimise their costs of transport, as any move out of the city centre will lead to an increase in commuting costs. However, residents do not 'use' space for gain as do the other users in a capitalist economy, so they are unlikely to be able to outbid other users.

Figure 5.6 shows a bid-rent curve.

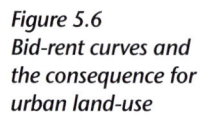

*Figure 5.6*
*Bid-rent curves and the consequence for urban land-use*

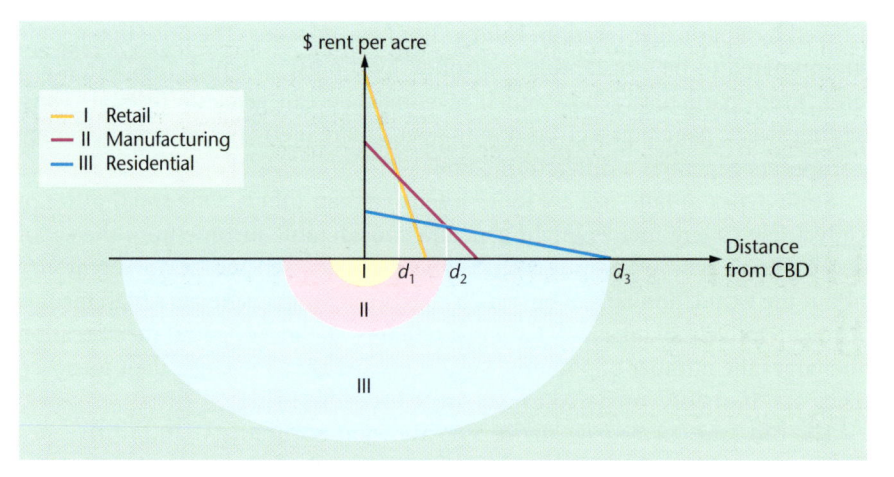

I Retail
II Manufacturing
III Residential

Recreational land use is not included in this model because it gives a low return. The existence of parks and open spaces in cities is not explained by bid-rent theory. Interestingly, parks and publicly accessible open spaces are rare in cities which have grown recently in an atmosphere of limited government regulation. Las Vegas is the obvious example.

There have been many changes in urban areas since bid-rent theory was first developed. The most obvious are described here.

- Almost all cities are now **planned**, with local governments taking a leading role in designating what can be built where. The land-market is no longer 'free'.
- **Retailing** has become more complex. The growth of supermarkets and the variation in site requirements for different types of retailers is widely acknowledged. A shoe shop needs centrality more than a specialist shop for model makers because something like 30% of shoe purchases are speculative purchases made without prior planning, while model airplanes tend to be

planned purchases. It is therefore important for a shoe shop, but not a model shop, to be located where it can be seen. Bid-rent theory suggests that all retail activity will locate centrally while low-order functions, such as newsagents, serve local markets.
- The growth of cities put greater and greater strains on **transport** systems designed to deliver people to the centre of these **monocentric** cities. During the twentieth century, the rapid increase in personal mobility made non-central locations more attractive, reducing the gradient of the bid-rent curves. Moreover, the car allowed cross-city journeys to be made. With the development of ring roads and urban motorways, cities tended to become more **polycentric**.
- **Negative externalities** became significant in the central city areas. For example, journey times grew with congestion, costs of running buildings rose and air pollution increased. The result was a reduction in demand for centrality.
- Cities are dynamic environments with rapidly changing costs. There is considerable **friction** in the movement of enterprises to reflect a new set of circumstances. The willingness to pay rent will vary over time, but location is subject to inertia. You cannot just change location; a move is not without costs. Thus cities reflect both old patterns and newer patterns at the same time.

None of these points invalidates the processes behind bid-rent theory. Rents are usually highest in central locations and the broad pattern holds true in most cities. As a consequence of the higher demand for space in the central city area, buildings here tend to be taller, if the planning authorities consent.

# Urban land-use models and theories

## The ecological model

The best-known urban model, even some 80 years after its original appearance, is that produced by E. W. Burgess. Burgess, working with Robert Park, in the School of Urban Sociology at the University of Chicago, developed a distinctive model known as an 'ecological model'. The key 'ecological' idea was that cities grew in response to processes that were common in 'nature'. Burgess and Park attempted to map the social composition of Chicago in great detail, and divided up the territory into a number of natural communities. At the core of this was a theory about the use of urban space which borrowed heavily from biology and ecology. The chain of reasoning was as follows.
- There was a shortage of space in the city, leading to **competition**.
- The space would be divided up into distinctive ecological niches, or **natural areas**, according to shared economic and sociological characteristics of the population such as race and income.
- Incoming migrant groups would **invade** the central area, as this is where they arrived when travelling into the area by train.

# Chapter 5

*Settlement patterns*

- They would displace their predecessors in a process of **succession** and, after a generation, would be more prosperous and more proficient at speaking the language.
- The city grows from the inside outwards, rather than by the 'hailstone' process of accretion whereby new arrivals are added to the outside of the city.

Thus a concentric arrangement develops, with successive generations of migrants pushing out their predecessors who, with increasing affluence, move towards the edge of the built-up area (Figure 5.7).

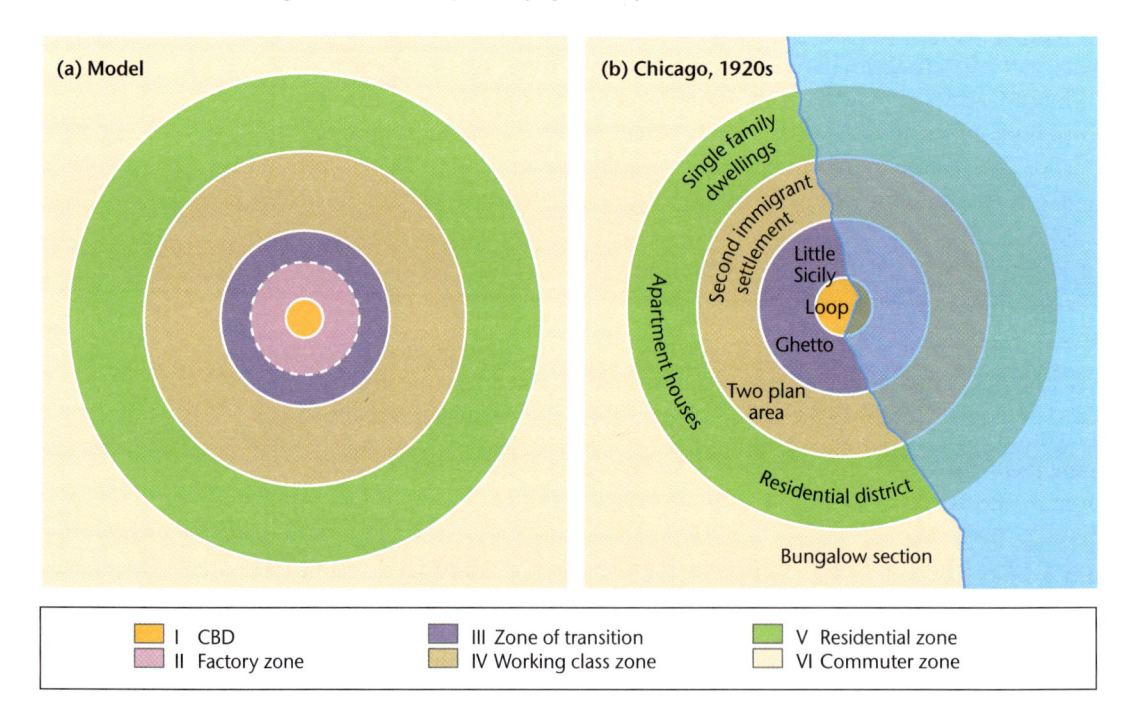

**(a) Model**

**(b) Chicago, 1920s**

Single family dwellings
Second immigrant settlement
Apartment houses
Little Sicily
Loop
Ghetto
Two plan area
Residential district
Bungalow section

- I CBD
- II Factory zone
- III Zone of transition
- IV Working class zone
- V Residential zone
- VI Commuter zone

*Figure 5.7 Burgess's concentric zone model as applied to Chicago in 1920*

- In the centre of Burgess's model was the central business district (CBD). In Chicago this is the Loop, surrounded by the elevated railway system. The train stations, which were the point of arrival for most migrants, were on the edge of the CBD adjacent to the next zone.
- Next was the twilight zone/inner-city slums, ethnic ghettos and industries and warehouses. In some cases a distinctive factory zone developed. This is sometimes known as the 'zone of transition'.
- Beyond this came the zone of 'working men's homes', stable if not affluent residential communities occupied by migrants who were now established.
- Furthest out was the zone of affluent suburban commuters, culminating in a commuting zone on the urban/rural fringe.

## The limitations of the theory and model

It is important to recognise that the driving force behind the competition for space reflected in this model was the extraordinarily rapid growth of Chicago (see Figure 5.5), which was where the urban ecology model was developed. This was a period of limited planning controls, almost anarchic development and considerable corruption involving organised crime. This is not a situation replicated in the development of all cities; indeed it is not replicated in many.

A further limitation is the very American cultural emphasis. Chicago was an enormously diverse city in terms of its ethnic mixture, including Italians, Irish, Germans, English, Scandinavians and, at the time of Burgess's work, a huge influx of black labour from the south of the USA.

Chicago had also grown so quickly that it had not developed the characteristic high-class residential areas near the centre of the city so typical of European and, indeed, some other US cities. Figure 5.8 is a map that shows the use of land in and around Chicago.

Modifications have to be made to the ecological model to deal with natural barriers, such as rivers and waterfronts, and to add major transport arteries, which are common in major cities. Sectors were introduced into the model to allow for patterns of development that were not in concentric circles. Thus in London, as in many British cities, the original industries and the associated working-class districts were concentrated to the east of the city centre. The more affluent residential areas are found to the west, stretching outwards to the suburbs, untroubled by fumes that the prevailing westerly winds carried away from them rather than towards them.

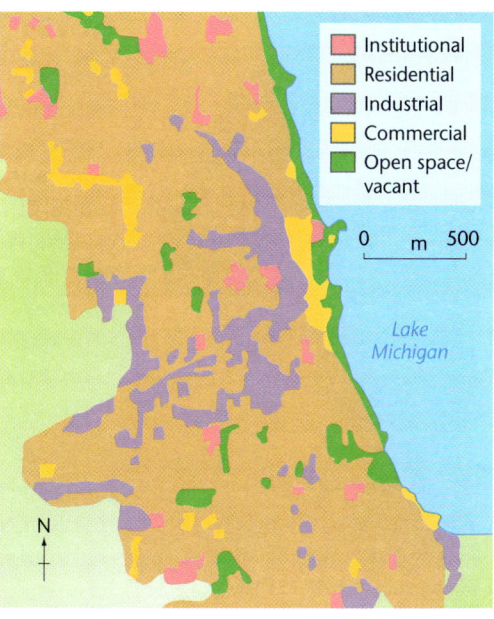

*Figure 5.8 Land use in Chicago*

## Case study 5.8  Invasion and succession in east London

Inner London, especially the area immediately to the east of the City of London, has long been an area of invasion and succession. Industry developed around the present CBD and around the docks along the river. Cheap housing developed there, especially from the eighteenth century onwards, and even today these are some of the poorest local authority wards in the UK.

### History of a building

A place of worship in the East End of London shows the process of succession very clearly.

- In the seventeenth century, Huguenots, French Protestant immigrants fleeing from repression in France, built a chapel in Brick Lane.
- The building was taken over by Methodists in the eighteenth century. The Huguenot community had moved out and in-migration from rural areas around London had began.
- Around 1900 the building became a synagogue for Jewish refugees.
- Later, the Jewish community moved out to Stamford Hill and Bethnal Green (working men's homes), then to Golders Green and Stanmore (affluent suburbia and commuting zone).
- By the 1980s, the building had become a mosque for local Bangladeshis. However, they are now beginning to move further eastwards to Newham, into property vacated by upwardly and outwardly mobile white East Enders.

# Chapter 5 — Settlement patterns

## Hoyt's sector model

Soon after Burgess published his work, Homer Hoyt re-worked the concentric ring model. After studying the rent levels in 142 US cities, he observed some consistent patterns. For example, he noticed that it was common for working-class households to be found in close proximity to railway lines and for commercial establishments to occur along business thoroughfares. He identified two forces that might modify the concentric pattern.

- Forces of **repulsion**, which are negative features in the natural or built landscape. Heavy industry is one, railway lines and noxious rivers might be others.
- Forces of **attraction**, which include areas of high ground, the CBD and attractive river frontages.

According to Hoyt, as cities grew outwards in response to the forces that Burgess identified, they created wedges or sectors spreading out from positive or negative features. The industrial area would have working-class housing attached to it and, as the city grew, that area would extend outwards. The high-class housing areas would be effectively cordoned-off from the working-class areas by lines of communication and, in all probability, middle-class housing.

In many respects, Hoyt's sector model is simply a concentric zone model modified to account for the impact of transportation systems on accessibility (Figure 5.9).

- Central business district
- Wholesale, light manufacturing
- Low-class residential
- Medium-class residential
- High-class residential

*Figure 5.9 Hoyt's sector model of urban land use*

## Other models

Many other land-use models have been developed to reflect economic, social and political conditions in particular countries at particular times.

- The Harris and Ullman model identifies that some cities have more than one centre; they describe these as '**multiple nuclei**'.
- Mann proposed a model for British cities that took into account the role of planning and postwar redevelopment of council estates on the urban structure, as well as the influence of the prevailing westerly winds.

Few of these models advanced upon the basic theoretical ideas outlined by Burgess and his colleagues in Chicago. It is productive to see cities as developing in response to several forces which overlap, creating a complex social space (Figure 5.10).

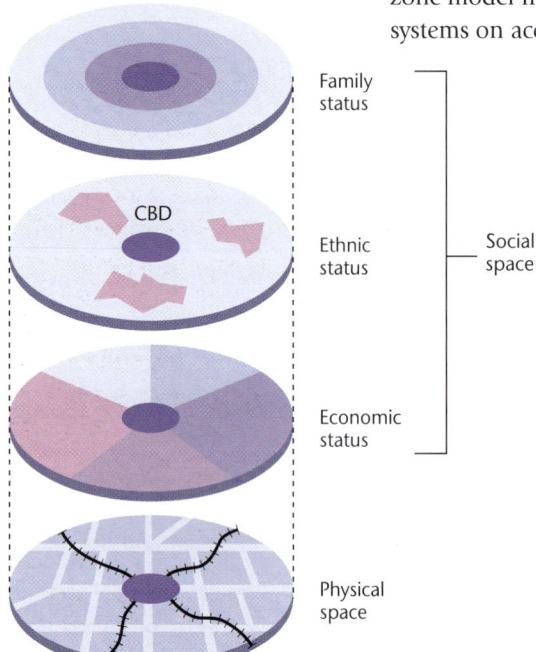

*Figure 5.10 The residential mosaic — another way of looking at how an urban area develops*

# Case study 5.9  Land-use patterns: Salisbury

The land-use pattern in Salisbury is typical of that of a medium-sized settlement in the south of England. With a population of about 40 000 and very few city-forming industries, Sailsbury's main functions are related to its central place activities. Figure 5.11 is an outline map of Salisbury that shows the land use.

## The CBD

The CBD is situated to the north of the cathedral in the old medieval core. The peak land-value intersection has shifted southwards in recent years, reflecting both the growing significance of tourist-related functions around the cathedral and the decline in importance of the bus station and railway station, both of which lie towards the north side of the CBD.

The rate of shop occupancy is very high (92% out of about 400 units, mostly quite small), reflecting the wealth of the town and the catchment area. Despite the growth of out-of-town shopping only 1 km away to the southeast at Bourne End on the Southampton road, the city centre has retained all the major high street chains and a large number of specialist functions. A market is still held here twice a week.

The CBD is also the location of a number of key tertiary employers, such as the cathedral and its diocesan administration, the local council and Friends Provident, an insurance company. There is residential use too — the second and third floors of a number of retail outlets are used as flats, something less common in larger towns with higher rent levels where financial services often out-bid residential use.

## The zone in transition

The next zone is certainly not a 'classic' area of in-migration, dilapidated buildings and urban blight. But there are changes here that reflect the processes of invasion and succession. The most obvious change is the use of what was once retail space as residential accommodation, especially in the Eastern Chequers area. In common with many counties in

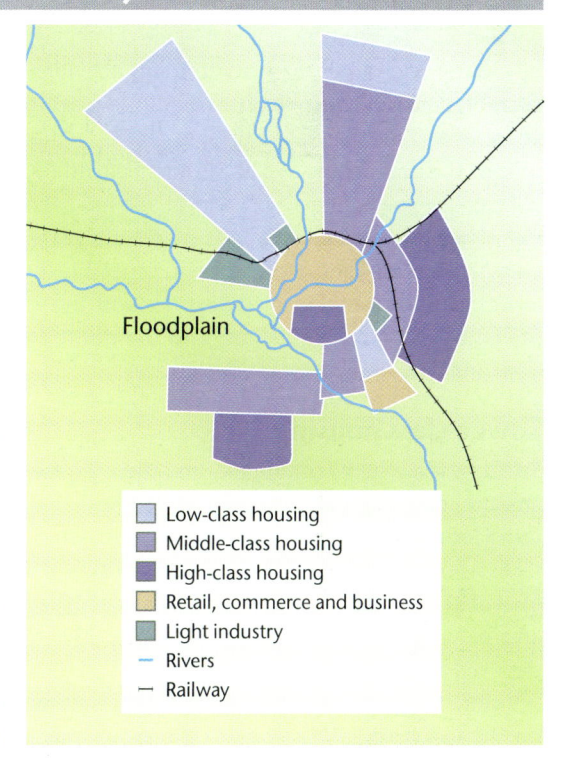

*Figure 5.11 Land use in Salisbury*

southern England, Wiltshire has to look for ways of accommodating a growing population fed largely by north–south drift. One of the preferred ways of doing this is to build or renovate brownfield sites to create small, usually two-bedroom, properties close to the city centre. In Salisbury, the land used in this way includes former car parks and small convenience retail units that have lost trade on the margins of the CBD, especially since the development of out-of-town shopping. The local authority ward data suggest that this is one of the poorer areas in the city.

## Industrial land use

Industrial land use is limited and quite concentrated. The oldest is in the Gas Lane area to the north of the station, a very small area of about 50 hectares with nineteenth-century terraced houses and a few older workshops, most now devoted to car repair and

# Chapter 5  Settlement patterns

re-spraying. The other two areas are both twentieth century industrial estates, which would be better entitled wholesale estates in that they are dominated by city-serving rather than city-forming activities. Examples are builders' merchants, double-glazing agencies and the ubiquitous garages and car rental offices. The larger of the two, the Churchfield estate, is located on the floodplain of the river and has been inundated on several occasions in recent years. Growth is unlikely and access is poor, especially for lorries. The distribution of industrial land use is driven by planning regulations and the desire to keep 'industry' away from residential areas.

## Lower-class housing

Salisbury is a wealthy community and there are few areas of social deprivation. The most deprived area, using local authority ward data, is Bemerton, which ranks as number 2120 in the national database of a total of 8140 wards. Thus, the most socially deprived area of Salisbury is not even in the bottom 25% of deprived areas in the country.

Victorian housing exists around the Gas Lane area on the low-lying land of one of the many valley bottoms. As the city expanded in the 1930s, cheaper housing developed along the Devizes Road to the northwest, along the ridge. This was also the zone where large estates were built from the 1960s onwards, and it contains the largest single planned extension of Salisbury onto greenfield sites beyond.

There is also an area of lower-class housing in The Friary, which is more or less adjacent to the CBD. Although it has been recently redeveloped, this was an area of 1960s development comprising blocks of flats on land only just above the floodplain. Salisbury lacks a heavy industrial base, and it would appear that the development of its lower-class housing has been strongly sectoral. This has as much to do with planning policy and municipal housing as it has to do with forces of repulsion and attraction.

## Higher-class housing

There are a number of areas of distinctly higher-class housing in Salisbury. The most notable are the houses in Cathedral Close, right in the middle of the city on the south side of the CBD. These are highly desirable, because of:

- their proximity to the attractive Gothic cathedral and the open space around it
- their proximity to the city centre
- the almost 'gated' nature of the close — walled on three sides and allowing very little vehicular traffic

These properties, many of which belong to the church authorities, are largely rented rather owner occupied, which is in itself unusual.

Other areas of high-class housing include new developments of executive style homes on the south side of the city on Harnham Hill and in the commuter village of Laverstock. There are also a number of large properties to the east of the city in Milford ward, but these have been subject to change in recent years. These large Victorian mansions built for the professional and property-owning class and their servants, on the high ground to the east of the city centre, are now too large for modern families. In a small example of invasion and succession, many of the houses either have been divided up into flats or are used as old people's homes or, in one case, a small school. In the meantime, the professionals have moved out to Laverstock and other surrounding villages.

## Middle-class housing

Much of the rest of Salisbury is middle-class housing, which takes on a largely sectoral form. Salisbury is at the meeting point of three chalkland rivers, and the floodplains dissect the higher ground. The constraints of the site limit development to the higher areas. The middle-class housing was built in two main periods.

- In the 1930s housing developments tended to be built along the main roads leading out of the city, especially to the north and northeast. These were served by bus routes, in a largely pre-car era, and by shopping parades built along these roads.
- Infill between these roads took place in the 1960s and onwards, with privately built estates offering houses with garages. Other than residence, there were no other functions provided in these areas.

Case study 5.10 is also about Salisbury, this time looking at the changes in the function of the city.

*Human environments* **AS Unit 2**

# Urban development in LEDCs

Cities in most LEDCs have grown more rapidly in recent years than cities in MEDCs. Figure 5.12 gives data for 1965 and 1990 and projected data for 2015 for seven LEDC cities.

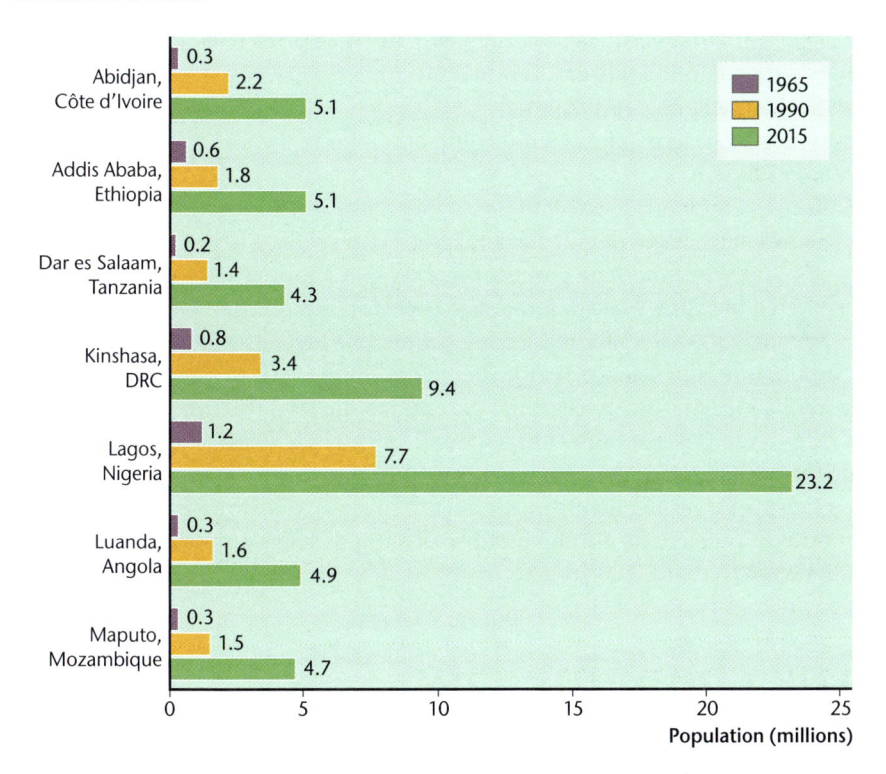

*Figure 5.12 Growth of cities in LEDCs*

| City | 1965 | 1990 | 2015 |
|---|---|---|---|
| Abidjan, Côte d'Ivoire | 0.3 | 2.2 | 5.1 |
| Addis Ababa, Ethiopia | 0.6 | 1.8 | 5.1 |
| Dar es Salaam, Tanzania | 0.2 | 1.4 | 4.3 |
| Kinshasa, DRC | 0.8 | 3.4 | 9.4 |
| Lagos, Nigeria | 1.2 | 7.7 | 23.2 |
| Luanda, Angola | 0.3 | 1.6 | 4.9 |
| Maputo, Mozambique | 0.3 | 1.5 | 4.7 |

Population (millions)

Cities in LEDCs have been growing in a different way from cities in MEDCs. The most obvious difference is that new arrivals in LEDC cities have not had jobs to go to in the inner-city area. Instead, they have arrived on the margin of the city, more in hope than in expectation of something better than the rural poverty from which they have escaped. Like hailstones, LEDC cities have mostly grown by accretion, with layers added to the outside.

In many countries in sub-Saharan Africa, Latin America and parts of Asia, urbanisation has taken place without any accompanying industrialisation. The result has been the staggering growth of slums, almost all on the periphery of the cities. This is what the American writer Mike Davis calls the 'urbanisation of poverty'. Only in a few places, such as Shanghai, does the urban growth reflect a rapidly growing manufacturing economy that provides the engine to soak up much of the urban population growth of recent years.

*The Challenge of Slums*, published in 1993 by the United Nations, places the blame for the growing urban problem on the IMF's structural adjustment schemes. These schemes, it argues, have stimulated major land-use changes in the countryside, because rural areas are now geared towards export crops that will help

### Examiner's tip

■ Don't forget that the rate of city growth is the percentage growth over time. The number of people added to the population is the absolute or numerical growth, which is frequently larger for larger cities.

## Chapter 5 Settlement patterns

pay back debt. This has resulted in an acceleration of rural–urban migration. The numbers involved are staggering.

- The 'official' number of slum dwellers globally is put at 921 million, much the same as the *total* global population in the early 1840s when the slum problem first attracted the attention of social reformers.
- The highest percentages of urban dwellers who live in slums are the astonishing 99.4% of Ethiopians, 98.5% of Afghans and 92% of Nepalese.
- In Delhi, planners talk of 'slums within slums' to describe the deterioration in the quality of accommodation.
- There may be as many as 250 000 different slums in the world, the vast majority in LEDCs.
- Slums are typically located on the city margins, and are known occasionally as 'slum sprawl'. Lagos doubled in land area in the space of 10 years, between 1990 and 2000, and the state governor reports that about 60% of the state area is covered in slums. Note that the state covers an area of over 3500 km$^2$ in total. No one knows for sure how many people live here; officially it is 6 million, but unofficially it is put at more than 10 million.
- About 85% of slum dwellers have no title to the land they live on.
- Fifty-seven per cent of the urban poor in Africa lack access to sanitation.
- In Mumbai, India, the poorer districts have 'one toilet seat per 500 people'.
- Construction is frequently on steep land or close to toxic emissions.
- Current predictions put the number of slum dwellers at more than 2 billion by 2035.

# The changing city

## Functional change

The functions of cities are what cities do, what they are there for. Cities are multi-functional, but those functions change over time, at least in their relative importance.

Functional change drives land-use changes in a city as areas once devoted to one usage convert to other types of land use. In recent years industrial warehouses have been converted to housing, as at Salford Quays (Manchester), and workshops that were engines of the Industrial Revolution have become retail 'outlet centres'. At Swindon Brunel's great locomotive workshops are now a shopping mall. County Hall in London, once the heart of London's city government, has become a hotel. The same processes go on all over the world as the functional role of a particular city in the world economy changes.

This type of change happens at all scales. For example, Corfe Castle in Dorset changed from a defensive position of national significance to a mining and quarrying centre, and now to a village that features on many tourist brochures and is a popular residential centre for retirees from urban areas. It is quite a short historical journey.

*Human environments* **AS Unit 2**

## Case study 5.10  *Changing city functions: Salisbury*

Salisbury is a city of about 40 000 inhabitants in southern England. It is best known today for its magnificent Gothic cathedral. It has experienced several changes of function and one dramatic change of site in its long history.

Originally the city was sited at Old Sarum, a defensive position of some military significance, occupying the site of an even older Iron Age fort slightly to the north of the existing city.

In 1220, the decision was taken to move the settlement down closer to the key bridging point on the River Avon, where the floodplain narrowed. The new city was planned and the central medieval core was laid out as a grid, much like US cities many centuries later. The site of the commercial and residential core lay to the north of the new cathedral, which, in those days, gave the city some significance as a point of administration and as a central place. The sheep grazing on the chalk lands that encircled the town generated local wealth, and it soon became a major industrial centre for the wool trade, with merchants' houses appearing in the city centre. The important central place function for the surrounding area was largely carried out in the large market place to the north of the cathedral.

*Photograph 5.5  Old Sarum*

By 1400 Salisbury was one of the most prosperous and largest settlements in the country, ranking fifth or sixth in the country in the years before the Black Death. This industrial function was not to last and, together with the decline in the significance of the Church as an administrative body, Salisbury was left with little more than its central place function by the nineteenth century.

As it lacked immediate natural resources, the city was passed by during the Industrial Revolution. It experienced only gradual population growth while northern towns and cities, such as Leeds (see Case study 5.7), experienced explosive growth.

What about today?

- Salisbury has continued to flourish as a market centre.
- It has very little manufacturing industry, with only 6% employed in this sector compared to a national average of over 17%. The largest single manufacturer is Dunlop Hiflex, located on the Churchfields Estate, which has only 80 employees.
- By contrast, Salisbury has experienced some growth as a service industry centre, with the insurance company Friends Provident becoming a significant local presence.
- There is a growing tourist industry in the town, as witnessed by an increasing number of food outlets and gift shops. The main focus is on the cathedral but, sadly for the town, the tourists tend to be day-visitors on coach trips that start or finish in London and which also take in Stonehenge and Bath.
- There is also an emerging quaternary sector and some hi-tech employment, related to the proximity of Porton Down and other research facilities in the immediate area.

Thus Salisbury has been a defensive and military city, an industrial city, an administrative centre and a tourist town. Throughout each of these periods it has also been a place to live and a central place.

*Chapter 5  Settlement patterns*

## Case study 5.11  Changing land use: Cairo

Cities in LEDCs do not share the history of growth associated with industrialisation that provides the basis of the Burgess and Hoyt models.

Cairo, the capital city of Egypt, has had four main periods in its 1000-year history.
- An Islamic phase when the city was founded by the Fatimid dynasty in 969, after which it became a key city in the emerging Islamic world.
- An Imperialist phase, which began formally in 1802 when Egypt became an important cog in the British empire.
- An Arab socialist phase after the eviction of the British by Colonel Nasser in 1952.
- A capitalist phase ushered in by a change of direction in 1987 and still in its early stages.

In each of these periods Cairo has changed both function and form. Building heights, street layout and the economic base of the city have all altered profoundly.

In the first of these four periods the main focal point of the city was an open parade ground, Bayn al Qasrayn, located between the east and west Fatimid areas. The focal point was the al-Azhar Mosque, which became a centre of Sunnî learning; the university which grew out of the mosque is the oldest university in the world. In addition to these military and religious functions, markets were developed to supply the city's residents. In the markets and residential areas a pattern of occupational segregation was established whereby particular trades occupied particular districts. This persisted in Cairo until the nineteenth century.

Rapid urban development took place in the eleventh century when famine struck the region and migrants arrived in the city. A port was built on the Nile and the strategic position of the city led to the development of a strong, trade-based economy; the city's market area and commercial importance grew, and the famous Khan al Khalili market was founded during this period. By the fourteenth century the city had a population of 500 000, making it one of the three largest urban settlements in the world. The large fortress complex

*Photograph 5.6  The al-Azhar Mosque, Cairo*

built by Saladin, situated on a hill, dominated the walled city. This complex, still in existence, is massive and contains many large mosques and royal buildings.

By the sixteenth century Ottoman rule from distant Istanbul and shifting trade routes that bypassed the Red Sea, followed by the discovery of the New World, meant that trade and economic power moved away from Cairo. As a result, Cairo experienced a period of decline in population as well as importance.

When the British arrived in the early nineteenth century, Cairo was restructured. A new administrative district was built, with a very European look. Elite areas, such as Heliopolis, were built specifically for the new European population, some of the old Arab quarters were removed and the city was effectively divided into two halves. The British also invested heavily in the development of cotton as an export crop, hastening rural–urban migration that initiated a period of rapid population growth in the city. New areas of housing were built to the north of the city.

The end of British rule brought yet another change of direction in the planning for Cairo. Under the new socialist government led by Colonel Nasser, Egypt shifted from an export, colonial style economy supplying European markets, to focus on the development of manufacturing industry, much of it built with Russian money and engineering help. Cairo served as the central control point for all the massive projects such as the Ministry of Agriculture and Ministry of Supply, being undertaken in rural areas. Huge ministerial departments were located in high-rise buildings to carry out the policies of the new regime. There was an enormous growth of public sector employment, from about 300 000 in 1952 to over 1 million in 1966–67. To house these technocrats, Nasser City was built at the northeast of the city.

Poorer people came in large numbers to the city. To house them the government created a public company which built thousands of apartments in poor areas of Cairo, such as Zeinhom, Helwan, Imbaba and Shubra al Khima. These were built on the Soviet model of high-rise apartment blocks. They differed totally from traditional housing in Egypt which emphasised a close relationship with the surrounding area. Rent controls were imposed elsewhere, making it difficult for landlords to afford to improve or even repair property. As a result, much property fell into a state of disrepair.

Meanwhile rural–urban migration was being encouraged by the government to provide the labour force for industry. Cairo became more and more the primate city in Egypt and the westward and northward expansion of the city was notable. In time this expansion became a concern as Cairo's infrastructure started to creak, so new towns were planned in the surrounding desert.

Towards the end of the socialist period, Nasser's successor, Anwar Sadat, began to encourage private enterprise in Egypt in general and Cairo in particular and a new generation of high-rise building began, this time for the wealthy. This has continued into the fourth and most recent period in which the Egyptian economy has been heavily controlled by outside forces, not least the World Bank and the International Monetary Fund.

The change is obvious in the built landscape. Fifteen years ago, few foreign enterprises, especially outside the tourism industry, could be found. Now there are, for example, 20 branches of Pizza Hut, as well as McDonald's, Taco Bell and many others. The food is only affordable by the new middle class, which has been able to make money during the economic transition. New shopping centres have been built, such as the up-market World Trade Centre. High quality imported goods are now available in the numerous shops that have appeared in notable focal points, such as the Gamiyyat al Duwal al Arabiyya (Arab Union) Street in Mohandesin. This street has developed into a strip of high-class retail stores and food outlets, and is cruised nightly by the relatively wealthy driving their imported cars. New housing developments have appeared on the edge of the city such as Al Rehab ('spacious'), which is being billed as the country's first privately built city. This 'new city', on the northeast fringe of Cairo, beyond Heliopolis, is planned for 150 000 residents and 30 000 jobs.

# Chapter 5

*Settlement patterns*

## Processes of change

### Suburbanisation

**Suburbanisation** is the decentralisation of people, employment and services from the inner part of the city towards the margins of the built-up area. The effects of suburbanisation are felt within the city and in the surrounding rural areas. Suburbanisation was driven by improvements in public transport and increases in income. In the UK it began in the 1830s, slowly at first, with the coming of the railways and the rise in disposable income that allowed a growing middle class to seek new places to live. For example, in Streatham, in south London, population growth was rapid.

- In 1792, Daniel Lysons reckoned that Streatham's population was about 1590 inhabitants occupying 265 houses.
- By 1850 when the railway arrived the population was 7000. The railway tied Streatham into the London economy and the population took off rapidly.
- Between 1871 and 1901, the population of Streatham rose from just over 12 000 to almost 71 000.

In the period between the First and Second World Wars, in both the USA and Europe, the rise of consumer industries led to the development of new sites on the arterial roads built on the urban fringes of many cities. Suburbs were not static communities and were increasingly invaded by upwardly mobile members of the working class who pushed first generation suburbanites out to the city margins in a process of invasion and succession similar to that described by Burgess.

### Counterurbanisation

**Counterurbanisation** is a change that extends beyond the city area. In terms of population concentrations, the process of urbanisation appears to be reversing in some MEDCs. In other words, people are leaving the city and moving to smaller towns and villages because of improvements in transport and ICT, government policy, perceived differences in quality of life and office relocations. This process affects both the shape and form of settlements. There are two distinct trends here:
- a movement of employment to rural areas
- a movement of people to rural areas who then commute

There is a good deal of evidence that there has been movement of industry to rural areas, but the definition of 'rural' needs qualifying. These areas are frequently within easy reach of a major urban centre but set in an attractive landscape. However, much of this movement is not significant economically. The 'counterurbanites' are still commuting into the cities or, like the inhabitants of Berkshire villages who commute to Reading or Newbury, they are still tied to the London economy.

### Re-urbanisation

**Re-urbanisation** is the return of people to the cities, especially inner cities, in recent years. This has an obvious effect within the city, both positive and negative. The return is particularly significant in large cities, especially global cities in MEDCs, where population has grown rapidly.

*Human environments*  **AS Unit 2**

## Redevelopment and change in the city

In most MEDC cities in the 1990s, it was recognised that there was an urgent need to revive and redevelop flagging city/central city areas. This was a response to the changing world economy: globalisation was having the effect of switching employment structure, with a shift in MEDCs from manufacturing industry to service industry, and inner cities were suffering unemployment and dereliction. Most plans aimed to stop the loss of population and employment, improve the housing stock (including removing the worst residences) and generally upgrade the city image. But, more importantly, plans had to attract the mobile investment created by globalisation. Many cities have responded to the challenge.

Birmingham's redevelopment of the city centre, for example, was aimed at the creation of a safe, profitable and pleasurable environment. Through a so-called 'growth coalition', several flagship schemes were promoted. The principal area of development was in the derelict areas in the northeast part of the city. The Heartlands initiative aimed to develop the office space within the city and to return some housing. Other city projects involved building an international convention centre, the National Indoor Arena and various developments in and around the Broad Street area.

### Case study 5.12 *City centre redevelopment: Nottingham*

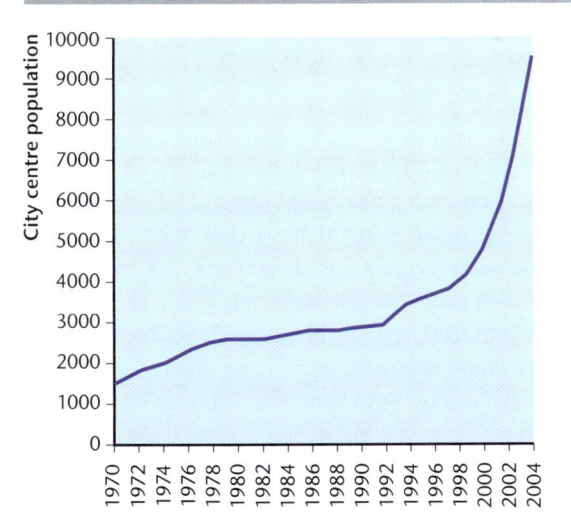

*Figure 5.13 Population change in Nottingham city centre*

Nottingham is the third richest city in the UK and has the fourth largest retail market, which contains over 1300 outlets that have limited out-of-town opposition. The city was once a significant centre for textiles, with a flourishing lace industry, but has now become an important centre for financial and business services with a burgeoning biomedical reputation that is linked to its highly rated university. The city centre, once a maze of small workshops, is being transformed by a series of development projects that are largely geared to creating both more office space and more space for houses.

The former Lace Market area, which had 90 residential units in 1998, will have 500 residential units by the end of 2005. The population of the inner city is forecast to rise significantly. Figure 5.13 shows the recent change in the city centre population of Nottingham.

The current plans include:

- renovation of the historic Lace Market
- redevelopment of the Broad Marsh and Trinity Square shopping areas, creating a further 77 000 m$^2$ of retail space
- doubling of the population living in the city centre from 3000 to 6000
- continued investment in key city centre sites
- development of national and international sports facilities: these include the National Ice Centre, the National Water Sports Centre and the Nottingham International Tennis Centre

# Chapter 5  Settlement patterns

## The changing CBD

The price and shortage of land in the CBD means that there has been an increase in the height of the buildings to cope with demand in many cities. There has also been a slow, but remarkable, change in the format of the CBD. Increasingly, it is going underground and into shopping malls. At street level, pedestrians are being separated from vehicle movements. Other social and economic changes have left the CBD to the banks and large department stores that can afford the location, with other businesses opting for more convenient and profitable sites away from the city centre.

### Case study 5.13  Changes in the CBD: London

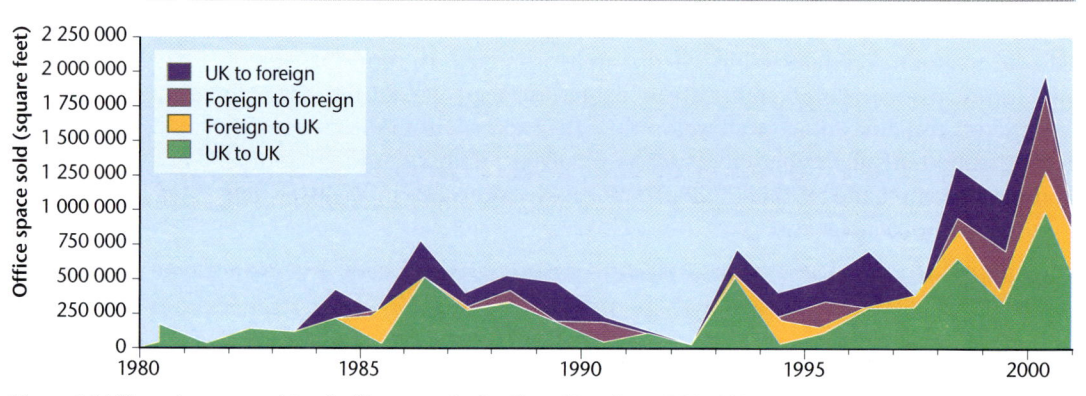

*Figure 5.14 Changing ownership of office space in the City of London, 1980–2001*

London is **polycentric** — it has **multiple nuclei** — although the various centres are close enough geographically to walk from one to the other without undue strain. One of these, the dominant financial centre, is still defined by the area of the City of London, 290 hectares dominated by office space, 7 million m² of it in all, which occupies 75% of the space in the City. A little further up the River Thames, Westminster is the political centre of the city and the West End is the social and cultural heart of the capital.

London has had an abundance of administrative, legal and financial services for at least 300 years, given its role as the centre of an empire that reached its geographical zenith in the early twentieth century.

Decolonisation temporarily threatened this global position, but the growth of the global economy and the leading role of London in the international organisation of that economy has placed renewed pressures on the city. These pressures have led to a rise in rental levels in the best areas to over £750 m$^{-2}$ for office space, among the highest rates in the world. In the 1980s the deregulation policies of the then Conservative government liberated the city from a whole series of regulations that had impeded its global role. Figure 5.14 shows the rise in transactions carried out in London since 1980.

The 2200 hectares of former docks have become the world's largest urban regeneration scheme and a major extension to London's CBD (see Figure 5.15). London is now a global city. It is the most important financial centre in Europe and, along with Tokyo and New York, is one of the three centres of global capital — operating as a centre for the raising of money, its re-circulation and the generation of many new ideas to help the flow of international capital. As such it generates 16% of GDP in the UK.

Dominant users of London's CBD are finance, insurance, law and accountancy (sometimes known as FIRE uses). The current London Plan assumes that another 600 000 jobs will be created in London over the next 15 years, many of them related to this global role. In the meantime the number of manufacturing jobs in London is forecast to fall from 319 000 in 2001 to 190 000 by 2016. Previously important as the centre of the newspaper and printing industry, central London now has a number of high-value-added and design-led manufacturing companies which will help the city retain an important role in the knowledge-driven economy.

Some 6800 people live permanently in the city, but over 300 000 work there in the daytime. Given the dominance of the office-based business sector in these employment projections, the availability of suitable office accommodation is a critical issue. This explains the urgent need to find more space and the pressure to expand the current CBD. The expansion eastwards into Docklands was little more than the first phase of a move that is now set to extend the estuary area, with the development of the Channel Tunnel rail link and the Gateway project.

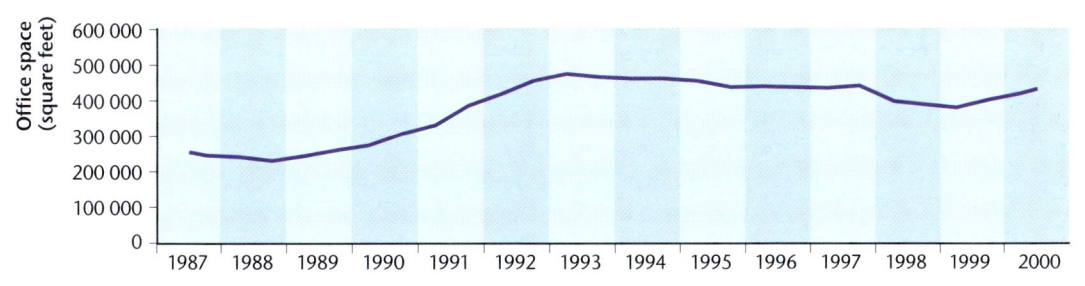

*Figure 5.15 Increases in London's stock of office space, 1987–2000*

## The transitional zone

For many decades, the transitional zone was characterised by a mix of housing and industry. Today, however, transitional areas are extremely diverse. There is pressure on this zone, particularly where it is in close proximity to higher-status areas of the CBD. On the whole, industry is being driven out, and overflow commercial activity from the CBD is moving in (so-called **active assimilation**); industry that is left tends to be the footloose IT and high-tech firms. Where the zone intersects with lower-status activity, it tends to retain much of its dilapidated housing and traditional small industrial units and warehouses. These areas are being passively assimilated into the CBD (e.g. in London and Nottingham).

### Case study 5.14 Changes in the transitional zone: Leeds

The historical development of Leeds was discussed in Case study 5.7.

Leeds city centre has been transformed in recent years and the next stage will be the development of the old industrial area along the Burley and Kirkstall roads, a kilometre-long strip of land, backed by the old Leeds–Liverpool Canal, that was once the industrial core of nineteenth century Leeds. This typical **zone in transition** was dominated by manufacturing and wholesaling but now, with companies such as Yorkshire Chemicals and Ala Foods relocating elsewhere, a substantial area has been released for development.

The city council, keen to establish Leeds as the major regional centre, wants to develop the area with a view to filling the gaps in what is one of the most important retail and service centres in the north of England. Thus a concert hall and a conference centre are in the embryonic plans, along with housing and some retail facilities, although without detracting from the Headrow, the retail heart of the city. This sort of mixed usage is a very long way from the segregated land-use plans of the 1960s and 1970s.

# Chapter 5

*Settlement patterns*

## Commercial activity in the suburbs and on the city edge

The hierarchically ordered sub-centres within the city (that is, the corner shop, shopping parades and small clusters of CBD-type activity) are changing. As cities have grown and the population has become more mobile, so have the numbers of district sub-centres serving specific areas of cities. These areas attract hypermarkets and supermarkets, and their emergence triggers the growth of residential areas. Corner shops in many areas are now a thing of the past. They have been overtaken by both district supermarkets and the forecourt shops attached to petrol stations.

## Out-of-town shopping

Out-of-town shopping first evolved in the USA and appeared in Europe when hypermarkets were built on the urban fringes in France. These are large single-storey warehouses of goods which sometimes operate at the heart of shopping malls and sometimes stand alone.

One of the first hypermarket developments has become a case study in retailing circles of 'what not to do'. The greatest density of out-of-town facilities in mainland Europe, in terms of the population served, is in Toulouse in southwest France.

- The first hypermarket to open in Toulouse was Mammouth in 1969 with 11 000 m$^2$ of selling space.
- It was followed soon after by Ouragan, which opened an 8000 m$^2$ store, 8 km out of the city just off the autoroute. Initially the store was a great success.
- One month later Carrefour opened a 25 000 m$^2$ store 3 km nearer the city centre. Within 2 months Ouragan had closed.

Superstores in the UK are typically smaller than those on the continent and in the USA, although store size is increasing. Britain had 457 supermarkets in 1986 but 1102 by 1997. Many of those built in the 1970s and 1980s, encouraged by the free-market spirit of that time, were in out-of-town locations. There were a number of reasons for this.

- Land on the city boundaries was cheaper and less constricted, allowing for expansion.
- Access by wealthy car owners was easier and car parking spaces were able to be provided at a ratio of about 2:1 to the area of the store itself. Motorway intersections were frequently chosen as sites.
- Access to a cheap female suburban workforce was straightforward.

Note that, in 1960, the UK had 5 648 000 cars and only 30% of UK households had a car. In 2001, the UK had over 28 000 000 private cars and vans.

As the planning atmosphere has changed and the supply of suitable out-of-town sites has dried up, each of the major companies has looked at ways of developing alternative formats and alternative means of growth. For example, Tesco has different types of store designed to meet the needs of different customers in different areas. These are:

- conventional superstore
- hypermarket format
- compact format (for smaller towns and suburban areas)
- city centre concept

Human environments  AS Unit 2

## Case study 5.15  Out-of-town shopping: Meadowhall, Sheffield

*Photograph 5.7 Meadowhall shopping centre*

Meadowhall out-of-town shopping centre in Sheffield was opened in 1990. It is situated 5 km north of Sheffield city centre. The site is an ideal location, as it has a catchment area of nine other cities, including Leeds, Nottingham and Manchester, all within an hour's drive. Meadowhall has 30 million visitors a year and is planning to expand along the lines of North American shopping complexes that have gone well beyond their original retail function. British Land, which owns Meadowhall, and MGM Mirage, which owns the MGM Grand in Las Vegas, are planning to build a £250m casino, hotel and leisure complex at Meadowhall.

In common with other such centres, Meadowhall offers:

- a purpose-built enclosed environment
- enormous choice with around 100 000 m² of retail floor area
- ready access, with over 10 000 free parking spaces
- largely comparison-goods shops and high-street 'brand name' shops, with one or two department stores that act as anchor stores and which provide the focus

Such regional shopping centres are event places – people go there for a day out because there are cafés, cinemas, crèches, restaurants and maybe casinos as well as shops.

There is a complex set of impacts.

- Meadowhall employs some 6000 people and the multiplier effect is not insignificant. However, most of these jobs are relatively low paid.
- Shops have closed in Sheffield city centre, creating something of a hollow city.
- The elderly and socially disadvantaged cannot reach Meadowhall easily.
- The development has forced some rethinking by the city government and stimulated long-overdue programmes to revive the inner city. These include substantial redevelopment of large parts of the centre, including Hallam University and the redevelopment of the John Lewis department store.

# Chapter 5

*Settlement patterns*

## Examination question

*Photograph 5.8*

Refer to Photograph 5.8 when you answer these questions.

**a** (i) Describe the shape of the village. (2)
  (ii) What evidence is there that this village is in an upland area? (3)

**b** (i) Define the term settlement function. (1)
  (ii) Name and outline TWO functional differences between urban and rural settlements. (4)
  (iii) Describe the differences between lower-class and higher-class residential areas. (4)

**c** With reference to a named urban settlement, describe and explain the changes that have taken place in the CBD in recent years. (6)

*Total = 20 marks*

# Synoptic link
# Influence of government policies on settlement characteristics and patterns

- Policies for managing changing urban and rural settlement.
- Issues, rationales and outcomes of policies.

## Managing urban sprawl in the USA

Most North American cities were built using technologies that assumed an inexhaustible abundance of cheap oil and land. These communities grew with low-density populations, compared to the crowded medieval origin of most European cities. Thus, they became increasingly dependent on transport systems. Cheap energy encouraged an infatuation with the automobile, and increased the separation of work places from homes.

Urban sprawl is therefore the legacy of abundant oil and assumes the right to unrestricted use of the private car, whatever the social costs and externalities. Per capita gasoline consumption in many cities in the USA and Canada is more than four times that of European cities and is over 10 times greater than that in high-density cities such as Hong Kong and Tokyo. These differences in consumption are partly explained by larger car sizes and low petrol prices in North America, but the most important factor is differences in the compactness of land-use patterns. Sprawling suburbs are arguably the most economically, environmentally and socially expensive pattern of residential development humans have ever devised.

- The negative local and regional level consequences of sprawl, such as congestion, urban air pollution and increased commuting distances between home and work are now widely recognised. Many of these external costs fall unevenly on the population. Less widely acknowledged are the global ramifications of North American land-use patterns. Largely because of low-density sprawl, the residents of US cities produce about twice as much carbon dioxide per capita as do Amsterdam residents.
- The argument about sprawl is one of the fiercest in the USA. Many people defend the civil liberty to live in whatever environment one chooses, and the development, building and real estate industries are powerful voices defending sprawl against environmentalist critics. The current US administration is unlikely to take a strong public view, although it is politically aligned with the anti-environmental lobby.

A study in San José, California, compared the environmental demands of 13 000 new residential units contained within an urban greenbelt with the same number if they were built in the usual sprawl pattern. The sprawl homes would require 320 000 more km of car commuting and 11.5 million more litres of water per day. The sprawl units would also require 40% more energy for heating and cooling than would their urban counterparts.

Sprawl is not confined to the USA. The current growth of Manila (Philippines) or Jakarta (Indonesia) is dominated by low-density outward spread, which is creating major problems of urban congestion. Here planners tend to react to such growth by improving the roads to lift congestion, which, in turn, generates more sprawl. This is an example of positive feedback. Urban sprawl can be contained by:
- setting planning limits on physical expansion
- favouring and promoting alternatives to the automobile
- limiting parking availability and automobile access to inner cities
- applying regional carbon dioxide taxes
- using traffic-calming street designs

## Chapter 5 Settlement patterns

## Planning in the UK

New towns were built in the period when the separation of functions into discreet neighbourhoods was almost a mantra of urban planning. The result was that the residential areas were poorly served with services and the town centres were completely dominated by retail and office functions. By the late 1980s this idea had become discredited as town centres were abandoned in the evenings and residential areas lacked vitality for much of the day.

Some of the designs put forward never made it to realisation. In Crawley New Town, for example, there was a plan to build a garage or parking plot for every home in the town. The idea was rejected on the grounds that the low probability of usage did not warrant the high costs involved. This may now seem somewhat short-sighted, particularly in comparison to the apparent forward thinking in creating pedestrianised town centres. At the time, the report explained that the likelihood of the increase in car ownership sustaining itself at the then current rate was improbable, and to spend such a large amount of money on an entirely car-orientated society would be unwise. This is one of the 'best' examples of how town planners got things dreadfully wrong, but at the time the decision was made, very few people could have expected that the disposable income of an average family would rise as fast as it did.

Planning of infrastructure is littered with similar examples; both the first Severn Bridge and the M25 reached their capacity in a far shorter time than was predicted. In the case of the M25, plans were being made to add more lanes within weeks of it opening.

## Planning and gender issues in LEDCs
### Urban planning

Since many women in LEDCs have to work in or near the home, they are the people most affected by failures in planning housing and settlement projects.

Examples of features of urban planning that fail to consider women's needs are:
- the absence of day-care facilities
- the inappropriate location of public water points in poor communities
- the lack of children's play parks
- inadequate lighting of streets and public areas
- costly and inconvenient public transportation facilities
- 'modern' housing designs that do not take into consideration women's traditional use of space
- zoning laws that prohibit economic activities and food-growing in residential areas, as these particularly damage women's income-producing strategies

LEDCs that have made major advances in ensuring women are housed include India and Namibia:
- In India, women have gained legal rights to inherit land and property, but gender inequalities still persist as women's shares are often smaller than those of men. In a major breakthrough, the government of Maharashtra state

## Case study 5.16  Urban planning: Britain's New Towns

Britain has a long history of urban planning. Case study 5.10 pointed out that Salisbury was a thirteenth-century planned city. By the late nineteenth century the idea was firmly embedded that planned settlements could influence social behaviour for the better and create economic wealth. This type of social engineering project reached its zenith with the New Town programme, which was launched in 1946 to create a series of new communities across the country. These were a planned reaction to concerns about urban sprawl (and the consequent risks to the countryside around the major conurbations) and to post-war housing demand.

The New Towns were also part of a wider plan to relocate enterprise and industry away from inner-city areas. Many of the New Towns were designed to encircle London's green belt. Others were later built round other major conurbations, such as Liverpool and Manchester.

Each of these New Towns was to work both independently and in cooperation with other towns. The project was seen as an opportunity to put all of Britain's architectural and environmental knowledge into a few 'perfect' settings, creating an urban utopia.

It is now more than 50 years since the New Town Act paved the way for what Lord Reith (the founder of the BBC) described at the time as 'an essay in civilisation'. People who moved to the New Towns in the early stages of their development have seen the towns grow.

Any assessment or judgment about the success of the New Towns of the 1940s and 1950s is highly subjective, although even their most loyal defenders wince a little at what has become of towns such as Cumbernauld, which exhibit very high levels of social deprivation and a decaying urban environment.

However, not all of the New Towns were built solely for the purpose of re-housing inner-city citizens. Although the majority were designed to cope with the overcrowding of the big cities, some had other purposes:
- Aycliffe was intended to act as an industrial magnet to attract business and commerce to County Durham.
- Corby in Northamptonshire was meant to house the influx of workers to the expanding steelworks nearby — the steelworks has now closed.

The last spate of New Towns, begun in the early 1960s, included Skelmersdale (near Liverpool), Telford (near Shrewsbury) and Milton Keynes (50 km northwest of greater London).

By 1985, the majority of the work of the Development Corporations in the New Towns was complete. The town centres of all the towns had long been established, as had the older neighbourhoods. The creation of the New Towns had led to many innovative ideas and unique architectural designs:
- pedestrianised town centres, such as those in Crawley and Stevenage
- development of neighbourhoods that were single function areas
- networks of cycle tracks in some of the towns, and specific public transport routes

---

recognised the right of slum and pavement dwellers to shelter. Through collective ownership, communities will be able to obtain security of tenure. Women's participation is an integral element of the programme.
- In Namibia, the laws that require a married woman to get her husband's signature for any official document are due to be repealed. However, the practice in human settlement policies has changed already. The government is active in designing and implementing housing programmes that are gender-sensitive. It encourages women to form saving and credit groups, and gives priority to households headed by women (41% of the total number of households) in credit schemes.

# Chapter 5

*Settlement patterns*

However, in many parts of Africa, when a man dies, his widow suffers a double tragedy: she not only loses her partner but also all the material wealth they owned. Her in-laws claim all the property and land, leaving her with none of the wealth she helped to create.

If the widow is childless or has no sons, she may be evicted from the land and forced to return to her parents. If she has children, particularly sons, she could continue to stay on the land because her sons will inherit it.

A study on widows of the Zambian national football team members who died in a 1993 air crash found that 24 of the 27 widows lost property to which they were entitled.

In Africa, deeply entrenched patriarchal traditions and values dictate that wealth, property and land belong to men. In such social systems, women do not inherit or own land or property because customarily these belong to husbands or fathers, and it is the sons who inherit. A woman is perceived as a 'temporary resident' in her parents' home until she gets married, and a resident in her husband's home so long as he is alive and satisfied with her.

Ironically, in most countries the law does not discriminate against women in land property ownership. Indeed, national laws in many African countries give women equal rights with men to own land or property. But the reality is very different because of traditions, customs and attitudes that have existed for centuries.

## Managing urban decay in the UK

The declining significance of manufacturing in most MEDCs and the increasing mechanisation of that manufacturing base posed severe problems for many of Britain's inner-city areas that lost employment and population from the 1950s onwards. This created a cycle of multiple deprivation which led to further loss of employment, loss of income and loss of social services.

Policy reaction to this decline evolved as follows.

- Attempts to improve the social well-being of the population. The Plowden Report, published in 1967, identified uneven educational achievement as a major factor in deprivation. This was followed by Community Development projects and the establishment of Development Agencies. This marked a change in policy; until then New Towns had been the focus of attention. Economic decline in the inner-city areas was recognised 'officially' by the government in 1977.
- Riots in Brixton, Toxteth and other inner-city boroughs stimulated the establishment of Urban Development Corporations, initially for London and Merseyside, but followed in the late 1980s by eight other UDCs.
- In the 1990s, a new set of policies emerged that stressed public–private partnerships and moved away from large-scale regional grants towards the improvement of the skill level of the population. A more flexible planning system gave priority to business over other competing land uses such as residence. The Social Enterprise Unit was set up in 2001 and tax credits for disadvantaged neighbourhoods began in 2003.

However, after almost 50 years of attention by government at all levels, the inner-city boroughs remain some of the poorest in Britain and the new initiatives have not been a total success. Nor has the new policy of community-centred action been wholly successful. For example, in Shoreditch, in northeast London, the local community wanted to repair council houses but keep them in council ownership. However, the government rejected the plan as 'unsustainable' and the community was asked to consider transferring the homes to a housing association. Once again local community wishes seemed to be contradicted by central government.

There are many issues here, not the least of which is the ethnic divisions in British society. Many of the most deprived inner-city areas are dominated by either Asian or Afro-Caribbean communities which have frequently expressed, through their community leaders, a sense of neglect and distance from decision-making processes.

In the view of almost all observers, Britain needs more houses. The current battle is, once again, between the regeneration of brownfield sites, many of which are in inner cities, and greenfield sites, often preferred by developers. The private sector is only producing a small fraction of the houses the nation needs as social change (e.g. more people living alone) rather than population increase fuels the demand. Fiscal measures and the planning system should be used to exploit sites that already have planning permission but which, for various reasons, are undeveloped.

The private sector has never been very good at providing cheap housing given that profit margins are larger if four 'executive' homes of £400 000 each are built instead of ten 'starter' homes of £160 000 each on the same area of land. The simple way of putting it is that you need fewer bricks! Thus the state has always had to devise systems to meet housing demand. If it does not, trailer parks will grow to serve the need, as in the USA.

The Thames Gateway has 12 000 ha of undeveloped land. Government has to be involved in projects of this magnitude, so that large-scale issues of decontamination, drainage and public transport can be dealt with efficiently, and whole communities can be created. There has to be a proper clean-up of contaminated land, either funded directly by government or subsidised through tax breaks. The brownfield sites that have been developed so far were the easy ones. Those that are left are the dirtiest sites, and therefore the most expensive to develop.

# Chapter 6

# Population movements

## Migration

Migration, especially international migration, is one the most controversial issues in the modern world. Despite this, only a handful of questions can be posed regarding migration.
- Who moved? How many moved?
- Why did they move?
- What were the patterns of origins, destinations and flows?
- What were the effects on the societies the migrants left and the societies they moved to?

### Who moved? How many moved?

It is a key characteristic of migrations that they tend to be selective. For example, it is only occasionally that everyone flees an erupting volcano. In almost all other cases it is a particular group that moves. Globally, migrants tend to be young and more often male than female, but there are many variations from place to place and from time to time. Selection might also take place according to religious belief or ethnic origin.

### Why did they move?

The motivation or **causes** of migration can be analysed at two levels, but in both cases they can be broken down into economic, social, political and physical factors.

Figure 6.1 gives some examples for migrants arriving in Italy and Spain. Note the gender differences.

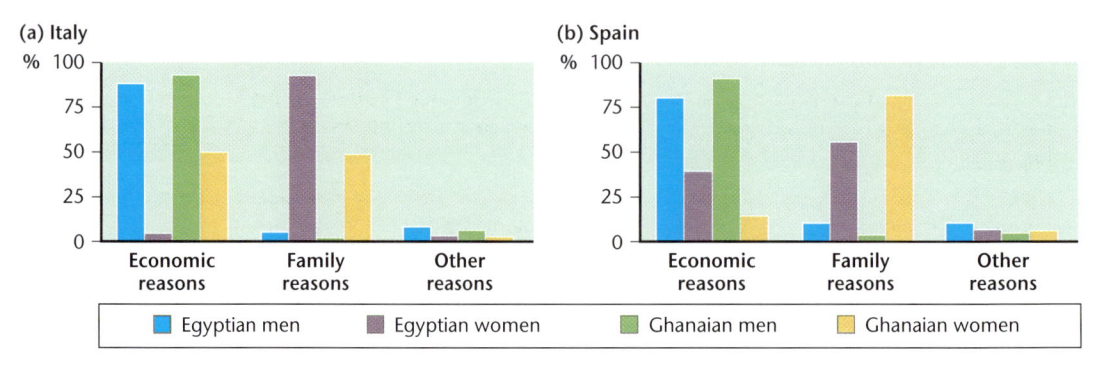

*Figure 6.1 Motives for migration to (a) Italy and (b) Spain*

The two levels at which reasons for migration occur are:
- The **individual decision**, which involves weighing a complex set of factors. Some factors make a new destination attractive (**pull factors**); some factors are negative about the current domicile (**push factors**).
- The greater forces that affect whole populations: most conspicuous of these is the growth in demand for labour in country or region A and the overabundance of labour in country or region B.

## What were the patterns of origins, destinations and flows?

This question concerns both the **source area** and the **receiving area** for migrants. It is important to recognise that for every flow there is a counter flow and the concept of *net* migration is critical. Even while large numbers are migrating into a particular region there are always some people migrating the other way. The rates of flow vary greatly over time.

## What were the effects on the societies the migrants left and the societies they moved to?

The impact or **consequences** of migration operate at both ends of the process. They can be negative or positive and their impact can be economic, social, political, demographic or physical.

## Definitions and categories

There is a distinction to be made between **population movement** and **migration**.
- Migration has been defined as a permanent or semi-permanent change of residence. This leaves ample scope for different interpretations. Some authors have restricted the term to changes of residence involving complete readjustment by the individual with a change of job, friends and neighbours. Such authors prefer to omit residential re-location within a town or city.
- **Population movement** covers all human movement which, of course, might be very short term and sometimes unimportant in terms of the economy. The term for short-term movements is **circulation**. This includes movement on a daily, weekly or even seasonal basis, in which the permanent place of residence does not change. Examples are shopping trips, holidays or commuting.

### Key terms

**Circulation** The movement of people involving no change in residential location.

**Internal migration** Movement of people within a country.

**International migration** Movement of people from one nation state to another nation state.

**Migration** The movement of people involving a change of residential location.

**Permanent migration** When there is no intention to return to the place of origin.

**Temporary migration** Migration for a pre-arranged fixed period.

# Chapter 6

## Population movements

## Internal and international migration

### Internal migration

**Internal migration** refers to movements within a nation state. Such movements can be classified as follows.

- Migration from a rural area to an urban area (rural–urban), which has been the dominant movement for many centuries in most countries. This is still the most significant global movement.
- Migration from a city to another city (urban–urban). In the highly urbanised societies of many MEDCs this is obviously an important flow.
- Migration from a city to a smaller urban area (urban–urban). This involves substantial numbers of people, especially in MEDCs. Large cities in some countries have lost population, but medium-sized cities and smaller towns have increased in size rapidly.
- Migration from a city to a rural area (urban–rural). This can happen when new industries create new jobs in areas outside the traditional city areas. Often known as counterurbanisation, this is a well-known social trend, although it is generally a movement from a large city to a smaller urban centre rather than to a village.

Most internal migration is **permanent migration**, as families move and never return to the place they left. However, it is hard to be precise about this definition for two reasons.

- People may have the intention of staying in the new area for the rest of their lives, but they cannot know for sure that they will. Thus **permanent migration** and **temporary migration** are quite complex and there is considerable overlap between them. For example, a British couple that retires to the Costa del Sol but keeps a property in the UK may very well return to the UK after a few years.
- The migration might only be to the next street or a nearby town. The vast majority of migrations in MEDCs are moves within a local area and usually involve upgrading property while leaving almost all other aspects of life, including schooling, employment and retail behaviour, untouched. In the USA and the UK, about one in seven of the population moves every year, but well over 50% of these movements are local and have no profound impact on the economy as a whole, except on the housing market.

All this internal migration can place costly demands on services (e.g. schools in one area might need to be expanded) and infrastructure (e.g. roads in one area becoming busier).

In the UK there is a clear mismatch between housing supply and housing demand, with the north–south shift (Figure 6.2) precipitating a serious housing crisis.

The most obvious forces for internal migration are economic, although social factors can play a significant role. For example, the most important reason for Indian women leaving their place of birth is marriage.

*Human environments* **AS Unit 2**

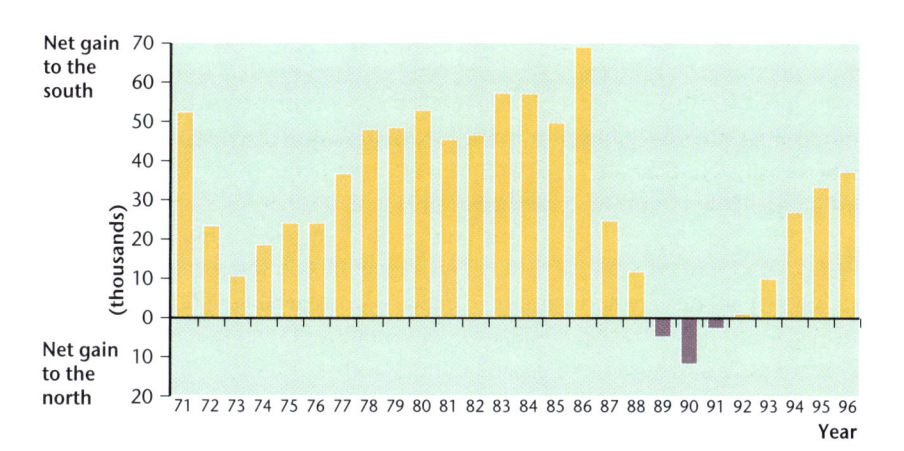

*Figure 6.2 Annual net migration between the north and the south of the UK, 1971–96*

## Case study 6.1 Migration: the drift to the south in the UK

Over the last 70 years there has been a significant shift in the population of the UK towards the south of the country. Despite brief periods of a reverse flow, most recently in the late 1980s, the trend has, if anything, gathered pace in recent years. The populations of southern regions grew at rates of 5.4% between 1991 and 2001 while all other regions, excepting Northern Ireland, grew at only half that rate.

Differences in natural increase, specifically birth rate, explain the anomalous Northern Irish figure, as the Catholic community has a much higher fertility rate than any other group in the country. The main reason for the growth of the population in southeast England and adjacent regions is a sharp rise in both internal and international migration (Table 6.1).

| Region | Population in 1996 (1000s) | Change 1991–96 (1000s) | Annual change rate 1991–96 (per 1000) Overall change | Natural change | Migration | Annual change rate 1981–91 (per 1000) Overall change | Natural change | Migration |
|---|---|---|---|---|---|---|---|---|
| Northern Ireland | 1 663 | 62 | 7.8 | 6.0 | 1.9 | 4.1 | 7.5 | −3.3 |
| Scotland | 5 128 | 21 | 0.1 | 0.0 | 0.1 | −1.4 | 0.5 | −1.9 |
| Wales | 2 921 | 30 | 2.0 | 0.5 | 1.5 | 2.8 | 0.9 | 1.9 |
| North | 3 091 | −1 | 0.0 | 0.3 | −0.4 | −0.8 | 0.6 | −1.4 |
| Northwest | 6 401 | 5 | 0.1 | 1.3 | −1.2 | −1.0 | 1.4 | −2.3 |
| Yorkshire and Humberside | 5 036 | 53 | 2.1 | 1.7 | 0.4 | 1.3 | 1.3 | 0.0 |
| West Midlands | 5 316 | 51 | 1.9 | 2.4 | −0.5 | 1.5 | 2.7 | −1.1 |
| East Midlands | 4 141 | 106 | 5.3 | 1.8 | 3.4 | 4.7 | 2.0 | 2.8 |
| East Anglia | 2 142 | 60 | 5.8 | 1.3 | 4.5 | 9.9 | 1.5 | 8.3 |
| Southeast | 18 120 | 484 | 5.5 | 3.5 | 1.9 | 3.7 | 2.7 | 1.0 |
| Southwest | 4 842 | 124 | 5.2 | −0.1 | 5.3 | 7.7 | −0.1 | 7.8 |
| UK | 58 801 | 993 | 3.4 | 2.0 | 1.4 | 2.6 | 1.8 | 0.8 |
| North | 29 556 | 219 | 1.5 | 1.5 | 0.0 | 0.4 | 1.6 | −1.2 |
| South | 29 245 | 774 | 5.4 | 2.5 | 2.9 | 4.9 | 2.0 | 2.9 |

*Table 6.1 UK population change by region (1981–96)*

# Chapter 6  Population movements

This migration process is driven by the continued decline of traditional employment in the primary and secondary sectors, which has led to a lack of opportunity in the old industrial heartland regions of, for example, south Wales. By contrast, the economy of the southeast has benefited from two trends.

- London's status as a 'global city' has led to the exponential growth in highly paid tertiary and quaternary jobs. These city-forming activities are typically financial services, law, insurance and the many related fields. This has stimulated construction and other city-serving activities, not least in restaurants, hotels and the leisure industry. Thus the labour demand has been at both ends of the labour market, with international immigration providing large numbers of those required in the poorer paid service sector.
- There has been development of new manufacturing industries in the city and the peri-urban fringe. Industries have tended to be hi-tech and low rise, but there have been manufacturing successes as well, for example the Honda plant at Swindon which has over 5000 employees.

The population of London grew by 500 000 between 1992 and 2004 and the growth rate of smaller towns and cities in the London region has been even more marked. Places such as Aylesbury, Newbury and Guildford have been among the fastest-growing settlements in the UK.

Much of London's increase is made up of international migrants fulfilling the less well-paid jobs. There is also a considerable influx of internal migrants, sometimes relocating to suburban or peri-urban counties such as Hampshire, Berkshire and Buckinghamshire. For example, in the 10 years since 1991, the population of Hampshire grew by 63 400 (or 4.0%) and in the last 50 years of the twentieth century Hampshire's population grew at over three times the rate of the UK as a whole, with inward migration contributing a significant proportion of the growth in every decade.

If the basic causes are economic, the consequences are more complex. Housing shortages and, partly as a result, rapidly rising property prices, traffic congestion and loss of habitat as agricultural land is lost to housing are negative consequences in the southeast. However, the migration does provide much-needed skills and services to one of Europe's most prosperous regions.

### Categories of international migrants

**International migration** refers to human movement across national boundaries. Note that not all migration patterns fit easily into this categorisation between internal and international. For example, the Kurdish people were nomadic and their migration paths were formed long before national boundaries existed. To call them 'international migrants' is therefore inaccurate. They are simply following traditional paths, much as the West African Fulani do. In these cases it is the nation state that creates the borders and so makes this distinction, not the migrants themselves.

International migrants can be separated into five main groups.

*Settlers*

**Settlers** are those people that intend to settle permanently in their country of destination. In the modern world there are really only four such receiver countries: the USA, Canada, Australia and New Zealand. These countries, which might be considered underpopulated, are keen to attract certain groups, but they remain selective. There are many criteria by which individuals are accepted as settlers, but the usual hurdles involve either skills or wealth. Thus $250 000 is one route for entry into Canada, and a university degree and a teaching diploma will help in New Zealand. Movement of labour is guaranteed within the European Union, which is increasingly becoming a super-state by removing barriers to entry between members.

### Contract workers

**Contract workers** are **temporary migrants**; they have fixed-period contracts. They are meeting a specific skill shortage and are expected to return to their homes at the end of their contracts.

The global distribution of contract workers tends to reflect the ebb and flow of the global economy; currently the largest numbers are employed in the Gulf countries where, for example, Kuwait has more than 300 000 foreign maids. Another example is young Pakistani men who supply the demand for construction labour in Saudi Arabia. The two main criteria here are poverty in the source country and the people having the same religious beliefs.

### Professionals

The global economy involves the movement of labour from place to place, albeit selectively. Among the most significant migrants are the **professionals** employed by **TNCs** (transnational corporations). These highly paid technicians and executives gain entry to host countries through their companies applying for temporary work permits. In most countries these are allowed if there are 'no suitable' local candidates for the job. In the USA there are about 500 000 such H-1B visas, of which half are for people working in computer-related industries.

### Illegal immigrants

More politely known as 'undocumented workers', the number of **illegal immigrants** is obviously difficult to count. From time to time they are legitimised by amnesty agreements and, on the basis of that sort of evidence, it is calculated that there are about 3 million of them in Europe today. They generally fill very poorly paid niches in the employment ladder. Attitudes towards such migrants are variable, given that they provide very cheap labour, especially in the textiles, catering and tourist sectors.

### Refugees and asylum seekers

The United Nations defines a **refugee** as 'Someone who has a well-founded fear of persecution for reasons of race, religion, nationality, membership of a particular social group, or political opinion'. Much of this is hard to judge and in recent years the term **asylum seeker** has gained currency, particularly with the popular press and some political groups in the receiving countries. Some asylum seekers are genuine refugees in fear of their lives, but others are 'economic migrants', fleeing poverty and seeking a 'better life' in the host country. Thus the term '**genuine refugees**' has become current.

# Chapter 6
## Population movements

*Photograph 6.1 Kurdish refugees*

# Theories of migration

## Ravenstein

E. G. Ravenstein's 'laws' of migration were formulated in the 1880s following his research using the 1871 and 1881 UK census data. They are not really laws at all but observations of consistencies based on the analysis of data, and can be summarised as follows.

- Most migrants move only short distances.
- Migration proceeds step by step.
- Migrants going long distances usually go to large cities.
- Each current of migration produces a compensating counter-current.
- The natives of towns are less likely to migrate than the rural population.
- Females are more likely to migrate within the county of their birth than males, but males are more likely to migrate further.
- Most migrants are adults: families rarely migrate outside their country of birth.
- Large towns grow more by migration than by natural increase.
- Migration increases in volume as industries and commerce develop and transport improves.
- The major direction of migration is from agricultural areas to centres of industry and commerce.
- The major causes of migration are economic.

Ravenstein's observations were a significant first step to understanding migration, but they are specific to that time and the place where they were written.

## Lee

Ravenstein's ideas were later refined by Lee (1969), who produced a series of hypotheses relating to:
- the volume of migration
- migration flows and patterns
- the characteristics of migrants

### Volume of migration
- The volume of migration within a given territory varies with the degree of variety of areas included in that territory. More contrasts in landscape mean more migration.
- The volume of migration varies with the diversity of people. More contrasts of age and ethnicity mean more migration.
- The volume of migration is related to the difficulty of surmounting intervening obstacles.
- Such obstacles might include family pressures, poor information, national policy, travel costs, lack of capital, illiteracy, military service and language.
- These obstacles might be perceived differently by different individuals, so some might welcome the opportunity to live in a large city, with all the facilities it might offer, whereas others might find it cramped, frightening and depressing.
- The volume of migration varies with fluctuations in the economy.
- Unless checks are imposed, both the volume and the rate of migration tend to increase with time.
- The volume and rate of migration vary with the state of progress in a country or area.

### Flows and patterns
- Migration tends to take place largely within well-defined streams.
- For every major migration stream, a counter-stream develops.
- The efficiency of stream and counter-stream tends to be low if areas of origin and destinations are similar and intervening obstacles help the efficiency of the system.

### Characteristics of migrants
- Migration is selective by age, sex and ethnicity.
- Migrants responding primarily to 'pull' factors from their destination tend to be positively selected in that they arrive in a place where they want to be.
- Migrants responding primarily to 'push' factors from their place of origin tend to be negatively selected in that they arrive in a place to which they are indifferent.
- The degree of positive selection increases with the difficulty of the intervening obstacles.
- The heightened propensity to migrate at certain stages in life is important in the selection of migrants.
- The characteristics of migrants tend to be intermediate between the characteristics of the population at their place origin and the population at their destination.

# Chapter 6

## Population movements

### Zelinsky

Zelinsky suggested that patterns of migration would change over time. He developed a model of 'migration transition' which can be run alongside the demographic transition model (DT). There are five stages in his model.

1. In a pre-industrial society in Stage 1 of the DT there is little residential migration and limited movement between areas.
2. As industrialisation begins in Stage 2 of the DT, then considerable rural–urban migration takes place. At the same time there is the colonisation of new lands, with the associated growth of longer distance migration (international migration).
3. In the third stage, rural–urban migration continues and there is a rapid rise in migration between cities. This corresponds with the later stages of industrialisation — Stage 3 of the DT.
4. Rural–urban migration continues, but at a markedly reduced rate. Residential migration remains high, but in the form of migration within and between cities — suburbanisation rather than emigration. There is some in-migration of unskilled workers from overseas, and highly trained professional workers may be exchanged between countries as a result of the operations of multinational companies. This is Stage 4 of the DT.
5. Advanced societies have almost exclusively inter- or intra-urban migration. New technology reduces the need for migration and there is less need for some types of circulation, such as long-distance journeys to work. Mobility between and within countries may be affected by state legislation. This too is Stage 4 of the DT, but at a post-industrial stage.

There is no doubt that Zelinsky's model is helpful in outlining something of the major flows, but it doesn't stand up to detailed examination. For example, we seem to have massively underestimated the mobility of pre-industrial societies.

### The application and limitations of gravity models in predicting migration flows

A gravity model (Figure 6.3) is based on two simple assumptions:
- the number of migrants is determined by the sizes of the places of origin and destination
- the number of migrants is determined by the distance between the places of origin and destination

Based on Newton's second law of gravitation, the gravity model creates a parallel between human behaviour and the behaviour of atomic particles. It should come as no surprise that the gravity model is rarely a good predictor of flows of migrants. At a basic level, the observation that two large cities which are close together will experience more migration between them than two distant smaller cities is hardly startling and doesn't do much to explain the motives for migration.

*Figure 6.3 The gravity model*

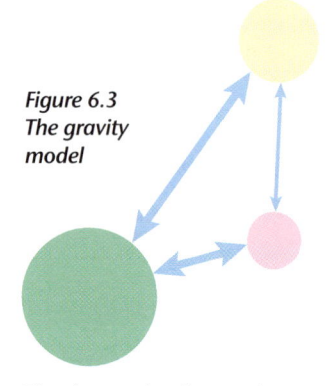

The shorter the distance between two objects, and the greater the mass of either (or both) objects, the greater is the gravitational pull between the objects

It is where there are divergences from model behaviour that motives for migration may be discovered. For example, two cities may have little flow between them because to get from one to the other involves crossing a national frontier, as with Tijuana (in Mexico) and San Diego (in California, USA). The reality is that there are too many variables for the gravity model to be useful. Thus the greatest value of the gravity model is to set up a theoretical flow, which may stimulate discussion as to why the real figure is different.

Much the same criticism can be levelled at Reilly's adaptation of the model for retail market areas. Distance and population size are not unimportant, but they do not allow us to make sensible comments about three of our opening questions, specifically:
- who moved?
- why did they move?
- what were the effects on the societies migrants left and moved to?

## Stouffer

Stouffer suggested that the rate of movement between two places is dependent on the number of **intervening opportunities** between the two places. Intervening opportunities are the nature and number of possible alternative migration destinations that may exist between place A (migration origin) and place B (migration destination). This radically affects the flow predicted by the gravity model. An essential feature of this model is that it is the nature of places, rather than the distance, that is more important in determining where migrants go. People will move from place A to place B based on the real, or perceived, opportunity at place B (e.g. work). According to Stouffer, therefore, the number of people moving over a given distance is directly proportional to the number of opportunities at that distance, and inversely proportional to the number of intervening opportunities.

## Recent developments

Recent developments in migration theory have concentrated on distinguishing between personal motives for migration and global economic changes that produce forces for large-scale migration, for example the movement of people from Mexico into the 'sunbelt' states of the USA.

It is also useful to identify a range of motives, or causes, for migration. These can be classified as:
- economic
- social
- political
- physical

These motives can operate as either **pull factors** or **push factors** — in other words, as negative aspects related to the origin of migrants (push factors) or as attractive elements in the planned destination. Needless to say, this is only relevant where people have a choice about migration — so-called voluntary migrants.

# Chapter 6

## Population movements

### Economic

Economic factors are mainly related to the **availability of work** and the **wage rates** involved, at least in the modern capitalist world. People move to improve their financial situations. Sometimes the migration is an inevitable part of an economic force that lies beyond the control of the individual, as with the relocation of a job. In our globalised economy such relocations may well be international as well as internal. In other cases migrants are pushed out by lack of employment and low wages. The greater the differential between the economic factors in the origin and the destination and the more open the borders, the greater will be the flow of migrants.

### Social

The dominant social motive for migation today, as when Ravenstein wrote, is **marriage**. This largely involves the movement of women when they marry. It is highly significant in rural regions of countries such as India where arranged marriages are still important. Women move relatively short distances to marry and this accounts for over 70% of female migration in states such as Bihar and Uttar Pradesh. Meanwhile the men the women marry may very well migrate either seasonally or permanently in the search for work.

Other significant social motives include **education** and **rejoining the family**. These factors are involved when one partner, usually the male, departs to look for work elsewhere. Having settled, he encourages his wife and children to follow him. **Racism**, **cultural isolation** or **religious beliefs** might also be powerful motives for leaving a country or region where individuals perceive themselves to be in a minority or unwelcome. When these social attitudes are formulated into laws, they become powerful political motives for migration.

### Political

The most obvious political motivation for migration occurs when the state takes a direct hand in legislating against particular political beliefs, ethnic groups or religious beliefs. The flight of the Huguenots from France through to the appalling ethnic cleansing in the Balkans in the 1990s suggest that we have made little progress in these matters. In fact, the twentieth century experienced some of the greatest political displacements, many of them forced. Examples include the Stalinist push for industrialisation, which forced millions of Russians from rural to urban areas, and the Cultural Revolution in China in the late 1960s, when Mao Tse-tung forced millions of Chinese back in the other direction.

Politics can also provide a whole series of intervening obstacles. Examples range from world wars cutting the flow of European migrants to the Americas, to the complex requirements for obtaining visas to enter certain countries. Immigration policy sets the tone and its changes over time are significant. Thus the fall of the Soviet empire in the late 1980s and the removal of restrictions on the flow of migrants changed the numbers of migrants in eastern Europe and the direction of their flow. What had been largely an internal movement, especially to the east, has become a much larger international movement to the west.

> **Key terms**
>
> **Economic causes and consequences**
> The reasons for and impacts of migration relating to employment and income.
>
> **Physical causes and consequences**
> The reasons for and impacts of migration relating to changes in the physical environment (lithosphere, atmosphere, biosphere and hydrosphere).
>
> **Political causes and consequences**
> The reasons for and impacts of migration relating to government policy and law making.
>
> **Social causes and consequences**
> The reasons for and impacts of migration relating to family, relationships and non-economic self-improvement.

Governments are major players in modern economies. For example, they are important in deciding the location of economic activity which then, indirectly, decides the location of the population. The building of a bridge, a dam or an airport are all obvious examples.

## Physical factors causing migration

The most obvious example of a physical factor causing migration is a sudden natural disaster that drives people away from their homes. These types of movements are unselective in terms of age, gender, income or ethnic background and are obviously forced migrations.

Physical factors can stimulate or at least be influential in other types of migration. In some cases the physical factors are direct causes of migration. Examples are:

- The migration southwards at retirement in the USA (see Case study 6.7 about the 'sunbelt').
- The similar flow from north to south in western Europe, with retirement to southern Spain being an obvious example. More than 50 000 British pensioners have made this move.

In both these examples there are factors other than climate that help explain the movement. Of course not all areas with the same climate attract in-migrants in equal numbers. The rate of these migrations is determined by the economic situation in both the country and region of origin and the country and region of destination, as well as being dependent on housing availability, the frequency and cost of air-flights and the nature of the health services in the receiving region. Political factors, such as terrorism in southern Spain, will also have a significant negative effect on the numbers choosing to settle in a region.

Physical factors can operate in a more complex way too. The choice of location of many quaternary and high-technology industries involves a process that often looks almost as closely at the attractions of the physical environment as it does at the economic environment and social and cultural facilities available. The growth of hi-tech clusters in Palo Alto, California (Silicon Valley) or Orange County, also in California, illustrates this. Case study 13.3 about Denver (Chapter 13, page 405) reinforces this idea.

# Chapter 6  Population movements

## Case study 6.2  Migration caused by physical factors: Montserrat

*Photograph 6.2 Montserrat, after the volcanic eruption in 1997*

Montserrat is a tiny Caribbean island of about 100 km². Before the devastating volcanic eruption in 1997 it had a population of about 9000, which has now reduced to under 1000. Throughout the twentieth century, Montserratians migrated from their island because of a lack of employment opportunities. Having found employment elsewhere, they often sent back remittance payments to relatives at home. Since the first eruption in 1995, approximately two-thirds of the island's population have migrated from the island for very different reasons, while 74% of the remaining population have had to relocate to the north of the island. Such drastic mass movements have had, and continue to have, serious social, economic, political and cultural effects upon all those involved. Montserrat is a particularly good place to study migration because it is also a good 'laboratory' of earlier migration patterns, including slavery and trans-Atlantic migration from Europe.

Montserrat was an Irish domain within the English empire, a place where Irish settlers were the dominant incoming group. This tiny island was one of only two regions where, before the creation of the Irish Republic, the Irish dominated.

The recent history of Montserrat is an example of a **forced migration**. The migration is not selective, in that almost everyone departs, and **pull factors** are not significant.

The existence of a resource is an obvious physical factor, as is the gradual reduction in the importance of that resource. The rapid rise in the significance of the south Wales coalfield attracted in-migrants to counties such as Mid-Glamorgan in large numbers, peaking in 1921. Since that time the fall in the demand for coal and the relative expense of extraction in south Wales have led to economic decline and population loss due to out-migration (Table 6.2). The same story can be picked up in Case study 14.1 about Potosi (Chapter 14, page 464).

Table 6.2 Population (thousands) of two contrasting counties in Wales

| Year | Powys (rural) | Mid-Glamorgan (urban industrial) |
| --- | --- | --- |
| 1891 | 117.2 | 330.3 |
| 1901 | 117.7 | 426.5 |
| 1911 | 118.5 | 585.0 |
| 1921 | 118.4 | 645.3 |
| 1931 | 111.9 | 590.7 |
| 1951 | 107.6 | 530.1 |
| 1961 | 102.3 | 519.6 |
| 1971 | 99.2 | 531.8 |
| 1981 | 110.6 | 538.5 |
| 1991 | 116.5 | 526.5 |

# The impact of migration on the physical environment

We have seen how the physical environment stimulates migration. It is equally obvious that the physical environment is affected by migration.

This can be seen at a local scale, for example in the debate about greenfield house-building in southeast England or in the desertion of central urban areas by the more mobile middle classes during the 1980s. The built environment has certainly been modified greatly by such processes.

On a larger scale, there are mass movements, such as migration to Amazonian territories from southern Brazil. This has been instrumental in the destruction of the rainforest through burning to make way for agriculture. However, the global forces involved in shifts like this go well beyond the power of individual peasant farmers who have been evicted to make way for agribusiness in southern Brazil.

Abandoned areas also change. The natural vegetation develops and climax communities can re-establish, as they have in states such as Georgia in the American south. Population decline through out-migration has led to the restoration of scrub oak woodland in areas previously used for wine-growing in many parts of southern France. Whether or not these changes are a 'good' thing is not possible to say in any objective sense. Most of the landscapes admired as 'natural' are anything but natural, so whether a landscape of scrub oak in the Tarn Valley is intrinsically more pleasing than the terraced vineyards that were common there in the 1870s is open to debate. Few would argue that the decline in mining has not improved the physical environment of the south Wales coalfield, even if the economic wasteland is rather less attractive.

## Case study 6.3 Physical impacts of migration: Indonesian transmigration

Indonesian transmigration was a government-funded programme, aided and supported by the World Bank, and was one of the most controversial of all migrations in the modern world. From the late 1970s, transmigration was the central policy of President Suharto's Indonesia. Despite widespread domestic and international criticism, the programme continued until the end of Suharto's reign in mid-1998. Transmigration involved the mass movement of millions of landless poor from the central Indonesian islands of Java, Madura, Bali and Lombok to the less densely populated outer islands. It was promoted as a humanitarian exercise with the primary goal of improving living standards. It was a voluntary, rural–rural migration programme.

Indonesia is an archipelago of islands and population density varies greatly from island to island. Out of a total population of about 200 million people, more than 110 million live on Java; this means that 55% of the population live on only 7% of the total land area. Rural population densities frequently exceed 1000 km$^{-2}$. As a consequence Java has small agricultural holdings. Despite its highly fertile volcanic soils, there are growing numbers of landless people that are swelling its towns and cities, most obviously the capital Jakarta. The outer islands have a large proportion of Indonesia's natural resources, a lower population density, and higher rural incomes, on average, than those in Java. Since the project began, more than 3.5 million people have been

# Chapter 6 Population movements

*Figure 6.4 Indonesia*

translocated to regions that were regarded by the government as being underpopulated and/or underdeveloped. These areas are frequently dominated by shifting cultivators whose rights are officially limited by central government:

> The rights of traditional communities may not be allowed to stand in the way of the establishment of transmigration sites.
>
> (Clause 17 of the Basic Forestry Act No 2823, 1967)

Ironically, one of the greatest catastrophes of transmigration occurred just before Suharto's fall from power, when the plan to turn 1 million ha of peat wetland in central Kalimantan into a major rice-growing area collapsed, leaving a legacy of expensive and pointless environmental destruction. The scheme was launched in 1995 as a means to guarantee rice self-sufficiency for the country, but quickly turned into an environmental catastrophe as peat forests were stripped of their timber, drained and rendered unusable. The native Dayak people were pushed aside and deprived of their livelihoods, while transmigrants brought in to grow rice soon found the land impossible to work. The dried out peat and debris left from the bulldozers also created ideal conditions for large fires, which took hold in 1997.

Placing transmigrants in areas where independence movements were strong, such as West Papua, Aceh and East Timor, was a key tool of Jakarta's internal colonialism — a process by which Suharto hoped to unify the country.

Suharto's 33-year dictatorship stored up huge problems for Indonesia's future. These problems included the annexation and brutal occupation of West Papua, the invasion and reign of terror in East Timor, and long-term military repression in Aceh. The Suharto regime made no attempt to create a government that reflected the ethnic diversity of the country. On the contrary, its highly centralised economic system led to growing disparities in the distribution of land and wealth. By resisting land reform, the government made transmigration the cornerstone policy to relieve tensions and to water down ethnic differences.

---

Desertification is often blamed on the in-migration of farmers, who exploit soils to such an extent that the nutrient levels decline, plants cannot become established, and surface runoff increases, causing rill and gully erosion and major landscape changes. The accusations are frequently ill-founded but clearly farming has an impact, whatever the primary causes for the movement of people. Case study 6.3, about Indonesian transmigration, illustrates that the impact on the environment, supposedly 'good' in both the areas from which people were moved and the areas in which they were placed, often failed to achieve what was planned.

Human environments    AS Unit 2

# Different types of migration: American case studies

Not all forms of migration involve personal choice. There is a fundamental distinction to be made between **forced migration** and **voluntary migration**. Even this apparently simple distinction is not as straightforward as it seems. The Atlantic slave trade is an obvious example of forced migration, while retirement to the Costa del Sol is obviously voluntary, but the range between these is considerable.

Today, 70% of the population of Mauritius are descendants of immigrants who left their country, India, as 'indentured labour'. This system, which replaced slavery, was almost as repressive. The method of recruiting Indian labour in the early days of the British Raj, to fill a gap created by the abolition of slavery, is not a happy chapter in Anglo-Indian relations. The 'coolies', as they were known (almost exclusively young men), were frequently tricked into signing agreements to work for 5 years without knowing where they were going. Once they realised that they were to be shipped overseas, they frequently resisted. This was forced migration in all but the fine detail. However, it continued to fill the gap left by the abolition of slavery for many years, until the early twentieth century. It has, like slavery, made a reappearance in the modern world with the growth of sex trafficking.

### Key terms

**Characteristics of migrants** How they can be classified according to age, gender, income, race, etc.

**Pattern of migration** Where migrants come from and go to.

**Volume of migration** The number of people involved.

### Examiner's tip

- There is a fine line drawn between some types of voluntary migration, such as escaping famine, and forced migration.

## Case study 6.4   Forced migration: the Atlantic slave trade

The forced movement of black Africans to the Americas was one of the most brutal episodes in world history. It was, of course, international (although nation states were of limited significance in the early period), rural–rural and mainly a movement of young men.

The African continent had been bled of its human resources for at least 600 years before the last and most extreme episode of forced migration began. The figures are disputed, but few deny the significance of the movement:

- 4 million slaves were exported via the Red Sea
- 4 million slaves were exported through the east coast Swahili ports on the Indian Ocean
- 9 million slaves were exported across the Sahara
- a staggering 11–20 million slaves were involved in the Atlantic slave trade from the sixteenth to the nineteenth centuries

This haemorrhaging of population, mostly young men, permanently weakened the continent, led to its colonisation by the Europeans in the nineteenth century, and gave birth to the racism and contempt from which Africans still suffer.

The causes for the slave trade are not hard to find. The 'newly discovered' American landmasses offered huge opportunities for the European powers that carved up the territory after the explorations by Columbus and others. The native population, numbering as many as 20–30 million, was not readily shackled to a system that was based upon pumping out as much wealth as possible from the land. South America offered mineral wealth, which was plundered by both the Portuguese and the Spanish, but North America did not seem as rewarding until the development of large plantations to grow crops which were not readily available in Europe, such as cotton and sugar.

With large quantities of land available to farm, voluntary European migrants to North and South America were not prepared to work for low wages in either the mines or the plantations. Labour was

# Chapter 6  Population movements

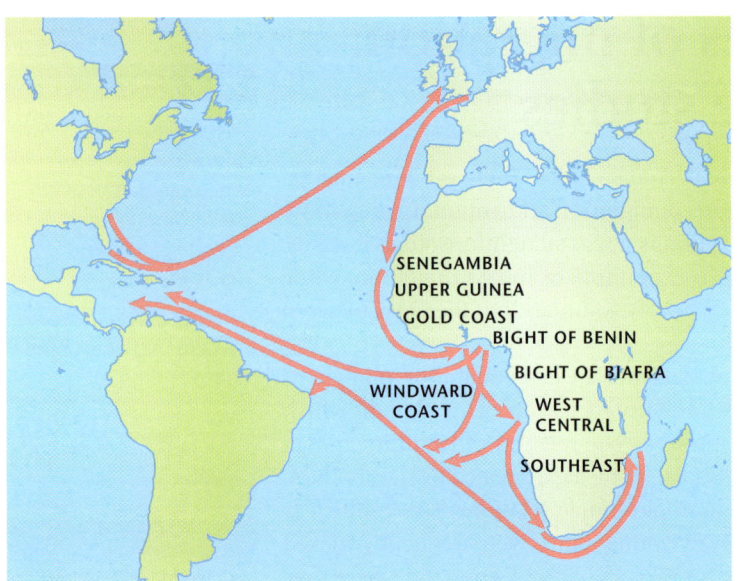

*Figure 6.5 The routes involved in the Atlantic slave trade*

The impact of the slave trade was immense. Many argue that the industrialisation of Europe was triggered by and, more importantly, funded by slavery. Profits from the plantations were recycled in the building of the railways and other infrastructure. The cotton that came to England was manufactured in Lancashire into cloth and clothing that was then re-exported.

In the early years of the USA the free labour provided by slaves allowed a rapid expansion of wealth and, of course, injected another ethnic group into the melting pot. It might be argued that the cultural footprint of the USA would not be the same without the input of the black population.

needed and, given that the native population simply ran away whenever the chance arose, the African slave trade became an 'obvious' response to this shortage. The African states fell into the trap set by the European slavers: trade or go under. All the states along the African coast were divided by the conflict between national interest, which demands that no resource necessary to security and prosperity be neglected (in this case the resource was people), and the basic function of a state, which imposes on rulers the obligation to defend the lives, property and rights of their subjects.

In Africa it is widely agreed that development was set back by many centuries owing to slavery. Another impact was the attitudes that were developed to 'legitimise' slavery. Enslaving another human being is hard to justify, especially perhaps for Christians, but the problem was side-stepped by labelling Africans as 'sub-human savages'. This allowed Europeans to treat them much as they would treat animals, or worse. From this history, it isn't hard to understand why 'racism' is so endemic in Western society.

| Region | Number of slaves accounted for | % of total |
| --- | --- | --- |
| Senegambia | 479 900 | 4.7 |
| Upper Gambia | 411 200 | 4.0 |
| Windward Coast | 183 200 | 1.8 |
| Gold Coast | 1 035 600 | 10.1 |
| Bight of Benin | 2 016 200 | 19.7 |
| Bight of Biafra | 1 463 700 | 14.3 |
| West Central | 4 179 500 | 40.8 |
| Southeast | 470 900 | 4.6 |
| Total | 10 240 200 | 100.0 |

*Table 6.3 Trans-Atlantic exports, by region 1650–1900*

| Region | Number of slaves accounted for | % of total |
| --- | --- | --- |
| Brazil | 4 000 000 | 35.4 |
| Spanish empire | 2 500 000 | 22.1 |
| British West Indies | 2 000 000 | 17.7 |
| French West Indies | 1 600 000 | 14.1 |
| British North America and USA | 500 000 | 4.4 |
| Dutch West Indies | 500 000 | 4.4 |
| Danish West Indies | 28 000 | 0.2 |
| Europe (and islands) | 200 000 | 1.8 |
| Total | 11 328 000 | 100.0 |

*Table 6.4 Trans-Atlantic imports, by region 1450–1900*

Human environments   AS Unit 2

The reasons why individuals move are often complex. A voluntary migrant may move for several reasons. For instance, individuals may move to support their families and to follow lifestyles that they cannot follow in their original location. It is often difficult to establish why they really want to move. A person may want to leave a place to escape family pressure to marry, or to follow a certain career, but will find it difficult to admit this.

Voluntary migration can also be analysed on a more general level, one that doesn't involve individual motives so much as basic causes.

Among the many millions who migrated across the Atlantic in the nineteenth and twentieth centuries, there would have been a multitude of individual reasons to leave Europe. Overall, the great driving force was the surplus labour in Europe. This was created by what some have called the 'great transformation'; agriculture became commercialised, releasing tens of millions of people for factory work. Some of these people were surplus to immediate requirements and so could be exported to the growing colonial economies overseas, which were later to become great economic powers in their own right, especially the USA.

### Case study 6.5  Voluntary migration: European migration to the USA

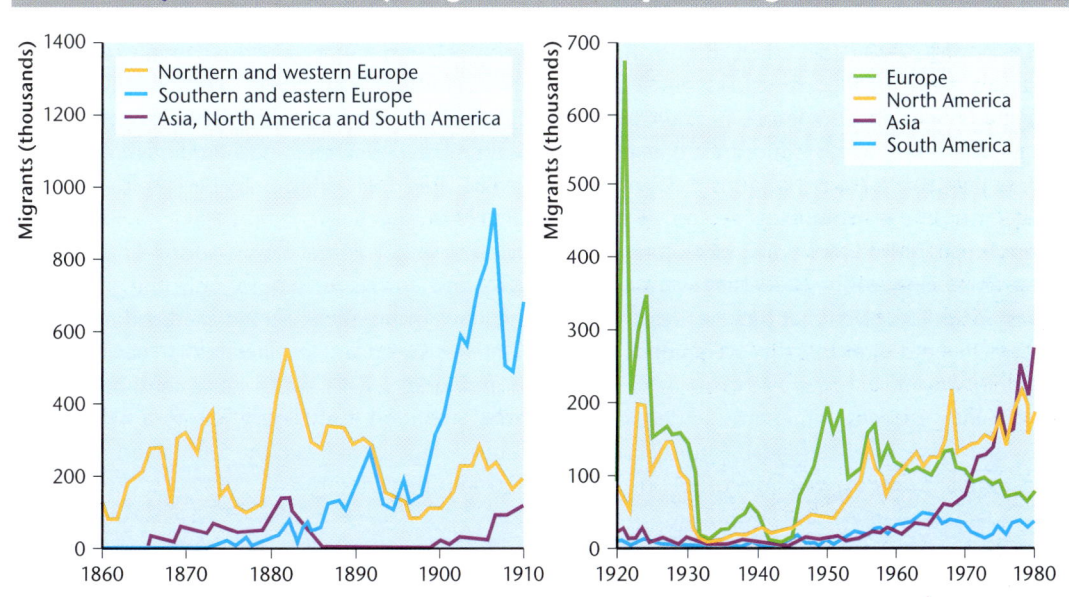

*Figure 6.6 The origin of migrants to the USA, 1860–1910*

*Figure 6.7 The origin of migrants to the USA, 1920–80*

The USA was made by migration. Strictly speaking, every part of the world was populated by migration, given that *Homo sapiens* spread out from north Africa to occupy the whole planet. However, the USA is a special case of very rapid migration in the recent past.

- The original inhabitants arrived some 10 000 years ago but are frequently overlooked in historical accounts.

- These accounts tend to begin with the waves of early settlers in the sixteenth and seventeenth centuries, who were supplemented by millions of enslaved Africans.
- The industrial workers who began to arrive in the late nineteenth century were also foreign-born.
- A total of 45 million migrants arrived between 1810 and 1930.

# Chapter 6: Population movements

- From the mid-1880s until the First World War began in 1914, they came from further south and east in Europe.
- The peak flow was reached when some 9 million migrants arrived during the period 1880–1900, and 13 million during 1900–1914.

Late-nineteenth-century migrants left Europe to escape economic problems, scarce land, poor harvests and the commercialisation of agriculture. They flooded into the rapidly growing cities of Europe where, as a 'reserve army', they posed a threat to social order and political stability much as the slum dwellers of Mumbai or Accra do today. They came to the USA in hope of economic gain or to escape political or religious persecution. Most settled in the USA permanently, but others came only to make their fortune and then return home.

Immigration dropped off during depressions, as in the 1870s and 1890s, and again during the First World War, with smaller downturns in between. Immigration was encouraged by new technology such as steamships, which reduced the time needed to cross the Atlantic from 3 months to 2 weeks or less. Almost all the migrants were rural in origin, although many had spent time in European cities before embarking for the USA. They were not the poorest in society, given that they needed the fare and something to show that they were not destitute when they arrived.

In the early decades immigrants settled in rural areas that were often ethnically quite specific. For example, Scandinavian immigrants used the Homestead Act, which gave land grants to start mid-western farms. Twenty years later, immigrants usually moved to industrial towns and cities. For example, they became unskilled labourers in New York.

Populations often grew very rapidly. For example, in Milwaukee, Wisconsin, the population increased tenfold from 1850 to 1890, with very large numbers of Poles and eastern Europeans finding work in rolling mills and blast furnaces. By 1910 immigrants and their families made up over half the total population of 18 major cities; in Chicago, eight out of ten residents were immigrants or children of immigrants.

The impact was enormous. The US economy is built on migrant labour, although given that it was always short of labour, it was also driven to become technically innovative. Immigrants' lives changed dramatically after they arrived. They usually came from rural areas in Europe, and had to adjust to industrial jobs, unfamiliar languages and urban life. They introduced ethnic foods, read foreign-language newspapers and celebrated ethnic holidays. Men outnumbered women in new immigrant communities because men often preceded their wives and families in migrating.

To native-born Americans, themselves often relatively new arrivals resident for only one generation, the new arrivals often seemed alien and more transient, less skilled and less literate than earlier groups of immigrants. This led to racial tension in many cities, although it also gave rise to a wholly new cultural impetus that laid the foundations for many traditions that endure.

---

A later example of the same sort of transformation came about with migration from the south of the USA, which had been a rather backward plantation economy, to the vigorous and rapidly growing northern 'Yankee' economy. That much more dynamic economy was starved of its labour supply by the First World War, which interrupted the supply of European migrants. There was therefore a huge pull factor, as northern factory owners set out to recruit cheap labour, almost always black, from the south. Many of these were poor, landless and illiterate as well as being subjected to racially motivated laws and discrimination. Thus there were both economic and social push factors.

Human environments  AS Unit 2

## Case study 6.6 Voluntary migration: south–north migration in the USA

This migration was voluntary, internal and rural–urban, and it represents the second stage of the movement of African Americans across the American landscape. The end of slavery brought few immediate benefits to the blacks of the southern states. They were 'free' but still needed to find a means of survival so, in practice, little changed for them and they still worked in the plantation system.

In 1910 80% of black Americans lived in the south, but by 1970 it was only 45%. In that period over 6 million black Americans moved out of the south, mostly from rural poverty to urban poverty, albeit relatively less poor.

In 1910 two-thirds of southern blacks were sharecroppers who worked on land that they did not own. Natural hazards such as the floods of 1915 and the boll weevil attack on the cotton crop in the 1920s made a desperate situation intolerable. They were extremely poor, discriminated against, politically disenfranchised by racist laws and frequently subjected to miscarriages of justice.

The 'Jim Crow' laws established as early as 1896 allowed states to segregate transport, education and housing throughout the south. This was most obvious in education. As late as 1950 the white schools of Clarendon county, South Carolina, received 60% of the education budget, although black children outnumbered whites by 3:1. This provided a powerful social impetus to migration as did the lynch mobs that routinely set upon black boys who cast as much as a glance at white girls.

The attractions of the north were not hard to see. As soon as the two regions connected up through railroads and the job opportunities in the north started to become known due to newspaper adverts and word of mouth, the great exodus began. The northern stockyards and factories were growing at a rate that outstripped the supply of European labour, especially during the periods when the supply line from Europe was cut off by war or depression. Wages in the north were around three times higher and the south, with its limited urban

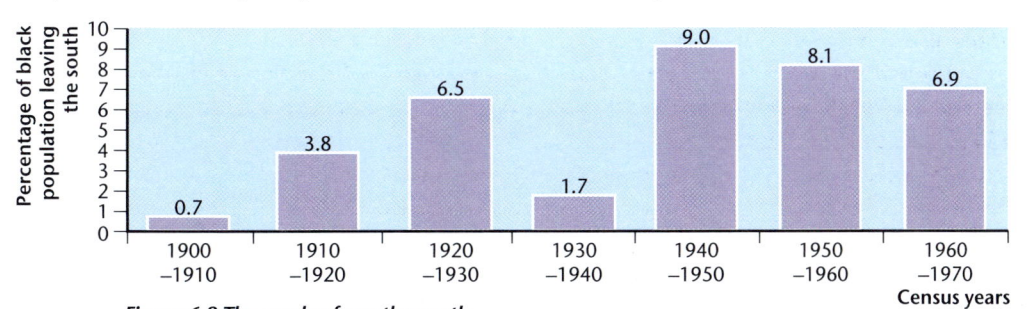

*Figure 6.8 The exodus from the south*

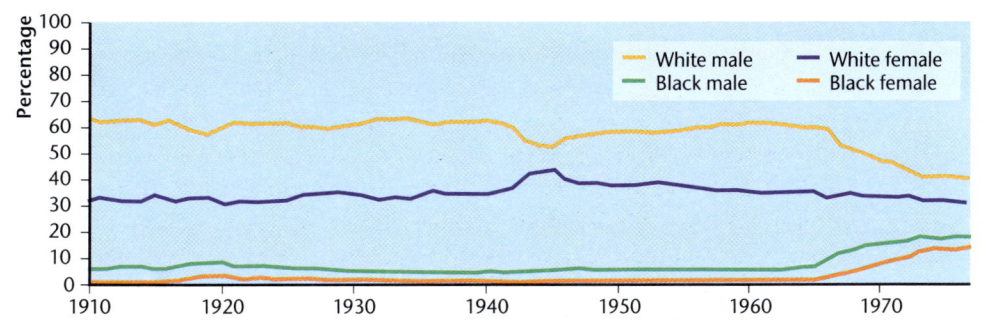

*Figure 6.9 Employment in the South Carolina textile industry, 1910–80*

# Chapter 6 Population movements

development, offered very few opportunities.

The southern economy had always been based on exporting primary products and importing the manufactured goods it needed, so factory jobs were in short supply. The recruiting companies from northern firms took risks because the southern landlords objected to losing their labour force and, potentially, their profit. However, this changed in the 1940s when the invention of the mechanical cotton picking machine meant that labour demand fell.

For the migrants, life in the northern cities was not always what they had expected. Although racist laws were not in place, racism certainly was and poor housing conditions were common. The in-migrants arrived in inner-city areas and occupied the tenement buildings previously occupied by European migrants who, a generation on, had moved out. It was this movement that gave rise to the Burgess model of urban growth (see Chapter 5). The migrants brought their culture with them and it was in these northern cities that the mixture of southern black gospel music and European folk ballads took place that gave rise to almost every brand of modern rock music.

---

It is inevitable that the USA provides some of the best case studies on migration for it is, of course, a country which has grown largely through migration. It is worth remembering, however, that, as in Case Study 6.6, not all of that movement is international; in a country the size of a continent it is unsurprising that there have been frequent shifts due to internal migration.

A more recent example of internal migration is the rise of the 'sunbelt'. The shift of population to the south has been a reversal of the historic flow from south to north in the USA (Figure 6.10). The population trends reflect this movement,

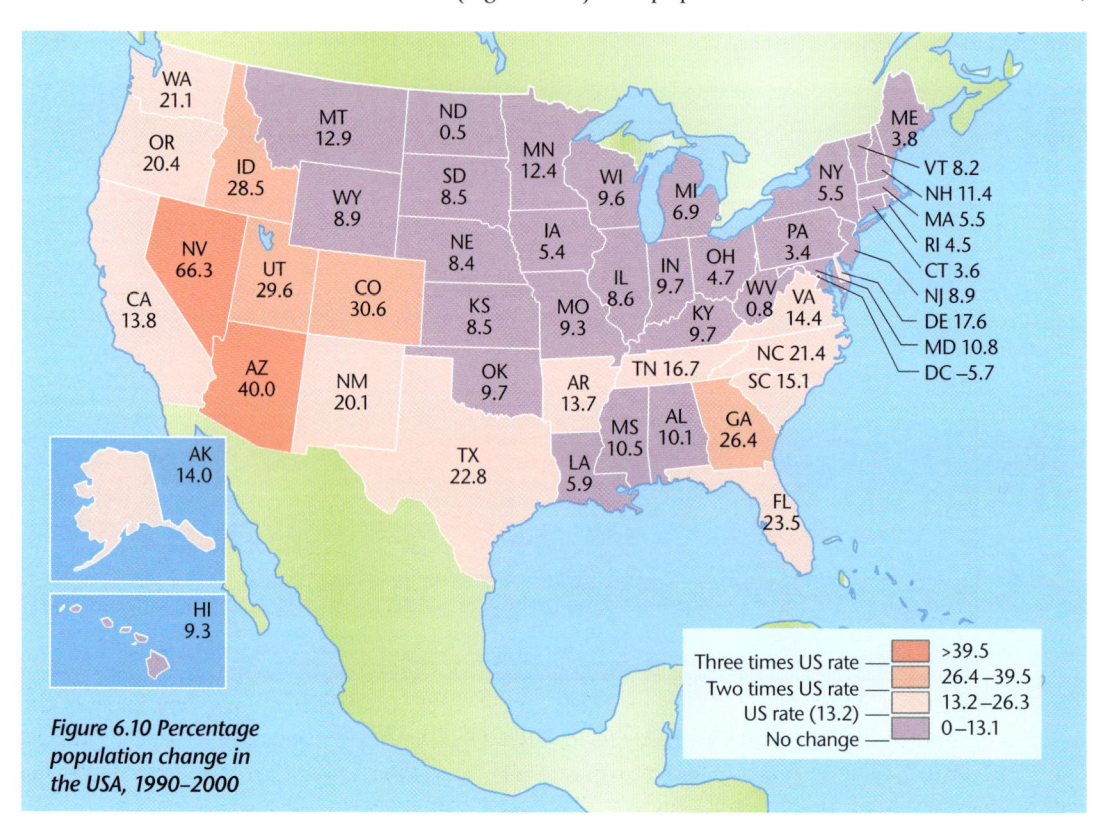

**Figure 6.10** Percentage population change in the USA, 1990–2000

State values:
- WA 21.1
- OR 20.4
- ID 28.5
- MT 12.9
- ND 0.5
- MN 12.4
- WI 9.6
- MI 6.9
- NY 5.5
- ME 3.8
- VT 8.2
- NH 11.4
- MA 5.5
- RI 4.5
- CT 3.6
- NJ 8.9
- DE 17.6
- MD 10.8
- DC −5.7
- PA 3.4
- OH 4.7
- IN 9.7
- IL 8.6
- WV 0.8
- VA 14.4
- KY 9.7
- NC 21.4
- TN 16.7
- SC 15.1
- SD 8.5
- WY 8.9
- NE 8.4
- IA 5.4
- KS 8.5
- MO 9.3
- NV 66.3
- UT 29.6
- CO 30.6
- CA 13.8
- AZ 40.0
- NM 20.1
- OK 9.7
- AR 13.7
- MS 10.5
- AL 10.1
- GA 26.4
- TX 22.8
- LA 5.9
- FL 23.5
- AK 14.0
- HI 9.3

Legend:
- Three times US rate: >39.5
- Two times US rate: 26.4–39.5
- US rate (13.2): 13.2–26.3
- No change: 0–13.1

with southern and western states showing much more growth than northern and eastern states (Table 6.5). The trends that lie behind these movements are complex.

*Table 6.5 The four tiers of sunbelt change, 1950–2000*

| Region | Population in 1950 | Population in 2000 | Numerical change | % change |
|---|---|---|---|---|
| Big three: Florida, southern California*, Texas | 16 414 474 | 57 439 121 | +41 024 647 | +250 |
| Booming four: Arizona, Georgia, Nevada†, New Mexico | 4 923 641 | 16 511 596 | +11 588 255 | +235 |
| Steady four: Louisiana, North Carolina, South Carolina, Tennessee | 12 154 190 | 22 219 584 | +10 065 394 | +83 |
| Lagging four: Alabama, Arkansas, Mississippi, Oklahoma | 9 383 519 | 13 415 812 | +4 032 293 | +43 |
| Sunbelt total | 42 875 824 | 109 586 413 | +66 710 589 | +156 |
| US total | 151 325 798 | 281 421 906 | +130 096 108 | +86 |

*California includes the ten counties south of the 36th north latitude. †Nevada includes only Clark County, which contains Las Vegas.

## Case study 6.7 Internal migration: the rise of the 'Sunbelt'

Figure 6.11 Movement into the sunbelt

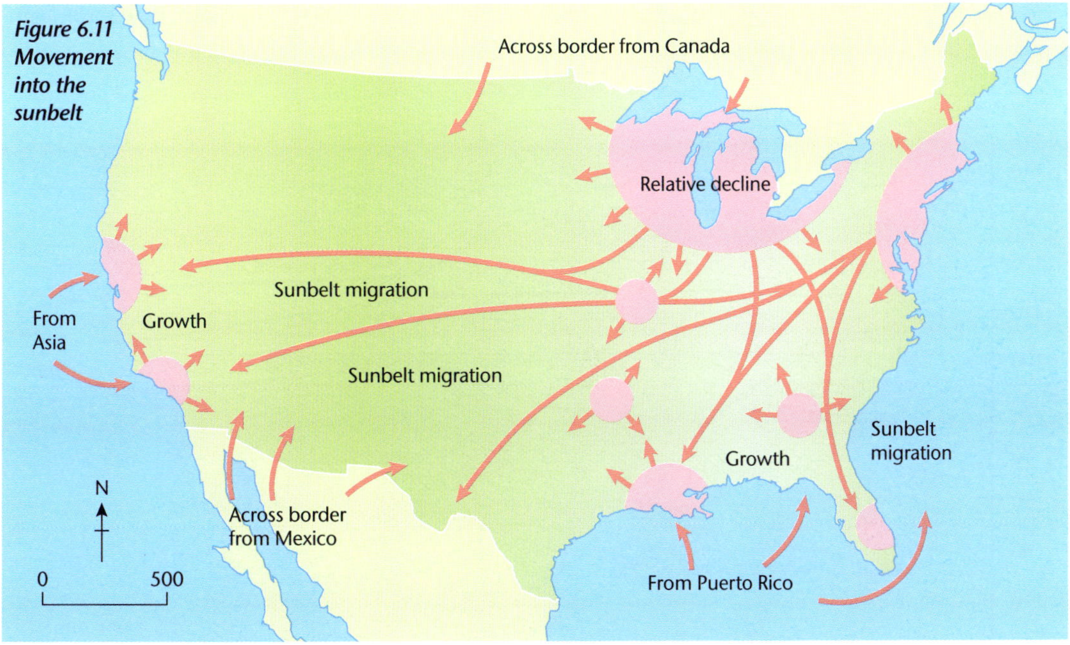

The term 'sunbelt' was first used in 1969 to describe a group of southern and western states stretching from Florida to California that contained a new constituency for the Republican Party. Traditionally the Republican Party had done very badly in the region, because of hostility dating back to the Civil War when the northern Yankees were led by the Republican President, Abraham Lincoln.

# Chapter 6
## Population movements

Sunbelt is now used to describe a region of growth in the US economy associated with modern manufacturing, hi-tech industries and retirement to the sun. The movement has been largely urban–urban, mixed in terms of gender, but with two distinct age groups:
- the young, looking for new economic opportunities
- more elderly people, looking to retire

The sunbelt's development required modern engineering, including:
- the provision of water in arid areas such as southern California, Arizona and Nevada
- the development of interstate highways (motorways) and later internal air communications to connect the region to the rest of the USA
- the development of air-conditioning systems to allow comfortable existence in desert regions such as the Mojave where summer temperatures frequently average above 35°C

The first two of these required substantial central government funds. The main push factor was the decline of traditional heavy industry in the northern and eastern states, which are now variously described as the rustbelt or frostbelt. Cities founded on import-substituting industries such as Cleveland, Ohio, or Detroit, Michigan, have lost population rapidly since the 1960s, as these industries have mechanised, rationalised production or simply closed as new areas of manufacture, both at home and abroad, have developed. Meanwhile, stimulated by the improvements in the infrastructure and encouraged by successive governments, new industries such as aircraft manufacturing and the associated defence industry have grown in southern locations:
- Houston became the centre of the American space effort
- Atlanta grew rapidly as nearby Marietta in Georgia became the centre of the US nuclear programme

This redistribution of the population has also had the consequence of narrowing the income differences between US regions (Figure 6.12). The idea of the 'poor' south is no longer valid.

There was and still is a significant flow southwards due to retirement, with Florida one of the preferred

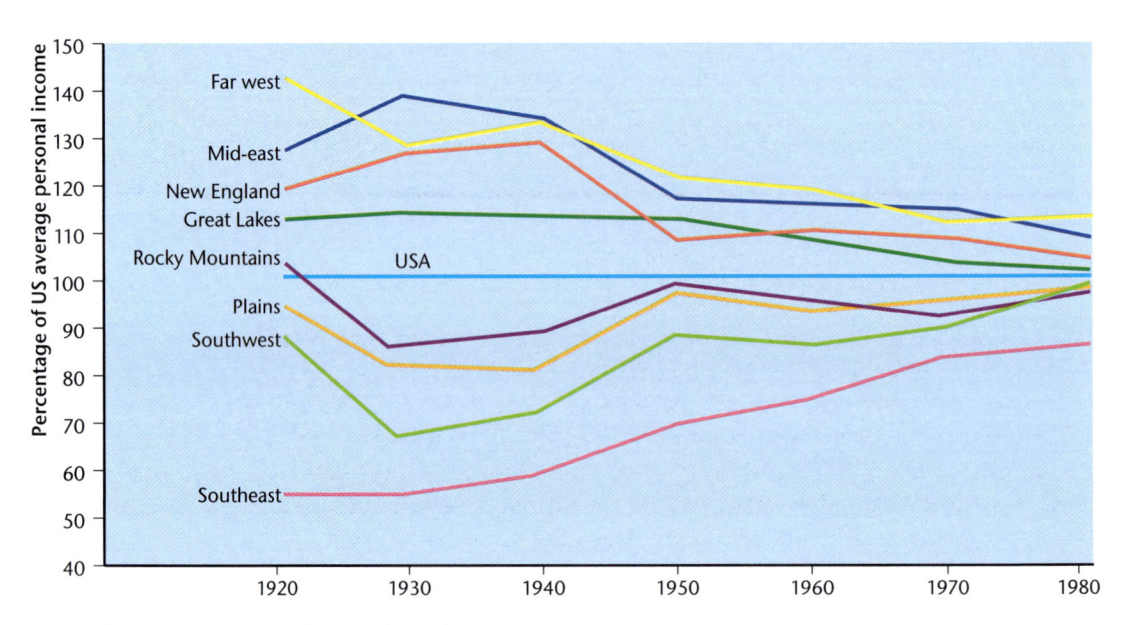

*Figure 6.12 US personal income by region, 1921–81*

*Human environments* **AS Unit 2**

destinations. The growth in population, which is variable from state to state, is sometimes remarkable:
- Florida contained less than 3 million people in 1950 but rose to nearly 16 million by 2000
- Clark county, Nevada, which includes Las Vegas, shot up a remarkable 2749% in just 50 years

The impact on the physical environment has been considerable. The benefits are hard to find, but the cleaning up of northern cities can be seen as a positive impact on the built environment. A reduction in pollutants entering large bodies of water such as the Great Lakes is also a positive impact, this time on the physical environment. However, there are considerable issues at stake in the south, including:
- loss of habitat in deserts such as the Mojave, which are increasingly built on or used for leisure such as golf courses or 4 × 4 trails; similar pressures for development exist in the wetlands of the Everglades in Florida
- the water demands made by cities such as Los Angeles and Las Vegas can only be met by taking more water from major rivers such as the Colorado, or by removing groundwater
- extraction from rivers has led to major environmental changes in the Colorado delta, where the river now scarcely reaches the sea
- removing groundwater has led to subsidence and building damage in many southern cities, such as Phoenix, Arizona
- the universal use of air-conditioning systems both raises urban temperatures by as much as 2°C and emits chlorofluorocarbons into the atmosphere, where they contribute to ozone layer damage

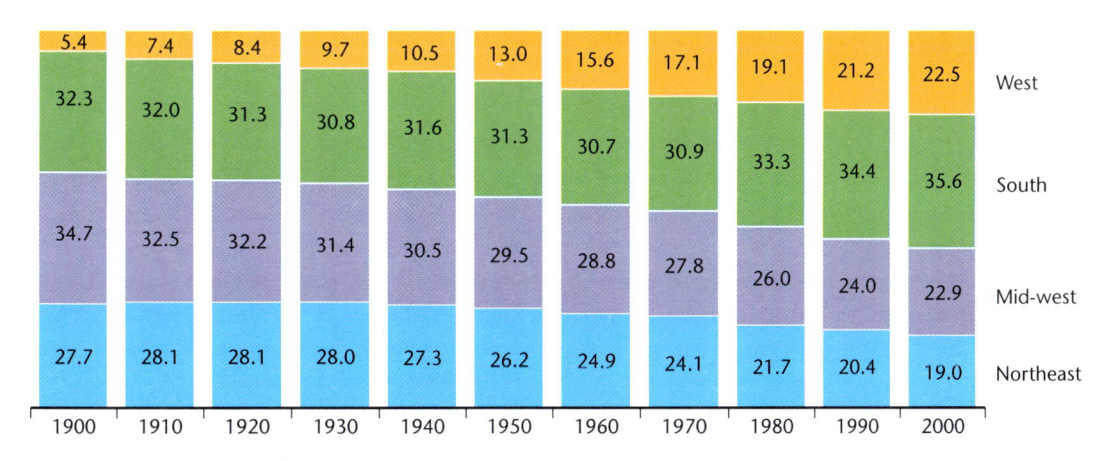

| Year | West | South | Mid-west | Northeast |
|---|---|---|---|---|
| 1900 | 5.4 | 32.3 | 34.7 | 27.7 |
| 1910 | 7.4 | 32.0 | 32.5 | 28.1 |
| 1920 | 8.4 | 31.3 | 32.2 | 28.1 |
| 1930 | 9.7 | 30.8 | 31.4 | 28.0 |
| 1940 | 10.5 | 31.6 | 30.5 | 27.3 |
| 1950 | 13.0 | 31.3 | 29.5 | 26.2 |
| 1960 | 15.6 | 30.7 | 28.8 | 24.9 |
| 1970 | 17.1 | 30.9 | 27.8 | 24.1 |
| 1980 | 19.1 | 33.3 | 26.0 | 21.7 |
| 1990 | 21.2 | 34.4 | 24.0 | 20.4 |
| 2000 | 22.5 | 35.6 | 22.9 | 19.0 |

*Figure 6.13 Population distribution in the USA (%), 1900–2000*

Not all sunbelt states have exhibited the same rates of growth. Parts of the deep south abandoned by so many black Americans in the first half of the twentieth century have yet to experience anything of the surge in population so characteristic of the rest of the south. And not all states that have grown have done so consistently. At the same time, although the areas of loss are identifiably in the north and the mid-west, there is patchiness here too, reflecting a complex picture that includes significant rural–urban and urban–rural movements.

The growth of the southern states of the USA has also been fuelled by in-migration from Mexico. This has resulted in a very different ethnic mix in the southern states compared with that found elsewhere in the USA. Many states are

# Chapter 6 Population movements

multi-ethnic, but the distribution of Hispanics (Spanish speakers who are often but not exclusively Mexican) is clearly related to the border with Mexico, as the numbers decline rapidly with distance from the frontier (Figure 6.14).

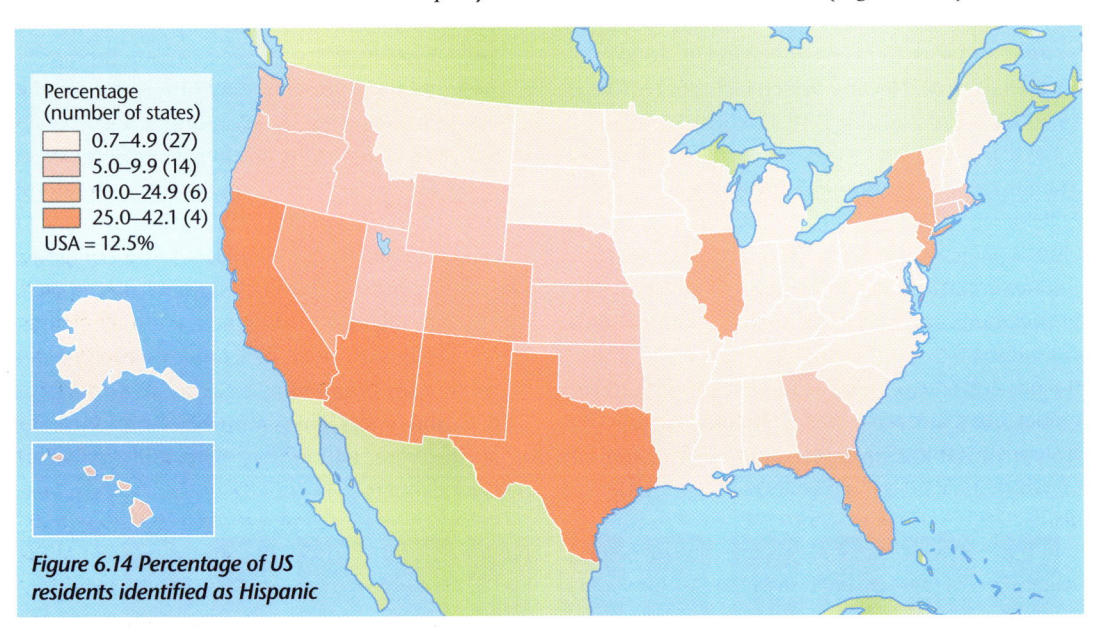

**Percentage (number of states)**
- 0.7–4.9 (27)
- 5.0–9.9 (14)
- 10.0–24.9 (6)
- 25.0–42.1 (4)

USA = 12.5%

*Figure 6.14 Percentage of US residents identified as Hispanic*

## Case study 6.8 Mexican immigration to the USA

- There are about 108 million people alive today who were born in Mexico.
- About 8 million of these have migrated to the US to settle. Half migrated in the 1990s, meaning that about 1 in 25 Mexicans moved to the USA in the 1990s.
- Some 250 000 to 350 000 Mexicans continue to settle in the USA each year, and many more, perhaps 1 million, work and live at least part of each year in the USA.
- Mexico's population grows by about 1.8 million a year; but with emigration of 300 000 a year, the population growth rate is 1.5%, not 1.8%.
- The Mexicans in the USA include about 2.5 million naturalised US citizens, about 3 million legal immigrants and about 2.5 million unauthorised persons.

The USA first approved the recruitment of Mexican workers for US farmers just after the First World War and again between 1942 and 1964. These Bracero programmes made many farmers dependent on Mexican workers. Seasonal agricultural workers are crucial; for example, 70% of fruit pickers in California are immigrants. This level of dependency has spread to other areas of employment:

- 80% of tailors
- 78% of waiters
- 75% of cooks
- 70% of taxi-drivers
- 55% of textile workers

Many of the above groups are dominated by illegal migrants. There is an ambivalent attitude to illegal immigration on both side of the border, especially since the US Congress forecasts that 50% of jobs available in 2050 will be in low-paid personal services such as construction, security, healthcare, fast food and other retail sales. Illegal migrants cannot complain about working conditions and are generally cheaper to employ because minimum wage legislation is not an issue. Mexico–US migration has diversified in terms of region of origin, but two-thirds of the Mexican migrants originate from 7 of Mexico's 31 states in the west central area of the country — dominated by the city of Guadalajara.

Mexico's population increased by 600% in the twentieth century, from 17 million in 1900 to 100 million in 2000. The preferred economic development model was state-owned industrialisation and the rapid commercialisation of agriculture, which produced widespread rural poverty. This, in turn, encouraged rural–urban migration. Policies have since changed to a more export-orientated growth model, but this too places an emphasis on commercial agriculture and concentrates on export crops to boost foreign currency earnings. The impact is to displace many small farmers and peasants.

In the 1980s millions of jobs were created in the industries of the sunbelt, driven by President Reagan's defence build-up. Many illegal migrants, mostly young men, were given legal status in order to enable them to be soaked up by these expanding industries. The impact on the 'donor' regions has been considerable. Mexico–US migration has become a key part of the socioeconomics of many rural areas in west central Mexico. In some areas, 80 to 90% of the young men make at least one trip to the USA before they are 40, usually as unauthorised workers. In many cases, families and villages have become international.

Some individuals divide their time between the USA and Mexico, some families have members living in both the USA and Mexico, and some Mexican areas are linked to US areas by a steady stream of migrants and visitors, and money flowing back in the form of savings (remittances). Mexican migrants send back approx. $6 billion per year to support their families, much of which is spent on consumer goods rather than saved. When this multiplier effect is accounted for, these remittances account for something like 5% of Mexico's GDP.

The cultural impact on California and other states is profound. With its slightly higher fertility rate, the Mexican population grows faster than most other ethnic groups. In San Diego over 40% of the population already speak Spanish as their first language. In June 1998 a cultural backlash occurred when California's voters approved Proposition 227. This ended bilingual education in Californian state schools and replaced it with 1 year of total immersion in English prior to participation in the regular curriculum.

---

The complex immigration history of the USA has had an obvious impact on the ethnic geography of the country (Table 6.6). Regional differences show up clearly:

- the black population is still over-represented in the deep south, despite years of out-migration
- the Hispanic population is over-represented in the south centre, especially Texas
- the west is the most ethnically diverse region
- Native Americans, always a tiny number, are found in larger than average percentages in the regions to which white immigration pushed them over a century ago
- the white population is still the dominant group in the north and east

| Country | Total number of migrants |
|---|---|
| Germany | 7 105 900 |
| Mexico | 5 400 105 |
| Italy | 5 353 200 |
| UK | 5 197 450 |
| Ireland | 4 780 891 |
| Canada | 4 348 541 |
| Russia (USSR) | 3 749 777 |
| West Indies | 3 372 716 |
| Austria | 2 644 728 |
| South America | 1 588 408 |
| Sweden | 1 398 578 |
| China | 1 232 740 |

*Table 6.6 Migrants to the USA by country of origin, 1820–1996*

## Chapter 6 Population movements

Table 6.7 Ethnic mixtures in metropolitan areas of the USA, 1998

|  | White | Black | Hispanic | Asian | Indian/Eskimo |
|---|---|---|---|---|---|
| Miami, FL | 42.0 | 17.7 | 38.5 | 1.7 | 0.1 |
| Los Angeles, CA | 43.1 | 7.4 | 38.5 | 10.6 | 0.4 |
| Fresno, CA | 44.0 | 4.3 | 42.4 | 8.6 | 0.7 |
| Salinas, CA | 44.3 | 5.6 | 40.6 | 9.0 | 0.5 |
| Merced, CA | 46.1 | 4.2 | 39.5 | 9.6 | 0.6 |
| Stockton, CA | 51.0 | 5.0 | 28.8 | 14.5 | 0.7 |
| Albuquerque, NM | 51.7 | 2.3 | 39.2 | 1.7 | 5.2 |
| Houston, TX | 53.1 | 17.3 | 24.6 | 4.8 | 0.2 |
| San Francisco, CA | 54.1 | 8.3 | 19.3 | 17.8 | 0.5 |
| San Diego, CA | 58.2 | 5.6 | 25.9 | 9.7 | 0.6 |
| Flagstaff, AZ | 58.2 | 1.5 | 11.8 | 1.1 | 27.4 |
| Santa Barbara, CA | 58.9 | 2.5 | 32.9 | 5.1 | 0.5 |
| New York, NY | 59.8 | 16.2 | 17.4 | 6.4 | 0.2 |
| Killeen, TX | 60.9 | 18.7 | 16.0 | 3.9 | 0.5 |
| Modesto, CA | 63.3 | 1.6 | 27.9 | 6.3 | 0.9 |
| Chicago, IL | 63.5 | 18.6 | 13.8 | 3.9 | 0.1 |
| Washington, DC | 64.1 | 25.4 | 5.3 | 5.0 | 0.3 |
| Yuba City, CA | 66.2 | 2.6 | 18.3 | 11.1 | 1.8 |
| Waco, TX | 66.9 | 16.0 | 15.8 | 1.0 | 0.3 |
| Dallas, TX | 66.9 | 13.5 | 15.8 | 3.4 | 0.4 |
| Sacramento, CA | 68.0 | 6.6 | 15.0 | 9.5 | 0.9 |

The regional picture is subject to many fascinating local variations, of which the most obvious is the urban–rural contrast.

# Rates and volumes of migration: contrasts in modern Europe

Rates of migration can vary greatly at a national level, depending upon national policies and economic forces. That is obviously so in the case of the USA, but it also applies to every other part of the world.

The changes that have taken place in central European societies in recent years, since the collapse of communism, are especially interesting. The combination of shattered economies, a fall in confidence and the opening of frontiers have combined to create a very different climate for migration than that which existed 20 years ago. In particular, the growing rate of urban–rural migration in central Europe is very different from that seen in the UK or France today.

*Human environments*  **AS Unit 2**

## Case study 6.9  Migration trends in Romania

*Figure 6.15 Romania and its neighbours*

With 22 million inhabitants, Romania ranks among the larger central and eastern European states. Between 1948 and 1989 Romania was a totalitarian state with a highly centralised government. In 1948 it was almost completely rural, with 80% of its population living in the countryside. The communist state then embarked on a programme of industrialisation that involved moving both people and capital from rural areas to towns and cities. Agriculture was 'collectivised' in that land was taken into national ownership, which initiated an out-migration of wealthier landowners while the borders still remained open. In the 1960s and 1970s, hundreds of thousands of villagers were encouraged or forced to move to cities and towns and become industrial workers in state factories.

Following the collapse of communism in 1989 and the exposure of these industries to competition, many state factories ceased to operate. Many of the workers who lost their jobs and were facing the cost of living in cities returned to the countryside, where they survive by subsistence farming. This is urban–rural migration of a quite different type than that experienced in western Europe.

In the years since the communist dictator Nicolae Ceauşescu was overthrown, Romania's population has declined by at least 415 000 people. The main factors behind the decline are the following.

- Higher mortality rates, especially among men, caused by increased rates of heart disease, alcoholism and drug abuse, exacerbated by the collapse of the health and welfare system.
- Lower birth rates, with fertility rates dropping from more than 2.2 children per woman in 1989 to 1.3 today. Ceauşescu pursued a pro-natal policy, which was abandoned in 1989. He made contraception and abortion illegal, leading to large numbers of unwanted children and an increase in female mortality due to back-street abortions.
- Large-scale emigration: lack of opportunity has led to population loss to neighbouring countries in western Europe. At its peak this was 100 000 people a year, but it has now declined to about 20 000. Unfortunately for Romania, those leaving are frequently the best-educated and the younger members of the population.

Ironically, Romania also faces serious problems of illegal immigration. Between 1993 and 1996, 45 000 foreigners were refused entry, but many more are living illegally in the country. Romania is on the most important transit route for illegal migrants into Europe from further east, with Turks, Afghans and Pakistanis among the largest groups.

# The problem of generalisation

One of the great difficulties when discussing migration as a topic is that it is very hard to make general comments without then having to qualify them.
- It is frequently claimed that counterurbanisation is a characteristic of MEDCs. However, there are many cases of MEDCs where this is not the case and almost

## Chapter 6  Population movements

- as many examples of LEDCs where urban sprawl and commuting are obvious features.
- There are many contrasts to be drawn between the UK and Iceland. They are both highly developed countries with more similarities of culture, philosophy and economic goals than differences, and yet their recent migration histories could hardly be more contrasting (see Case studies 6.10 and 6.11).

In the past 50 years the growth of cities and major improvements in the transport system have led to the increase in population in the areas immediately adjacent to large cities, the peri-urban fringes. This change is almost universal, although the impact on the population of the city itself is variable, depending on the levels of prosperity. Thus New York has lost population to the rural fringes without losing much in the way of economic dominance, while Chicago, Detroit and Manchester have lost both population and economic significance.

### Case study 6.10  Comparing MEDCs: counterurbanisation in the UK

The term **counterurbanisation** was first coined by the American geographer Brian Berry. He identified an urban–rural migration pattern associated with the dispersal of industry into areas beyond the city limits, and depopulation of major conurbations as their industries declined. It was not so much a shift in industrial location, but more the death of some industries and the birth of newer ones in more rural locations that drove this process. The retirement to rural areas of an increasingly large elderly population also made a contribution.

The most obvious weakness in the concept is that it becomes stereotyped as the movement of people from inner-city suburbs to isolated rural cottages, probably with roses growing around the front door. Most counterurbanisation is not at all like this, but is instead the growth of small cities and towns on the periphery of major city regions. So Aylesbury, Swindon and Bournemouth have all grown rapidly in recent years, while many villages, especially in protected landscape areas such as the Chilterns, have scarcely grown at all.

Berry also recognised that rural *depopulation* remains important in many areas. Approximately one in five counties in the USA (each state is divided into counties) lost population between 1990 and 1995. The vast majority of these were rural counties, especially in the mid-west, with a high dependency on agriculture and a lack of alternative employment. Similarly, remote rural areas in the UK still show marked rural decline. For example, in a recent survey only 4% of schoolchildren in Ullapool (northwest Scotland) expressed a wish to stay in the area after leaving school. In the UK counterurbanisation probably peaked in the 1970s when the economic decline of heavy industries was at its height and the new types of footloose employment were emerging strongly. The key factors for the industries involved were accessibility, availability and attitudes.

- **Accessibility:** many of the firms that locate in rural areas have national or even international markets, so it is not vital for them to be close to any one city. They adopt 'interim locations'.
- **Availability:** the availability of relatively cheap building land is a major attraction of rural areas for companies, as is the cheaper housing for the labour force.
- **Attitudes:** residents of rural areas often have fewer choices of employment available. With fewer options to change employment, they seem to be better motivated than urban employees.

*Human environments* **AS Unit 2**

However, for every five people leaving the cities of the UK there are four people arriving. Many of the arrivals are young adults moving into higher education, who use cities as 'social escalators'. They then move out of the cities when they start families, even if they continue to commute, as many do.

Counterurbanisation is important, but the recent growth of London, Paris and many other cities suggests that the importance of cities is not radically diminished.

Case study 6.11, about Iceland, is instructive in that it is a sharp reminder that generalisations are dangerous. Iceland is one of the wealthiest nations on Earth, one of the most developed and one of the happiest, yet much of its recent history fits poorly into the stereotypes of MEDCs.

## Case study 6.11  Comparing MEDCs: rural–urban migration in Iceland

Iceland is a small country of about 100 000 km$^2$. It was the last place on the planet to be permanently occupied by human beings, who arrived there in the form of Viking explorers in the later part of the ninth century. Only 40% of Iceland is available for human use; the rest is lava flows, ice sheets or great outwash plains of sandur. Cold and damp winters and short summers, also damp, prevent arable agriculture apart from a few potatoes and turnips. Pastoral agriculture developed, based on sheep and horses, supplemented by fishing from the very productive seas around the country.

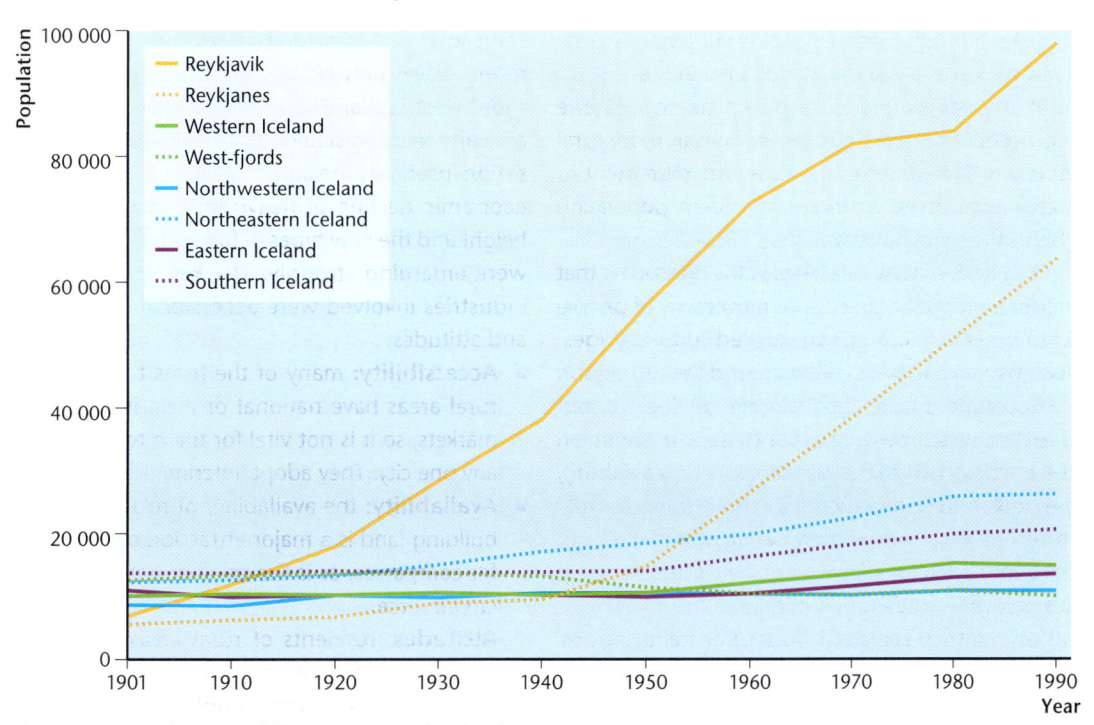

*Figure 6.16 Population in different regions of Iceland, 1900–90*

235

## Chapter 6   Population movements

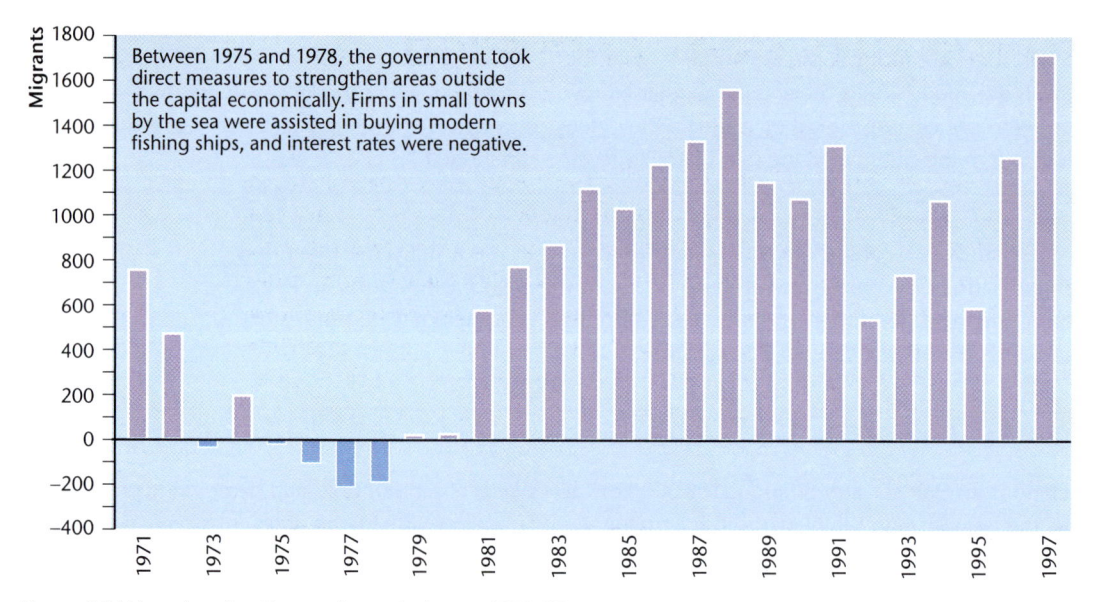

*Figure 6.17 Net migration flow to the capital area, 1971–97*

Between 1975 and 1978, the government took direct measures to strengthen areas outside the capital economically. Firms in small towns by the sea were assisted in buying modern fishing ships, and interest rates were negative.

Remote and isolated communities developed, largely made up of independent small farmers, who met up only rarely at the world's first parliament to settle the odd boundary dispute. The population was probably around 50 000 for much of Iceland's history until it became, slowly but inevitably, integrated into the rest of the world economy.

Only then did the urbanisation process begin. The capital city, Reykjavik, a fishing village blessed by hot thermal spring water, had a population of about 6000 in 1900. It has grown to 110 000 today, with the most rapid growth coming in the past 30 years.

The economy depends heavily on the fishing industry, which provides 75% of export earnings and employs 12% of the workforce. The area around the capital, the city region, has also grown, so that about 75% of the whole population of Iceland (283 000) live within an hour of the capital. Most of this growth is rural–urban migration from Iceland's increasingly depopulated regions. The following are the driving forces for this migration.

- Lack of employment opportunities in the regions. Very few towns exist and the remoteness, difficulty of travel and small labour market make it unlikely that enterprises will establish in the regions.
- The negative multiplier effect is more pronounced because the most likely migrants are the young, who are also the most likely to be potential employees.
- The old fishing industry created processing and packing jobs in Iceland's fishing ports. This was helped by government assistance that briefly stemmed the outflow of people in the 1970s. Today, processing is done at sea on the much larger fishing vessels that operate out of Reykjavik.
- Education is available in the capital. Icelanders are among the most literate people on the planet and the most educated. The only university in Iceland is in Reykjavik.
- Almost all Iceland's social and cultural renaissance is located in the capital. The remote regions offer little for the young.
- Reykjavik is the centre of a lively tourist industry and there have been significant developments in quaternary employment in diverse fields ranging from geothermal energy to off-road tourism and fishing.

# Consequences of migration

## Demographic consequences

Migration may have profound effects on the size, structure and growth patterns of populations. It affects both the populations of the places that people leave and the populations of the places in which they settle. These effects vary with different types of migration, the age structure of the migrants and the length of time the migrants stay. The absence of large numbers of either men or women may have a limited impact on the sending society in the short term, but if they are absent for longer periods of time, their absence will have significant effects on population growth rates in the medium and longer terms.

## Social consequences

Social consequences include the impact of migration on individuals, families and communities. They can be either positive or negative and they can change over time. Disruption of family life is probably the most significant consequence of modern population movement.

## Economic consequences

Migration almost always has significant effects on economies. These effects vary with different types of migration, the skills of the migrants and the lengths of time involved. Most observers believe that migration has a largely positive impact on the economy of receiving countries, which is hardly surprising, since the demand for labour is the most important driving force behind most major migrations.

## Political consequences

Migration can have an impact on politics in both the places which people leave and those to which they move. Governments might have to make policies to attract migrants, to persuade migrants to return, or to limit migration and ensure that migrants have access to the skills they need. These political effects vary with different types of migration.

## Cultural consequences

Closely related to social consequences, cultural consequences include the impact of migrants on art, music and ideology. There is no doubt that ethnic diversity is stimulating in all these fields, although stimulation is not always welcomed by every section of society. The US case studies illustrate that many of the modern icons, from Madonna through to Eminem, are directly descended from a musical tradition that is a product of migration.

> **Examiner's tip**
>
> ■ Using the case studies, make sure you know at least two examples of both causes and consequences of migration from each category, i.e. economic, social, political, cultural and demographic.

# Chapter 6

*Population movements*

## Examination question

Study Table 6.8, which shows the number of migrants arriving in Florida (USA) as a percentage of the total population.

Table 6.8

| Year | International migrants | Internal migrants |
|---|---|---|
| 1990–91 | 0.36% | 1.24% |
| 1991–92 | 0.39% | 0.76% |
| 1992–93 | 0.40% | 0.77% |
| 1993–94 | 0.44% | 1.03% |
| 1994–95 | 0.60% | 0.72% |
| 1995–96 | 0.66% | 0.82% |
| 1996–97 | 0.59% | 0.89% |
| 1997–98 | 0.50% | 0.77% |
| 1998–99 | 0.54% | 0.58% |

**a** (i) Identify the year when the smallest number of migrants arrived. (1)
(ii) Describe the relationship between international and internal migration. (3)
(iii) Suggest reasons why coastal regions, such as Florida, often have relatively large numbers of migrants. (4)

**b** (i) Identify *two* political causes of out-migration from a country. (2)
(ii) Describe the impact on the physical environment of the out-migration of a population. (4)

**c** With the use of a named migration, describe and explain the variations in the number of migrants over time. (6)

*Total = 20 marks*

# Synoptic link
# Government policies influence migration patterns

- Reasons for, effects of and issues associated with government policies to influence migration into, out of and within a country.
- The causes and consequences of recent migrations and refugee movements.

## Social and cultural impacts of migrants

In many cases, the economic benefits of migration are obvious to both communities, except during periods of economic recession. The reception that migrants encounter in the receiving society depends on a number of factors. Migration usually brings together people with different world views and lifestyles, so it is not surprising that there is often some suspicion on the part of both the hosts and the migrants. Where the world views and lifestyles are similar, there are typically lower levels of suspicion, as with migration within Europe. Where groups understand little of each other's world views and lifestyles, there is typically the following.

- A greater dependence on social stereotypes (that is, generalised views of others based on limited personal contact and very few observations).
- A higher level of prejudice, based on social stereotypes rather than personal experience.
- A higher probability of ethnic discrimination (that is, when a person's beliefs about a migrant group influence their behaviour towards that group and lead them to treat members of that group differently from others). While this discrimination occasionally leads to more favourable treatment of members, it usually results in less favourable treatment.

Social stereotypes, prejudice and discrimination often combine to produce 'self-fulfilling prophecies', which seem to 'confirm' social stereotypes and to justify prejudice and discrimination.

Members of the host society are not alone in this conduct. Migrants also discriminate against their own people. For example, experienced migrants may take advantage of a new arrival's lack of knowledge to discriminate against him or her. This is well documented in cities in the USA (e.g. loan-sharking).

# Arguments about immigration: politics

There are few subjects more likely to engender heated debate and dissension than migration. The issue is especially controversial in the wealthiest MEDCs where the tensions that arise between economics and nationalist politics are most evident. This presents a major set of opportunities and challenges for potential migrants.

The opportunities are largely economic, although for political refugees they might also be expressed in terms of political freedom and freedom from oppression.

The challenges are readily identified as language, religion and culture.

## Language

Adoption of the host language is a key element in integration of migrants into a new culture. Great emphasis is placed on this in many MEDCs. However, the arguments are complicated by the rights given to older ethnic minorities to teach and learn in their own language. Welsh is the obvious example of this.

## Religion

Religion plays a major part in countries that have a state religion, such as Saudi Arabia. Even in the UK, religion plays a part, although the formal adoption by migrants of the state religion is not an expectation. Religion has become an issue in France where, in line with policy disallowing any form of religious symbols in schools, the government banned the hajib, the traditional headscarf worn by many

## Chapter 6
## Population movements

Muslim women. This move provoked an enormously varied response, from those who claimed that it liberated students from any form of religious oppression to those who saw it as a fundamental intrusion on civil liberties. Similar arguments have surrounded the problems of the Sikh community with the wearing of turbans and the difficulties they have in complying with the law on the wearing of crash helmets on motorbikes.

This type of debate is often very complex. For example, the ritual slaughter of foxes might be defended by some as a 'fundamental human right' that is deeply embedded in the 'traditions' of the English countryside, while the same people might wince at the slaughtering techniques adopted by Jewish or Islamic butchers.

### Culture

A recent Conservative prime minister of the UK, John Major, invoked an image of his native land which was 'the country of long shadows on county grounds, warm beer, invincible green suburbs, dog lovers and pool filters'. This is not an image of Britain that resonates with all its citizens. It is not even national, in that it fits poorly in rural Wales or the Scottish Highlands.

Similar views about British culture and its 'contamination' by migrants are deeply held and frequently pronounced. 'Great waves of immigration by people who do not share our culture, our language, our way of social conduct, in many cases who owe no allegiance to our country' was the summary of the Conservative Lord Tebbit. He sits in the same intellectual tradition as Enoch Powell who, in 1968, used the phrase 'River Tiber foaming with much blood' as his prediction for the future of the UK if immigration was allowed to continue. Quite what either of them would have made of the chant 'I'd rather be a Paki than a Turk', one of the many racist chants invented by England supporters during the Euro 2000 football tournament, is not known.

Whether or not chants like these are 'English culture' and a proper way of showing allegiance to a shared set of values, they do highlight the problem that this 'shared set of values' is elusive. If we cannot define the values of a country, then it is hard to argue that the values are being threatened.

## Arguments about immigration: economics

80 million people entered the UK in 1998, most of them visitors or British citizens returning to the UK — nearly double the figure of a decade before.

Migrants are not the same as members of an ethnic minority. This is frequently confusing for nationalists.
- The majority of migrants are white.
- Most members of ethnic minorities are not migrants, but were born in the UK.

There is no one main source country of migration to the UK. The largest single group is UK nationals, mostly returning emigrants, though some will have been born away from the UK.

*Human environments* **AS Unit 2**

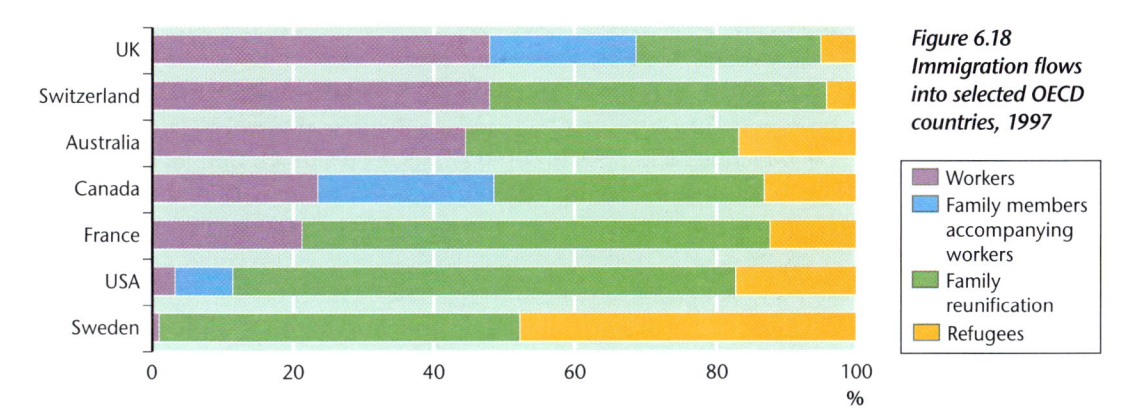

*Figure 6.18 Immigration flows into selected OECD countries, 1997*

- Workers
- Family members accompanying workers
- Family reunification
- Refugees

The argument that 'native born' candidates for jobs are disadvantaged by in-migration is almost entirely without statistical support, although it still plays well politically. The German politician Jurgen Ruttgers fought a recent state election on the slogan 'Kinder Staat Inder' or 'educate German children, don't import Indians', thus adding his name to the long list of European politicians who have played the 'race card'.

If migration of workers into a particular employment sector is restricted, this will not stimulate the greater supply of, and wages for, native British workers, at least not in the foreseeable future. The obvious example would be in the IT sector. Rather than wait for the local labour supply to appear, which would take many years of changes in education systems and training, IT businesses would simply shift elsewhere. You only need look at the multi-ethnic society of Palo Alto (Silicon Valley, in the USA) to question the ideas of the most hardened opponents of immigration, unless, of course, they are uninterested in economic performance. Figure 6.19 shows this clearly in the case of Switzerland.

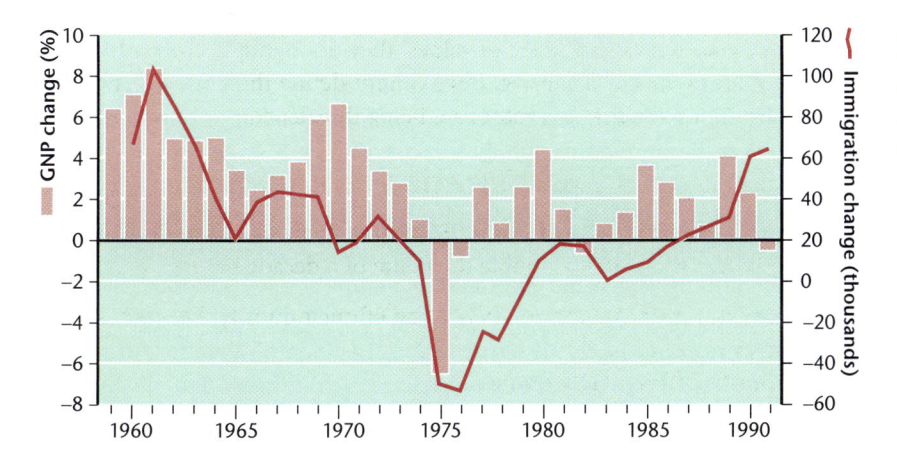

*Figure 6.19 The relationship between immigration and % change in GDP in Switzerland, 1959–91*

There are significant skills shortages in the UK labour market. The fact that many migrants are concentrated in the industries and sectors where there are

## Chapter 6
*Population movements*

particular labour or skill shortages is clear both anecdotally and from the available data (Figure 6.20). Here are some examples.
- 31% of doctors and 13% of nurses in the NHS are non-UK born.
- Half the expansion of the NHS in the last decade of the twentieth century, that is, 8000 of the additional 16 000 staff, qualified abroad. Despite this, a Royal College of Nursing survey reports 78% of hospitals with medium to high recruitment difficulties.
- A growing number of London education authorities are recruiting staff directly from abroad to address staff shortages in schools.
- In 1995–96, the Higher Education Statistics Agency showed that non-British nationals made up 12.5% of academic and research staff.
- The increase in demand for people with specialist IT skills has been spectacular, and is expected to continue. Projections suggest that the IT services industry alone will need to recruit another 540 000 people between 1998 and 2009.

*Figure 6.20 Number of migrants needed per year to maintain the workforce in a range of regions*

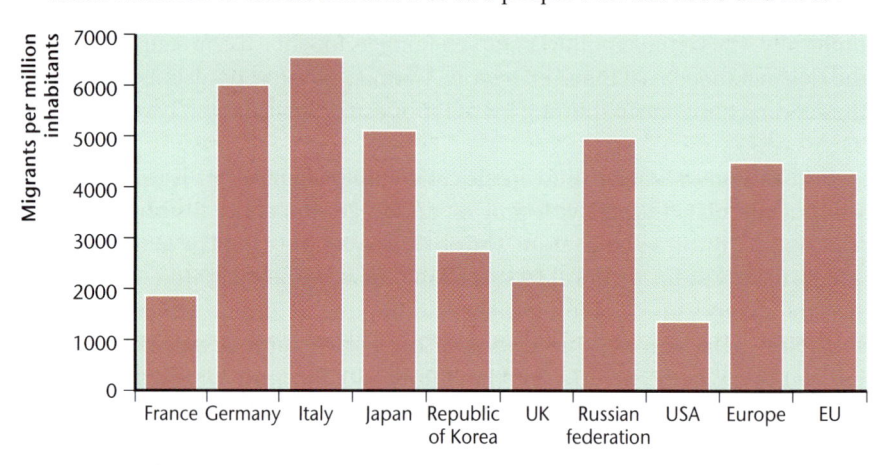

### Arguments about immigration: the costs?

An initial analysis for the UK suggests that migrants contribute more in taxes and National Insurance than they consume in benefits and other public services. It is estimated that the immigrant population contributes 10% *more* to government revenues than it receives in government expenditure, equivalent to perhaps £2.6 billion in 1998/99. Thus, it is calculated, if there were no foreign-born people in the UK, taxes (or borrowing) would have to rise, or expenditure would have to be cut, by £2.6 billion. This is the equivalent of adding about 1p to the basic rate of income tax.

Migrants are more likely to be in receipt of unemployment and housing benefits, but less likely to be receiving sickness or disability benefits or a state pension. Overall, they claim about as much, per person, as the 'native' population.

British eating habits have been transformed by immigration. In 1996, there were 10 000 curry houses in Britain. These had 60 000–70 000 employees and a turnover of £1.5 billion, which, startlingly, is more than the steel, coal and ship-building industries put together.

## Government policies

Few issues arouse such strong emotions as migration. Governments are obliged to take a view and to be seen to be active in pursuit of a well-defined policy or they risk losing significant support. The argument about immigration policy is complex and clouded by misconceptions and racial myths.

In almost all MEDCs one can find many people to argue that immigration should be more restricted. Some of the diverse reasons given for this are listed below. Note that these arguments are sometimes mutually contradictory and not all will be held by any single person or interest group.

- The view that a country should remain predominantly of one race, and that continued immigration threatens the numerical and cultural predominance of that group, usually white.
- The view that the present immigrant flow (at least at the low-wage end) is a 'low-quality' stream — many immigrants arriving in MEDCs today have few skills or assets to contribute.
- The view that the present immigration flow (or certain parts of it) is a 'high-defect' stream — many immigrants arriving in MEDCs today bring problems with them, including diseases and habits of poverty, drugs, illiteracy and a tendency to crime.
- The view that continued immigration will exert a downward pressure on wages and working conditions of workers already in the country, especially those toward the bottom of the economic ladder who are most likely to be in direct competition with low-skill migrants.
- The view that a country such as Britain welcoming high-skill immigrants encourages a brain-drain from the poorer countries which need to keep their native talent at home for the work of national development.
- The view that there is simply not enough to go around, that the receiving country cannot afford to share its wealth with the hordes of people who want to come in from poorer neighbours.

On the other hand, some people argue that immigration should be more generously permitted than it is. Among the diverse reasons advanced for greater liberalisation are the following.

- The view that past immigration restrictions operated with a strong racial bias whereas it is good for a country, both economically, socially and culturally, to have a wide ethnic mix.
- The view that the present immigrant flow (including the low-wage end) is a 'high-quality' stream, that is, that many immigrants arriving in MEDCs today have characteristics that make them particularly valuable to the receiving economy. Various commentators see different characteristics as desirable, but some examples include the following:
    – a strong work ethic
    – an admirable devotion to family
    – deep religious convictions
    – entrepreneurial skills

# Chapter 6
## Population movements

– strong traditions of organising labour and community improvement
– a willingness to perform work despised by indigenous workers
– an ability to throw themselves into a job without complaint

- The view that increased immigration flows will help discipline indigenous workers and encourage them to give up their short-sighted insistence on high wages and soft working conditions, thus making them readier to pitch in and make their employers more efficient and productive competitors in the global economy.
- The view that increased immigration flows will help indigenous workers learn about and make common cause with their fellow workers around the world, with whom they share many important interests and experiences, despite divisions of race, culture, language and income.
- The view that present low-wage immigrant workers are likely to be politically active fighters against injustice and will bring important knowledge and energy to the task of winning greater democracy and economic justice for those toward the bottom of the economic heap.
- The view that freedom of movement is a basic human right that should be recognised and protected in a global society, either in all instances or at least in instances where capital and goods are flowing freely across borders.
- The view that all MEDCs have a responsibility to accept (certain) immigrants and refugees because the MEDCs have played a powerful role in creating the conditions that have produced migration flows, sometimes as a result of military activity, sometimes as a result of economic policies forced on various 'sending' countries.
- The view that attempts to keep immigrants out of the MEDCs are, in any case, doomed to failure given the social, economic, geographical and technological realities of today's global economy, and that continued insistence on such restrictions will only lead to further removal of fundamental rights for all citizens.
- The view that immigrants are more likely to vote for one political party than another.

It is hard to find much middle ground here. The possibility of consensus is remote. Migration, like fox hunting, arouses powerful emotions and governments have to be responsive to these. In democracies it is likely that newspaper campaigns and anecdotal information will carry at least as much weight as official statistics or logic. If one believes a country to be 'overpopulated', then attitudes to immigration are likely to be negative, whereas if you believe that your nation is short of skills and will profit from cultural diversity, then you are likely to be positively inclined to higher levels of immigration.

## Immigration history of the USA

The USA was built upon in-migration, yet it has a long history of regulating immigration. Early legislation tried to limit entry but was strongly biased in favour of northern Europeans.

The Immigration and Nationality Act Amendments of 1965 established important principles that are still significant. They abolished the national origins quota system as the basis for immigration and replaced it with a preference system for visas. Numbers of migrants were increased from 154 000 to 290 000, of which 120 000 were reserved for immigrants from the western hemisphere. In this period, the USA started to experience a more varied entry geography, with significant contributions from Asia and, especially, Latin America.

In later developments increasingly severe treatment of illegal immigration was put in place. In an attempt to 'close the back door while opening the front door', the US government granted amnesty to illegal immigrants who had lived in the USA for a significant period of time but also made illegal immigrants harder to employ and increased the fines for doing so. The government stepped up border controls too, especially at the border with Mexico.

## Impact of the new migrants

Over the past 50 years the number of foreign-born people in the USA has increased dramatically. In 2000, 28.4 million foreign-born people (excluding illegal immigrants) lived in the USA, representing about 10% of the entire population. This is a record for absolute numbers. The estimated number of illegal immigrants is thought to be 5 million, so the total number of foreign-born residents is over 33 million, or about 11% of the population. As a proportion this remains lower than the figure of 15% that occurred in both 1890 and 1910.

The attitude to illegal immigrants in the USA is best described as ambivalent. On the one hand there are serious concerns about the impact of a large number of untraceable and undocumented immigrants, especially since 9/11, but on the other hand they provide the cheapest and least vocal sources of labour.

According to US Census Bureau 2000 Current Population Survey (CPS) data, some 51% of these foreign-born persons originate from Latin America (including Central America, South America and the Caribbean), 25.7% from Asia, 15.2% from Europe and 8.1% from other regions of the world, such as Africa and Oceania.

Migrants from Central America (including, for data purposes, Mexico) account for nearly one-third of the entire foreign-born population. Mexicans, the largest single group, now comprise 27% of all those born outside the USA. Mexicans also account for the largest proportion of the illegal immigrant population by far, with El Salvador and Guatemala running a distant second and third. Hispanics now comprise the largest ethnic minority group in the US, at 12.5%.

## Refugees

In common with many other MEDCs, the USA has a refugee resettlement programme. The ceiling 'height' varies.
- In 1999 admissions were increased from 78 000 to 91 000 to accommodate Kosovan refugees.
- The ceiling for 2002 was set at 70 000.
- There is a 1994 agreement with Cuba that allows 20 000 Cuban immigrants through legal admissions channels.

## Chapter 6 Population movements

The attacks of 9/11, in 2001, altered public attitudes, and in turn altered government perception of immigration. In broad terms, the 'war on terrorism' has encouraged opponents of immigration to step up their campaign to restrict entry into the USA.

Increasingly broad measures to prevent or intercept terrorist activities passed since 2001 have been a source of concern for some immigrant and civil liberties groups. In the coming decade, the relationship between the threat of terror, public fears and the civil liberties that are challenged by more restrictive legislation will be closely examined. Some observers believe the threat of terrorism is being overplayed in order to strengthen the hand of those who oppose further immigration.

The economic prospects of the USA and many other MEDCs with more or less full employment and a shortage of skills will remain strongly tied to immigration forces. In the USA, immigrants comprise 13.5% of the population aged 16 and over, and account for nearly the same percentage of the labour force. During the second half of the 1990s, immigrants of all legal statuses contributed a net 35% to total growth in numbers, while the number of foreign-born workers increased by nearly 25% compared with just 5% for all native-born workers.

# AS Unit 3

## Personal enquiry/ Applied geographical skills

# Chapter 7

# Personal enquiry

## The sequence of events

1. Candidate chooses and researches the viability of an appropriate topic.
2. Candidate completes a GeogA1 (coursework proposal) form, which is sent to the moderator allocated to the centre.
3. Candidate responds to any recommendations or requests on the return of the GeogA1, if necessary.
4. Data are collected and the coursework written up. The deadline for this is set by the centre.
5. The work is marked internally within the centre. All the work within the centre is moderated by the teaching staff under the supervision of the head of department.
6. A sample of work is sent to the moderator for checking. The deadline for this is set by Edexcel.

## Timing

The timing will, to a certain extent, be determined by the centre and is reflected in Table 7.1. Note that there is no opportunity to submit coursework for assessment in the January of either year of the course.

*Table 7.1*

|  | Submission at end of first year | Submission at end of second year |
|---|---|---|
| Deadline appropriate for | ■ 1-year AS students<br>■ 1-year A-level students | ■ 2-year AS students<br>■ 2-year A-level students |
| GeogA1 to Edexcel moderator by | 1 February | 1 February in the 2nd year of the course |
| Feedback/approval from the moderator | Within 2 weeks of receipt of GeogA1 | Within 2 weeks of receipt of GeogA1 |
| Realistic research and write-up period | Mid-February to mid-April | Early summer of Year 1 to mid-April of Year 2 |
| Date by which marks must be with the moderator | 1 May of Year 1 | 1 May of Year 2 |

## Suitable topics

Topics can be chosen from any part of the specification. Successful enquiries tend to be those with a clear focus and with a clearly stated aim. Enquiries with a specific question or clear hypothesis to test tend to allow easier analysis, interpretation and conclusion.

Once a topic has been selected, a question can be posed that forms the basis for one or more hypotheses to be tested. For example:
- Topic: soils
- Question: how do soil characteristics change downslope?
- Hypothesis: soil depth increases with distance downslope.

A list of possible topics is given below.

## Settlements: urban
- Land-use changes from the PVLI towards the edge of the CBD
- Land values changes with distance from the PVLI
- Building height changes with distance from the PVLI
- Clustering of shop types
- Building age changes with distance from the CBD
- The application of land-use models to the urban structure of a town
- Differences in traffic flow, pedestrian flow and land use between the CBD and an out-of-town retail centre
- Differences in spheres of influence of different towns
- Variation in shopping habits between centres of different sizes
- Levels of crime related to social, environmental and economic characteristics of different areas
- Park-and-ride schemes and their effect on a city
- Park-and-ride buses and their uses
- The impact of supermarkets on local retailers

## Settlements: rural
- The extent to which a settlement is a commuter village
- How a village's development has changed its population structure and amenities
- Housing style and density changes as a village has developed
- The economic factors affecting agricultural land use
- Variations in the intensity of farming with distance from the farm
- Agricultural land-use changes as a result of government policy
- The impact of recreational activities on a rural area
- Visitor patterns in a country park

## Weathering
Note that a major problem here is the very slow rate of processes.
- Differing rates of weathering on gravestones of different rock types
- Differing rates of weathering on buildings of different rock types

## Rivers
- Downstream changes in hydrological characteristics
- Relationships between hydrological characteristics
- Impact of human activity on river channel variables or water quality
- Changes in load with distance downstream
- Comparing Bradshaw's model to a specific river
- Comparing characteristics of streams of different order

## Chapter 7

*Personal enquiry*

### Coasts

- Vegetation changes across a sand dune complex or a salt marsh
- Impact of human activity on sand dunes or salt marshes
- Factors influencing rates of marine erosion
- Factors affecting rates of longshore drift
- Variations in characteristics of beach material along a beach or up a beach

### Climate

Note that only microclimatic studies are really suitable for enquiries in this specification.

- Relationship between microclimate and vegetation characteristics in a woodland
- Relationship between microclimate and building characteristics in an urban area
- Microclimate variations during the day
- Variations in microclimate in a valley
- Comparison of microclimates in urban and rural areas

### Glacial landscapes

Note that there are only limited possibilities here and accessibility is often a problem.

- Analysis of glacial sediment
- Analysis of fluvioglacial sediment

### Ecosystems

- Variations in vegetation across a transect in sand dunes, salt marsh, fresh water or recently colonised bare rock
- Soil profile changes downslope
- Relationships between human activity and soil characteristics
- Microclimatic differences between woodlands of different types
- Variations in soil characteristics on a slope or in an area

## How to complete the GeogA1

- Complete *all* the details at the top of the form, including the candidate number and centre details.
- Each aim must be clearly linked to specific data sources and means of analysis.
- It is a good idea to separate each aim with its linked data and techniques from the next one by a horizontal line across the table.

### Title

- Is the title defined as a question, problem or issue with a clear focus?
- Is the title unique at the centre? If not, are there clear distinctions in the plan that will ensure individuality?
- Does the topic fall within the specification?
- Is the proposal manageable within the word limit?

*Personal enquiry/Applied geographical skills*  **AS Unit 3**

## Context statement
- You need to give about three sentences that explain the detail and give background information to the title. Is it testing reality against a theory? Why have particular sites been selected?
- Is there a clear reference to the location where data will be gathered? A map locating the sites may be included. (The location must be appropriate in terms of accessibility, safety, time spent in the area, etc.)

## Aims, key questions or hypotheses
- Is there potential for useful data collection and analysis within each aim?
- Are the aims listed in a logical and sequential way? This will help in the organisation of the write-up.

## Plan of data collection and analysis
- Is this clearly and directly linked to specific aims, key questions or hypotheses?
- How many sites will be used for data collection? Are there enough sites for statistical purposes? Are the sites suitable for the purpose?
- Are 'hard' primary data being collected (not just value-laden views)?
- Have realistic sources of data and equipment been identified?
- What are the sample sizes? Are they realistic for the task? Is the sampling technique clearly stated?
- Are some of the data collected by a group? If so, which elements are group data and which are individual?
- Remember that some data collection is quantitative and some is qualitative.
- Examples of quantitative data
    - pedestrian flows
    - soil pH
    - traffic counts
    - land-use surveys
    - river velocity
    - temperature/wind speed
- Examples of qualitative data
    - field sketches
    - environmental quality index
    - questionnaires seeking views and opinions
- Are secondary sources specifically detailed? Terms such as 'library' or 'textbook/notes' suggest poor planning and little thought.
- Secondary data sources include:
    - media, including newspapers, television and radio and geographical sources such as *Geography Review* magazine
    - organisations and action groups, such as the Environment Agency, the Geographical Association and English Heritage
    - charities, such as Oxfam
    - environmental pressure groups, such as Friends of the Earth and WWF
    - the Meteorological Office
    - office for National Statistics
    - Ordnance Survey
    - estate agents
    - Thompson's Directory
    - data collected by students during coursework in previous years

## Chapter 7 *Personal enquiry*

- Is a sufficient variety of methods proposed to provide for a range of presentation and analytical skills? Does the proposal suggest that adequate data are likely to be collected for this task?
- Have appropriate techniques been chosen and linked to specific objectives and data sources, rather than just being listed? A simple list is not good enough and suggests poor planning.
- Remember that annotated photos and diagrams can save a lot of words. Field sketches are rarely used, yet they are incredibly useful.

## Checking work

### General

- Present the enquiry in a standard, soft-backed folder.
- Provide a title and a properly structured contents page, number all the pages and check all cross referencing.
- Check grammar and spelling.
- Differentiate between group work and individual work; make sure that anything done *individually* is clearly flagged.

### Investigation design and planning

- Locate and describe (possibly with annotated photos) all the data collection sites, and justify their selection.
- Provide aims and objectives as a series of bulleted, or numbered, points and check that they are followed up in a logical sequence and mentioned in the conclusion.

### Data collection

- Decide on and state the sample size. Ensure that the sampling technique is discussed, evaluated and justified.
- Clearly outline and justify the equipment used.
- Say whether questionnaires were piloted and whether they were altered in the light of experience — if they were, say why.

### Data presentation

- Be selective about maps. Create your own rather than relying on a photocopy or something from a piece of routefinder software. Maps must have a scale, title and orientation. The best maps are used and referred to in the text and may be annotated with further information — this also helps to reduce the word count.
- Tabulate and summarise data to avoid lengthy text and to make the data intelligible.
- Organise the presentation of your data. Combine presentation with summary statistics for the site to help give a general feel for the data. Graphs which need to be compared should be on the same page, or on overlays. Make sure that your written analysis is close to the presented material, rather than further on in another section. A wide range of methods may be used; many of these are dealt with in Chapter 8.

- Remember that simple calculations like means, medians, modes and inter-quartile ranges are perfectly valid statistical techniques and their use can also help description and analysis.
- Remember to use scattergraphs and to use significance in correlation exercises.
- Use clear, well-labelled and properly scaled graphs. Annotation of graphs, with commentary and analysis, is a good technique and makes the study easier to read.
- Use *annotated* field sketches and photographs as they save a lot of words.
- Use a variety of methods and consign repetitive techniques to the appendix, with a clear reference to them.

## Analysis and interpretation

- Describe the results obtained and how the collected data are presented.
- Suggest reasons for the results that have been obtained.
- Apply the principles of established theories and concepts.
- Seek explanations of any anomalies, which are likely to be due to the influences of specific local factors.
- Consider to what extent the results are as expected or are surprising.

## Conclusion and evaluation

- Refer back to the original aims and objectives; to what extent have they been met?
- Ensure that the question or hypothesis posed has been answered or tested.
- Identify possible extensions to the work; could it be taken further?
- Could other data be collected?
- Could improvements be made to the way in which data were collected?
- Who might be interested in the results of the enquiry?

## Other points

- Do not go over the word limit; this will seriously affect the marks that can be awarded.
- Do not include large amounts of unprocessed raw data.
- Do not use a word processor without doing a spellcheck.
- Do not rely on a spellcheck to pick up mistakes. Read through your work and check it yourself.
- Do not rely on an appendix — it rarely has creditable material and is often totally divorced from the project without a useful cross-reference.
- Do not include unstructured background information. All information should be closely linked into the aims.
- Do not include all questionnaires/rough work in an appendix.
- Do not include photos or diagrams without referring to them. What are they for? Why were they chosen? What do they represent? Annotations are *essential* for geographical photos.
- Do not give an unsupported list of 'what I could/should have done'. There should be cross linkage to data and some appreciation of the feasibility of suggestions.
- Do not use weak excuses such as 'if more time had been available…' or 'if the weather had been better…'.

# Chapter 8

# Applied geographical skills: Section A

## Introduction

In the Unit 3b examination, Section A is a compulsory question with a number of sub-sections. It will require the manipulation and organisation of a range of resource materials and the application of practical skills in an unfamiliar context.

The aim of the examination is to test knowledge, understanding and application of geographical skills. Questions are, therefore, designed to assess the candidates' ability to:

- know the main sampling methods used in geography and the importance of sampling to obtain information about a population; appreciate the meaning of sampling error
- understand the need for classification of data
- use a variety of data presentation methods and be able to interpret data presented in these ways
- sketch a best-fit line from an array of points on a scattergraph and identify anomalies
- compute measures of central tendency and dispersion (variation)
- apply simple statistical tests such as the Spearman rank correlation coefficient and the chi-squared test
- make geographical interpretations of a variety of resources relating to a question or hypothesis
- draw conclusions
- comment on the validity of such conclusions
- recognise opportunities for further investigation

A range of resource materials will be provided, and these will relate only to topics from Units 1 and 2 of the specification. The material will be based on data collected during real or imaginary fieldwork investigations and might relate to such issues as:

- river investigations: downstream changes in river channel variables, relationships between river channel variables, load analysis…
- coastal investigations: vegetation changes across sand dunes or salt marshes, analysis of beach material, influence of human activity…
- weathering investigations: rates of weathering on gravestones or buildings…
- urban area investigations: land-use patterns, clustering of services, pedestrian flows, shopping patterns, spheres of influence, functional hierarchy, environmental quality…

*Personal enquiry/Applied geographical skills* **AS Unit 3**

- rural area investigations: changing village character, population density, distribution and structure, patterns of commuting, nearest neighbour analysis, shopping patterns, spheres of influence…

## Resources

The data may be presented to you in a range of different ways. In addition to the graphical methods detailed below, you may also be provided with:
- maps: including Ordnance Survey 1:25 000 and 1:50 000 scales, land-use maps, often to provide a locational context
- tabulated data: perhaps summarising results prior to their presentation
- extracts of text from books, magazines, reports or newspapers: perhaps providing background information
- photographs: taken at ground level and from the air
- field sketches: possibly annotated with detail and explanatory comments
- sketch maps: also possibly annotated with detail or explanatory comments
- sketch sections: cross-sections and long-sections

## Sampling

### Why sample?

Most 'populations', i.e. the data set from which information is being obtained, are too large to be measured and/or handled. A sample is a portion of the full population that should be representative of the characteristics of that population.

*Table 8.1 Types of sampling method*

| Type | Meaning | Method | Suitability | Weaknesses |
|---|---|---|---|---|
| Random | All items in the population have an equal chance of being selected in the sample | Use of random number tables, dice, coins, etc. | When the area or length being surveyed is fairly uniform, e.g. a residential area of a town | A subjective element is often included. For example, stopping people in the street 'at random' is seldom completely random |
| Systematic | Ensures even coverage of the area or length | Sample points at regular intervals along a length or equally spaced over an area | When the area or length has much variation and it is desirable to ensure that all parts contribute to the sample, e.g. a length of river | There may be practical problems, e.g. access or ownership issues with points along a river |
| Stratified | To replicate the structure of the population in the sample | Identify the population characteristics and select points in the sample to produce a similar profile | When the population characteristics can be easily established, e.g. the age/sex profile of a town's population | A lot of prior knowledge is needed and information needs to be gained about potential sample points before they are included, e.g. people need to be asked their age before they can be selected |

255

# Chapter 8

*Applied geographical skills: Section A*

## What size of sample?

The sample should be large enough that it is representative of the total population but not so large that unnecessary amounts of time and energy are spent collecting and handling the data.

Another influencing factor is what you are intending to do with the results. If you are planning to use the Spearman rank correlation coefficient, for example, then a sample of ten is generally thought to be the minimum number for the test to be reliable.

## What type of sample?

There are three types of sample generally used by geographers working at this level: random, systematic and stratified. Careful consideration should be given to the choice of sampling method to ensure that the most appropriate type is used and that its use can be justified. Even after careful selection, it should be appreciated that there are potential problems and weaknesses with the type of sampling chosen.

## Methods of data presentation

In Section A, data may be presented in a wide range of different ways. Candidates should be familiar with the range of different techniques that might be used and be prepared to add further data to a partially completed presentation. The suitability of each method should be recognised and advantages and disadvantages appreciated. Candidates should also be able to extract data from presentations, interpret the data shown in terms of patterns and trends, and draw conclusions from them.

*Figure 8.1 Plotting annual data*

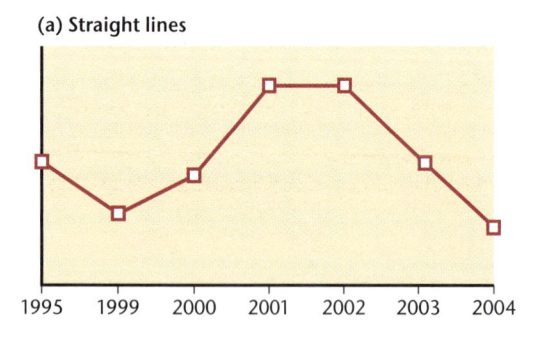

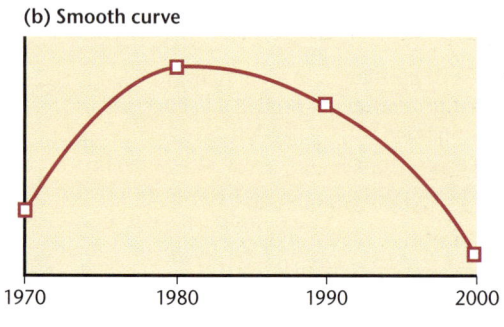

(a) Straight lines

(b) Smooth curve

### Line graphs

**Line graphs** are used to show absolute changes in a set of data. If the data record changes over time, the time period should be shown on the horizontal ($x$) axis. The dependent variable (i.e. the one that changes depending on the other) is plotted on the vertical ($y$) axis. Different lines may be shown on the same graph, in which case different colours or symbols (e.g. dots, dashes) are likely to be used. If two different variables are plotted on the vertical axis, one scale is usually provided on each side of the graph. If the horizontal axis shows continuous data, then the graph should have a smoothly drawn curve that passes through each point. If the axis has discrete data, each point should be joined up to the next with a straight line. If you are not sure whether the data are continuous or not, a good way to tell is to see if it makes sense to interpolate values from the graph between those plotted. For instance, if the graph is showing annual data, it does not make sense to try and estimate a value between 1998 and 1999, as there is no year between the two (Figure 8.1a). In that case, the points should be joined up with straight lines. However, if the data were only given for 1980 and 1990, it would be reasonable to estimate values for the years in between — so a curve should be used (Figure 8.1b).

## Cumulative line graphs

Data are converted into percentages for this method and added cumulatively until 100% is reached. The points are often plotted in rank order, with the largest plotted first.

One common application of this type of graph is to form a **Lorenz curve**. Lorenz curves are used to illustrate the extent to which a geographical distribution is even or concentrated.

- The category with the highest percentage is plotted first.
- The second plot consists of adding together the percentages for the highest and the second highest categories.
- The third plot consists of adding together the first three and so on.
- When all the categories have been plotted, 100% will have been reached.

In Figure 8.2, the data for energy sources in Italy have been plotted cumulatively. The bold line indicates a situation in which the various energy sources would each be making an equal contribution to the total. The figure clearly shows the dominance of oil and the relative unimportance of nuclear and other sources (which include wind, solar and geothermal).

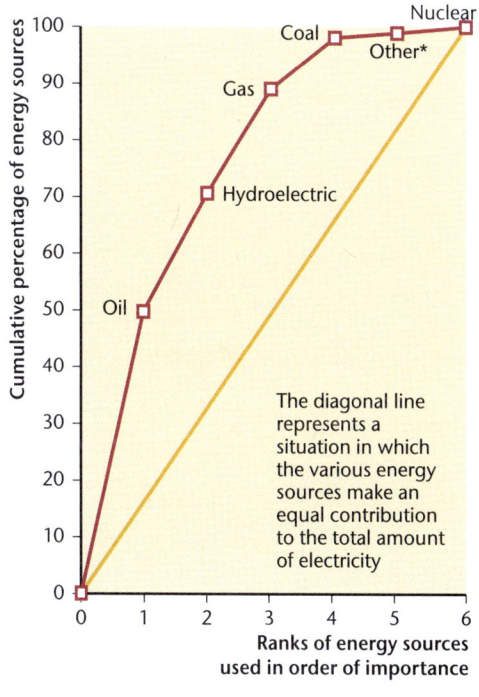

The diagonal line represents a situation in which the various energy sources make an equal contribution to the total amount of electricity

\* Other energy sources include wind power, solar power and energy derived from heated rocks in the Earth's crust

*Figure 8.2 A Lorenz curve showing energy sources used to generate electricity in Italy, 1995*

## Logarithmic graphs

**Logarithmic graphs** are drawn as line graphs, but the scales are non-linear. Each division on the scale represents a 10-fold increase in the range of values. Lines may therefore be numbered 1, 10, 100, 1000 and so on. The special graph paper used may be fully logarithmic, in which both *x*- and *y*-axes have logarithmic scales, or semi-logarithmic, with one axis linear and one logarithmic.

Logarithmic graphs can be useful for showing rates of change — the steeper the graph the faster the rate of change — and for plotting data with a wide range of values, such as populations in very different types of settlement, from village to city. These graphs cannot be used to show negative values and the origin point does not have a value of zero. It can be challenging to plot points accurately, but these graphs are very useful for recognising settlement hierarchies using the rank–size rule.

## Chapter 8

*Applied geographical skills: Section A*

*Figure 8.3 A semi-logarithmic graph showing primary fuel production*

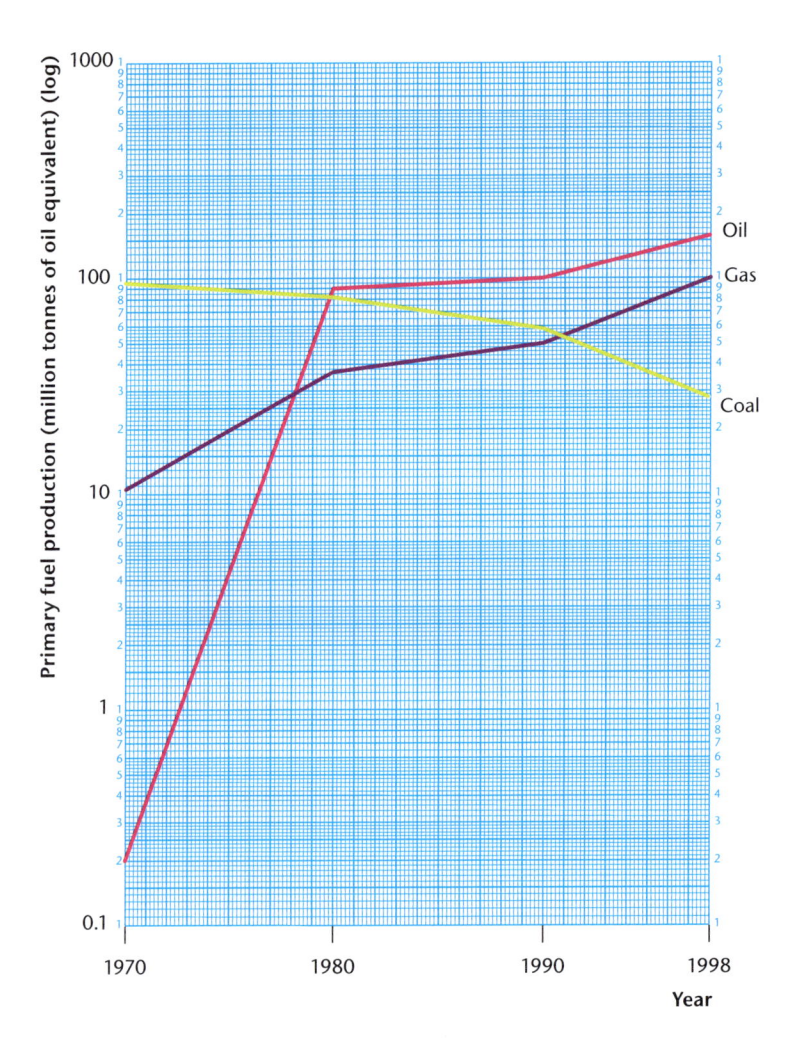

### Bar charts and histograms

A **bar chart** has vertical columns rising from a horizontal base line. The height of the column is proportional to the value it represents. A scale on the vertical axis enables the heights to be compared. The vertical scale can be absolute or shown as a percentage.

Bar charts are particularly useful when the horizontal scale is a location or type rather than a numerical value such as time. Data for a number of different countries or settlements can be shown in this way. They can be shown in rank order or perhaps alphabetically, but the key is that the order is not predetermined, as it is with time-based scales.

Bar charts are easy to construct and use, and they ensure that comparisons are easy to make. They can be used for data with both positive and negative values, such as rates of change, if a scale is drawn through zero on the *y*-axis.

**Histograms** are similar to bar graphs, but technically different. They are used to show the frequency distribution of data (i.e. how many of each item is in the

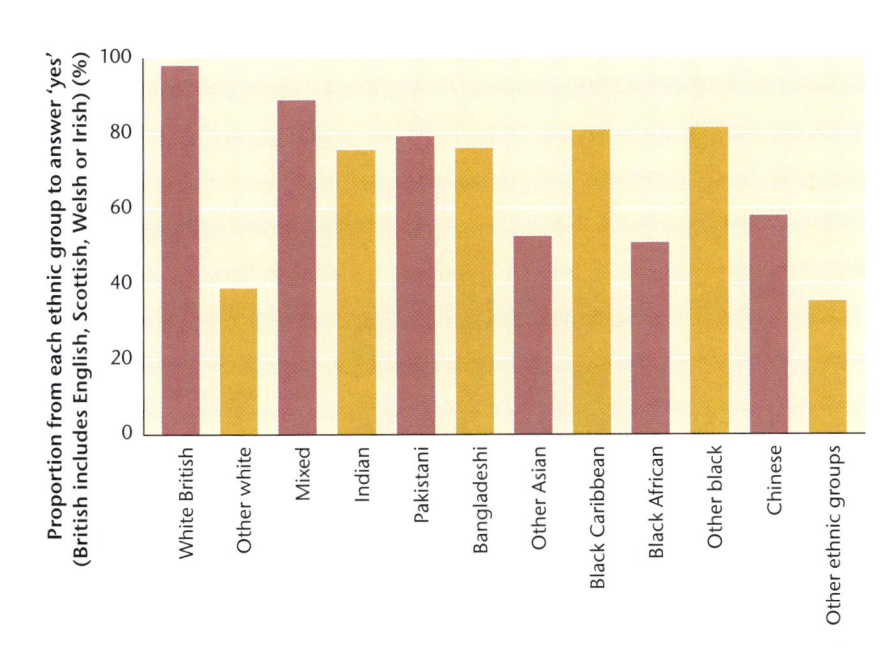

*Personal enquiry/Applied geographical skills*  **AS Unit 3**

*Figure 8.4 A sample bar chart showing responses to the question: 'Do you feel British?'*

data set) and are ideal for presenting data by groups or classes. As the size of the groups is not necessarily equal, it is the *area* within each bar that is proportional to the value being plotted rather than its height. However, if equal-sized groups are used, the height will, of course, be proportional to the value. Histograms can reveal important features of frequency distributions. The distribution is *normal* if the histogram is symmetrical (Figure 8.5a), but *skewed* if the histogram is asymmetrical (Figure 8.5b and c).

*Figure 8.5 Normal and skewed distributions shown by histograms*

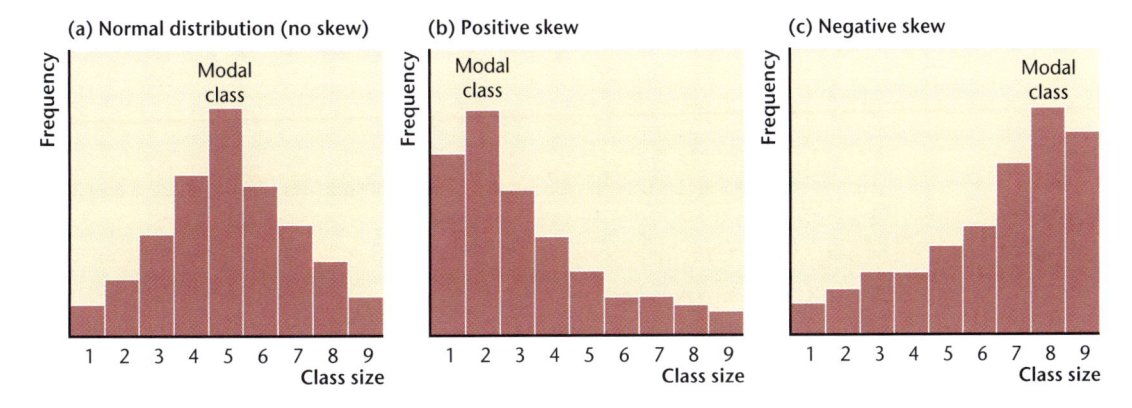

## Pie graphs and divided bar charts

A **pie graph** is a circle divided into segments according to the share of the total value represented by the segment. This is a useful and visual method to use, as relative sizes can be readily recognised. The lack of a scale may make extracting specific values difficult, but these may be written onto the segments. There is also a danger in having too many segments, especially if a lot of them are small. To

## Chapter 8

*Applied geographical skills: Section A*

calculate the angle of the sector, the sector value should be divided by the total value and multiplied by 360°. To calculate the value represented by a sector, the angle should be divided by 360° and multiplied by the total value.

*Figure 8.6 A pie graph showing the origin of migrants to Australia, 1999*

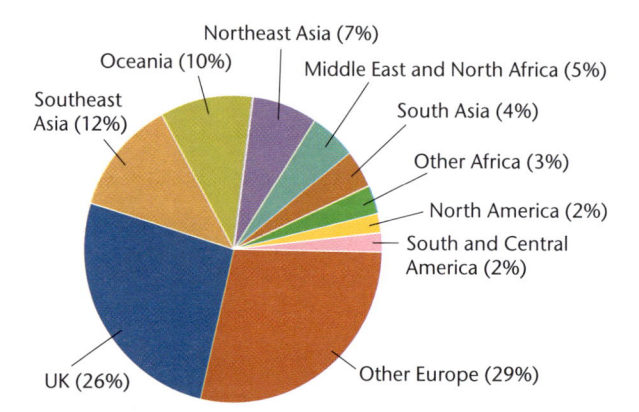

Northeast Asia (7%)
Oceania (10%)
Middle East and North Africa (5%)
Southeast Asia (12%)
South Asia (4%)
Other Africa (3%)
North America (2%)
South and Central America (2%)
UK (26%)
Other Europe (29%)

A series of pie graphs can be used to show changing proportions over time or to compare different situations and places. However, accurate comparison is again difficult because of the lack of a scale. In such cases it may be more effective to use a **divided bar chart**. Each bar is divided into sections proportional to the value of each component. By plotting these alongside each other, direct comparison is straightforward and the scale allows actual values to be read off easily and accurately.

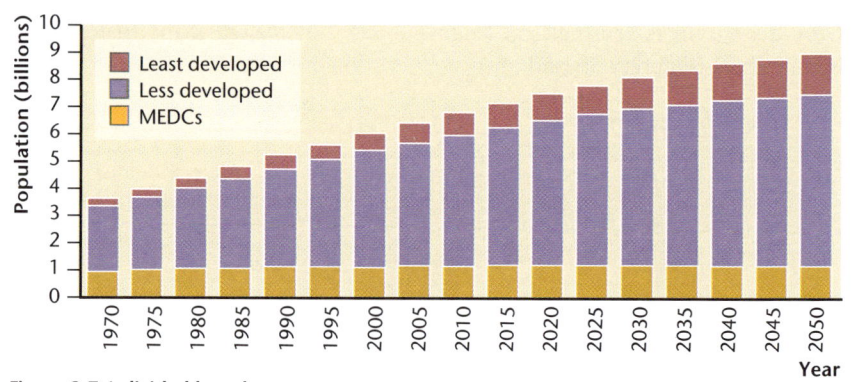

*Figure 8.7 A divided bar chart*

## Proportional symbols

Symbols can be drawn that are **proportional** in area or volume to the value they represent. This is a similar principle to that used in bar graphs, but shaped or pictorial symbols can be used.
- Circles, squares, cubes and spheres are commonly used shapes and these can, like pies and bars, be divided to reflect the relative size of the components that make up their total.
- Suitable symbols might include cars to show traffic flow, stick people for pedestrian flows or houses for house prices.

*Personal enquiry/Applied geographical skills* **AS Unit 3**

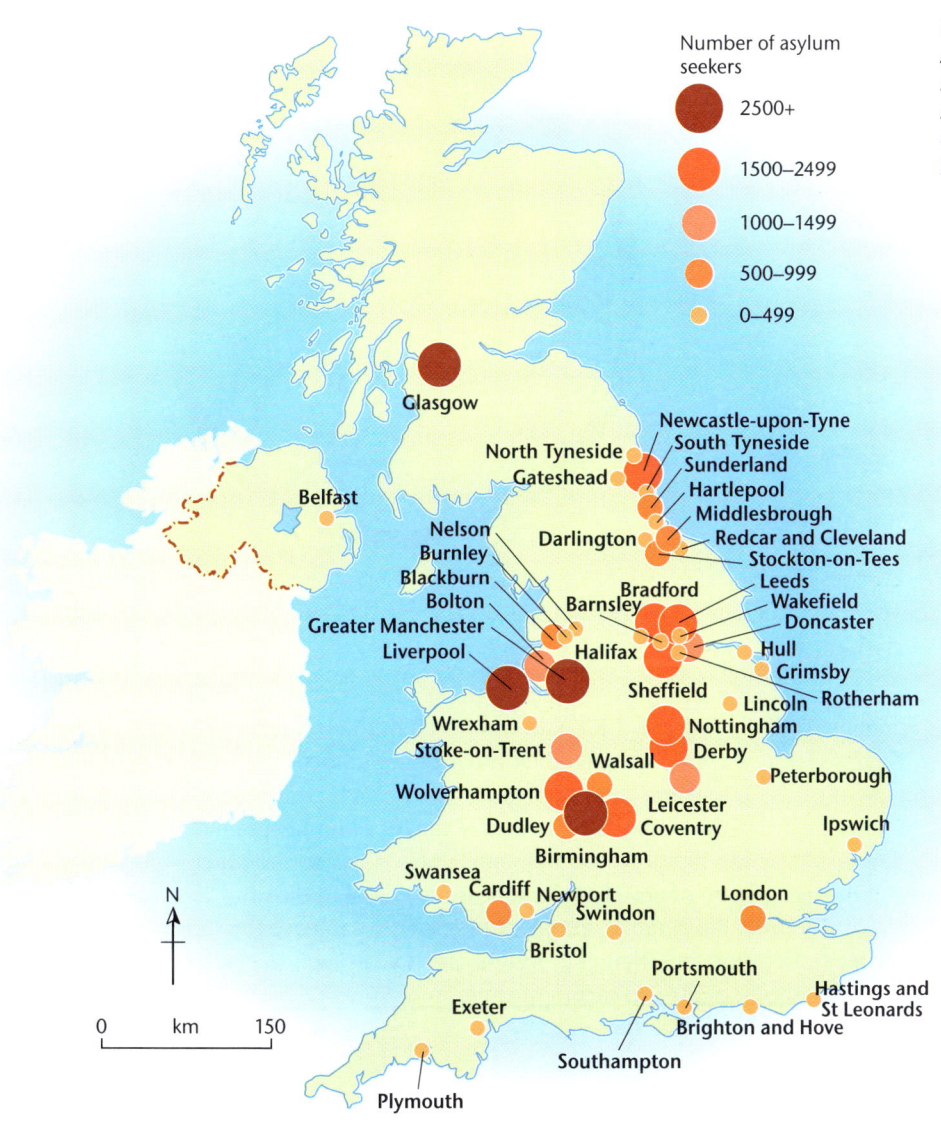

*Figure 8.8
A proportional symbol map showing location of asylum seekers in Britain*

Number of asylum seekers
- 2500+
- 1500–2499
- 1000–1499
- 500–999
- 0–499

The symbols can be drawn independently. Often they are placed on maps in order to show spatial patterns, which can be very useful. If possible, the symbols should not overlap and they should be centred on the location to which they refer. Choosing a suitable scale is therefore important.

## Flow lines

**Flow lines** are a specific type of proportional symbol. They use an arrow to represent the volume of a movement. The width of the arrow is proportional to the size of the movement and the arrowhead indicates the direction of movement. Flow lines are particularly useful for showing traffic and pedestrian flows. Routes can sometimes be drawn that link an origin and a destination, such as a shopping journey or a commuting journey. These are then known as **desire lines**.

# Chapter 8

*Applied geographical skills: Section A*

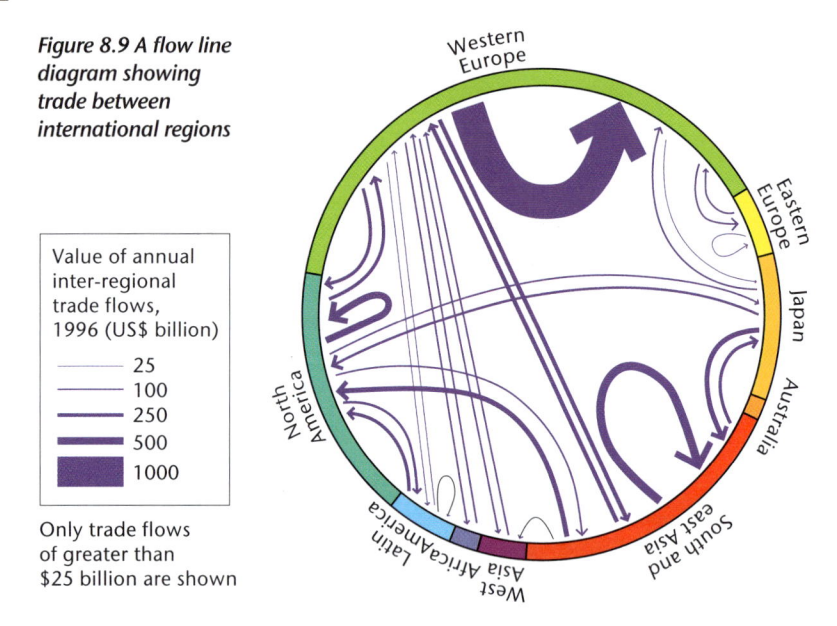

*Figure 8.9 A flow line diagram showing trade between international regions*

Value of annual inter-regional trade flows, 1996 (US$ billion)
- 25
- 100
- 250
- 500
- 1000

Only trade flows of greater than $25 billion are shown

## Triangular graphs

**Triangular graphs** are drawn on special graph paper in the form of an equilateral triangle. They are only used for data composed of three components that can be expressed as percentages, and so their use is rather limited. The advantage of this type of graph is that the varying proportions of each component can be seen, indicating the relative importance of each. Data for a number of places or different points in time can be shown on the same graph. This enables direct comparison and the recognition of any clustering, possibly enabling classification.

For example, a triangular graph can be used to analyse the occupations of people from different settlements. In this case the occupation type would be split into three components and each settlement would provide one entry on the graph (Figure 8.10).

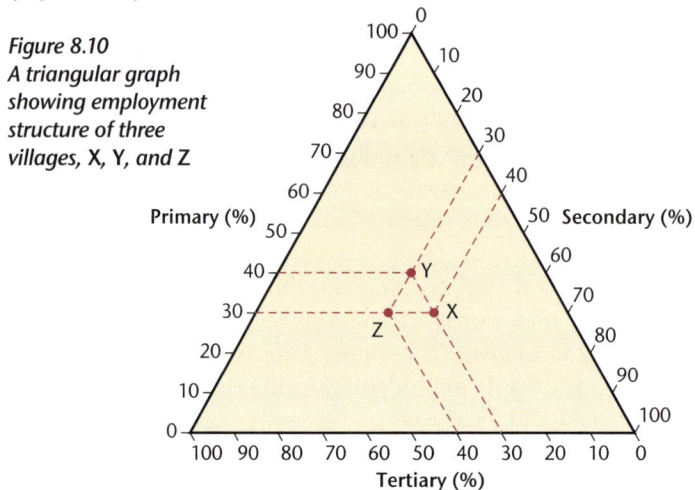

*Figure 8.10 A triangular graph showing employment structure of three villages, X, Y, and Z*

## Scattergraphs

Drawing a **scattergraph** can show the relationship between two sets of data. Although scattergraphs are a way of presenting data, their interpretation is really the first step in the analysis of such data. Scattergraphs should only be used when a relationship is thought to exist, and this might be the basis of a hypothesis that is being tested. As with line graphs, the dependent variable, if one can be established, should be plotted on the $y$-axis and the independent variable on the $x$-axis.

Once the graph is plotted, a pattern may be visible that suggests that a relationship does exist. To highlight this, a best-fit line can be drawn on the graph. This should be drawn 'by eye' rather than calculated and it should pass through the spread of the points, minimising the total distance of points from the line and with roughly equal deviations both above and below the line. The graph will then indicate some important points about the relationship between the two data sets.

- If the line has a positive gradient, i.e. it runs from bottom left to top right, there is a positive relationship (Figure 8.11a).
- If the line has a negative gradient, i.e. it runs from top left to bottom right, there is a negative relationship (Figure 8.11b).
- If you cannot establish a position for a best-fit line because the points are too widely or unevenly scattered, then there is no relationship (Figure 8.11c).
- The closer the points are to the best-fit line, the stronger the relationship.
- Points lying some distance from the best-fit line are anomalies.

Note, however, that any such conclusions will be fairly tentative at this stage. Further investigation, perhaps using the Spearman rank correlation coefficient, needs to be undertaken before firmer conclusions can be drawn. Plotting the scattergraph is a good starting point and can help in the decision about whether or not it is worth proceeding any further with data analysis.

Best-fit lines can be curves as well as straight lines, so it can be difficult to establish whether a relationship is non-linear or whether it is linear but with an anomaly or two. It should be remembered, however, that the Spearman rank correlation coefficient is a measure of linear relationships and does not recognise non-linear ones.

## Choropleth maps

On **choropleth maps** the density of shading used in different areas is proportional to the value of a variable for that area. Differences in shading therefore indicate differences between areas. Suitable data include, for example, population density or crime statistics.

In order to shade the map, the data need to be grouped into classes. Graduated shading is used, with the highest values represented by the heaviest or darkest shading and the lowest by the lightest. Choropleth

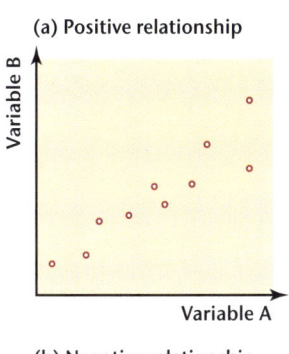

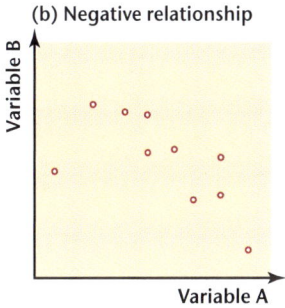

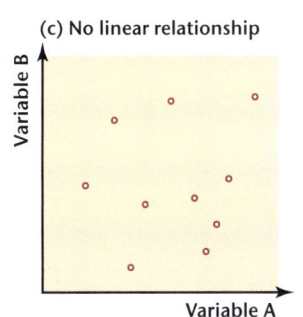

*Figure 8.11 Examples of the possible conclusions that can be drawn from a scattergraph*

### Examiner's tip

- Examination questions will often require the justification of the choice of a technique of data presentation or an awareness of the disadvantages of a particular method.

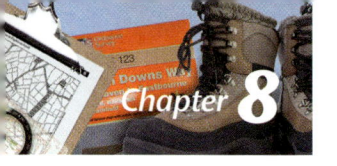

## Chapter 8 Applied geographical skills: Section A

*Figure 8.12*
*A choropleth map showing population density according to ward in a city*

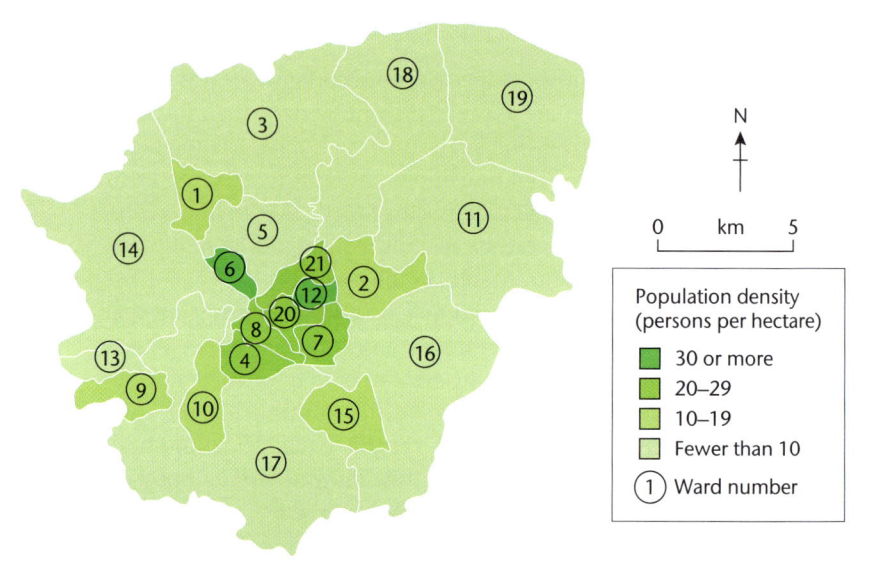

maps may be shaded in black and white or by using different shades of the same colour. Care should be taken over the choice of class sizes so that not too many different classes are used and the differences are visible.

Maps of this type can give a clear visual impression of the pattern of variations across an area. However, they do have limitations. They suggest that everywhere within a shaded area has the same value and this may be far from the case. They also suggest abrupt changes at boundaries, whereas in reality the changes are likely to be more gradual.

### Isopleth maps

The prefix 'iso' means equal; **isopleth maps** are constructed by drawing lines to join up places that have equal values for the variable under consideration. **Contour maps** are, therefore, a type of isopleth map, as are **isotherm maps** (temperature) and **isobar maps** (pressure) showing weather data.

Drawing these maps allows patterns to be seen across a study area. In fieldwork investigations they can be used for pedestrian flows, traffic flows, land/property values, environmental quality indices, etc. When you have a series of sample points across an area, drawing the isolines can make the pattern much clearer and easier to see. If the lines pass through the points, they are simple to construct. However, when the points fall between the chosen isoline values, care must be taken to draw the line the correct side of the point. The analogy with contour lines can be helpful to keep in mind when you are determining the pattern of high values (hills) and low values (valleys).

*Figure 8.13*
*A contour map — an example of an isopleth map*

### Summary

Table 8.2 provides a useful summary of the data-presentation techniques considered in this section.

264

*Table 8.2 Comparing methods of data presentation*

| Technique | Use | Advantages | Disadvantages |
|---|---|---|---|
| Line graph | Continuous data; often change over time/distance | Easy to see trends | Can lead to unrealistic interpretations between points |
| Cumulative line graph | To show effect of adding to previous frequencies | Normal distributions should have S-shaped curve | Distributions are often skewed and so the curve is unbalanced |
| Logarithmic graph | Non-linear scales used on special graph paper; can show trends over time and relationships | Allows a wide range of values to be plotted; allows accuracy at the lower end of the range | Limited accuracy at the upper end of the range |
| Bar chart | One quantitative scale, frequency of categories | Enables easy comparison (blocks should be separated) | – |
| Histogram | Two quantitative scales, continuous horizontal scale | Easy to see variations as continuous (blocks should be connected together) | Area of block gives the frequency, but not a problem if equal intervals are used |
| Pie graph | To show relative contributions of components to a total | Easy to see relative importance, more suitable than divided bars for more than four or five categories | Actual values hard to determine |
| Divided bar chart | To show a number of components of a category | Easy to see changes over time | May be harder to see changes in components themselves, so may be better to use percentages rather than actual values |
| Scattergraph | To show relationship between two variables | Pattern may be easy to see | Independent variable should be on *x*-axis, but often isn't |
| Best-fit line | Line drawn by eye that passes as near as possible through all points on a scattergraph | Highlights anomalies, predictions of unknown values possible | Cause–effect sometimes wrongly inferred; straight line or curve? |
| Triangular graph | To show three variables that are components of a total | Shows dominance of a variable; values should always total 100% so easy to check | Can be hard to see which axis to read from |
| Flow lines | To show volumes of flow in a given direction by arrows of proportional width or length | Easy to see relative magnitude and direction; good when used on maps | Only really suitable for one type of information, scales can be difficult and exact values hard to establish |
| Proportional symbols | Size of symbol proportional to total, e.g. population of settlements on a map | Very visual | Can be hard to construct accurately, especially with large amounts of data |
| Choropleth map | Shaded map (colours/density shading) to show variations in data across an area | Can see spatial patterns, easy to construct | Boundaries may be artificial, gradual changes not shown, care must be taken with categories chosen |
| Isopleth map | Map with lines joining points of equal value | Not affected by boundaries, individual data points can be left on as well | Lots of data needed, can be hard to construct if patterns are not obvious, assumes gradual change between values |

## Chapter 8

*Applied geographical skills: Section A*

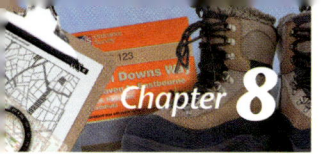

# Methods of statistical analysis

The use of statistical analysis is a common feature of geographical investigations. Statistical analysis can be used to test subjective conclusions about the results of an investigation or to test hypotheses. It is important to know why particular techniques are appropriate and what they are designed to reveal about a data set. Statistical analysis can be especially useful in determining the significance of a result or a conclusion.

All calculations should be performed accurately and it is essential that you take a calculator into the examination room for this unit. However, marks are also given for the interpretation of results, not just for their calculation.

## Averages

There are three different types of average that you are expected to know. All three are statistical measures of the central tendency of a data set.

The **mean** is the most commonly used measure. It is found by adding together all the values in a data set and then dividing the total by the number of values:

$$\text{mean } (\bar{x}) = \frac{\Sigma x}{n}$$

where $n$ is the number of values in the data set.

The **mode** is the most frequently occurring value in a data set. There may be more than one mode. If there are, two the data set is bi-modal; if there are more than two, it is multi-modal.

The **median** is the middle value in the data set once all the values have been arranged in rank order. If the data set has an odd number of values, there will be a value which is exactly in the middle of the set with an equal number of values above and below. If there is an even number of values in the data set, the median will be the mid-point between the two values that occupy positions either side of the middle.

In all cases, the median value is located in the $(n + 1)/2$-th position in the ranked data set. For example:
- if there are 11 values in the data set, the median will be the one in the $(11 + 1)/2$-th or 6th position
- if there are 12 values in the data set, the median will be the $(12 + 1)/2$-th or 6.5th position, half-way between the 6th and 7th values

It is possible that the mean, median and mode for a data set will all be the same. In this case, the distribution is described as *normal*. However, this is rather unlikely with sets of geographical data collected from the real world. It is more likely that the three values will be different and the distribution *skewed* (see Figure 8.5).

## Measures of dispersion (variation)

**Dispersion** is the spread or variability in a data set around the point of central tendency. There are two measures of dispersion that you need to know about.

The **range** is the simpler of the two measures and is found by subtracting the lowest value in the data set from the highest. The larger the range, the greater is the spread of values in the data set.

*Personal enquiry/Applied geographical skills*  **AS Unit 3**

The **inter-quartile range** is a more sophisticated measure and is the difference between the upper quartile and the lower quartile.
- The upper quartile (UQ) is the $(n + 1)/4$-th term in the ranked data.
- The lower quartile (LQ) is the $3(n + 1)/4$-th term.
- The inter-quartile range (IQR) therefore = UQ − LQ

This measure of dispersion has the advantage of not being influenced by extreme values in the data set as only the middle 50% of the values are involved in its calculation.

### Example
A group of students measured the diameter of a sample of 20 pebbles on a beach.

The results (mm) were:
15, 39, 31, 28, 10, 21, 19, 28, 33, 8, 40, 26, 36, 24, 17, 38, 20, 14, 35, 23

They put the data in rank order:
8, 10, 14, 15, 17, 19, 20, 21, 23, 24, 26, 28, 28, 31, 33, 35, 36, 38, 39, 40
mean = (8 + 10 + 14 + 15 + 17 + 19 + 20 + 21 + 23 + 24 + 26 + 28 + 28 + 31 + 33 + 35 + 36 + 38 + 39 + 40)/20 = 525/20 = 26.25 mm
mode = 28 mm (this value appears twice; all other values appear once)

The **median** is found in the $(20 + 1)/2$-th position: 21/2 = 10.5, so the median is found at the midpoint between 24 (the 10th value) and 26 (the 11th value) = 25 mm.
range = 40 − 8 = 32 mm
inter-quartile range = $3(20 + 1)/4$-th value − $(20 + 1)/4$-th value
= 15.75th value − 5.25th value = 34.5 − 17.5
= 17 mm

> **Examiner's tip**
> - Always take a calculator into the examination room. Even simple tasks can be carried out quickly and reliably with a calculator.

## Summary of measures of averages and dispersion (variation)

Table 8.3 summarises the techniques considered in this section and points out their advantages and disadvantages.

*Table 8.3 Comparing various methods of statistical analysis*

| Technique | How to use | Advantage | Disadvantage |
|---|---|---|---|
| Mean | Add all values, divide by number of values | Simple to calculate as a measure of central tendency, good for large data sets | All values included, extreme values may have a large effect on small data sets |
| Median | Put all values in rank order, find the middle value | Easy to use if data already ranked or small data set, unaffected by extremes | Only based on a single value, data have to be ranked first |
| Mode | Value with highest frequency | Can be unreliable in small data sets | Data sets often have more than one mode |
| Range | Highest value minus lowest value | Simple measure of dispersion in the data set | Only based on two values, affected by extremes |
| Inter-quartile range | Range of the middle 50% of values in a data set | Unaffected by extremes | Data have to be ranked initially, only half the data set considered, rank positions are often not whole numbers |

# Chapter 8

*Applied geographical skills: Section A*

## Measuring clustering: nearest neighbour analysis

The nearest neighbour statistic ($R_n$) analyses the distribution of individual points in a pattern. It can be applied to the distribution of any items that can be plotted as point locations and so it is often used for analysing the distribution of shop types in town centres or settlement types in an area.

The basis of the statistic is the measurement of the distance between each point in the pattern and its nearest neighbour of the same type. Once all the measurements have been made, the mean distance between each pair of points should be calculated ($\bar{d}$). The nearest neighbour statistic can then be calculated:

$$R_n = 2\bar{d}\sqrt{N/A}$$

where $N$ is the number of points and $A$ is the area of study.

The value of the statistic can only lie between 0 and 2.15.

- 0 represents a pattern that is perfectly clustered. In this case, all of the points would be at exactly the same place with no spaces between them. This, of course, could not happen in reality and so it is the proximity of the statistic to 0 that indicates the degree of clustering: the closer to 0, the greater the degree of clustering.
- 2.15 indicates a pattern displaying perfect regularity; all points lie at the axes of equilateral triangles. All distances between the points are exactly the same. Again, this is most unlikely in reality and it is the proximity to 2.15 that indicates the degree of regularity in the spacing.
- 1.0 is taken as an indication of a random pattern.

Care should be taken when using this technique to ensure that the units for distance and area are equivalent, for example metres and square metres or kilometres and square kilometres. Accuracy in making the measurements and calculations is also important if the result is to be reliable.

A further problem, often experienced in studying settlement patterns, is in the delimitation of the study area. It can be useful to establish a buffer zone around the study area. If the nearest neighbour to a point within the study area falls within the buffer zone, then the distance should be measured to this point. Points in the buffer zone, however, are not themselves included in the survey. In other words, distances can be measured to points in the buffer zone, but not from them.

### Example

Data for the distribution of post offices in Salisbury, Wiltshire are given in Table 8.4.

$$\text{mean distance} = \frac{\text{total distance}}{\text{number of points}}$$

$$= \frac{10.5}{9}$$

$$= 1.17$$

*Table 8.4 Nearest neighbour data for post offices in Salisbury*

| Post office number | Number of nearest neighbour | Distance between points (km) |
|---|---|---|
| 1 | 3 | 1.1 |
| 2 | 1 | 1.1 |
| 3 | 3 | 0.4 |
| 4 | 7 | 1.3 |
| 5 | 2 | 1.0 |
| 6 | 4 | 1.2 |
| 7 | 9 | 0.4 |
| 8 | 8 | 1.9 |
| 9 | 7 | 2.1 |
| | Total distance | 10.5 |

area = 36 km²

$$R_n = 2 \times 1.17 \sqrt{\frac{9}{36}}$$
$$= 1.17$$

The distribution is, therefore, almost random.
There are some points of caution to be aware of when using this technique.
- It does not distinguish between clustering around a single point and clustering around several points.
- Values indicating a random pattern may hide patterns based on unmapped factors, for example settlements located at springs.
- The outcome may be influenced by the size of the area being studied. A pattern may appear quite clustered within the context of the immediate locality, but much more widely dispersed when looked at in terms of a larger area.

It is, therefore, best to use this statistic to make comparisons between patterns, for example different types of shop in the same area.

## Predicting interaction between places: gravity models and breakpoint analysis

Reilly's law of retail gravitation suggests that the amount of movement of people between two settlements is influenced by the population of each settlement and the distance between them.

The law is used to calculate the breakpoint between two settlements; this fixes a notional distance beyond which people would be expected to travel to the other settlement to use its functions and services.

$$d_{jk} = \frac{d_{ij}}{1 + \sqrt{\frac{P_i}{P_j}}}$$

where $d_{jk}$ = the distance of the breakpoint from town j
$d_{ij}$ = distance between towns i and j
$P_i$ = population of town i
$P_j$ = population of town j

This, in effect, is a measure of the sphere of influence of a town. The larger the population of one town compared to the other, the nearer to the smaller town the breakpoint will be. This suggests that people between the two towns are more likely to travel further to the larger one than to the smaller one.

### Example
Grimsby has a population of 131 000 and Lincoln has a population of 75 000. The distance between them is 71 km. Calculate the distance of the breakpoint from Lincoln:

# Chapter 8

*Applied geographical skills: Section A*

$$d = \frac{71}{1+\sqrt{\frac{131\,000}{75\,000}}}$$

$$= \frac{71}{1+1.32}$$

$$= 30.58 \text{ km}$$

The breakpoint is 30.6 km from Lincoln (the smaller city) and 40.4 km from Grimsby.

Note that there are weaknesses in this model, mainly due to the fact that people do not always behave rationally and that other factors apart from distance and size have an influence:
- car parking may be cheaper and better in one town
- public transport links may be better to one town
- one town may be cleaner and safer
- one town may advertise better
- specialist shops may exist in one town

## Measuring correlation: the Spearman rank correlation coefficient

Two sets of numerical data can be compared to each other in order to see if there is a relationship between them. It should be appreciated, however, that even if a relationship is shown to exist, this does not prove that there is a causal link. In other words, it does not show that one variable is *causing* the other to change, only that as one variable changes so does the other.

Initially, in order to try and identify whether a relationship exists or not, a scattergraph can be drawn (see page 263). This will show whether a relationship is likely to exist and, if so, the Spearman rank correlation coefficient can then be calculated in order to measure the strength of the relationship. Once calculated, the coefficient can be tested to establish whether it is statistically significant or not. To be reliable, the test should only be used when there are at least 10 pairs of data, although too many can make the calculation complex and prone to error.

The method of calculation is as follows:
- rank one set of data from highest (rank 1) to lowest ($R_1$)
- rank the other set of data ($R_2$)
- if ranks are tied, the average values of the positions that they occupy should be given: for example, if two values are sharing positions 3 and 4 because they are equal, rank them each as 3.5
- calculate the difference in the ranks $R_1 - R_2$: this gives $d$
- square each difference ($d^2$)
- add up all the squared differences ($\Sigma d^2$)
- substitute $\Sigma d^2$ and $n$ (number of pairs of data) into the formula:

$$r_s = 1 - \frac{6\Sigma d^2}{n^3 - n}$$

*Personal enquiry/Applied geographical skills* **AS Unit 3**

If the calculation produces a positive value, then there is a positive (or direct) relationship between the two sets of data. In other words, if one variable increases, so does the other. However, if the result is a negative number, the relationship is negative (or inverse). If one variable increases, the other decreases.

The statistical significance of the result should then be tested. When comparing sets of data, there is always a possibility that the relationship has occurred by chance; the figures may coincidentally follow a similar or inverse pattern of change. The result should be compared to a table of critical values. If the coefficient equals or exceeds the critical value, the result can then be deemed significant. Two levels of significance are used. At the 0.05 (5%) level, there will be a 5% possibility that the result has occurred by chance, but it can be concluded with 95% confidence that there is a relationship between the data sets. At the 0.01 (1%) level, there will be a 1% possibility that the result has occurred by chance, but it can be concluded with 99% confidence that there is a relationship between the data sets. If the result is below the 0.05 level, then the relationship should be interpreted as not significant, even though a relationship exists. The greater the number of pairs of data being compared, the more reliable the evidence will be. This means that for larger $n$ values, slightly lower values of the coefficient can be accepted as being significant.

Table 8.5 shows some critical values.

| $n$ | 0.05 (5%) significance level | 0.01 (1%) significance level |
|---|---|---|
| 10 | ±0.564 | ±0.746 |
| 12 | 0.506 | 0.712 |
| 14 | 0.456 | 0.645 |
| 16 | 0.425 | 0.601 |
| 18 | 0.399 | 0.564 |
| 20 | 0.377 | 0.534 |
| 22 | 0.359 | 0.508 |
| 24 | 0.343 | 0.485 |
| 26 | 0.329 | 0.465 |
| 28 | 0.317 | 0.448 |
| 30 | 0.306 | 0.432 |

Table 8.5 Critical values for testing the significance of $r_s$ scores

As an example, a coefficient value of 0.6 from 12 pairs of data would be significant at the 0.05 level as it exceeds the critical value of 0.506, but not at the 0.01 level as it is less than the critical value of 0.712. A value of 0.4 would not be significant at the 0.05 level. (There is no need to say that it is not significant at the higher level as well, as this must be the case.)

Values of the Spearman rank correlation coefficient can only lie between −1 and +1.

- A value of −1 means that there is a perfect negative correlation, and so the ranks of the two sets of data would be the exact opposite of each other. On a scattergraph this would be shown by all the points being on or very near to a line of best-fit with a negative gradient.

## Chapter 8 — Applied geographical skills: Section A

- A value of +1 indicates a perfect positive correlation, with the ranks of the two sets of data being exactly the same. On a scattergraph this would be shown by all the points being on or very near to a line of best-fit with a positive gradient.
- A value of 0 means that there is no relationship. It should be remembered that this coefficient only measures linear relationships and so the two data sets may be related in a non-linear manner even if the coefficient is equal to or close to 0.

### Example

In September 2004, a group of students collected data on the River Bottle in Wiltshire. From the data collected at a sample of 10 systematically selected sites, they calculated the discharge and hydraulic radius at each site (Table 8.6).

The students developed a null hypothesis:

> There is no significant relationship between discharge and hydraulic radius on the River Bottle.

To test the hypothesis, they first calculated the Spearman rank correlation coefficient.

Table 8.6 Hydraulic radius and discharge values for 10 sites on the River Bottle

| Site | Hydraulic radius (A) | Rank A | Discharge (B) | Rank B | A − B (d) | d² |
|------|----------------------|--------|---------------|--------|-----------|----|
| 1 | 0.19 | 8 | 0.08 | 10 | −2 | 4 |
| 2 | 0.13 | 10 | 0.11 | 9 | 1 | 1 |
| 3 | 0.20 | 7 | 0.23 | 7 | 0 | 0 |
| 4 | 0.21 | 6 | 0.18 | 8 | −2 | 4 |
| 5 | 0.16 | 9 | 0.32 | 5 | 4 | 16 |
| 6 | 0.28 | 5 | 0.31 | 6 | −1 | 1 |
| 7 | 0.38 | 4 | 1.10 | 4 | 0 | 0 |
| 8 | 0.45 | 3 | 1.14 | 3 | 0 | 0 |
| 9 | 0.56 | 2 | 1.44 | 2 | 0 | 0 |
| 10 | 0.77 | 1 | 4.46 | 1 | 0 | 0 |

$\Sigma d^2 = 26$

$$r_s = 1 - \frac{6 \Sigma d^2}{n^3 - n}$$

$$= 1 - \frac{6 \times 26}{1000 - 10}$$

$$= 1 - \frac{156}{990}$$

$$= 1 - 0.158$$

$$= 0.842$$

The value can then be compared to the table of critical values for $n = 10$:

| n | 0.05 (5% level) | 0.01 (1% level) |
|---|-----------------|-----------------|
| 10 | 0.564 | 0.746 |

Table 8.7 Critical values

As the calculated value of 0.842 is greater than the critical value of 0.746 at the 0.01 level, there is a positive correlation which is significant at the 0.01 level. The null hypothesis is therefore rejected in favour of the alternative hypothesis, that there is a significant relationship between discharge and hydraulic radius on the River Bottle.

> *Examiner's tip*
>
> - In examination questions, the table of critical values will be provided. However, it may include figures for *n* values other than the *n* in the question. Ensure that you use the correct figures.

## Measuring differences between observed and expected data: the chi-squared test ($\chi^2$)

The chi-squared test is used to compare data collected in the field or from secondary sources with the results that would be expected according to the theoretical hypothesis being tested. It is usual to formulate a null hypothesis, stating that there is no significant difference between the observed and expected values. The test is then applied to determine whether there is enough evidence to reject the null hypothesis in favour of the alternative hypothesis that there is a significant difference between them.

To calculate $\chi^2$:
- list the observed value of each item: $O$
- list the expected value of each item: $E$
- subtract the expected value from the observed value: $O - E$
- square the result for each item: $(O - E)^2$
- divide each of these by the expected value:

$$\frac{(O-E)^2}{E}$$

- add up all of these values:

$$\chi^2 = \Sigma \frac{(O-E)^2}{E}$$

As with the Spearman rank correlation coefficient, the statistical significance of the result then needs to be established. This time, however, the critical value corresponds to the number of degrees of freedom, which is equal to the number of categories of data plus one, i.e. $(n + 1)$. The procedure for establishing whether the result is significant or not is the same: compare the calculated value to the 0.05 and 0.01 levels and if it is greater or equal to either of them, the result is significant and the null hypothesis can be rejected in favour of the alternative value. Critical values for the chi-squared test are available in books of statistical tables.

One other difference between $\chi^2$ and $r_s$ is that $\chi^2$ can be used for data that are divided up into categories, such as rock types, rather than just being used for numerical data. This is shown in the worked example below.

### Example

Students collected data on the degree of weathering of gravestones in a churchyard. Fifty gravestones of each of three different rock types were inspected and the degree of weathering was identified using Rahn's index. Their results are summarised in Table 8.8.

# Chapter 8

*Applied geographical skills: Section A*

Table 8.8

| Rahn's index | Granite | Sandstone | Marble | Total |
|---|---|---|---|---|
| 1–2 | 14 | 9 | 40 | 63 |
| 3–4 | 31 | 26 | 5 | 62 |
| 5–6 | 5 | 15 | 5 | 25 |
| Total | 50 | 50 | 50 | 150 |

A null hypothesis was developed:

> There is no significant difference between the degree of weathering and the different rock types of the gravestones.

To calculate $\chi^2$, Table 8.9 is used for each rock type.

Table 8.9 Calculating $\chi^2$ for granite

| Rahn's index | O | E | O – E | (O – E)² | (O – E)²/E |
|---|---|---|---|---|---|
| 1–2 | 14 | 21 | –7 | 49 | 2.33 |
| 3–4 | 31 | 20.6 | 10.4 | 108.16 | 5.25 |
| 5–6 | 5 | 8.3 | –3.3 | 10.89 | 1.31 |
|  |  |  |  | $\sum(O-E)^2/E =$ | 8.89 |

The expected values (E) are determined by looking at the average number of gravestones of all three rock types which have a particular category of weathering. As there were 63 gravestones with an index of weathering of 1–2 (Table 8.8), this is divided by 3 to give an expected value of 21. Thus if there were no difference in the way the different rock types were weathered, each rock type would have 21 gravestones with an index of 1–2. The result, 8.89, is then combined with the results for sandstone and marble (calculated in the same way as the result for granite in Table 8.9 but not shown here) to give a total of 45.99. This is then compared to the critical values for $(n + 1) = 4$ (three rock types + 1) (see Table 8.10).

Table 8.10 Critical values

| Degrees of freedom | 0.05 (5% level) | 0.01 (1% level) |
|---|---|---|
| 4 | 9.49 | 13.28 |

As the result exceeds the critical value at the 0.01 level, the difference is statistically significant and the null hypothesis can be rejected in favour of the alternative hypothesis. In other words, there is a significant difference between the degree of weathering and gravestones of different rock types.

### Drawing conclusions

You should be able to draw conclusions from the data provided and from the interpretation of those data. Such conclusions should always be based on evidence from the data and so concluding comments should be supported with specific detail and figures. Where an initial question or hypothesis is provided, the conclusion should explicitly refer to it, either giving a direct answer to the question or indicating whether the hypothesis was proven or not. It should be remembered that hypotheses are not disproved, only proven or unproven, based on the evidence of the data. With small samples of local fieldwork data, there may well be insufficient evidence for a hypothesis to be proven. That does not mean it was wrong or inappropriate as a starting point for an investigation.

> **Examiner's tip**
>
> ■ When statistical calculations are asked for in the examination paper, the formulae will be provided.

The conclusions drawn may well need to be tentative given the nature of an investigation with a small sample size and/or from a small area. The explanations you offer may be incomplete, because other variables may be at work for which data were not collected. It may, therefore, be appropriate to use phrases such as 'it is possible that…', 'it might be true to say that…' or 'there is some evidence to suggest that…'.

## Opportunities for further investigation

Many sets of fieldwork data presented in Section A of the examination paper may, by definition, be limited in their coverage of a topic or issue. If the data are from actual or imaginary AS investigations, they will have been collected with a fairly narrow focus, with limited time and equipment. It is essential, therefore, to be able to suggest how an investigation could be extended. This may include the collection of further data (see below).

## Evaluation

Any investigation should be evaluated upon completion. This should involve:
- returning to the original aims and objectives to establish if they have been achieved
- identifying areas of the investigation that were completed successfully
- identifying areas of the investigation that could be improved upon
- suggesting who might find the investigation of use

Improvements to fieldwork data collection can broadly be seen as:
- collect more data
- collect the same data more efficiently and effectively

When collecting more data, some or all of the following possibilities might be appropriate:
- collect the same data but at a greater number of sample points from a wider area/longer time period to increase the reliability of statistical calculations
- collect the same data but repeat the data collection at a later time so that comparisons can be made and variations over time identified
- collect the same data but also at another place so that spatial comparisons can be made

When collecting data more efficiently and effectively, the following ideas may be relevant:
- use different/better equipment
- measure to a greater degree of accuracy
- take a number of measurements of the same item and then calculate an average
- use a more appropriate sampling method
- take a bigger sample within the designated location/timeframe to increase the reliability of any statistical calculations

In each case it should be made clear how the additional or improved data would help produce a better investigation with a more reliable conclusion.

# Chapter 8

## Applied geographical skills: Section A

### Examination question

In June 2004, a group of students investigated the impact of tourism on the town of Swanage in Dorset. They collected data from 26 sites in the centre of the town on a dry, sunny weekday. At each site they carried out a pedestrian count, a street-quality survey and a land-use survey. They also collected data at a number of car parks; the origin of the cars was obtained from a survey of tax discs.

**a** Suggest what factors the students should have considered when identifying their data collection sites. (4)

**b** Table 8.11 shows the results of the tax disc survey.

Table 8.11

| Name of car park | Swanage tax disc | Other Dorset tax disc | Other county tax disc |
|---|---|---|---|
| Broad Road | 1 | 13 | 22 |
| The Parade | 3 | 1 | 6 |
| Short Stay | 6 | 9 | 24 |
| Short Stay | 13 | 5 | 13 |

(i) Suggest a suitable method for presenting these data. (1)
(ii) Justify your choice of method. (3)
(iii) What disadvantages or limitations does this method have? (3)

**c** The students used the results of the land-use survey to classify the land uses according to their relevance to the tourist industry. They then wanted to test a null hypothesis that there was no significant difference between tourist related and non-tourist related land uses. To do this, they decided to use the chi-squared test.

(i) Complete Table 8.12 and calculate the chi-squared value. (7)

Table 8.12

| Land use | Observed number | Expected number | O – E | (O – E)² | (O – E)²/E |
|---|---|---|---|---|---|
| Tourist related | 13 | 34 | | | |
| Mainly tourist related | 37 | 34 | 3 | 9 | 0.26 |
| Neither | 14 | 34 | –20 | 400 | 11.76 |
| Mostly non-tourist related | 34 | 34 | 0 | 0 | 0 |
| Non-tourist related | 72 | 34 | | | |

$\Sigma(O - E)^2/E =$

$$\chi^2 = \Sigma \frac{(O-E)^2}{E}$$

(ii) Test the significance of the result using the critical values in
Table 8.13. (2)

| n | 0.05 (5% level) | 0.01 (1% level) |
|---|---|---|
| 5 | 11.08 | 15.09 |
| 6 | 12.59 | 13.28 |
| 7 | 14.07 | 15.09 |

Table 8.13

(iii) Should the null hypothesis be accepted or rejected? (1)

**d** The street-quality survey was based on students' judgements of six criteria, each being scored on a scale from 1 to 5, with 1 being the worst and 5 the best. The results are shown in Table 8.14.

Table 8.14

| Site no. | Street quality (max. 30) | Pedestrian flow (per min) |
|---|---|---|
| 1 | 10 | 1.0 |
| 2 | 9 | 6.0 |
| 3 | 14 | 12.5 |
| 4 | 15 | 9.1 |
| 5 | 14 | 2.5 |
| 6 | 13 | 1.3 |
| 7 | 18 | 1.3 |
| 8 | 15 | 4.6 |
| 9 | 13 | 5.2 |
| 10 | 14 | 6.1 |
| 11 | 16 | 9.2 |
| 12 | 15 | 9.8 |
| 13 | 14 | 8.7 |
| 14 | 14 | 7.5 |
| 15 | 13 | 3.2 |
| 16 | 14 | 5.0 |
| 17 | 14 | 5.2 |
| 18 | 14 | 2.9 |
| 19 | 13 | 9.9 |
| 20 | 16 | 9.1 |
| 21 | 17 | 15.2 |
| 22 | 11 | 7.7 |
| 23 | 9 | 11.0 |
| 24 | 19 | 7.5 |
| 25 | 8 | 2.8 |
| 26 | 8 | 1.2 |

(i) What are the limitations of a street-quality survey as a method of
data collection? (3)

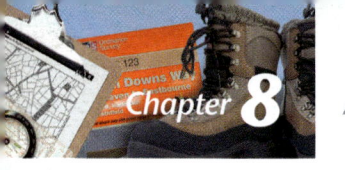

*Applied geographical skills: Section A*

(ii) Complete the scattergraph below (Figure 8.14) by adding results for sites 25 and 26. (2)
(iii) Draw a best-fit line on the graph. (1)
(iv) Describe the relationship between street quality and pedestrian flow. (3)

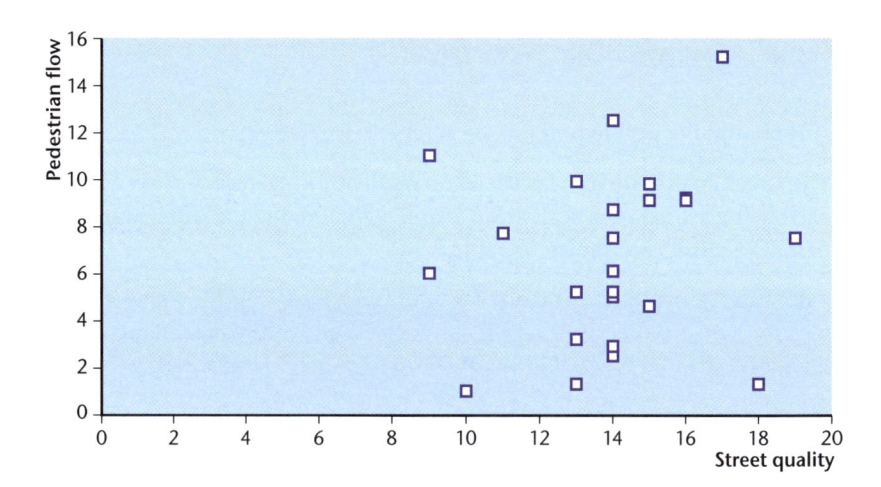

*Figure 8.14*

**e** Using all the evidence provided and suggesting what other data could have been collected, evaluate this investigation. (10)

*Total = 40 marks*

# Chapter 9

# Applied geographical skills: Section B

## Introduction

Question 2 on the Unit 3 examination paper is always about your fieldwork experience. There are three types of question that you must be prepared for:
- a question about your physical fieldwork
- a question about your human fieldwork
- a question about the fieldwork techniques and methods you used, which might be drawn from either your physical or your human fieldwork trips

There are a limited number of questions that can be asked about your fieldwork, so you can prepare for this section of the examination beforehand in the expectation that at least some of the following will be asked.
- Where did you carry out your fieldwork?
- What were your aims?
- What were your hypotheses?
- How did you carry out your primary data collection?
- Did you use sampling? If so, what sort and why?
- Did you use secondary data? If so, what sort and why?
- How did you present your data and why did you choose those methods?
- What were your conclusions?
- Were your conclusions reliable? If not, why not?
- How would you improve your work? More data, different data?

## Where did you carry out your fieldwork?

Can you draw a locational sketch map showing the sites where you collected data and the general situation of the area in which you conducted the fieldwork? All such maps should have:
- a scale
- a north point
- clear labels

If you are asked to annotate this map, then some explanation is also required, probably justifying the choice of sites.
   You should be able to sketch two maps:
- one should show the general location of the fieldwork exercise in relation to surrounding places, i.e. its situation
- the second (to a larger scale) should show the sites themselves

You can practise drawing these maps well ahead of the examination date.

## Chapter 9 — Applied geographical skills: Section B

## What were your aims?

The aim of your fieldwork is the broad objective that you had. Your aim will not be 'To find out more about Whereville' or 'To look at the River Riddle'. You will have focused on particular aspects of the area, such as 'To investigate the impact of tourism on the CBD of Whereville' or 'To investigate downstream changes in channel characteristics in the River Riddle'.

## What were your hypotheses?

Hypotheses are testable statements that will have been designed to deliver some answers to the question outlined in the aim of your fieldwork. You should have more than one hypothesis but fewer than five, so that the focus on the central question or aim is maintained. Thus:

- channel depth will increase downstream
- channel width will increase downstream
- river velocity will remain constant downstream

## How did you carry out your primary data collection?

You need to know where, when and why you collected your data. This is the justification of site selection.

### Where?

A number of criteria that have little to do with geography are pretty much constant features here:

- safety
- accessibility (which may be related to safety)
- time and opportunity

The geographical criteria will obviously depend on your fieldwork but, once again, there are a number of recurring ideas:

- Are there enough sites to cover the range of possible environments studied?
- Are there enough sites to allow you to apply statistical tests such as the Spearman rank correlation coefficient?

### When?

As well as the actual timings you used in your data collection, you need to be aware of the time of day and the time of year, as these are almost certain to have an impact on your results. You need to say more than 'pedestrian counts were conducted for 5 minutes at hourly intervals'. You may also need to consider the day of the week: results obtained on a Sunday may differ from those obtained on a weekday, while a market day may introduce even more variation.

### Why?

You need to be able to justify the methods chosen in terms of the hypotheses. Thus, for example, 'the pedestrian counts were designed to show whether or not pedestrian densities increased as one approached the peak land value intersection'.

## Did you use sampling? If so, what sort and why?

There are three possible sampling techniques you could have used:
- systematic
- stratified
- random

These techniques are explained in Chapter 8. You need to be able to say why you used a particular technique in a particular location. For example, on a survey of a sand dune you may wish to pick sites where a change is expected, such as a break of slope or a change in drainage characteristics. In these circumstances a transect which sampled the sand dune systematically, say every 10 m, might miss significant changes in the environment, whereas a stratified sample would deliberately pick these out.

Not all types of fieldwork will involve sampling. For example, a land-use survey will sometimes map every usage in an area.

## Did you use secondary data? If so, what sort and why?

Primary data are the data you collect in the field. Secondary data cover everything obtained from other sources. Usually this means published material found in books, magazine articles or on the internet, but it also includes data from previous year groups that have done the same fieldwork. Secondary data are not more or less reliable than your own data. In some cases, especially if you use published material, the information may have been gathered from a larger sample size and thus be more likely to reflect the real world. This is particularly true of census material, for example.

## How did you present your data and why did you choose those methods?

The methods used should include a number of graphical and tabular methods but also a few statistical tests. Once again, you need to be able to justify the techniques used and suggest why they were appropriate. The various methods that you might have used are presented in Chapter 8. You will need to be able to say why one method was chosen rather than another.

## What were your conclusions?

The conclusions should relate to the original hypotheses. Thus you may have found that width, depth and velocity in a river all change downstream. However, your conclusion would be much stronger if you could say that width changed most, depth changed by a smaller amount and, shall we say, velocity changed hardly at all.

You need not learn exact data in this context but it would help greatly if any anomalies in your data were noted. As with descriptions of data in answering the 'skill' sections of Unit 1 and 2 questions, you should describe the overall trends, the variations in those trends and pick out any anomalies. You could say, for example, 'Depth decreases downstream but at a decreasing rate in the lower stages, except at Site 7 where there is a significant change in channel profile and the channel is markedly shallower than at Sites 6 or 8'.

## Were your conclusions reliable? If not, why not?

The reliability of the conclusions is the degree to which they conform to the 'real' world. Reliability is dependent upon two main sets of variables:
- the quality of the 'design' of the fieldwork used to collect the original data
- the quality of the execution of that design: i.e. how well it was carried out

### Design of the fieldwork

Questions to do with the design of the fieldwork are more complex. The most obvious design 'fault' may be an insufficiency of data. There are several aspects to this:
- It is sensible to have at least 12 sites if correlation tests such as the Spearman rank correlation coefficient are to be carried out effectively.
- The greater the number of pieces of data, the greater is the 'sample' size of the population as a whole, and thus the more reliable the results will be.
- The choice of sites may well have proved to be potentially distorting. For example, if 3 of the 12 data collection sites for a questionnaire happened to be outside nursing homes for the elderly, the results might well be unrepresentative of the population as a whole. Similarly, a stratified sample of sites downstream on a river or across a sand dune may well miss significant areas such as meander bends or slacks.
- The choice of timing for data collection can also affect the results. Both time of day and time of year are key variables in controlling the results.
- The criteria applied may affect results. For example, in a land-use mapping exercise of retail usage in a town centre, any of the following could be used to measure changes in the retail pattern of a town:
   - the number of retail units devoted to convenience goods
   - the length of shop frontage devoted to retail goods
   - the floor area devoted to retail goods
   - the retail turnover per m$^2$ of convenience goods

To take the last example, it is most unlikely you, as a student, will be able to obtain data about retail turnover. Similarly, you are likely to find that measuring the floor area would take far too much time and effort. In reality, you can record either the raw number of units or the length of shop frontage. It is obvious that this may distort the results. The number of units, for example, simply 'stands-in' or proxies for what we really want to find out and may not do so effectively. Imagine that one very large restaurant has closed down but two much smaller ones have opened since the last survey. From this, it could appear that 'restaurants' have become more important in the town. If the turnover of the two 'new' eating places is smaller than that of the now defunct older restaurant, then the opposite is actually the case.

On the closely related theme of the quality of the mapping, it might well be that only ground-floor usages were mapped. In other words, sampling has taken place, albeit unconsciously, and this may very well have a large impact on the reliability of your conclusions.

Perhaps the most complex potential error in conclusions is that of drawing false conclusions because a key variable has been missed out altogether. In an (in)famous piece of research conducted some years ago into the incidence of mental disease, a large amount of data were accumulated to show that the mental illness of schizophrenia was far more commonly found in inner-city areas than in any other geographical location. From this it was concluded that 'inner-city' living, with its stress, noise levels and pollution, might well be causing the problem. Unfortunately, although this may be the case, the data had omitted one vital ingredient, which was an analysis of the previous histories of these mentally ill people. It transpired that many of them were in inner-city areas because they were schizophrenic, not the other way round. They had developed schizophrenia in their 'home' area. As a consequence of this, perhaps linked with the loss of their jobs or their relationships falling apart, they left home and found lodgings in the poorer districts of inner cities. The original conclusion was, if you like, back to front.

### Execution of fieldwork

Factors to do with the execution of fieldwork are more straightforward. The rather less interesting reason for the unreliability of conclusions is that the data were simply of poor quality. There may be plenty of data and the initial design may have been excellent, but the data are just weak. There are two possible reasons.

- **Equipment quality.** Measuring river flow with submerged oranges may not give the same quality of results as using a flow meter. However, your flow meters might malfunction on the day…
- **Operator error.** Not all fieldwork is carried out scrupulously. Some problems might be unavoidable, such as the incompetence of one of the party (the 'Tim dropped the ranging pole' critique of conclusions). But, more seriously, even supposedly eminent academics have occasionally been found guilty of manufacturing their data to help prove their point. This is the equivalent of one fieldwork group sitting in a café making up their questionnaire responses, although at least in this case the only motive is laziness.

## How would you improve your work? More data, different data?

This is the evaluation part of your fieldwork, during which the various and inevitable flaws can be addressed. You should aim to make at least three evaluative comments. One of these can be based on the idea that the reliability of your data could be improved by gathering more data, perhaps at different times and at different places. But you must include at least one, preferably two, evaluative comments for each piece of fieldwork that reflects on the design flaws and makes suggestions of how to improve them by gathering different data. For example, the comment 'We only looked at ground floor usage using Goad maps and therefore

# Chapter 9　Applied geographical skills: Section B

we might have missed significant office usage on first floors' mentioned in your conclusion can be followed by 'This could be improved by mapping the usage of the first floor and even the second floor'.

## Examination question

**a** (i) What were the aims of your fieldwork exercise? (2)

(ii) Copy and complete the table below for *two* different types of data collected.

Table 9.1

| | Data collected | Sampling method used | Justify your choice of sampling technique |
|---|---|---|---|
| 1 | | | |
| 2 | | | |

(6)

(iii) Comment on the effectiveness of the methods in which you analysed your data. (3)

**b** (i) In what ways might your data collection have been unreliable? (3)

(ii) What other data could you have collected? (3)

**c** Summarise the conclusions you have reached in this investigation. (3)

*Total = 20 marks*

# A2 Unit 4

## Physical systems, processes and patterns

# Chapter 10

# Atmospheric systems

## The dynamic atmosphere

### The Earth's heat budget

**Insolation** is short-wave radiation emitted by the sun due to its temperature of approximately 6000°C. This radiation passes relatively easily through the Earth's atmosphere, but it does suffer a significant amount of loss and dilution en route.

Imagine 100 units of solar energy reaching the Earth's atmosphere
- 31 units are reflected back into space
- 69 units are absorbed by the Earth and the atmosphere.

Of the 31 units that are reflected back into space:
- 8 units are reflected by the atmosphere
- 17 units are reflected by clouds
- 6 units are reflected by light-coloured surfaces, such as ice, on the ground

Of the 69 units absorbed by the Earth's atmosphere and surface:
- 23 units are absorbed by the atmosphere (by water vapour and other gases)
- 46 units reach the Earth's surface

Of the 46 units of radiation that reach the surface of the ground, which is short-wave radiation:
- 14 units are re-radiated as long-wave radiation
- 10 units pass through the atmosphere by conduction (contact heating)
- 22 units are transferred by latent heat, which is energy used during evaporation, for instance

Of the 14 units that are re-radiated as long-wave radiation:
- 7 units are absorbed by gases such as carbon dioxide, water vapour and methane in the atmosphere
- 7 units are of such a wavelength that they cannot be absorbed by any of the atmospheric gases; this energy escapes back into space through the 'radiation window'

> **Examiner's tip**
> 
> - When referring to global warming, you must refer to the 'enhanced' greenhouse effect.

The 7 re-radiated units that are absorbed by gases in the atmosphere give rise to the **greenhouse effect**. The long-wavelength radiation does not pass through the atmosphere as easily as the incoming short-wavelength insolation and so it is readily absorbed. The greenhouse effect is an entirely natural phenomenon; it causes the Earth's temperature to be about 30°C higher than it would otherwise be and allows the Earth to support life.

*Physical systems, processes and patterns*  **A2 Unit 4**

The current concerns about **global warming** relate to the *enhancing* of the greenhouse effect by the increased concentration of 'greenhouse gases' such as carbon dioxide and methane in the atmosphere, making temperatures on Earth even higher.

The term **albedo** refers to the proportion of radiation scattered and/or reflected by the atmosphere and the Earth's surface. The global average is 31%. However, the albedo varies enormously with the type of surface:
- fresh snow, for example, may reflect as much as 95%
- dark soil may only reflect 5%

Another variable which affects albedo is the angle of the sun:
- oceans may reflect 3% of the energy reaching the water's surface when the sun is directly overhead
- oceans may reflect 80% when the sun is at a very low angle

The situation is complicated by the fact that much of the heat gained by the lower atmosphere is radiated back to the Earth's surface. However, a useful summary of the **heat budget** is shown in Figure 10.1.

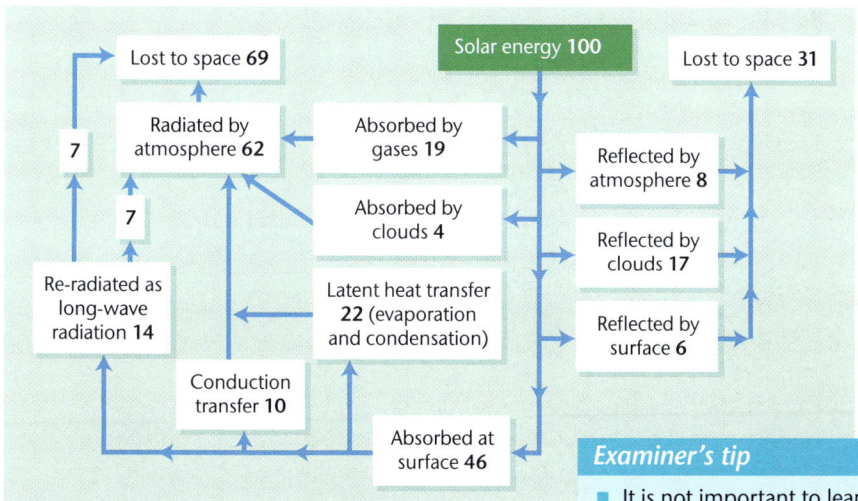

*Figure 10.1 The Earth's heat budget*

Heat budget diagram:
- Solar energy 100
- Lost to space 69
- Lost to space 31
- Radiated by atmosphere 62
- Absorbed by gases 19
- Reflected by atmosphere 8
- Absorbed by clouds 4
- Reflected by clouds 17
- Re-radiated as long-wave radiation 14
- Latent heat transfer 22 (evaporation and condensation)
- Reflected by surface 6
- Conduction transfer 10
- Absorbed at surface 46

### Examiner's tip
- It is not important to learn all the figures in Figure 10.1. However, you should be aware of the relative differences between them and be able to explain them. A few selected figures could provide useful evidence in an essay answer to support or illustrate a point that you are making.

The above figures all relate to the overall global situation, so it is important to appreciate that there are spatial and temporal variations in the amount of insolation actually received at the surface of the Earth. The main variables are as follows.
- **The solar constant:** this is estimated at $2.35 \times 10^{28}$ joules per minute, but output is in fact variable (due to sunspot activity, for instance). Recent studies of ice cores from Greenland suggest that these variations may well be linked to long-term climate changes.
- **Angle of the sun:** the high angle of the sun above the horizon in the tropics has a more intense heating effect than the low-angle sun experienced near the poles. Nearer the poles the radiation passes through a larger depth of

# Chapter 10 Atmospheric systems

atmosphere and it also has to heat a larger surface area. At a latitude of 60° north (for example the southern tip of Greenland) or 60° south, the sun's rays are spread over twice as large an area as they are at the equator.
- **Length of day/night:** this varies with the position of the overhead sun. In the northern hemisphere there are long days and short nights in summer, so more insolation is received than in winter when the pattern is reversed.
- **Aspect:** in the northern hemisphere south-facing slopes receive more direct insolation than north-facing slopes, which are often in shadow. This is clearly a factor of local importance.
- **Cloud cover:** the greater the extent and thickness of cloud cover, the greater the amount of incoming solar radiation that is absorbed and reflected.

Figure 10.2 shows the latitudinal and seasonal variations in the amount of insolation that result from the changing position of the overhead sun. The key features of the pattern are:
- total insolation is greater at the equator than at the poles
- seasonal differences are greater at higher latitudes
- the seasonal pattern in the northern hemisphere is the reverse of the pattern in the southern hemisphere

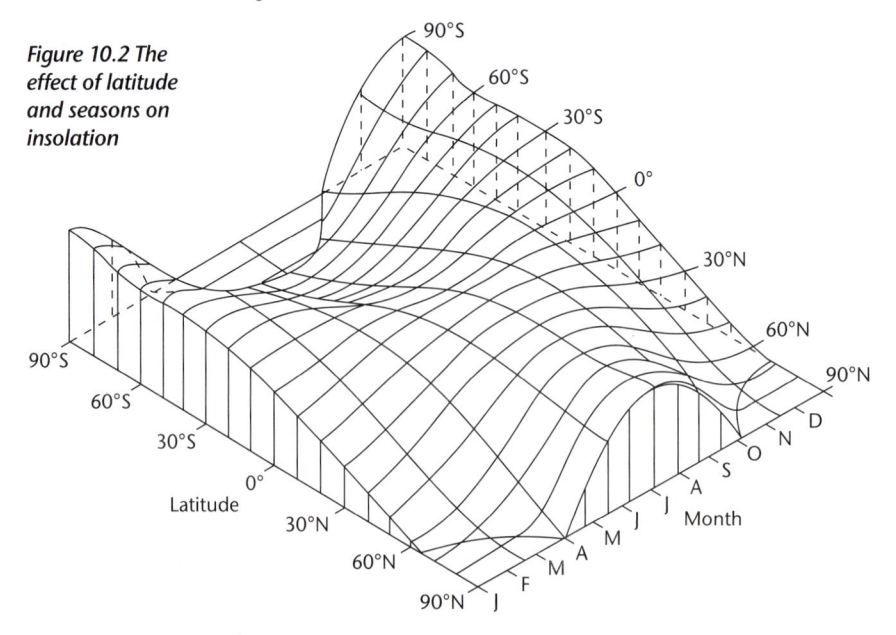

*Figure 10.2 The effect of latitude and seasons on insolation*

### Examiner's tip

- There is no need to learn Figure 10.2, as you would not be expected to draw it in an essay answer. However, you should be aware of its key features and be able to explain the main reasons for the variations that exist.

### Key terms

**Albedo** The proportion of insolation reflected and scattered by the Earth's surface and atmosphere.

**Earth's heat budget** The relationship between insolation and terrestrial radiation.

**Insolation** Incoming solar radiation.

# The global surface and upper air circulation

The major factor that causes atmospheric circulation is the uneven distribution of insolation, which has been described above. Between 38°N and 38°S (i.e. either side of the equator) there is surplus heat; more energy is received from insolation than is lost from terrestrial radiation. Nearer the poles there is an energy deficit; more energy is lost from terrestrial radiation than is gained from insolation. Energy is therefore transferred from low latitudes to high latitudes to balance this unequal heating. These transfers are in the form of:

- planetary winds such as the westerlies
- depressions and hurricanes
- ocean currents such as the Gulf Stream

The energy transferred within the atmosphere is known as the **general circulation**. A simple model of the general circulation, known as the tri-cellular model, formed the basis for our understanding of atmospheric motion between 1856 and 1941. It was noted that there were three cells of atmospheric movement (Figure 10.3). You need to know that the 'tropopause' is effectively the top of the lower atmosphere.

The processes involved in generating the first of these cells are as follows.

- At the equator the sun is always at a high angle in the sky and so its heating effect is intense and concentrated. The air that is heated rises, creating an area of low pressure.
- When the warmer air reaches the tropopause, having lost heat while rising, it moves towards the poles in both directions.
- In the sub-tropical areas the air sinks because it is colder and denser. In addition, there is less space in the atmosphere as the tropopause becomes lower with increasing latitude; it is about 16 km above the equator and only 8 km above the poles.
- This sinking air produces high pressure at the ground surface. As a vacuum cannot exist, air returns towards the tropics to replace the air that has risen. This creates the **trade winds**.
- Near the equator the trade winds converge at the **inter-tropical convergence zone** (ITCZ). The rising air creates low pressure and gives conditions of great instability and high thunderstorm activity.

The cell of atmospheric motion that is created by these processes is known as the **Hadley cell** and there is one in each hemisphere.

At the poles, cold dense air sinks and spreads out towards the lower latitudes. Here it is warmed and rises, spreading out at the tropopause and moving polewards again to replace the air that has sunk. This produces the **Polar cell**. The Polar cell results in high pressure at the poles under the sinking air and relatively low pressure at about 60°N and 60°S, where the warmed air rises again.

A third cell develops between the other two cells, driven by their movements. This is called the **Ferrel cell**. It consists of air moving polewards at the surface and towards the equator below the tropopause.

# Chapter 10  Atmospheric systems

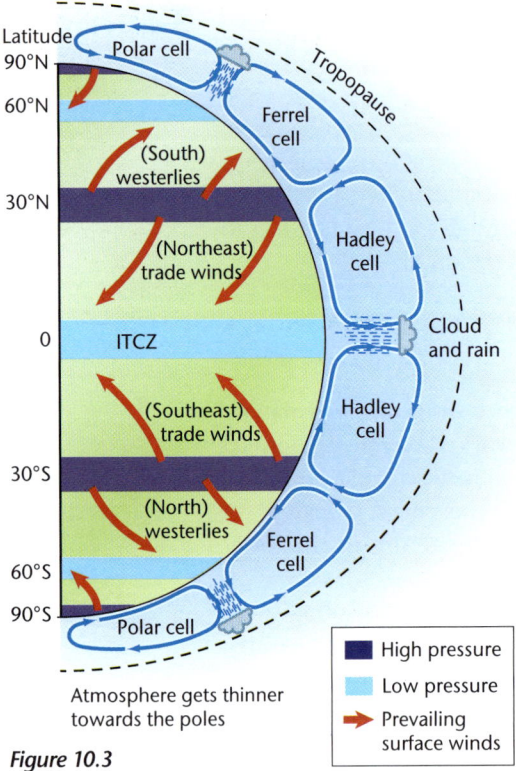

**Figure 10.3** The tri-cellular model

It should be noted that the positions of the Hadley cells do not remain constant. The position of the overhead sun shifts with the seasons because the Earth is tilted. This leads to a shift in the area of greatest heating and results in the movement of the ITCZ. In summer in the northern hemisphere, the sun is overhead at or near the tropic of Cancer and the ITCZ shifts about 6° northwards. In winter in the northern hemisphere, the sun is over the tropic of Capricorn and the ITCZ shifts by a similar amount to the south. These movements result in changes in the global pressure belts and planetary winds, giving rise to the seasonal variations in the climates of places in the sub-tropical areas of Africa in particular.

Figure 10.3 shows the tri-cellular model of atmospheric circulation. More recently it has been recognised that the tri-cellular model is inaccurate and therefore limited in its usefulness. Research by radiosonde balloons and laboratory experiments with dishpans provided important information about possible east–west movements and energy transfers in the middle latitudes. Two important new considerations came to light.

- **Jet streams**: these are strong and regular winds that blow in the upper atmosphere, about 10 km above the surface. Air moves at between 100 and 300 km h$^{-1}$. There are two major streams: one between 30° and 50° in each hemisphere moving eastwards, which is known as the **polar jet**; and the other

**Figure 10.4** The global air circulation

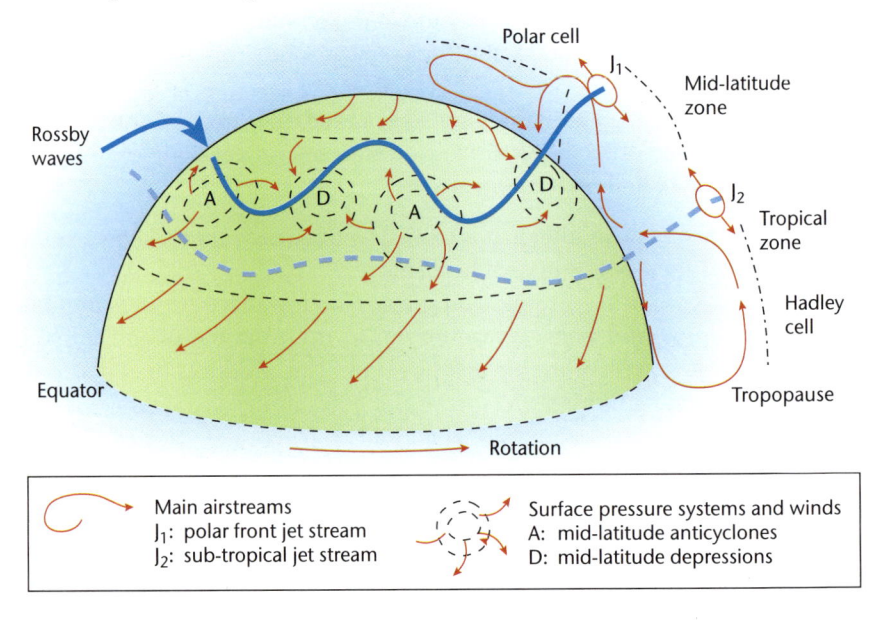

Main airstreams
J$_1$: polar front jet stream
J$_2$: sub-tropical jet stream

Surface pressure systems and winds
A: mid-latitude anticyclones
D: mid-latitude depressions

290

moving westwards between 20° and 30°, which is known as the **sub-tropical jet**. The jet streams result from the temperature differences between the tropical air and sub-tropical air (the sub-tropical jet) and between the polar air and the sub-tropical air (the polar jet). The greater the temperature difference, the stronger is the stream.
- **Rossby waves**: these are meandering rivers of air formed when westerly winds are influenced by major relief barriers such as the Rocky Mountains. A wave-like pattern is created which may last for about 6 weeks. There are 3–6 waves in each hemisphere and they are also influenced by land/sea differences.

These two phenomena greatly disrupt the north–south air movements in the Ferrel cells. As a result, the simple tri-cellular model is now seldom used at high academic levels. The model shown in Figure 10.4 gives a much more accurate representation of the global air circulation.

> **Examiner's tip**
> - You should be able to describe and explain the tri-cellular model and discuss why there are limitations to its use in studying atmospheric motion.

## Global patterns of pressure and wind

The global pattern of pressure is closely linked to the atmospheric circulation discussed above. The rising and sinking of air produces low and high pressure areas respectively. These are named according to their latitudinal position. It should be noted that the pattern is repeated in each hemisphere and that the pressure belts shift position by a few degrees of latitude as the overhead sun changes position seasonally.

The global winds are controlled by a combination of the following forces:
- pressure gradient force
- centripetal force
- Coriolis force
- friction

### Pressure gradient force

Pressure gradient force is the main driving force and it causes movement of wind from high-pressure areas (where there is a build-up of surplus air) to low-pressure areas (where there is a deficit). In effect, the winds are equalising pressure differences across the Earth's surface. The steeper the pressure gradient (shown on synoptic charts by closely spaced isobars), the greater is the wind speed that results. Thus in mid-latitude depressions wind speeds are generally high, whereas in anticyclones they are low.

### Coriolis force

Coriolis force is the deflection of wind due to the Earth's rotation. It is to the right in the northern hemisphere and to the left in the southern hemisphere. It does not exist at the equator but it increases with distance towards the poles. The Coriolis force is often likened to the effect of a person sitting on a revolving roundabout trying to throw a ball to a stationary target. Because of the rotation of the roundabout, the thrown ball will follow a curved path and, in order to hit the target, it would have to be thrown before the target was reached. It is the Coriolis force that generates the westerly direction of the mid-latitude winds and the easterly element of the trade winds. The trade winds would blow from north to south towards the equator in the northern hemisphere if the Earth was not

# Chapter 10 Atmospheric systems

rotating. However, the Coriolis force causes the air to be deflected to the right and the trade wind to blow from a northeasterly direction. In the southern hemisphere, the trade wind on a non-rotating Earth would blow from south to north, towards the equator. However, the Coriolis force causes a deflection to the left and so the trade winds blow from a southeasterly direction. In the mid-latitudes, the pressure gradient force and the Coriolis force are directly balanced. This leads to air moving, not from a high-pressure area to a low-pressure area, but between the two, parallel to the isobars. This is called the geostrophic wind.

## Centripetal force

Air moving towards a low pressure area in a deflected path accelerates towards the centre of low pressure; this is due to the centripetal force. The reverse effect, centrifugal force, causes air to accelerate out of high-pressure areas. However, the effect of a centrifugal force is less significant as the air is spreading across an increasingly large surface area. In a low-pressure area the inward-moving air is converging into an area of decreasing size, which enhances the acceleration, producing rapidly moving, spiral winds.

## Friction

Frictional drag from the Earth's surface not only decreases wind speed but also modifies direction, crossing isobars as it moves from high-pressure areas to low-pressure areas. With increasing height through the atmosphere, the effects of friction decrease.

## In summary

The effects of these forces are that air moves from high pressure to low pressure on a deflected path, crossing isobars and accelerating into the area of low pressure. Figure 10.5 provides a diagrammatic summary.

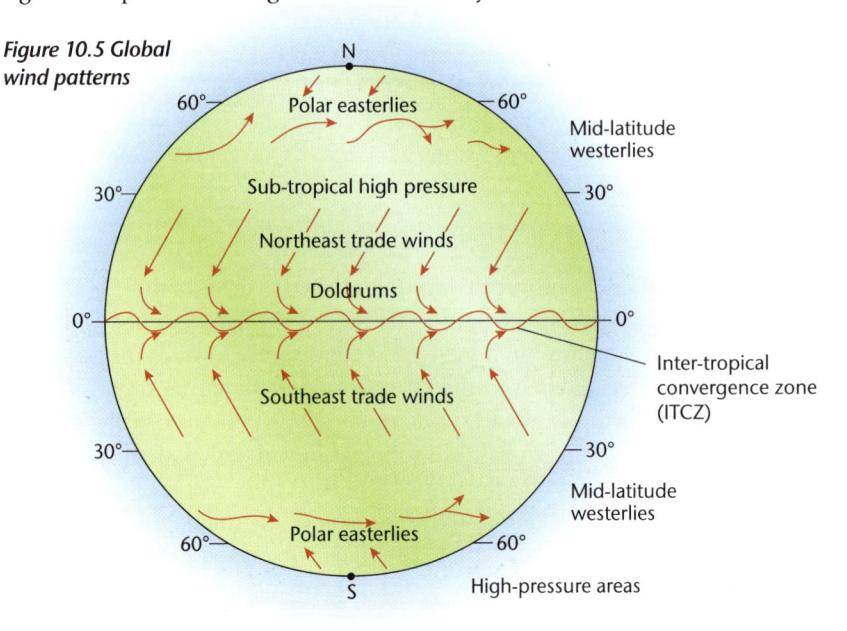

*Figure 10.5 Global wind patterns*

> **Examiner's tip**
>
> ■ Figure 10.5 is a good summary of global winds. You should be familiar with the names and locations of the main pressure belts and global winds.

292

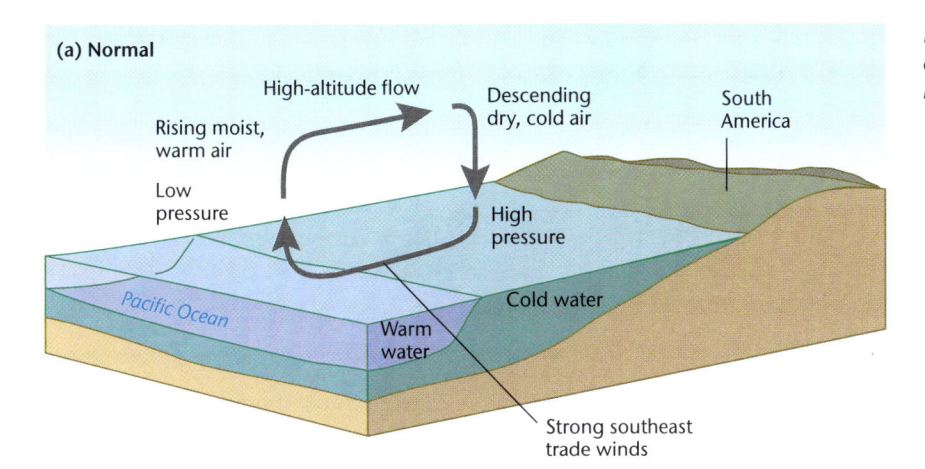

*Physical systems, processes and patterns*  **A2 Unit 4**

*Figure 10.6 The changes caused by El Niño*

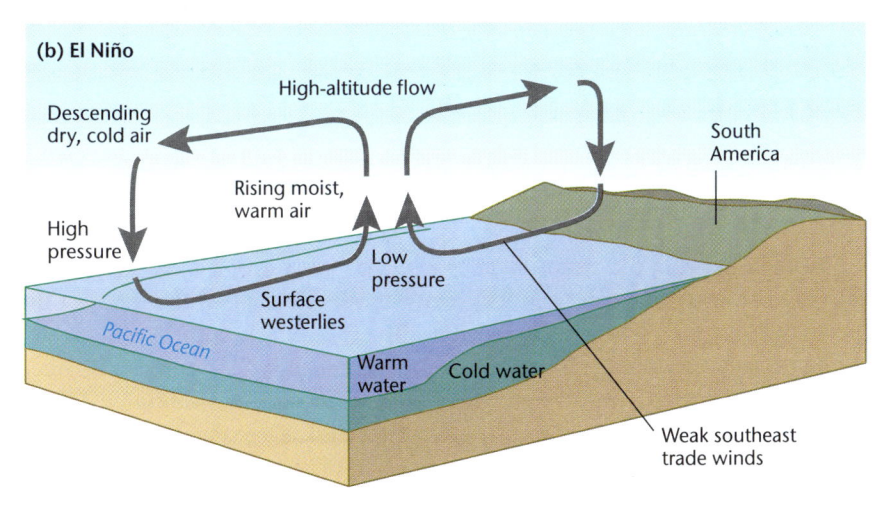

# El Niño southern oscillation (ENSO)

Variations occur in the global patterns of wind and pressure. These variations may, at least in part, be associated with human activity and global warming. However, variations also occur naturally; one example of this is the phenomenon known as **El Niño**, or the El Niño southern oscillation (ENSO).

The Walker circulation is the east–west circulation that occurs in low latitudes. In the Pacific near South America, winds blow offshore, causing up-welling of cold water. By contrast, the warm surface water is pushed into the western Pacific. Normally sea-surface temperatures in the western Pacific are over 28°C, causing an area of low pressure and producing high rainfall. Over the coast of South America, sea-surface temperatures are lower, high pressure exists and the weather is generally dry (Figure 10.6a).

Every 4–7 years, and lasting for up to 2 years, there is a shift in the temperature structure in the Pacific Ocean, which results in warm water occurring in the eastern Pacific off the coast of South America. This produces low pressure, as the

# Chapter 10  Atmospheric systems

air above the warm water rises, and so the trade wind circulation is weakened. This leads to high rainfall along the west coast of South America but much drier conditions than usual in the western Pacific, as far west as Indonesia (Figure 10.6b). It is likely that, through teleconnections (linkages), the wind pattern in other parts of the world is modified as well. A reverse alteration also exists, known as La Niña.

In July 1997 the sea temperature in the eastern tropical Pacific was 2.0–2.5°C above normal, breaking all previous climatic records. This continued into early 1998 and was thought to be the cause of:

- a stormy winter in California
- above average rainfall in southern USA
- worsening drought in Australia, Indonesia and the Philippines
- lower rainfall in northern Europe
- higher rainfall in southern Europe

> **Examiner's tip**
>
> - El Niño is an important topic. However, here it needs to be viewed in the correct context: it is a relatively small-scale variation in the global pattern of pressure and wind.

## Moisture in the atmosphere

### Orographic, frontal and convectional mechanisms of uplift

Moisture is found in the atmosphere as a liquid (water droplets), as a solid (ice crystals) and as a gas (water vapour). Only about 0.0009% of the water in the global hydrological cycle exists in the atmosphere, but the proportions in each of the three states vary and the formation of rainfall depends upon changes between them. Water droplets are formed when cooling air becomes saturated (100% relative humidity) at its dew point temperature. This leads to the condensation of water vapour into water droplets as cold air can hold less water vapour than warm air. The dew point temperature is not a fixed figure; it varies, particularly depending upon how much water vapour the air is holding.

The cooling of air can occur in three ways:

- when air comes into contact with a colder surface
- when a relatively warm air mass meets a colder air mass
- when air rises through the atmosphere, expands and cools. This is known as **adiabatic cooling** and occurs because rising air encounters decreasing pressure with increasing height. As a result, the air expands and uses up energy, and the temperature falls. There is no loss of heat between the parcel of rising air and the surrounding air

There are three main mechanisms by which air is uplifted, leading to adiabatic cooling and, potentially, rainfall. They are **orographic**, **frontal** and **convectional** (Figure 10.7).

These trigger mechanisms lead to air rising, cooling and, if the dew point temperature is reached, condensing. Rainfall, however, only occurs when water droplets in the air are heavy enough to fall through the rising air currents beneath them. When droplets first form through condensation, they are usually microscopic

*Physical systems, processes and patterns* **A2 Unit 4**

### Convectional

When the land becomes very hot it heats the air above it. This air expands and rises. As it rises, cooling and condensation take place. If it continues to rise, rain will fall. Convectional rainfall is very common in tropical areas. In Britain it is quite common in the summer, especially in the southeast.

(3) Further ascent causes more expansion and more cooling; rain takes place

(2) The heated air rises, expands and cools; condensation takes place

(1) The Earth's warmed surface heats the air above it

### Frontal or cyclonic

Frontal rain occurs when warm air meets cold air. The warm air, being lighter and less dense, is forced to rise over the cold, denser air. As it rises, it cools, condenses and forms rain.

Warm air rises over cold air; it expands, cools and condensation takes place; clouds and rain form

This line represents the plane separating warm air from cold air

Warm air is forced to rise when it is undercut by colder air; clouds and rain occur

### Relief or orographic

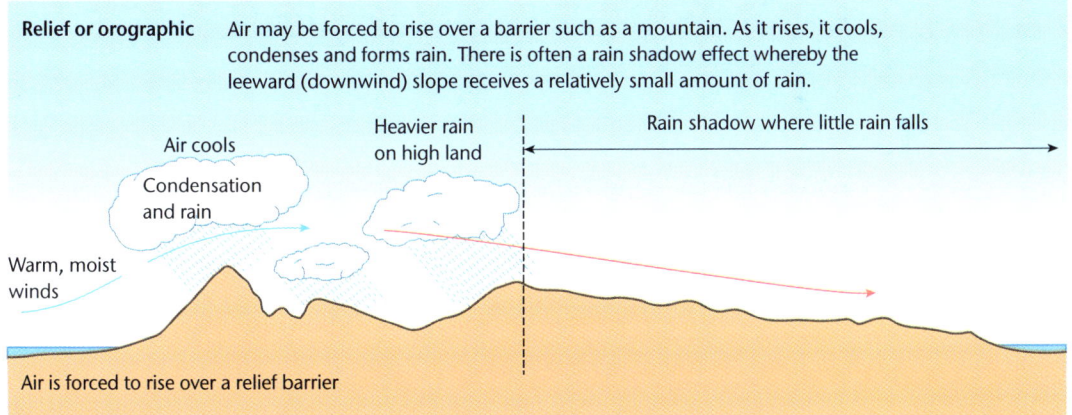

Air may be forced to rise over a barrier such as a mountain. As it rises, it cools, condenses and forms rain. There is often a rain shadow effect whereby the leeward (downwind) slope receives a relatively small amount of rain.

Air is forced to rise over a relief barrier

and seldom large enough to fall. They need to grow first. This can happen due to the collision of small droplets with other droplets in turbulent air, known as the coalescence process, or due to the growth of ice crystals at high altitudes. In this case water droplets are lifted by air currents and grow as sublimation takes place around them. Eventually they become large enough and heavy enough to fall through the rising air currents, usually melting as they fall and enter warmer air, becoming raindrops. If they remain frozen, they fall as hailstones.

*Figure 10.7 The three types of rainfall*

### Examiner's tip

- Do not forget that you covered this material in Unit 1 of AS, as part of your study of the hydrological cycle. However, at A2 you would be expected to have a greater depth of understanding of the mechanisms involved.

### Key terms

**Convectional uplift** When air that has been heated rises.

**Frontal uplift** When warm air rises on meeting cold air.

**Orographic uplift** When air is forced to rise by relief barriers.

# Chapter 10 Atmospheric systems

## High- and low-pressure systems and their associated weather

### Anticyclones

**Anticyclones** are large masses of subsiding air that create high pressure at the ground surface. Subsidence is linked to the general global circulation discussed earlier, with air descending at the poles, creating the polar high-pressure belts and, at the descending limb of the Hadley cell, creating the sub-tropical high-pressure belts. In the mid-latitudes these are often extensions of the sub-tropical high pressure area, particularly in the summer when they shift polewards due to the changing position of the overhead sun. Small extensions of the larger body are called ridges, and often occur between depressions. Anticyclones typically have the following characteristics:

- they are areas of high atmospheric pressure
- they are represented on synoptic charts by a system of widely spaced isobars with pressure increasing towards the centre
- they tend to be fairly static once developed and they sometimes persist in the same location in the mid-latitudes for a few weeks
- the subsiding air warms adiabatically as it descends, producing a decrease in relative humidity and a lack of cloud cover as little, if any, condensation will take place
- winds are weak and flow gently outwards, clockwise in the northern hemisphere, due to the Coriolis force

The typical weather conditions associated with anticyclones over the British Isles vary enormously between summer and winter.

In winter an anticyclone will tend to bring:

- clear skies by day and night due to the lack of condensation from sinking, warming air
- cold daytime temperatures, from sub-zero to a maximum of about 5°C, because even though the skies are clear the sun is at a low angle and its heating intensity is limited
- very cold night-time temperatures, below freezing with frosts, as the lack of cloud cover means that terrestrial radiation is lost to space readily and little heat is trapped in the lower atmosphere
- radiation fog in low-lying areas caused by rapid heat loss at night and condensation occurring at low altitude (if the air has enough moisture)
- temperature inversions (where the air at the ground surface is slightly colder than the air immediately above), which can trap pollutants over urban areas — the resultant dull weather is known as anticyclonic gloom

### Key terms

**Anticyclone** An area of high pressure.

**Depression** An area of low pressure.

**Hurricane** An area of extreme low pressure in the tropics.

*Physical systems, processes and patterns* — A2 Unit 4

## Case study 10.1  Winter anticyclone: Britain, 1962/63

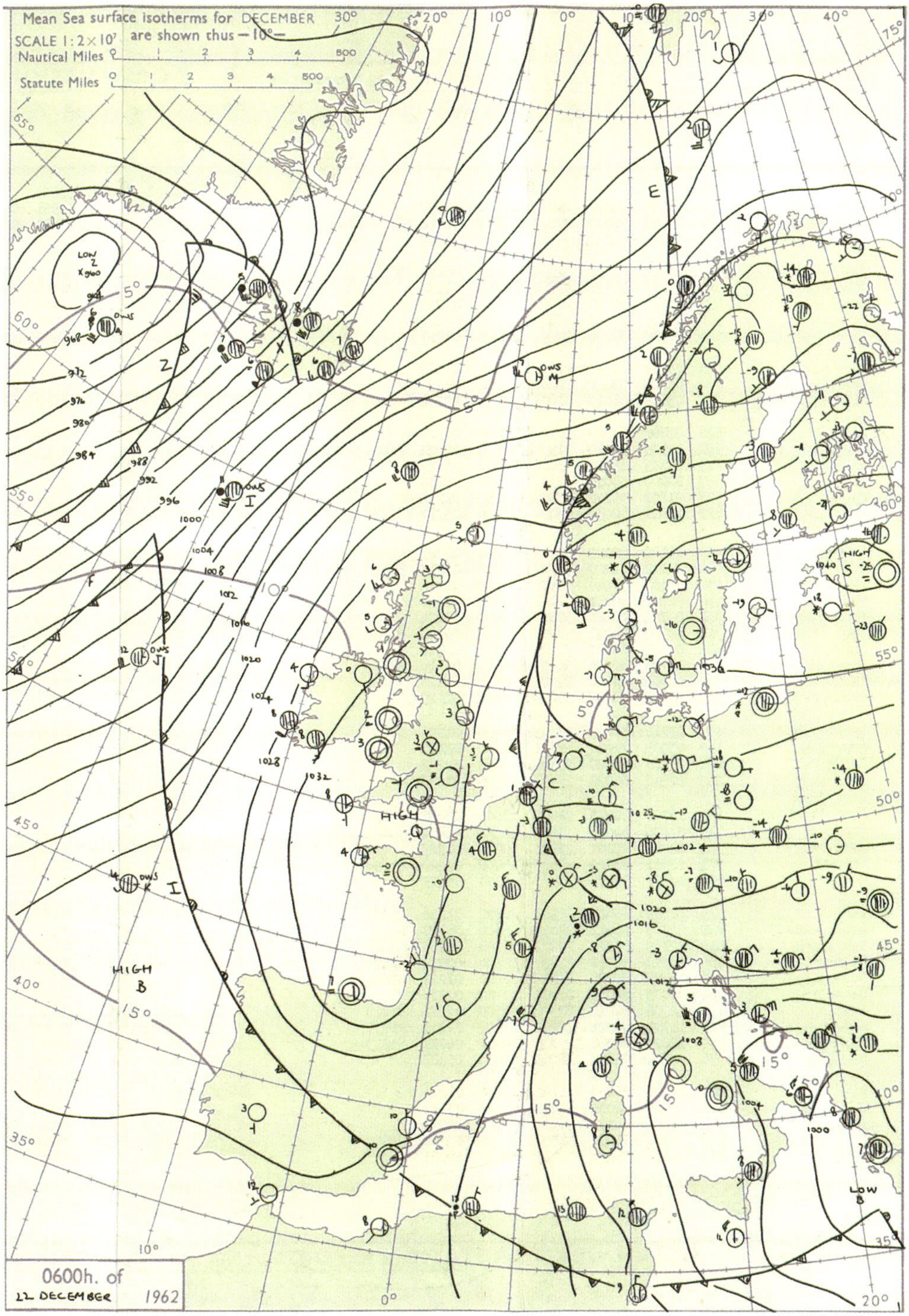

Figure 10.8 The synoptic chart for 22 December 1962

# Chapter 10  Atmospheric systems

The winter of 1962/63 was the coldest in England and Wales since 1740. Anticyclones to the north and east of the British Isles brought bitterly cold winds from the east day after day.

The mean maximum temperatures for January 1963 were more than 5°C below average over most of Wales, the Midlands and southern England and in some places more than 7°C below average.

On 23 December 1962, the pressure in the Scandinavian high reached 1050 millibars, with high pressure extended all the way from the southern Baltic to Cornwall (Figure 10.8). This brought cold easterly winds to much of England and Wales.

A belt of rain over northern Scotland on 24 December turned to snow as it moved south, giving Glasgow its first white Christmas since 1938. The snow belt reached southern England on 26 December and became almost stationary. The following day, snow lay 5 cm deep in the Channel Islands and 30 cm deep in much of southern England.

- Lakes and rivers froze.
- Ice formed on harbours in the south and east of England.
- Patches of ice formed on the sea.
- Huge blocks of ice formed on beaches where waves broke and the spray froze.
- Coastal marine life suffered severely.

In summer anticyclones typically result in:
- clear skies by day and night due to the lack of condensation from sinking, warming air
- hot daytime temperatures, often 25°C or more, as the clear skies and high-angle sun mean intense solar heating
- warm night-time temperatures, often 15°C or more, as the sinking air warms
- early morning mists and dew; these form near the ground as it loses heat at night, but burn off rapidly in the morning
- occasional thunderstorms due to convection created by the intense solar heating under clear skies
- sea fret or haar (advection fog) on the east coast caused by onshore winds giving relatively high moisture levels and an increased likelihood of condensation as the air moves across land surfaces cooled by the loss of heat by terrestrial radiation at night

> **Examiner's tip**
>
> - It is important that you recognise the temporal (summer/winter and day/night) variations in the weather conditions as well as the spatial (coastal/inland and possible north/south) differences.

### Blocking anticyclones

**Blocking anticyclones** can establish themselves over the British Isles and remain stationary for many days or even weeks. These force the mid-latitude depressions that normally track across Britain from the west to pass further north, or even occasionally south. Extreme weather conditions often result, including heat waves and droughts in the summer and dry, freezing spells in the winter.

*Physical systems, processes and patterns* **A2 Unit 4**

### Case study 10.2  Summer anticyclone: Britain, 1994

A blocking anticyclone over northwest Europe led to a long spell of settled, fine weather for Britain in 1994. This led to the second-hottest summer since records began, with July having 197 hours of sunshine. Temperatures were as high as 20°C from 6:00 a.m. and a maximum of 37°C was recorded at Cheltenham in Gloucestershire. These high temperatures had both positive and negative effects on human activity.

- On the positive side, the tourism industry benefited, with many people holidaying in the UK rather than going abroad. Ice cream and soft drink manufacturers experienced increased demand; for some the demand was double their average for the time of year. The high temperatures also helped vine growers in southeast England, who experienced 50% higher yields and were able to produce a large and good quality vintage.
- On the negative side, rail lines buckled in Worcestershire, tarmac roads melted and there was a 20-fold increase in hospital admissions for people with breathing difficulties. Thunderstorms caused 29 London underground stations to be flooded.

### Examiner's tip

- You should try to support your description of human impact with evidence in the form of dates, names of places, facts and figures. It is good practice to refer to both positive and negative impacts.

## Depressions

**Depressions** are low-pressure systems of the mid-latitudes and are associated with the meeting of warm and cold air masses. The traditional model is depressions that form as a wave on the polar front (the area of mixing between cold polar air and warm tropical air, Figure 10.9a). As the warm air moving northeast meets the colder, polar air, the warm air rises over the colder air due to the lower density of

*Figure 10.9 The stages in the development of a depression*

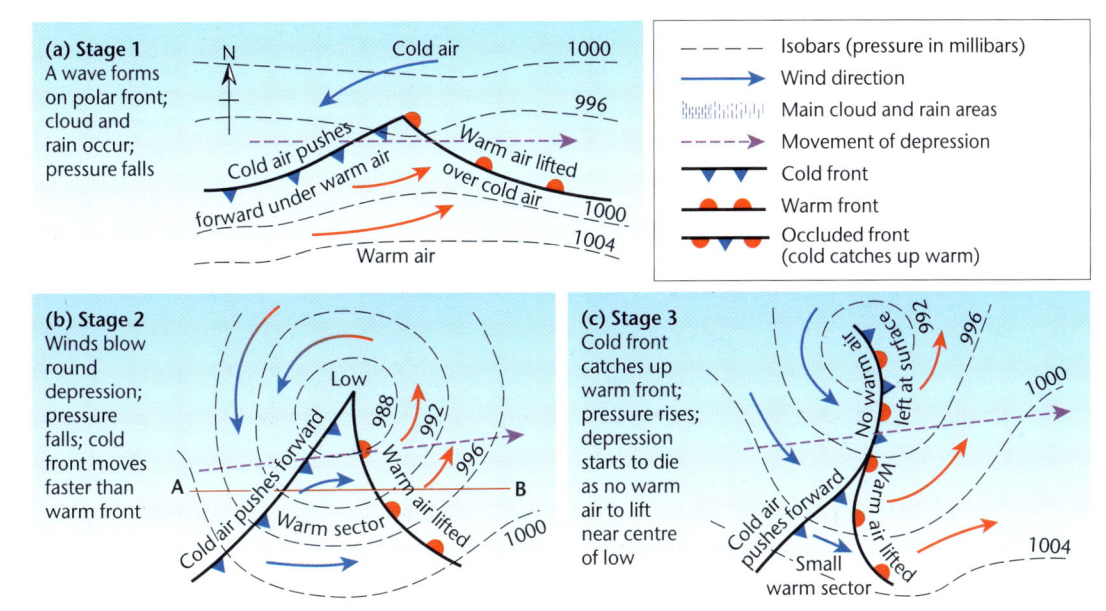

299

# Chapter 10 Atmospheric systems

the warm air, leading to low pressure at the surface. A more recent model is the conveyor belt model, which suggests that cold dry air near the tropopause divides as it sinks and meets the warmer air rising from beneath.

Once formed, these depressions are carried eastwards in the northern hemisphere along the Rossby wave system (Figure 10.9b). They initially deepen and develop distinctive warm and cold fronts. The cold front moves more rapidly than the warm front, eventually catching up with it and producing an occluded front (Figure 10.9c). The warm air between the fronts is then lifted away from the surface, leaving more uniform air at the surface and resulting in the decay of the system.

The typical sequence of weather associated with the passage of a non-occluded depression over the British Isles is summarised in Figure 10.10 and Table 10.1.

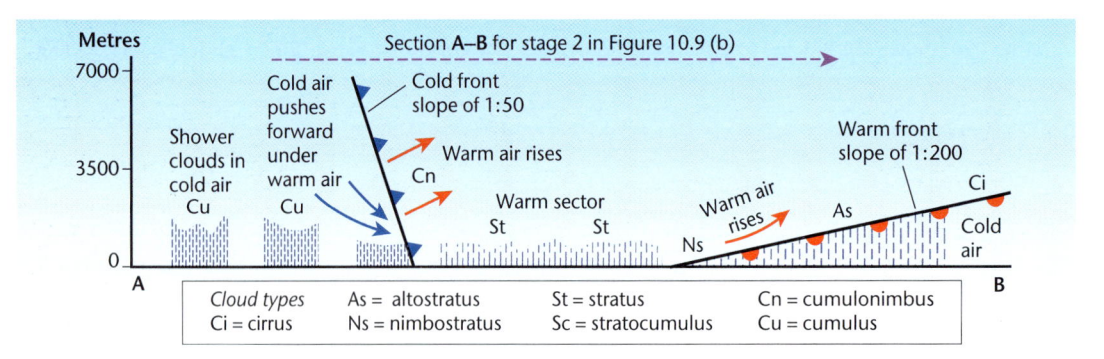

*Figure 10.10 Weather associated with a depression*

Cloud types:
As = altostratus    St = stratus          Cn = cumulonimbus
Ci = cirrus         Ns = nimbostratus     Sc = stratocumulus
                                          Cu = cumulus

| Weather element | Cold front — In the rear | Cold front — At passage | Cold front — Ahead | Warm front — In the rear | Warm front — At passage | Warm front — Ahead |
|---|---|---|---|---|---|---|
| Pressure | Continuous steady rise | Sudden rise | Steady or slight fall | Steady or slight fall | Fall stops | Continuous fall |
| Wind | Veering to northwest, decreasing speed | Sudden veer, southwest to west; increase in speed, with squalls | Southwest, but increasing in speed | Steady southwest, constant | Sudden veer from south to southwest | Slight backing ahead of front; increase in speed |
| Temperature | Little change | Significant drop | Slight fall, especially if raining | Little change | Marked rise | Steady, little change |
| Humidity | Variable in showers, but usually low | Decreases sharply | Steady | Little change | Rapid rise, often to near saturation | Gradual increase |
| Visibility | Very good | Poor in rain, but quickly improves | Often poor | Little change | Poor, often fog/mist | Good at first, but rapidly deteriorating |
| Clouds | Shower clouds, clear skies and cumulus clouds | Heavy cumulonimbus | Low stratus and stratocumulus | Overcast, stratus and stratocumulus | Low nimbostratus | Becoming increasingly overcast, cirrus to altostratus to nimbostratus |
| Precipitation | Bright intervals and scattered showers | Heavy rain, hail and thunderstorms | Light rain, drizzle | Light rain, drizzle | Rain stops or reverts to drizzle | Light rain, becoming more continuous and heavy |

*Table 10.1 How the weather changes*

*Physical systems, processes and patterns*  **A2 Unit 4**

The actual weather experienced in Britain depends upon whether polar maritime or tropical maritime air is dominating at the time.

- Polar maritime air brings average temperatures for the season in winter (5–8°C in January) but cooler than average temperatures in the summer (16–18°C in July). Showers of rain are common in both seasons, with sleet likely in winter.
- Tropical maritime air brings humid and mild weather in winter, with temperatures well above the norm (12–14°C in January). Summers are warmer than average (24–27°C in July), with thunderstorms possible.

## Case study 10.3  Depression: the Great Storm, 1987

*Photograph 10.1 Tree and structural damage in west London following the 1987 Great Storm*

During the afternoon of 15 October 1987, winds were very light over most parts of the UK. The pressure gradient was slack. A depression was drifting slowly northwards over the North Sea off eastern Scotland. A col lay over England, Wales and Ireland. Over the Bay of Biscay, a depression was developing.

- The first gale warnings for sea areas in the English Channel were issued at 06:30 on 15 October and were followed, 4 hours later, by warnings of severe gales.
- At 12:00 on 15 October, the depression in the Bay of Biscay was 970 mb (millibars). By 18:00, it had moved northeast and deepened to 964 mb.
- At 22:35, winds of Force 10 were forecast.
- By midnight, the depression was over the western English Channel, and its central pressure was 953 mb.
- At 01:35 on 16 October, warnings of Force 11 were issued.
- The depression now moved rapidly northeast, filling a little as it did, reaching the Humber estuary at about 05:30, by which time its central pressure was 959 mb. Dramatic increases in temperature were associated with the passage of the storm's warm front.

In southern England, 15 million trees were lost, among them many valuable specimens. Trees blocked roads and railways, and brought down electricity and telephone lines. Hundreds of thousands of homes in

# Chapter 10   Atmospheric systems

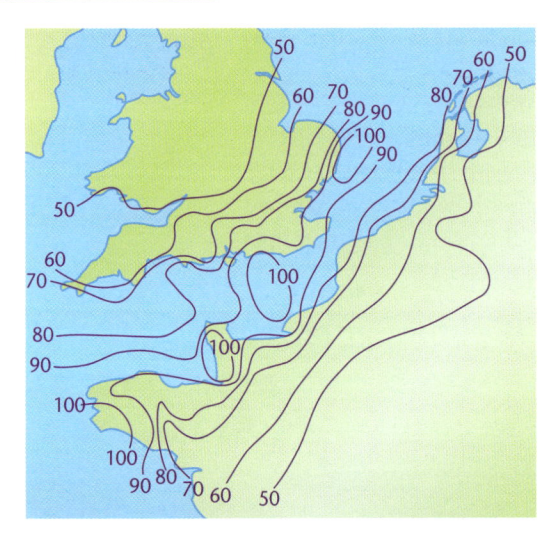

*Figure 10.11 Maximum wind speeds (knots) recorded during the 1987 Great Storm*

England remained without power for over 24 hours. Falling trees and masonry damaged or destroyed buildings and cars. Numerous small boats were wrecked or blown away. A ship capsized at Dover, and a Channel ferry was driven ashore near Folkestone.

The storm killed 18 people in England and at least four more in France. The death toll might have been far greater had the storm struck in the daytime.

- The highest wind speed reported was an estimated 119 knots in a gust soon after midnight at Quimper coastguard station on the coast of Brittany (48° 02' N, 4° 44' W).
- The highest measured wind speed was a gust of 117 knots at 00:30 at Pointe du Roc (48° 51' N, 1° 37' W) near Granville, Normandy.
- The strongest gust over the UK was 106 knots at 04:24 at Gorleston, Norfolk.
- A gust of almost 100 knots occurred at Shoreham on the Sussex coast at 03:10.
- Gusts of more than 90 knots were recorded at several other coastal locations.
- Even well inland, gusts exceeded 80 knots: 82 knots was recorded at the London Weather Centre at 02:50, and 86 knots at Gatwick Airport at 04:30 (the authorities closed the airport).

Figure 10.11 shows the maximum wind speeds recorded.

## Hurricanes

**Hurricanes** are intense low-pressure systems of the tropics. They develop under certain specific conditions:

- a warm tropical ocean (min. 27 °C) to provide a continuous source of heat which generates rising air currents
- a minimum water depth of 60 m as the moisture provides enough latent heat, released by condensation, to drive the system
- a location of at least 5°N or at least 5°S of the equator with northeast and southeast trade winds converging at the surface and, due to the Coriolis force, air spiralling as it converges; hurricanes cannot, therefore, form on the equator itself where the Coriolis force does not exist
- a rapid outflow of air at the tropopause

The spiralling, rising air develops around a calm central eye that is 10–50 km in diameter (Figure 10.12). Wind speeds are at least 118 km h$^{-1}$ and can be as much as 300 km h$^{-1}$, and the whole system may be 200–700 km in diameter. High rainfall occurs, often over 100 mm per day.

Hurricanes can also produce storm surges as a result of the piling-up of wind-driven waves and the heaving-up of the ocean surface under very low pressure. As a hurricane moves westwards it will eventually cross land, lose its source of heat and moisture (the warm sea) and so decline. The land surface also provides friction, which slows down the air movements.

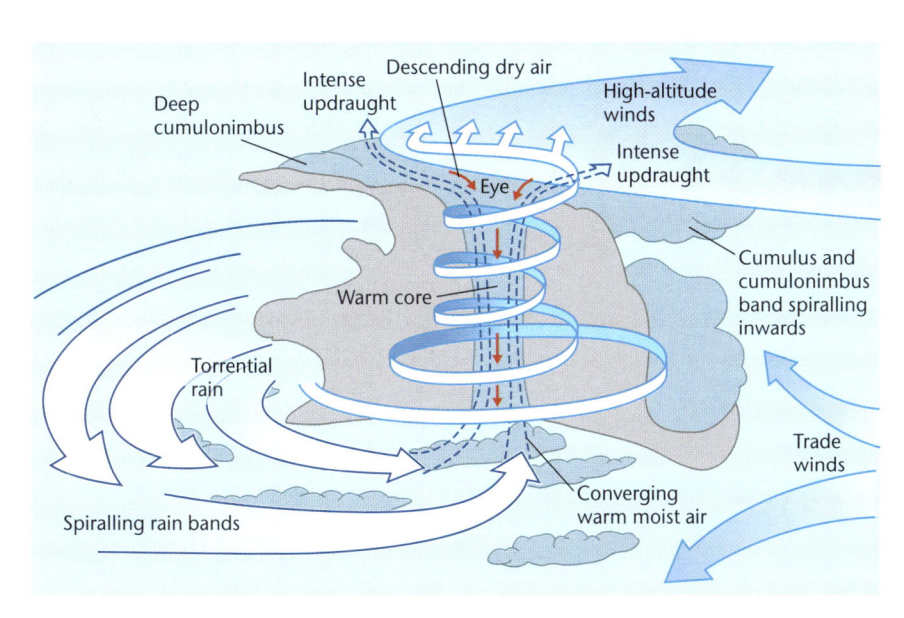

*Physical systems, processes and patterns*  **A2 Unit 4**

*Figure 10.12 The structure of a hurricane*

Tropical storms are known as hurricanes in the Caribbean, as cyclones in south Asia, as typhoons in southeast Asia and as willy-willies in Australia.

## Case study 10.4  Hurricane Mitch

Hurricane Mitch started as a tropical storm in the southern Caribbean Sea on 22 October 1998. It moved slowly westwards, on a typically unpredictable track. The wind speeds gradually increased until by the end of October it was officially classified as a category 5 hurricane — the most severe type.

*Photograph 10.2 An example of the devastation caused by Hurricane Mitch*

# Chapter 10   Atmospheric systems

After moving further westwards it turned southwards, hitting the north coast of Honduras before moving through Nicaragua, El Salvador and Guatemala (Figure 10.13). The winds were as high as 300 km h$^{-1}$, but what made the hurricane so devastating was the fact that it moved so slowly and produced over 1000 mm of rainfall in 5 days, which led to severe flooding. All aspects of life and economy in the countries it passed through were affected: transport, settlement, agriculture and industry.

- Over 19 000 people were killed or were missing.
- Many settlements were destroyed, particularly on the north coast of Honduras.
- Transport links were severely disrupted. Over 50 bridges were destroyed in Nicaragua and the capital, Managua, was cut off.
- 2.7 million people were left homeless.
- Crime rates rose due to looting of abandoned homes and businesses.
- Coffee and banana plantations were severely affected. Even when the crops could be harvested, they could not be sold due to the transport disruption. Losses were put at $1.5 billion. The limited export base in Nicaragua and Honduras meant serious economic implications for their servicing of foreign debt.

Honduras and Nicaragua were already among the poorest countries in the Americas with a GDP of under $700 per person. Their economies suffered enormously and large amounts of foreign aid had to be supplied.

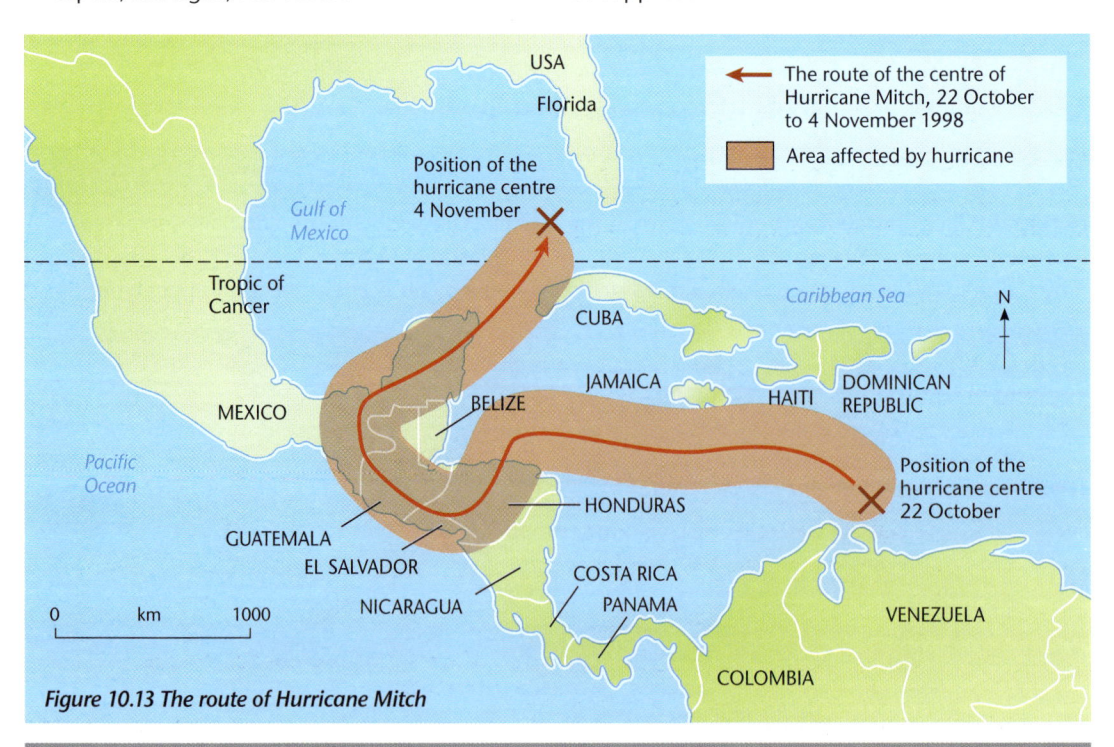

Figure 10.13 The route of Hurricane Mitch

### Examiner's tip

- You should be able to consider the formation, movement and change of each type of system (anticyclone, blocking anticyclone, depression and hurricane). You need to know the weather associated with them and be able to detail their impact on human activity. Case studies 10.1–10.4 provide the information you need.

*Physical systems, processes and patterns*  **A2 Unit 4**

# Weather associated with different air masses

An air mass is a large body of air of fairly uniform temperature and humidity. These masses tend to form in areas of permanent high pressure, such as the sub-tropical high around the Azores, and in polar regions, such as Canada and Siberia. In these areas air can stagnate for several weeks, allowing the air mass to take on the characteristics of the source region.

The British Isles is affected by a variety of different air masses. Each one tends to bring its own typical weather conditions and these are classified according to the **source region** and the **track**.

- The source region is where the air mass originated. This tends to control the temperature of the air mass — hence sub-tropical air masses bring warm air and polar air masses bring cold air. The source regions for air masses affecting Britain are generally anticyclonic areas, out of which air moves towards low-pressure areas that often cross the British Isles.
- The track is the route taken by the air mass. It affects temperature, so passing over a cold land mass lowers the temperature, for instance, making the air mass more stable. The track also affects the moisture content. Air masses following oceanic tracks increase in humidity whereas those following continental tracks tend to experience little change in their moisture content.

### Examiner's tip
- You only need to study this section in relation to the British Isles.

The air masses affecting the British Isles are shown in Figure 10.14. Information about the air masses is given in Table 10.2. The abbreviations used in Figure 10.14 are explained in Table 10.2.

Air masses vary both temporally and spatially in the impact they have on the weather in the British Isles. Table 10.2 indicates the differing impacts that air masses have at different times of the year (temporal variation). Not all masses are common throughout the year; tropical continental air masses are not, for instance, received in the winter. The effects of polar continental air are very different depending on the season, whereas the impact of returning polar maritime air is similar throughout the year.

Table 10.3 reveals some important differences in the spatial pattern of air masses. For example, Stornoway in the north of Scotland has a different frequency of air mass types from Kew in the southeast of England. One difference is that Stornoway experiences Arctic air masses 11.3% of the time, while Kew experiences them only 6.5% of the time. When locations are not experiencing passing air masses, they are under the direct influence of depressions or anticyclones. Table 10.3 also reveals that Kew experiences anticyclonic conditions much more frequently than Stornoway.

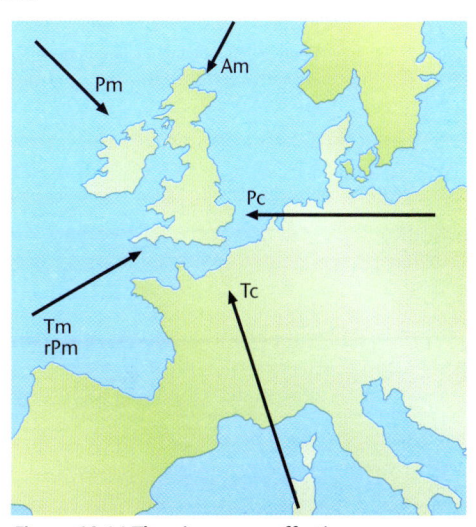

*Figure 10.14 The air masses affecting the British Isles*

### Key terms

**Air mass** A large body of air of uniform characteristics.

**Source region** Where the air mass originates.

**Track** The type of surface the air mass passes over.

# Chapter 10 Atmospheric systems

| Air mass | Winter | Summer |
|---|---|---|
| Arctic maritime (Am) | A very cold air mass<br>Reaches the British Isles on a northerly airstream<br>Temperatures well below average (Jan. max. −1 to 2°C)<br>Unstable as it is warmed by the North Atlantic (8–10°C)<br>Wintry showers, especially in northern Britain and on exposed north-facing coasts<br>Brings our most severe winter weather | Brings a cool northerly airstream with temperatures well below average (July max. 12–13°C)<br>Unstable<br>Rain showers, more especially in the north |
| Polar maritime (Pm) | Brings average temperatures in winter on a northwesterly airstream<br>Unstable after crossing the North Atlantic<br>Showers of rain and sleet, particularly in western areas<br>Instability dies down at night<br>There may be clear skies and frost | Cool conditions with daytime maxima in July only 16–17°C<br>Unstable<br>Showers in the west with possibility of thunderstorms |
| Polar continental (Pc) | Arrives from the Continent on an easterly airstream<br>A cool air mass, warmed as it crosses the North Sea<br>Unstable on reaching eastern Britain as it is warmed by the North Sea (6–8°C)<br>Wintry showers that die out towards the west<br>Daytime temperatures only just above freezing<br>Frost at night | The air mass is warm in its source region, but cools and becomes stable as it crosses the cool North Sea<br>Advection fog common along the east coast north of the Humber in spring and early summer<br>Inland clear skies and temperatures of 20–25°C |
| Returning polar maritime | Similar to Tm air in both winter and summer<br>A polar air mass forced further south than usual; it approaches the British Isles from the southwest | |
| Tropical maritime (Tm) | Source region is the Azores<br>Comes to the British Isles on a southwesterly airstream<br>A humid, stable and mild air mass<br>Temperatures well above average (Jan max.10–13°C)<br>Low stratus cloud and fog on high ground | Stable in summer after being cooled by the Atlantic<br>Some advection fog around southwest coasts<br>This quickly evaporates inland<br>Clear skies and temperatures around 25°C |
| Tropical continental (Tc) | Does not usually occur in winter | Occasionally affects the British Isles<br>The air mass comes from north Africa and is hot<br>Very low humidity and largely cloudless<br>Heat wave conditions (28–30°C)<br>Poor visibility (haze) because of pollution from the continent |

(Above) Table 10.2 Air masses and the seasons

| Air mass | Kew | Stornoway |
|---|---|---|
| Am | 6.5 | 11.3 |
| Pm | 24.7 | 31.5 |
| rPm | 10.0 | 16.0 |
| Pc | 1.4 | 0.7 |
| Tm | 9.5 | 8.7 |
| Tc | 4.7 | 1.3 |
| Anticyclones | 24.3 | 13.8 |
| Depressions | 11.3 | 11.8 |

(Left) Table 10.3 Percentage frequency of air masses at Kew (London) and Stornoway (northwest Scotland)

*Physical systems, processes and patterns* **A2 Unit 4**

# Local effects of human and physical factors on the atmosphere

## Lapse rates and atmospheric stability

**Lapse rates** are rates of temperature change with altitude. **Adiabatic** means an internal change without external gains or losses. A knowledge of lapse rates is important in understanding whether the air is unstable (rising) or not, as this is closely linked to rain formation (this has been discussed earlier).

The key is to investigate the relationship between individual 'parcels' of air and the air around them — the environmental air (Table 10.4).

The **environmental lapse rate** (ELR) is not a fixed rate. It varies with local conditions such as:
- height — it is lower near the ground than further up in the troposphere
- humidity — it is lower in humid air than in dry air
- season — rates are lower in winter than in summer
- time — there are variations between day and night depending on rates of terrestrial radiation

The **saturated adiabatic lapse rate** (SALR) also varies a little; it is usually between 4°C per 1000 m for warm air and 9°C per 1000 m for cold air. This is because warm air holds more moisture than cold air and so there is more heat released during condensation as it cools.

A rising parcel of air will cool at the **dry adiabatic lapse rate** (DALR) until it becomes saturated. When air is saturated it is holding all the water vapour that it can (100% relative humidity). The temperature at which this happens is called the **dew-point temperature** and the height at which this happens is the **condensation level**. The dew-point temperature is not a fixed figure. If the air is lacking moisture, it will need to cool to a lower temperature to become saturated compared to very moist air, which will soon become saturated as the air cools.

If the parcel of air continues to rise, condensation takes place (water vapour is converted into microscopic water droplets). Clouds form, as the air cannot hold more water vapour than it has when it is saturated. Condensation releases latent heat and so the rising air does not cool so quickly when condensation is occurring. This is why the SALR is lower than the DALR.

*Table 10.4 Lapse rates*

| Name | Abbreviation | Rate | Application |
|---|---|---|---|
| Environmental lapse rate | ELR | Average 6.5°C per 1000 m | The general change in air temperature with height through the troposphere at any given time |
| Dry adiabatic lapse rate | DALR | 10°C per 1000 m | The rate at which a rising parcel of air cools before saturation, or the rate at which it warms when sinking |
| Saturated adiabatic lapse rate | SALR | Approximately 5°C per 1000 m | The rate at which a rising parcel of air cools after saturation, or the rate at which it warms when sinking |

# Chapter 10  Atmospheric systems

At any given height, if the temperature of the parcel of air is higher than the temperature of the environmental air, then the parcel will rise as it is less dense and so the air is **unstable**.

If the temperature of the parcel of air is lower than the temperature of the environmental air, then the parcel will not rise as it is denser and so the air is **stable**. The parcel of air will actually sink as it is denser than the air around it.

If the parcel of air is at the same temperature as the environmental air, then conditions are described as **neutral**. The parcel of air will neither rise nor sink as it has the same density as the surrounding air.

If the parcel of air is cooler than the environmental air at a low level but warmer than the environmental air at a high level, then it requires a trigger to uplift it. In these cases it is said to be **conditionally unstable** as the instability is conditional upon there being an uplift trigger. This is likely to be in the form of an orographic (relief) barrier or the meeting of warm and cold air at a front.

These situations of stability and instability are often shown graphically, as in Figure 10.15.

*Figure 10.15 Variations in atmospheric stability*

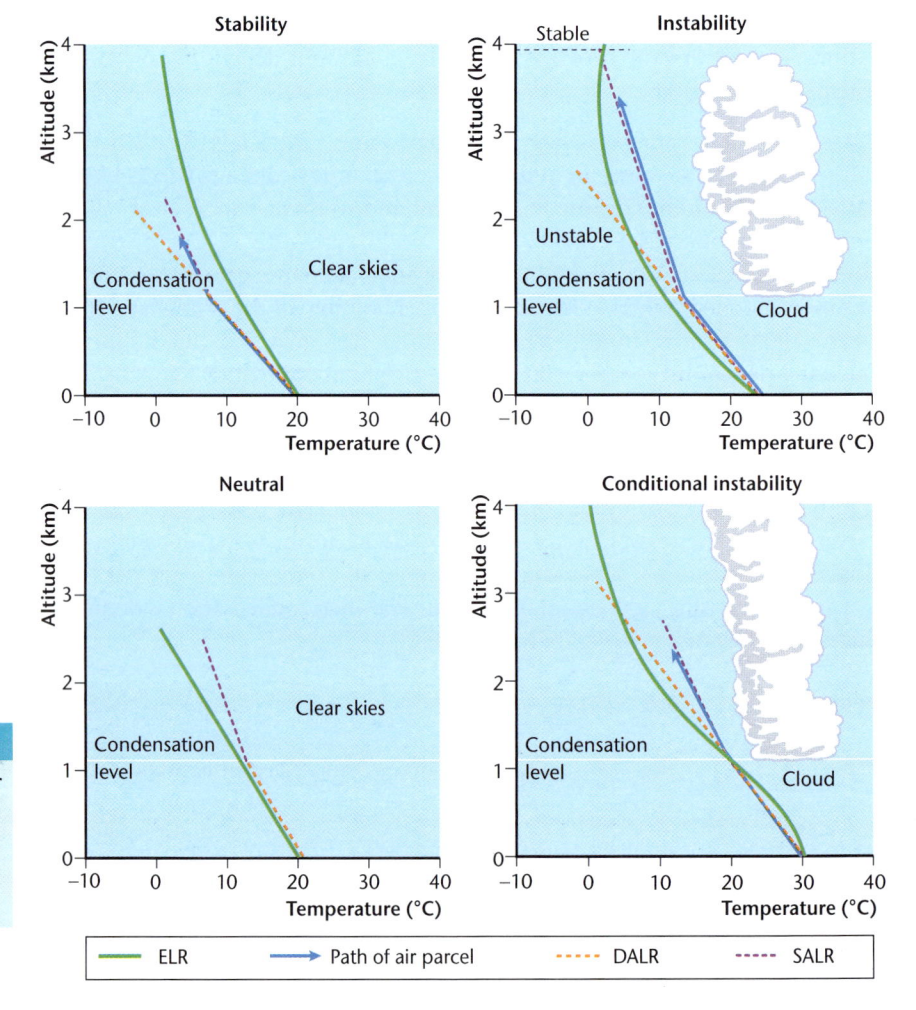

> **Examiner's tip**
> 
> ■ Do remember that the ELR and the SALR are not fixed rates; they vary.

ELR ——  Path of air parcel ——▶  DALR ----  SALR ----

308

*Physical systems, processes and patterns* **A2 Unit 4**

### Key terms

**Conditional instability** When air rises due to a trigger mechanism.

**Instability** When air is rising, as it is warmer and less dense than the surrounding air.

**Neutral stability** When there is no rising or sinking of air.

**Stability** When air is descending, as it is colder and denser than the surrounding air.

## Characteristics of weather associated with different states of stability

There is a simple but strong link between stability and high pressure and also between instability and low pressure. The weather associated with these two states is therefore generally comparable with that described earlier for low pressure and high pressure systems (Table 10.5).

| Stability | Pressure system | Weather |
|---|---|---|
| Stable | High (anticyclone) | Calm, clear, dry, settled |
| Unstable | Low (depression) | Cloudy, wet, windy, changeable |

*Table 10.5 Stability, pressure and weather*

The key reason for the differences shown in Table 10.5 is that stable, sinking air is warmed and so has a decreasing relative humidity. The air will not therefore become saturated and so no condensation, cloud or rain can occur. However, it should be noted that these are generalisations and the specific weather experienced is influenced by a number of different variables. The particular situation has to be looked at closely before a full, accurate explanation can be offered.

## Measurement and recording of local weather

Local weather can be recorded using relatively simple instruments. Your own school or college may have a weather station of its own. This can be particularly helpful in carrying out investigations into microclimates, a topic that some students use as a base for a fieldwork investigation in Unit 3. Some centres may even have access to a remote station linked via satellite technology, which enables fuller records to be kept over a long period of time. Even so, knowledge of the instruments is still necessary and is summarised in Table 10.6.

| Weather variable | Instrument | Units |
|---|---|---|
| Temperature | Thermometer | °C |
| Max and min temperature | Six's thermometer | °C |
| Relative humidity | Hygrometer | % |
| Precipitation | Rain gauge | millimetres |
| Pressure | Barometer | millibars |
| Wind speed | Anemometer | $m\ s^{-1}$, $km\ h^{-1}$, knots |
| Sunshine | Campbell Stokes recorder | hours |
| Cloud cover | Observation/estimation | oktas |

*Table 10.6 Recording local weather*

# Chapter 10 Atmospheric systems

You should be aware of the limitations of any measurements made. Some methods may be judgmental (e.g. cloud cover), some may only give data for a particular point in time (e.g. temperature), while some may require an estimate of an average value (e.g. wind speed). Others are based on the weather experienced over a period of time (e.g. precipitation during a 12- or 24-hour period).

Issues such as the design and siting of the weather station should also be considered. Most of the instruments should be housed in a Stephenson screen which has the following design:

- it is white, to reflect sunlight and prevent overheating
- it is louvred, to allow air to circulate around the instruments but to protect them from direct sunlight
- it is 1 m off the ground to ensure that it is air temperature and not ground temperature which is recorded
- it has a sloping roof to allow water and snow to move off easily and prevent insulation

The Stephenson screen should be sited away from buildings and trees: these can provide shelter from wind, shade from sun and buildings can release stored heat energy.

The use of local data as an aid to forecasting can be attempted, particularly if local meteorological data are also available for comparison. Recording the changing weather during the passage of a depression can be a very worthwhile exercise in helping to describe and explain the typical sequence. The effect of blocking anticyclones can also be determined by recording the weather characteristics during such an event.

> **Key terms**
>
> **Urban heat island effect** The increasing temperature with distance towards the centre of an urban area.
> **Urban microclimate** The distinctive climate of an urban area compared to its surrounding rural climate.

> **Examiner's tip**
>
> - Any data collected can be useful as evidence in support of ideas you are conveying in an essay.

## Urban microclimates

**Urban microclimates** are the local climatic conditions of a built-up or urban area and they tend to differ noticeably from those of the surrounding rural areas. Urban microclimates tend to exhibit distinctive features of temperature, wind, visibility and precipitation and, particularly in large urban areas, these may vary spatially across the area. There is also a temporal dimension here, in that some differences are more pronounced at particular times of the year and even at certain times of day.

The impact may extend to 250–300 m upwards and as much as 10 km downwind. Some geographers refer to this as the **climatic dome**. Within this, two levels can often be recognised. Below the average roof level there is an **urban canopy** where processes act within the spaces between buildings. Above this is

# Physical systems, processes and patterns — A2 Unit 4

the **urban boundary layer**, whose characteristics are governed by the nature of the urban surface. The dome extends downwind as a plume into the surrounding rural area. These areas are shown in Figure 10.16.

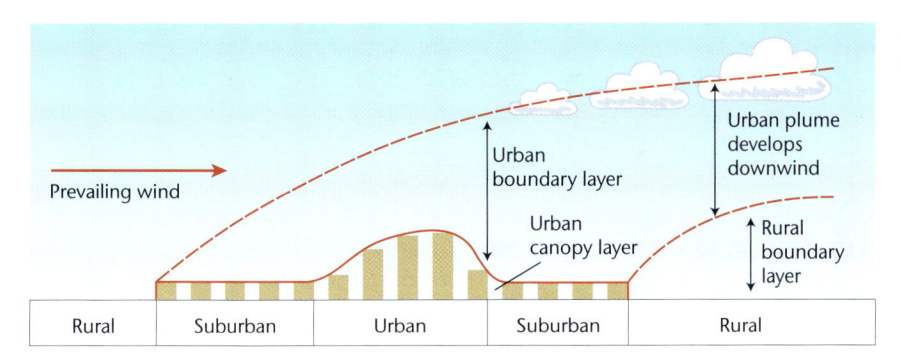

*Figure 10.16 The urban climate dome*

## Temperature

Urban areas tend to have higher mean annual and higher winter minimum temperatures than the surrounding rural areas and the differences are generally greater towards the centre of the urban area. If temperatures are mapped, the isotherms tend to appear as an island on a contour map and so the term **urban heat island** is often applied. The differences are due to the following.

- Heat is given off by factories, vehicles and homes, all of which burn fuel and produce anthropogenic heat.
- Urban surfaces, such as concrete and tarmac, absorb large amounts of solar radiation during the day and release it slowly at night. Some surfaces, such as glass, also have high reflective capacities, and multi-storey buildings in particular tend to concentrate the heating effect in the surrounding streets by reflecting energy downwards.
- Smog and pollutants form a **pollution dome** which allows short-wave insolation to enter but which traps outgoing terrestrial radiation as this is of a longer wavelength.
- Precipitation falling in urban areas is quickly removed from the surface by drainage systems. Reduced evapotranspiration means more energy is retained by the atmosphere.

Variations occur over time. The greatest differences between urban and rural areas happens under calm, anticyclonic conditions, especially if a temperature inversion forms in the boundary layer. Heat islands are also more significant in winter when there is a greater addition of heat energy due to the use of domestic heating systems. Differences are more distinct at night too, when the impact of insolation is absent and the urban surfaces are slowly releasing the heat they stored during the day.

Differences may well occur spatially across the urban area. Industrial zones may produce more heat than retail areas, for instance. Tall buildings may provide lots of shade in certain streets. In general the edge of the urban heat island is

# Chapter 10  Atmospheric systems

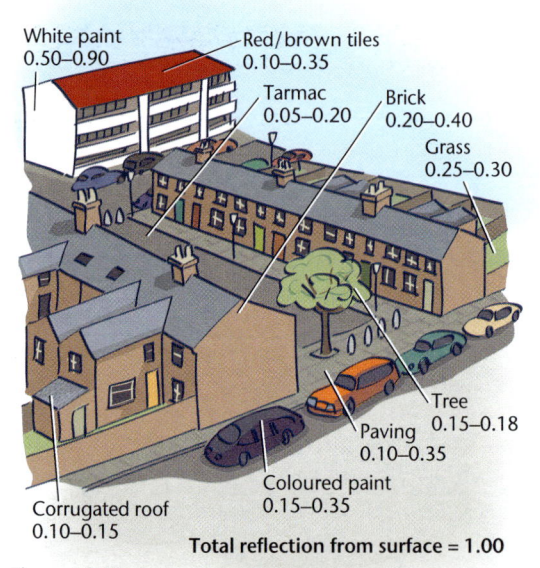

**Figure 10.17** Urban albedos

- White paint 0.50–0.90
- Red/brown tiles 0.10–0.35
- Tarmac 0.05–0.20
- Brick 0.20–0.40
- Grass 0.25–0.30
- Tree 0.15–0.18
- Paving 0.10–0.35
- Coloured paint 0.15–0.35
- Corrugated roof 0.10–0.15
- Total reflection from surface = 1.00

usually well defined, with significant increases in temperature occurring. From here to the city centre temperatures typically rise by 2–4 °C km$^{-1}$. However, the specific pattern will vary according to the albedo of the surfaces found. Grass-covered areas within the city will reflect much more insolation than concrete and tarmac surfaces and so temperatures here would be relatively lower.

The release of heat from buildings is slow and so urban temperature change often lags behind the general diurnal (daily) and seasonal patterns.

Examples of different urban albedos are shown in Figure 10.17. Compare the values in Figure 10.17 with the values for typical rural surfaces, such as pine forest (0.14) and deciduous forest (0.17).

## Wind

On average, urban areas tend to have lower wind speeds than surrounding areas, as tall buildings provide a frictional drag on air movements. This creates turbulence, giving rapid changes in both direction and speed. There is a general decrease in wind speeds towards city centres from the suburbs.

However, maximum wind speeds are often higher due to the channelling effect of very tall buildings. The windward side of a building may well have a relatively high pressure as the air pushes against it. On the leeward side, pressure is lower and air is drawn form the windward high pressure to the leeward low pressure across a steep, localised pressure gradient. Such winds may be strong and gusty. The spacing of buildings is also a factor. Closely packed buildings tend to act in combination, whereas more widely spaced buildings act individually to disrupt air flow. Figure 10.18 shows how the air flow in urban areas is modified by more than one building.

In some cases the urban heat island effect may change the local wind patterns completely. A localised low pressure area may develop as warm air rises over the urban area, drawing in air from the surrounding area. Typically, winds are 20–30% lower in urban areas. However, it is important that air flow does occur so that pollutants can be dispersed.

(a) Widely spaced buildings act like single buildings

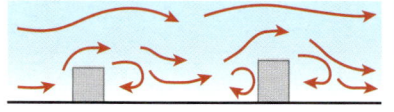

(b) Narrower-spaced buildings – flows interfere with each other

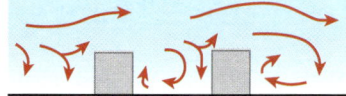

(c) Very close spacing causes winds to skim over the top

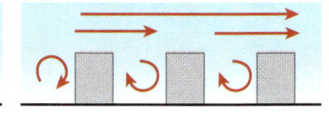

**Figure 10.18** Urban air flow

*Physical systems, processes and patterns*  **A2 Unit 4**

## Case study 10.5  Urban temperature patterns: Montreal, Canada

Figure 10.19 shows the temperature pattern over Montreal at 07:00 hours on 7 March 2000. Winds were cold and from the north at an average speed of 0.5 m s$^{-1}$. Several distinctive features can be seen.

- The lowest temperatures are in the north of the city, facing the direction of the cold winds.
- The edge of the urban area and the −15°C isotherm correspond closely.
- There is a general incidence of slightly higher temperatures towards the centre of the city, typically 5°C warmer than at the edge.
- The temperatures are 1–2°C lower over the park area, which is an open space with plants.

However, the pattern is far from perfect and there must be variables at work in the urban area that the map does not show, such as building heights and materials.

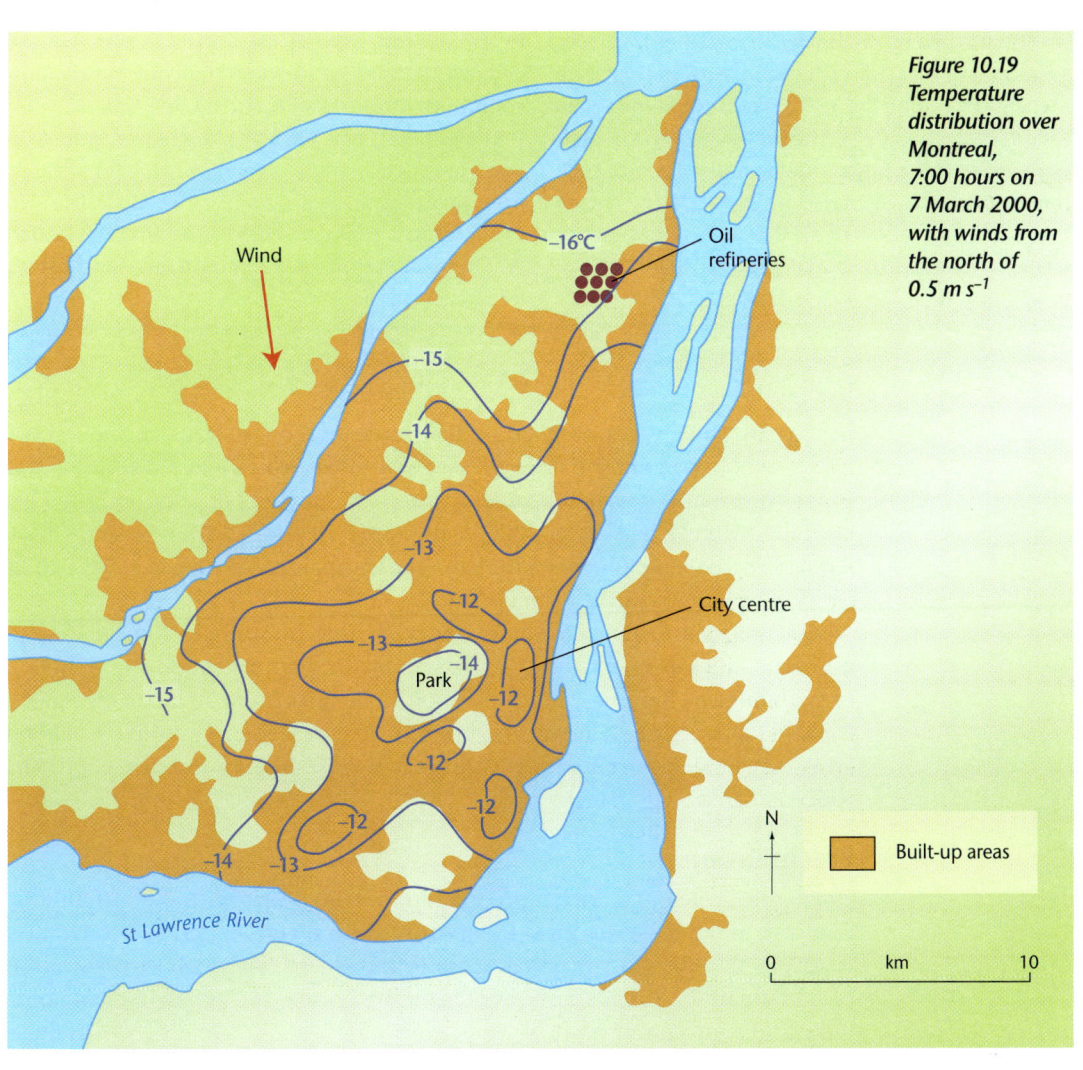

Figure 10.19 Temperature distribution over Montreal, 7:00 hours on 7 March 2000, with winds from the north of 0.5 m s$^{-1}$

# Chapter 10 Atmospheric systems

## Case study 10.6 Urban wind patterns: London

In 1935 winds over 38 km h$^{-1}$ were recorded in suburban Croydon for a total of 371 hours over the year. However, in more central and more built-up Kensington, only 13 hours of such winds were recorded. In 1961/62 studies were carried out to investigate the diurnal differences in wind speeds. At Heathrow airport, beyond the western edge of the built-up area of London, mean annual wind speeds at 13:00 hours were 10.2 km h$^{-1}$ compared to 3.4 km h$^{-1}$ in central London. At night, however, the situation was different; Heathrow averaged 7.8 km h$^{-1}$ and the central area 9 km h$^{-1}$. The reason for this is the greater turbulence created in the urban area as stored heat is released and temperatures rise.

## Case study 10.7 Urban rainfall: Greater Manchester

Recent research by the University of Salford has suggested that the building of high-rise tower blocks in Manchester during the 1970s has brought more rain to some parts of the city. The prevailing wind is westerly, and rainfall in areas downwind of Salford, such as Stockport, has increased by as much as 7% over the past few decades. This is the result of turbulence created by the micro-scale effect of tall buildings forcing air upwards. The heat island effect, which leads to temperature being as much as 8°C higher in the city centre than the surrounding countryside, also contributes, causing air to rise by convection.

In the north and east of Manchester, pollution levels are particularly high and particulates released from factories and cars appear to reduce the size of raindrops, making them heavy enough to fall without coalescing, creating more of Manchester's trademark drizzle.

### Visibility

In the past, smog was a problem in many industrial urban areas, mainly caused by the combustion of fossil fuels, the emissions from which reduced visibility and caused respiratory problems. In December 1952 a smog in London lasted for several days and caused over 4000 deaths. Most developed countries now have legislation that has ended the problem. In particular, zones have been created in which only smokeless fuels can be burnt.

However, photochemical smog, caused by intense sunlight combined with nitrogen oxides, is now creating similar problems. Photochemical smog is often compounded by the temperature inversion sometimes found in urban areas, with cold air trapped beneath a layer of warmer air at or near condensation level. Urban areas may have 100% more fog in winter and 30% more in summer, compared with rural areas.

### Precipitation

Urban areas often have higher precipitation totals compared to the surrounding rural area as tall buildings provide a micro-scale orographic effect and convection is very active due to the urban heat island effect. The high concentration of particulate pollutants also provides more condensation nuclei, and some industrial sources and power stations emit water vapour as steam. On average, there are 10% more rainy days and 5–30% more rainfall in total in urban areas.

However, the lack of water and vegetation may mean less evapotranspiration and so lower humidity levels in urban areas; humidity can be 2% lower in winter

*Physical systems, processes and patterns*  **A2 Unit 4**

and about 10% lower in summer. The higher temperatures also mean that snow is less likely than in rural areas and if it does fall, it thaws relatively quickly. It is estimated that there are 14% fewer snow days in urban areas than rural areas. Thunderstorms, however, are more common due to the intense convection that can occur, particularly during hot summer evenings.

> **Examiner's tip**
>
> - It is important that you can illustrate and support your theoretical observations with evidence from at least one located example.

## Examination questions

**1** Figure 10.20 shows the relationship between air temperature and moisture content.

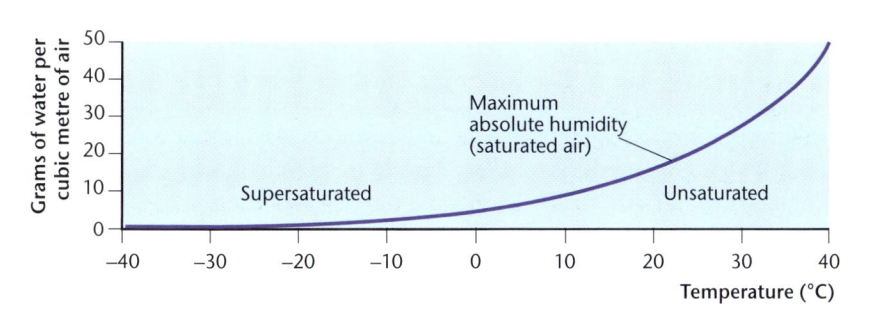

*Figure 10.20*

(a) Define the terms 'saturation' and 'dew point'. (5)
(b) Describe and explain the characteristics of weather associated with different states of stability. (20)

*Total = 25 marks*

**2** The synoptic chart (Figure 10.21) shows a depression over the British Isles.

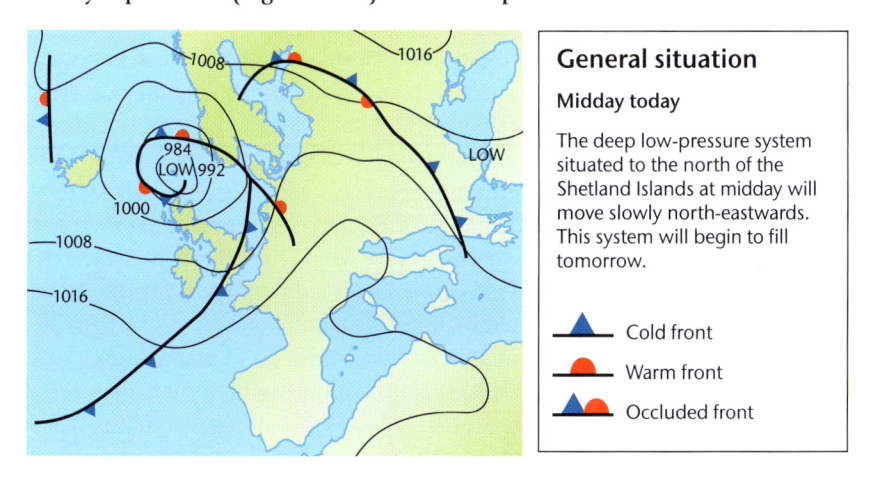

*Figure 10.21*

**General situation**

**Midday today**

The deep low-pressure system situated to the north of the Shetland Islands at midday will move slowly north-eastwards. This system will begin to fill tomorrow.

▲ Cold front
● Warm front
▲● Occluded front

(a) Outline the conditions leading to the formation of a depression in the mid-latitudes. (5)
(b) Describe and account for the weather associated with the passage of a depression across the British Isles. (20)

*Total = 25 marks*

315

# Chapter 10 Atmospheric systems

# Synoptic link
## Weather and human activity are interdependent

- Possible impacts of humans on weather and climate, including pollution, ozone destruction, global warming, cloud seeding.
- Weather hazards and their impact on human activity, to include hurricanes, tornadoes and drought, with an emphasis on recent events.

## The dominant view

The Earth's climate has always been changing. Only 20 000 years ago, much of northern Europe was still covered in an enormous glacier that was up to 3 km thick. Sudden climate shifts happened quite frequently during the Pleistocene, and made the area covered by ice expand or contract. In the cold climate south of the ice-covered areas, small groups of people hunted reindeer, wild horses and bison. Over thousands of years, the Earth's orbit around the sun changed so that the summers became warmer, and the ice began to melt. The ice age ended around 10 000 years ago. Since then, the climate in the northern hemisphere has been warmer and far more stable. Over the last 10 000 years of milder climate, people have developed agriculture, then cities, and most recently industries. The dominant view is that this has caused increasing emissions of carbon dioxide and other greenhouse gases. As a result many experts believe that over the next 100 years we will see the fastest warming of the Earth since the end of the ice age.

Man-made climate change occurs because we emit greenhouse gases to the atmosphere. These emissions come from many sources, including the factories and agriculture that supply us with food and other material goods, power stations that provide us with electricity, and cars and planes that take us where we want to go.

The greenhouse effect is a natural phenomenon: water vapour, carbon dioxide and other gases in the atmosphere allow sunlight to pass through, but then absorb much of the heat from the Earth that otherwise would have escaped to outer space. Without the natural greenhouse effect, the mean temperature on Earth would be about −18°C and the Earth would be uninhabitable.

Emitting greenhouse gases in larger amounts through human action increases the concentration of these gases in the atmosphere, which then increases the greenhouse effect so that more heat is trapped by the atmosphere. This can increase the temperature of the atmosphere and change the climate on Earth. This man-made enhancing of the greenhouse effect is known as **global warming**.

## Evidence of changing climate

- Global mean surface temperatures have increased by 0.5°C since the late nineteenth century (Figure 10.22).

*Physical systems, processes and patterns*  **A2 Unit 4**

- The twentieth century's ten warmest years all occurred since 1985.
- Of these, 1998 was the warmest year on record.
- The snow cover in the northern hemisphere and floating ice in the Arctic Ocean have decreased.
- Globally, sea levels have risen by 100–120 mm over the past century.
- Worldwide precipitation over land has increased by about 1%.
- The frequency of extreme rainfall events has increased throughout much of the USA.

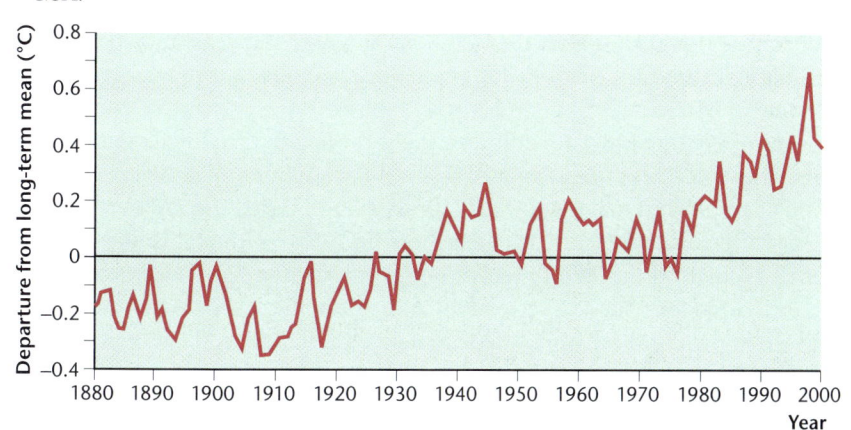

*Figure 10.22 Global temperature change, 1880–2000*

Increasing concentrations of greenhouse gases are likely to accelerate the rate of climate change. Many scientists expect that the average global surface temperature could rise by 0.6–2.5°C in the next 50 years, and 1.4–5.8°C in the next century, with significant regional variation. Evaporation will increase as the climate warms, and this will increase average global precipitation. Soil moisture is likely to decline in many regions, and intense rainstorms are likely to become more frequent. Sea level is likely to rise about 60 cm along most of the US coast.

**Examiner's tip**

- It should be appreciated that one of the difficulties with the evidence about climate change is that it is from a relatively short time period. Accurate records of global climate are only about 200 years old and, of course, there is much spatial variation that makes the reading of global trends very difficult indeed.

## Linkages between ozone and global warming

Ozone (the gas $O_3$) is produced and found in a thin layer of the upper atmosphere where it absorbs harmful ultraviolet radiation from the sun. When stratospheric clouds form, reactions take place between ozone gas and chlorine, bromine and other halogen gases on the surface of ice or water droplets in these clouds, depleting the ozone. Most of those halogens are from chlorofluorocarbons emitted into the atmosphere by industrial processes. Thus a 'hole' has appeared in the ozone layer, at both poles. The loss of this protective layer has led to an increase in harmful radiation and is widely blamed for increases in the occurrence of skin cancers.

Because the number of particles that form in polar stratospheric clouds is extremely sensitive to changes in temperature, the reactions that lead to ozone depletion are also very sensitive, occurring only below about −78°C.

# Chapter 10  Atmospheric systems

> **Examiner's tip**
> 
> - The 'hole' in the ozone layer is often confused with greenhouse gases and this linkage between them might help tie together two very different products of human behaviour.

Greenhouse gases cause atmospheric warming at the Earth's surface, in the troposphere. However, they cool the higher level stratosphere, where the ozone occurs, and thus are a likely cause of the increased ozone depletion.

## Global warming: the sceptics' view

Given that climate is subject to many variations over time, the global air temperature increase of about 0.45°C over the past 100 years may be no more than normal climatic variation, and several biases in the data may be responsible for some of this increase.

Some of the satellite data indicate a slight cooling in the climate in the last 18 years. These satellites use advanced technology and are not subject to the heat island effect around major cities that affects ground-based thermometers.

Projections of future climate changes are uncertain. Although some computer models predict warming in the next century, these models are very limited. The effects of cloud formations, precipitation, the role of the oceans, or the sun, are still not well known and are often inadequately represented in the climate models, although all play a major role in determining our climate. Scientists who work on these models are quick to point out that they are far from perfect representations of reality, and are probably not advanced enough for direct use in policy implementation. Interestingly, as computer climate models have become more sophisticated in recent years, some predicted increases in temperature have been lowered.

### Human impact

In fact 98% of total global greenhouse gas emissions are natural (mostly water vapour); only 2% are from man-made sources. By most accounts, man-made emissions have had no more than a tiny impact on the climate. Although the climate has warmed slightly in the last 100 years, 70% of that warming occurred prior to 1940, before the upsurge in greenhouse gas emissions from industrial processes.

A recent survey indicated that only 17% of the members of the American Meteorological Society and the American Geophysical Society thought the warming of the twentieth century was the result of an increase in greenhouse gas emissions.

### Impacts of global warming

Larger quantities of carbon dioxide in the atmosphere and warmer climates would be likely to lead to an increase in vegetation growth. During warm periods in the past, vegetation flourished in some areas, at one point allowing the Vikings to farm in Greenland and the English to have a successful wine industry. The impacts of warming could be positive rather than negative.

The idea that global warming would melt the ice caps and flood coastal cities is highly controversial. An increase in global temperatures, whether due to human activity or not, may not lead to a massive melting of the Earth's ice caps, as sometimes claimed. In addition, sea-level rises over the centuries relate more to expansion of oceans as the water has warmed, not to melting ice caps.

Some have argued that climatic warming will increase the spread of diseases but, the opponents argue, there may be more important issues affecting the spread of vector-borne diseases:

- deterioration in public health practices, such as rapid urbanisation without adequate infrastructure
- forced large-scale resettlement of people
- increased drug resistance
- higher mobility through air travel
- lack of insect-control programmes

## Weather hazards

It is well known that hazards have become both more frequent and more damaging in recent years. This is certainly true of climatic hazards. Part of the reason for the increase in damage is, of course, that there are more of us living in the path of destructive hurricanes or tornadoes (Figure 10.23).

There are some who argue that the frequency of these events has also increased and links have been made with global warming:

- hurricanes are generated in areas where the sea water temperature is above 26°C
- global warming has increased the areas where this condition is met
- hurricanes have become more frequent

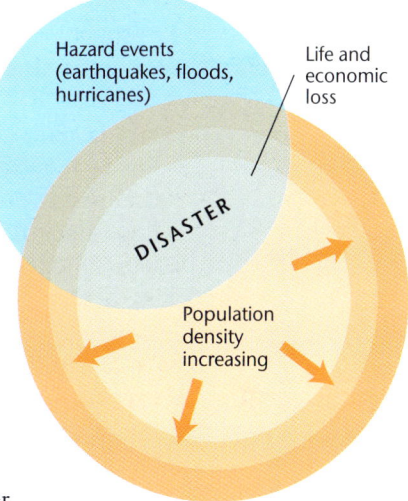

*Figure 10.23 The relationship between population density and natural events*

The extreme climatic episodes during the late 1960s and early 1970s were assessed as a switch to a more variable and dangerous climate epoch, although the reason for this is uncertain. Whether or not the global climate system has recently shifted to a more variable state, the perception that it has has stimulated both studies of the impacts of extreme events and greater awareness of them. But in MEDCs, at least, the other obvious truth is that there has been a decline in loss of life but rapid increase in economic losses from such events (Figure 10.24).

The capacity to survive and recover from the effects of a natural disaster is the result of two factors:

- the physical magnitude of the disaster in a given area
- the social and economic conditions of individuals or social groups in that area; vulnerability varies between social groups in almost all natural disasters

The various natural disasters that occurred in 2002, the vast majority of which were climate or weather related, cost countries and communities an estimated $70 billion. Many of the disasters were related to the high level of rainfall. One-third of the 526 natural catastrophes in 2002 were floods. In total, there were more wind-related natural disasters. But floods killed more people and cost far more than windstorms, earthquakes or other natural catastrophes. It is estimated that

# Chapter 10 Atmospheric systems

*Figure 10.24 Financial costs and deaths from natural disasters*

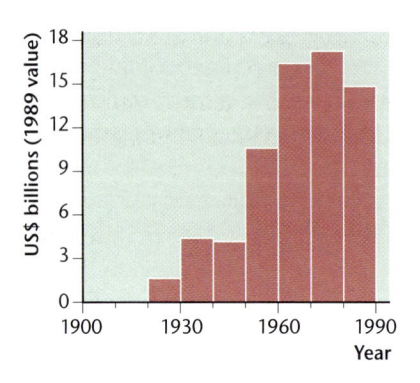

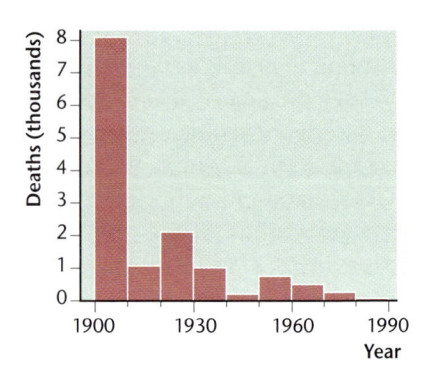

42% of fatalities, 66% of the economic losses and 64% of insured losses were due to floods. Windstorms, including hurricanes and tornadoes, accounted for 13% of fatalities, 23% of economic losses and 34% of insured losses.

It is estimated that 90% of victims and 75% of all economic damage occur in LEDCs, although it is worth remembering that the LEDCs also contain some 80% of the world's population.

The relationship between socioeconomic conditions and the impact of natural disasters can be expressed as follows.

- Due to economic constraints, the poor are forced to live in precarious homes, made of flimsy, non-durable materials.
- These homes are located on the least-valued plots of land, for example on steep hillsides, on floodplains, in fragile ecosystems and watersheds, on contaminated land, rights-of-way and other inappropriate areas.
- There is inadequate management of rainwater and wastes on these plots — during disasters, poor services and infrastructure further complicate survival efforts.
- Health risks are similarly accentuated.

This works at an international scale but also applies at a national level, so poorer areas are most significantly affected. For example, in the case of Nicaragua the municipalities affected by Hurricane Mitch were those with the highest levels of poverty in the country, especially in rural zones. Of the 58 poorest municipalities in the country, 40 are located in the provinces worst affected by Hurricane Mitch.

There are many ways of dealing with climatic hazards, such as:
- accepting them and living with the losses
- accepting them but insuring against the losses
- avoiding/reducing the hazard by intelligent siting of settlements
- attempting to manage their impacts — either by hard management (e.g. dams, dykes) or soft management (e.g. land-use changes such as reforestation)

## Drought: El Niño gets the blame

Papua New Guinea (PNG) seems to be adversely impacted by El Niño each time the event occurs. It is affected by La Niña (cold events) as well. When a cold event is in progress, PNG suffers from heavy rains and flooding. This is typical of normal

*Physical systems, processes and patterns* **A2 Unit 4**

to extremely cold sea surface temperatures in the central and eastern Pacific. The effect is in marked contrast to the impact of El Niño in 1997/98 in PNG.

Severe drought conditions began in mid-1997 and extended into the early months of 1998. By the end of 1997, an estimated 700 000 inhabitants had been adversely affected by food shortages, with 40% of the country's rural population suffering severe food shortages.

Despite a 6-month relief effort, hundreds died of drought-related causes (food shortages, lack of water, an increase in the incidence of infectious disease). PNG's Highland provinces were doubly affected because the lack of cloud cover (i.e. drought conditions) produced frost in the Highlands and that too reduced crop output, making a bad situation worse. This was similar to the situation in PNG's Highlands that occurred during the 1972/73 El Niño and the intense 1982/83 El Niño. But El Niño-related climate anomalies were not the only problems that PNG leaders faced during 1997/98 and at least some of the problems were caused by poor distribution of food supplies, corruption and neglect of rural regions. However, it is likely that blaming a catastrophe on 'natural' forces is tempting for governments who might otherwise have to accept some blame themselves.

> **Examiner's tip**
>
> - The following incident can be used to provide a pertinent comment about governments and their responsibilities. In August 1998, the Chinese government changed its story on the devastating floods of 1997. The government had originally blamed El Niño for the floods but now believes that they were caused in large measure by deforestation in the Yangtze River's watershed. This is a bold admission — a government taking blame (and responsibility) for human activities, activities most likely allowed, if not encouraged, by decades of government policy or foreign neglect.

## Hurricanes

For more information about hurricanes, see Case study 10.4.

## Tornadoes

One of the most terrifying weather events is a fully developed tornado. The strong air currents within a storm cloud can create a high speed vortex (spiral) or funnel of winds. Wherever the end of the vortex touches the ground, it creates a path of concentrated destruction that has no equal in nature. Near the core of a tornado, winds may spiral around at more than 480 km h$^{-1}$.

### How are tornadoes created?

Tornadoes develop near the boundary between the up-currents and down-draughts in a storm cloud. A 'funnel cloud' develops first from the cloud base and this may then extend down to ground level. The destruction caused by tornadoes is due mainly to the violence of the winds. There is very low pressure at the centre of the vortex. If a tornado goes over a building, the building can explode outwards because of the sudden drop in pressure as the vortex passes over

# Chapter 10  Atmospheric systems

it. Although tornadoes are usually less than 250 m across, they can travel a long way, sometimes more than 200 km across the land surface. They pick up material from the ground as they go, including cows, cats, dogs and other animals — even humans.

'Waterspouts' are simply tornadoes that form over water rather than land. They tend to lose their energy as soon as they cross from water to land. Waterspouts are usually confined to shallow waters during warmer seasons.

Tornadoes are classified on the Fujita scale according to the level of damage they cause (Figure 10.25).

*Figure 10.25 The Fujita damage scale for tornadoes*

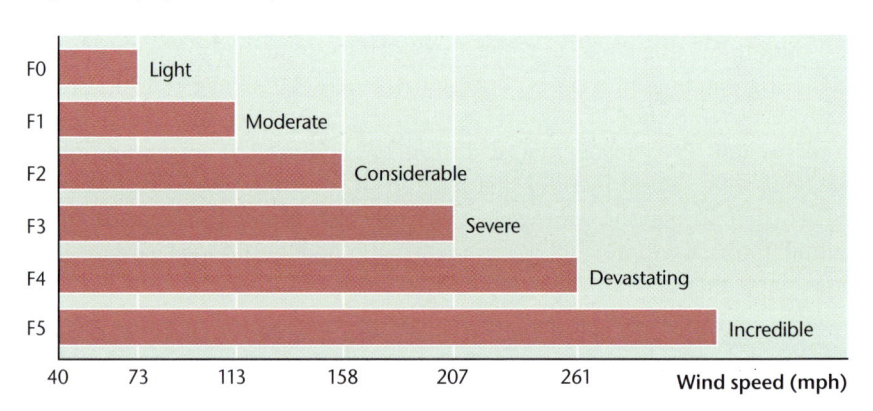

| Category | Damage | Wind speed (mph) |
|---|---|---|
| F0 | Light | 40–73 |
| F1 | Moderate | 73–113 |
| F2 | Considerable | 113–158 |
| F3 | Severe | 158–207 |
| F4 | Devastating | 207–261 |
| F5 | Incredible | 261+ |

## Case study 10.8  Tornadoes: Kansas, 2003

The late afternoon of Sunday 4 May 2003 produced one of the deadliest outbreaks of tornadoes in many years, responsible for 37 deaths.

Initial reports indicated that as many as 80 tornadoes touched down in eight states, putting the outbreak in the top ten of all time. Many were wide, long-track tornadoes that were well-documented by stormchasers.

- West and north of Kansas City, one member of a damaging family of tornadoes caused a death in northern Wyandotte County, KS.
- In southeast Kansas and southwest Missouri, a swarm of large, intense tornadoes produced three major damage tracks. Hundreds of homes, dozens of farms and several entire towns were wiped out.
- The northern track began near McCune, Crawford County, Kansas, and moved northeast, killing three people before crossing into Barton County, Missouri, where one person was killed at Liberal. Three more people were killed as the town of Stockton, Cedar County, Missouri, was virtually destroyed.
- The middle storm track caused three deaths near Columbus, Cherokee County, Kansas, and two deaths near Carl Junction, Missouri.
- The southernmost track began southwest of Pierce City, and devastated that small town, with three deaths in the area. Further to the east north-east, death and destruction hit near Clever, Marionville and Battlefield. The massive tornado mercifully spared the city of Springfield.

# Chapter 11

# Glacial systems

## Glaciers are dynamic systems

There are several types of glacier. Although valley glaciers are the focus in this specification, it is worth being aware of the range of different types.

- **Ice sheets** are the largest accumulations of ice, defined as complete and continuous cover of more than 50 000 km². There are currently only two — in Antarctica and Greenland. Today these possess 96% of the world's ice. During the last glacial period huge ice sheets also covered much of Europe.
- **Ice caps** are smaller and tend to cover only the highest relief in high latitude areas, each being less than 50 000 km². The largest ice cap in Europe is the Vatnajökull glacier in Iceland.
- **Ice shelves** are floating masses of ice attached to the coast at one edge. The Ross shelf off Antarctica is one of the best known and is the size of the state of Texas.
- **Valley glaciers** are confined by valley sides. These may be outlet glaciers from ice caps or sheets, or fed by snow and ice from one or more cirque glaciers. They follow the course of existing river valleys or corridors of lower ground. They are typically between 10 km and 30 km in length, although in the Karakoram Mountains of Pakistan they are as long as 60 km.
- **Cirque glaciers** are relatively small in area and occupy hollows in upland areas. They may be up to hundreds of metres in diameter.
- **Piedmont glaciers** occur where valley glaciers spread onto lowland areas, often joining together.

Glaciers exist today in areas of high latitude, such as Antarctica and Greenland, and places of high altitude, such as the Andes, Himalayas, Alps and Rockies. Note that the Arctic area is a mass of floating ice, not a glacier.

In the past, when our climate was much colder, glaciers also existed in Britain, mainly in Scotland, the Lake District and North Wales. They occurred in Ireland, Scandinavia and much of northern Europe as well, although cover in the Alps was relatively limited in comparison. Britain has experienced several glacial periods during the last 2 million years of the Quaternary. The main advances have been:

- the Anglian glaciation between 425 000 and 380 000 years ago
- the Wolstonian glaciation between 175 000 and 128 000 years ago
- the Devensian advances between 26 000 and 15 000 years ago, with a maximum advance 18 000 years ago
- the most recent between 12 000 and 10 000 years ago

A more recent Loch Lomond re-advance was confined to western Scotland.

### Examiner's tip

- Although you are not likely to be questioned directly about types of glacier or the existence of glaciers in the present and past, this is useful background knowledge and can provide a context in which to place your understanding at the start of an essay. Most of Britain's glacial landforms are the result of the most recent advance (12 000–10 000 years ago).

# Chapter 11 Glacial systems

## The glacier system

A **glacier system** is the balance between inputs, stores/transfers and outputs.
- **Inputs** include the accumulation of precipitation, avalanches, rock debris, heat energy and meltwater.
- **Outputs** include sublimation of ice into water vapour, ablation or melting, and sublimation of snow and ice, sediment and heat energy.
- **Stores** are mainly snow, ice and debris in, on and under the ice and meltwater.
- **Transfers** are the movements of snow and ice through the glacier.

Glaciers form when temperatures are low enough for the snow that falls in one year to remain frozen throughout the year. This means that the following year, fresh snow falls on top of the previous year's snow. Fresh snow consists of flakes with an open, feathery structure and a low density. Each new fall of snow compresses and compacts the layer beneath, causing the air to be expelled and converting low-density snow into higher-density ice. Snow that survives one 'summer' is known as **firn** (or **neve**) and has a density of 0.4 g cm$^{-3}$. With further compaction by subsequent years of snow fall, the snow becomes glacier ice with a density of between 0.83 g cm$^{-3}$ and 0.91 g cm$^{-3}$. This process is known as **diagenesis**. It may take between 30–40 years and 1000 years for this to happen. True glacier ice is not encountered until a depth of about 100 m and is characterised by a bluish colour rather than the white of fresher snow (the white colour is due to the presence of air).

The majority of inputs occur towards the top of the glacier and this area, where accumulation exceeds ablation, is called the **accumulation zone**. Most of the outputs occur lower down where ablation exceeds accumulation, in the **ablation zone**. The two zones are notionally divided by the **firn** or **equilibrium** line, where there is a balance between accumulation and ablation. This can be shown diagrammatically (Figure 11.1).

*Figure 11.1 The glacier system*

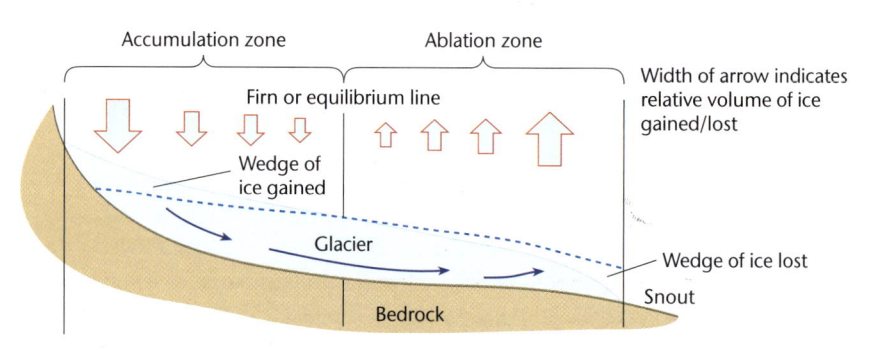

## Glacial budget or mass balance

The **annual budget** of a glacier can be calculated by subtracting the total ablation for the year from the total accumulation.
- A positive figure indicates a net gain of ice through the year, i.e. there is **net accumulation** and so the glacier will **advance** or grow. The firn line will, in effect, move down the valley.

*Physical systems, processes and patterns*  **A2 Unit 4**

- A negative figure indicates a net loss of ice through the year, i.e. there is **net ablation** and there will be a **retreat** or contraction of the glacier and an up-valley movement of the firn line.
- If the amount of accumulation equals the amount of ablation, the glacier is in **equilibrium** and therefore will remain **stable** in its position.

There will often be seasonal variations in the budget, with accumulation exceeding ablation in the winter and vice versa in the summer. It is therefore possible that there will be some advance during the year even if the net budget is negative and some retreat even when the net budget is positive (Figure 11.2). An idea that many fail to appreciate is that even when in retreat, the ice in a glacier may move forwards across the firn line under gravity — it can appear that a retreating glacier is actually advancing.

Rates of advance and retreat vary enormously. Table 11.1 shows the mean rate of retreat for selected glacial regions.

> **Key terms**
>
> **Ablation** The loss of ice to processes such as melting and evaporation.
>
> **Accumulation** The addition of ice by processes such as snowfall and avalanches.
>
> **Annual budget** The difference between the total accumulation and the total ablation for a year.

| Region | Period | Mean rate of retreat (m yr$^{-1}$) |
|---|---|---|
| Iceland | 1850–1965 | 12.2 |
| Norway | 1850–1988 | 28.7 |
| Rockies | 1890–1974 | 15.2 |
| Alps | 1874–1980 | 15.6 |
| New Zealand | 1894–1990 | 25.9 |

*Table 11.1 Rates of glacial retreat*

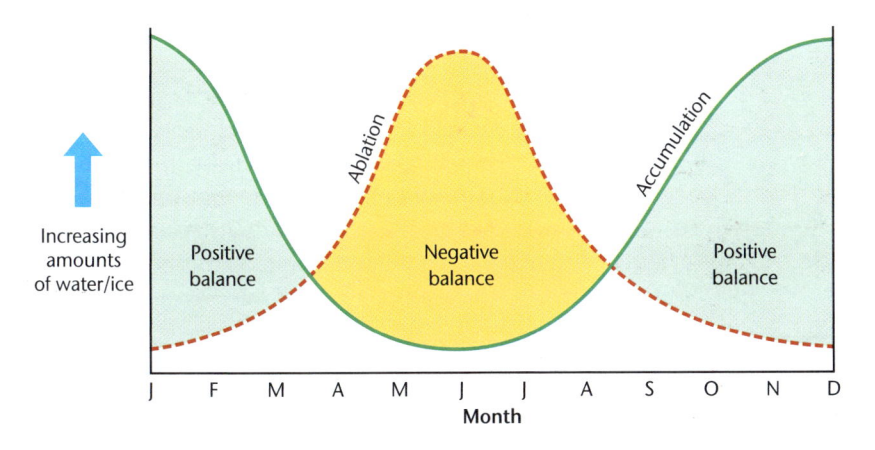

*Figure 11.2 Net budget in a northern hemisphere glacier*

> **Examiner's tip**
>
> - Don't forget that in southern-hemisphere glaciers the pattern shown in Figure 11.2 would be reversed.

## The movement of glaciers

The movement of ice is important as it allows the glacier or ice sheet to pick up weathered debris and erode at its base and sides, as well as transport and alter the materials it is carrying. Rapidly moving ice in dynamic glaciers is capable of eroding and shaping landscapes significantly. However, stagnant ice, as often found in ice sheets, can protect the landscape.

### Why glaciers move

Glaciers move, fundamentally, due to the force of gravity. Ice moves downslope from areas of high altitude to lower areas. In a valley glacier this involves moving from the accumulation zone, across the firn line, and into the ablation zone.

# Chapter 11

## Glacial systems

There are a number of factors that influence the movement of glaciers:
- **gravity**: the fundamental cause of the movement of an ice mass
- **gradient**: the steeper the gradient of the surface, the faster the ice will move, if other factors are excluded
- **the thickness of the ice**, which influences basal temperature and the pressure melting point
- **the internal temperature of the ice**, which can allow movements of one area of ice relative to another
- **the glacial budget**: a **positive budget** (net accumulation) causes the glacier to advance

The pressure melting point is discussed in more detail later.

### How glaciers move

Ice moves in different ways. When it is solid and rigid it will tend to break apart, as shown by the existence of crevasses. However, when under steady pressure it will deform and behave more like a plastic. There are typically two zones in glaciers where these different movements occur (Figure 11.3):
- an upper zone where the ice is brittle and breaks
- a lower zone where, under pressure, the ice deforms

As a result of research on the Mer de Glace glacier in the French Alps in 1842, James Forbes concluded that the sides and base of a glacier tend to move more slowly than the top and middle (Figure 11.4). This is because the ice may be frozen onto the rocks of the valley floor and sides. There may be obstructions as well that slow down movement. It is also due to the accumulative effect of laminar flow in which each lower layer of ice not only moves itself but carries the layers above with it.

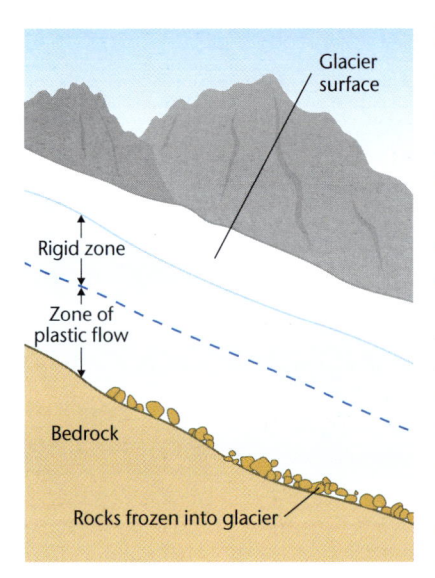

**Figure 11.3** The zones within a glacier

Glaciers move differently depending on the temperature of the ice at their base. **Cold (polar)** glaciers are characterised by:
- high-latitude locations
- low relief
- basal temperatures below the pressure melting point
- very slow rates of movement, perhaps a few metres a year

In such areas as Antarctica and Greenland, even summer temperatures are below freezing and precipitation is low. This means that both accumulation and ablation are limited and there are no great seasonal differences. The glaciers are not very dynamic and there is not only limited movement but limited landscape impact as well; little erosion, deposition and transportation take place.

**Warm (temperate)** glaciers usually have:
- high-altitude locations
- steep relief
- basal temperatures at or below the pressure melting point
- rapid rates of movement, typically 20–200 m yr$^{-1}$

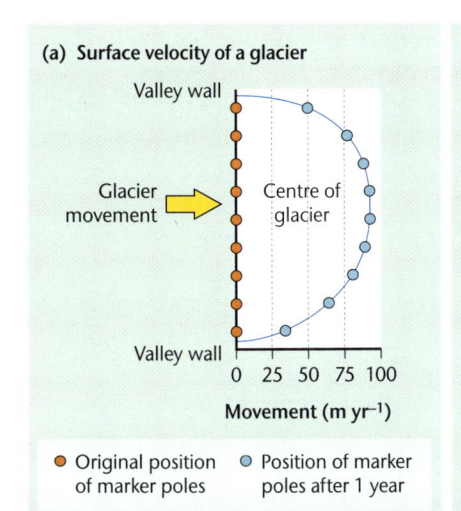

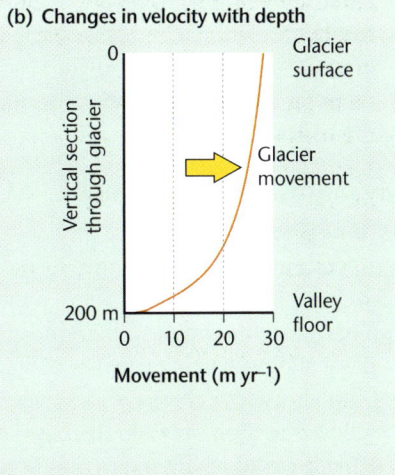

*Physical systems, processes and patterns* **A2 Unit 4**

*Figure 11.4 Differential rates of flow in a glacier*

Locations such as the Alps and the Rockies experience high rates of accumulation in the winter and, due to significantly warmer, above-zero temperatures, high rates of ablation in the summer. This makes the glaciers in these locations very active, with large volumes of ice being transferred across the firn line, and significant seasonal advance and retreat. Not only do the rapid movements cause erosion and produce erosional landforms, but the ablation also produces lots of meltwater and so landforms of fluvioglacial deposition are common.

The most important difference between the two types of glacier is their basal temperature, as this largely determines the mechanism of movement. Figure 11.5 contains temperature profiles for the two types of glacier.

The **pressure melting point** is the temperature at which ice is on the verge of melting. At the surface this is at $0°C$, but within an ice mass it is fractionally lowered by increasing pressure. Ice at its pressure melting point deforms more easily than ice at a temperature below its pressure melting point.

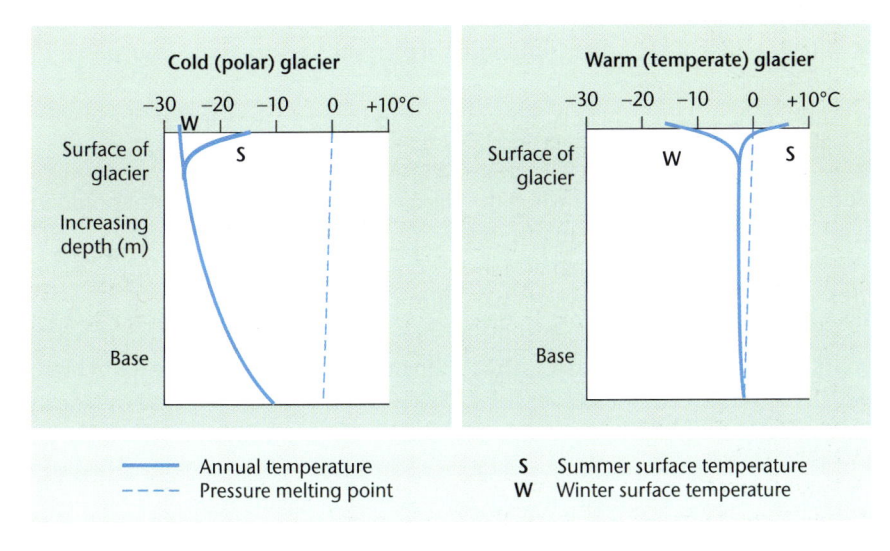

*Figure 11.5 Glacial temperature profiles*

# Chapter 11 Glacial systems

- Most temperate glaciers are at their pressure melting point at the base and sometimes within the glacier itself. Movement is facilitated by the production of meltwater.
- In polar glaciers temperatures are below the pressure melting point and movement is limited.

## Mechanisms of movement

### Basal sliding

Temperate glaciers mainly move by **basal sliding**. If the basal temperature is at or above the pressure melting point, a thin film of meltwater exists between the ice and the valley floor and so friction is reduced.

Basal sliding actually consists of a combination of different mechanisms.

- **Slippage**, where the ice slides over the valley floor because the meltwater has reduced friction between the base of the glacier (and any debris embedded in it) and the valley floor. Friction between the moving ice/debris and the valley floor can also lead to the creation of meltwater.
- **Creep or regelation**, which occurs as ice deforms under pressure when encountering obstructions on the valley floor. This enables the ice to spread around and over the obstruction, rather as a plastic, before re-freezing again when the pressure is reduced.
- **Bed deformation**, when the ice is carried by saturated bed sediments moving beneath it on gentle gradients. The water is under high pressure. This movement is often compared to the ice being carried in roller skates.

The Franz Josef glacier in New Zealand moves approximately 300 m yr$^{-1}$ by basal sliding. Basal sliding accounts for 45% of the movement of the Salmon glacier in Canada, but as much as 90% in extreme cases.

### Internal deformation

Polar glaciers are unable to move by basal sliding as the basal temperature is below the pressure melting point. Instead they move mainly by **internal deformation**. This has two elements:

- intergranular flow, when individual ice crystals re-orientate and move in relation to each other
- laminar flow, when there is movement of individual layers within the glacier, often layers of annual accumulation

Both these movements occur when the glacier is on a slope. They do not occur on level surfaces; here the ice remains intact.

The Meserve glacier in Antarctica moves only 3–4 m per year at its firn line and 100% of this movement is by internal deformation.

When ice moves over a steep slope it is unable to deform quickly enough and so it fractures, forming crevasses. The leading ice pulls away from the ice behind, which has yet to reach the steeper slope. This is **extending flow**. When the gradient is reduced, **compressing flow** occurs as the ice thickens and the following ice pushes over the slower-moving, leading ice. The planes of movement, called **slip planes**, are at different angles in each case, as seen in Figure 11.6.

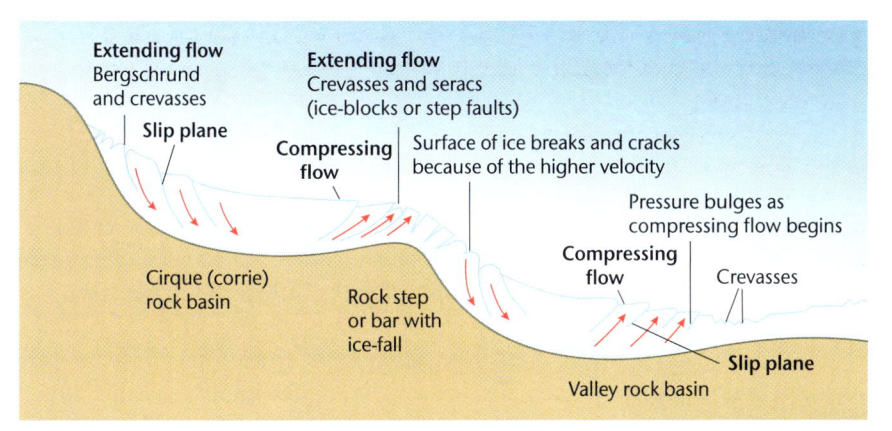

*Physical systems, processes and patterns* **A2 Unit 4**

*Figure 11.6 Extending and compressing flow*

Some glaciers may occasionally **surge** at rates of up to 100 m per day. This is only likely on relatively steep gradients in temperate glaciers after large inputs have been received. A glacier on Discko Island, Greenland, was found to have moved 10 km in 4 years between 1995 and 1999, with a peak movement of 30 m per day. Some surges may be triggered by tectonic activity such as earthquakes.

### Examiner's tip

- One detailed case study is not as useful here as a range of different examples for which you have specific data. You need to be able to illustrate variations. Learn the name, location, rate and type of movement of at least one polar glacier and one temperate glacier.

# Glacial processes and landforms

One application of the 'systems approach' is to consider the shaping of the landscape as being part of a system. The inputs into the system are the controlling factors that influence the rate and type of process occurring, the throughputs are the processes such as erosion and deposition, and the outputs are the landforms that are formed by the processes (Figure 11.7).

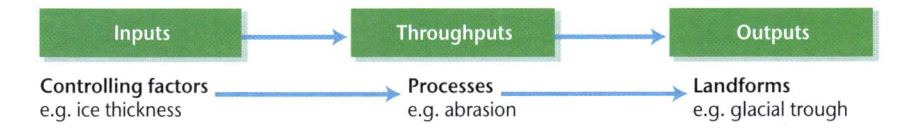

*Figure 11.7 A systems approach to glaciation*

Many glacial landforms are visible today in places that were glaciated in the past, such as the Lake District. In order to see how glacial processes have produced these landforms, it can be helpful to study locations that are currently being glaciated and then make linkages between the active processes and the relict landforms.

## Processes of erosion

Glacial erosion occurs as glaciers move forward; this largely happens in upland areas. There are two main processes of **erosion** by glaciers: plucking and abrasion.

# Chapter 11  Glacial systems

**Weathering** processes also help, by producing some of the glacial debris used in abrasion and by enlarging joints in the rock to assist in plucking. The two main processes involved in weathering are freeze–thaw and dilatation.

## Plucking

**Plucking** mainly occurs when meltwater seeps into joints in the rocks of the valley floor/sides. This water then freezes and becomes attached to the glacier. As the glacier advances it pulls pieces of rock away. A similar mechanism takes place when ice re-freezes on the down-valley side of rock obstructions that it is moving over by the process of creep. Plucking is particularly effective at the base of the glacier as the weight of the ice mass above may produce meltwater due to pressure melting. It is significant when the bedrock is highly jointed, as this allows meltwater to penetrate. Plucking is also known as quarrying.

## Abrasion

As a glacier moves across the ground surface, the debris embedded in its base/sides rubs against surface rocks, wearing them away — **abrasion**. The process is often likened to the action of sandpapering. The coarse material scrapes, scratches and grooves the rock. The finer material tends to smooth and polish the rock. The glacial debris itself is also worn down by this process, forming a fine rock flour that is responsible for the milky white appearance of glacial meltwater streams and rivers.

## Freeze–thaw

When water freezes it expands by between 9% and 10% of its volume. In glacial environments this may happen as temperatures fluctuate between day or night. In areas when temperatures are generally below freezing during the day as well as at night, it may happen seasonally. The pressure exerted by expanding water as it freezes breaks down jointed rocks, particularly those exposed above the level of the ice. Beneath the glacier, temperatures tend to vary less and so freeze–thaw is thought to be less significant there.

## Dilatation

Rocks fracture when the overlying pressure is released. This happens when the glacier is melting, losing weight and exerting less downward pressure. The rocks of the valley floor, and to a lesser extent the sides, expand and fracture as the weight is removed. The rocks tend to fracture parallel to the surface.

### Examiner's tip

- Do not confuse weathering and erosion: this is a common mistake.
- Good essays often recognise the different locations and situations in which different processes occur. For example, plucking is particularly effective in eroding well-jointed rocks.

### Key terms

**Erosion**  The wearing away and removal of material by a moving force; in glacial erosion the force is the moving ice.

**Weathering**  The breakdown and decay of rock in situ, with no movement involved.

## Nivation

A further process, which is not easily classified as erosion or weathering, is **nivation**. This is complex and includes a combination of freeze–thaw action, solifluction, transport by running water and, possibly, chemical weathering. Nivation is thought to be responsible for the initial enlargement of hillside hollows as part of the formation of corries and cirques.

## Rates of glacial abrasion

A number of factors influence rates of glacial abrasion. These are summarised in Figure 11.8.

- **Presence of basal debris**: pure ice is unable to carry out abrasion of solid rock and so basal debris is an essential requirement. The rate of abrasion increases with the amount of basal debris up to a point where the debris produces increased amounts of friction, which reduces the movement of the glacier.
- **Debris size and shape**: particles embedded in ice exert a downward pressure proportional to their weight, and so larger debris is more effective in abrasion than fine material. Angular debris is also more effective as the pressure is concentrated onto a smaller area of debris–bedrock interface.
- **Relative hardness of particles and bedrock**: abrasion is most effective when hard, resistant rock debris at the glacier base is moved across a weak, soft bedrock. If the bedrock is more resistant than the debris, then little abrasion will be accomplished.
- **Ice thickness**: the greater the thickness of overlying ice, the greater is the pressure exerted on the basal debris and the greater the rate of abrasion. This is, however, only true up to a point. Beyond a certain thickness the pressure becomes too great and there is too much friction between the debris and the bedrock for much movement to happen. This is not a fixed depth, as it depends upon ice density and the nature of the debris, but it is typically 100–200 m.
- **Basal water pressure**: the presence of a layer of meltwater at the base of a glacier is vital if sliding, and therefore abrasion, is to take place. However, if the water is under pressure, perhaps because it is confined, the glacier can be buoyed up and so the debris is not under so much pressure.
- **Sliding of basal ice**: this is important as it determines whether abrasion can take place. Abrasion requires basal sliding to move the embedded debris across rock surfaces. The greater the rate of sliding, the more potential there is to erode as more debris is passing across the rock per unit of time.
- **Movement of debris to the base of the glacier**: abrasion does not only wear away the bedrock, it also wears away the basal debris. This therefore needs to be replaced and replenished over time if abrasion is to remain effective. Debris may be added by plucking and by the downward movement of weathered debris from the surface under its own weight and localised pressure melting.

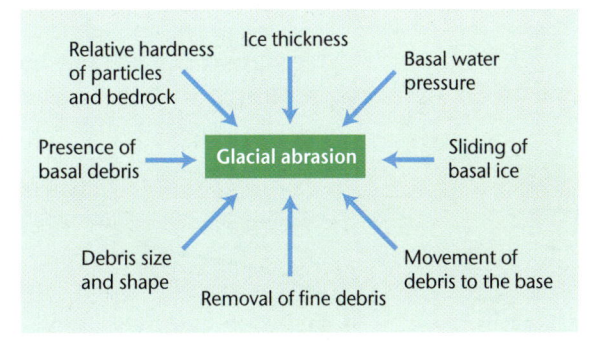

*Figure 11.8 Factors influencing rates of glacial abrasion*

# Chapter 11 Glacial systems

- **Removal of fine debris**: to maintain high rates of abrasion, rock flour (fine debris) needs to be removed so that the larger particles can rub against the bedrock. The is mainly done by meltwater.

> **Examiner's tip**
>
> - As with any set of influencing factors, essay questions about factors afflecting glacial abrasion are likely to require you to:
> - categorise the factors (the debris, the ice, other factors...)
> - assess their relative importance (which are the most important factors?)
> - appreciate that factors vary spatially and temporally (from place to place and over time)
> - identify interrelationships (e.g. sliding of basal ice is influenced by basal water pressure)
> - illustrate at a range of scales (spatially: within a valley, in polar regions/temperate regions; temporally: from season to season, from year to year)

Rates of glacial abrasion vary enormously both spatially and temporally.
- Embleton and King (1968) suggest that mean annual abrasion for active glaciers is between 1000 and 5000 m$^3$.
- Boulton (1974) measured abrasion on rock plates placed beneath the Breidamerkurjökull glacier in Iceland and found that under ice 40 m thick, basalt abraded at 1 mm yr$^{-1}$ and marble at 3 mm yr$^{-1}$. The ice had a velocity of 9.6 m yr$^{-1}$. However, if the velocity increased to 15.4 m yr$^{-1}$, the rate of abrasion of marble increased to 3.75 mm yr$^{-1}$, even though the ice was 8 m thinner. This would suggest that velocity is more important than ice thickness here.
- In comparison, ice 100 m thick flowing at 250 m yr$^{-1}$ in the glacier d'Argentière abraded a marble plate at up to 36 mm yr$^{-1}$.

> **Examiner's tip**
>
> - Some examples of rates of abrasion are important to provide evidence supporting your argument in an essay. For example, the importance of the bedrock type can be illustrated using Boulton's data for basalt and marble.

## Erosional landforms

Many erosional landforms, particularly those in lowland areas, become hidden by later glacial and fluvioglacial deposits laid down on top of them. Other glacial landforms may be modified by post-glacial processes of weathering and erosion. However, landforms that are etched into the solid rock of upland areas may remain distinctive long after the end of the glacial period.

### Cirques (corries)

Cirques or corries are armchair-shaped hollows found on upland hills or mountainsides. They have a steep back wall, an over-deepened basin and a lip at the front, which may be solid rock or made of morainic deposits.

Cirques vary in size and shape, but the length to height ratio (from the lip to the top of the back wall) is usually between 2.8:1 and 3.2:1. Some are only a few hundred metres across, but they can be over 15 km wide. For example, the Walcott Cirque in Antarctica has a 3-km high back wall.

The formation of a cirque is the result of several interacting processes.

- It starts with the nivation of a small hollow on a hillside in which snow collects and accumulates year on year.
- Over time the hollow enlarges and contains more snow, which eventually compresses into glacier ice.
- At a critical depth, the ice moves through the hollow in a rotational manner under its own weight. This helps enlarge the hollow further and it becomes a cirque.
- The rotational movement causes plucking of the back wall, making it increasingly steep.
- Debris is gained by plucking and by rock fragments formed by weathering above the hollow falling into the bergschrund or crevasse at the back. This enables abrasion to take place in the hollow, causing it to deepen.
- Once the hollow has deepened, the thinner ice at the front is unable to erode so rapidly and so a higher lip is left. This lip may also contain moraine deposited by the ice as it moves out of the cirque.
- In the post-glacial landscape the cirque may become filled with water, forming a tarn lake.

These processes are summarised in Figure 11.9.

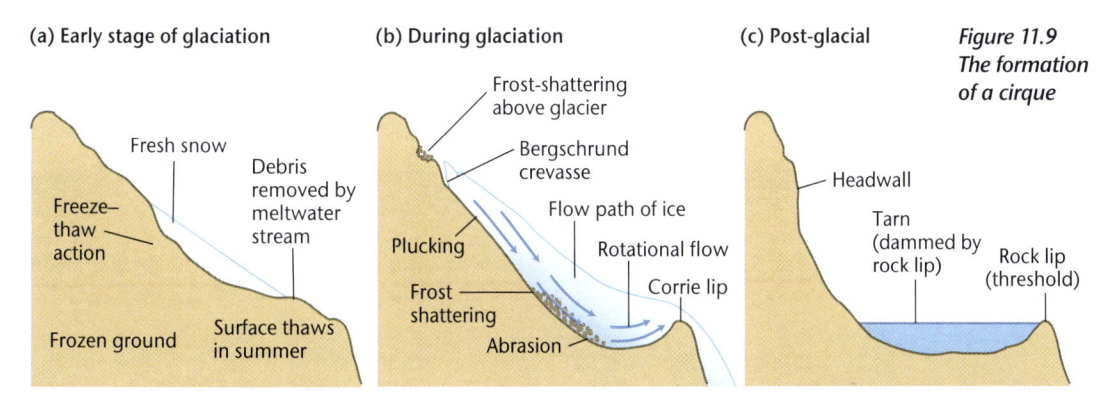

*Figure 11.9 The formation of a cirque*

## Arêtes and pyramidal peaks

An **arête** is a narrow, steep-sided ridge found between two cirques (Figure 11.10). The ridge is often so narrow that it is described as knife-edged. They form from the erosion, steepening and retreat of cirques that are back-to-back or alongside each other. A good example is Striding Edge in the Lake District, which has slopes that are 200–300 m high and almost vertical in places on both sides. The arête itself is so narrow that it is just wide enough for one person to walk along the footpath that runs along it towards the peak of Helvellyn.

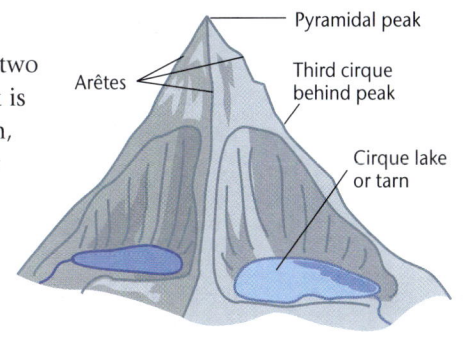

*Figure 11.10 Arêtes and pyramidal peaks*

# Chapter 11 Glacial systems

*Photograph 11.1 Red Tarn and Striding Edge, Lake District*

Where three or more cirques around a hill or mountain top have been eroded and their back walls retreat, the remaining central mass will itself be steepened to form a **pyramidal peak**. Weathering of the peak may further sharpen the shape. The Matterhorn in the Swiss Alps is an excellent example and is over 4500 m high.

## Glacial troughs (U-shaped valleys)

Glaciers flow down pre-existing river valleys under gravity. As they move they erode the sides and floor of the valley, causing the shape to become deeper, wider and straighter. The mass of ice has far more erosive power than the river that originally cut the valley. Although the valleys are usually described as being 'U-shaped', they seldom are. Rather they are parabolic, partly due to the weathering of the upper part of the valley sides that goes on both during the glacial period and in the subsequent periglacial period as the glacier retreats. The resultant scree slopes that accumulate at the base of the valley sides lessen the slope angle.

There are often variations in the long profiles of glacial troughs. When compressing flow occurs, the valley is over-deepened to form **rock basins** and **rock steps**. This process may be particularly evident where there are alternating bands of rock of different resistances on the valley floor. The weaker rocks will be eroded more rapidly to form the basins.

## Hanging valleys and truncated spurs

Tributary valleys do not contain as much ice as the main valley and so they are not deepened as much. When the ice melts, the tributary valley floor may be left perched high above the main valley floor as a **hanging valley**. In the post-glacial

*Physical systems, processes and patterns* **A2 Unit 4**

landscape, waterfalls may form here as tributary streams fall over the steep sides of the main valley to reach the much lower floor. In the Lake District, Fisher Gill flows into the valley of Thirlmere from the slopes of Helvellyn. A series of waterfalls is found above Fisher Place as the gill drops 200 m in less than 1 km.

The original river valley would have contained interlocking spurs, which the river could not erode. During the straightening of the river valley, the interlocking spurs are removed by the more powerful glacier. The spurs are chopped off or 'truncated' to leave a near vertical cliff or **truncated spur** on the valley side (Figure 11.11). These occur either side of a hanging valley at the point it meets the main valley, as is the case at Fisher Gill.

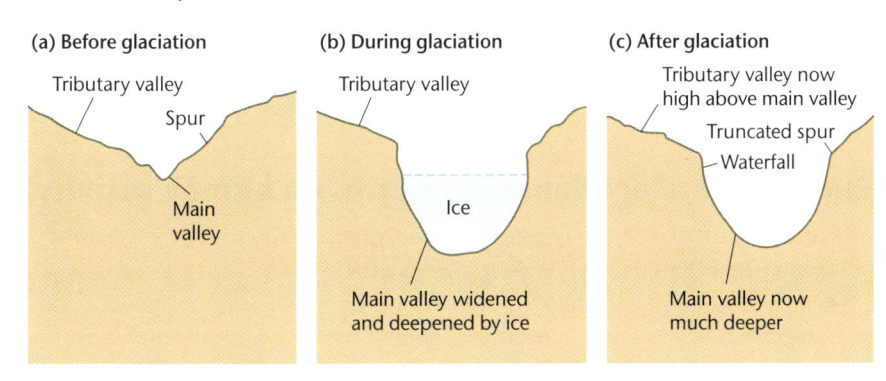

*Figure 11.11 The formation of a hanging valley and truncated spur*

## Roches moutonnées

Projections of resistant rock are sometimes found on the floor of a glacial trough. As advancing ice passes over them, there is localised pressure melting on the up-valley side. As a result, the projections are smoothed and streamlined by abrasion; the final landform is known as a **roche moutonnée**. Often these have **striations**, which are scratches or grooves made by embedded debris. When the ice has passed over the projecting rock, the pressure is reduced and the ice re-freezes. This leads to plucking and steepening on the down-valley side, so these landforms can indicate the direction in which the ice moved through an area (Figure 11.12). Roches moutonnées vary in size but are typically 1–5 m high and 5–20 m long; they are numerous in the Coniston area of the Lake District.

*Photograph 11.2 A roche moutonnée in Snowdonia. Deep striations are visible on the right, indicating that the ice flowed from right to left.*

# Chapter 11 Glacial systems

*Figure 11.12 Formation of a roche moutonnée*

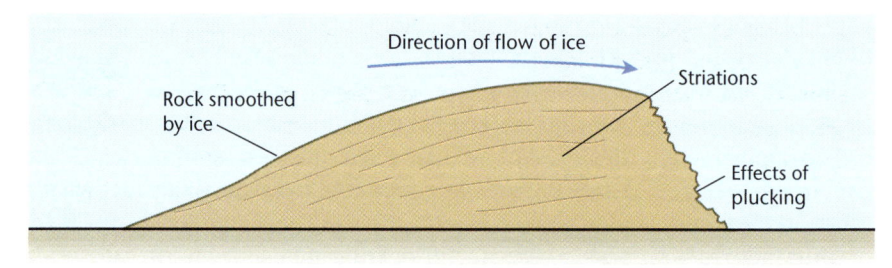

### Examiner's tip

- Examination questions are frequently based on the specific wording found in the specification. In Edexcel (A) Unit 4, the relevant bullet point refers to the 'range, variety and location' of erosional landforms. You should be able to address these key words by selecting appropriate landforms that illustrate these characteristics.

## The impact of erosional landforms on human activity

As with all impacts of the physical landscape on human activity, erosional landforms have both positive and negative effects.

Positive impacts may include:

- **transport**: glacial troughs may provide natural routeways through upland areas, but they can make communication between parallel valleys very difficult
- **agriculture**: flat-floored troughs may provide flat land in an otherwise hilly area, but glacial stripping means very thin soils on the hillsides
- **industry**: deeply eroded troughs and the stripping of the surface by ice may expose rocks and minerals of economic value, but accessibility may be limited
- **settlement**: valley floors may encourage linear settlement, but steep relief means that hilly areas have widely dispersed settlement with a low density
- **tourism**: walkers and climbers may be attracted by the sharp relief features and lakes, aiding the local economy but putting pressure on the environment

## Transportation processes

Moving ice is capable of carrying huge amounts of debris. The material transported by ice sheets and glaciers comes from a wide range of sources:

- **rockfall**: weathered debris falls under gravity from the exposed rock above the ice down onto the edge of the glacier
- **avalanches**: these often contain rock debris within the snow and ice that moves under gravity
- **debris flows**: in areas of high precipitation and occasional warm temperatures, melting snow or ice can combine with scree, soil and mud
- **aeolian deposits**: fine material carried and deposited by wind, often blowing across outwash deposits
- **volcanic eruptions**: often a source of ash and dust
- **plucking**: large rocks plucked from the side and base of the valley
- **abrasion**: smaller material worn away from valley floors and sides

*Physical systems, processes and patterns*  **A2 Unit 4**

While being transported, the material may be classified according to its position in the glacier (Figure 11.13):
- **supraglacial** material is carried on top of the glacier; this will most often come from weathering and rockfall
- **englacial** is material within the ice; this may be supraglacial material that has either been covered by further snowfall, fallen into crevasses or sunk into the ice due to localised pressure melting beneath larger stones and rocks
- **subglacial** is material embedded in the base of the glacier which may have been derived from plucking and abrasion, or that has continued to move down through the ice as former englacial debris

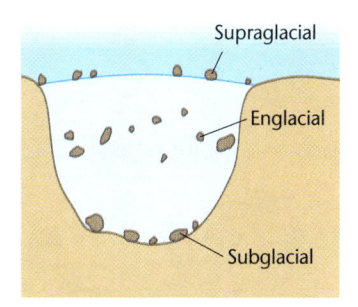

*Figure 11.13 Locations of debris transportation in a glacier*

## Depositional processes and landforms

Glaciers deposit their load when their ability to transport material decreases. This usually occurs as a direct result of ablation (e.g. melting, sublimation) during periods of retreat or during de-glaciation. However, material can also be deposited during advance or when the glacier becomes overloaded with debris.

All material deposited during glaciation is known as **drift**. This can be subdivided into **till**, which is material deposited directly by the ice, and **outwash**, which is material deposited by meltwater. Outwash is also known as 'fluvioglacial' material and is dealt with later in this chapter (page 343).

It is estimated that glacial deposits currently cover about 8% of the Earth's surface. In Europe they cover almost 30%, mainly in the form of material left by earlier ice masses that have since retreated. The East Anglia area, for example, has deposits that are 143 m thick while in the Gulf of Alaska they are 5000 m thick in places. In active glacial areas, it is possible that rates of deposition are in the order of 6 m per 100 years.

### Till

There are two types of glacial till.
- **Lodgement till** is material deposited by advancing ice. Due to the downward pressure exerted by thick ice, subglacial debris may be pressed and pushed into existing valley floor material and left behind as the ice moves forward. This process may be enhanced by localised pressure melting around individual particles that are under significant pressure. **Drumlins** are the main example of landforms of this type.

### Key terms

**Deposition** The dropping or laying down of material.
**Drift** Material deposited during glaciation.
**Englacial** Within a glacier.
**Fluvioglacial deposition** The dropping of material by glacial meltwater.
**Glacial deposition** The dropping of material by ice.
**Glacial transportation** Movement by ice.
**Outwash** Material deposited by glacial meltwater.
**Subglacial** Beneath a glacier.
**Supraglacial** On top of a glacier.
**Transportation** The movement of material.

# Chapter 11 Glacial systems

- **Ablation till** is material deposited by melting ice from glaciers that are stagnant or in retreat, either temporarily during a warm period or at the end of the glacial event. Most other depositional landforms are of this type.

Whichever type it is, till material can be recognised by three distinctive characteristics. These can be compared to the charateristics of material deposited by moving water.
- Till is jagged and angular in shape. This is because it has been embedded in the ice and has not been subjected to further erosion processes, particularly by meltwater which would make it smooth and rounded.
- Till is unsorted. When glaciers deposit material, all sizes are deposited together. In contrast, when moving water deposits material, the material loses energy progressively and is deposited in a size-based sequence.
- Till is unstratified. Glacial till is dropped in mounds and ridges rather than in layers. Water-borne deposits, on the other hand, are typically deposited in layers.

Photograph 11.3 shows students investigating till deposits exposed on a river bank in Kingsdale, Yorkshire Dales.

*Photograph 11.3 Investigating till deposits*

## Case study 11.1 Human activity in a glacial area: Keswick, Lake District

### Transport
The main A591 road through the Lake District from north to south follows the glacial trough that contains Thirlmere, with the road running along the eastern edge of Thirlmere. The Penrith to Cockermouth railway, now dismantled, followed valleys from east to west through Threlkeld and Keswick. This route is now followed by the A66 trunk road.

### Settlement
Keswick, the largest settlement in the northern Lake District, lies at the confluence of three glacial valleys in the biggest expanse of flat land in the area. Many small villages, such as Legburthwaite and Dale Bottom, lie on the valley floor and form a linear pattern. Only a few isolated farms are found on the upland areas, in a very dispersed pattern.

### Industry
In the past, mining for graphite and quarrying for slate were important in this area.

The area has limited industry now, apart from tourism. Tourists are attracted to the spectacular glacial landscape and the lakes. Ribbon lakes such as Derwent Water and tarns such as Watendlath are popular sites, as are waterfalls from hanging valleys such as Lodore Falls, which is fed by Watendlath Beck and has a hotel near its base.

### Agriculture
Dairy farming takes place on the flat valley floors; an example is Yew Tree Farm in St John's in the Vale. Sheep farming is the only option on hillsides that have been stripped of their soil by glacial erosion. Low Bridge End Farm, situated over 150 m above sea level below Yew Crag, is an example.

*Physical systems, processes and patterns* **A2 Unit 4**

*Figure 11.14 Ordnance Survey map showing the Keswick area*
Reproduced by permission of Ordnance Survey on behalf of HMSO, © Crown copyright 2005. All rights reserved. Licence no. 100027418.

# Chapter 11 Glacial systems

**Till sheets** are formed when large masses of unstratified drift, deposited at the end of a period of advance, smother the underlying surface. Till sheets may not be conspicuous in terms of relief, but they are significant landforms because of their extent. The till itself is variable in composition, depending greatly on the nature of the rocks over which the ice moved. If there is a high clay content, compaction by the weight of the overlying ice may lead to relatively hard deposits.

**Drumlins** are generally recognised as a feature of lodgement, although there is much debate about their origin. They are mounds of glacial debris that have been streamlined into elongated hills. They range in size from a few metres long and high to over 1 km in length and 100 m in height. They have a pear-shaped long profile, which is aligned to the direction of ice advance. The higher and wider **stoss** end faces the ice flow, while the **lee** side is more gently tapered.

Drumlin formation is thought to be the result of one, or a combination, of the following:
- lodgement of subglacial debris as it melts out of the basal ice layers
- reshaping of previously deposited material during a subsequent re-advance
- the accumulation of material around a bedrock obstruction — these are known as **rock-cored drumlins**
- the thinning of ice as it spread out over a lowland area, reducing its ability to carry debris, with the continuing forward movement streamlining the feature

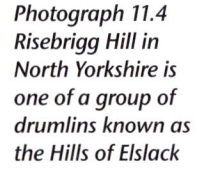

*Photograph 11.4 Risebrigg Hill in North Yorkshire is one of a group of drumlins known as the Hills of Elslack*

Drumlins tend to occur in groups or **swarms**, also known as basket of eggs topography. Good examples can be seen in central Ireland, New York State and in the Eden Valley in Cumbria.

## Moraines

**Moraines** are accumulations of glacial debris, piled into ridges or mounds. The name is sometimes used to describe the material being transported by a glacier, but technically it should only be used when referring to the resultant landforms.

*Physical systems, processes and patterns* **A2 Unit 4**

## Case study 11.2 Till deposition: East Anglia

In East Anglia, the till is quite chalky in content, due to the rocks over which the ice passed. It is typically 30–50 m deep, although it can be as much as 143 m deep in places.

Several different ice advances are thought to have been responsible. In places the chalky deposit overlies an earlier deposit known as Norwich Brickearth.

## Case study 11.3 Drumlins: Eden Valley, Cumbria

A wide range of drumlins, hundreds in total, can be found in the Eden Valley area. The smallest are 2 m high and 10 m long, while the largest are 100 m high and 20 km long. There are concentrations around Appleby and Penrith. Some of these are rock-cored while others are composed entirely of till. Fieldwork research in the area has shown that they have an average elongation ratio (long axis:short axis) of 3:1.

**Lateral moraine** is a ridge of till running along the edge of a glacial valley. The material accumulates on top of the glacier, having been weathered from the exposed valley sides above the level of the ice and falling onto the ice by rockfall. As the glacier melts or retreats, this material is gradually lowered down to the ground and deposited.

- A lateral moraine left by the retreating Athabasca glacier in Canada is 1.5 km long and 124 m high.

**Medial moraine** is formed when two glaciers meet. The debris on the edges of the two adjacent valley sides converges and is subsequently carried on top of the middle of the enlarged glacier. As the glacier melts or retreats, this predominantly supraglacial material is deposited as a ridge running down the middle of the valley floor. These landforms are seldom distinctive.

- An example is the Kaskawash glacier in the USA, which has a ridge that is 1 km wide at the point of confluence of the two glacial valleys.

**Terminal moraine** is a ridge of till extending across a glacial valley. These moraines are usually steeper on the up-valley side and tend to be crescent-shaped, reaching further down the valley in the centre. Such landforms mark the position of the maximum advance of the ice and they would have been deposited at the snout when the ice melted and the glacier retreated.

Their crescent shape is due to the position of the snout; the centre of the glacier advances further because friction with the valley sides causes more rapid melting at the sides of the glacier. The steeper up-valley side is the result of the ice behind supporting the deposits and making them less likely to collapse.

- Franz Joseph glacier in New Zealand has left a terminal moraine 430 m high.

**Recessional moraine** forms a series of ridges running broadly parallel to each other and to the terminal moraine, found further up the valley than the terminal moraine. They form during a standstill in the glacier's retreat. For significant ridges to develop, these pauses need to be sufficiently long, and so recessional moraines seldom exceed about 100 m in height.

# Chapter 11  Glacial systems

**Push moraine** is really a modified terminal or recessional moraine. It has the same outward appearance, but stones in the till are tilted upwards. Push moraine formed during periods of lower temperature when the glacier re-advanced, pushing the ridge of debris forward in front of the snout and tilting the stones upward. The faster the advance, the greater is the angle of tilt seen in the orientation of the stones.

- In front of the retreating snout of the Sòlheimajökull in Iceland there is a ridge of moraine about 7 m high which was originally deposited during a period of retreat, but which was pushed forward during a period of re-advance in the 1960s.

> **Examiner's tip**
> 
> - It should be remembered that glacial deposits are often re-worked by further advances and retreats of the glacier. They will be subjected to weathering and erosion in the post-glacial period, which may further modify them. They can also be colonised by vegetation, making them less visible.

**Ground moraine** is typically found all over the valley floor, often in hummocky mounds. Ground moraine may have been deposited during advances and retreats of the glacier. Later movements in both directions may well mould them into more streamlined shapes.

- The Glen Torridon area of northwest Scotland is the largest area of hummocky ground moraine in Britain.

**Erratics** are individual pieces of rock, varying in size from small pebbles to large boulders. What makes them distinctive is that their geology is different from that of the area in which they have been deposited. They were eroded, most likely by plucking, or added to the supraglacial debris by weathering and rockfall, in an area of one type of geology and then transported and deposited into an area of differing rock type.

- A good example is the erratic blocks of Silurian shale deposited on carboniferous limestone at Norber in the Yorkshire Dales (Photograph 11.5). The erratics have protected the underlying rock from carbonation weathering, leaving the erratic perched on a pedestal of the underlying rock.

*Photograph 11.5 Erratic block at Norber, Yorkshire Dales*

It is worth noting that many of these landforms are difficult to recognise in the field as each successive advance and retreat alters and modifies the appearance of features such as ridges of moraine. During one glacial period a single valley will have experienced several such advances and retreats. You also need to bear in mind that the valley may have suffered several glacial periods over the last 2 million years.

> **Examiner's tip**
> 
> - As with erosion, you should be able to describe the appearance of each landform (a diagram can help), explain its formation (link processes to shape) and refer to a named example.

## The impact of depositional landforms on human activity

The impact of these landforms and of depositional landscapes in particular is less striking than the impact of erosional landforms on human activity.

The major impact is a positive one, because in many places glacial till encourages the formation of fertile soils. The mix of particle sizes aids drainage and the presence of fine clays helps retain moisture. The stones present retain heat and the varied origins of the material tend to provide a range of minerals and nutrients.

- Valley floors in upland areas are therefore generally much more suitable for dairying than the surrounding hills. This is true of the Lake District in general and of the Eden and Ribble Valleys in particular.
- Larger, flatter glaciated expanses, such as those in East Anglia, have such fertile soils that they have become key areas for arable farming.
- In areas of high rainfall such as Northern Ireland, drumlin fields are poorly drained and are subject to waterlogging, making them unsuitable for commercial agriculture.

Ridges of moraine on the valley floor sometimes contribute to the formation of **ribbon lakes**, damming the down-valley flow of post-glacial streams and rivers. These lakes provide benefits for recreation and leisure as well as water supply.

> **Examiner's tip**
>
> - As with the impact of erosional landforms on human activity, try and find some evidence of both positive and negative impacts of landforms of glacial deposition and provide details of location as evidence in an essay.

# Fluvioglaciation

The meltwater produced by glaciers may carry out erosion, transportation and deposition, leading to the formation of distinctive landforms.

Fluvioglacial deposits tend to be very different from glacial deposits. They are deposited by meltwater streams and rivers, which flow under and then beyond the snouts of glaciers. Meltwater streams deposit in the same manner as any other streams — as they lose energy due to a loss of velocity and or volume, they deposit sequentially. The largest material is deposited first and, as energy continues to be lost, smaller particles are deposited. The material deposited, known as **outwash** or fluvioglacial deposits, tends to be:

- generally smaller than glacial till as meltwater streams typically have less energy than valley glaciers and so only carry finer material
- smooth and rounded by contact with water and by attrition
- sorted horizontally, with the largest material found furthest up the valley and progressively finer material found further down the valley due to the sequential nature of the deposition mechanism
- stratified vertically, with distinctive seasonal and annual layers of sediment accumulation in many of the landforms

This description is a simplification of reality, however. Meltwater rivers may have extremely high discharge in summer months, enabling them to carry very large rocks.

# Chapter 11  Glacial systems

- The glacier d'Argentière in France, for instance, has a mean winter discharge of 0.1–1.5 cumecs, but a summer discharge of 10–11 cumecs. (1 cumec is $1 \, m^3 \, s^{-1}$.)

In Iceland, jökulhaulps occur: these are extreme glacial outbursts caused by geothermal or volcanic activity beneath glaciers, causing massive and sudden melting.
- On 2 October 1996, a volcanic fissure of Grimsvötn erupted beneath the Vatnajökull ice cap in southeast Iceland. Discharge on the Skeidara River that flows from the ice cap rose from 70 cumecs to 90 000 cumecs, enabling it to transport boulders the size of houses, which it then deposited on the coastal plain.

A further, more sophisticated division may be made between outwash material and ice-contact drift.
- Outwash material is carried relatively long distances, possibly well beyond the snout and any terminal moraine, and so becomes smooth, rounded and highly sorted.
- **Ice-contact drift** is material deposited under or against the ice, which tends to be less rounded and less well sorted.

## Landforms of fluvioglacial erosion

**Meltwater channels** are common features, occurring anywhere that meltwater flows in discrete channels. However, many subglacial channels are shallow, especially if the ground is frozen, and they are not always seen in the post-glacial environment. When meltwater is dammed, perhaps behind a ridge of moraine or even when meeting an advancing glacier, its level increases. If it overflows through a relative low point in the local relief, an **overspill channel** may be formed.
- Newtondale, in North Yorkshire, was formed when an ice sheet from Scandinavia blocked the passage of meltwater through the Esk Valley. The water level rose and eventually spilled over a low col, eroding a deep gorge in the landscape.

**Sichelwannen** are crescent-shaped scour marks made by meltwater streams, sometimes flowing supraglacially on the edge of a glacier. The marks are eroded into the rocks of the valley side. When the glacier has retreated, these marks are exposed high up on the valley side, well above any present-day river channel erosion.

**Tunnel-valleys** are thought by some to have been eroded by subglacial meltwater streams, and then infilled by deposition. A number of them have been found in East Anglia, especially in the Stour Valley near Long Melford. However, there is much debate about their exact origins.

## Landforms of fluvioglacial deposition

Meltwater deposition is particularly significant where there is variable discharge. During times of high flow, streams and rivers have sufficient energy to carry large amounts of load. As the discharge decreases, energy is gradually and progressively lost and so sequential deposition takes place and numerous landforms are formed.

**Outwash (sandur)** is the name given to a flat expanse of sediment in the proglacial area. Meltwater streams gradually lose energy as they enter lowland areas and deposit their material. The largest material is deposited nearest the snout and the finest further away.
- On the south coast of Iceland there is an extensive area of sandur fed by numerous meltwater streams from glaciers such as Gigjökull and Sòlheimajökull. From the edge of the upland area to the present position of the sea is a distance of some 5 km.

*Photograph 11.6 Sandur in Thorsmork nature reserve, southern Iceland*

**Varves** are layers of sediment found in the bottom of proglacial lakes. Sediment carried by meltwater streams is deposited when the water enters a lake, as the energy of the stream is lost.
- In summer, when large amounts of meltwater are available, the sediment is coarse (sand and gravel) and plentiful, depositing a wide band of sediment of relatively large material.
- In winter, with little meltwater present, sediment is limited in amount and size (silt and clay) and so bands are thin and fine.

Each pair of bands is therefore representative of 1 year's deposition (Figure 11.15). Rather like tree rings, the age of the sediment can be determined from the number of bands. The thickness of the bands also varies over time and so studying the varves can give indications of climate change; thicker bands indicate warmer climates when there would have been more meltwater and more load.

The colour of the bands may also be relevant. Darker colours may be the result of more organic matter and indicative of warmer climatic conditions.

**Kettle holes** are small circular lakes in outwash plains. During ice retreat, blocks of **dead ice** can become detached. Sediment builds up around them and when the ice eventually melts, a small hollow is formed in which water accumulates to form a lake. Although it is generally thought that the deposits that build up around the dead ice are fluvioglacial, some argue that they may be supraglacial deposits.

# Chapter 11  Glacial systems

Figure 11.15 Varves

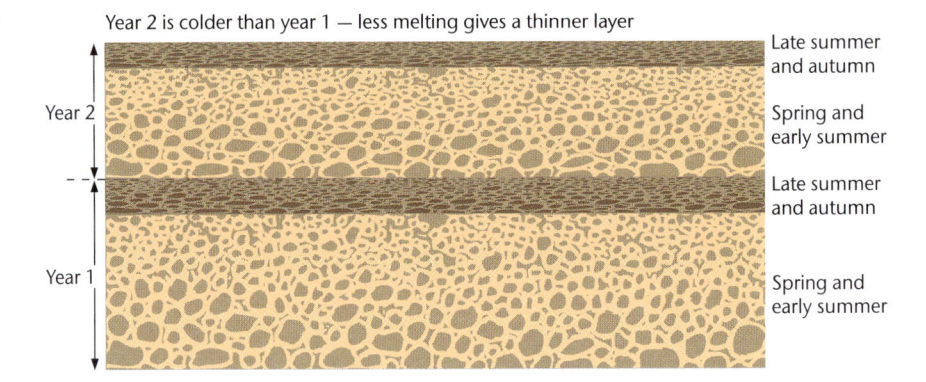

*Year 2 is colder than year 1 — less melting gives a thinner layer*

- Year 2 — Late summer and autumn / Spring and early summer
- Year 1 — Late summer and autumn / Spring and early summer

These lakes vary in size enormously.
- Around Ellesmere in Shropshire are a number of large kettle lakes with diameters of over 100 m.
- In front of the Sòlheimajökull in Iceland there is a collection of kettle holes, none more than 15 m or so in diameter.

Photograph 11.7 Kettle lakes at Sòlheimajökull

**Eskers** are long, sinuous ridges on the valley floor. Material is deposited in subglacial tunnels as the supply of meltwater decreases at the end of the glacial period. Subglacial streams may carry huge amounts of debris under pressure in the confined tunnel in the base of the ice. Some argue that deposition occurs when the pressure is released as the meltwater emerges at the snout. With a retreating ice snout, the point of deposition would gradually move backwards. This may explain why some eskers are **beaded**. The ridge can show significant variations in height and width, with the beads of greater size representing periods when the rate of retreat slowed or halted. However, some argue that the beads are simply the result of the greater load carried by summer meltwater.
- The Trim esker, near Dublin, is one of a group of 12 in the area. It is 14.5 km long and between 4 m and 15 m high.

**Delta kames** are small mounds on the valley floor. There are differing circumstances under which they can form. Some are produced by englacial streams emerging at the snout of the glacier and falling to the valley floor. The streams

346

then lose energy and deposit their load. Others are the result of supraglacial streams depositing material on entering ice-marginal lakes. It is also possible that debris-filled crevasses collapse during ice retreat.
- There are many kames in East Lothian, Scotland. They tend to be a few hundred metres long and tens of metres high.

**Kame terraces** are ridges of material running along the edge of the valley floor. Supraglacial streams on the edge of the glacier pick up and carry lateral moraine that is then deposited on the valley floor as the glacier retreats. The streams form as a result of the melting of ice that is in contact with the valley sides, due both to friction and the heat-retaining properties of the rocks. Although kame terraces may look similar to lateral moraines, they are composed of fluvioglacial deposits that are more rounded and sorted than the glacial debris of the moraines.
- In the Kingsdale valley of the Yorkshire Dales, a kame terrace extends for about 2 km along the north side of the valley. It is approximately 2 m high for most of its length.

**Braided streams** are river channels subdivided by numerous islets and channels. Debris-laden streams lose water at the end of the melting period and so can carry less material. This is deposited in the channel, causing it to divide and, possibly, then rejoin. Braiding begins with a mid-channel bar which grows downstream. Discharge decreases after a flood or a period of snow melt, causing the coarsest particles in the load to be deposited first. As discharge continues to decrease, finer material is added to the bar, increasing its size. When exposed at times of low discharge, the bars can become stabilised by vegetation and then exist as more permanent features. The river divides around the island and then rejoins. Unvegetated bars lack stability and often move, form and re-form with successive floods or high discharge events. They are common in outwash areas due to the seasonal fluctuations in the discharge.
- On the outwash plain of southern Iceland, braiding occurs widely across the areas of Sòlheimasandur.

Figures 11.16 and 11.17 summarise much of the information about fluvioglacial landforms.

*Figure 11.16 Fluvioglacial features (a) during glaciation and (b) after glaciation*

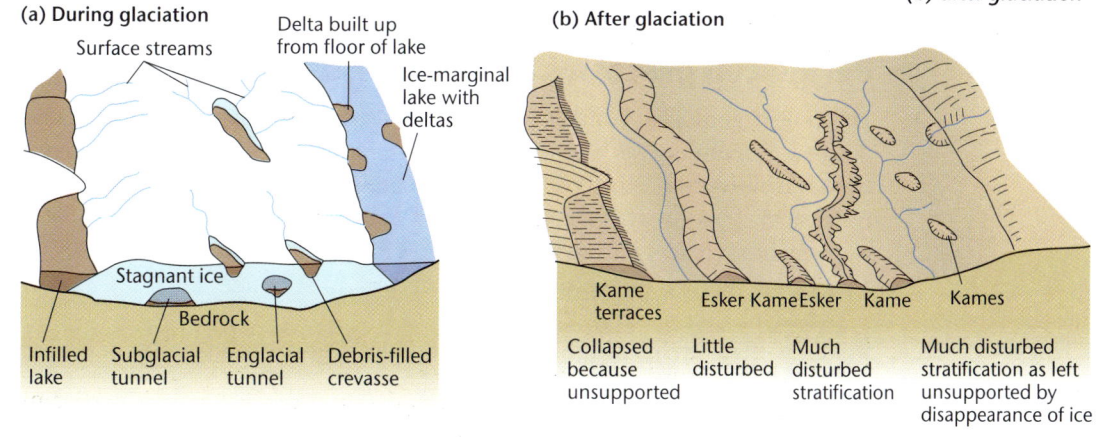

# Chapter 11 Glacial systems

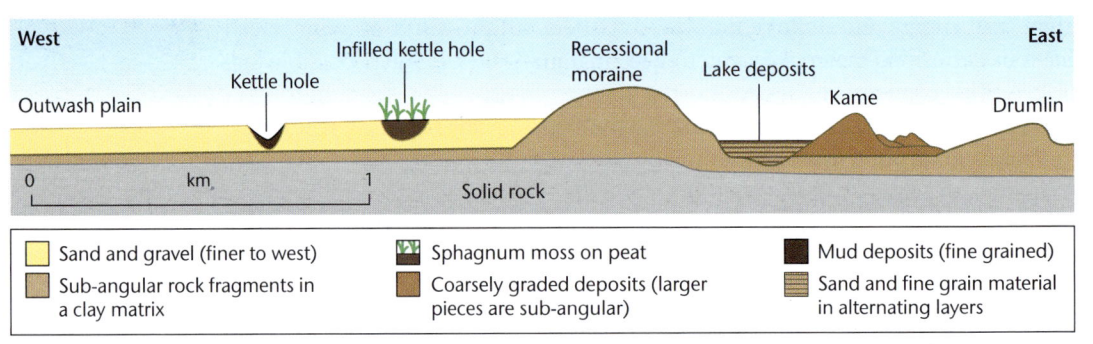

**Figure 11.17** Features of lowland glaciation

*Diagram labels: West — Outwash plain, Kettle hole, Infilled kettle hole, Recessional moraine, Lake deposits, Kame, Drumlin — East; Solid rock; scale 0–1 km.*

*Legend:*
- Sand and gravel (finer to west)
- Sub-angular rock fragments in a clay matrix
- Sphagnum moss on peat
- Coarsely graded deposits (larger pieces are sub-angular)
- Mud deposits (fine grained)
- Sand and fine grain material in alternating layers

> **Examiner's tip**
> - As with glacial landforms, for each fluvioglacial landform you know, you should be able to describe its appearance (a diagram can help), explain its formation (link processes to shape) and refer to a named example.

As with glacial deposits, fluvioglacial deposits are often difficult to identify in the field. Again, repeated advance and retreat modify and alter the appearance of landforms, and they are also subject to weathering, erosion and colonisation by vegetation in the period since the end of the last glacial episode. The proglacial area, where most of the landforms are to be found, is often **chaotic** in appearance.

## Impact of fluvioglacial landforms on human activity

Once again, the impact of fluvioglacial landforms on human activity is varied, and includes both positive and negative effects.
- Outwash deposits, for instance, often lead to the development of poor, infertile soils with little agricultural potential, but they can be a source of sand and gravel for use in the construction industry.
- The kame terrace in the Kingsdale Valley in the Yorkshire Dales provides a route for the only road through the valley, safely above the wet valley floor and any risk of flooding.

## Periglacial processes and landforms

The polish geologist Von Lozinski first used the term **periglacial** in 1909. Periglacial environments have traditionally been referred to as being 'at or near ice sheets'. However, the work of Hugh French and others has meant that they are increasingly thought of as areas with:
- permafrost (perennially frozen ground overlain by an active layer)
- seasonal temperature variations (above zero in summer, albeit for a short period)
- freeze–thaw cycles dominating geomorphological processes (frost heave, freeze–thaw)
- distinctive flora and fauna

*Physical systems, processes and patterns*  **A2 Unit 4**

Such areas exist in a variety of locations, including high latitudes (e.g. Alaska and northern Canada), continental interiors (e.g. Siberia) and high altitudes at lower latitudes (e.g. the Tibetan plateau, the high Andes and the Alps). These areas make up 25% of the Earth's land surface. It is estimated that another 25% has been periglacial in the past. Much of southern England was periglacial during the most recent glacial period, as it was not permanently covered by ice during that time, but experienced severe winters.

Another important issue is that the landforms of periglacial areas are very varied and include not only those which are distinctive and unique to periglacial areas (pingos, patterned ground, ice-wedge polygons) but also some found widely elsewhere in other climatic regions (braided streams, scree, loess).

> **Key terms**
>
> **Active layer** The layer near the ground surface in permafrost that is frozen in the winter but thaws each summer.
>
> **Periglacial** Literally means near or almost glacial.
>
> **Permafrost** Perennially frozen ground.

## Permafrost

**Permafrost** is defined as perennially frozen ground. In other words, it remains frozen for at least two consecutive summers. It is estimated that for permafrost to develop, the mean annual temperature needs to be at most −4°C. It exists as three types (Figure 11.18):

- **Continuous** occurs where the upper limit of the permafrost effectively remains at the ground surface throughout the year with little, if any, surface melting during the summer. The whole of the ground is frozen with no unfrozen patches. Continuous permafrost is only found within the Arctic Circle and continental areas of Eurasia and North America. It is commonly over 300 m in depth in these areas, although it may be much deeper in Siberia.
- **Discontinuous** contains noticeable areas of unfrozen ground or **talik** within the permafrost. It tends to be shallower than continuous permafrost, typically 10–50 m deep. The surface also shows a significant depth of melting in the summer, forming an **active layer**.
- **Sporadic** occurs where there is more talik than permafrost. Here, the mean annual temperature may be around 0°C.

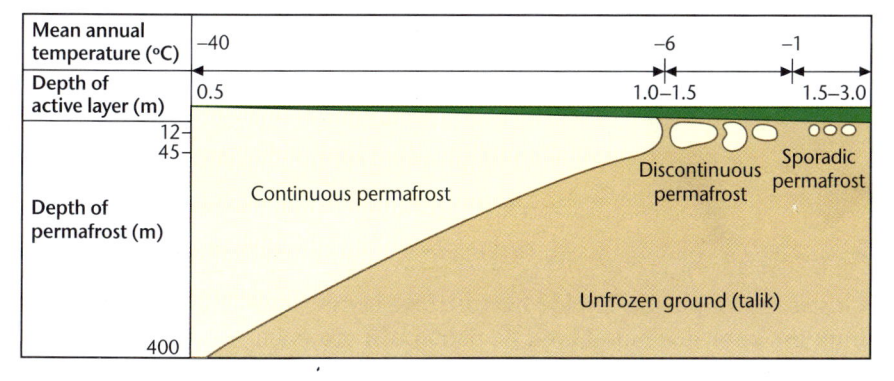

Figure 11.18 Types of permafrost and variations in depth

The depth and continuity of the permafrost layer diminishes with decreasing latitude and decreasing altitude. Its maximum depth is limited by the existence of geothermal heat.

349

# Chapter 11  Glacial systems

### Active layer
The active layer is the upper part of the ground that thaws during the warmer summer temperatures. The active layer is a significant feature of the permafrost landscape and is critical to the process–landform relationships. As it thaws it produces meltwater, which saturates the upper layer, partly due to the impermeable nature of the permafrost beneath, making it mobile. By contrast, the permafrost is inert and plays only a limited part in landscape shaping.

## Periglacial landforms
**Pingos** are rounded ice-cored hills that can be as much as 90 m in height and 800 m in diameter. They are essentially formed by **ground ice** which develops during the winter months as temperatures fall. There are two types of pingos.

**Open-system pingos** are formed in valley bottoms where water from the surrounding slopes collects under gravity, freezes and expands. This forces the overlying surface material to dome upwards (Figure 11.19).

*Figure 11.19 The formation of (a) open-system and (b) closed-system pingos*

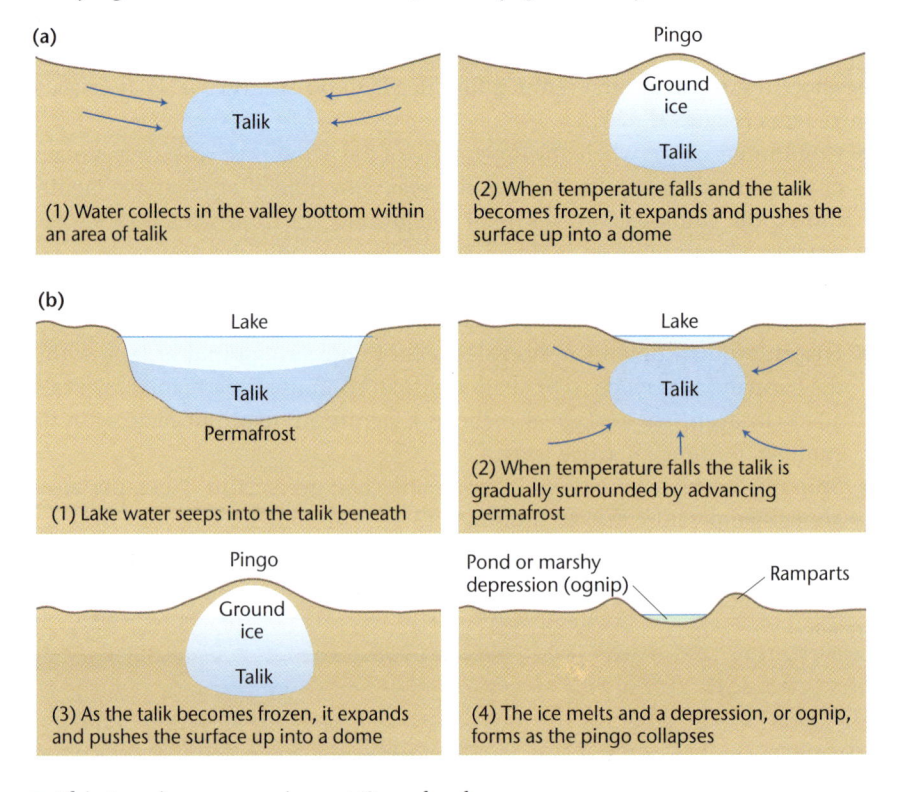

(a)
(1) Water collects in the valley bottom within an area of talik
(2) When temperature falls and the talik becomes frozen, it expands and pushes the surface up into a dome

(b)
(1) Lake water seeps into the talik beneath
(2) When temperature falls the talik is gradually surrounded by advancing permafrost
(3) As the talik becomes frozen, it expands and pushes the surface up into a dome
(4) The ice melts and a depression, or ognip, forms as the pingo collapses

- This type is common in east Greenland.

**Closed-system pingos** develop beneath lake beds where the supply of water is from the immediate local area. As permafrost grows during cold periods, the groundwater beneath the lake is trapped by the permafrost below and the frozen lake above. The saturated talik is compressed by the expanding ice around it. When it eventually freezes itself, it forces up the overlying sediments.
- Over 1400 pingos of this type are found in the Mackenzie delta of Canada.

*Physical systems, processes and patterns*   **A2 Unit 4**

*Photograph 11.8 A pingo on the Arctic coast of Canada*

When the ice core thaws, the top of the dome collapses, leaving a **rampart** surrounding a circular depression called an **ognip** (Figure 11.19). Relict features can be found in Britain, although due to the thawing of the permafrost they are collapsed features and only the remains of the rampart may be seen.
- These are common in Ireland and a good example, about 15 m in diameter, can be seen at Llanberis in north Wales.

**Patterned ground** is the collective term for a number of fairly minor features of periglacial environments. Their origin is uncertain, although they are thought to be the result of **frost heave** — a process that leads to a vertical sorting of material in the active layer (Figure 11.20).

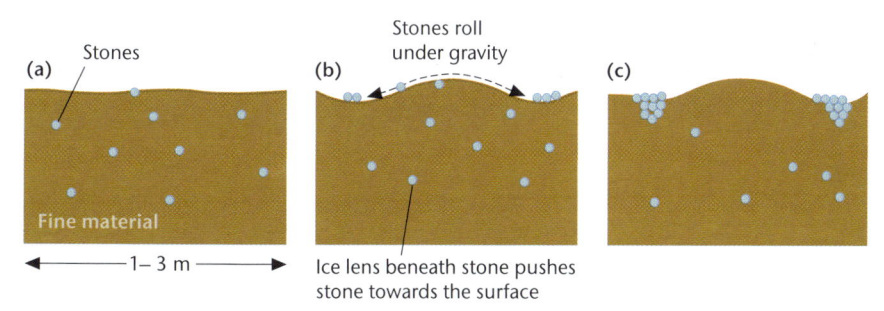

*Figure 11.20 Frost heave*

Stones within fine material heat up and cool down faster than their surroundings as they have a lower specific heat capacity. As temperatures fall, ice lenses develop beneath the stones and they are forced upwards as the freezing and expansion occur. The larger stones are moved upwards through the fine material towards the surface. They also push some of the overlying material upwards, so the surface is domed slightly and the large stones eventually force their way out onto the surface. This process is known as **frost push**.

It also likely that **frost pull** plays a part. As temperatures drop, an advancing ice front forms in the ground, extending from the surface progressively downwards the lower the temperature falls. When this ice front comes into contact with a large stone, it freezes onto it and lifts it upwards as the freezing ground expands.

When the large stones reach the surface, they move off the dome under gravity. This forms an arrangement of large stones around the domed area of fine material. From above, this appears as a network of **stone polygons**, typically 1–2 m in diameter.
- A particularly distinctive occurrence of these can be seen in the area around Barrow in Alaska.

# Chapter 11 Glacial systems

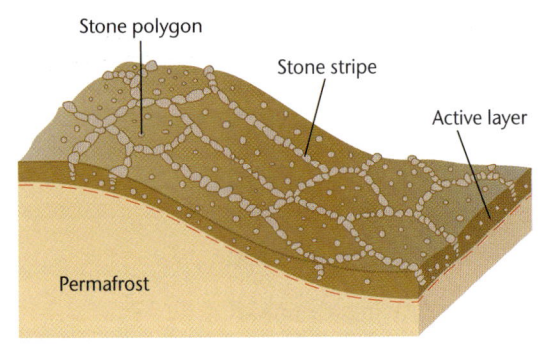

*Figure 11.21 Stone polygons and stripes*

If there is a slope angle of 3–5°, the larger stones move greater distances downslope and the polygons become elongated into **stone garlands**. On slopes of 6° and over, the effect is even more significant. The polygons lose their shape and **stone stripes** are more usually seen (Figure 11.21).

**Ice-wedge polygons** give rise to a distinctive surface appearance in many periglacial landscapes. They form due to the process of **ground contraction**, which occurs at very low temperatures (below −15°C) and results in the formation of cracks and fissures in the ground. Over many years, these are enlarged by the freezing and thawing of summer meltwater. The cracks eventually form a wedge shape, becoming wider and deeper at a rate of 1–20 mm per year. At the end of the periglacial period they fill with sand and gravel to form fossil ice wedges, which are sometimes exposed on river banks and in quarries.

*Figure 11.22 Patterned ground*

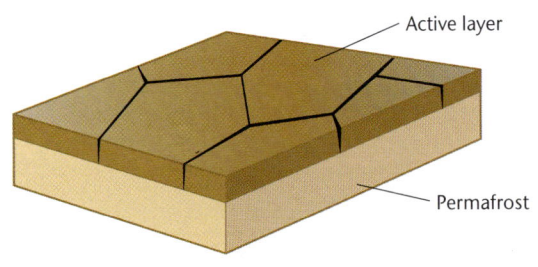

(a) Thermal contraction causes cracks

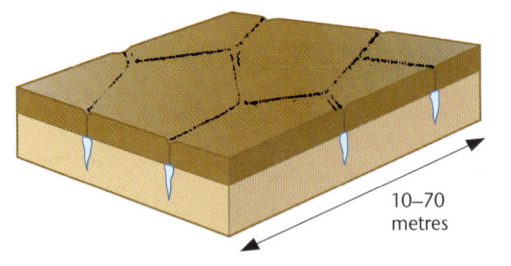

(b) Ice wedges develop in fissures, which reopen with each winter freeze

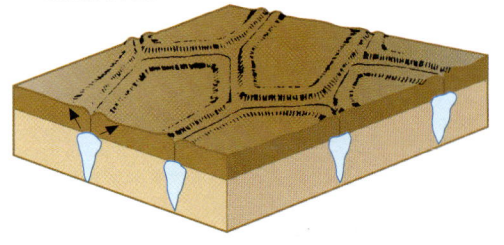

10–70 metres

(c) Wedges expand laterally, pushing the rims up

*Photograph 11.9 Stone polygons in upland Wales*

Janet Baxter/SPL

*Physical systems, processes and patterns* **A2 Unit 4**

*Photograph 11.10 Ice-wedge polygons in northern Canada*

- The Long Hanborough Carrot near Oxford is over 2 m deep, although fossil ice wedges can be as deep as 10 m.

When ice-wedge polygons are still forming, the expanding ice forces the surface upwards and so the wedges are slightly higher than the central area they surround. Surface water often collects in the depressions. The polygonal pattern of these depressions is much larger than those of the stone polygons, often 20–30 m in diameter.

- Ice-wedge polygons are clearly visible from the air in many parts of Greenland, Canada and Alaska.

### Examiner's tip

Students often confuse stone polygons and ice-wedge polygons. They are different in size, in cross-sectional structure and in method of formation. In summary:

- stone polygons are small, raised in the centre and formed by frost heave
- ice-wedge polygons are large, low in the centre and formed by ground contraction

**Solifluction** is a type of mass movement process. The thawed active layer becomes mobile in the summer months as melting reduces friction both between the particles in the layer and between the layer and the permafrost beneath. Due to this mobility, solifluction can take place on slopes of only a couple of degrees. Tongue-like features called **solifluction lobes** form at the base of slopes as the material spreads out on the less steep gradient.

- As melting was greater on south-facing slopes in Britain during periglacial times, valleys tend to become **asymmetrical** in cross-section, such as in the Cheviot Hills of Northumberland.
- In southern Britain, the deposit of rock and other material left by solifluction is usually known as **head**. A layer of this can be seen on top of the cliffs on the Isle of Portland in Dorset.

All the above processes and landforms only occur in periglacial environments. There are others that are important, but which also occur in other, rather different, environments.

353

# Chapter 11  Glacial systems

*Photograph 11.11 The blockfield near Loch Torridan, Scotland*

**Freeze-thaw weathering** is a dominant process in periglacial environments due to the distinctive seasonal variations in temperature. **Scree slopes** develop at the base of exposed cliffs as weathered fragments drop under gravity in a process called **rockfall**. In flatter areas large boulders tend to accumulate as **blockfields** or **felsenmeer** (sea of rocks).

- A blockfield of Torridan sandstone can be found near Loch Torridan, Scotland.

**Braided streams** are also common, as meltwater streams tend to have a highly variable discharge and there is a plentiful supply of debris.

**Aeolian** (wind) processes are important too, but mainly in terms of transportation and deposition rather than erosion. For much of the year the ground is relatively dry, as any water present is frozen. This, combined with the low amount of vegetation cover, means that fine material can be picked up by the wind and deposited often great distances away as **loess**.

- Silt deposits from Siberia have accumulated in China in this way.

This wide range and great variety of landforms may be seen collectively in periglacial landscapes, each tending to occur in a particular type of location (Figure 11.23).

*Figure 11.23 A summary of periglacial landforms*

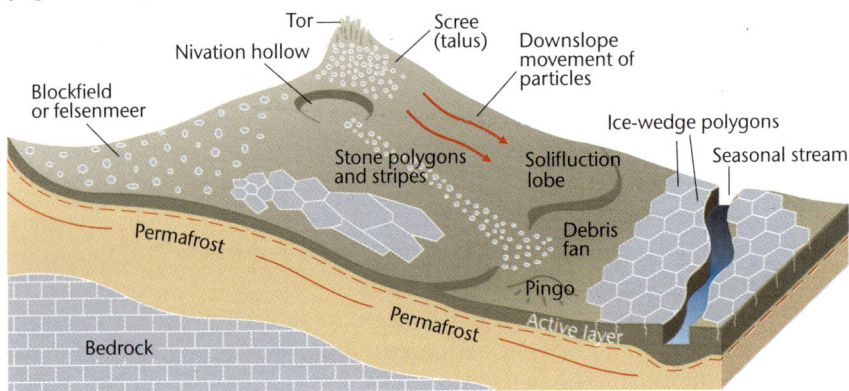

## Impact on human activity

Periglacial areas have a particularly significant impact on human activity, especially when we attempt to overcome the challenges of a periglacial environment in order to exploit it. Areas such as Alaska and Siberia have substantial reserves of raw materials.

*Physical systems, processes and patterns* **A2 Unit 4**

## Case study 11.4 Human activity in a periglacial landscape: oil exploitation in Alaska

**Photograph 11.12** The Alyeska pipeline crossing the Tanana River

Alaska has huge oilfields, including those around Prudhoe Bay on the north coast. Permafrost is susceptible to melting due to the heat produced by human activity, including building construction and the transport of hot oil through pipelines such as the 800 km Alyeska pipeline which runs from Prudhoe Bay to the ice-free port of Valdez on the south coast. As a result of this environmental challenge:

- buildings are raised off the ground and constructed with telescopic piles to counter frost heave in the active layer
- piles have to be sunk deep into the permafrost for stability
- the pipeline is raised on vertical support members 11 m deep which cost over $3000 each when built in the 1970s
- roads are constructed on 1–2 m high gravel pads to reduce physical damage and heat transfer
- the use of air cushion vehicles by companies such as Exxon helps to avoid damage to the fragile ecosystem, which has slow rates of recovery

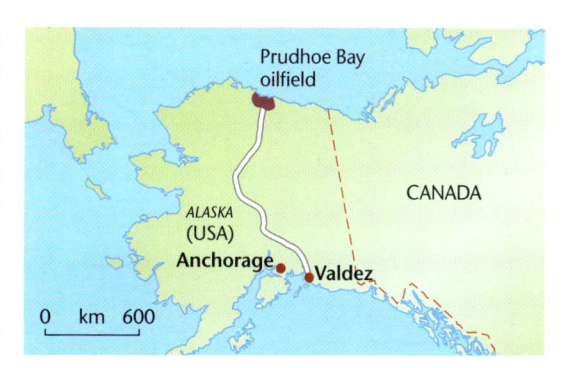

**Figure 11.24** The route of the Alyeska pipeline

- domestic water and sewage pipes are raised off the ground and heavily insulated with liquid ammonia and fibreglass in utilidors
- pipeline suspension bridges are used to cross rivers with highly seasonal discharge, such as the 700 m wide Yukon River and the Tanana River
- the pipeline is laid on the bed of smaller rivers, but 2 m of backfill and a 12.5 cm concrete jacket have to be used to prevent scour by the debris-laden rivers

### Examiner's tip

- The synoptic link in this chapter deals with challenges and opportunities of periglacial environments. You may be able to extend Case study 11.4 to cover this topic more fully.

# Chapter 11 Glacial systems

## Examination questions

**1** Table 11.2 shows the climate data for Ruskoye Ust'ye in Siberia.

Table 11.2

| Month | J | F | M | A | M | J | J | A | S | O | N | D |
|---|---|---|---|---|---|---|---|---|---|---|---|---|
| Temperature (°C) | −39 | −38 | −32 | −24 | −8 | 4 | 9 | 7 | 0 | −14 | −27 | −35 |
| Precipitation (mm) | 5 | 5 | 5 | 3 | 10 | 20 | 28 | 28 | 18 | 8 | 8 | 8 |

(a) Outline the main characteristics of periglacial environments. (5)
(b) Account for the range and variety of periglacial landforms. (20)

*Total = 25 marks*

**2** Figure 11.25 shows ice-sheet movement and erosional intensity in Britain during the last glacial advance.

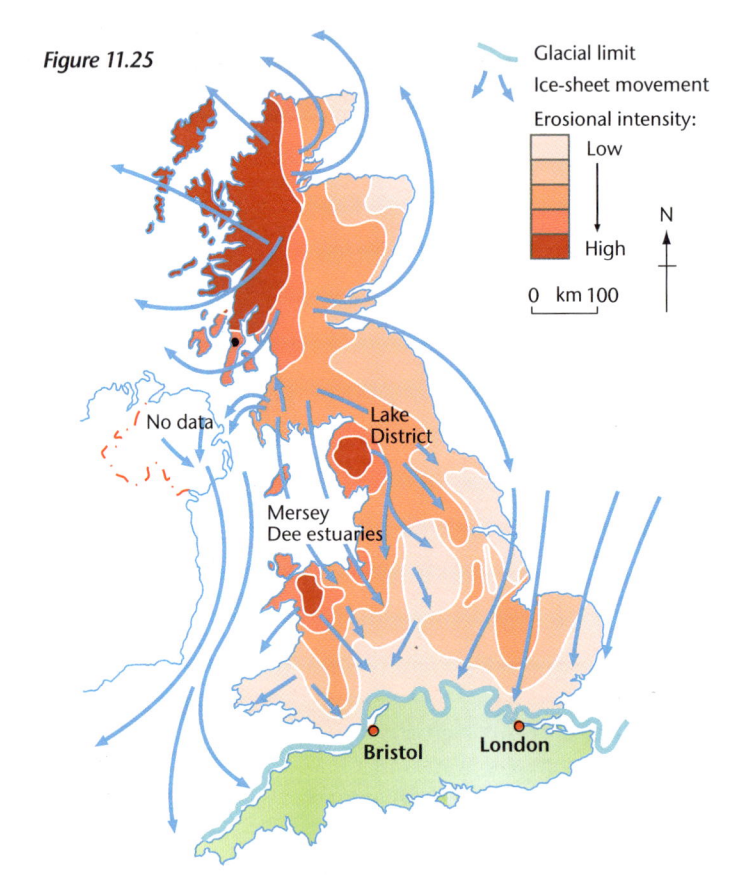

*Figure 11.25*

(a) Distinguish between the erosional processes of abrasion and plucking. (5)
(b) Examine the impact of processes of glacial erosion on the landscape. (20)

*Total = 25 marks*

*Physical systems, processes and patterns*    **A2 Unit 4**

# Synoptic link
## Opportunities and challenges associated with glacial and periglacial areas

- Opportunities and challenges exist in upland areas (either glaciated in the past and/or currently active) for tourism, energy production, quarrying, transport, agriculture, settlement, etc.
- Periglacial and permafrost environments present their own challenges and opportunities.

### Upland glacial areas

Mountainous regions are, in most cases, inaccessible and environmentally fragile environments. They are frequently marginalised or neglected by political and economic decision-making, because of their inaccessibility and their low population densities, which make them electorally insignificant. Thus they are often poor regions, peripheral and remote.

Major physical constraints (challenges) exist in upland areas:
- relatively more extreme climatic parameters so a reduction in the thermal growing season
- steepness of slopes
- thin soils
- inaccessibility, and consequent time taken to travel

To this list can be added distinctive human challenges:
- low population density, leading to shortage of skills and labour
- low electoral importance, leading to neglect and remoteness from the centre/core
- vulnerability to economic change

Upland areas that have been glaciated may have especially restrictive physical environments given that over-deepening of river valleys and the stripping of the topsoil make them even more inaccessible, increase slope angle and reduce the carrying capacity through further limitations on the availability of fertile soil. In extreme cases, such as in Norway, communication between communities was only possible by boat until very recent times.

These remote rural societies are characterised by:
- relatively few people
- dependence on natural resources
- limited educational opportunities after secondary school
- limited job demand and opportunities for skilled and educated labour
- absence of urban commodities such as cafés and restaurants, theatres and galleries

As a consequence, mountainous regions have been particularly exposed to population decline and a loss in vitality as soon as they become integrated into a

# Chapter 11    Glacial systems

national economy. Many of the mountain regions of Europe have experienced long periods of depopulation and decline, although there is some evidence that those furthest from the core (e.g. southern Italy) have suffered greater decline than those closer to the economic centre of Europe (e.g. Switzerland).

Mountains are home to about 10% of the global population. Nearly 30% of this population depends directly on the resources flowing from mountain regions. Functionally, mountains play a critical role in the environment and economic process of the planet. The economic opportunities in such regions include forestry, mineral extraction, power generation through hydroelectric schemes, livestock rearing, tourism and recreation.

## Mineral extraction: the Andes

*Photograph 11.13 Altiplano, Peru*

The Andes Mountains stretch for some 7500 km along the western edge of South America from the equator to Tierra del Fuego. As a result, they contain the most extreme variations in landscape, vegetation and climate of any mountain group in the world. The central Andes is characterised by large, high plateaus above 3500 m, known as puna or altiplano, which are the most densely populated areas of the whole range. In Peru 50% of the population lives in the mountains, 33% in the puna. Residents of these high plateaus grow frost-resistant crops, especially potatoes, and in common with mountain populations all around the world, are heavily dependent on the grazing of hardy animals. In Peru these are llamas, alpacas and sheep. The central Andes was also the heartland of the highly sophisticated Inca empire, which has become a focus for tourist activity in recent years.

In the area around Cerro de Pasco, over 4000 m above sea level, the traditional herding lifestyle has been eroded by the introduction of mining into the region. The Volcan company has become one the world's largest zinc-mining companies and has a large open pit here. The Cerro de Pasco Copper Corporation has been a major investor and reserves of silver also exist. Although this has brought employment and some infrastructure to the region, some argue that environmental issues have been neglected. These include:

- acidic waste from the mines, which seeped into the water supply
- pollution of the springs, which fed the meadows
- destruction of ecosystems in the lakes, endangering the bird species such as Junin rail
- atmospheric pollution
- deteriorating health of people and livestock

> **Examiner's tip**
>
> ■ Peru is well known in Europe for its growing tourist industry, which is also largely based in the mountains. Attention should be given to the obvious conflicts between the development of a tourist industry built around Inca culture, such as at Macchu Picchu, and the extractive industries described above.

## Tourism: Iceland

Icelanders have long been dependent on a harsh environment for their survival. The fisheries surrounding Iceland supply 80% of the country's export earnings. Both in the past and today this has brought wealth to many. The fishing industry has generated the finance to drive Iceland's service economy, which has recently developed a distinctive quaternary sector specialising in tourist-related information and exploiting its knowledge of geothermal power. Fishing is not without its costs. In one small town on Haemay Island, 500 fishermen out of 5000 inhabitants have lost their lives since 1860.

The glacial environment of Iceland has provided both a challenge and, more recently, opportunities. The development of the glaciers, especially Vatnajökull and Mýrdalsjökull, as tourist locations has led to a spin-off quaternary industry based largely in Reykjavik that specialises in off-road adventure holidays and snowmobiling.

Iceland's scenery is among the most beautiful and dramatic in the world. Iceland is also a society with an extraordinary history of isolation, in which a literary, story-telling culture has developed. After years of isolation, tourism has begun to have a significant impact. The numbers will never rival those visiting the Alps in proportion to the total population, partly on account of the uncertain summers (which are frequently cold and wet). However, the recreational impact is highly significant.

*Photograph 11.14 Off-roading in Iceland*

# Chapter 11 Glacial systems

Cultural changes and exposure to outside influences have led the younger generation of Icelanders to adopt different attitudes toward material assets, as has happened in many parts of the world (from the west of Ireland to the Andes). Reykjavik, the capital, has grown from a small village with 1200 people in 1850 to more than half the country's total population of 295 000 in 2004.

Icelandic tourism is not well organised — limited accommodation and a rather repetitive cuisine are drawbacks — but the number of tourists has grown dramatically over the last 20 years and, as is common with tourism, a few localities receive concentrated attention. One such locality is the Skaftafell National Park.

- Skaftafell is a small community close to the 2119 m summit of Öraefajökull and looking out across the outwash plains and volcanoes.
- It dates back to the earliest settlement of Iceland and has been farmed by the same family since AD 1400.
- Until the 1950s Skaftafell could only be reached on horseback across dangerous glacier meltwater rivers. Visitors were few in number.
- In 1969, with help from WWF, the Icelandic government bought the farms and made Skaftafell the centre of the country's first National Park. In 1974 the river was bridged and Skaftafell became accessible from Reykjavik by bus.
- A park visitor centre and campsite were open by the time the road was completed.
- In the first year (1974) there were 13 000 overnight stays. This increased steadily to exceed 24 000 by 1985.
- An increasing proportion of visitors were from overseas; foreign visitors exceeded 50% by 1987. This growth of interest in Skaftafell was demonstrated by a 1987 hotel association questionnaire, which indicated that it was the most popular destination in Iceland.

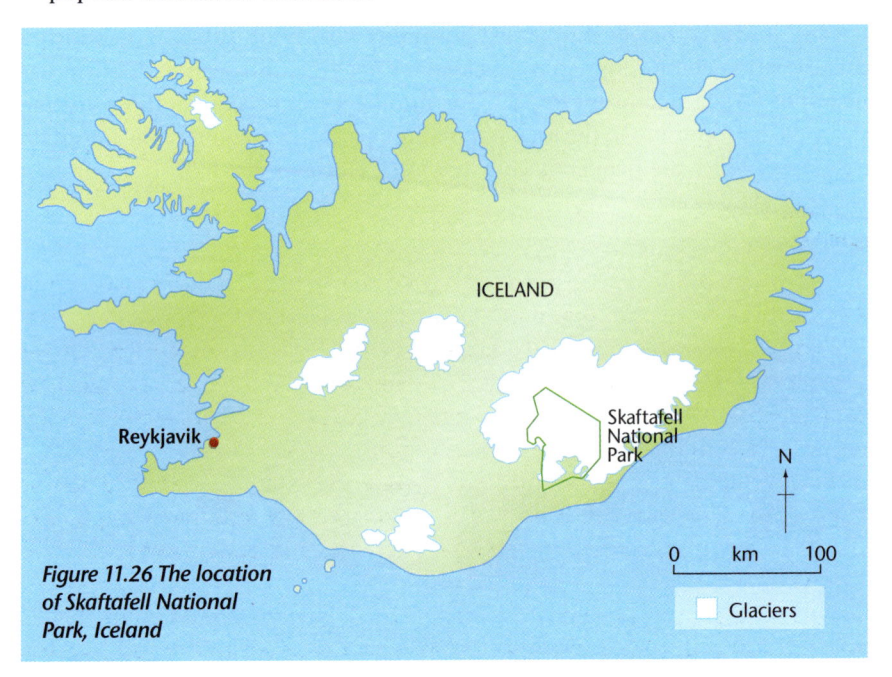

Figure 11.26 The location of Skaftafell National Park, Iceland

There are are two important themes to consider in growth of tourism at Skaftafell:
- visitor impact
- type of management policy

The landscape is the main attraction. Visitor numbers are quite modest, but the landscape is very fragile. Short walks bring visitors to dramatic glacier views with ice avalanches. The physical impact of these walks is significant and spreading. The Nature Conservatory has so far resisted pressures to permit park entry by four-wheel drive vehicles, but these pressures continue to mount as Iceland becomes both more affluent and more dependent on tourist revenues.

Mangement initiatives include the Nature Conservatory's attempt to restore the natural vegetation by eliminating sheep grazing from the surrounding hill and mountain slopes. Sheep are not 'native' to Iceland, but this policy has ended the farming in the area. It is based upon the US National Park ideology, which assumes an incompatibility between 'natural' mountain landscape and human occupation. This is in marked contrast to the UK tradition of National Parks which embrace human occupation as a key element in the formation of landscape. Without sheep farming, Icelanders would certainly have died out during the Little Ice Age 200 years ago, as their cousins in Greenland did.

This raises many questions about our understanding of 'natural' landscapes and the role of humans in their creation.

> **Examiner's tip**
>
> - This is a useful example of the complexity of the interface between the 'natural' landscape and the 'man-made' landscape. Synoptic geography often divides the 'physical' environment and the 'human' environment as though they are uncontroversial and easily determined categories; in fact the division is not always easy, possible or even desirable.

## Living with permafrost

Permafrost underlies an estimated 20–25% of the world's land surface; it occurs in more than 50% of Russia and Canada, 82% of Alaska, 20% of China, and probably all of Antarctica. Permafrost in northern Siberia is 1600 m thick and it is 650 m thick in northern Alaska.

The active layer of permafrost thaws every summer and freezes in winter. A frozen sub-layer keeps the water on the surface. The active layer is extremely vulnerable to environmental damage. For example, tracks from a passing vehicle will tear up the fragile insulating tundra, allowing the soil to thaw into scars that may remain for hundreds of years.

Permafrost regions of the planet are cold, remote and frequently inaccessible. Almost all of them have supported indigenous peoples. For example, the Inuits of North America and Greenland have, by and large, survived despite the permafrost rather than because of it. In recent years some of these regions have been integrated into the global economy more closely, sometimes with a complex set of impacts on the landscape and the people.

As development pressures increase in arctic and sub-arctic areas, planning approaches and techniques needed to offset the environmental impacts of development are not yet fully developed, especially in poorer regions such as Siberia.

As a general principle, it is becoming increasingly recognised that the terrain around building sites in the Arctic should be disturbed as little as possible.

# Chapter 11  Glacial systems

Another important issue to keep in mind is that the impacts of development on permafrost soils can occur very slowly over long periods of time. It may be too early to tell how well we have adapted our land-use practices to the Arctic lands. Development during the 1970s and 1980s in Alaska (and the Alyeska pipeline itself) that occurred as a result of the oil boom may not be sustainable over the long term in permafrost areas.

There are several general approaches to dealing with permafrost when developing housing or supporting infrastructure.

- Ignore it. If there is virtually no water in the soil, thawing of the permafrost will not significantly affect its bearing strength.
- Thaw the permafrost. This can be done by removing vegetation from the site; often peat and other permafrost-susceptible layers are dug out of the building site and replaced with gravel or sand. Such an approach is more appropriate for the discontinuous zone of permafrost, where it is thinner and unlikely to refreeze after thawing. (Soil replacement has been carried out extensively in Fairbanks, Alaska.)
- In the continuous zone, the best approach is to keep the permafrost from thawing at all. There are several common ways that this is accomplished.
  - Ventilation involves providing for the circulation of air between the bottom of the structure and the soil surface. Gravel pads fitted with ventilation ducts have been used beneath buildings. Insulating matting is also sometimes used in conjunction with the pads.
  - More commonly piles are used to raise the building several metres above the surface. Piles are usually driven down to twice the thickness of the active layer, to avoid frost-heaving problems. The pile placement is often surrounded with material that readily freezes the piles in place. Increasingly, the piles being used are telescopic so that they can adjust to the expansion and contraction of the ground during freezing and thawing cycles.

*Photograph 11.15 A modern building with telescopic piles*

*Physical systems, processes and patterns*   **A2 Unit 4**

- On airport runways and other dark surfaces, white paint has been used to reflect solar energy that would otherwise be transmitted down to the soil.
- The use of coarse gravel roadbeds helps stop water moving upwards under roadways that could form ice lenses and frost heave.
- The supply of utilities to buildings on permafrost is problematic because the pipelines for water and sewerage can thaw the soil (or the pipes can freeze). Utility corridors (utilidors) raised on low pilings and with insulated casings are one solution.

*Photograph 11.16 A utilidor carrying a domestic water supply*

There is increasing concern that global warming will lead to the thawing of vast areas of existing permafrost. If this does occur, it will have serious implications for human settlement. Stable soils may become more prone to subsidence. Increase in solifluction and landslides may occur as well. In several communities in the interior of Alaska, general thawing of the permafrost has been blamed for widespread undermining of buildings and roads.

A larger issue related to the thawing of permafrost is the potential release of gases from the decomposition of now frozen organic soils — greenhouse gases that would likely exacerbate the warming phenomenon.

**Examiner's tip**

- You should be able to comment on the fact that periglacial environments such as Alaska have a delicate and sensitive ecosystem. Any damage to natural vegetation or the ground surface may have very long-term consequences.

# Chapter 12

# Ecosystems

## Ecosystems are dynamic systems

### The components of ecosystems

**Ecosystems** have been defined as 'functioning, interacting systems comprising one or more living organisms and their effective environment, both physical and biological' (Fosberg).

Ecosystems can exist at a variety of different scales, from a single tree to a small pond, an area of woodland, the Amazon rainforest or even the Earth itself (sometimes known as the biosphere).

**Biomes** are global-scale ecosystems. For example, the tropical rainforest is a biome; the tropical rainforest of the Amazon basin is an ecosystem.

Ecosystems are **open systems** in that both energy and materials cross ecosystem boundaries. Figure 12.1 shows the inputs and outputs to and from an open ecosystem.

*Figure 12.1 Inputs and outputs in an open ecosystem*

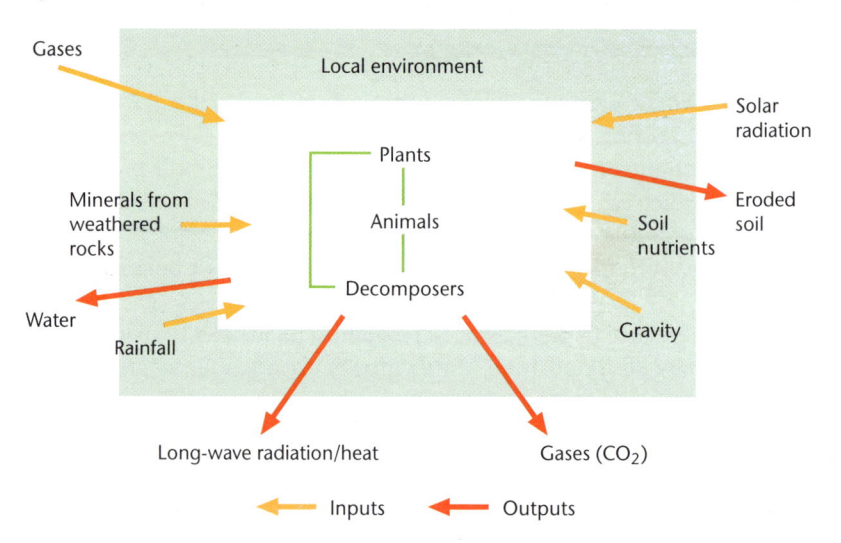

The main **function** of an ecosystem is the transfer of food energy in order to support the life forms in the ecosystem. The interactions (or **interrelationships**) are the linkages that exist between one component and another. These interactions show where one component has an influence upon another, as in Eyre's diagram (Figure 12.2). This indicates that climate is the most important component in an ecosystem, as it not only has *direct* influences but also *indirect* influences.

*Physical systems, processes and patterns*  **A2 Unit 4**

An example of a direct influence of climate is that it provides rainfall for soil moisture. An example of an indirect influence is the temperature pattern, which determines the length of the growing season and what plants can grow. This in turn influences what organic matter is added to the soil via leaf fall.

The structure of ecosystems can also be considered in terms of the types of components that they contain. These may be either **biotic components** (living matter) or **abiotic components** (non-living elements). Biotic components include plants, animals, decomposers and humans. Abiotic components include insolation, gravity, water, soil, gases, relief, rocks, temperature and wind. The biotic components can be further subdivided into feeding groups: autotrophs and heterotrophs.

Autotrophs are **producers** capable of converting solar energy into food energy for their own use and for consumption by others. They are mainly green plants that produce food energy by photosynthesis. There are some exceptions, such as the venus flytrap, which also derive energy from other organisms.

Heterotrophs are **consumers** that cannot produce their own food energy but have to feed on other organisms in order to obtain energy. They include:

- **herbivores** (plant eaters, such as rabbits)
- **carnivores** (meat eaters, such as foxes)
- **omnivores** (eaters of both plants and meat, such as humans)
- **detritivores** (decomposers, such as fungi and bacteria)

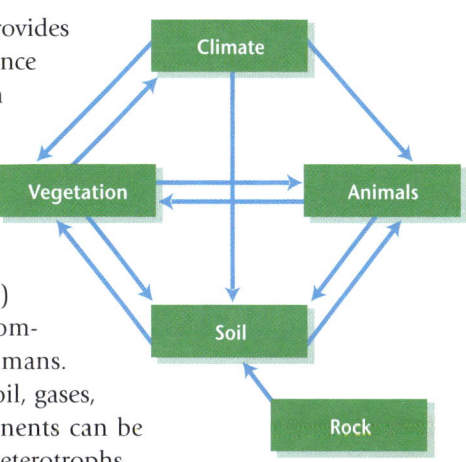

*Figure 12.2 Interrelationships in ecosystems*

### Key terms

**Biome** A type of ecosystem on a global scale, classified according to its dominant vegetation type and climatic zone.

**Ecosystem** The interrelationship between plants and animals and their living and non-living environment.

**Food chain** The linear, unidirectional flow of energy from one trophic level to the next.

**Food web** A series of interconnecting food chains.

**Trophic level** A feeding level in the transfer of energy through the ecosystem.

**Trophic pyramid** The structure of the trophic levels in an ecosystem with decreasing biomass and number of organisms with increasing trophic level.

## Energy flow

The **food chain** shows the largely unidirectional, linear transfer of food energy between trophic levels (Figure 12.3). It is a simplistic view of the process, but it is a good starting point to see the nature of energy flow.

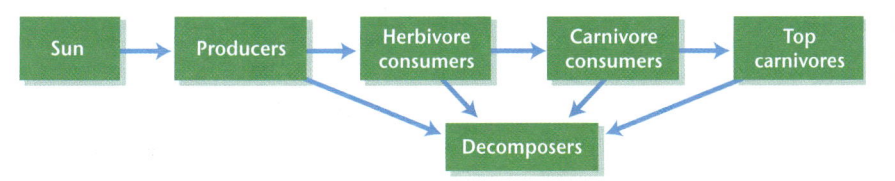

*Figure 12.3 A generalised food chain*

# Chapter 12 *Ecosystems*

**Trophic levels** are the feeding levels of the organisms in each component:
- tropic level 1 = autotrophs
- trophic level 2 = herbivores
- trophic level 3 = carnivores
- trophic level 4 = top carnivores

*Figure 12.4 A trophic pyramid*

| Top carnivores |
| Carnivores |
| Herbivores |
| Autotrophs |

Only about 1% of the energy from insolation is converted by photosynthesis, in which the green plants convert sunlight, water and carbon dioxide into carbohydrates such as glucose. This provides not only energy for the producers themselves that they use in growth, reproduction, movement and respiration, but also food energy that can be consumed by herbivores.

At each trophic level, energy is used by the organisms and so at each successive level there is less energy available; hence the amount of living organic matter that can be supported is reduced. This reduction in energy leads to a pyramid structure known as a **trophic pyramid** (Figure 12.4). The width of each bar in the pyramid can be regarded as being proportional to the amount of energy stored (the **biomass**). It is also generally true that the number of organisms at each successive level decreases, partly because of their increased size but also because of the reduced energy available on which they can feed.

The concept of a food chain is obviously a simplistic one. In any particular ecosystem there are complex feeding patterns between individual species that are more accurately summarised by the use of a **food web**. This is a much more useful approach than the food chain when investigating energy flows within a specific ecosystem. Figure 12.5 shows the typical food web of a deciduous forest ecosystem in Britain.

### Examiner's tip

- The position of omnivores and detritivores in the food chain can be difficult to identify. Omnivores obviously don't fit easily into a trophic level, as they feed on more than one of the levels beneath them. Detritivores derive food energy from decomposing organisms found in all of the levels. You may need to explain this difficulty in an essay to clarify your answer and to show that you understand the complexity of the topic.

*Figure 12.5 A typical food web*

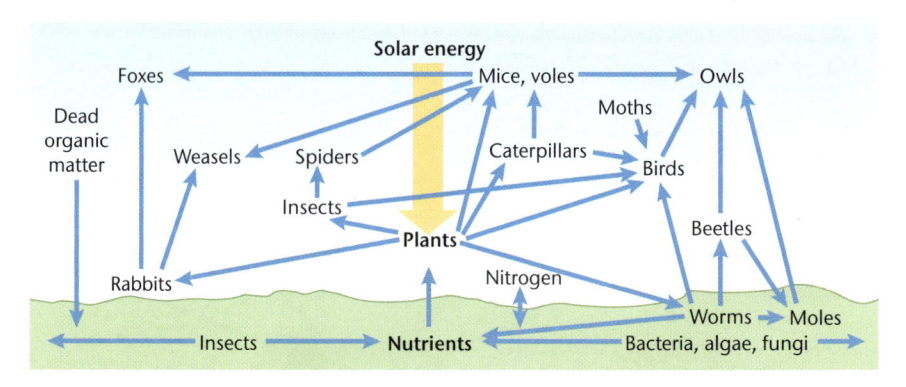

# Productivity

**Productivity** refers to the rate of energy production, which is normally given as an annual figure.

**Net primary productivity** (NPP) is the amount of energy from plants made available to herbivores, usually expressed in $g\,m^{-2}\,yr^{-1}$. It is, in effect, the amount of energy produced by photosynthesis (or gross primary productivity) minus the amount used by the plants themselves in respiration. It is generally estimated and found by trying to quantify the difference between the maximum and the minimum biomass in an area over a year. The net primary productivity of the major global biomes is shown in Table 12.1.

| Biome/ ecosystem | Net primary productivity ($g\,m^{-2}\,yr^{-1}$) |
|---|---|
| Tropical rainforest | 2200 |
| Deciduous woodland | 1200 |
| Tropical grassland | 900 |
| Coniferous forest | 800 |
| Mediterranean woodland | 700 |
| Temperate grassland | 600 |
| Lakes and rivers | 400 |
| Tundra | 140 |
| Oceans | 125 |
| Deserts | 90 |

*Table 12.1 Biome productivity*

The main factors influencing the size of NPP are:
- temperature
- nutrient availability
- vegetation type
- vegetation health
- moisture
- competition
- vegetation density
- human activity

It should be appreciated that NPP will vary not only from biome to biome but within each biome as well. In addition to these spatial variations, there will be temporal variations. In general terms, NPP increases towards the equator, as long as moisture is available, and decreases towards the poles.

**Biomass** is the amount of organic matter, plant and animal, per unit area, usually expressed as dry weight in $g\,m^{-2}$. It can also be seen as the amount of stored energy. Biomass may be stated for the ecosystem as a whole, or for each of the trophic levels. It is generally measured by sampling a number of areas in the ecosystem, and then drying the organic matter slowly in an oven (to avoid burning) before weighing. This removes any water so that only the organic matter itself is being weighed.

## Key terms

**Biomass** The amount of living matter or stored energy in an ecosystem or in a particular trophic level.

**Gross primary productivity (GPP)** The rate at which energy is produced by photosynthesis.

**Mineral–nutrient cycle** The stores and transfers of minerals and nutrients in an ecosystem.

**Net primary productivity (NPP)** The rate at which energy is produced by plants and made available to animals: GPP minus respiration.

## Examiner's tip

- Good-quality essays invariably recognise spatial and temporal variations. The very best answers tend to illustrate this at a range of scales. On a small spatial scale, this may be the difference between a low-lying part of the area and a position with higher relief. On a large scale it may be the difference between one ecosystem in one part of the world and another in a very different part. Temporally, the differences may be shown seasonally and in the longer term during succession.

# Chapter 12  Ecosystems

## Nutrient cycling

Ecosystems cycle nutrients as well as transferring food energy. Nutrients are the chemical elements and compounds that are needed by plants and animals.

- **Macronutrients** are the nutrients needed in large quantities: oxygen, carbon, hydrogen and nitrogen.
- **Micronutrients** are trace elements that are only required in very small quantities, such as sulphur, phosphorus and magnesium.

There are three main sources of nutrients:

- **rocks:** these release nutrients when weathered, providing a soluble form of elements such as sodium, potassium and calcium that can be absorbed by plant roots
- **atmosphere:** elements such as carbon and nitrogen can be obtained directly from the air, and others may be dissolved in rainwater
- **dead organic matter:** nutrients that are recycled by decomposers are made available as humus in the soil

All **nutrient cycles** involve interactions between the soil, the atmosphere and organisms. A useful summary of mineral–nutrient cycling is provided by Gershmehl's model, which shows both stores and transfers of nutrients (Figure 12.6).

*Figure 12.6 Gershmehl's model*

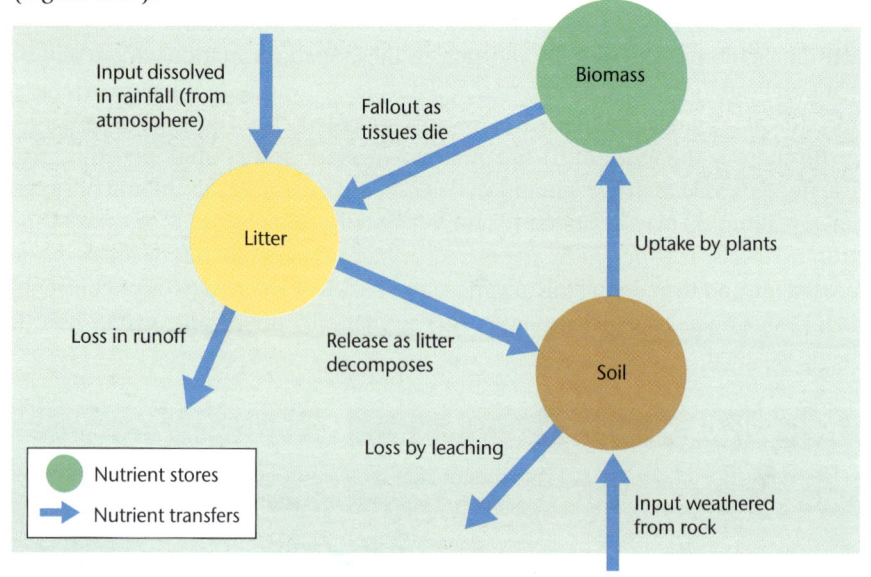

> **Examiner's tip**
>
> - The size of the stores and flows varies with the type of ecosystem. It is useful to know the details of the nutrient cycles for the forest and grassland biome case studies that are discussed later in this chapter.

The main factors that influence the stores and transfers are:

- rate and type of weathering of rocks
- the amount of surface runoff and soil erosion
- the rate of removal of nutrients from the soil (leaching)
- the amount of rainfall
- the length of the thermal growing season
- the rates of decay and decomposition by decomposers
- the nature and type of vegetation cover

*Physical systems, processes and patterns* **A2 Unit 4**

- the density of vegetation cover
- the incidence of fire
- the age and health of the plant species
- the degree of competition between species
- human activity

Explaining the specific features of the mineral–nutrient cycle in any one particular ecosystem or type of ecosystem may involve consideration of many, if not all, of these factors. Significant changes to the natural cycle can be made by human activities such as deforestation or agriculture.

# Grassland and forest biomes

It is a requirement of Edexcel (A) that you study *one* grassland biome and *one* forest biome. The focus of the study should be on their:
- **distribution:** in which parts of the world is the biome located
- **structure:** what the biome consists of and how it is organised
- **functioning:** how the biome operates

For each of the three features of the biome, you need to be able both to describe and to explain. The factors that influence the features will be physical and human and will therefore mainly include climate, soil, relief and human activity.

### Case study 12.1  *A forest biome: tropical rainforest*

#### Distribution

Tropical rainforests occur in a broad band around the Earth between about 15°N and 15°S of the equator (Figure 12.7). The three main areas are the Amazon basin, the Congo (Zaire) basin and Indonesia. The band is incomplete as there is an absence of tropical rainforest in East Africa, mainly due to the influences of relief and soil in the area. However, in southeast

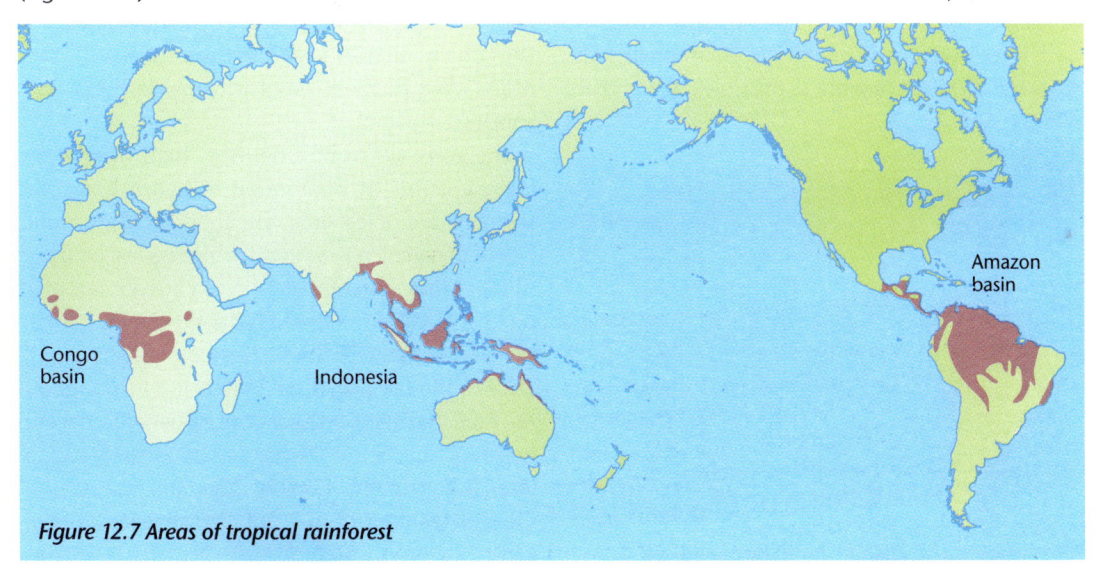

*Figure 12.7 Areas of tropical rainforest*

# Chapter 12  Ecosystems

Asia the climate is warm and wet enough for the forest to extend into the sub-tropical parts of India and Bangladesh.

## Structure

The **climatic conditions** in tropical rainforest are characterised by:
- high mean annual temperatures of 27–30°C due to their equatorial location and the high angle of the sun
- low seasonal temperature ranges of only 2–3°C due to the lack of variation in the sun's high angle
- high rainfall, typically 2000 mm per year or more, which is largely intense, convectional rainfall falling in the early afternoon after strong heating of the ground during the morning
- high humidity levels, often 100% near the ground surface
- a 12-month thermal growing season

Figure 12.8 shows the climate data for Singapore. The total rainfall is 2413 mm, the mean temperature is 27°C and the annual temperature range is 2°C.

The **vegetation** is dominated by evergreen trees, which retain their leaf cover throughout the year to take advantage of the 12-month growing season. Species diversity is very high with as many as 200 species of tree per hectare, including mahogany, balsa, teak and the emergent kapok.

The vegetation has a layered vertical structure, each layer having a different type of crown in order to receive as much sunlight as possible (Figure 12.9).

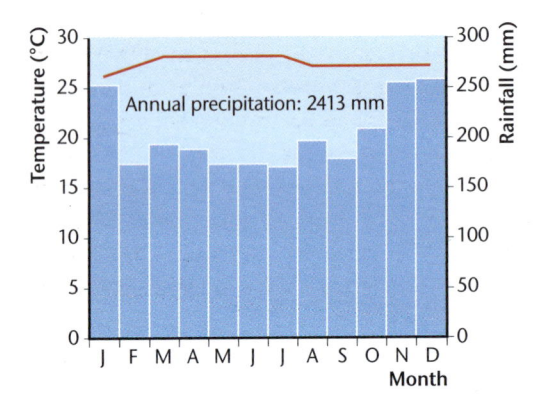

*Figure 12.8 Climate graph for Singapore*

*Photograph 12.1 Tropical rainforest in Costa Rica*

There is usually little undergrowth because hardly any light penetrates to the forest floor. Similarly, most trees lack branches on the lower part of their trunks. Lianas (creepers) use the tree trunks to climb their way upwards in search of light. The trees grow quickly and reach heights of over 30 m. As their main roots are shallow (most of the nutrients are in the organic matter near the top of the soil), they often have buttress roots to provide them with support.

The **soils** are tropical latosols that, despite the intense chemical weathering of bedrock and the regular input of leaf litter, are not particularly fertile. The high rainfall results in quite intense leaching and nutrient loss. Iron and aluminium are not removed by leaching and so the soil has a red colour, particularly in the middle of the soil profile (Figure 12.10). Despite the regular inputs of leaf litter throughout the year, the rate of breakdown and decay by decomposers is high, about 1% per day, and so little surface litter accumulates. (Soil profiles are discussed in more detail later in this chapter, and described in Table 12.2 on p. 386.)

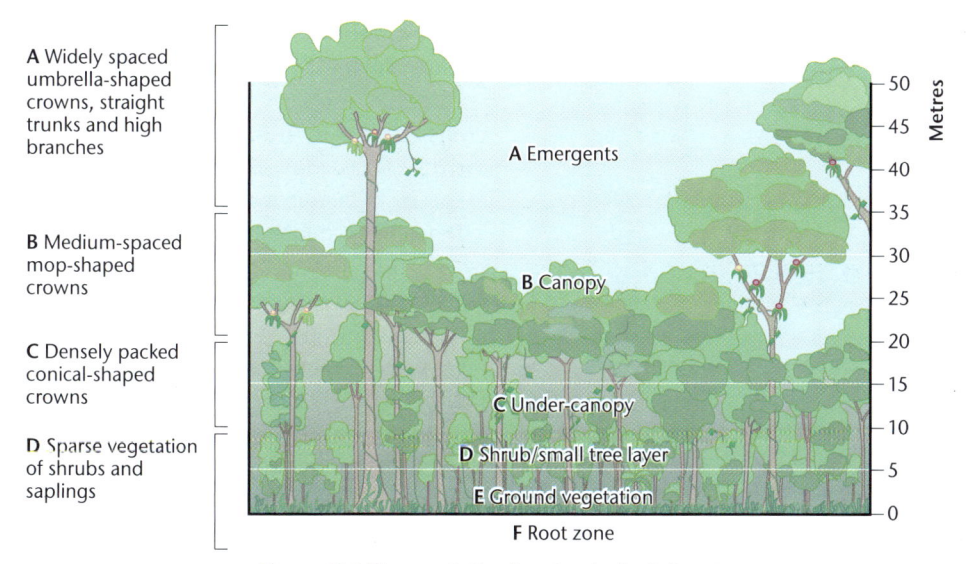

*Physical systems, processes and patterns*  **A2 Unit 4**

**A** Widely spaced umbrella-shaped crowns, straight trunks and high branches

**B** Medium-spaced mop-shaped crowns

**C** Densely packed conical-shaped crowns

**D** Sparse vegetation of shrubs and saplings

- A Emergents
- B Canopy
- C Under-canopy
- D Shrub/small tree layer
- E Ground vegetation
- F Root zone

*Figure 12.9 The vegetation in a tropical rainforest*

The **animals** of the rainforest are great in number and variety. Although ground animals are limited, the rainforests of Brazil are thought to have 2000 species of birds, 600 species of insects and mosquitoes and 15 000 species of fish.

## Functioning

Rainforests are productive systems and their NPP is typically 2200 g m$^{-2}$ yr$^{-1}$. There is a great deal of stored energy, with the biomass averaging 45 kg m$^{-2}$. The climate is the key reason for this, with the year-long thermal growing season, high temperatures and abundant rainfall all allowing high levels of photosynthesis and growth.

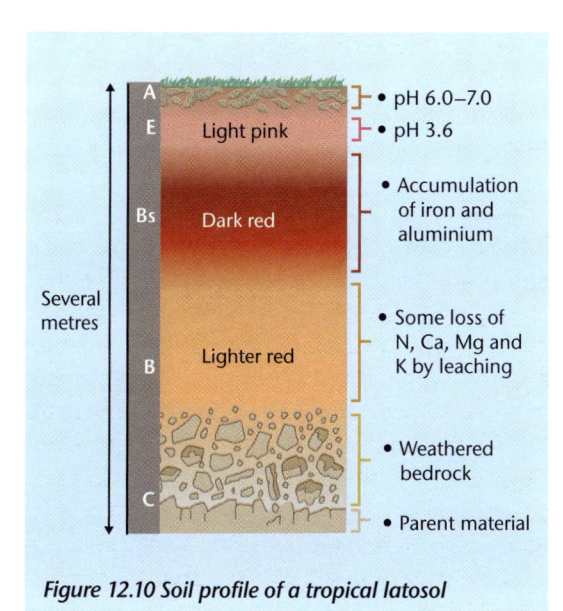

The mineral–nutrient cycle has a very large biomass store and very large, rapid transfers. Decomposition of dead organic matter is rapid (averaging 1% per day) as the climatic conditions permit plenty of extremely active soil fauna. Uptake by plants is rapid due to the ideal growing conditions. However, losses can also be high due to both runoff and leaching, particularly once vegetation has been cleared (Figure 12.11).

## Human activity

Increasingly high levels of human activity in the tropical rainforest in the last 30 years or so have had significant effects on the distribution, structure and functioning of this biome.

Deforestation for plantation agriculture, timber, mining or as a result of the pressure from an increasing population dependent on shifting cultivation has led to a reduction in the distribution of the forest. The current rate of loss is about 40 million ha yr$^{-1}$ (which works out at around 1 ha s$^{-1}$). Malaysia is losing 2% of its forest per year and Brazil is losing 8.5 million ha yr$^{-1}$. It is estimated that, so

Soil profile labels:
- A, E: Light pink — pH 6.0–7.0, pH 3.6
- Bs: Dark red — Accumulation of iron and aluminium
- B: Lighter red — Some loss of N, Ca, Mg and K by leaching
- C: Weathered bedrock, Parent material
- Several metres

*Figure 12.10 Soil profile of a tropical latosol*

371

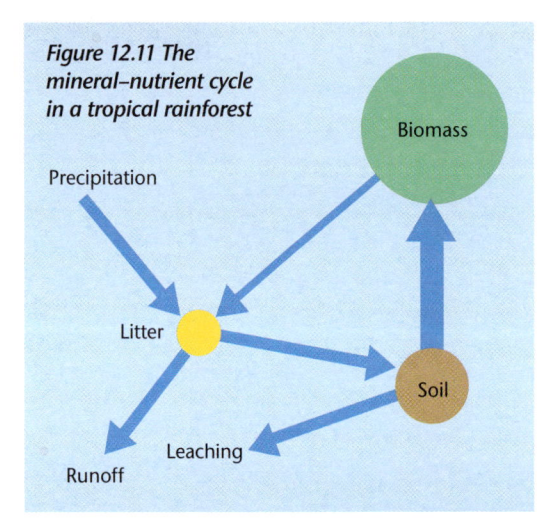

# Chapter 12  Ecosystems

**Figure 12.11 The mineral–nutrient cycle in a tropical rainforest**

Size is proportional to the amount of nutrients stored

Width equals nutrient flow as a percentage of the nutrients stored in the source

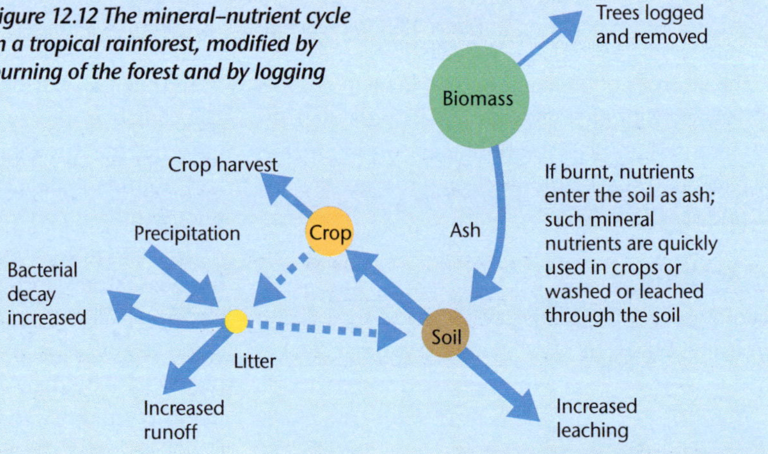

**Photograph 12.2 Clearing rainforest by burning**

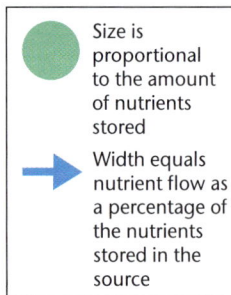

far, the world has lost half of its original cover of tropical rainforest.

There has also been a great reduction in species diversity, with an estimated 75 species being lost each day. In Ecuador, 90 species of plants were made extinct by farmers in just 8 years.

The mineral–nutrient cycle is particularly affected when crops replace trees. The crop biomass is removed at harvest and so very little organic matter is transferred to the store. The exposed soil is also more heavily leached and this quickly negates any short-term benefit from the ash produced if the forest is burnt (Figure 12.12). As a result of these changes, NPP has fallen from 2200 g m$^{-2}$ yr$^{-1}$ to as low as 100 g m$^{-2}$ yr$^{-1}$ in parts of Costa Rica.

**Figure 12.12 The mineral–nutrient cycle in a tropical rainforest, modified by burning of the forest and by logging**

372

*Physical systems, processes and patterns*     **A2 Unit 4**

## Case study 12.2  A grassland biome: temperate grassland

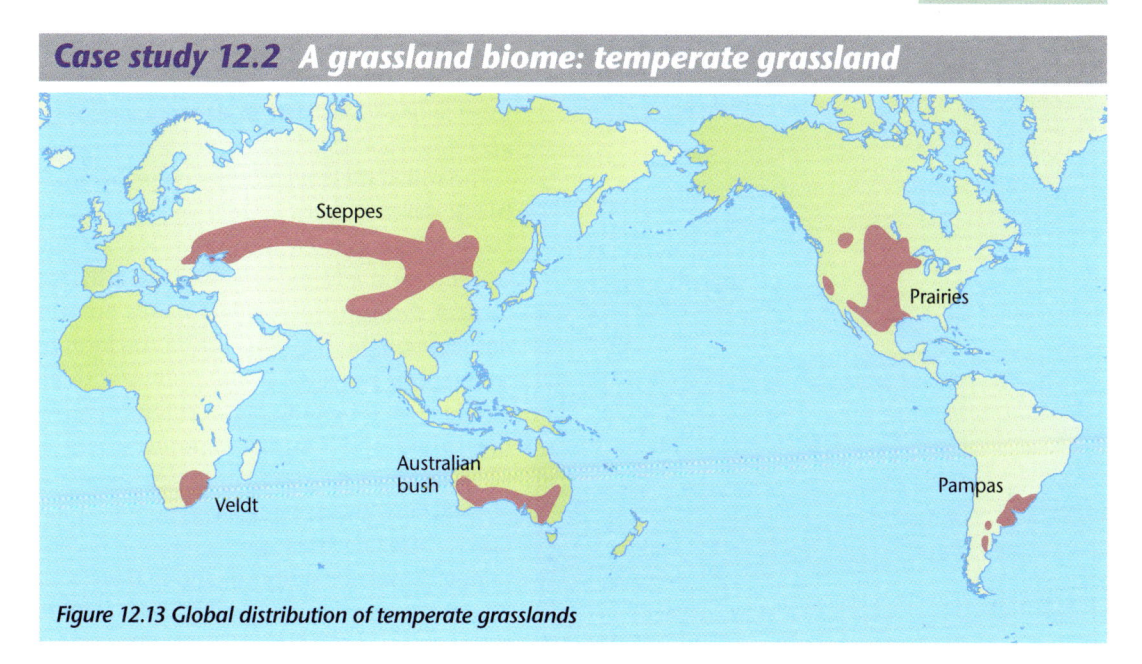

*Figure 12.13 Global distribution of temperate grasslands*

### Distribution

Temperate grasslands lie between 40°N and 55°N and between 40°S and 55°S of the equator. They are concentrated on the continental interiors of large land masses, and so the two main areas are the prairies of North America and the steppes of Russia (Figure 12.13). The pampas of South America are of a smaller extent, given the lesser distance from the sea in this area.

### Structure

The **climate** of these areas is described as cool temperate. Temperatures range from −15°C in the winter to +20°C in the summer. The thermal growing season is 5–6 months. The temperature range is high due to the lack of moderating influence of the sea in the continental interior locations. Rainfall is quite low, ranging from 250 mm in the heart of the continental areas to 600 mm or more towards the coasts. Mountain barriers such as the Rockies and the Urals provide a rain-shadow effect. In winter the precipitation is mainly in the form of snow, while in summer the rainfall is generally due to convection as a result of the high temperatures. Figure 12.14 summarises the climatic data for Omsk, Russia. The total precipitation is 318 mm, the mean temperature is −1°C and the annual range is 40°C.

The dominant **vegetation** cover is grass, as there is insufficient moisture for trees to survive, except in river valleys. The winter frosts also slow the

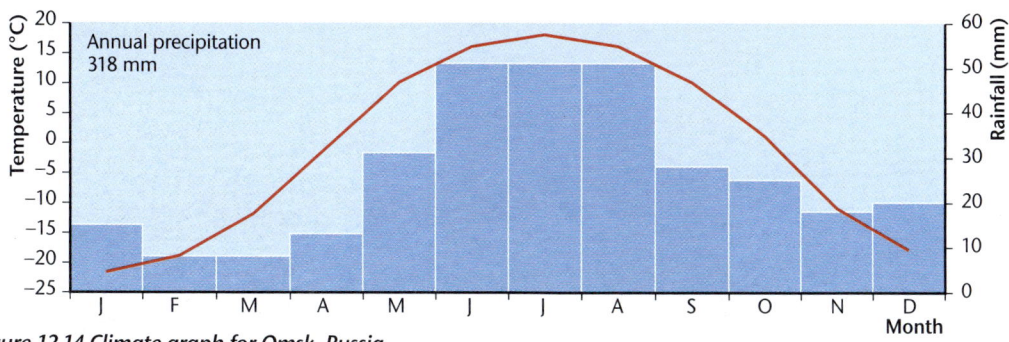

*Figure 12.14 Climate graph for Omsk, Russia*

# Chapter 12 Ecosystems

**Photograph 12.3**
**The Steppes in Siberia**

growth of trees. Feather grasses grow to about 50 cm and give fairly even coverage, while tussock grass reaches 2 m or more and has a very deep root system.

The **soil** in these areas is the **chernozem** or **black earth** (Figure 12.15). The thick grass cover dies back each winter due to the extreme cold and the effects of the **physiological drought** (precipitation falls but it is not available to plants, as it is in the form of snow). This provides a mull humus with an abundance of biota to aid decomposition in the spring and summer months. There is some downward movement of base minerals, but leaching is very limited as evapotranspiration is greater than precipitation, except during convectional storms, and so there is an upward movement of water and soluble minerals by **capillary action**. This helps maintain a neutral pH and **calcification** occurs as calcium is precipitated out of solution, leaving white nodules or threads of calcium in the B horizon. Thus the soil is deep, rich and fertile. In drier areas, where only shorter grasses grow, this is less so and here the lighter-coloured chestnut soils are found.

The **animal** life is dominated by grazing herbivores such as bison and buffalo in the prairies and the siga in the steppes. Birds such as the

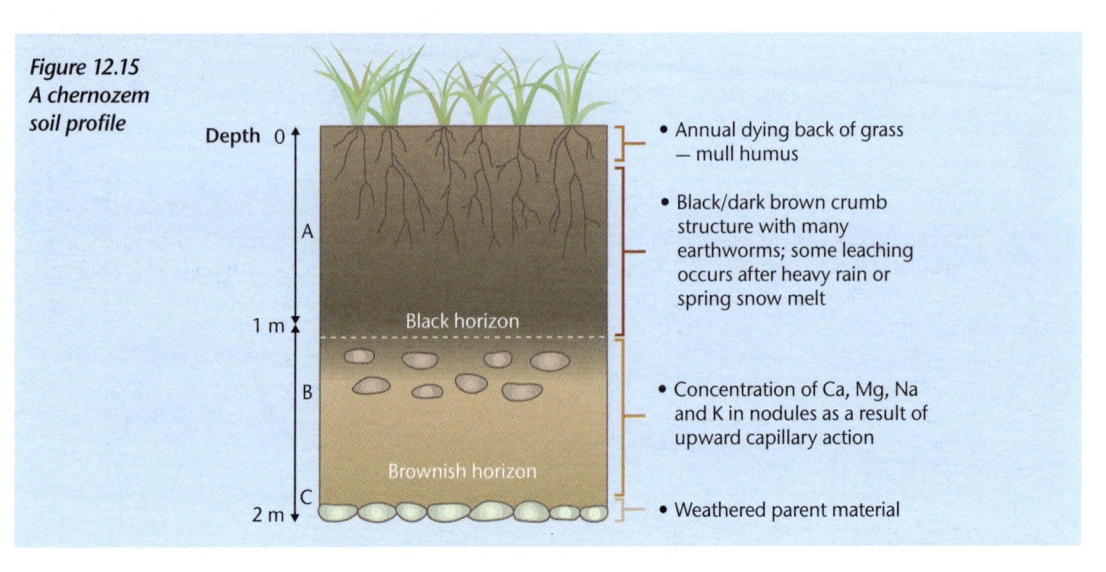

**Figure 12.15**
**A chernozem soil profile**

- Annual dying back of grass — mull humus
- Black/dark brown crumb structure with many earthworms; some leaching occurs after heavy rain or spring snow melt
- Concentration of Ca, Mg, Na and K in nodules as a result of upward capillary action
- Weathered parent material

Black horizon
Brownish horizon

*Physical systems, processes and patterns*  **A2 Unit 4**

prairie chicken, actually a grouse, feed on shoots. Prairie dogs live underground, where in winter they can survive by feeding on the roots of the grass. Rattlesnakes hibernate during the cold winter months. However, in large parts of the prairies in particular these species were replaced by domesticated herbivores as people started to use the area for cattle ranching. Wolves and coyotes are major top carnivores in these locations.

## Functioning

The NPP of the temperate grasslands averages 600 g m$^{-2}$ yr$^{-1}$, although this can be as high as 700 g m$^{-2}$ yr$^{-1}$ in the wetter, long grass prairies and as low as 500 g m$^{-2}$ yr$^{-1}$ in the dry, short grass areas. Biomass is low at 1.6 kg m$^{-2}$ due to the lack of trees. The mineral–nutrient cycle is dominated by the large store of nutrients in the chernozem soils and the significant transfer of nutrients from the biomass to the litter store that occurs in the winter die-back of the grasses (Figure 12.16). Losses due to leaching are limited due to the relationship between evapotranspiration and precipitation. Decomposition occurs quite quickly in the spring and summer due to the abundant soil biota.

## Human activity

Today, over 90% of the original grassland of the prairies and 20% of the steppes have been lost to agricultural activity. Initially this was for cattle ranching, as the early settlers looked to take advantage of the extensive grassland vegetation. Intensive cattle production still takes place in Nebraska and South Dakota. More recently the emphasis has shifted to cereal growing, with wheat and soya the main crops. Both North America and Russia have over 100 million hectares of arable cropping. However, harvesting the crop removes the litter input from the soil, and huge amounts of fertiliser need to be used in order to maintain soil fertility. The effect of agriculture on the mineral–nutrient cycle is similar to that in the tropical rainforest (see Figure 12.12) if remedial action is not taken.

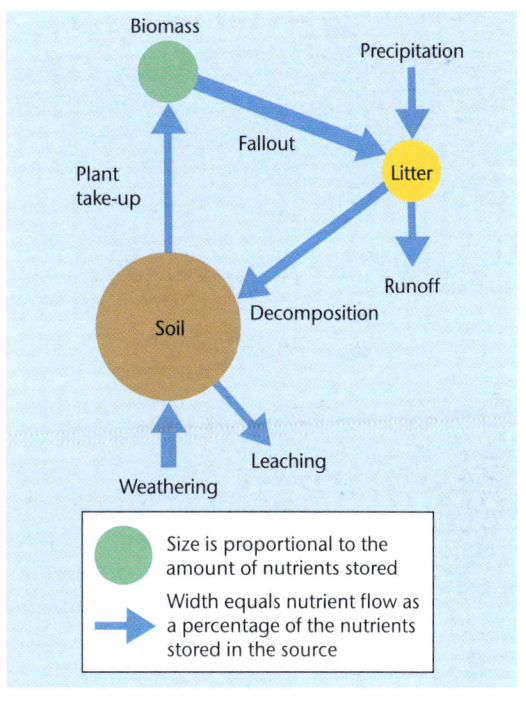

*Figure 12.16 The mineral–nutrient cycle in a temperate grassland*

### Examiner's tip

For each of the case studies you need to be able to:
- describe and explain its distribution — draw this on a world map and explain the role of climate and other lesser factors (soil, relief, human activity, fire) in its distribution
- describe and explain its structure — this could refer to the species found in each trophic level or the physical structure of the vegetation layers (canopy, under-canopy, etc.)
- describe and explain its functioning — the main functions of ecosystems are to transfer energy through the trophic levels and to cycle nutrients, both of which enable the ecosystem to support life

To ensure that you are able to use your examples effectively, you should be able to provide supporting data and evidence in the form of:
- names of places
- climatic data
- species names
- soil type

# Chapter 12   Ecosystems

## Plant succession

The composition of a **plant community** depends on the interaction between all of the components that make up an ecosystem. Plant communities vary from place to place and, as a general rule, become more complex over time. This is because the physical environment tends to become more favourable to plant growth. The reason for this is that there is a progressive rise in nutrient and energy flows, and so biodiversity and NPP also increase over time.

**Succession** is the long-term change in a plant community as it develops from a bare inorganic surface to a **climax community**. The main changes that occur are that, in general, the vegetation becomes taller, denser, more varied and more woody over time until the climax community is reached. At this time the variety of species tends to fall and level off as dominant species emerge. Each stage in the sequence is a **sere**, which alters the local environment and allows another group of species to dominate. It may increase the depth and fertility of the soil and allow for improved storage of moisture. Larger species may provide protection for smaller ones while plants may trap and stabilise sand or sediment with their roots. The climax community is the group of species best able to develop under the prevailing environmental conditions. This is usually controlled by climate and is then known more specifically as the **climatic climax community**.

The **mono-climax theory** argues that this is the only outcome of succession. However, the **poly-climax theory** argues that the climax community may be determined by other factors such as soil or relief:

- **edaphic climax:** when soil conditions determine the climax community
- **topologic climax:** when the relief is too steep for many species
- **biotic climax:** when grazing or trampling is the key control
- **hydrologic climax:** if the availability of water is the main factor
- **plagioclimax:** when human activity determines the climax community

### Key terms

**Climatic climax community** The group of species that is in equilibrium with the climatic conditions.

**Climax community** The group of species that is best able to exploit the prevailing environmental conditions.

**Hydrosere** Plant succession in a freshwater environment.

**Lithosere** Plant succession on a bare rock surface.

**Plagioclimax community** The group of species that exist in an ecosystem as a result of the influence of human activity.

**Plant community** The group of plant species living together in an ecosystem.

**Plant succession** The long-term evolution of a plant community from initial colonising species to the climax community.

**Sub-climax community** The group of species found in an ecosystem that have not yet reached a state of equilibrium with the environmental conditions or which have been prevented from doing so.

If there are no further changes to the environment, the climax community will persist and achieve a state of equilibrium (Figure 12.17). This will be maintained until a change does occur.

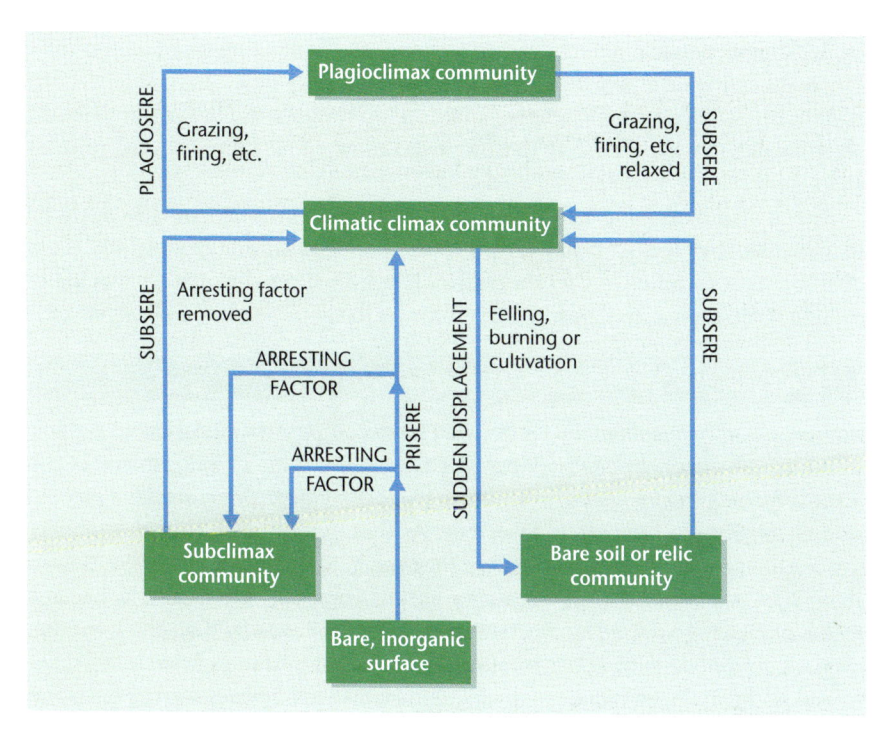

*Figure 12.17 The long-term evolution of plant communities*

**Primary succession** occurs on bare ground that was previously unvegetated. It includes bare rock, lava flows, sand and mud. The changes that take place are mainly the result of internal or **autogenic factors** — changes in the environmental conditions brought about by the plant community itself. The processes of invasion, colonisation, competition, domination and decline tend to operate in sequence to influence the composition of the plant community.

When plants first invade a bare, inorganic surface, perhaps by wind dispersal of seeds or spores, groups of a particular species or colonies of two or more species become established. These are known as **pioneer species**. They are very hardy, tolerant species that can survive in the harsh conditions because they are well adapted. They are successfully able to compete for the limited water, light, nutrients and space, partly by having a low demand for these essentials. They also help to improve the conditions over time. They may contribute to rock weathering by releasing organic acids or by root penetration into cracks and crevices. As they die they add organic input to the thin veneer of soil that starts to develop.

This enables other, less hardy species to colonise the improved surface and to change the composition of the community. Each stage in the succession produces a better environment for a greater number of species. The increasing soil depth allows for taller species that need more stability and these slower growing species, trees in most cases, eventually become dominant as they are successfully able to compete with other, smaller species. Eventually the dominants kill off some of the other species and the biodiversity is often slightly reduced once stability has been achieved in the climax community.

**Secondary succession** involves an external or **allogenic factor** causing a change to the primary succession that occurs naturally. Factors such as human activity, fire

# Chapter 12  Ecosystems

or flooding may occur, which lead to a change in the existing plant community. Once the allogenic factor is removed or ceases to have an influence, the community will continue to evolve and develop towards its climax state, but the seral stages followed are likely to differ from those of an uninterrupted primary succession (Figure 12.17).

After a disturbance has happened and the plant community embarks again on its successionary route, it may return to the previous climax community if the disturbance was minor or very short term. However, if the disturbance was major or long term, then a new climax community is likely to be ultimately achieved.

## Case study 12.3  Succession: lithosere, Isle of Arran

A **lithosere** is succession on a bare, consolidated rock surface. On the Isle of Arran, this has happened due to the gradual exposure of a rocky shore platform during a period of relative sea level fall, mainly due to isostatic readjustment after the end of the last glacial period about 9000 years ago.

The newly exposed rock was initially colonised by blue-green algae, which have no root systems and can survive with few mineral nutrients. Lichens and mosses then followed — these have similar characteristics but can obtain nutrients from rainwater, which the moss absorbs. As these species are autotrophs, they provide their own energy through photosynthesis. Gradually, they helped to break up the rock, both physically and chemically, and formed a very thin layer of soil. As they died, they contributed to soil formation by adding litter, which is decayed by decomposers to provide an organic input. Seeds were then able to lodge in cracks and crevices in the rock surface, giving grasses a place to take root. As these were taller than the pioneer species, they began to dominate. Gradually, slower-growing shrubs, bracken and bramble became established, as the soil was by then thick enough to support their root systems. Eventually, slow-growing dominating tree species such as birch and ash replaced smaller species of alder and rowan until a climax community of oak woodland became established.

Due to the restrictive, limiting conditions at the water's edge, succession is only partially complete in many places. With increasing distance from the sea there is an increasingly mature and well-developed plant community (Figure 12.18).

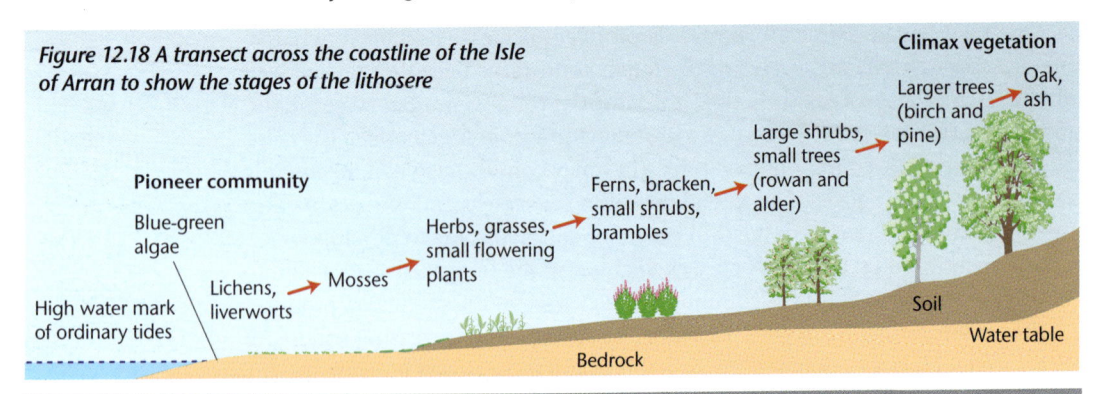

Figure 12.18 A transect across the coastline of the Isle of Arran to show the stages of the lithosere

### Examiner's tip

- You do not need to know a case study of secondary succession, nor of each of the different types of climax community. You do need to understand the concepts, however.
- You do need a case study of a lithosere and of a hydrosere.

*Physical systems, processes and patterns*  **A2 Unit 4**

## Case study 12.4 Succession: hydrosere, Sweetmere, Shropshire

A **hydrosere** is succession in a freshwater environment. Sweetmere is one of a series of small kettle lakes formed at the end of the last glacial period.

As the ice retreated about 10 000 years ago, the climate slowly started to warm and algae and water lilies, floating aquatic plants, started to colonise the lake. The dead organic matter created by these plants provided fertility in the lakebed sediment and this enabled species such as reeds and bulrushes to take root. The roots of these plants helped to trap more sediment and the dead plants added further nutrients, creating a genuine soil. Sedges, willow and alder were then able to take root and develop, further adding to the trapping of sediment and the increasing depth of soil. As the sediment was trapped, so the surface level rose and became drier, making it more suitable for a wider range of species. Birch eventually came to dominance and by this stage the ground level was above the water table. Terrestrial species of grasses and shrubs quickly developed below the trees, and the acidity level of the soil increased, enabling greater rates of nutrient exchange. Finally, with the aid of artificial drainage (an allogenic influence), the oak/ash climax community was able to develop (Figure 12.19).

Human activity may continue to play a part here as the area to the southeast of the lake is a park. Some cutting back and clearance of species by the water's edge takes place in order to allow access for recreation and leisure. If this continues, it will prevent the lake from following its natural succession, which would ultimately result in the whole lake becoming oak/ash woodland.

*Figure 12.19 Succession at Sweetmere, Shropshire*

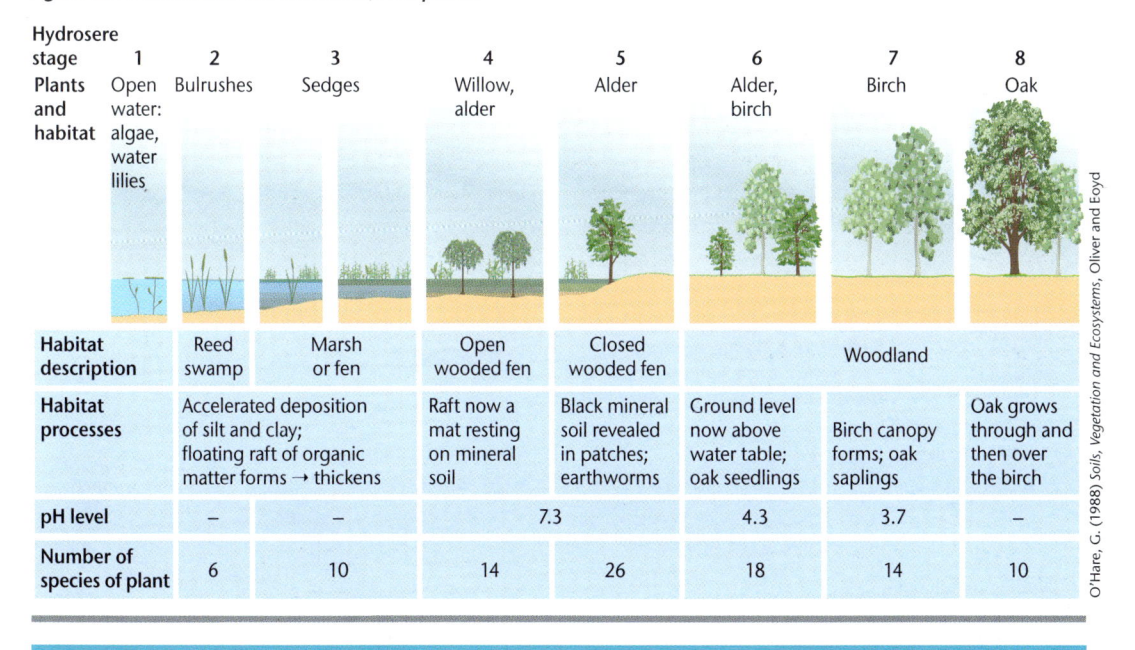

| Hydrosere stage | 1 | 2 | 3 | 4 | 5 | 6 | 7 | 8 |
|---|---|---|---|---|---|---|---|---|
| Plants and habitat | Open water: algae, water lilies | Bulrushes | Sedges | Willow, alder | Alder | Alder, birch | Birch | Oak |
| Habitat description | | Reed swamp | Marsh or fen | Open wooded fen | Closed wooded fen | Woodland | | |
| Habitat processes | | Accelerated deposition of silt and clay; floating raft of organic matter forms → thickens | | Raft now a mat resting on mineral soil | Black mineral soil revealed in patches; earthworms | Ground level now above water table; oak seedlings | Birch canopy forms; oak saplings | Oak grows through and then over the birch |
| pH level | – | – | | 7.3 | | 4.3 | 3.7 | – |
| Number of species of plant | | 6 | 10 | 14 | 26 | 18 | 14 | 10 |

O'Hare, G. (1988) *Soils, Vegetation and Ecosystems*, Oliver and Boyd

### Examiner's tip

- Examination questions often ask for this topic to be tackled with reference to one named and located example. Specific details, such as species names, pH values and number of species, help to support an essay answer. If you use a transect diagram to show different species, remember that you must be able to describe and explain the changes that have happened over time. Try to link the conceptual aspects of the topic to the located example.

# Chapter 12  Ecosystems

## Plagioclimax communities

**Plagioclimax communities** are those determined by human activity. They may have been stopped from reaching their full climatic climax or deflected towards a different climax by activities such as:

- cutting down the existing vegetation
- burning as a means of forest clearance
- planting trees or crops
- grazing and trampling by domesticated animals
- harvesting of planted crops

In each case, the human activity has led to a community which is *not* the climax community expected in such an area. If the human activity continues, the community will be held in a stable position and further succession will not occur until the human activity ceases.

### Case study 12.5  Plagioclimax: heather moorland, North Yorkshire Moors

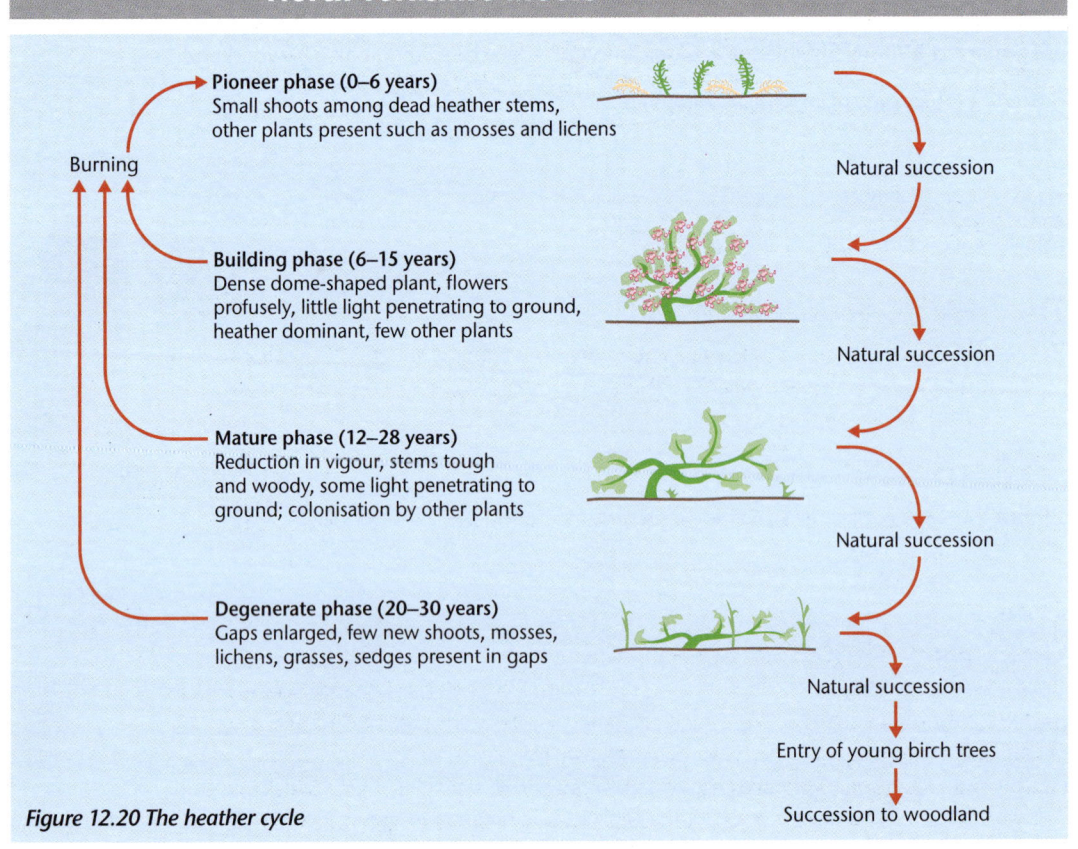

**Pioneer phase (0–6 years)**
Small shoots among dead heather stems, other plants present such as mosses and lichens

**Building phase (6–15 years)**
Dense dome-shaped plant, flowers profusely, little light penetrating to ground, heather dominant, few other plants

**Mature phase (12–28 years)**
Reduction in vigour, stems tough and woody, some light penetrating to ground; colonisation by other plants

**Degenerate phase (20–30 years)**
Gaps enlarged, few new shoots, mosses, lichens, grasses, sedges present in gaps

Burning → (cycles back to Pioneer phase)

Natural succession → Natural succession → Natural succession → Natural succession → Entry of young birch trees → Succession to woodland

Figure 12.20 The heather cycle

The uplands of northern England were once covered by deciduous woodland. Some heather (*Calluna vulgaris*) would have been present, but in relatively small amounts. Gradually the forests were removed during the early Middle Ages for timber and fuel and to create space for agricultural activities. The soils deteriorated as a result and heather came to dominate the plant community, usually up to 500–600 m above sea level. Sheep grazing was the major type of agriculture and the sheep prevented the re-growth of woodland by destroying any young saplings.

In more recent times the process of **muirburn** has been taking place. This involves the controlled burning of the heather. In any 1 km² area, about six patches of 1 ha are burnt, resulting in a patchwork of heather at various stages of development. The heather is burnt after about 15 years of its life cycle, once it has reached the end of the building stage and before it becomes mature and eventually senile. The ash adds to the soil fertility and the new growth that results increases the productivity of the ecosystem and provides the sheep with a more nutritious diet than that provided by the older heather. Some fast-growing species such as bilberry can become established before the heather, but the heather will eventually choke out any invaders and re-establish itself as the dominant species (Figure 12.20). The burning does have to be carefully controlled so that the plants are not killed. New shoots have to be able to grow from the existing rootstock just below the ground surface.

The burning, therefore, maintains a plant community that is a plagioclimax.

> **Examiner's tip**
>
> - Some examination questions may be general ones on succession and require you to draw on your knowledge of a range of case studies. Read the question carefully! You may also be able to use some of your succession case studies to illustrate other ideas, such as variations in productivity.

# Soil characteristics

## The development of soils

The formation, development, profile and characteristics of any soil are determined by the influence of a range of factors. These were first suggested by Professor H. Jenny in 1941 who stated that:

$$\text{soil} = f(c, o, r, p, t \ldots)$$

where f means 'is a function of', $c$ = climate, $o$ = organisms, $r$ = relief, $p$ = parent material and $t$ = time.

### Climate

Climate is generally the major factor, which explains why world soil maps correspond very closely to maps of climatic zones. Climate has both direct and indirect influences on soil. The indirect influence is mediated via plants and animals. The two main influences are temperature, because it affects rates of chemical and biological processes, and precipitation, because it provides soil moisture and, in conjunction with temperature, determines evapotranspiration rates. One of the key determinants of soil processes is the relationship between the amount of precipitation received by the soil and the rate of moisture loss due to evapotranspiration. Climate also affects rates of weathering, which provides the soil with its main source of minerals.

# Chapter 12 Ecosystems

## Key terms

**Azonal soil** An immature soil without distinctive horizons, not determined by one of the soil-forming factors.

**Intrazonal soil** A mature soil determined by a factor other than climate.

**Soil-forming factors** The set of factors that, according to Professor Jenny, interact to produce soil profiles: they are climate, organisms, relief, parent material and time.

**Soil horizon** A layer in the soil with distinctive physical and chemical characteristics.

**Soil profile** A two-dimensional vertical section through the soil to show the horizons of varying depth.

**Zonal soil** A mature soil determined by climatic factors.

## Organisms

Organisms include vegetation, bacteria, earthworms, burrowing animals and humans. Vegetation is the main source of dead organic matter (leaves, twigs, etc.), which is decomposed by bacteria and other decomposers and incorporated into the soil by earthworms and water. Plant roots also help stabilise the soil and reduce rates of erosion. Of course the roots also take nutrients out of the soil in solution. Earthworms mix the soil and aerate (provide oxygen). Burrowing animals aerate too and provide matter for decomposition (excreta, dead animals). Humans have a range of possible influences and increasingly these influences may have significant impacts on soil characteristics. They may be positive:

- crop rotation and fallow periods to allow the soil to replenish its store of nutrients
- adding fertilisers to increase the soil fertility
- liming to raise the pH of acidic soils, which aids nutrient uptake by plants and encourages organisms to develop
- improving drainage in saturated soils, which helps increase the oxygen content
- ploughing to aerate the soil and to improve the soil structure

However, human activity also has negative impacts:

- harvesting crops removes a natural source of organic nutrients
- removing vegetation exposes the soil to wind and water erosion
- monoculture depletes the soil of specific nutrients
- heavy farm machinery and animals can compact the soil, which impedes drainage

## Relief (topography)

Slope angle is important as it influences drainage and mass movement of soil. Soils at the base of slopes tend to be thicker, as soil particles move downslope under gravity, and more moist, as water drains to the bottom of the slope, than those on the slope. Aspect and altitude can both affect the microclimate and cause small but important differences in temperature, which affect evaporation and organic activity.

In northern Britain this often leads to:
- hill peat on the flat hilltops
- brown earths on the lower slopes
- podzols on the upper slopes
- gley soils at the base of slopes

## Parent material

The parent material is important in the initial formation of a soil as it provides the mineral input from weathering. It may be solid rock, such as chalk or granite, or superficial material, such as glacial drift or river alluvium. As the soil becomes deeper and more mature, the parent material is usually less important, only really influencing the 'C' horizon (see Table 12.2 on p. 386 for an explanation of the term 'C'). The exception is calcareous parent materials such as chalk and limestone. They lead to well-drained, alkaline soils. The more resistant or highly soluble the parent material, the thinner the soils tend to be.

> **Examiner's tip**
> - You need to understand the principles by which these factors influence soils. When you come to write about an example of a soil later in this chapter, you should be able to apply these principles to the specific soil and make clear cause–effect links.

## Time

Time is not really a factor itself but the medium through which the other factors operate. Soils develop at different rates, primarily depending on the climate. In the mid-latitudes it might take 10 000 years for the development of a mature soil, but the process is much more rapid in tropical locations and even slower in high latitudes and altitudes.

# The zonal concept

The **zonal concept** is a means of classifying soils. It is a simple but well-known and widely used classification system. It recognises three types of soil: zonal, intrazonal and azonal.

## Zonal soils

**Zonal soils** are those in which climate is the dominating factor affecting their formation and development. On a global scale, a map of climatic zones and a map of soil zones are very similar, with brown earths in temperate zones, chernozems in continental climates, latosols in tropical climates and podzols in cool temperate climates. Zonal soils are mature soils that are in equilibrium with their environmental conditions.

## Intrazonal soils

**Intrazonal soils** are determined by another factor, such as local geology, and can therefore occur in different climatic zones. A good example is rendzina, which develops on limestone. These soils are mature and in equilibrium, not with the climatic conditions but with another factor. Gleyed soils are also intrazonal soils as they occur due to waterlogging, which may be caused by poor drainage rather than high rainfall. Solonetz is an intrazonal soil dominated by a high salt content.

Although intrazonal soils can develop in different climatic zones, in reality the zones in which any one intrazonal soil develops are often not that dissimilar.

# Chapter 12 Ecosystems

> **Examiner's tip**
>
> - You are required to study one zonal soil and one intrazonal soil. You should be able to describe and explain the factor that dominates its development and recognise the impact of other factors as well. Podzol (Case study 12.6) and rendzina (Case study 12.7) are the ones used in this book.

## Azonal soils

**Azonal soils** are young, immature and poorly developed soils characterised by a lack of distinctive horizons. They have not been forming for long enough for any one factor to become dominant. However, due to their lack of age they often retain many of the characteristics of the parent material on which they are developing. In many cases their development is restricted by local conditions. For instance, soils developing on floodplains — alluvial soils — are repeatedly flooded and in receipt of new mineral particles.

## Soil processes

A wide range of different soil processes operates. Their importance varies spatially and, to a lesser extent, temporally.

Most of the processes involve **translocation** (the movement of water and material downwards or upwards through the soil profile) and/or **transformation** (a physical or chemical change in state).

### Weathering

**Weathering** involves the physical and chemical processes of rock breakdown and decay that provide the soil with its mineral content. It is the weathering of the underlying bedrock that generally gives a soil its **parent material** (C horizon). It also influences the **structure** (the shape and size of the individual grains or peds in the soil) and **texture** (literally the feel of the soil, determined by the relative proportions of sand, clay and silt) of the soil. Environments in which rapid chemical weathering occurs, usually when it is hot and wet, generally have soils of significant depth.

### Humification

**Humification** is the process by which dead organic matter (leaves, excreta, dead organisms) is broken down by bacteria, fungi and algae into a dark, amorphous mass called **humus**. This is then incorporated into the soil by the mixing action of earthworms and by the downward movement of soil moisture. In the top 30 cm of 1 ha of soil there may be 25 tonnes of organisms, including 10 tonnes of bacteria and 4 tonnes of earthworms. In one day the worms are able to ingest and pass through 18–40 tonnes of soil, enough to produce a layer 5 mm deep.

The decaying organic matter releases nutrients and organic acids, which increase the acidity of soil moisture and can be strong enough to break down clays and other minerals. The resulting chelates (organic-metal compounds) are soluble and are easily transported down the soil profile — a process known as **cheluviation**.

### Leaching

**Leaching** is the downward movement of soluble nutrients, such as calcium, in solution in soil water. It occurs anywhere that has more precipitation than evapotranspiration and so is very common in Britain and other mid-latitude locations. **Eluviation** is often associated with leaching, but in this case the downward movement of material is purely physical. Particles are not taken into solution but are physically carried by the soil water as it drains through the soil.

## Podzolisation

**Podzolisation** is often described as an extreme form of leaching. Some soils have a high degree of acidity, particularly when coniferous forest litter leads to the formation of **mor**, an acidic form of humus. In such soils the downward movement of soil water has significant effects. As the soil water passes through the litter and mor, it takes organic acids into solution and becomes quite acidic itself. It is then able to take into solution not only the readily soluble bases such as calcium, but also **sesquioxides** of iron and aluminium. This leaves the upper part of the soil primarily with silica, giving it a sandy texture. Lower down in the soil profile the water reduces in acidity. The iron and aluminium come out of solution and are deposited in distinctive horizons — a process known as **illuviation**. In extreme cases, a layer of concentrated iron, or hard **iron pan**, is formed.

## Gleying

**Gleying** involves the soil becoming waterlogged and producing **anaerobic** (oxygen-deficient) conditions. Iron is chemically reduced because of the lack of oxygen, and the soil develops a blue-grey colour. Blotches or **mottles** of orange or red result from dry periods when the iron oxidises. Waterlogging may occur for two main reasons, giving rise to different types of gleyed soil.
- **Surface-water gleying** involves saturation from the top of the soil downwards. This is usually caused by an excess of precipitation and/or impeded drainage. In podzols the iron pan may restrict water movement and so the horizons immediately above it are the ones that show the evidence of gleying.
- **Groundwater gleying** is the result of saturation from the bottom of the soil upwards, usually due to a high water table. In this case it is the lowest horizons in the profile that show the evidence of gleying.

Gleyed soils are not necessarily, therefore, waterlogged permanently. It is the colouring in the soil rather than the presence of water which is the key indication that the process is taking place.

## Capillary action

**Capillary action** is the upward movement of soil water caused by evaporation rates at the surface being higher than precipitation rates. Water is drawn up through the pores in the soil before being evaporated. The minerals that were in solution are left behind in the upper part of the soil as solids. This particularly affects readily soluble minerals, such as calcium. Concentrations of calcium in the form of white nodules may be found in the soil. This is known as **calcification**. If salt is drawn up through the soil and accumulates at or near the surface, this is called **salinisation**, which may lead to the development of salt pans and cause serious problems for agriculture in semi-arid environments. Leaching is largely ineffective under such conditions, although the excess of evapotranspiration over precipitation may only be seasonal, as in the temperate grasslands.

> **Examiner's tip**
>
> ■ You should be aware that in any particular soil a combination of processes is likely to be taking place. Some processes will be more important than others and there may well be variations both temporally (particularly seasonally) and spatially within the local area, perhaps due to local relief, geology or human activity variations. You should have a particularly good understanding of the processes that take place in your examples of zonal and intrazonal soils.

## Soil profiles

A **soil profile** is a two-dimensional, vertical section through a soil. It is divided into **horizons**, which are layers with distinctive physical and chemical characteristics. The boundaries between horizons may be blurred in immature soils or due to earthworm mixing. The main horizons (Table 12.2) are:

- O — the organic horizon consisting of litter, decomposing litter and humus
- A — the mixed organic/mineral horizon, usually dark in colour due to the incorporated humus
- E — the eluvial horizon, depleted of nutrients by leaching and so usually pale in colour
- B — the illuvial horizon in which nutrients are deposited, making it darker in colour
- C — the parent material, usually weathered bedrock or superficial deposits such as alluvium or glacial deposits

Within each of these categories, further subdivisions can be applied. For example, in the B horizon, Bh indicates deposition of humus and Bfe accumulation of iron.

*Table 12.2 Soil horizons*

| Main category | Sub-category | Description |
| --- | --- | --- |
| O | L | Undecomposed litter; leaf layer |
|   | F | Partially decomposed; fermentation layer |
|   | H | Well decomposed; humus layer |
| A | Ah | Dark-coloured humic horizon |
|   | Ap | Ploughed layer in cultivated soils |
| E |    | Eluvial horizon from which clay/sesquioxides removed |
|   | Ea | Bleached (albic or ash-like) layer in podsolised soils |
|   | Eb | Brown eluvial layer, depleted of clay |
| B | Bt | Illuvial clay redeposited (textural B horizon) |
|   | Bh | Illuvial humus layer |
|   | Bfe | Illuviated iron layer |
|   | Bs | Brightly coloured layer of sesquioxide (iron/aluminium) accumulation |
| C |    | Weathered parent material |

*Physical systems, processes and patterns* **A2 Unit 4**

### Examiner's tip

You are required to know a case study of a zonal soil (Case study 12.6) and an intrazonal soil (Case study 12.7). In each case you should be able to:

- give each horizon its appropriate letter label
- describe its appearance and characteristics (colour, pH, texture, etc.)
- explain how the soil profile has been influenced by physical and human factors
- explain how the soil processes operating have led to the features of the profile

An effective way of doing this is by using an annotated diagram.

It is also useful to know the soil profiles found in your case studies of a grassland (Case study 12.2) and a forest biome (Case study 12.1). You then have a range of profiles that you can draw upon in a general essay question about process–profile linkages.

### Case study 12.6  A zonal soil: podzol, Churchill, northern Canada

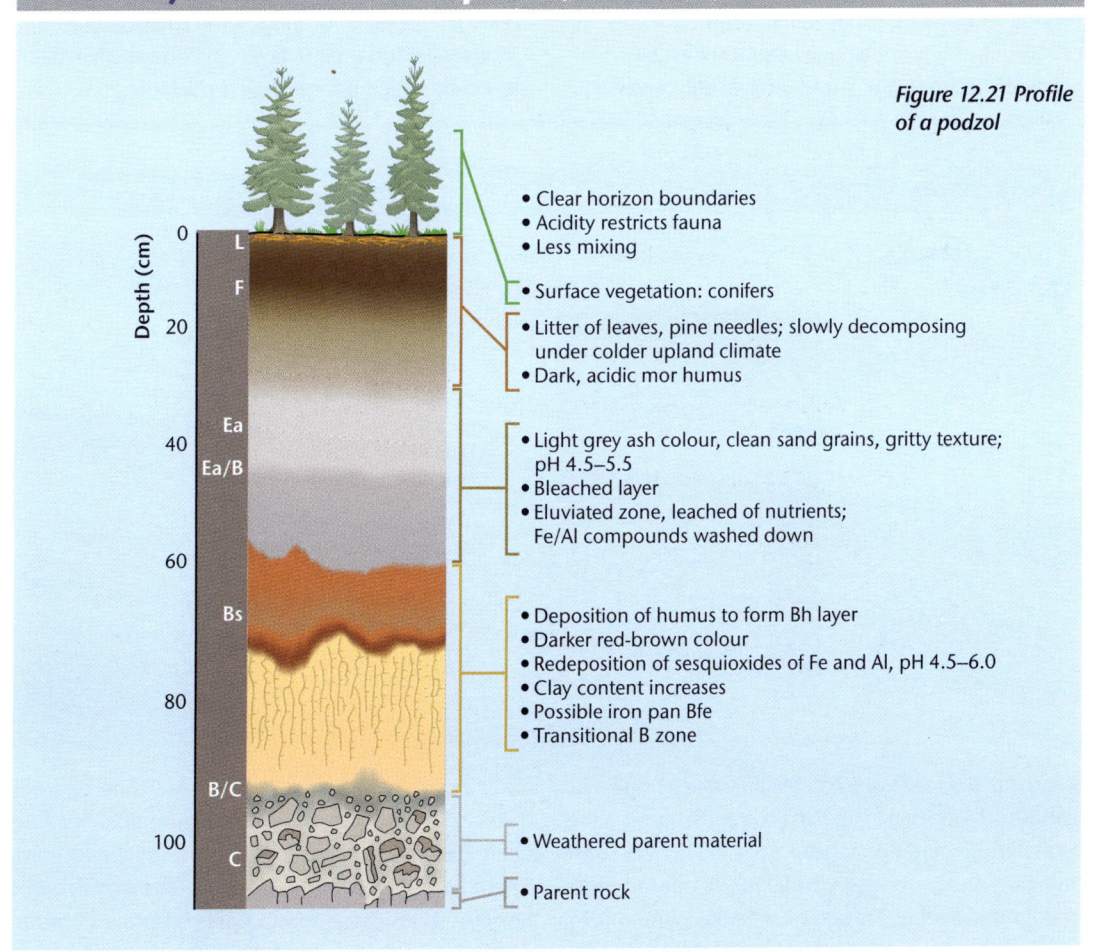

**Figure 12.21** Profile of a podzol

- Clear horizon boundaries
- Acidity restricts fauna
- Less mixing

- Surface vegetation: conifers

- Litter of leaves, pine needles; slowly decomposing under colder upland climate
- Dark, acidic mor humus

- Light grey ash colour, clean sand grains, gritty texture; pH 4.5–5.5
- Bleached layer
- Eluviated zone, leached of nutrients; Fe/Al compounds washed down

- Deposition of humus to form Bh layer
- Darker red-brown colour
- Redeposition of sesquioxides of Fe and Al, pH 4.5–6.0
- Clay content increases
- Possible iron pan Bfe
- Transitional B zone

- Weathered parent material
- Parent rock

This area is dominated by coniferous forest, mainly pine, with very little undergrowth. The climate is cool and relatively dry, with an annual precipitation total of 402 mm. Temperatures peak in July at 12°C and reach a minimum of −28°C in January. Much of the precipitation in the winter falls as snow and there

# Chapter 12 Ecosystems

is significant snowmelt in May, giving surplus water to drain through the soil.

The needles dropped throughout the year by the coniferous trees are high in lignin and are difficult to decompose. The low temperatures for the majority of the year inhibit the presence of bacteria and so decomposition is slow and the mor humus produced is very acidic. Water draining through the soil is, therefore, able to eluviate soil particles and take minerals into solution.

Because of the high acidity, this affects not only the soluble bases such as calcium, sodium and potassium, but also the sesquioxides of iron and aluminium that are produced by reaction with the acidic water. All that remains in the upper Ea horizon below the organic layer is a sandy, grey structureless layer.

Lower down in the profile the acidity levels rise slightly and some of the dissolved nutrients come out of solution sequentially. This gives a series of distinct, darker B layers which may include an iron pan. The soil profile is shown in Figure 12.21.

In some of the low-lying areas of this region the podzols are very poorly drained. This leads to the formation of a layer of **peat** on the surface. Here, decomposition is even slower due to the anaerobic conditions produced by waterlogging. A thick, fibrous layer of partially decomposed organic matter develops. It is thought that some of the tree trunks and branches in the peat have been there for centuries.

In other areas the iron pan has developed to a significant thickness of 3 cm, which has impeded drainage and led to gleying in the Ea horizon.

Podzols with a peat layer that have also been gleyed are called **peaty gleyed podzols**.

---

## Case study 12.7 An intrazonal soil: rendzina, Isle of Purbeck, Dorset

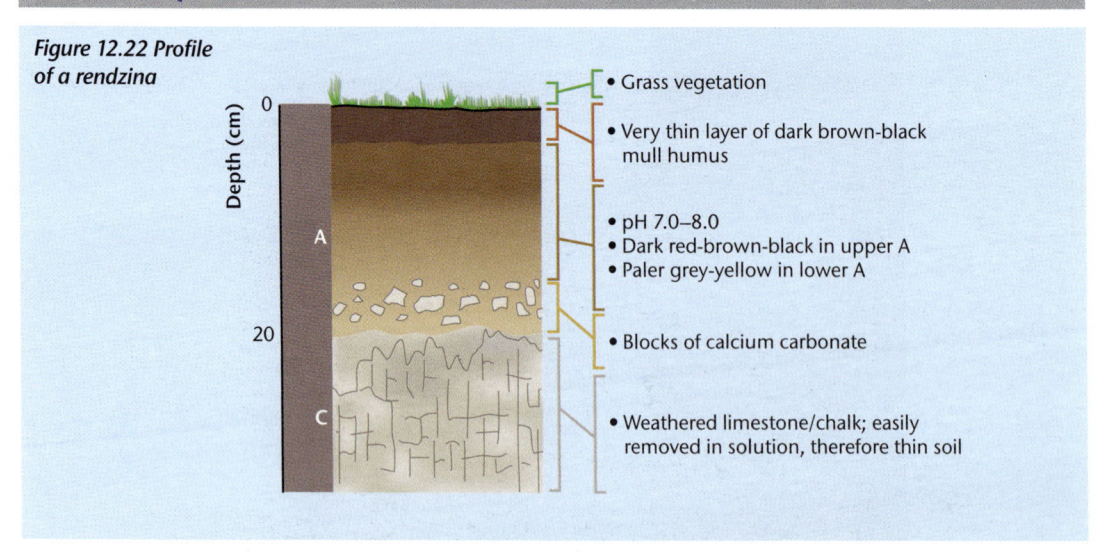

**Figure 12.22 Profile of a rendzina**

- Grass vegetation
- Very thin layer of dark brown-black mull humus
- pH 7.0–8.0
- Dark red-brown-black in upper A
- Paler grey-yellow in lower A
- Blocks of calcium carbonate
- Weathered limestone/chalk; easily removed in solution, therefore thin soil

Rendzina soils are found on the limestone ridge near Langton Matravers. The surface vegetation is short grassland, yielding a litter that is rich in base minerals and produces a **mull** type of humus which has limited acidity. This gives suitable conditions for microbiological activity and so the litter is readily decomposed. The presence of calcium-rich clays and calcium carbonate from the limestone bedrock restricts the movement of soil constituents because chemically stable calcium compounds are formed. There is, therefore, no B horizon. The rich organic content gives these soils a dark red-brown colour. The calcareous nature of the parent material gives the soil a neutral or even slightly alkaline pH. The soil is fairly shallow; in places on the ridge top it is only 10 cm deep, because the parent material produces soluble residues that are readily removed. However, on the lower slopes of the ridge towards Corfe Castle the soil is about 30 cm deep. The soil moisture content is low due to the permeability of the limestone beneath and the area is therefore used mainly for sheep rearing.

This soil is intrazonal because of the dominating influence of the parent material over climate as a soil-forming factor. However, the climatic conditions do allow for weathering and eluviation. The annual precipitation total is 900 mm and temperatures range from 5°C to 19°C. Precipitation will still generally exceed evapotranspiration, giving a net downward movement of water through the soil.

*Photograph 12.4 A rendzina soil exposed on a worn footpath on the Isle of Purbeck*

## Examination questions

**1** Photograph 12.5 shows an area of deciduous woodland in the British Isles.

*Photograph 12.5*

(a) Distinguish between the terms biome and ecosystem. (5)
(b) Examine the factors that influence variations in the productivity of ecosystems and biomes. (20)

*Total = 25 marks*

# Chapter 12

*Ecosystems*

**2** Figure 12.23 shows the zonal classification of soils.

|  | Warm climates | Cool climates |
|---|---|---|
| **Moist climates** | Yellow and red podsols<br>Red tropical soils<br>Lateritic soils | Podsols<br>Brown earths |
| **Dry climates** | Tropical pedocals | Tundra soils<br>Brown soils<br>Chestnut soils<br>Chernozems |

*Figure 12.23*

(a) Distinguish between zonal and intrazonal soils. (5)
(b) Describe and explain the profile of a named zonal soil. (20)

*Total = 25 marks*

# Synoptic link

# Opportunities and challenges associated with ecosystems

- The management opportunities and challenges associated with grassland and forest ecosystems.
- The causes and management of soil erosion.

## Grasslands: the tropical grasslands of the Serengeti

The Serengeti National Park is one of the best-known game reserves in Africa; with its herds of ungulates and their associated predators, it is also one of the most complex ecosystems in Africa. The park, in combination with the adjacent Ngorongoro Conservation Area and Maasai Mara National Park, is sufficiently large to ensure the survival of this tropical grassland ecosystem.

### Management issues in the Serengeti

- The human population to the west of the park has expanded rapidly over the past 30 years, wildlife and livestock populations have grown, and demand for land is high.
- Grazing land is becoming scarce as pastureland is converted into cropland.
- Local people are vulnerable to external development and large-scale agricultural schemes, which do not benefit local communities.

## Physical systems, processes and patterns — A2 Unit 4

- Open land ownership has resulted in local people over-exploiting common resources.
- Agricultural encroachments have appeared on park boundaries, and former subsistence poaching has now become large-scale and commercial.
- An estimated 200 000 animals are killed annually, resulting in large declines in the numbers of some species.
- The rise in demand for meat has partly been driven by the growing local population and in-migration as wildlife and fuel wood is depleted elsewhere.
- The need for bushmeat has also been exacerbated by the relatively low contribution tourism has made to the local economy.
- At one time the Serengeti was not inhabited by elephants, but cultivation and settlement outside the park resulted in changes in distribution.
- The combination of elephants, uncontrolled fires, and subsequent browsing and stunting of regrowth by giraffe has caused a decline in woodlands.
- Approximately 30 000 domestic dogs live in the area, most of which are not vaccinated, thus creating a large reservoir for diseases.

*Photograph 12.6 The Serengeti National Park*

In 1951, the original boundary of the National Park included land to the south and east of the present park and the Ngorongoro Highlands. Pastoralism and cultivation by the Maasai were allowed to continue until 1954 when it was felt that this was incompatible with resource conservation, and the park was divided into the present day Serengeti National Park and the Ngorongoro Conservation Area. The National Park was set aside strictly for wildlife conservation and tourism, and human access was restricted. This preservationist approach to protected area management has slowly been changing throughout the 1980s to a conservation and development programme in the Serengeti region. The overall goal is to change the approach of the management and utilisation of the Serengeti from the traditional approach, which has excluded and alienated local communities, to one in which the needs of human development in the region are reconciled with natural resource conservation.

### Examiner's tip

- It is worth noting that much of this discussion involves the conflict that frequently arises between communities that live and survive in areas of international 'environmental' significance and the governments and development lobbies that wish to exploit the areas quite differently.

# Chapter 12  Ecosystems

- It has been recognised that wildlife is an important economic resource for rural communities around the park.
- It is hoped that creating schemes whereby local communities are given legal rights to manage the wildlife around their villages will prevent the present illegal and unsustainable levels of wildlife poaching from the park.
- Areas suitable for development as buffer zones to the park have been identified where wildlife can be managed by the local people, and village Wildlife Committees are supervising conservation activities.
- The Serengeti Regional Conservation Strategy also includes programmes to stabilise land use, and plans to channel more money earned from tourist activities within the park back into the community.
- The park administration works with the village authorities to resettle encroachers and re-mark the boundary. The Grumeti Game Controlled Area has been incorporated into the park as greater control of the area was thought to be necessary.

## Forests: the coniferous forests of Canada

It is estimated that 0.4 ha of Canadian forest is felled every 30 seconds. The forestry industry is vital to the Canadian economy, providing 3% of its GDP, 13% of its trade and 6% of its employment. In 1997 it was estimated that 63% of the

*Photograph 12.7 Logging in Canada*

country's coniferous forest had either been felled or was under threat of felling. The forests are mainly used for timber, with over 6 million m$^3$ of wood felled per year for this purpose, and about half the area is owned and operated by a few large-scale companies. Harvesting rates were unsustainable and so in the 1990s the government, the timber industry and environmental groups became actively committed to sustainable management of forests.

The Model Forest Program applies the principles agreed at the 1993 Helsinki Conference on Forestry. Strategies include:
- the development of multi-purpose forestry including recreation and hunting with zoning to avoid conflict
- locally agreed working plans for felling cycles and management operations that can sustain yields
- conservation and maintenance of natural forests on sensitive watersheds
- the avoidance of clear felling to assist regeneration and to reduce soil erosion
- marketing and certification of sustainable timber products by the Forest Stewardship Council (FSC)

North America now has over 100 sites certified by the FSC — over 25% of the sites in the world.

## Soil erosion

All soil systems can be evaluated in terms of their:
- stability — the ability of a system to return to its previous 'equilibrium' state after temporary disturbance
- resilience — the capacity of a system to absorb change without significantly altering the community composition
- spatial significance (scale)
- human and geological time scales
- long term = non-reversible or permanent change in terms of human life time, i.e. it takes more than 50 years to fix
- short term = reversible or temporary change, i.e. it can be readily fixed in 5–10 years

The magnitude and frequency of events and activities can also be considered:
- Low magnitude/high frequency versus high magnitude/low frequency: which does the most (geomorphological) work?
- In deciding management strategies it is important that we consider the risk we are prepared to underwrite, both financially and as a community.

### Soil erosion by water

Soil erosion by water involves entrainment, transport and (consequently) deposition of fine-grained materials (organic and mineral) by water.
There are various types of erosion process:
- raindrop splash
- sheetwash (rainwash) — overland flow
- rill and gully erosion

# Chapter 12 Ecosystems

- piping and tunnelling — sub-surface flow
- bank erosion and river bed scour

A number of factors control the amount of erosion that takes place:
- rainfall erosivity (intensity, duration)
- soil erodibility (texture, structure, organic content)
- gradient or slope steepness
- land surface cover type (vegetation, crusts, stones etc.)
- land surface cover density
- land management practices (both present *and* past)

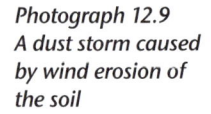

Some of the consequences of soil erosion are:
- topsoil and (sometimes) subsoil loss
- loss of nutrients and organic matter
- downslope and downstream siltation
- lowering of water table
- damage to infrastructure

*Photograph 12.8 Gully erosion*

## Wind erosion

Wind erosion involves the entrainment, transport and (consequently) deposition of fine-grained materials (organic and mineral) by wind moving over a surface. The controlling factors are:
- soil erodibility (texture, structure, organic matter content, moisture content)
- land surface cover type and density
- land management practices — overgrazing is a primary cause
- meteorological conditions (wind speed, dryness)

*Photograph 12.9 A dust storm caused by wind erosion of the soil*

*Physical systems, processes and patterns* **A2 Unit 4**

Some of the consequences of wind erosion of soil are:
- topsoil loss
- loss of nutrients and organic matter
- reduced visibility
- contamination of water bodies
- burying of cropland and infrastructure, especially fences, by wind drifts

## Soil structure decline

The structure of a soil can be damaged as a result of the reduction in continuity and distribution of soil pores caused by the repeated use of machinery, e.g. tractors, ploughs and trucks. The controlling factors are:
- soil characteristics (binding agents, clay content and type, organic matter content, soluble and exchangeable nutrients)
- rainfall characteristics (intensity)
- land surface cover (vegetation)
- land management practices (machinery, animals)

The consequences are:
- compacted upper subsoil
- discontinuous pores
- reduced porosity
- increased soil strength

## Summary of the impact of soil erosion

- Soil erosion is a big problem. It is estimated that globally 3 million $km^2$ of arable land have been strongly degraded over the past 100 years (an area equal in size to western Europe).
- Another 9 million $km^2$ of arable land have been moderately degraded.
- A further 1.5 million $km^2$ may be degraded in the next 20 years unless changes are made.
- Land degradation rates are exceeding the opening of new lands for agriculture; thus there is a net loss of productive land for food production.

Processes of soil formation operate on geological timescales and average less than 0.1 mm per year. This means that the loss of topsoil exceeds its replacement by as much as 1000 times.

## Soil erosion in Iceland

When Iceland was first settled in 874, it is estimated that at least 60% of the island was vegetated. Birch woodland covered at least 25% of the country and these trees, as well as smaller shrubs, protected the soil beneath from wind erosion. Since the initial settlement, vegetation cover has been reduced to 25% and tree cover to 1%. Much of the woodland was cleared for firewood and building material, and the land was overgrazed by the sheep that had been introduced. Climatic deterioration from 1200 to 1800 hastened the rate of loss, which increased by 400% in some places.

# Chapter 12  Ecosystems

The effects are highly visible. Huge areas of virtually bare surface with the occasional *rofabard* are all that remain from the original soil and grass-covered surface. High winds of over 25 m s$^{-1}$ in October 2004 were estimated to have removed 15 million tonnes of surface material from the centre of the island.

*Photograph 12.10 Rofabards near Hagavik, Iceland*

There have been many attempts to slow the rate of erosion:
- birch and pine trees have been extensively planted since 1990
- eroded areas have been fenced to keep sheep off
- lyme grass coated with fertiliser has been spread by air
- motorists have been given packets of seeds at petrol stations to scatter from their cars as they drive

The main management objectives are to minimise disturbance, to retain as much soil as possible on site, and to promote stability in the long term by changing farming practices. This will often imply a reduction in farming intensity.

> **Examiner's tip**
>
> - It is worth noting that increased soil erosion and a growing world population put more pressure on farmers to increase yields. This in turn increases the problem of soil erosion as soils become exhausted and lose structure.

# A2 Unit 5

## Human systems, processes and patterns

# Chapter 13

# Economic systems

## The classification of economic activity

Economic activity can be classified in a number of ways. One of the most useful is the division into four categories based on the nature of the product. This classification is artificial but useful, in that the organisation of production can be seen as a flow.

- Iron ore is excavated.
- Iron ore is one of the raw materials used in the production of steel ingots, which are then turned into sheet steel.
- Sheet steel is rolled and pressed into shape.
- The pressed steel parts become parts of a car body.
- Cars are sold by garages.
- Garages service and maintain the vehicles.

At each stage of this chain of processes, more value is added in the form of both labour and/or capital (in the form of machinery). Low value-added processes offer fewer opportunities for profit than those later in the chain.

### Primary activities

**Primary activities** are fundamental — this sector produces the raw materials on which all other economic activity and human survival depend. Even in pre-modern societies there was some sub-division of tasks, but if some people undertook non-productive roles, other members of the tribe had to catch or grow the food so the 'non-productive' members could survive.

Primary activity involves the exploitation and extraction of raw materials (digging coal or drilling for oil), the growth of food crops and textile crops in agriculture, fishing and any other activity in which very little processing of the basic material is involved. Quarrying is another example. In these extractive processes little is added to the value of the product because little is done to it. These are low value-added industries.

In the modern world the production of many primary products takes place at a substantial distance from their point of consumption. A walk down a supermarket aisle in any MEDC readily demonstrates that we are no longer supplied by locally produced food and basic necessities.

*Human systems, processes and patterns* **A2 Unit 5**

## Secondary activity

**Secondary activity** is the sector that adds value to the raw materials by processing them in some way or, more usually, both processing them and combining them to create goods. Thus the growing of cotton is a primary industry but the making of cloth and clothing is secondary. Humans have made goods for a very long time, from flint arrowheads through to mobile phones. To make something requires labour and, almost always, some equipment (otherwise known as capital).

The word 'manufacturing' literally translates as 'handmade'. Handmade or 'craft' goods have undergone something of a revival in recent years in those sections of society that want individuality and design above all other things, but the majority of manufactured goods are produced using machines. Sometimes there is very little human intervention in the production process itself.

It is useful to draw a distinction between those secondary industries that produce essential equipment or material for other industries (e.g. machine tools, iron and steel), which are often known as **capital** or **basic industries**, and those that produce goods for direct sale to consumers. It is **consumer industries** that have taken a leading role in economic growth in the twentieth century.

In the last 30 years, a distinctive group of **hi-tech industries** has emerged, characterised by high levels of technological inputs and thus a heavy reliance on the development of new products that are faster, cleaner and more efficient. Of course the term 'hi-tech' is a relative one. There was a time when cars, and even flint arrowheads, would have qualified as hi-tech.

## Tertiary activity

The **tertiary sector** is otherwise known as the **service sector**. It produces no physical product, in other words nothing that you can touch. All societies have had some tertiary activity, ranging from Aztec priests and shamans many centuries ago through to merchant bankers and checkout assistants today.

Historically, the largest single group in the tertiary sector was servants or slaves. The largest categories today include fast-food operatives and healthcare workers for elderly people. In neither case do these workers enjoy the high wages that many associate with the tertiary sector. There are, of course, some highly paid tertiary workers such as lawyers and accountants, but they are in the minority in all societies. Tertiary employment always depends upon primary and secondary activity for the provision of the basic necessities of life. In recent years this dependency has been on a global scale, given that many primary and secondary products are imported into MEDCs.

## Quaternary activity

**Quaternary activity** is a sub-set of tertiary activity. It is a new category that recognises the importance of information technology and research in the creation of wealth. It includes employment in the gathering and exchange of information, research and development (R&D) activities and senior management roles. So-

# Chapter 13  Economic systems

called hi-tech industry, which is the manufacturing of goods using a highly advanced, usually electronic system of production is not necessarily in this sector. The production of microprocessors is a hi-tech industry belonging to the secondary sector. However, there will be people involved in the development of new products in this industry who undertake quaternary activities.

### Examiner's tip

- Don't confuse hi-tech and quaternary industries. Hi-tech industries produce something tangible whereas quaternary industries produce ideas.

### Key terms

**Primary industry** Extractive industries with limited value.

**Quaternary industry** A subset of tertiary industry, producing information and ideas through ITC.

**Secondary industry** Manufacturing materials either for other industries or for consumers.

**Tertiary industry** Service industries with no tangible product.

## Using this classification

National and international statistics do not use these sub-divisions. Look at Table 13.1. At face value, it appears that 21.1% of the working population of Worcestershire is in the manufacturing sector. However, this is too simplistic — in reality many of the people who work in the manufacturing sector for companies that make goods, are secretaries, canteen workers, security guards or managers. None of these people are working on the assembly line itself. This complication is true in all sectors.

There are a number of other significant factors in understanding employment structure. These include:

- full-time as opposed to part-time employment
- permanent as opposed to temporary employment
- the growth of female employment

In the past 50 years there has been a significant rise in the number of part-time employees in many industries. From the point of view of the employer this has been a cost-saving response to falling profits. The advantages for employees are less clear, although it has allowed some flexibility for people such as those with young children.

Table 13.1 Employment structure in Worcestershire, UK, 2000

| Sector | Number of employees | Share of all employment (%) |
|---|---|---|
| Mining, energy supply and agriculture | 941 | 0.4 |
| Manufacturing | 47 053 | 21.1 |
| Construction | 8 644 | 3.9 |
| Distribution, hotels and restaurants | 58 727 | 26.3 |
| Transport and communications | 11 070 | 5 |
| Banking, finance and insurance | 33 681 | 15.1 |
| Public administration, education and health | 54 226 | 24.3 |
| Other services | 8 701 | 3.9 |
| Total | 223 043 | 100 |

Source: Annual Business Inquiry 2000

*Human systems, processes and patterns* **A2 Unit 5**

Flexibility in the workforce has also been gained by a growing use of fixed-term (often 1 year) renewable contracts. These allow employers to respond to changing market conditions. Once again, the advantages for employees are less obvious. Reduced job security makes long-term planning of their own lives more difficult (anything from getting a mortgage to planning a family).

The rise of female employment in many MEDCs can be seen as positive, in that it has increased the financial independence of women. However, it can also be seen as a response to financial need in many households. The decline in well-paid manufacturing jobs has meant that both partners need to work to maintain their overall income.

Figure 13.1 shows how the nature of UK employment changed between 1977 and 1997.

*Figure 13.1 Changes in UK employment, 1977–97*

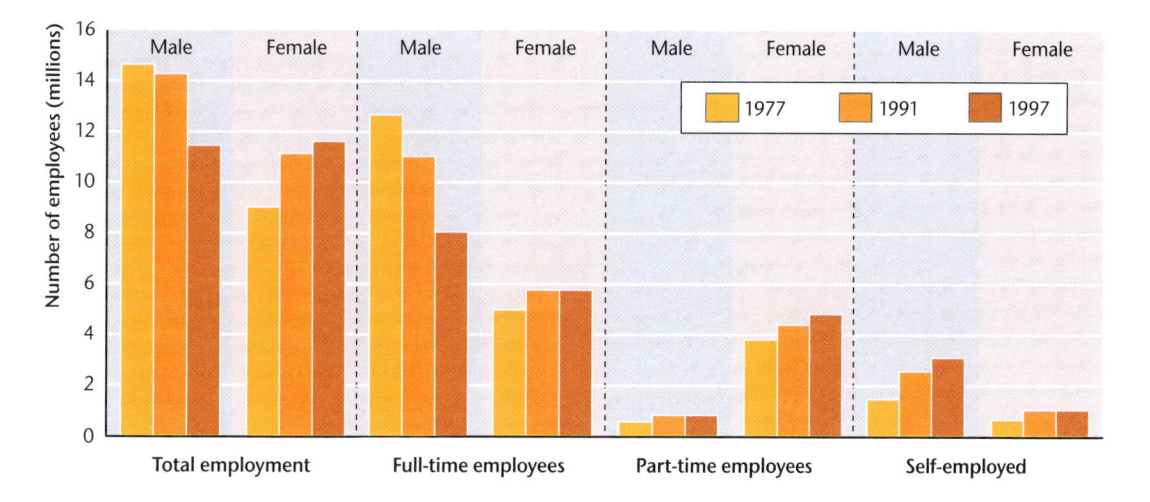

Taken together, these changes have made the organisation and structure of production harder to interpret. With many, especially the young, drifting in and out of employment, and with that employment frequently being both part-time and temporary, it has become much harder to monitor changes in the workforce.

Employment categories such as primary, secondary, tertiary and quaternary do not necessarily reflect the importance of particular sectors in an economy.

- The commonest confusion here is to mix up employment with output. The agricultural sector is insignificant in terms of the numbers employed in the UK (less than 2%), but it is far more important in terms of output and contribution to GNP, quite apart from its broader social and political importance.
- Similarly, the numbers employed in the automobile industry in the UK have been in sharp, although not consistent, decline since the 1970s. However, more cars were produced in the UK in 2000 than in any other recent year. This apparent contradiction is easily resolved, as this is an industry that has replaced labour with machinery, reducing employment but increasing output.

# Chapter 13  Economic systems

## Industrial location

Location has two elements, site and situation.

The **site** of a factory, or group of factories, is the actual physical location — the area of land on which the factory is built. There are some obvious constraints on the location of a site, such as:
- availability of flat land
- ability to expand
- accessibility
- water supply
- ability to dispose of waste
- access to transport, power and water

The situation of a factory is where it is located in relation to other places. Almost all regions in MEDCs possess appropriate industrial sites, although not all of these sites may be available because of planning limitations by the local council or regional government.

Physical factors influence the choice of location for individual plants or factories. For some basic manufacturing industries, especially those that require very large areas of flat land, need to extract water for use and need to discharge waste into water, there is a limited range of possible sites. Large iron and steel plants and chemical works, such as the Port Talbot steelworks in south Wales or the ICI chemical complex at Billingham on Teesside, tend to be coastal. In the past they were often sited on estuaries, which would be swept 'clean' by the tide on a twice-daily basis. (The development of stricter environmental controls now prevents this form of 'cleaning'.)

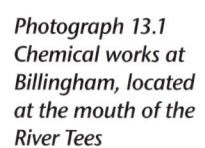

*Photograph 13.1 Chemical works at Billingham, located at the mouth of the River Tees*

*Human systems, processes and patterns* **A2 Unit 5**

At a more basic level, it is obvious that primary industries are strongly controlled by the location of their raw materials. To put it at its simplest, coal mining takes place where coal is found — an obvious physical factor. However, even in these apparently simple cases the full explanation is more complex: coal mining does not take place in all areas where coal occurs. Economic, social and political factors determine whether or not it is actually mined. Thus the decline of mining in south Wales is only partly explained by exhaustion of the resource. It had much more to do with the price of imported fuel and the decline in demand for coal as energy usage altered in the last half of the twentieth century.

The term 'natural environment' encompasses anything from the attractiveness of the landscape to the climate. The development of cleaner industries that create less waste and use less space has meant there is now a much wider range of possible sites in almost every corner of a country, especially given falling transport costs.

### Key terms

**Site** The precise physical characteristics of a location.

**Situation** The relationship of a location to other places.

## Case study 13.1 Industrial location: the Ruhr valley

The Ruhr valley in Germany is famous for its industry, once based on coal mining and steel production and now benefiting from its mix of energy production, environmental technologies and modern service industries.

Since the eighteenth century, mining has been the main economic activity in a region that, like south Wales, was relatively poor before the Industrial Revolution. The coal was easily accessible and was initially extracted by strip mining. Eventually deep mines were dug and many of the impressive industrial buildings emerged. Collieries, steelworks and the tall chimneys of the Industrial Revolution shaped the face of the region.

One of the largest mines, known as the Zollverein, was once the centre of life in Katernberg and Stoppenberg. This area was home to the highest concentration of mines in the entire Ruhr area (and also in Germany). The mine began operations in 1847; by 1932 it was the most up-to-date mine in the world and employed more than 5000 miners.

Ecological and economic problems, including a fall in demand, concerns about pollution both locally and nationally, and cheaper imports, destroyed confidence in the mines. Most of them were shut down, despite the large reserves of coal that still existed — albeit in deeper and thus more expensive seams. The numbers of people employed fell rapidly. In common with many other regions that once

*Photograph 13.2 Zollverein: pit head XII*

depended upon raw materials, the Ruhr valley has experienced many problems, among which unemployment and social unrest are just some of the more pressing. Meanwhile the Zollverein, with its pit head XII, is considered to be one of the most important industrial monuments in Europe. Recently, it was declared a World Cultural Heritage site.

# Chapter 13  Economic systems

In the early stages of industrialisation, by contrast, high transport costs and primitive technologies which incurred much wastage of raw materials tied a large number of basic industries to the sources of both power and materials.

One of the most important of the basic industries is the iron and steel industry. Its international location has changed dramatically in recent years, but a number of common factors still guide the choice of site.

## Case study 13.2  Industrial location: Port Kembla steelworks, Australia

BlueScope Steel's Port Kembla steelworks is a 5 million tonne a year integrated steelworks located 70 km south of Sydney, NSW. It produces a wide range of finished and semi-finished flat steel products for Australian and international customers. The plant encircles the harbour and is located on about 800 ha (8 km$^2$) of flat land. There has been steel-making at this site since 1921. The site was initially chosen by Hoskins Bros so that they might expand their output in order to enjoy economies of scale in their competitive battle with the BHP steelworks at Newcastle, also in NSW (Figure 13.2).

The reasons for the choice of Port Kembla to replace the older and smaller Lithgow works were:
- a large area of flat land with room for expansion
- good access to power and water — 96% of the 1 billion litres of water consumed per day is recycled seawater
- local supplies of coal and limestone
- the deepwater harbour
- cheap rail links with Sydney — the main market (Lithgow is 150 km from Sydney)

The factors that influence the location of steel-making plants have changed over time. For example, new bulk-handling technology means that raw materials can be transported greater distances at very much lower costs. Over time these factors change in significance, but industrial inertia often prevents a plant being shifted to another location. The scale of steel-making operations (i.e. the large size of the equipment) means that it is not practical to a move an existing plant to a new location. Thus the distribution of steelworks reflects a mixture of locational variables, some of which are no longer operative.

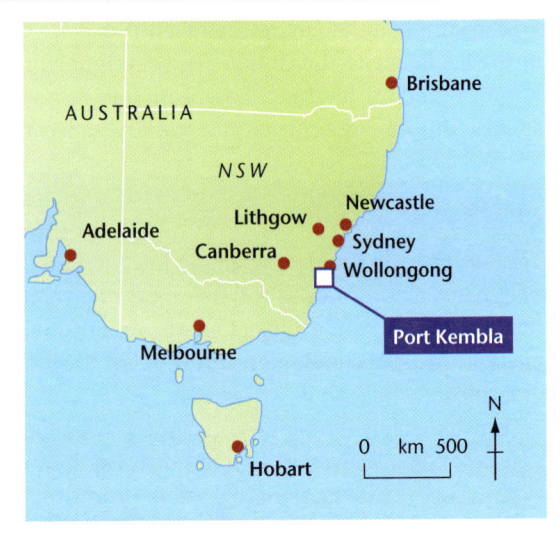

*Figure 13.2 The location of Port Kembla*

The natural environment is of particular significance in those industries that are more footloose — in other words, they have low transport costs and could locate in a wide range of places with little impact on their total costs or their efficiency. Particular attention has been paid to quaternary sector and hi-tech industrial growth in areas perceived to be physically and climatically attractive, such as southern California and Mediterranean Europe. The location of Microsoft

*Human systems, processes and patterns* **A2 Unit 5**

and Amazon in the US city of Seattle can be used as examples of this trend. Seattle was a preferred place to live for a large number of university graduates, long before the founding of Microsoft.

The choice of location for such enterprises is driven by the need to attract the highest quality labour supply. This high-quality labour is relatively highly paid, geographically mobile and tends to seek the best possible environment in which to live and work. The persistent appeal of southern England, despite the higher costs of living, is attributed to a number of factors, but, the environment is one of these. At a more local scale, the development of a significant insurance sector in the southern English town of Bournemouth can be partly explained by the attractive local landscape and the mild climate.

## Case study 13.3 *Location according to environment: Denver, Colorado*

**Photograph 13.3** Downtown Denver looms behind Ferril Lake in City Park, a 126 ha preserve east of the city

Denver advertises itself to prospective investors, whether they be companies looking for new locations or potential migrants, using the following 'facts'. Some of these are clearly related to its physical environment.

- Denver is one of the most vibrant and youngest cities in America.
- It offers a landscape that varies from flat plains to high mountains.
- It has clean air at an altitude of over 1500 m — the 'mile high' city.
- The city's business and residential areas are intermingled with over 200 parks. In fact, Denver has the largest city park system in the USA.
- Its animal and plant life, as well as the city's population of 2.4 million, enjoy sunshine more than 300 days a year.
- Denver is noted as being the USA's most highly educated city, with the highest percentage of high-school and college graduates.
- Denver has the USA's second-largest performing arts centre, is home to major league sports teams (the Denver Broncos, Colorado Avalanche, Colorado Rockies and Denver Nuggets), and has a multitude of well-known shopping areas, such as the Sixteenth Street Mall.
- Denver has a lot to offer families, from art, nature and science museums to the Butterfly Pavilion and Insect Center, the Denver Zoo, the Botanical Gardens and the US Mint.
- For even more family fun in the city, there is Six Flags Elitch Gardens Theme Park.
- Skiing, hiking, fishing, rafting, camping and other activities are all within a 90-minute drive of the city.

405

## Chapter 13 Economic systems

# Classical location theory

**Location theory** was introduced into the teaching of A-level geography in the 1970s. This was some 10 years after it made an impact in universities, along with a whole range of quantification techniques (essentially the use of mathematics and statistics) and tools of analysis. Before then, most locational work in geography had involved the detailed description of places and an attempt to recognise distinctive regions. At the time, location theory was of considerable interest, especially in a country such as Britain with an extensive empire. Prominent among the adopted theories were those that came from the German school of economic geographers; they attempted to build explanatory models and theories to better explain the world rather than concentrating on simply classifying and describing it.

*Models* are attempts to simplify reality and can be physical, diagrammatic or mathematical. For example, maps are models of the real world. By contrast, *theories* are both predictive and testable.

- A theory is predictive in that it allows you to forecast what is going to happen in the future. For example, if you let go of this book while reading it in the bath, you can use the theory of gravity to predict that it will fall into the water. You could then use a number of other theories to predict what will happen next to the book, depending on how quickly you rescued it from the water.
- A theory is testable in that we can test its validity (or otherwise) by experimentation. You could drop the book into the bath and observe what happened.

A criticism of all theories is that they have to operate within a series of assumptions. These assumptions are simplifying, in that they remove aspects of the world that may produce too much 'background noise' for the theory to be properly testable. One common assumption in economic geography theories is that 'human behaviour is rational'. Although this is obviously not always the case, it is a necessary assumption because the alternative — assuming all human behaviour is irrational — would prevent us building any predictive theories.

## Weber

**Alfred Weber** developed the dominant theory of industrial location, which he published in *Über den Standort der Industrie* (*Theory of the Location of Industries*) in 1909. The book was only translated into English in 1929. His model took into account several spatial factors in order to find the optimal location for manufacturing plants at the point of minimal costs. Weber also applied the model to service industries, such as investment firms, and applied it more broadly to specific political systems, thus recognising the role of government.

The problem of explaining the location of industry was especially relevant at the end of the nineteenth century:

- the Industrial Revolution was in progress
- there was rapid development of rail and road transport
- there were major changes in energy supply with the development of electricity

- the development of the telephone and rapid urban growth were concentrating markets in a manner that was unprecedented

These changes provided many more options for distributing firms and the manufacturing process as a whole. Weber's theory was developed at a time when local markets and manufacturing were becoming national markets. It predicts that industries will locate where they can minimise their transport costs. It holds the other variables constant, much as one would in the chemistry laboratory by making various assumptions. The most important of these assumptions are:
- inputs are available in unlimited supply
- there are numerous competitive single-plant firms — this is what economists call 'perfect competition' — and so prices are not fixed by firms but determined by the market
- all firms have the same available technology
- transport costs are proportional to weight and distance on an isotropic (even, flat, unvarying) plane and movement is possible in any direction
- there is infinite demand at a fixed price concentrated at a number of limited locations
- no economies of scale exist
- there are two types of material used by manufacturing industry, localised and ubiquitous; ubiquitous material can be found anywhere

Although Weber was not specific about it, he also assumed that firms were independent enterprises producing a single product. Choices made about location would be rational. He recognised that labour-cost variations and the possible savings to be made from co-location (agglomeration) might pull firms away from 'ideal' cost minimisation points.

The first variable Weber considered was whether to locate an industry where the raw material is located or where the market is located. If there is no change in weight during production, the factory can be sited at either location, because the transport costs will be the same for both raw materials and manufactured goods.

As this is rarely the case, a relative weighting must be calculated. Weber used a material index to calculate relative weight gain or loss:

$$\text{material index} = \frac{\text{total weight of materials used to manufacture the product}}{\text{total weight of the finished product}}$$

- If the product is a pure material its material index will be 1.
- If the index is less than 1 the final product has gained weight in manufacture, thus favouring production at the market place. The weight gain is most likely to come from the addition of ubiquitous materials, like water. Examples of these products are a soft drink or beer, in the manufacture of which a small quantity of usually dried materials is added to water and the liquid bottled to make a much heavier and more fragile final product.
- Most products lose weight in manufacture, for example a metal being extracted from an ore. Their material index will be more than 1, thus favouring location at the raw material site. This is the case with most of the basic industries, such as manufacture of iron, steel or chemicals.

# Chapter 13 Economic systems

The significance of the material index is in calculating the difference between the unit transport costs of raw materials and finished products. As well as considering the weight loss or gain, the material index and weighting of transport costs can also take account of losses or gains in transport and features such as perishability, fragility and hazard.

Table 13.2 Various location decisions in a Weberian world

| Industry | Ubiquitous materials | 'Pure' localised materials | 'Weight-losing' materials | Cheap labour source | Savings from external economies | Likely location |
|---|---|---|---|---|---|---|
| A | Yes | No | No | No | No | Market |
| B | Yes | Yes | No | No | No | Market |
| C | Yes | No | Yes | No | No | Materials |
| D | Yes | Yes | No | Yes | No | Uncertain |
| E | No | Yes | No | Yes | No | Labour |
| F | Yes | Yes | No | Yes | Yes | Co-located |
| G | Yes | Yes | Yes | No | No | Uncertain |

### Key terms

**Cost minimisation** The processes whereby companies seek to minimise their costs.

**Profit maximisation** The processes whereby companies seek to maximise their profits.

## Weaknesses of location theory

Weber himself rejected the simplified view that he offered up for analysis, pointing out that real locations frequently reflect variations in labour costs and/or agglomeration savings.

**Agglomeration** is the term used to describe the concentration of similar firms in an area. It occurs when there is sufficient demand from the companies and the labour force for support services, including schools and hospitals. Suppliers, such as those that build and service machines or offer specialist financial services, will then tend to locate near the cluster of industry, as will auxiliary industries, such as those making component parts. Companies which supply specialised machines or services used only occasionally by larger firms tend to locate in agglomeration areas, not only to lower their own costs but also to find sufficient customers to allow them to operate. This is now frequently referred to as **clustering** (Figure 13.3).

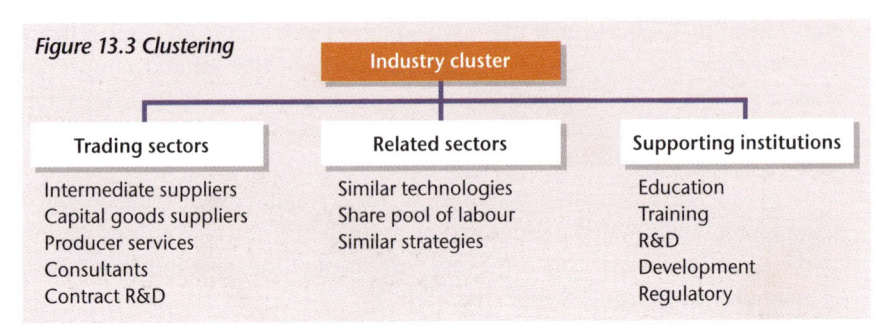

**Figure 13.3 Clustering**

Industry cluster

- **Trading sectors**
  - Intermediate suppliers
  - Capital goods suppliers
  - Producer services
  - Consultants
  - Contract R&D
- **Related sectors**
  - Similar technologies
  - Share pool of labour
  - Similar strategies
- **Supporting institutions**
  - Education
  - Training
  - R&D
  - Development
  - Regulatory

*Human systems, processes and patterns*  **A2 Unit 5**

A factory may also grow locally because of the cost savings arising from agglomeration when there is local cheap labour, or a labour pool that can adapt to fluctuations of markets for products in the same industry.

You should appreciate that much has changed since Weber wrote his book on location in 1909. He wrote in a different economic world from that of today — one which was dominated by relatively small firms, with each factory typically producing a single product.

The most obvious changes since 1909 are the following.

- There has been a revolution in transport, which has greatly reduced the importance of transport costs as a proportion of the total cost of producing most manufactured goods (Table 13.3). This has rather colourfully been described as the 'death of distance', and it has a marked effect on any theory built around the minimisation of transport costs.

|  | 1930 ($) | 1950 ($) | 1990 ($) |
|---|---|---|---|
| Mean cost of air travel per km | 0.43 | 0.12 | 0.07 |
| Mean cost of shipping freight (per tonne) | 60 | 34 | 29 |
| Mean cost of transatlantic phone call (per minute) | 80 | 16 | 1 |

*Table 13.3 International transport costs (adjusted to 1990 values)*

- There have been fundamental changes in the organisation and structure of industry, with the emergence of large corporations and global markets. Weber developed his theory to deal with the radical changes taking place as economies integrated on a national level. The international market and international trade were in their infancy when he wrote.
- There has also been an increase in the significance of government in the management of national economies and the location of industry. Not only do governments employ large numbers of people directly, they also control many contracts — defence goods are an obvious example — and by either building the infrastructure or determining the location of that infrastructure they have an enormous influence on the location of economic activity.

Many of the criticisms of Weber's and other related models are directed at specific generalisations they make. In many cases the models can be rescued by adapting them to a new set of conditions. However, the problem of the models' relevance to modern industry is much more difficult to overcome because of fundamental changes in the structure of production. In recent years a new set of terms has emerged and a new, less deterministic view of location has grown which is less preoccupied with distance.

- **Sector (or industry)**: a group of enterprises that manufacture similar products, as defined in national statistics. Thus textiles and clothing might be one sector, automobiles another.
- **Industry cluster**: a group of business enterprises and non-business organisations in which each member needs to belong for its individual competitiveness. In other words, they need to be in a cluster. The cluster is held together by a complex set of 'buyer–supplier' relationships, or shared technologies, buyers or distribution channels, and labour pools.

# Chapter 13 Economic systems

- **Regional industry cluster**: a cluster that shares a common regional location. The automobile industry in the USA is an obvious example.
- **Potential industry cluster**: a situation in which a cluster could occur if there were some changes. It might not have quite enough companies or not quite the correct specific weight to lead to growth. Companies might not be trading with one another to a high level.
- **'Italianate' industrial district**: a highly concentrated group of companies that work directly or indirectly for the same end market, share values and knowledge and dominate the cultural environment. They are linked to one another in a mix of competition and cooperation. Elements of trust, solidarity and cooperation exist between firms, a result of a close intertwining of economic, social and community relations.
- **Innovative environment**: not a specifically geographic concept but rather a complex of relationships, which is capable of initiating a synergetic process by which firms depend on one another. It will extend into local universities, colleges and schools, lending institutions and local government. Silicon Valley is frequently cited as the blueprint for this type of environment.

## Key terms

**Agglomeration** The coming together of economic activity to enjoy external economies of scale.

**Clustering** The co-location of economic activity to provide mutual benefit for all companies through external economies but also a shared client base.

**Optimum location** The best possible location given the information available.

**Sub-optimal location** A location that may not minimise cost or maximise profit.

Industry clusters vary in their level of geographical concentration and their independence. Early theorists such as Weber recognised that interdependence between enterprises would be important and allowed for it in his treatment of agglomeration. Some of the linkages that bind together automotive parts suppliers in Ohio and Kentucky are created by their mutual dependence on the relatively distant final market assemblers in Michigan and South Carolina, without which they could not exist. At one scale, an automotive cluster may appear to be concentrated in the southern Ohio/northern Kentucky region, but in reality this is linked to the traditional vehicle heartland in Michigan. Some clusters are harder to explain, as shown in Case study 13.4.

There are some examples of mutual location which are far from benign. Seattle has been Boeing's production base for a long time. The growth of Microsoft in the Seattle area was quite unrelated and happened completely by chance. But from Boeing's standpoint it has been a most unwelcome development in the local labour market.

As a rapidly growing firm taking advantage of high levels of innovation, Microsoft has offered high wages to attract the people it wants, and needs, to Seattle. This has pushed up house prices which, in turn, is having a negative

*Human systems, processes and patterns*  **A2 Unit 5**

## Case study 13.4 Industry clusters: boat-building in Kentucky

It is hard to imagine a more unlikely cluster of manufacturing than that of the specialist, custom houseboat industry that is concentrated in the US state of Kentucky. Kentucky is land-locked and is 800 km on average from the nearest ocean. Nonetheless, it is the centre of this small industry.

It turns out that water is not significant at all in this particular branch of boat-building. The boats are not ocean-going vessels and even though the products end up on water they are neither constructed in, nor transported on, water. Most customers do not live near navigable rivers and water transport would be slow and expensive. Road delivery is arranged and included in the price of the boats.

Construction takes place indoors in large sheds strung out along Route 90 just outside the town of Monticello. The cluster of builders is readily explained as an example of co-location that has developed in order to maximise the number of possible clients. A potential buyer comes to visit one builder that he has seen advertise, only to find another six or seven in the same location. According to one company, 'drop-in' clients are frequent; perhaps six or seven a week who are determined to buy a boat for $200 000 and are able to visit all the builders in the area to get the best deal. A company that located in another area would not benefit from this co-location and would limit its client base to those who had happened to see its advertisements. The same logic applies to DIY stores in edge-of-town retail locations. Co-location maximises client numbers.

---

impact on recruitment at Boeing. Potential Boeing employees want higher wages so they can afford housing in Seattle.

Boeing is in a tight competitive fight with Airbus, made worse by the events of 9/11 which led to a decline in demand for passenger aircraft, its core business. The company is resisting the pressure to raise wages. Boeing is trying to reduce its dependence on labour in the Seattle area by subcontracting work to firms in lower wage areas of the Far East, particularly China. Recently, amid great consternation, Boeing announced that it was moving its head office to Chicago (Figure 13.4).

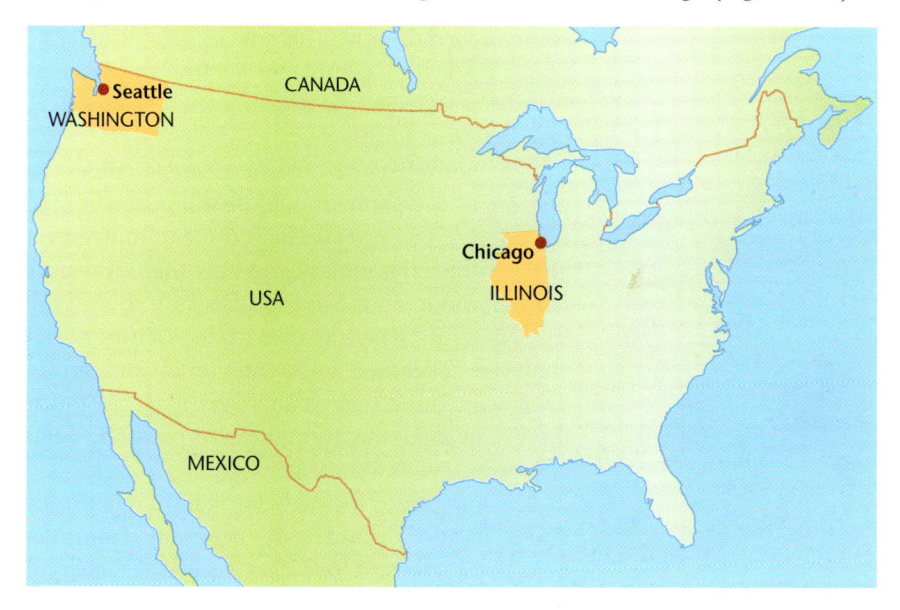

*Figure 13.4 The locations of Seattle and Chicago*

# Chapter 13 Economic systems

## Factors influencing present-day industrial location

The study of present-day factors that influence industrial location involves the concepts of optimal and non-optimal locations, and behaviourist and structuralist explanations.

The interaction between people and their environments has of course been a focus of geographical inquiry for a long time, both at a descriptive level and in offering explanations for those relationships.

- Many geographers adopted a normative approach and described the behaviour that should result under specified environmental conditions. This is particularly true of classical location theory, where it is frequently assumed that entrepreneurs and consumers act as rational economic individuals.
- Other researchers have preferred to study mass behaviour in the hope of highlighting the main features of 'statistically average behaviour'.
- Some geographers more or less ignored people and concentrated on the geometry of landscape. This is most clearly demonstrated in the spatial analysis school of thought which dominated human geography in the 1960s and which ought to dissect the environment into nodes, networks and surfaces.

Table 13.4 shows the results of a survey on the importance of various factors when a firm was choosing its location: they include social, economic and political factors.

Weber might help us understand and explain factors such as accessibility, and might also help us interpret why it is a more significant factor in cities than rural areas. Classical theory might also help us unravel the labour advantages that are surprisingly more of a factor for the remote rural firms than the urban ones. However, these theories do not help us understand the behavioural factors (such as nearness to founder's home), nor the political ones (government grants), nor those that clearly didn't have much relevance to Weber (such as company acquisition). This type of complexity gave rise to a reaction against the classical models and theories, and led to the development of a different perspective on location.

**Behaviouralist** explanations argue that the purpose of geography should be to understand the way in which we *as individuals* come to understand and use the environment and how that influences our behaviour. Thus behaviouralists try to shift the focus of locational geography towards a micro-level analysis, concerned with individual and group behaviour within society, and away from too much theorising about the distribution of power between elite groups within the state.

Behaviouralists insisted that valid conclusions about location could only be made from the objective analysis of hard data, such as statistics on how people make decisions, as opposed to abstract

*Table 13.4 Reasons for location (figures show the percentage of firms that mentioned the factor as significant)*

|  | Remote rural locations (% of firms) | Urban locations (% of firms) |
| --- | --- | --- |
| Nearness to founder's home | 37.0 | 40.9 |
| Environmental attractiveness | 35.4 | 37.5 |
| Labour advantages | 40.5 | 15.6 |
| Premises advantages | 24.1 | 46.9 |
| Local market or materials | 35.4 | 37.5 |
| Good communications | 40.5 | 15.6 |
| Government grants | 24.1 | 46.9 |
| Historic reasons | 35.4 | 37.5 |
| Company acquisition | 40.5 | 15.6 |
| Access to clients and suppliers | 24.1 | 46.9 |

Source: Keeble and Tyler, 'Entrepreneurship, business growth and enterprising behaviour in rural west England', *The South West Economy — Trends and Prospects 1996*.

## Human systems, processes and patterns — A2 Unit 5

speculation about the nature of the state. Many companies are situated where the original entrepreneurs lived. (Dyson, Case study 13.5, is an example of this.) These locations may be sub-optimal in strictly economic terms but may 'satisfy' the owners and managers (so the motive is satisfaction).

**Structuralist** explanations in geography are those whereby location is explained by the underlying structures of society. This covers almost any approach that sees the meaning of something as subordinate to its position in the structure. As the economic system develops in particular ways, so it requires particular types of location. Thus location is driven by changes in the national and the world economy and, in recent years, the need to preserve profits in the face of a crisis. Individual motivations and decisions will make a difference, but within very clear constraints imposed by the structure.

### Case study 13.5  Location and structuralism: Dyson's move from Wiltshire to Malaysia

James Dyson, who owns the Dyson company and is its sole shareholder, founded his company in his adoptive home town of Malmesbury, in Wiltshire. This is an example of behaviouralism. Other similar examples are Laura Ashley's clothing factory in Newtown, Wales, and the historical accident of a family marriage that brought the process of vulcanisation to M. Michelin in the remote Auvergne region of France. Michelin proceeded to manufacture tyres from imported rubber, even though the location was about as far from the sea as it is possible to be in France.

After investing £32 million in his factory in Wiltshire, in 2001 Dyson decided to move part of the production line to Malaysia. This meant the loss of some 800 jobs in Wiltshire. Dyson felt that the strong pound, which will continue while the UK stays outside the eurozone, the global recession and the need to be near to supply chains meant that a new location was essential.

As the market leader in the UK vacuum cleaner market, Dyson argued that costs and reliability of spares, and other after-sales factors (some of which feature highly in the non-price competitive perceptions of consumers), would be best controlled from the UK. However, Dyson had been forced to shift production to Malaysia because of soaring manufacturing costs in Britain. Labour costs had doubled in 10 years, partly because of the need to pay high wages in an area around Swindon at a time of virtually zero unemployment. Dyson also argued that he needed to increase his profit margin. This was not to increase his personal wealth but to increase the spending on research and development (R&D) which was vital to the survival of the company (as it is to most electronics companies where product refinement is vital to market share). This R&D section of the Dyson company was to remain at Malmesbury.

Two years after the move, Dyson was on course to double its profits, justifying the decision. Washing-machine production was moved to the same part of southeast Asia, with the loss of a further 65 jobs. At the same time the R&D budget had increased by 50% over the £12 million spent in 2002.

Dyson suffered much criticism, both local and national, for his decision, but he insisted that the site had taken on 100 more employees and that by retaining R&D in Wiltshire the company was offering higher paid jobs:

> We employ 1300 at Malmesbury — engineers, scientists and people running the business. The decision to shift production to Malaysia was not good for Britain in one sense because we don't employ manual labour any more, but we are taking on more [people] at higher pay rates and more value-added levels and that's what the government is always asking us to do.

# Chapter 13 Economic systems

> **Key terms**
>
> **Behaviouralism** Explanations of economic decisions that concentrate on individuals and the psychology of their decision making.
>
> **Marxism** The view that the most significant structures are economic and best understood through the medium of class.
>
> **Structuralism** The view that deep-seated structures in societies rather than individual action drive economic and social change.

A central theme here is that the factors that control the location of plants (factories) for a large transnational corporation (TNC), such as General Motors or Toyota, are not the same as the factors that explain the location of a smaller, owner-managed company. The differences in the structural organisation of companies will have a large influence on location. This is clearly illustrated by the case of Dyson. The best-known facet of structuralism in geographical thinking was the development of radical, **Marxist geography** in the 1970s. Marxist geography called into question the racism, ethnocentrism and sexism in geographical texts and teaching and attacked conventional geography as being largely uncritical of the 'bourgeois' state. It pointed out that geographers and, by extension, geography teachers, had collaborated with the state to ensure the continuity of the dominant ideology, that is to say the support of the capitalist state that rewards the owners of capital.

The method of location selection used by a major TNC such as Sony illustrates the complex set of both economic and political factors involved in the selection of new sites in a world where transport costs have reduced in significance. The method also illustrates significant differences in the type of labour required at different stages of the production process.

## Case study 13.6 A location decision: Sony

Sony Corporation started production in Tokyo in 1955, with the introduction of the transistor radio. Five years later, the company established its first overseas operation, Sony Corporation of America, in an abandoned warehouse in Manhattan. Today, Sony is a world leader in the consumer/industrial electronics and entertainment industries. As with many companies in the electronics sector, innovation is vital to survival. Sony has marketed some of the world's most distinctive technological products and introduced breakthroughs such as the Walkman personal stereo.

One of the company's most successful and well-known products is the television set manufactured by Sony Electronics, a subsidiary based in Park Ridge, New Jersey. In 1972, Sony became the first Japanese company to establish a television manufacturing plant in the USA. At that time California was the preferred location because it had available labour and, most importantly, allowed the shipment of components from Japan across the Pacific. With a strong yen the West Coast location, as close as possible to Japan, was ideal; hence the establishment of Sony's San Diego manufacturing centre.

A plant was later set up in Tijuana to supply components. This took advantage of the much cheaper labour of Mexico. The rapid growth in the market led to the search for a suitable location for another production plant in the USA. The stages in finding this location were broken up as follows:

- identify the preferred region
- identify the preferred city or sub-region (the situation)
- identify the preferred site

In this case Sony was determined to pursue a policy of vertical integration in the manufacture of television sets. It needed a site that would provide adequate suppliers, especially of glass, and skilled labour, as well as offering factors such as flat land and accessibility. Vertical integration meant that Sony would manufacture some of its own components, especially the glass screens. The choice of region was determined by the following:

- the main market was in the eastern USA
- the main suppliers of glass were in Ohio and Pennsylvania
- transport and utility costs (water and electricity) were much lower in these states than in California

Sony then appointed consultants to look at the secondary data for 80 cities and sub-regions. From these data, which included a wide range of economic, social and demographic information, Sony drew up a short-list of eight semi-finalists.

Armed with their secondary data, two Sony executives set out on a 5000 km 'road trip' to find the ideal location. The reasons for some of their decisions are fascinating and remind us how stochastic (i.e. random) factors are often influential. For example, on arriving in Kentucky after a long day's drive the two men stopped for some food. Having ordered their steaks, they asked for a beer. The restaurant, unable to serve alcohol on a Sunday, said 'No beer', so Sony said 'No' to Kentucky.

One of their final choices, Pittsburgh, was received with little enthusiasm in corporate circles for it had the reputation of being a decaying 'steel' city, a typical 'rust-belt' location. Even the consulting firm had cut Pittsburgh from its list. Nonetheless, the two finalists were Pittsburgh and Cincinnati. Pittsburgh was finally chosen because:

- state support was good
- the local labour force was skilled

This last point was vital; in common with all modern electronics companies, Sony was planning to use highly automated processes, with sophisticated machinery that needed maintaining. This required high levels of skill, albeit from a relatively small workforce. This is known as 'total equipment productivity'.

To choose between the final two locations, Sony asked the following question of companies in both cities that ran highly automated production lines:

- How long would it take to recruit 100 employees with associate or advanced technical degrees?

The answers were '1 year' for Cincinnati and 'a few days' for Pittsburgh.

Local educational facilities were also evaluated to establish that training was available. The results combined with what one of Sony's travelling executives called his 'gut instinct' and Pittsburgh was chosen.

Only at this stage did the choice of *site* become an issue. It was determined by:

- flat land (not easy in hilly Pennsylvania)
- adequate water and sewerage facilities
- rail access

A helicopter tour of the region passed over a disused factory at Mount Pleasant in Westmoreland county, previously owned by Volkswagen of America and before that by the Chrysler Corporation.

- The factory had a floor thickness of 5 m.
- It had a 3.5 million litre per day sewage capacity.
- The rail line ran beside the plant.
- Energy costs were competitive and lower than those of a competing site in neighbouring Allegheny county.

The final assessment looked at local housing, educational and cultural facilities for Sony executives. The high ratio of golf courses to population was a factor noted by several!

Sony executives decided the Mount Pleasant site met their criteria, including an adequate supply base. Eager to land the Sony plant, Pennsylvania officials offered incentives of about $27 million through loans, customised job-training funds and a demolition grant.

Sony began producing rear-projection colour televisions at the site in 1992. The company relies on more than 500 suppliers in the region to provide components for the various products built at the facility. It outsources 100% of the mechanical parts and 80% of the components locally.

# Chapter 13 Economic systems

## Political processes influencing the location of industry

The role of government was marginal in the early days of industrial capitalism. Governments had a crucial role in the location of defence industries, such as the naval dockyards, but their role in managing the economy was small. This is no longer true, even in those countries where politicians make election promises about reducing control of enterprise. In practice government has a central and direct role in the management of the economy as an employer, as an awarder of contracts and in developing infrastructure. In fact, government intrudes in every decision about location to some degree.

This influence can be approached on several levels:
- global
- national
- local

The questions we need to answer are:
- How do political processes affect global location?
- How do political processes affect location within a nation state?
- How do political processes affect location within an area?

Case study 13.7 looks at these questions in relation to the location of a Honda car plant.

### Case study 13.7 Location and political processes: the Honda car plant, Swindon

The Honda factory at Swindon is a good example of the location decisions made at a number of scales, and the influence of political policy on each of these.

#### Why Europe?

Honda, in common with the other volume car manufacturers from Japan, wanted to expand into both the other major car markets, the USA and Europe. This could be achieved by:
- establishing branch plants in these regions using imported parts
- establishing full production facilities with local sourcing
- exporting vehicles manufactured in Japan

This last option was not possible given the tariffs and quotas imposed by both the USA and the EU on foreign-made vehicles entering their territory. Thus Honda had to produce in Europe if it wished to compete there.

*Photograph 13.4 The Honda works at South Marston, Swindon*

*Human systems, processes and patterns* **A2 Unit 5**

## Why the UK?

The UK was seen to be the softest of all the European markets. The decline of the British car industry and a history of higher levels of import penetration than those found in France or Germany made Britain a good choice. However, the British government also played a role in facilitating the move, encouraging investment by companies such as Honda in a way that certainly was not true in France.

## Why Swindon?

Local politics plays a role here, in the attitude of local government to foreign investment, the attitude of local officials to questions of planning and the flexibility of those officials. Broader, more indirect influences include the existing infrastructure which is, of course, provided by government, both central and local, and promises for future improvements in that infrastructure. Government has sometimes had a role in providing a skilled workforce through training schemes in local schools and colleges.

## Why South Marston?

The choice of site is usually the last part of this process. Political processes may intrude not just in the provision of essential services such as sewerage and water but also in facilitating land purchase and, in some cases, compulsory purchase. Land-use planning zones may need to be adjusted, not just for the factories themselves but also in order to accommodate the extra housing needed for the workforce.

---

Political processes at a number of levels often come together in quite a complex manner, as in the creation of the London Docklands Development Corporation. This corporation was a response by central government to global competitive forces that were seen to threaten London's status as a global city. The national government's response to those forces can be seen in the context of a local area — the London Docklands.

It is important to distinguish carefully between direct and indirect influence of government. Regional policies which offer incentives and tax breaks are obviously direct government action. The building of a new motorway or a new bridge is indirect government action, in that it is not targeted exclusively at industry. The new motorway or new bridge does not determine that firms will move into an area, although it may help. The building of the second Severn crossing, for example, has reduced journey times to south Wales and made the area a little more attractive to potential investors.

*Photograph 13.5 The new Severn Bridge should help to regenerate the region of South Wales*

# Chapter 13  Economic systems

## Case study 13.8  Micron's search for a new site

Micron is one of the world's largest manufacturers of computer chips. Founded in 1978, it operated in the low-profile location of Boise, Idaho, USA, for many years. It was originally founded by a couple of Idaho twins in the basement of a dental studio, and has a board of directors dominated by local businessmen, not the usual hi-tech engineers characteristic of such companies.

In the autumn of 1994 Micron announced that it was to build a new $1.3 billion factory and was actively looking for a site. The new site would probably not be in Boise, where resistance to expansion was obvious in the local community. Political support for the expansion was felt to be lukewarm at best.

Micron's decision to expand was based on the huge success of its main product, the tiny pieces of silicon that act as the link between a computer microprocessor and the main memory.

The company was looking for a community that fulfilled six basic criteria and in its global public announcement it set a 2-month deadline for proposals from any community, in any country, wanting the business. The six criteria were:

- a population of at least 100 000 within 72 km of the site
- at least 200–300 hectares to develop
- 7.5 million litres of water a day from a groundwater source
- 25–30 megawatts of electricity
- sewage discharge of 4000 litres a minute
- 500 000 therms of natural gas

The company received 400 proposals, including several brought to Boise by hand by elected politicians who were desperate for the business. The intention was to narrow down the field to ten semi-finalists within a few weeks but, given the larger than expected 'entry', Micron ended up with 13, all in the USA. Four were in the home state, including Boise itself, but none were in any of the traditional hi-tech hotspots such as Oregon, California, Arizona, Texas or Colorado.

The next step was to investigate these 13 communities in more detail but quickly and stealthily, given that the self-imposed deadline was early March 1995. The team did a tour of all 13 sites in a week. After inspecting the proposed site they investigated the local community, largely by wandering into local coffee houses and bars and chatting. Meanwhile the 13 communities were sent a set of questions, broken into groups, about local taxes, physical site, potential government assistance, various community factors and workforce. Micron wanted information about every possible aspect of these communities, from the quality of the schools to the percentage of graduates in the local population.

At the same time the 13 communities were working hard to assemble packages that would be attractive to Micron. Local political leaders were at the forefront of this, not least back home in Idaho where the general feeling was against expansion in Boise itself. Resistance was also evident in the three other Idaho locations where people didn't want taxpayers to have to provide the funds to offer 'incentives' for Micron to choose them.

Another problem Micron had with Idaho was the educational structure of the state, which allocates university subjects to different universities; the nearest engineering faculty to Boise is 480 km to the north. Back in 1988 the company had agreed to expand its Boise site, rather than open an alternative plant in Portland Oregon, in a $90 million investment programme on the express understanding that the state education authorities would deliver an engineering programme for Boise University. By 1994 that had hardly got off the ground. Micron even offered $5 million to get the course moving, but it was accused of bribery and an inter-university dispute broke out that made matters even more complicated. Even the Idaho-loyal board of Micron was forced to admit defeat and the final three on the second short-list did not include an Idaho site. They were:

- Payson, Utah
- Omaha, Nebraska
- Oklahoma City

*Human systems, processes and patterns* **A2 Unit 5**

The three key criteria applied to make the final decision were:
- engineering education
- water availability
- road infrastructure

Even when making this selection the panel knew that, although they liked Utah in general, they didn't much like any of the eight sites offered in Payson. They were too close to railway tracks, which caused vibrations, suffered from a high water table and the town was too far south. Thus the Utah state authorities were asked for an alternative. Ultimately one was found at Lehi, close to Mount Timpanogos in the required area between the Brigham Young University at Provo and the University of Utah in Salt Lake City.

At this point all three finalists were eager to provide state government tax incentives. One of the three, Oklahoma City, offered a $780 million package including paid relocation of executives and free job training for employees. Omaha, Nebraska promised $200 million off the cost of water and electricity, while Utah put together a $86 million package.

The final decision was announced on 13 March 1995. Lehi in Utah was the choice.

The state had cooperated fully in the process, right down to helping with the land sale. Utah had spent years developing its attractions for new business, including spending on education — in contrast to the problems back home in Idaho. As a complicating factor, it should be noted that at least one of the board members who made the choice was an active Mormon, a religion which has its homeland in Utah.

# The rise and decline of consumer industries in MEDCs

Consumer industries are those that sell goods directly to the public. These industries are almost entirely twentieth century in origin and their rise is commonly associated with the rise of 'Fordism'. Prior to this, most manufacturing industries had a very small market of potential customers. Few people were wealthy enough to purchase consumer goods.

*Photograph 13.6 A car assembly line in the 1960s*

# Chapter 13  Economic systems

**Fordism** involved:
- the mass production of standardised products using standardised parts
- the manufacturing of goods using assembly-line techniques
- the division of tasks into smaller, repetitive jobs that require little skill
- the use of time and motion studies to improve productivity

From the point of view of employers, Fordism offered the possibility of cutting costs and increasing productivity. However, so large were the increases in productivity that the workforce, which saw the value of their labour being downgraded from skilled to semi-skilled, pressed for higher wages. Thus in the car industry, followed by many others, the last and crucial part of Fordism was:
- higher wages paid to a workforce which could, as a result, become consumers themselves

## Case study 13.9  Fordism: the Ford Motor Company

Henry Ford's main contribution to revolutionising car manufacturing was to standardise all parts of the production process. In order to do this, parts needed to be perfectly interchangeable. By exploiting advances in machine tools and manufacturing processes he was able to achieve this. To exploit the interchangeability of parts, he introduced a moving assembly line at Highland Park in Michigan in 1914, which increased labour productivity by a stunning 1000% and thus enabled him to introduce price cuts for the Model T Ford; in 1910 the price was $780, but by 1914 it was $360.

Assembly-line production was deeply unpopular. It was physically demanding, required high levels of concentration and was exceptionally boring. The assembler had only one task to perform, perhaps to put two nuts on two bolts, again and again for the whole working day. He never saw the finished product, didn't supply his own tools and had no part in the production process other than his task. Unsurprisingly absenteeism, labour unrest and turnover of staff were all high. Ford believed 'men work for only two reasons; one is for wages, one is for fear of losing their jobs'. Given his considerable increase in profits, Ford dealt with the high and costly labour turnover by doubling pay to $5 a day — a policy rapidly copied by other manufacturers. In effect, the manufacturers were paying their workers for putting up with mass production's monotonous and degrading work process.

The effect was that for the first time working people had high enough wages to buy the products they made. This system hugely expanded demand, thus maintaining high profits for the companies through high turnover and economies of scale. This process came to be known as Fordism.

### Key terms

**Fordism** The application of Henry Ford's approach of mass production, high division of labour and less skill in the workforce, but with high wages that allow the employees to become a new generation of consumers.

Fordism dominated the world economy for 50 years, between the 1920s and the 1970s. In that time Japan joined Europe and the USA as a major industrial power.

Since the 1970s many consumer industries have looked for new locations both within MEDCs and beyond. They have done this in an attempt to reduce costs and, in some cases, to find new markets. Economists vary in their interpretation of this. Some see it as the end of another Kondratieff cycle (Figure 13.5). This is a theory based on a study of nineteenth-century price behaviour, which included wages, interest rates, raw material prices, foreign trade, bank deposits and other data.

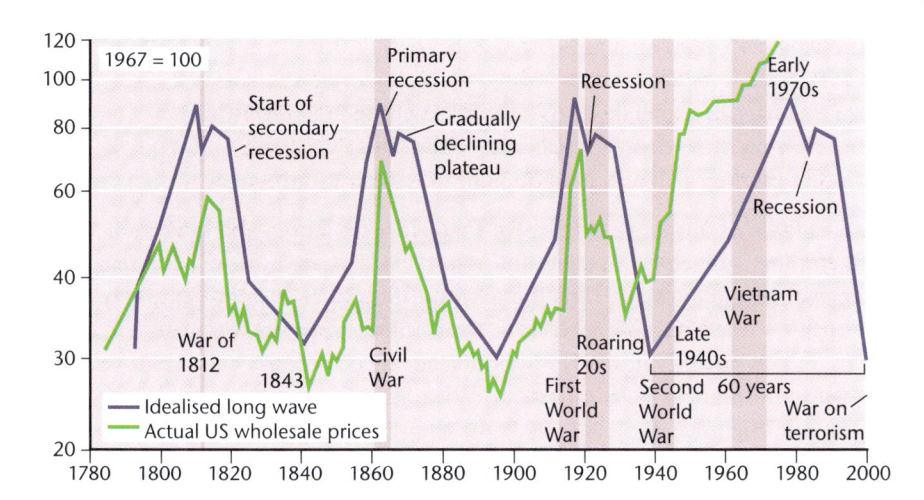

*Figure 13.5 The Kondratieff cycle*

Kondratieff was convinced that his studies of economic, social and cultural life proved that a long-term pattern of economic behaviour existed and could be used for the purpose of predicting future economic developments.

Others regard the recessions and depressions in the Kondratieff cycle as major crises of production in the capitalist system, with too much factory capacity for the demand. The rapid growth of postwar economies involved rises in demand for goods as the Western world stocked up with refrigerators, hi-fi systems and, above all, cars. Inevitably demand slowed down as almost all the people who could afford such things acquired them. The factories that had been turning out goods to meet a rapidly growing demand from new customers were now faced with largely 'replacement' customers who simply sought to change their old fridge. This is over-capacity, whereby factories are designed for an output higher than the demand. This inevitably leads to higher costs and a crisis.

## The 'Fordist' crisis

After a long period of growth that was particularly noticeable after 1945, the world economy started to slow down in the 1970s. This slowdown came about as the relatively small number of global consumers, perhaps 1 billion, most of whom lived in North America, Europe and Japan, came to the natural end of their collective spending spree. They now had in their possession a wide range of domestic goods from cars to washing machines.

The firms who produced these consumer durables had, up until this point, been expanding output (and employment) to meet a high annual growth rate in sales as new consumers bought their products. Now they found themselves faced with falling demand as market saturation was reached.

There are several possible responses by a manufacturer to falling demand and over-capacity. They include:
- a change of location, often to an NIC (newly industrialised country), to reduce costs
- mechanisation of production to reduce costs

# Chapter 13 Economic systems

- merging with other companies to reduce costs — this often leads to the closure of factories
- finding new customers (which also links with the rise of NICs)
- diversifying into other products and businesses
- working on research and development to create new products

The first three of the above list involve a reduction in labour and thus an increase in unemployment. Unfortunately, this in turn leads to a reduction in money spent on consumer durables and thus the negative multiplier ensures that a recession (downturn) turns into a depression. This situation was apparent by the early 1970s, most obviously in the world's most advanced economy, the USA.

At this moment a political crisis in the Middle East stimulated two large increases in the price of oil (in 1973 and 1979), which massively increased costs for almost all products (as virtually everything needs to be transported). This made a difficult situation worse. There were various impacts.

- Consumer confidence, already low because of high unemployment and fears about job security, was reduced even further. In an atmosphere of uncertainty, future spending plans such as on houses or a new car were postponed.
- Companies and corporations also put off any spending plans; certainly planned expansion at this time given the obvious fall in demand for their products and services. This pessimism further reduced consumer confidence.
- Banks were caught in a serious trap. Banks have to 'recycle' the savings deposited with them by lending money. They make their money by charging borrowers more than they pay depositors through manipulating interest rates. The oil states accumulated huge surpluses in their current accounts. (In other words, these countries couldn't spend their money fast enough on 'big' projects and personal consumption.) This money *had* to be recycled or the banks faced catastrophe. A bank that has to pay out interest on the money saved with it but which is not making money from loans is technically bankrupt.

The banks had two potential sets of customers:
- business and industry borrowing for new projects
- individuals borrowing for house purchase, new cars and consumer durables

Neither of these groups was in the least bit interested in borrowing at this time. They lacked either the confidence or the potential market to contemplate borrowing money. As a result, the banks were obliged to seek new 'markets' for their money — hence their enthusiasm for lending to the NICs and others. These loans went both to governments and to businesses.

This, combined with companies searching for new, cheaper manufacturing locations and new customers, was among the driving forces behind the emergence of the NICs. They were a necessary development of the world system, which was otherwise in crisis. The companies that pursued this route became TNCs (trans-national corporations) as a consequence.

*Human systems, processes and patterns*  **A2 Unit 5**

The lending policy of the banks was ultimately to lead to an increasing growth of debt in both NICs and, more generally, the peripheral LEDCs. This led to a series of financial crises (Mexico 1982, East Asia 1993–, Argentina 2000–) that have destabilised the world economy.

This debt crisis is not resolved, although the banks have worked hard to cover their losses (real and potential) by massively increasing lending to individuals in MEDCs. This has led, in turn, to a crisis of personal debt: in the UK this amounts to £151 187 000 000 in outstanding unsecured debt — about £3000 for every individual (2002 figures).

Common mistakes in the treatment of this topic include a tendency to generalise overmuch about MEDCs. The speed of change has been much more noticeable in some than in others. For example, the decline in consumer industries has been limited in Germany, but has been sharp in the UK. However, in Spain, for example, there have been recent increases in manufacturing employment. Similar variations can be found in the new market economies (NMEs) of eastern Europe. For example, while part of the Soviet Empire, Romania was initially discouraged from developing a manufacturing industry base but was given the role of grain supplier for the region by Moscow, with much of this taking place on large collective farms. However, in the 1970s and 1980s rapid industrialisation took place, led by large state-financed projects, often of doubtful efficiency. Since the fall of communism in Russia, this influence has disappeared and there has been a rise in employment in the primary sector in Romania, particularly in agriculture as people have returned to the land, and a rapid decrease in manufacturing employment (Figure 13.6).

### Key terms

**Region** An area that constitutes a distinctive sub-division of a country, characterised by geographical, economic, social and cultural facets that make areas within the region more like each other than they are like anywhere else (e.g. East Anglia).

**Global region** The same concept applied at a global level (e.g. Europe).

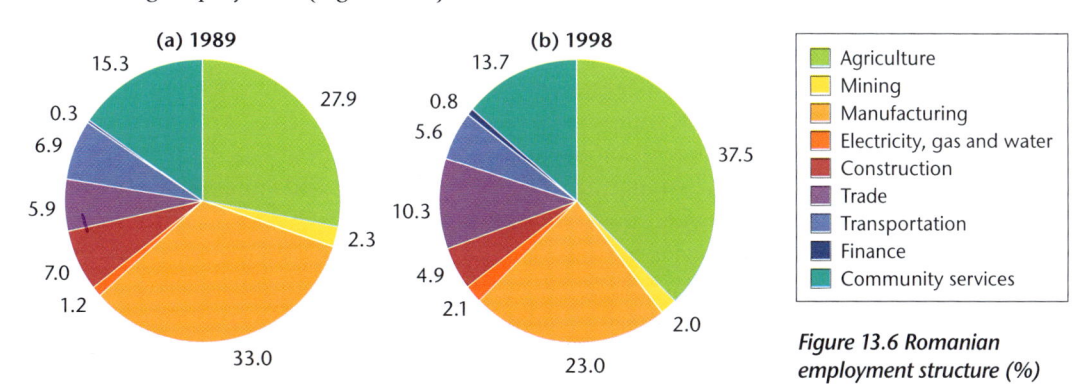

*Figure 13.6 Romanian employment structure (%)*

You also need to recall that, although new areas of industry have emerged in the world, most of the MEDCs still have a positive trade balance in manufactured goods. That is to say, they export more than they import.

## Changes in the employment structure in the UK

Figure 13.7 shows the main trends in UK employment since 1945:
- continuing falls in primary employment — there are now three times as many public relations consultants as there are coal miners

# Chapter 13  Economic systems

*Figure 13.7 Employment, output and investment in manufacturing in the UK, 1985–1997 (at constant 1990 figures)*

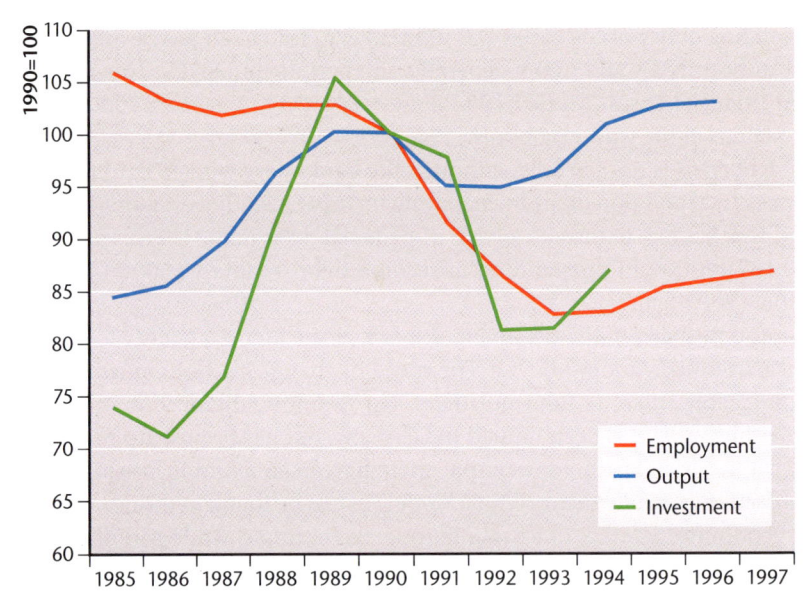

- an initial rise in secondary employment, followed by a long period of decline since the 1950s with a particularly steep fall in the 1980s
- a rapid growth in tertiary employment, especially since the 1980s — more people now work in Indian restaurants than in shipbuilding and steel manufacture combined
- the emergence of a quaternary sector in the past 30 years

The reasons for these trends vary in their complexity and are not without controversy. The decline in primary and secondary employment can be attributed to some of the same causes, although there are significant variations both between these sectors and within them. These trends have been accompanied by shifts in the age and gender composition of the workforce as well as significant changes in the contractual arrangements between employees and employers.

The background theme to change is **mechanisation**, which is systemic, and therefore inevitable and universal (at least in MEDCs). In other words, there is a natural tendency for companies that are seeking profit to replace labour with machines wherever possible because machines are cheaper to run, do not take holidays and do not require pensions. Unless labour is exceptionally cheap or even free, as in slave societies, the tendency to mechanise is always there. This is sometimes known as positive deindustrialisation in that it does not always lead to a fall in output. It was particularly pronounced in those industries in which assembly-line production and later the use of robots were possible. Societies with shortages of labour have always tended to mechanise early, introducing 'labour-saving' devices. The best-known example of this is the USA, with its deserved reputation for a love of gadgetry. However, it is not as easy to introduce in some industries. Those which have small product runs or deal with non-resistant materials, such as the clothing industry, have proved resistant to mechanisation. It has proved difficult here to advance much beyond technology that was well-known 100 years ago.

*Human systems, processes and patterns*  **A2 Unit 5**

*Photograph 13.7 Robot assembly in the Nissan plant, Washington UK*

The profit crisis of the 1970s also stimulated **rationalisation**. To rationalise is to make more rational or logical. In practice this involves mergers, plant closures and redundancies.

One recent case involved Waterford Wedgwood, which was badly hit by a downturn in US demand for their luxury glass and ceramic products. Having moved towards outsourcing for many years, they have now reverted to production in their under-capacity factories. Unless demand recovers, further rationalisation will take place in the form of short-time working, job-sharing and other cost-saving exercises.

Perhaps the most obvious feature of rationalisation is merger, in which one plant is closed in order to concentrate production at a more modern but underused second facility. Thus a change of location occurs, but not always for specifically geographic reasons. The plant that is closed might simply be the oldest.

The same profit crisis in the 1970s also stimulated the search for sources of cheap labour and new markets, which stimulated **globalisation** and the rise of the **TNCs**. A revolution in transport facilitated this process. Globalisation is dealt with separately in this book (Chapter 15) and is best understood as a consequence of a crisis rather than a cause in itself. No firm has ever moved its production out of the UK *because of* globalisation. Many firms, including Dyson, have moved their production facilities out of the UK to reduce costs by finding sources of cheap labour. This is globalisation.

These three factors (mechanisation, rationalisation and globalisation) had an impact in most MEDCs, including the UK, from the 1970s onwards. However, the impact was varied in timing and in extent. It also varied within countries, according to regional dependence on secondary-sector employment.

# Chapter 13  Economic systems

## Factors particular to the UK

In the UK the impact was made worse by a number of other factors (some of which are controversial):

- the Thatcher years
- poor labour relations
- the 'first to industrialise' argument
- the loss of empire
- inappropriate education
- involvement of banks
- government priorities

The **Thatcher years**, especially 1979–84, were a period during which high interest rates led to higher costs for many exporting manufacturers and a loss of their markets as the pound rose in value against other currencies. When Margaret Thatcher came to power in 1979 her clearly stated intention was to 'defeat' inflation. The method employed by her government, swiftly followed by the Reagan administration in the USA, was to raise interest rates. This dealt a double blow to some sectors of British industry.

- It made the cost of overdrafts and borrowing very much higher. Given that many companies were in difficulties because of the world recession, this was the last straw for them.
- The high interest rates attracted capital into the UK, which pushed up the value of the pound. This made British goods expensive in foreign markets, which was especially damaging for companies that had a large export market in the USA, such as Wedgwood, the pottery and ceramics company.

Poor **labour relations** and antiquated systems for settling disputes led to many days lost in strikes during this period and thus higher costs. British trade unions were frequently based on particular skills or crafts. For example, a shipyard might have workers from eight or nine different unions. This made wage bargaining particularly difficult and also impeded the efficiency of the yard when disputes between different groups of workers held up production. Industrial unions that represented all the workers in a particular industry were not unknown in the UK (the National Union of Mineworkers was an example). Ironically, the British TUC (Trades Union Congress) had a key role in establishing the postwar German trade union system, which was widely praised for its efficient and non-confrontational industrial relations.

Because the UK was the **first to industrialise**, it was thought that it would be the first to deindustrialise: capital equipment was outdated, the infrastructure was old and attitudes were complacent. The first part of this argument is quite appealing on the 'first in, first out principle', but makes simplistic assumptions about economic development. The second part of the argument has much more validity. Old rails, old machines and old factories are likely to lead to lack of competitiveness when these industries are exposed to foreign competition. This in turn leads to loss of employment as factories close.

In a similar vein, the **loss of empire** has been blamed for our uncompetitive industries. The argument is that UK industry did not need to be competitive because it had both highly protected domestic markets and the added benefit of the empire, also highly protected from foreign competition. Many British industries had no need to be troubled about technical deficiencies in these circumstances. With substantial import duties imposed on foreign machine tools, cars and many other products, design deficiencies and manufacturing faults were only likely to be revealed in exceptional times. One example of this was the Battle of Jutland in 1916. 'There seems to be something wrong with our bloody ships today...' remarked Admiral David Beatty to his Flag Captain. And he was right. Poor design and poor management sealed the fate of three battle-cruisers and four armoured cruisers. German engineering skills had long been a preoccupation of the British, but very little had been done to improve their own since the first alarm bells had been rung in the 1870s.

It has been suggested that Britain has never given a proper stress to **education** in engineering and design skills. As a result, British firms have failed to innovate and the quality and reliability of products has declined. When the protection of tariffs was eventually removed with the end of empire, these problems were fatal. There is a good deal of evidence supporting this view. As late as the end of the nineteenth century, science was regarded as a distraction, taking valuable time away from playing games or studying Latin and Greek. Even at the height of our industrial power, in the middle of the nineteenth century, much of our wealth came from overseas trade and investment. Thus UK firms became less innovative and fell behind in technical competence, which was exposed as soon as foreign competition began after the removal of tariffs in the 1950s.

Similar criticisms are directed at other institutions, above all the **banks and investment** institutions. It was believed that more money was to be made elsewhere, especially in property, so manufacturing was starved of funds. Proponents of this argument point out that other competing MEDCs do not have the same inflated property values, which, it is claimed, have sucked potential investment funds away from industry. Furthermore in some countries, and Germany is an example, there are investment banks that specifically lend money to manufacturing industry. In countries such as Sweden the quick capital gains offered by the property market are not possible. The high profits generated by property development in the UK are real enough; the view is that this prevents adequate funds reaching other borrowers, so that firms fail to modernise effectively and thus become more and more uncompetitive, ultimately leading to job losses.

Successive **British governments** have also been accused of being unenthusiastic about manufacturing, preferring to see the future of the UK in terms of high-level tertiary and quaternary activities, which are encouraged through interest rate and macroeconomic policies that have a negative impact on manufacturing. Only a few years ago the idea that countries do not 'need' a strong manufacturing base would have been inconceivable. The law of comparative advantage (see Chapter 15) was not used to support the idea of free trade in, for example, tanks. Defence and security arguments were used to defend a country maintaining a number of strategic industries, including iron and steel and chemicals, without which it would be unable to sustain many other industries. Dependency on imports is a problem, it was argued, in times of war.

# Chapter 13  Economic systems

## Case study 13.10  The changing fortunes of the British car industry

The car industry is intimately associated with the rise of modern consumer society. To own a car is almost the definition of modern consumerism. To own two cars implies a certain status in the middle classes. The second phase of capitalism was ushered in with the development of Fordism in the 1920s.

Early industry concentrated in engineering areas in Europe and the USA. Typical locations were close to the small market of wealthy clients for customised vehicles.

In 1896 speed limits in the UK were lifted to 14 mph and the embryonic car industry expanded rapidly. Daimler became established in Coventry, using German designs, so the car industry had an 'international' flavour very early. The period between 1896 and 1914 is best described as the 'experimental period' with few standard designs, lots of innovation, and a high rate of turnover (this offers parallels with the computer industry today). A typical firm produced a few hundred cars that were custom built for a luxury market. This required high levels of skill in the labour force. In the UK 393 firms were founded in this period, of which 280 had closed by 1914. Only ten of these companies produced more than 1000 cars per year. The largest UK plant was at Trafford Park (Manchester) where Ford (another early example of international linkage) reconstructed Model T Fords imported from the USA in a kit form, four to a crate.

Most firms in the UK were concentrated in the West Midlands in towns and cities such Coventry, Birmingham and Wolverhampton. These places had a tradition of metal working and employment in sectors such as bicycle making, and an unorganised labour force unlikely to oppose the use of new types of machinery. During the First World War engineering firms grew in these regions. Local capital helped, as local entrepreneurs had money to invest in 'venture' projects that were risky. This early history was dominated by frequent mergers; there was a high 'birth rate' and a high 'death rate'. Companies remained clustered in traditional areas.

Car production rose sharply between the wars, with Morris, Austin and Ford leading the way. A flurry of mergers occurred as part of the rationalisation of production to reduce costs. In 1922 there were 88 car companies in the UK, in 1929 there were 31 and in 1932 there were 22.

After years of falling sales, Ford finally accepted the need to replace the Model T with a new British production plant at Dagenham, just outside London. This plant was a smaller version of the giant River Rouge plant in Detroit and included its own steel-production facility as part of a strategy of vertical integration.

By the end of the 1930s, the three leading companies had some 60% of the UK market, but many firms had disappeared as they were unable to compete on either price or quality. The strategy of all the major companies was to concentrate on the existing market rather than working to expand the customer base. One of the most significant factors in this period was the large 33.3% tariff wall that had been erected as a 'temporary' measure during the First World War but in fact continued until 1956. This protected UK firms from all foreign competition. As the richest country in Europe and with an empire in which UK firms had a virtual monopoly, the British industry was the largest in Europe. By 1938 there were some 500 000 workers in the car industry in the UK. Large numbers were employed in plants located at some distance from existing industrial zones, where land was cheap and space for expansion was available. Location of car factories in the UK became more concentrated as peripheral firms disappeared:

- Scottish firms disappeared
- Ford moved to Dagenham
- Vauxhall opened a new plant in Luton
- Rolls Royce moved to Derby

Market locations had become increasingly attractive; Dagenham, Luton and the Morris factory at Oxford are all close to London.

After the Second World War, the UK car industry continued much as before until it entered a critical period in the 1960s. Despite a brief success in the US market in the late 1950s, supplying small 'compacts', often as second cars for increasingly affluent Americans, a number of problems began to emerge. Empire markets were opened up to competition as former colonies gained their independence, and the reduction of the tariff wall in 1956 saw foreign cars enter the UK market in larger numbers. Meanwhile UK firms had continued their path of rationalisation by merger, with Austin and Morris combining as BMC. There was new and much more dynamic management in foreign companies such as Renault and Volkswagen, both of which had reorganised production. For example, while VW was producing 400 000 Beetles a year and Renault produced 200 000 Dauphines a year, the largest production run for any UK model was just under 100 000 a year.

The UK industry had too many models and an increasing technological backwardness when compared to competitor countries. Poor industrial relations, frequent strikes, some catastrophic management decisions, weak design, which saw only the Mini emerge as a successful design concept, and inept government policies contributed to the decline. The entry of the UK into the EEC (forerunner of the European Union) in 1973 saw a dramatic leap in the number of cars imported to Britain, with numbers reaching 34% in 1974 and 57% by 1980.

Foreign producers were not immune from a crisis of falling sales and over-capacity that affected all car companies in the 1970s and 1980s. However, Volkswagen, for example, managed a reduction in the labour force of 33 000, which was achieved without a government takeover (as happened in Britain), and the company was saved by cooperation between government, unions and management.

When the Japanese first started exporting cars to the West in the early 1970s it became apparent that their products were relatively reliable and cheap compared to domestically produced vehicles. The new era of 'Toyotaism' had arrived. This corresponded with a complete change of political climate. Under Thatcher, government help for domestic industries from steel to shipbuilding to cars was stripped away. She also positively encouraged foreign direct investment. BMC had become British Leyland, but the problem of uncompetitive models, poor industrial relations and low profitability persisted. The rise in value of the pound hurt export markets and the future for large volume British car manufacture began to look bleak.

Meanwhile the Japanese, eager to open up the European market, chose the UK as their bridgehead with its 'softer' domestic market, a common history of left-hand drive, a language with which the Japanese were familiar and, above all, significant government encouragement at both the local and national levels. The management of Nissan (Sunderland) followed by Toyota (Burnaston) and Honda (Swindon) chose their locations carefully. They avoided traditional centres of union militancy but looked carefully at communications, labour skills and potential supply chains. These were important in an industry that was adopting just-in-time production techniques and required close contact between main assembly plants and component suppliers. The arrival of the Japanese has seen a revival in UK car production, although the extent to which this is 'British' is open to debate. No British-owned volume car producer has survived into the twenty-first century. This is in contrast to all our traditional competitors: Germany, France, the USA and Japan.

## The rise of tertiary employment

There has been a rise in both the total numbers and the percentages in tertiary employment, which is a highly varied category. Many factors are involved in this rise, as detailed below.

**Rising income levels** are important: household incomes rose rapidly from 1945 until the early 1970s. This rise generated spending on consumer goods and thus generated a growth in employment in retail activity.

# Chapter 13 Economic systems

The fall in the length of the working week allowed more time for **leisure activities**, which, in turn, generated a demand for those activities and a rise in their provision. Leisure-related employment covers a wide range of activities, from lifeguards at leisure centres through to tour guides and personal 'fitness' trainers for the more affluent.

The **'greying'** of the population has led to the growth of a specialist sector of healthcare in almost all MEDCs. Geriatric healthcare is seldom well paid, but it often provides part-time employment for a dominantly female workforce.

**Social changes**, allied to rising incomes, have led to a growth in restaurants, especially fast-food outlets (which have increased rapidly in the past 30 years). Over three times as many people are employed in restaurants as in agriculture, and the total income from this sector is 80% higher than in agriculture. The first McDonald's opened in the UK in 1974 in Greenwich. Since then, despite the continuation of traditional and usually independent operators that have well over 60% of the market, chain-outlets have become common features of every high street. Burger King, Kentucky Fried Chicken (KFC) and Pizza Hut quickly followed McDonald's. They are frequently co-located, in that the decision by one company to open a store in a town or city is rapidly followed by one or more competitors. For example, in places where McDonald's and Burger King are both found, the mean distance between them is only 260 m.

New types of employment have developed, with the rise of **teleworking**, especially the growth of call centres, being an obvious example. The rapid growth in consumption and personal incomes has increased the need for support services. Call centres have played a key role in the restructuring of the banking and insurance industries, particularly in the development of 'direct' banking and insurance services. There has been a distinct shift from a customer service ethic towards a sales approach. It is estimated that there are over 10 000 call centres in the UK, employing more than 400 000 people, 60% of whom are women. This is by far the largest number in Europe, partly because of the dominance of English as a global language, partly because of government incentives, and partly because of flexible working practices and the importance of the UK as a headquarters location for TNCs. Preferred locations in the UK are the large northern cities such as Leeds, Newcastle and Glasgow (Figure 13.8). Clustering is a well-known

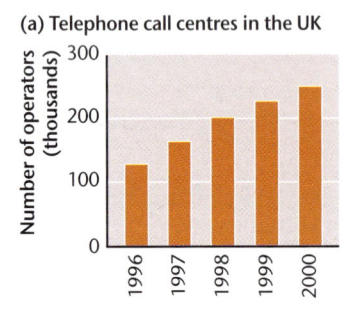

*Figure 13.8 The growth of call centres in the UK, 1996–2000*

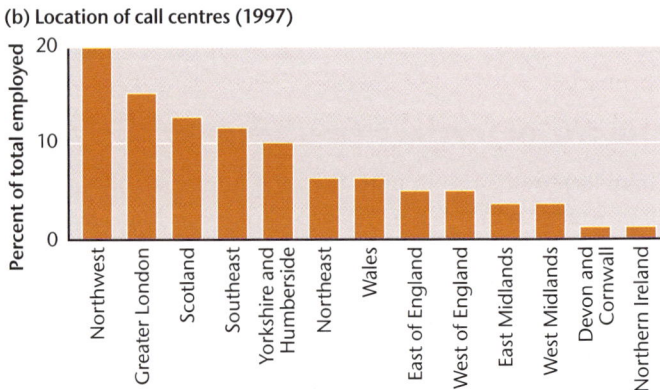

phenomenon in the industry because of the costs of training. The ideal premises are about 5000 m² on a single storey, although many centres are found in far from ideal, often shared premises. However, the most recent developments in the industry have been far from encouraging for the UK, with the transfer of large call-centre companies to cheap labour locations overseas. Both India and Malaysia have been popular choices.

The growth of the internet and the rapid decline in the cost of communication has stimulated a growth in home-working. In some cases this is so-called 'back-office' work. It allows people to work at home, either spending less time in the office or dispensing with it altogether. A recent example has been the motorist services provided by the AA (now part of Centrica). The organisation made the decision to close its call centre in Leeds and shut the vehicle recovery section of its facility in Newcastle. Instead, the AA developed its home-working section. Home-workers need specialist equipment installed, and the cost incurred has to be clawed back through higher productivity, measured by 'calls per hour'.

> **Examiner's tip**
>
> ■ Don't assume that all tertiary employment is well paid. Most of it is poorly paid and insecure. Thus the growth of tertiary industry very seldom causes a decline in other types of industry.

## The rise of manufacturing in NICs

The terms 'newly', 'industrialised' and 'countries' raise many problems. Look at Table 13.5.

Table 13.5 suggests a very different story from that usually told, with China and India experiencing rapid de-industrialisation between 1750 and 1900, while a much older generation of NICs, the European countries, rose to world dominance at the end of the nineteenth century. Thus the term NIC can itself be subject to critical scrutiny.

|  | 1750 | 1800 | 1830 | 1860 | 1880 | 1900 |
|---|---|---|---|---|---|---|
| Europe | 23.1 | 28.0 | 34.1 | 53.6 | 62.0 | 63.0 |
| UK | 1.9 | 4.3 | 9.5 | 19.9 | 22.9 | 18.5 |
| China | 32.8 | 33.3 | 29.8 | 19.7 | 12.5 | 6.2 |
| India | 24.5 | 19.7 | 17.6 | 8.6 | 2.8 | 1.7 |

*Table 13.5 Shares of world manufactured output (expressed as a percentage of the total), 1750–1900*

- 'Newly' is hardly appropriate for China or India, for example, given their long history as industrial powers.
- 'Industrialised' really means 'modern manufacturing industry' rather than 'industry' (which is a much more general term).
- 'Countries' may no longer be the best tool of analysis when looking at the global economy, as the rising importance of TNCs has diminished the significance of national boundaries.

The modern set of NICs needs identifying. There is no 'official' list and no agreed definition, but they commonly include:
- the 'Asian Tigers' that emerged in the late 1970s and 1980s
    - South Korea
    - Taiwan
    - Hong Kong
    - Singapore
- a number of 'tiger cubs' which emerged in the late 1980s and 1990s, including:
    - Malaysia
    - Thailand
    - Indonesia
    - the Philippines

> **Key terms**
>
> **Asian Tigers** The first four NICs in south and east Asia — Taiwan, Hong Kong, Singapore and South Korea.
>
> **NICs** Newly industrialised countries.
>
> **RICs** Recently industrialised countries — a wider group of countries that have industrialised since the 1980s.

# Chapter 13  Economic systems

*Figure 13.9 The Asian Tiger and 'tiger cub' economies*

The Asian Tigers and the tiger cubs can be located on Figure 13.9.

There are a number of more complex and speculative cases including India, Brazil, China and Mexico.

The idea to grasp is that many of these countries, especially the Asian Tigers, experienced very rapid growth in GNP (gross national product) of up to 10% per annum sustained over several years. They did this by developing a manufacturing sector. This is in contrast to the equally rapid growth experienced by countries such as Uruguay earlier in the twentieth century, or many oil states today, which have grown by exporting primary products. Reliance on exporting primary products exposes countries to the falling prices of such products, increasing competition and in some cases (such as copper and zinc) the invention of substitutes. Falling demand may be an issue as well.

Figure 13.10 shows how global manufacturing production is spread around the world and the substantial variation of its significance.

The growth of the NICs took place when the core economies were in crisis, a crisis exacerbated by the oil shocks of the early 1970s. This was probably no coincidence. The banks had to find new customers and the NICs fulfilled this role.

*Human systems, processes and patterns* **A2 Unit 5**

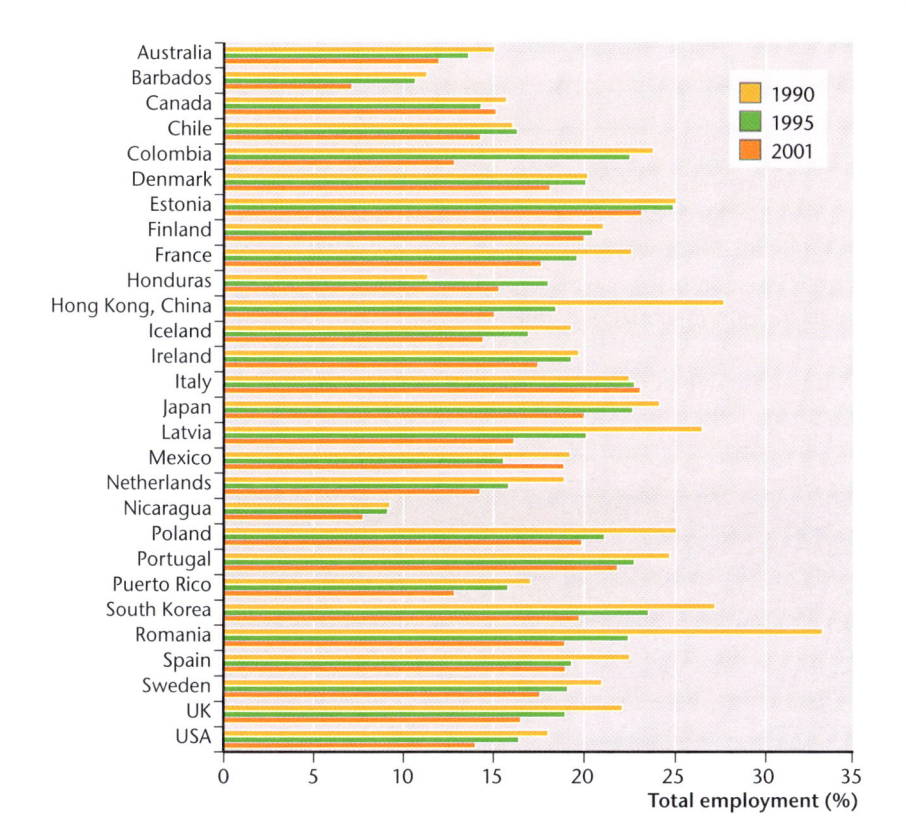

*Figure 13.10 Manufacturing as a percentage of total employment for selected countries*

NICs are not a uniform group and they must not be treated as such. They adopted different policies and applied them differently. Any use of the phrase 'the NICs...' requires immediate qualification. What they do have in common is spreading the benefits of economic growth to parts of the world outside the core countries. For much of the last century industrialisation was restricted to the core of Europe and North America. The rest of the world was a periphery, much of which was utilised by the core countries to provide raw materials through one system or another (colonialism, neo-colonialism or dependency — see Chapter 15). NICs broke through this, with considerable implications for theories of development.

Dependency theory argues that countries are locked into unequal relationships in which dependent economies are systematically underdeveloped by the core countries. The causes and consequences can be approached as a series of questions.
1 How did they do it?
2 Who, if anyone, helped?
3 What problems have been caused?
4 What are the implications for theories of development and economic growth?

The answers depend on the choice of country or countries used as examples, but here are some guidelines.

# Chapter 13 Economic systems

## How did they do it?

### Route 1

The first route was by **export-led growth**, sometimes known as export valorisation. This is the preferred route of the major world institutions such as the IMF (International Monetary Fund) and the World Bank, which funded some of this development. Export valorisation is the process whereby resources are directed towards those industries with export markets (Table 13.6). Little attention is paid to the domestic market, which is often only weakly developed. It is argued that once a sufficiently large export market is established then, and only then, will domestic demand start to rise as wages rise. The theoretical justification comes from the theory of comparative advantage (see Chapter 13, page 442 or Chapter 15, page 569).

*Table 13.6 Shares of world manufacturing trade by global region ($billions)*

| Country | Exports Value | Exports Share (%) | Imports Value | Imports Share (%) | Total Value | Total Share (%) |
|---|---|---|---|---|---|---|
| USA | 781.1 | 15.7 | 1257.6 | 23.9 | 2038.7 | 19.9 |
| EU | 858.9 | 17.3 | 965.9 | 18.3 | 1824.6 | 17.8 |
| Japan | 479.2 | 9.6 | 379.5 | 7.2 | 858.7 | 8.4 |
| Canada | 276.6 | 5.6 | 244.8 | 4.6 | 521.4 | 5.1 |
| China | 249.3 | 5.0 | 225.1 | 4.3 | 474.1 | 4.6 |
| Hong Kong | 202.4 | 4.1 | 214.2 | 4.1 | 416.6 | 4.1 |
| Mexico | 166.4 | 3.3 | 182.6 | 3.5 | 349.0 | 3.4 |
| Korea | 172.3 | 3.5 | 160.5 | 3.0 | 332.8 | 3.2 |
| Taiwan | 148.3 | 3.0 | 140.0 | 2.7 | 288.3 | 2.8 |
| Singapore | 137.9 | 2.8 | 134.5 | 2.6 | 272.4 | 2.7 |
| Malaysia | 98.2 | 2.0 | 82.2 | 1.6 | 180.4 | 1.8 |
| Switzerland | 81.5 | 1.6 | 83.6 | 1.6 | 165.1 | 1.6 |
| Russian Fed. | 105.2 | 2.1 | 45.5 | 0.9 | 150.7 | 1.5 |
| Australia | 63.9 | 1.3 | 71.5 | 1.4 | 135.4 | 1.3 |
| Thailand | 69.1 | 1.4 | 61.9 | 1.2 | 131.0 | 1.3 |

Source: World Trade Organization

### Key terms

**Export valorisation**
A set of policies that seek economic development through concentrating on exports.

**Import substitution**
A set of policies that seek economic development through developing industries to produce goods that are currently imported.

### Route 2

The second route was by **import substitution**. This occurs when countries begin to manufacture products that they previously had to import. This was clearly the route to development pursued by Japan. Many people think that development in Japan revolved around the existence of successful cities that provided the right levels of skill, entrepreneurial activity and education to allow the process to take place. However, import substitution does require some protection from foreign competition; the relevant explanatory idea here is economies of scale. A country cannot hope to develop its own industries if it faces competition from cheaper imported goods from countries with large established industries.

*Human systems, processes and patterns*  **A2 Unit 5**

## Who, if anyone, helped?

**Autarkic development** is development that takes place without any 'outside' help. It is uncommon historically (although someone had to be first) and more or less impossible in the modern world because trade rules disallow protectionist policies. The role of TNCs and foreign investment is central to the development of the NICs.

- Foreign direct investment (FDI) has been important in some NICs, such as Singapore, where foreign companies established production centres.
- 'Indirect investment' involves movements of money, with lending from other countries being critical in the establishment of industries. This was the main route employed by South Korea; as its TNCs developed they subsequently become important in investing in other countries.
- Support has also been provided for what are best described as 'geopolitical' reasons. All four of the first wave of NICs were and are important strategic pieces in the political chessboard of the Pacific area and billions of dollars of US money has been directed towards them.

Figure 13.11 shows the change in GDP per person in the period 1965–95.

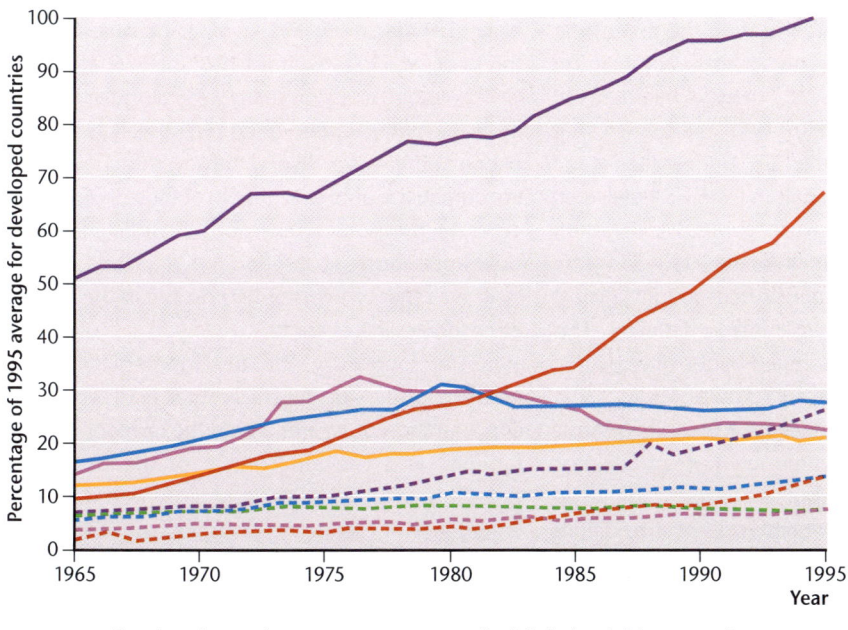

*Figure 13.11 Real GDP per capita in selected world regions*

Legend:
— Developed countries
— Western hemisphere excluding major oil exporters and Chile
— Major oil exporters excluding Iraq
— Middle East and Europe excluding major oil exporters
— Asian newly industrialised economies: Hong Kong, South Korea, Singapore, and Taiwan
-- Rapidly industrialising countries: Chile, Indonesia, Malaysia and Thailand
-- Africa excluding major oil exporters
-- Developing countries excluding Asian newly industrialised economies
-- Asia excluding China, Hong Kong, Indonesia, South Korea, Malaysia, Singapore, Taiwan and Thailand
-- China

# Chapter 13  Economic systems

## Case study 13.11  Asian Tigers: South Korea

Of all the Asian 'economic miracles' of the post-Second World War era, that of Korea most caught the attention of the USA and the rest of the industrial world. Companies such as Hyundai, Daewoo, Samsung and Lucky Goldstar became household names, like the Japanese companies before them, Honda and Sony. The expansion of companies such as these pushed Western balances of trade with countries such as Japan and South Korea into the red, and many economists were obliged to redefine their theories about industrialisation and development.

Korea was occupied by the Japanese between 1905 and 1945. Koreans were kept out of education and prohibited from most significant positions in industry and government. The Japanese developed heavy industries to the north of Korea and left light manufacturing in the south. After the Japanese were defeated in the Second World War, the peninsular nation was divided into north and south; South Korea was briefly occupied by the USA before being given its independence in 1948. Nonetheless it remained important for the USA, both militarily and as a shop-window of capitalism. The USA delivered massive amounts of aid (up to 18.5% of South Korea's GNP) in the aftermath of the 1950–53 war in which US-led United Nations forces fought against North Korea and its Chinese and Soviet allies. This inflow of capital was distributed by the government and became a major source of growth.

The main agency behind the Korean economic policy is the chaebol, a uniquely Korean business organisation. The chaebol are large, family-run companies with extensive networks of subsidiaries and political connections. Political favour is especially important to the chaebol, and extensive government intervention in the economy has been embodied in Five Year Plans, which were begun in 1962. These sought to organise the nation's productive factors and reduce competition among South Korean firms, in order to improve their ability to compete in the global market. Cooperation between government and industry was pursued in Korea between 1950 and 2000 and took the country's per-capita GNP through a more than 20-fold increase between 1965 and 1985. Much of the money came from foreign investment, initially Japanese and then increasingly global.

Foreign capital continued to be a vital source of growth. About 60% of the capital used during the first Five Year Plan came from abroad and the government actively encouraged the inflow with the Foreign Capital Inducement Promotion Act of 1960.

The third Five Year Plan was probably the most influential; it called for the development of Korea's heavy manufacturing and chemical industries. During this plan, which lasted from 1972 to 1976, growth of GNP averaged an astonishing 9.7% per year. Imports of heavy manufactured goods were seen as holding back on Korea's growth and the government chose to encourage the domestic growth of those industries in order to promote a sense of self-sufficiency. This policy of import substitution was pursued behind a tariff wall that would not be possible in the current atmosphere of 'free trade'.

In the long run, the development of heavy industry under the third Five Year Plan led to the concentration of wealth within the chaebol-owning families and the emergence of the dominant chaebol, which essentially crowded out the emergence of competing firms well into the 1980s. It was this emphasis on heavy industry and the resulting dominance of a small cadre of chaebol that led to the precipitous fall of the Korean economy in the late 1990s.

### What problems have been caused?

All economic development incurs some sort of costs. The rise of the NICs is no different in this respect. The costs are both economic and environmental, but might also be broadened to include the political and cultural impact of rapid growth. Some of that damage may well have been inflicted outside the NICs. For example, any global shift of employment to the NICs has frequently left a trail of unemployment in the original country of production. Thus the rise in Korean incomes has to be off-set against the reduction of incomes of workers in MEDCs who have been forced to take lower-paid jobs in the service sector.

Rapid growth has frequently been at the expense of political freedom and expression, while cultural values have been heavily subverted by Western-influenced values of conspicuous consumption, hedonism and individualism. This has occasionally led to a violent reaction. However, the best-documented negative impact has been environmental; rapid economic growth has taken place with only limited recognition of the need to protect the environment.

> **Examiner's tip**
>
> ■ Don't suggest that the NICs all have the same history of development. There is no single message to be drawn from their progress.

## What are the implications for theories of development and economic growth?

South Korea is often quoted, understandably enough, as a case study of economic success, and the example of South Korea certainly suggests that countries can break through the barriers to development, as suggested by the dependency theorist Andre Gunder Frank. However, for other countries the route to development is not, as some would have it, quite so obvious. For the neo-liberals, who are critics of too much government, South Korea is a poor example. For the same neo-liberals, who advocate the virtues of free trade, it again represents a poor choice given its early development history of protecting its industries. It can also be added that this sort of growth does not represent a real *expansion* of world output and markets if the only impact of the growth of Hyundai, for example, is that consumers buy their cars and not Fords. It does not lead to a growth in the world economy but simply shifts production from place to place.

### Case study 13.12  Problems of NICs: environmental issues in Taiwan

The process of economic growth in Taiwan, the country that was once called Ilha Formosa, the 'beautiful island', has made it one of the most polluted countries on the planet.

Some sections of Taiwanese society are now insisting that their quality of life, and that of future generations, should receive priority, even if this reduces the rate of economic growth. In the past, political protest against any government-sponsored development was difficult, punishable by imprisonment and worse. However, in a world of changing attitudes and a country determined to show itself as modern and freer than its large neighbour, China, this is no longer possible. In 1945 75% of economically active Taiwanese were farmers. Between 1952 and 1987 annual economic growth averaged an impressive 9%. As a result, income per person climbed from $145 in 1951 to $9805 in 1993 and $23 400 in 2003; it is now more than the income per head in Greece or Portugal.

The benefits of this growth have been widely shared among the Taiwanese people. In 1992 unemployment stood at just 1.5%. Life expectancy, at 74 years, is similar to that in established MEDCs. Nutritional levels are among the best in Asia, while almost universal educational opportunities have allowed the sons and daughters of peasants to become engineers, scientists, doctors, lawyers and business executives. In contrast to Japan and Korea, where large conglomerates predominate, enterprises with 30 workers or fewer account for 80% of industrial jobs.

# Chapter 13  Economic systems

But there has been a high environmental price to pay for all this. Among the negative entries in Taiwan's environmental account are the following:
- only 1% of human waste receives primary treatment
- the water in streams near the petrochemical complex in Hou Chin in southern Taiwan is combustible
- there is one factory for every km$^2$ of Taiwan
- air pollution in Taipei is among the worst in Asia
- Taiwan's farmers are among the world's heaviest users of agricultural chemicals — many consumers now look for fruit with insect bites to indicate that it is pesticide-free
- Taiwan has 15 times as many motor vehicles per km$^2$ as the US
- cancer is a leading cause of death — the rate doubled between 1960 and 1990

In 1986 the government approved plans by the US-owned TNC DuPont to build a $160 million plant making dye from titanium dioxide in the port town of Lukang. At the time it was the single largest investment project in Taiwan's history, but it was given the go-ahead without community consultation and without an environmental impact study. The community led a vigorous campaign that was influenced by the disasters at Chernobyl and Bhopal; ultimately the government and DuPont agreed to move the venture elsewhere and to conduct environmental and community opinion assessments.

In recent years, Taiwan's environmental movement has devoted much attention to the petrochemical industry, which accounts for about 20% of the island nation's GNP. Farmers and fishermen living near plants argue that hazardous wastes are destroying their livelihoods. Hou Chin residents began 3 years of sit-ins at the state-owned China Petroleum Company (CPC) complex in mid-1987 to protest against plans for a new 'naphtha cracker' (a plant to make the raw materials for plastics). In 1990 the government agreed to spend $55 million on additional pollution controls and community health benefits, and appointed a panel to monitor environmental practices.

Nuclear power is also a major environmental issue. The government chose nuclear energy as the solution to the 1973 oil crisis. Taiwan now relies on nuclear energy for 38% of its electricity, the highest dependence on nuclear power in the world after France. However, highly publicised safety problems, including serious accidents at existing plants, have led to widespread opposition to further expansion. The risks of nuclear power on a densely populated island are obvious. Two plants are located within 19 km of Taipei. Taipei has a population of 6 million, is in an area prone to earthquakes and typhoons, and is on the edge of a semi-active volcano.

Taiwan's authorities have looked for easy answers to the problem of radioactive waste. Until recently they shipped it to Orchid Island, off Taiwan's southeast tip. Orchid Island is home to 3000 Yami aborigines; residents have reported high rates of leukaemia and cancer due to radioactive leaks.

Taiwan's radioactive waste is now destined to be buried in a disused coal mine at Pyungsan in Hwanghaedo, North Korea. Pyungsan is hardly suited to the purpose: the area is geologically unstable and earthquake-prone. Furthermore, the mine has an underground lake, increasing the chances of nuclear containers corroding and of radioactive materials, such as caesium and tritium, seeping into the surrounding area. Its not hard to imagine the likely effect on the health of local people.

---

Two of the NICs (Hong Kong and Singapore) are rather odd examples, in that they are really city states. They have no real territory other than the city itself — they are trading centres that are not easily comparable with countries such as South Korea and Taiwan, both of which have had considerable American influence in the last 60 years.

Education is often seen as pivotal to the rise of the Tiger economies. Taiwan and South Korea undeniably 'educated ahead of demand', even at the risk of substantial educated unemployment, and it is widely assumed that the other fast-growing economies of southeast Asia (Singapore, Malaysia, Thailand and Indonesia) did the same. In fact, recent research suggests that the course of educational development in these four countries has been very different from that in Taiwan and South Korea.

## Case study 13.13 Singapore and hi-tech industry

Singapore is an important trading centre for the southeast Asian region. It has a highly developed and successful free market economy and strong service and manufacturing sectors. Singapore has a tradition of authoritarian governments that have taken a central role in the planning of its economy. The economy has always depended on international trade and on the sale of services. Singapore's major industries include petroleum refining, electronics, oil drilling equipment, rubber products, processed food and beverages, ship repair, financial services and biotechnology.

### An educated labour force

Other than its location as a port, the only resource that could be a basis for the economic development and prosperity of Singapore was its labour force, more specifically a trained and educated labour force. Singapore was unable to compete, for any length of time, on the basis of labour costs, given the presence of neighbouring countries with far lower costs. Instead, Singapore chose the strategy of creating technical skills that were unavailable elsewhere in southeast Asia and the Pacific rim.

Table 13.7 shows the changes in employment in Singapore compare with OECD (Organization for Economic Cooperation and Development) countries.

### Foreign investment

Local industry in Singapore was limited to trade and did not have the capability of creating export industry given the very small population and high start-up costs. Singapore, under the leadership of Lee Kuan Yew, sought to bring in foreign industry. His first aim was to make potential employers aware of the stable political climate. TNCs do not want to risk investing millions of dollars in facilities and factories in a country where, with a change of political climate, assets can be taken. The laws in Singapore might not be exactly to the liking of foreign companies, but they would be fairly enforced and are clearly committed to corporate capitalism. This proved to be a highly attractive feature of Singapore.

### Political stability

Singapore is a rare example of an 'administration state' run by one party since its independence. The political mood is apathetic with a form of 'self-

|  | Change in number (thousands) | | Percent change | |
|---|---|---|---|---|
| Industry | Singapore | OECD countries | Singapore | OECD countries |
| Total | 503 | 35 940 | 46.8 | 15.5 |
| Manufacturing | 112 | −1 646 | 34.6 | −2.9 |
| Construction | 32 | 1 339 | 45.1 | 7.4 |
| Commerce | 128 | 9 093 | 55.9 | 20.1 |
| Services | 244 | 31 404 | 57.6 | 34.6 |
| Transport and communications | 39 | 2 157 | 32.3 | 16.1 |
| Finance and business | 87 | 9 731 | 104.0 | 62.4 |
| Community, social and personal | 118 | 19 515 | 53.6 | 31.6 |

Source of basic OECD data: ILO Yearbook of Labour Statistics

Table 13.7 Employment changes in Singapore compared with OECD economies, 1980–92

# Chapter 13  Economic systems

censorship' imposed through an apparatus of institutions that allows people to channel opinions direct to government. The government is seen to be responsive to the population while heavily discouraging the development of opposition parties, the leaders of which often end up living abroad. The tax system was also attractive to foreign companies, often giving lower tax rates for foreign investment than for local residents.

Another key element of Singapore's development strategy was the upgrading of infrastructure — streets, roads, an airport and port facilities. The upgrading was financed not primarily by borrowing but by a special infrastructure tax.

## Raising income levels

The intention was to raise income levels of Singaporean labour. When the programme of attracting TNCs was proving to be successful, the Singaporean government increased the minimum wage levels, which provoked a serious downturn in the economy as investment stalled. The higher wages were an obvious discouragement to foreign companies considering Singapore as a potential location, but, more seriously, the move threatened Singapore's carefully developed reputation as a politically stable environment.

## Raising skill levels

Having learnt from its mistake, the government adopted an alternative strategy to increase wages. The plan was to improve the training of Singaporean workers through government training institutes. A typical training programme would meet twice a week for 3-hour sessions over a 2-year period. The training was voluntary and free and it was geared to the needs of the companies operating in Singapore at that time.

Apple Computers was one TNC that located facilities in Singapore. Initially Apple just produced electronic boards in Singapore for shipping to the USA where they were assembled into computers. In response to the increasing supply of skilled personnel in Singapore coming through the government training institutes, Apple decided to produce its entire computer in Singapore. The newly trained and highly motivated Singaporean workers not only replicated the old production process successfully but also made improvements that further lowered costs. This was an example of an innovative culture that could improve and modify rather than simply copy. The government training programme proved so beneficial to employers that they made little fuss over the tax imposed on them to help pay for it.

## Singapore today

Singapore now has:
- science parks to share research between government and industry
- a national 'computer board' to promote the computerisation of schools, offices and homes
- a 300% increase in student capacity in the two engineering universities
- the creation of a $50 million venture capital fund to encourage Singaporean start-up companies

Table 13.8 shows the link between wages and GNP for a range of countries. Note that the countries are listed in order of increasing GNP.

Singapore is one of several possible examples of a country that defies immediate classification. As the figures in Table 13.8 suggest, it is perverse to call it an LEDC and yet aspects of its political system prevent most analysts from placing it firmly in the box labelled MEDCs.

| Country | Hourly wage (US$) | GNP per capita (US$) |
| --- | --- | --- |
| Mexico | 1.099 | 3 970 |
| South Korea | 4.774 | 7 970 |
| Spain | 9.565 | 14 080 |
| Canada | 11.594 | 20 020 |
| UK | 15.073 | 21 400 |
| France | 9.454 | 24 940 |
| Germany | 15.219 | 25 850 |
| USA | 13.490 | 29 340 |
| Singapore | 7.945 | 30 060 |
| Japan | 11.522 | 32 380 |
| Norway | 16.629 | 34 330 |
| Switzerland | 20.091 | 40 080 |

*Table 13.8 Wages and GNP in US dollars, 1998*

*Human systems, processes and patterns*   **A2 Unit 5**

Strong government has been a unifying feature of most of the Tiger economies. Several of them were under military rule for many years, although these days Taiwan, South Korea and Thailand are among the most democratic countries in Asia. The government has had a key role in planning, demographic management and investment in all of these countries. Governments have also used national identity and loyalty as political weapons to ensure the compliance of their people.

### Examiner's tip
- Avoid using terms like MEDC and LEDC without qualifying them. They are very broad categories and many countries cannot be easily placed in either because there are wide variations of wealth within them.

### Key terms
**Hi-tech industry** Manufacturing industries involving a high level of sophistication both in research and development and in the finished product.

# The emergence of large trading blocs

The emergence of large trading blocs in recent years is part of the wider process of globalisation. In some ways globalisation is a natural development from large trading blocs into a unified world market. At the simplest level, free trade associations have been set up between countries, such as Mercosur in South America. At the other extreme is the overarching trend towards full political integration which is occurring, for example, in the European Union (Table 13.9).

Globalisation involves the removal of barriers to movements of capital, goods and labour. There is much resistance, both political and cultural, to freeing up the movement of labour because this would imply no restrictions on immigration.

|  | No internal trade barriers | Common external tariff | Mobility of labour, capital and investment | Common currency | Common economic policy |
|---|---|---|---|---|---|
| Free trade area | ✓ | | | | |
| Customs union | ✓ | ✓ | | | |
| Single market | ✓ | ✓ | ✓ | | |
| Monetary union | ✓ | ✓ | ✓ | ✓ | |
| Economic union | ✓ | ✓ | ✓ | ✓ | ✓ |

*Table 13.9 The stages of development of economic cooperation between nation states*

- The origins of trading blocs are found in the postwar economy, although the imperial system of free trade was an obvious precursor.
- The original super-state was the USA, which provides one model of integration into a continental market. The American civil war was a battle between two

441

# Chapter 13 Economic systems

groups with very different economic interests: the south had an export economy based on plantation crops and the north had import-substituting manufacturing industries.
- The theoretical advantages of free trade are rooted firmly in classical economic doctrine, laced with neo-liberalism. The eighteenth-century theories of Adam Smith and David Ricardo have been recycled and modified by twentieth century economists, such as Milton Friedman, who are advocates of the 'free' market.
- The economic theory behind the implementation of free trade is based on the law of comparative advantage and the principle of economies of scale.

### The law of comparative advantage

The **law of comparative advantage** states that countries should specialise in producing and exporting only the goods that they can produce at a lower relative cost than other countries. Where a country is more efficient in the production of two commodities than another country, it should specialise in the commodity in which its comparative advantage is greater. The logic of this was first expounded by David Ricardo and has been used ever since as the justification for free trade with no tariff barriers. It is suggested that this allows countries to specialise in those things that they do best or, more properly, least badly.

Until the recent past this argument met resistance on the grounds of national security. In times of war, the interruption of trade might leave a country exposed to shortages of food or other critical goods such as steel or machine tools if it was not making these products itself. In the new world that has emerged since the end of the Cold War this no longer seems a realistic objection. However, many critics argue that Ricardo and the other classical economists did not intend a world dominated by TNCs, which are the main drivers of globalisation and integrated markets. On the contrary, they assumed a world dominated by small firms in perfect competition with one another. This is not what happens in free trade areas.

### Economies of scale

The principle of **economies of scale** states that the greater the scale of production the lower the average costs of producing each unit. Fixed costs such as heating and lighting are not likely to increase if production goes up within the same factory; nor will more security or cleaning staff be needed. However, the variable costs will rise as more labour is employed and more components are used. Overall, the cost of production is the fixed cost added to the variable cost; hence the average cost of production reduces as production units (factories) increase their output. Such economies of scale give a huge advantage to large countries with large markets that allow big production runs. It explains why, in an era of tariffs and quotas, only large countries with large domestic demand maintain a domestic car industry.

The application of this theory which results in free trade areas is based on the presumption that if all barriers are removed, production will become more efficient, bringing benefits to all. Large trading blocs, according to this theory, are in everyone's interest.

*Human systems, processes and patterns*  **A2 Unit 5**

Practical free trade policies *always* include:
- the removal of tariffs
- the establishment of trade rules, e.g. anti-dumping procedures
- the deregulation of services, e.g. the USA frequently considers state services to be a subsidy to employers, and therefore unfair
- the treatment of foreign investment as though it were national, thus removing local content provisions such as the 60% rules that exist in the car industry
- the maintenance of intellectual property rights, ensuring copyright regulations are strictly enforced

### Key terms

**Comparative advantage** The principle that by trade all countries will benefit even when they are comparatively less efficient in the production of commodities than their neighbours.

**Economies of scale** The principle whereby the greater the output of a company or factory the lower is the average cost of producing each unit of production because fixed costs such as heating or lighting costs are spread more thinly.

### Case study 13.14  A free trade area: NAFTA

NAFTA, the North American Free Trade Agreement, came into effect in 1994. Under NAFTA, the USA, Canada and Mexico became a single, giant, integrated market of almost 400 million people with $6.5 trillion worth of goods and services traded annually. Mexico is the world's second largest importer of US-manufactured goods and the third largest importer of US agricultural products. NAFTA is a trading bloc or customs union, and there is not a free flow of labour across the national borders; these remain intact.

Before NAFTA, Mexican tariffs averaged about 250% compared to US duties. After the pact, about half the tariffs on trade between Mexico and the USA were eliminated, and the remaining tariffs and restrictions on service and investment are being phased out over a 15-year period. The USA and Canada have had a free trade agreement since 1989.

A number of factors led up to NAFTA. By far the most important was the relative decline in US international competitiveness. Central to this was the rise of Japan and the recovery of Europe as serious rivals in the US domestic market as well as the global market. By the 1970s it was clear that the international trade system the USA had constructed after the Second World War was no longer working solely in the interest of American corporations. This was despite the USA emerging from the Second World War as the dominant power.

Multilateralism, which in the 1950s and 1960s favoured the expansion of US corporations across the globe, also favoured their foreign rivals. Free trade meant that the USA now faced competition from the revitalised economies of western Europe and Japan. It was losing out on its traditional bedrock industries: cars, consumer electronics, and textiles and apparel.

Officials in all three countries argue that the strength of NAFTA markets has been one of the brightest spots in the economies of all three countries in recent years, especially for US farmers, agricultural exporters and the industries that support them. Together, the NAFTA partners of the USA, Canada and Mexico now purchase 27% of US agricultural exports. Farmers in all three countries, it is argued, benefit from NAFTA. Two-way agricultural trade between the USA and Mexico has increased more than 55% since 1994, reaching more than $11.6 billion in 2004 (Figure 13.12). Although US imports have grown under NAFTA, so have US exports. Without NAFTA, it is argued that the USA would have lost these expanded export opportunities.

# Chapter 13 Economic systems

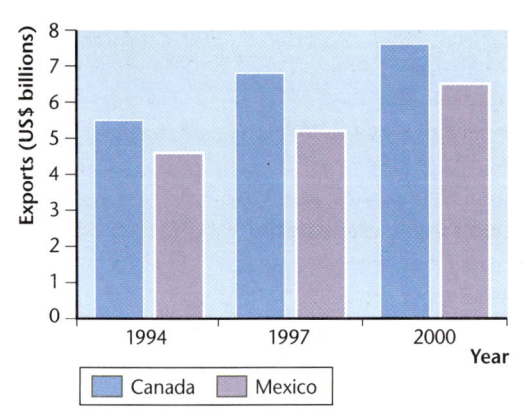

*Figure 13.12 US agricultural exports to NAFTA trading partners, Canada and Mexico*

On the other hand, NAFTA is blamed for a wide range of economic ills that have befallen the North American economies. Its proponents argued that it would create jobs in the USA. By July 2002, an official 400 000 people had claimed for relief on the special programme of benefits the government had allowed, because it expected some job losses. This figure is almost certainly an underestimate of job losses given that by no means all those who lose jobs either claim the benefit or are able to do so. Service workers are not eligible, nor are workers in a car parts plant who lose their jobs because the main plant they serve moves to Mexico.

Employment grew in the US through the 1990s, but this was largely in the service sector. One and a half million jobs were lost in manufacturing, while service employment grew by over 10 million. Average wages in the service sector are only 77% of those in manufacturing. Not all of these problems can be laid at the door of NAFTA, but there is no doubt that NAFTA has made it much easier for companies to flee to the maquiladoras.

Maquiladoras are US-owned assembly plants in Mexico that employ Mexican labour to make products mostly for export back to the USA. In 1965 the Mexican government set up the Border Industrialisation Programme, which created export platforms for US companies on favourable terms. The word maquiladora is derived from *maquilar*, which means 'to submit something to the action of a machine'. When US companies opened factories in Mexico, the name evolved to refer to the process of assembling parts manufactured elsewhere. Approximately 2100 maquiladora plants produce electronic goods, auto parts, chemicals, furniture, machinery and other goods. The number has increased more than fourfold since the mid-1980s. About 600 000 workers are employed in the maquiladoras and 90% of the plants are US-owned. Many of the largest US corporations have maquiladora plants, including AT&T, Ford Motors, General Electric, General Motors, DuPont, Eastman Kodak, Emerson Electric, IBM, ITT, PepsiCo, United Technologies and Xerox. Maquiladoras receive government subsidies such as preferential tariffs and taxation. They pay no tariffs on materials and semi-finished products imported into Mexico. When they ship finished products back to the USA, they pay tariffs only on the value added in Mexico, not the value of the entire product. Maquiladoras have continued to grow since the formation of NAFTA in 1994. The reduction in employment in US manufacturing has been significant, as has the growth of manufacturing employment in Mexico.

The Mexican government has often proclaimed NAFTA a success and for some Mexicans it clearly has been. However, growth overall has been disappointing at only 1% per annum through the 10 years of NAFTA, and the export figures hide the fact that most of the profits go to US-owned companies. Three of the five largest export companies in Mexico are US car companies and 54% of Mexico's exports are either maquiladora production or petroleum. There has been a disappointing 'trickle down' of these benefits to the nation as a whole.

The impact in agriculture has been even more dramatic. Mexico now imports much of the huge US corn surplus: 8.8 million tonnes in 1993, 20.3 million tonnes by 2000. Agriculture is heavily subsidised in the USA while Mexican farm aid has been reduced. Critics say that the failure of NAFTA to deliver on its promise of a better life for Mexicans represents more than just a misplaced faith in free trade.

Behind the laissez-faire rhetoric, Mexico's neo liberals, mostly US educated, were pursuing a large-scale programme of government social engineering aimed at forcing Mexico's rural population off the land and into the cities, where it could provide cheap labour for the foreign investment that the new open economy would attract. Of course this is not, and was not, admitted to be a policy. On the contrary, the Mexican government promised that as tariffs on US agricultural products fell, generous financial and technical assistance would enable small farms to increase their productivity in order to meet the new competition. But, after the treaty was signed, the reformers reneged on this. Funding for farm programmes dropped from $2 billion in 1994 to $500 million by 2000. Rural depopulation has continued, as has the pressure to migrate to the USA.

The hope that NAFTA would bring a better balance to Mexico, with a growing middle class generating wealth at home, has not been realised. For the first 7 or 8 years the maquiladoras boomed, but in the last 2 years an estimated 200 000 maquiladora jobs have left Mexico for China, where workers are paid one-eighth the Mexican wage. In a deregulated world, there is always someone who will work for less.

# The emergence of a new international division of labour

## The old world order

**Colonialism** is the subjugation by physical force of one culture by another, generally through military conquest of territory. Beginning in the early sixteenth century, European adventurers, sailors and ultimately colonists travelled along African coasts to the New World and across the Indian Ocean and the China Seas seeking fur, precious metals, slave labour, spices, tobacco, cocoa, potatoes, sugar and cotton. The European powers, initially Spain and Portugal, but followed by the Netherlands, France, and Britain, exchanged manufactured goods such as cloth, guns and implements for these products and for the African slaves they transported to the Americas. In the process they reorganised the world.

One aim of the colonists was to establish specialised extraction of metals, especially gold and silver, in the colonies. These metals were not found in North America. Here the production of primary agricultural products which could not be grown in Europe or were in short supply there, such as sugar, cotton and tobacco, was encouraged instead. European manufacturing was able to expand using these products, which became both industrial inputs and foodstuffs for its labour force. On a world scale, this specialisation between European economies and their colonies came to be termed the colonial, or first international, division of labour.

The colonial division of labour had two basic effects: it stimulated European industrialisation and it forced non-Europeans into primary commodity production. Such specialisation disorganised the non-European cultures, undermining local crafts and mixed-farming systems as well as artificially dividing up their lands. Not only did non-European cultures surrender their own local industries in this exchange, but local farmers were often forced to become producers of a single crop for export.

# Chapter 13 Economic systems

## The new international division of labour

In the twentieth century the colonies gained their political independence, and at the same time increasing tensions in the postwar economy culminated in the crisis of Fordism. From this a new division of labour emerged.

- The core MEDC countries dominated the development and production of more sophisticated products with high research and development costs and relatively high skill requirements in the labour force — examples are aeronautics, automobiles and pharmaceuticals.
- A semi-periphery of NICs emerged that took over capital industries such as iron and steel and some consumer industries that had high labour costs and were not readily mechanised, such as clothing, footwear and toy production.
- A periphery of raw-material-producing economies remained, by and large still locked into the system of economic colonialism. They depended on large MEDC-based TNCs for the expertise and capital to develop export crops or extract raw materials.

Table 13.10 shows the international division of labour at the end of the twentieth century.

Table 13.10 International division of labour at the end of the twentieth century

| Global region | Primary (%) | Secondary (%) | Tertiary (%) | Quaternary (%) |
|---|---|---|---|---|
| Sub-Saharan Africa | 60 | 10 | 28 | 2 |
| Arab countries | 36 | 20 | 41 | 3 |
| South Asia | 56 | 19 | 22 | 3 |
| East Asia | 64 | 18 | 14 | 4 |
| Southeast Asia | 50 | 21 | 25 | 4 |
| Latin America | 22 | 23 | 52 | 3 |
| North America | 8 | 23 | 60 | 9 |
| Eastern Europe | 18 | 36 | 43 | 3 |
| European Union | 6 | 25 | 62 | 8 |

In more recent times this picture has become much more complex.

- Quaternary services, especially research and development, are found in the core world regions of Japan, Europe and the USA. So too are some of the higher-order manufacturing sectors.
- A growing level of manufacturing activity is found in the NICs, where labour costs are cheaper. However, both Singapore and Taiwan have developed significant higher-level manufacturing sectors, particularly in electronics. Examples are Apple in Singapore and Acer (a Taiwanese TNC) in Taiwan.
- Primary production is still dominated by peripheral countries, but this is complicated by the fact that, for example, the USA is the world's major exporter of wheat.
- The new information-led economy works on a global scale, but markets are far from fully integrated. Governments in MEDCs and their associated regional trading blocs (e.g. the EU) still play crucial regulatory roles, keeping the most important elements under their control.

*Human systems, processes and patterns* **A2 Unit 5**

- There are regional differences within the three areas of influence: North America, the European Union and the Asia Pacific region (dominated by Japan). There is competition between them, but there is an increasing blurring of the significance of the 'national' origins of investment (Figure 13.13).

The following quote from Castells (1996) sums this up:

> Around this triangle of wealth, power and technology, the rest of the world becomes organised in a hierarchical and asymmetrically interdependent web, as different countries and regions compete to attract capital, human skills and technology to their shores.

The global capitalist economy is also divided according to dominant technology:
- producers of high-value goods based on 'knowledge' and skills: this is the quaternary and hi-tech sector
- producers of high volumes based on lower-cost labour, such as trainers and T-shirts
- producers of commodities such as coffee, cotton and cocaine
- producers of materials based on natural resources, such as diamonds and timber
- producers who remain outside the global capitalist system, such as subsistence farmers in Somalia

This division does not coincide with particular countries: in all countries there is this type of division, but the distribution and density of the various categories varies greatly.

Two other factors need to be considered to give this picture a little more detail.
- The emerging significance of global cities that are often the control centres of this system of production. The dominant cities are New York, Tokyo and London — these are the locations of transnational headquarters, including banks, and of the financial markets that control the flow of capital.
- The central role of the TNCs in both the development of this system and its further refinement.

There are, of course, major generalisations involved in this sort of analysis.

The quaternary sector and hi-tech industries tend to be concentrated in particular areas. For example, in Europe there are obvious concentrations of hi-tech industries along the M4 corridor and in 'silicon fen' (around Cambridge

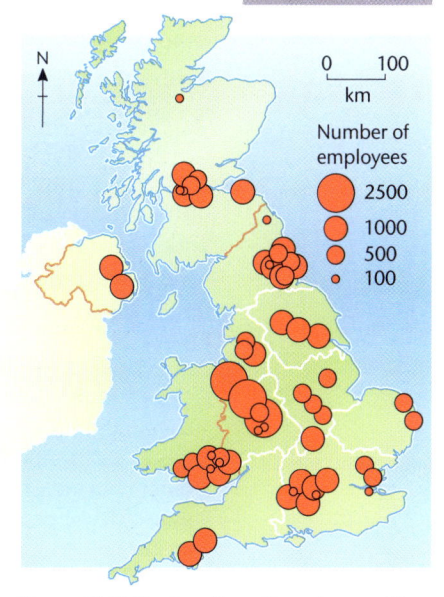

*Figure 13.13 The location of employment in Japanese manufacturing companies in the UK, 1998–99*

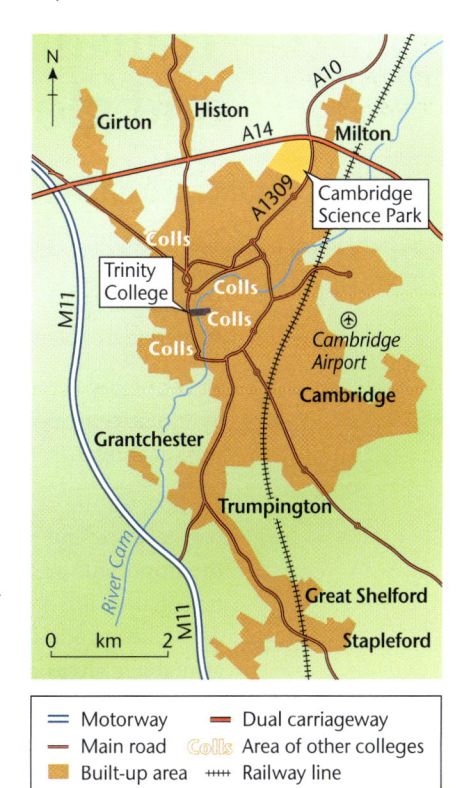

*Figure 13.14 The location of the Cambridge Science Park*

# Chapter 13 Economic systems

with its Science Park) in the UK (Figure 13.14), in Toulouse in southwest France (with strong co-locational ties with the aircraft and space industry), in Grenoble in eastern France, and in other distinctive clusters in northern Italy and Germany.

There are pockets of cheap labour and thus sweatshop-style industry in global cities. For example, there are significant clothing and textile sectors, sometimes only semi-legal in their operation, in the inner-city areas of London and New York. These are sometimes associated with the growth of home-working and the revival of a 'putting-out' system, in which companies deliver material to be made up into clothing at the same time as they collect finished garments from workers' homes.

### Examiner's tip
- Don't confuse decline in employment with decline in output. An industry might employ fewer people but enjoy an increase in output. The UK car industry is a case in point.

### Key terms
**Colonialism** When one nation state absorbs fresh territory by force of arms or by treaty.

**Neo-colonialism** When one nation state retains economic control of another nation state despite the latter having political independence.

# Transnational corporations and globalisation

## Transnational corporations

Transnational corporations (TNCs) are:
- large
- managed by professionals who do not own the company
- operate in several different countries

In his highly eccentric *Devil's Dictionary*, Ambrose Bierce defined a corporation as 'An ingenious device for obtaining individual profit without individual responsibility.'

The key feature of the modern business corporation, which distinguishes it from partnerships and sole proprietorships, is that it gives the owners of the business the important legal protection of 'limited liability'. This means that the owners' personal wealth is not at risk should the company go bankrupt. During the Middle Ages, the corporate form of ownership for business organisations was generally available only as a special privilege awarded by the government to a handful of royal favourites. Changes in the legal codes of Great Britain, the USA, France and other economically progressive countries as capitalism advanced in the nineteenth century saw the removal of most obstacles to the corporate form of ownership. This made it much easier to raise capital for new business and industry.

TNCs have been around for more than 100 years, but their recent growth has been remarkable. As an indication of the early origins of TNCs it is worth recalling that the first cars made in the UK were produced by the German company Daimler. Although many modern TNCs bear the name of a country on their logos (for example, British Petroleum or BP) and they frequently have their headquarters in their country of 'birth', they owe no special loyalty to those countries. Only recently British Airways contemplated a name change, which involved dropping 'British'. TNCs are driven by the need to maintain profitability to reward shareholders, so they are constantly searching for more 'efficient' ways of production and cheaper places for that production.

A mere 500 TNCs produce 30% of planetary output and account for 70% of trade (Table 13.11). Some people believe TNCs are so powerful that they threaten the sovereignty of national governments.

*Table 13.11 The world's largest companies, 2002*

| Rank | Company | Home country | Total revenue (US$ million) |
|---|---|---|---|
| 1 | Wal-Mart | USA | 219 812 |
| 2 | Exxon Mobil | USA | 191 581 |
| 3 | General Motors | USA | 177 260 |
| 4 | BP | UK | 174 218 |
| 5 | Ford | USA | 162 412 |
| 6 | DaimlerChrysler | Germany | 136 897 |
| 7 | Royal Dutch/Shell | UK/Netherlands | 135 211 |
| 8 | General Electric | USA | 125 913 |
| 9 | Toyota | Japan | 122 814 |
| 10 | Citigroup | USA | 120 022 |
| 11 | Mitsubishi | Japan | 105 813 |
| 12 | Mitsui | Japan | 101 205 |
| 13 | Chevron Texaco | USA | 99 699 |
| 14 | Total Fina Elf | France | 94 311 |
| 15 | Nippon Tel/Teleph | Japan | 93 424 |
| 16 | Itochu | Japan | 91 176 |
| 17 | Allianz | Germany | 85 929 |
| 18 | IBM | USA | 85 866 |
| 19 | ING | Netherlands | 82 999 |
| 20 | Volkswagen | Germany | 79 287 |
| 21 | Siemens | Germany | 77 358 |
| 22 | Suminoto | Japan | 77 140 |
| 23 | Philip Morris | USA | 72 944 |
| 24 | Marubeni | Japan | 71 756 |
| 25 | Verizon Communications | USA | 67 190 |

# Chapter 13  Economic systems

## Case study 13.15  TNCs: Levi Strauss

The clothing industry is one of the oldest manufacturing industries and also one of the most geographically dispersed. It produces a huge range of products. The diversity of human size, taste and need means that it has small product runs and has proved remarkably immune to technical innovation. Textiles are not resistant materials — they cannot be bent into shape — and clothing is three-dimensional. Only T-shirts cut from flat cloth can be readily mass-produced by machines. Most production methods are much as they have been for centuries, millennia even, with women working at sewing machines, albeit now powered by electricity rather than hand. As a consequence, labour costs represent anything between 40% and 60% of the total cost of production. It is thus unsurprising that this industry has a long history of searching out sources of cheap labour. Initially this occurred within national boundaries, but falling transport costs mean it has now moved beyond them.

### Levi Strauss

Levi Strauss is one of the best known clothing manufacturers. Given the huge market for clothing in the world, no single company, however well-known, has a large market share of clothing, unlike the position of Ford or Toyota in the car market. However, some do have a significant part of certain niche markets. In the case of Levi Straus that is the jeans market.

The San Francisco-based company was founded during the 1850 California gold rush by Levi Strauss. The company made denim trousers for miners and other workers. Jeans became fashionable in the 1950s, and even more so in the 1960s, when they became a kind of uniform. As Levi Strauss enjoyed this period of rapid growth in sales both at home and abroad, it kept many of its plants on US soil, unlike some rival companies which took advantage of lower production costs overseas.

### Changing fortunes

In the late 1980s and 1990s, Levi jeans lost their appeal. Market share among younger consumers had plummeted. Levi was slow to catch on to the taste for wider legs and a lower cut on the hips. The company reported a fall in sales of 13% in 1998 alone as its market was eaten away by cheaper rivals at one end and the entry into the market of designer jeans marketed under Tomy Hilfiger, Calvin Klein and other 'designer' labels.

The company responded by finally accepting the need to move production overseas and closed ten of its eleven US factories: four factories in Texas, two in Tennessee, and one each in Georgia, North Carolina, Virginia and Arkansas, leaving Levi with just a tiny US manufacturing presence, a plant in San Antonio, Texas, which was kept to produce quickly in response to sudden shifts in demand or taste.

### Job losses

One of these closures was near Blue Ridge, a town of nearly 2000 people in north Georgia, just south of the Tennessee and North Carolina state lines. Residents are mostly of Scots-Irish descent. The factory had been producing at full capacity for 50 years and was the economic heart of the town, employing 400 workers, mainly female. For most of these women, job prospects are grim as the town becomes another abandoned 'industrial' settlement.

These 400 workers are a small fraction of the hundreds of thousands of American garment workers who have lost their jobs in the last few years as the industry has deserted a country of high wages for cheaper alternative locations. In 1950, 1.2 million Americans were employed in apparel manufacturing. By 2001, that figure had fallen to 566 000. In 1989, the USA imported $24.5 billion worth of clothes; in 2001, the figure was $63.8 billion. In the last quarter of 2001, 83% of all apparel sold in the USA was imported. The Levi Strauss example is particularly symbolic because this company had a reputation for good management and strong relations with employees.

### The race to the bottom

As soon as wages rise in one country, work can be moved to another. Some observers have called this a 'race to the bottom' of the wage scale. Wage rates in LEDCs vary: in Guatemala they are 37 cents an hour; in China, 28 cents; in Nicaragua, 23 cents; and in Bangladesh, 13–20 cents. In addition to low wages, manufacturers in many countries 'benefit' from child labour and long working hours as well as the absence of health plans, environmental protection, workplace safety standards and union efforts to organise workers.

Levi now calls itself a 'marketing company', and has stated that future production in almost all cases will be by contract manufacturers. It will take place in 50 countries, including Mexico, Bangladesh and China. This is often taken to mean that a company can use cheap labour without the attendant bad publicity because it doesn't directly employ these people. To its credit, Levi Strauss has been a pioneer in its management of overseas contracting, creating a corporate code of standards for every manufacturer with which it contracts. Levi Strauss also pays inspectors to enforce the standards, although it is not possible for a corporation to enforce its code in a foreign state even if it has direct ownership of the factories and sweatshops.

## The last stop?

So in the race to the bottom, is Mexico the last stop? In the sand-blown Mexican border town of Piedras Negras, 2 hours southwest of San Antonio, they are only too well aware of the answer: no. In 1998 the Levi jeans factory closed and the jobs moved to Central America and the far east. It is claimed that 4500 garment jobs have disappeared from that small city in the past 3 years and wages have gone from $4 per hour 10 years ago to an average of 80 cents today. Further south, the town of Tehuacán, in the state of Puebla, southeast of Mexico City, was known principally for its mineral waters until the 1970s, when it began to attract workshops that produced garments primarily for the domestic market. But with the coming of the North American Free Trade Agreement (NAFTA) in 1994, many US firms sought locations here. There are now 60 registered maquiladoras that pay taxes, and up to 500 illegal sweatshops in the town. Maquiladoras range in size from a handful of workers to over 1000. Sixty per cent of the industry is denim, with the material usually produced and cut in the USA, then shipped to Tehuacán for sewing. US-based companies, like Guess? Inc, the VF Corporation (Lee and Wrangler), Calvin Klein, Sun Apparel, Ditt and Kellwood contract with local plants to sew their jeans. Despite this, Mexico is falling well behind.

A better candidate for last stop might be Bangladesh. In 2000, Bangladesh companies made 924 million garments for US companies, including Levi Strauss, with a wholesale customs value of $2.2 billion. Recently, however, the Bangladesh minister of commerce complained that wages in other countries, such as China, were undercutting workers in his nation. This carousel is not surprising, given the logic of globalisation.

## Case study 13.16  TNCs: Toyota

The Toyota Motor Corporation is the fastest growing auto manufacturer in the world. The company had a 16% increase in revenues between 1998 and 1999, and was the ninth largest corporation in the world with revenues of $122.8 billion in 2002. Toyota manufactures on six continents and has become increasingly global in its manufacturing operations in the past 30 years, shifting much of its production capacity from Japan to a number of LEDCs and, above all, the USA in order to develop new markets. Most production takes place in Japan, the USA, Canada, Brazil, New Zealand and Thailand. Toyota also produces at Burnaston in the UK. It is not known for technical innovation in its design, but it is renowned for its production methods, sometimes described as lean production, sometimes as just-in-time production (*kanban* in Japanese).

### History

The Toyota organisation was founded by Sakichi Toyoda. As with many car companies, it began in an older branch of mechanical engineering, in this case the manufacture of automatic looms. It made its first cars in 1935 as a division within the loom company.

# Chapter 13  Economic systems

The Japanese government adopted strongly protectionist policies for these infant industries and established tariffs and quotas to keep out foreign imports. This allowed Toyoda to form an independent Toyota motor company in 1937, with the subtle change of name.

After the devastating defeat of Japan at the end of the Second World War, the Toyota company escaped the loss of management that was imposed on some of the Japanese zaibatsu by the occupying Americans. It was thus able to resume car production almost instantly.

### Just-in-time

The Japanese studied US production methods carefully, paying particular attention to Fordism. Toyota began to incorporate Fordism into its production, but also to adapt it significantly, evolving a just-in-time system that recognised the central role of stock control. It was stimulated to do this because of falling sales — thus the development of just-in-time production was a response to a crisis, just as Fordism was a response to an earlier crisis.

The philosophy of just-in-time production is best described as continuous improvement. This had to involve the workforce, which was given a role in production not apparent in Fordist methods. Fordism strongly dehumanised the workforce, giving it simple and very repetitive tasks. Just-in-time production was led by demand; thus it had to become flexible, with parts and components flowing into the factory as they were needed and vehicles flowing out as they were sold. This kept to a minimum stocks of parts and hectares of space occupied by vehicles awaiting sale. It also meant that the inflow of money for sales kept up with the outflow of money used to purchase components. This had considerable geographical implications for suppliers, which had to be able to guarantee quality (100% defect free is the rule) and supply of the exact number of parts where and when they were needed. Suppliers had to develop close relationships with the mother company. Just-in-time production led to clustering.

### Foreign investment

Toyota gradually expanded its domestic market through the 1950s, and opened its first foreign subsidiary in California in 1957. This attempt at establishing an American market failed as early models were both too small and too under-powered for US roads and driving conditions. In 1959 Toyota's first foreign production began in Brazil with Land Cruisers. Toyota began exports to the middle east and Africa during the 1960s, and local production in South Africa in 1964. Political factors intruded, because of Japan's opposition to apartheid, and foreign direct investment (FDI) was limited. Toyota also established local assembly in Ghana in 1969, partly to escape the imposition of tariffs on imported vehicles.

Global sales grew throughout the 1970s and 1980s, and Toyota became the third largest automobile company in sales after Ford and General Motors. Having established a market for its models in the USA, Toyota restarted full production there by establishing a joint venture with General Motors. While other US car manufacturers were closing plants during the 1980s, Toyota opened an $800 million plant in Kentucky and introduced a smarter Camry in an attempt to steal market leadership away from the Honda Accord. In 1990, between its luxury line, trucks and inexpensive models, Toyota sold over 1 million vehicles. Toyota has now expanded production into Europe and into the luxury car market, seeking to boost revenues to even higher levels.

## Globalisation

**Globalisation** is the process whereby the whole planet is being incorporated into the capitalist system. It involves an increasingly free flow of capital (international investment), reduction of tariffs and the growth of free trade. It is not entirely new, for at the height of the British empire a similar system prevailed, albeit within the empire rather than between the empire and other nation states. In some respects the world economy was more integrated before the First World War. There was even a single currency — gold. But today, globalisation has gone further in some areas. For example, the ratio of exports to total output has reached 17%, financial markets are highly integrated and technology is being transferred rapidly from country to country.

What is not generally included in this free flow is labour. Although the shifting of labour from areas of surplus to areas of demand would be a perfectly logical extension of free trade, the concept of open international borders poses serious challenges to politicians faced by hostility within their own country to high rates of immigration (see Chapter 6).

The principal agents of globalisation are the large corporations and the majority of governments in MEDCs which perceive there to be major benefits for their populations. It has been facilitated by rapid falls in the cost of transport. Technology makes globalisation feasible.

- Between 1930 and 2000 the average cost per kilometre of air travel fell from 43 cents to 5 cents.
- Between 1930 and 2000 the average cost of a 3-minute phone call between London and New York fell from $240.65 to $4.00.
- Between 1960 and 2000 the cost of one unit of computing power fell by 99.9%.

## Glocalisation

**Glocalisation** is a system of production pioneered by Japanese automobile producers. Production takes place outside the home country but, in contrast to production in assembly branch plants, local sourcing of parts is involved (hence **glocal**isation). This allows producers to minimise the holding of stocks of parts and to take daily delivery from local suppliers as part of the just-in-time production system.

An alternative definition of glocalisation that is occasionally used is when a TNC adapts the product that it is offering to meet local tastes, incorporating aspects of the local culture into its marketing strategy. The variation in McDonald's menus from country to country might be quoted as an example.

## The case for globalisation

The case for globalisation can be considered by looking at the alternative. The states that have not followed the path of integration into the world economy are not numerous but they are instructive. They include, Cuba, North Korea, Iran, Iraq and Libya.

Some of these countries have suffered from lack of development and even famine. Indeed, the real progress made by some of them in recent years has come about because they have begun to liberalise their policies and open their borders.

The countries that have grown fastest in recent years, the NICs, have all pursued policies that have encouraged trade,

### Key terms

**Globalisation** The trend towards a world in which the movement of goods, services and investment is increasing, making national boundaries less distinct.

**Glocalisation** The development whereby TNCs either use local components in their manufacturing process in countries outside their 'home' base or where companies gear their products towards local tastes and cultural preferences.

# Chapter 13 Economic systems

and each has developed export markets for their goods, whereas the worst-performing economies have tried for 'autarkic' development without the benefit of trade.

Proponents of globalisation state that consumers in the richer MEDCs have benefited from the falling prices of almost all commodities and services. We are all better off. Even the poorest 20% in the UK are relatively better off than the same group was 50 years ago. Admittedly the benefits have been uneven, but the theory is that, in time, the 'trickle down' process will benefit even the poorest groups.

Those in favour of globalisation argue that supposed job losses in MEDCs because of cheap production elsewhere, especially the NICs, are an illusion. Only 9% of imports into European Union countries, for example, come from LEDCs (including the NICs). The main reason for job losses is mechanisation and rationalisation, not globalisation.

Advocates like the World Bank and the IMF argue that the problem with globalisation is that it hasn't been properly applied. They argue that the existence of trading blocs and customs unions which impose unfair and restrictive tariffs are the real barriers to the benefits of globalisation reaching the poorest countries of sub-Saharan Africa. There is not, they argue, too much globalisation but too little.

## The case against globalisation

Opponents of globalisation point out that, of the 200 largest TNCs in the world, 172 have their world headquarters in the USA, Germany, the UK, France or Japan. They argue that these headquarters remain in MEDCs because TNCs benefit from close ties to their own national governments. Such benefits include tax breaks, favours from political leaders, links to education and training institutions, use of infrastructure, and the possibility of using the state military or police in times of crisis.

Opponents also point out that research and development departments of TNCs typically remain in the home country. This is because the knowledge and skills of new technology are central to the power of TNCs and certain states. If research and development were spread throughout the world, people in all countries would develop the capacity to make the latest computers or cell phones and would then be able to compete in the world market. Research and development is a carefully guarded secret in most TNCs. For example, it is alleged that only a handful of people know the whole chemical formula and process for making Coca-Cola or the recipe for Kentucky Fried Chicken.

You have studied at least one major TNC and have some details about its locational decisions and a history of its development. These can be used to illustrate the increasingly global nature of production and some of the issues outlined above. This is a controversial topic and there is an increasingly noisy opposition to the apparent dominance of TNCs on the world stage, the policies that are driven by them and those governments that collaborate in the growth of this world system.

*Human systems, processes and patterns*  **A2 Unit 5**

Employment structures have changed as a consequence of the development of the global economy. The most obvious features of this are:
- the growth of secondary employment in some countries and regions (especially in the NICs)
- the decline of secondary employment in some core countries, although not in all industries (see the UK experience)
- the changing pattern of primary employment in some peripheral countries and the increasing importance of commercial agriculture for the production of export crops
- the growth of quaternary employment in a few key regions of core countries

*Examiner's tip*

- The word 'industry' means work. If an essay title doesn't narrow this down for you, it is perfectly acceptable to discuss a wide range of industries from retailing through to research development.

## Examination questions

**1** (a) What is meant by the term 'footloose industry'? (5)
(b) Assess the view that the search for cheap labour is the most important factor in determining the location of modern manufacturing industry. (20)

*Total = 25 marks*

**2** (a) Outline the main characteristics of the global economy. (5)
(b) Examine the view that the growth of the global economy is a consequence of the rise of TNCs (transnational corporations). (20)

*Total = 25 marks*

# Synoptic link

# Industrialisation and de-industrialisation have an impact on the physical and human environment

- The interrelationships between the physical environment and industrial location and the management of environmental impact.
- The role of government in the control and modification of the relationship between industry and the environment, and the response to changing values and attitudes within society as a whole.

## Industry and environment

The natural environment and physical factors clearly have some impact on the choice of industrial location, although this varies greatly from industry to industry. Remember that 'industry' is just another word for 'work' and it is a term that applies to economic activity covering any area of employment from primary to quaternary.

*Key term*

**De-industrialisation** The loss of manufacturing jobs and the reducing significance of manufacturing industries.

# Chapter 13 Economic systems

Physical factors are the small-scale *site* factors that are influential in the choice of location for individual plants or factories. The natural environment is a broader term that might embrace anything from the general attractiveness of the landscape to the climate.

Site factors include:
- availability of flat land
- ability to expand
- accessibility
- water supply
- ability to dispose of waste

The relationship between the physical environment and industry is especially obvious in regions of the world that have industrialised quickly in recent times and developed a heavy industrial base built around potentially high-polluting industries such as chemicals, iron and steel, and energy extraction. This development frequently shows the same lack of concern about despoliation of the environment that occurred in the UK during the Industrial Revolution. Among the most notorious of these 'late developers' were the east European countries that formerly belonged to the Warsaw Pact and were pushed very rapidly to develop by their Soviet advisers.

## Poland

### Economy and industry

Despite the fall of the Soviet Union and the building of a capitalist economy in the former eastern bloc countries, large-scale industry in Poland remains largely in state hands. Since 1990, inflation has declined. Manufacturing industry accounts for 25% of total employment. In the agricultural/food sector, production is rising rapidly and many of Poland's largest companies are in the food industry. The chemical industry consists mainly of large state companies and a few private enterprises. There are 23 steel mills in Poland; their production has dropped since 1986, but has now stabilised. Modernisation programmes are planned, but it will take a lot of time to solve the technical, commercial and management deficiencies. The environmental problems connected to the steel industry are large.

### Environment

The environmental situation in Poland has improved since 1990 due to a decline in heavy industry and increased government concern for the environment. When the economy started to grow, industrial pollution did not increase at the same rate, because of a large investment programme, industrial restructuring, and the changing attitude of the state, especially since joining the EU.

Poland still has severe environmental problems. Acid rain has taken its toll on Polish forests, and health standards for the population are threatened because of the excessive amounts of sulphur dioxide ($SO_2$), nitrogen oxides ($NO_x$), carbon dioxide ($CO_2$), dust particulates, heavy metals (lead, cadmium, arsenic, mercury) and carcinogenic hydrocarbons being placed into the atmosphere by the Polish energy sector, which still relies on coal as its primary fuel source.

While all areas of Poland have experienced some of the negative effects associated with high sulphur and nitrogen dioxide emissions, the effects have been unevenly distributed. The most affected region, the southwest corner of the country, Upper Silesia, was also the primary emissions source.

During the 1980s, Poland's Katowice district, which is part of the Silesia region and makes up 2.1% of the country, accounted for as much as 20–25% of the country's total emissions of sulphur dioxide ($SO_2$), oxides of nitrogen ($NO_x$) and dust. In the region now known as 'the black triangle', home to the largest basin of brown coal in Europe, approximately 200 million tonnes of coal were produced each year, leading to the emission of 3 million tonnes of sulphur dioxide and approximately 1 million tonnes of oxides of nitrogen each year.

Environmental pollution in Silesia resulted from years of industrial activity concentrated in a relatively small area. Emission of excessive amounts of pollutants led to severe acid rain, practically destroying the mountain forests and acidifying the soils in the Karkonosze and the Izerskie Mountains. The death rate for men in Katowice between the ages of 30 and 59 exceeds the national average by 40%, children are usually born underweight, and the occurrence of birth defects in the region is up to 60% above average.

It is expected that the recent accession of Poland to the OECD and the EU will provide a stimulus for improvement in the levels of pollution and specifically the management of hazardous waste. Since 1989, significant resources have been channelled for upgrading some 300 wastewater treatment plants. Further investments in this sector are, however, required. In 1990 the government required power plants to start reducing sulphur dioxide, nitrogen oxide and dust emissions by approximately 30% (depending on the particular fuel used in a specific type of furnace) by 1997. For plants put into operation after the law took effect, the levels have to be much lower than the 30% reduction rate. An environmental policy was adopted in 1991, and in 1995 a detailed review of policy implementation was published, setting detailed targets for the year 2000. Using environmental fees and fines, Poland has been quite successful in obtaining financial resources to invest in reducing industrial and water pollution.

Pollution of the environment is costly, but many of these costs fall on people not responsible for the original damage. These are *external costs* and in most economic systems companies are not keen to pay for this sort of damage. In other words, governments often have to intervene to insist that companies do not pollute or, if it is really unavoidable, that they pay for some of the damage done — known as 'internalising the externalities'. This is especially difficult if the externalities are exported beyond national boundaries, as is the case with acid rain.

The Polish example also illustrates how problems that are frequently perceived as national are, in reality, much more localised.

## Generation of waste

Rising incomes and economic growth favouring domestic consumption are both factors likely to increase waste. For some 20 years industrialised countries have been disappearing under a rising heap of rubbish, which they need to dispose of.

# Chapter 13  Economic systems

- In the early 1990s the USA produced 870 kg of waste per person each year, which is equivalent to 2.5 kg a day.
- In 1997, about 64% of municipal waste in the USA was put into landfill sites, 18% was incinerated and 18% was recycled.
- For every person employed to dump waste in a landfill site, ten jobs could be created in the recycling industry. The new raw materials produced by recycling will also reduce our demand for new resources.
- Babies born within 3 km of landfill sites taking hazardous waste are 40% more likely to have chromosomal anomalies, such as Down's syndrome. Babies born within 2 km of a landfill site are more likely to have birth defects than babies born elsewhere.
- Recycling aluminium cans brings energy savings of up to 95%, and so produces 95% less of the emissions that cause climate change than when aluminium is produced from raw materials.
- Productive use of waste has increased in industrialised countries, but increasing production of waste means that there is still a growing amount to dispose of.
- OECD forecasts show that production of municipal waste, an estimated 540 million tonnes in 1997, will increase by 43% by 2020.

In an attempt to slow down this development, the European Union has issued a number of directives designed to encourage the productive use of waste. These regulations deal with all types of waste, from the largest, for example motor vehicles at the end of their lives and electrical and electronic products, to the smallest, such as batteries and packaging.

The short-term benefits of landfill as a means of disposing of waste may be heavily outweighed by the long-term costs. We do not yet have a complete picture of this cost, which, in any event, is highly controversial. Long-term health problems arising from methane emissions, housing subsidence and toxicity in groundwater are among the best-known effects.

### Are the French setting the trends?

France was one of the first European countries to take the management of its waste seriously. After 10 years' experience, the picture is generally encouraging. Eighty per cent of French people say they are worried about the environment and most are quite prepared to sort their household waste. According to the latest statistics from IFEN (the French Institute for the Environment), the production of waste (excluding waste soil and rock) in France is estimated at 600 million tonnes, most of it consisting of agricultural waste (350 million tonnes) and building-site waste (110 million tonnes).

Every French person throws away an average of more than 1 kg of waste material each day. France lies in second place for sorting waste among the European countries, behind Germany, with 45 million sorters in 2001. This means that 1.7 million tonnes of household packaging each year is sent to 250 approved sorting centres, through the officially approved household packaging waste management body. This organisation set itself the target of recycling 65% of household packaging by 2002.

*Human systems, processes and patterns*     **A2 Unit 5**

There has been a shift towards better waste management in recent years in the wealthiest countries. These changing values reflect greater prominence afforded to green issues by politicians. In France and Germany, political advances by various green parties have more or less forced mainstream political parties into action.

### Do MEDCs export their rubbish?

Between 1994 and 1996 a German company sent 95 000 tonnes of household rubbish to Pyongyang (North Korea) for recycling. This story was very sensitive for Germans, who see themselves as world leaders in waste recycling. The German spokesman for Greenpeace commented that Germany was the world master in exporting rubbish overseas and went on to list over 30 countries, including China, Indonesia, Pakistan, Vietnam and the Philippines, to which German companies made 'legal or illegal' rubbish shipments. North Korean workers could have been affected by fumes and toxins released from the waste. It may be that the health of the workforce is not a primary consideration in North Korea, which may be one reason why it is cheaper to send waste many thousands of kilometres for recycling.

## Environmentalism

Concern about the environment is not new, but it is largely a product of industrialisation. It occasionally includes a certain amount of nostalgia about a pre-industrial world. Concern for the environment has grown alongside high mass consumption and is seen by critics as a hypocritical indulgence. It is easier to worry about environmental damage when you have achieved some degree of material wealth than it is when you are still striving for developmental goals.

It is worth thinking about the following questions.
- Who benefits most from slowing down the rate of deforestation in Amazonia?
- Who benefited most from the deforestation of Europe?
- Do Bengal tigers threaten the life or livelihood of any Europeans?
- Did bears and wolves threaten the livelihoods of any north Europeans and what did we/they do about it?
- What gives us the right to lecture poorer societies about the environment?

You need to note that there isn't any one 'correct' attitude towards nature, any more than there is one universally accepted idea of what nature is. Some include the human species as an integral part of nature, and thus exonerate us from any blame, while others tend to place 'human' and 'nature' in opposition.

The view that the natural world is there to be exploited ruthlessly in the pursuit of personal gain has dominated Western culture until very recently. In Britain different conceptions of 'nature' can be seen to have affected both values and policy. Hill walkers and bird watchers, preservationists and green-spacers all have their own idea of 'nature', and they have had varying levels of success in getting their differing visions incorporated into planning regulations.

The following are examples of the kind of questions that arise when it comes to protecting the natural world.
- Which definition of 'nature' is being protected?
- Are humans in or out of the 'natural' world?

# Chapter 13  Economic systems

- In medicine, does anything 'organic' count as natural? Do we let cancer cells or malaria have their way, as part of 'nature'? Do head lice deserve our protection?
- Is the environment we wish to protect some imaginary world without the impact of human history — without landscapes transformed, species eradicated and plant varieties cultivated that we want to protect?
- Is what we want to preserve the world *with* the human interventions we like, whether they are English hedgerows and dry stone walls, or the deforested and depopulated Scottish moors?

The trouble with declaring one's reverence for a system, be it a market, a culture or an ecosystem, is that people disagree strongly about what the 'natural' state of that system is.

But whatever one's beliefs, the dominant scientific view about the environment makes gloomy reading. Environmental problems have gone from bad to worse in the past 50 years.

- 15–20% of all species have already been lost, threatening the fifth great extinction (the previous great extinction involved the dinosaurs). Bird and mammal species are disappearing at a rate that is estimated to be 100–1000 times the rate at which extinctions naturally occur.
- Tropical deforestation has occurred at rates of around half a hectare a second.
- An area of about 100 000 km$^2$ is made barren each year by desertification.
- Growing energy use has led to a 2–3 °C rise in mid-latitude temperatures and to significant changes in rainfall patterns over the past 100 years.
- Half of the world's mangroves and wetlands have been destroyed.
- Industry and agriculture are fixing nitrogen at rates that exceed nature's, and among the many consequences of the resulting over-fertilisation are 50 dead zones in the oceans.
- In 1960, 5% of marine fisheries were either fished to capacity or over-fished. Today, 70% are in this condition.

Not only have governments failed to reverse these trends, but they have laid a poor foundation for progress in the future. Many years of international environmental negotiations have been disappointing. The agreements made, for example, in the conventions on climate, desertification and biodiversity are sound enough, but most of it is simply advisory. Governments can sign up to agreements which they support in principle, but then fail to work vigorously towards their implementation.

These weaknesses should not be a surprise; the agreements were forged with procedures that gave maximum leverage to countries with an interest in thwarting international action.

- The USA successfully weakened the Kyoto Protocol.
- Brazil has worked against a convention on deforestation.
- Japan and other major fishing countries watered down the international marine fisheries agreement.

There is no international consensus. Indeed, there is no consensus between different interest groups within countries.

# Chapter 14

# Rural–urban interrelationships

## The world's major urban areas

### Location and distribution

**Urbanisation** can be defined as either:
- the process by which there is an increase in the number of people living in urban areas; or
- the increase in the land area occupied by towns and cities.

Of these definitions, the first is by far the most useful, but it does contain some difficulties. The number of people living in cities can be expressed as an absolute figure, but is more commonly given as a percentage of the total population of a country. If rural populations are also growing, then the relative rate of increase of urban populations may not appear as significant as it would do if rural populations were stagnating or declining.

We need to define our terms. Here are some common methods used to define towns.
- **Population size**: highly arbitrary and likely to vary from country to country.
- **Population density**: always going to require the use of an arbitrary division as modern cities merge with the surrounding countryside.
- **Function**: what do cities do? This was much easier to answer in the past than it is today.
- **Level of administration**: e.g. rural or urban district. This is a common way of escaping the need to define by adopting the existing local government designations.

The UN does not classify settlements as towns or rural areas. Instead it has chosen to classify settlements by size. This causes variations depending on the census methods of individual countries, which make the figures for city size unreliable in detail, but the 'big' picture emerges clearly enough.

Tables 14.1 and 14.2 give some statistics about the world's largest cities. Figure 14.1 shows how the size of the largest cities has grown.

| Region | 1800 | 1900 | 1950 | 2000 |
|---|---|---|---|---|
| Africa | 0 | 0 | 2 | 34 |
| Asia | 1 | 3 | 26 | 136 |
| Europe | 1 | 9 | 30 | 61 |
| Latin America | 0 | 0 | 7 | 39 |
| North America | 0 | 4 | 14 | 36 |
| Oceania | 0 | 0 | 2 | 5 |
| Total | 2 | 16 | 81 | 311 |

*Table 14.1 Number of cities with more than 1 million people, by region, 1800–2000*

| Region | 1800 | 1900 | 1950 | 2000 |
|---|---|---|---|---|
| Africa | 4 | 2 | 3 | 6 |
| Asia | 64 | 23 | 32 | 44 |
| Europe | 29 | 51 | 37 | 19 |
| Latin America | 3 | 5 | 8 | 16 |
| North America | 0 | 16 | 18 | 13 |
| Oceania | 0 | 2 | 2 | 2 |

*Table 14.2 Distribution of the world's 100 largest cities, 1800–2000*

# Chapter 14   *Rural–urban interrelationships*

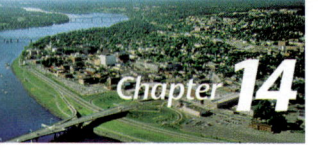

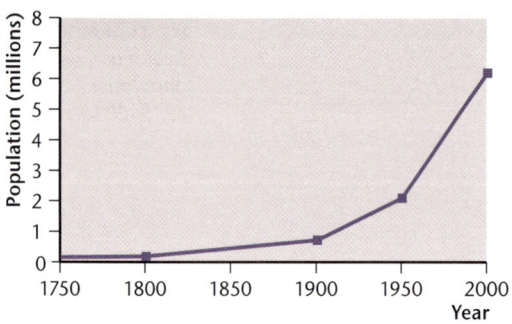

*Figure 14.1 The average size of the world's 100 largest cities, 1800–2000*

Figure 14.1 shows that the growth of large urban areas (cities) has been especially notable over the last 50 years. The distribution of the largest cities has also changed significantly (Table 14.2). The dominance of Europe and North America in 1900 has been challenged and then overtaken by Asia (2000). Africa has also emerged as a continent of rapid urbanisation in the last few decades; significantly, Africa is the continent that remains the least urban.

It is worth remembering that there is no pre-determined path here. Urban growth has causes and if those factors are not in place, then growth will not occur. The history of the world is not pre-ordained. Africa may or may not reach the same levels of urbanisation as other continents, but beware of 'explanations' that suggest a known outcome. This is what is known as **teleological reasoning** and it is thoroughly misleading.

A relatively recent feature of urbanisation has been the emergence of mega-cities with populations of over 5 million (Figure 14.2). These were once almost exclusively found in the highly developed economies of Europe, the USA and Japan and were associated with enormous concentrations of wealth and power. Although these cities persist and remain economically dominant, a new breed of mega-city has emerged. These are far more widely distributed and are largely a consequence of forces in rural areas rather than positive pull factors in the urban areas concerned.

*Figure 14.2 World mega-cities*

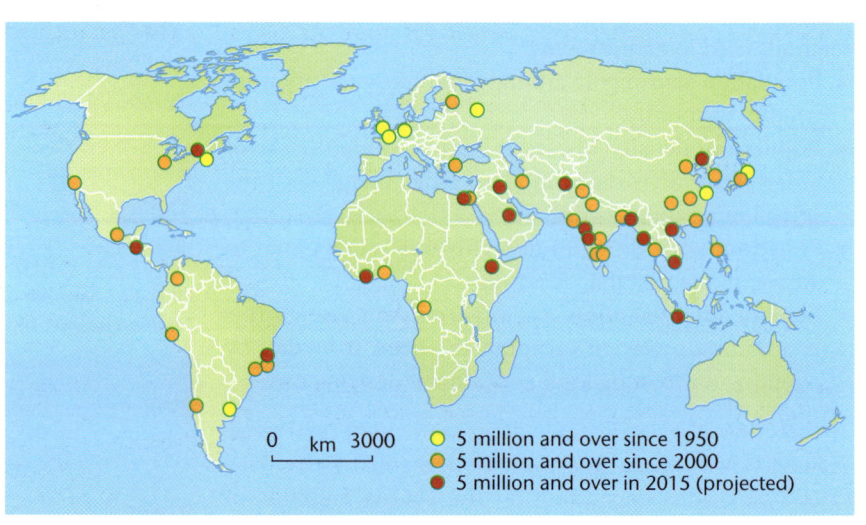

○ 5 million and over since 1950
○ 5 million and over since 2000
● 5 million and over in 2015 (projected)

Source: United Nations, World Urbanization Prospects

## Urbanisation as a process

Figure 14.3 shows how the urban and rural populations have changed between 1950 and the present day. The graph also shows projected values to 2030.

*Human systems, processes and patterns* **A2 Unit 5**

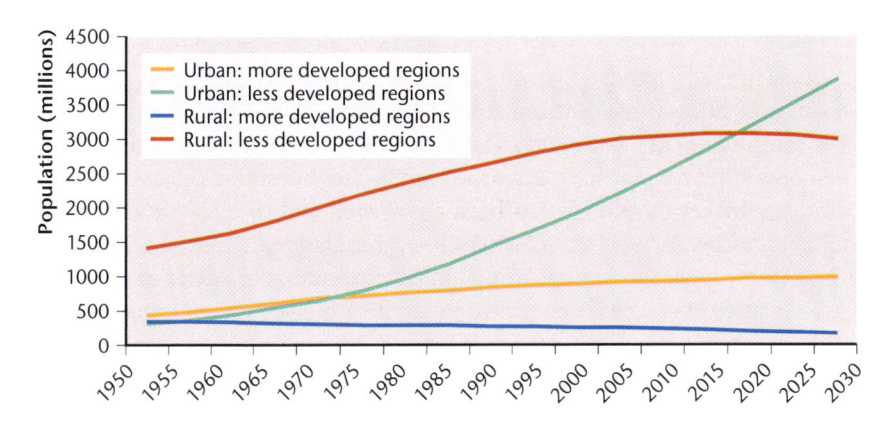

Figure 14.3 Urban and rural population, 1950–2030

At the demographic level, urbanisation involves two processes:
- natural increase
- migration

Rates of natural increase in cities are often significantly higher than those in more rural areas, with higher fertility rates in urban areas reflecting the lower age profiles of the urban population. These differences in age structure are often the consequence of migration. In many cases it is younger people who arrive in cities, and they drive up the birth rate. This is most common in situations where having more children is likely to increase family income, as it frequently does if work is available in what is known as the 'informal' sector.

**Rural–urban migration** has often been the dominant force in the development of cities and you will remember that there are likely to be factors operating at both ends of this process: there are **push factors** from rural areas and **pull factors** from urban areas.

Not all types of society will develop an urban structure. There was, for a long period, a marked contrast between the economies of the northern and southern regions of the young USA. These differences are summarised in Table 14.3.

Table 14.3 Comparing north and south in the USA

| Location | Economy | Social and political system | Urban development |
| --- | --- | --- | --- |
| Northern states of the USA | Small-scale homestead farms, none especially wealthy | Independent farmers with artisans and craft industry developing with an emerging middle class; democratic | Towns and cities grow as market centres and points where local industry develops to serve wide market area |
| Southern states of the USA | Plantation system: cotton is the dominant crop, almost all of which is exported to the UK | A dominant elite property-owning class with slave labour, subsequently replaced by poorly paid indentured labour | All luxury goods are imported and all the cotton is exported with no value added, as is typical of a supply region; only a few ports develop |

## Key terms

**Counterurbanisation** A trend that involves the movement of people and enterprises out of urban areas to more rural areas.

**Urbanisation** The growth of urban areas both physically and as an absolute or relative share of total population.

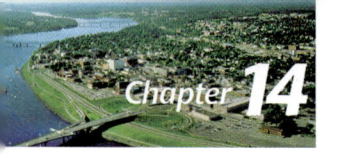

# Chapter 14  Rural–urban interrelationships

## Pre-industrial cities

There are a number of forces that tend to lead to the agglomeration of populations in towns and cities. These **centripetal forces** include the following.

- The exploitation of a raw material such as coal or iron ore. Some of the world's first urban centres may have developed around such resources. Catal Huyuk in modern Turkey is thought to have developed into a 'city' of over 6000 inhabitants because of local obsidian (volcanic glass).
- The development of manufacturing industry at sources of power or resources. As industries developed they sucked in labour from the surrounding countryside. In an era before mass transport, this workforce was obliged to live close to the place of work.
- Military control and the development of a centralised state. The large cities of the pre-modern world were usually centres of administration such as Edo (in Japan), Beijing and ancient Rome. These centres of power brought together the wealthy and politically powerful together with their servants and slaves into one large urban area.

### Case study 14.1  A pre-industrial city: Potosi, Bolivia

The Spanish began their extraction of precious metals from their South American colonies by simply taking the artefacts made by the Aztec and Inca peoples. Just as this source of wealth was running out, the Spanish discovered the silver amalgam method of silver extraction. This allowed them to mine the silver that existed in such abundance in the mountains of Cerro Rico in Bolivia.

The city of Potosi was founded in 1545 by the Spanish to exploit huge reserves of silver in the nearby hills. Excavation of the ore began using slave labour drawn from the local population, and the silver was sent to Spain.

Potosi is now a small, poverty-stricken town in the Bolivian highlands. At its peak, around 1600, it was as large and perhaps as rich as any city in western Europe, with a population of about 200 000. At this time, Potosi was producing half the world's supply of silver.

Once the original resource ran out there was very little to replace it, so the population declined rapidly to less than 10 000 by the time of independence in 1825. Some historians calculate that as many as 8 million Indians died during the 300 years that Potosi was a mining city. Most of them were killed by disease or accidents or by the brutal conditions in which they lived and worked, either as slaves or as forced labour.

*Photograph 14.1 Potosi from Cerro Rico*

*Human systems, processes and patterns*  **A2 Unit 5**

The growth of cities in more modern times has frequently been driven by industrialisation. This process concentrates production at one place, replacing a manufacturing system which relied on 'putting-out' work (sometimes known as **proto-industrialisation**).

The proto-industrial system relied on merchants delivering raw materials and picking up finished products from producers who worked in their homes (sometimes known as cottage industry). This 'putting-out' was common in the textile industry, as it allowed poor peasant farmers to supplement their income by spinning or weaving. The merchants lived in the local towns and as the trade grew, so did the town which 'organised' it.

### Examiner's tip

- Remember that even when urbanisation is taking place at a significant rate through rural to urban migration, rural population might still be increasing, fuelled by a rapid rate of natural increase.

## Case study 14.2 *Industrialisation and city growth in the UK*

During the nineteenth century only the growth of London's population could compare with the rapid growth of Glasgow, Birmingham, Leeds, Sheffield, Manchester and Bradford. These towns grew both by natural increase, for birth rates were still high, and from internal migration. The migration was dominated initially by short-distance moves from rural areas to the nearest town. Manchester, for example, drew on north Lancashire, the Lake District and north Wales. It was not easy to recruit labour for the textile mills and factories and there was much local resistance to the appalling conditions in these factories. Right from the outset much of the market for the goods produced was the overseas market, with two-thirds of cotton cloth being exported as early as 1820 and no less than 85% by the 1880s. While this rapid urbanisation was taking place there was, by and large, no rural depopulation given high rates of natural increase, except in a few remoter upland areas such as the Scottish Highlands where farm enlargement was taking place.

|      | Liverpool | Glasgow | Manchester | Leeds   |
|------|-----------|---------|------------|---------|
| 1801 | 82 000    | 77 000  | 70 000     | 53 000  |
| 1831 | 202 000   | 193 000 | 238 000    | 123 000 |
| 1851 | 376 000   | 329 000 | 303 000    | 172 000 |

*Table 14.4 Population of industrial towns, 1801–51*

**Pre-industrial cities** are those urban settlements that existed before the early nineteenth century or cities in those countries that have yet to show signs of a transition to an industrialised economy. The functions of pre-industrial cities varied, but they were often dominated by administration, government, religion and trade. They were frequently military strongholds and centres of authority. Of course all of them had some industry in the sense that things were made there (for example, Edo in Japan was a centre of weapon manufacture), but it was not the primary function of these cities, nor the cause of their development. This manufacturing was not **city-forming**, in that it was not the fundamental cause of the growth of the city. On the contrary, this manufacturing was **city-serving**. The inhabitants needed clothing and household articles and an industry of sorts developed to serve these needs.

### Key terms

**Industrial cities** Cities that developed due to the coming together of a working population involved in manufacturing.

**Post-industrial cities** Cities that have developed in recent times without a major input from developing industries.

**Pre-industrial cities** Cities that grew without the development of a concentrated manufacturing sector.

## Chapter 14 Rural–urban interrelationships

### Location quotients

This distinction between city-forming and city-serving activities is useful today and can be explored by looking at the location quotients for various towns and cities. The relative concentration of a particular industry or group of industries can be measured by comparing location quotients. A location quotient is a measure of the degree to which any two quantitative characteristics (in this case employment in a particular industry) are distributed between any two areas.

$$\text{location quotient} = \frac{\text{regional employment in industry I in year T}/\text{national employment in industry I in year T}}{\text{total regional employment in year T}/\text{total national employment in year T}}$$

City-forming activities have a location quotient of greater than 1, suggesting that the industry is 'over-represented' in the town or city. A location quotient of less than 1 suggests 'under-representation'. In Figure 14.4, it is clear that Coventry has several industries that are highly significant as city-forming activities, specifically those involving aircraft and rubber products.

*Figure 14.4 Location quotients for selected trades in Coventry and Birmingham*

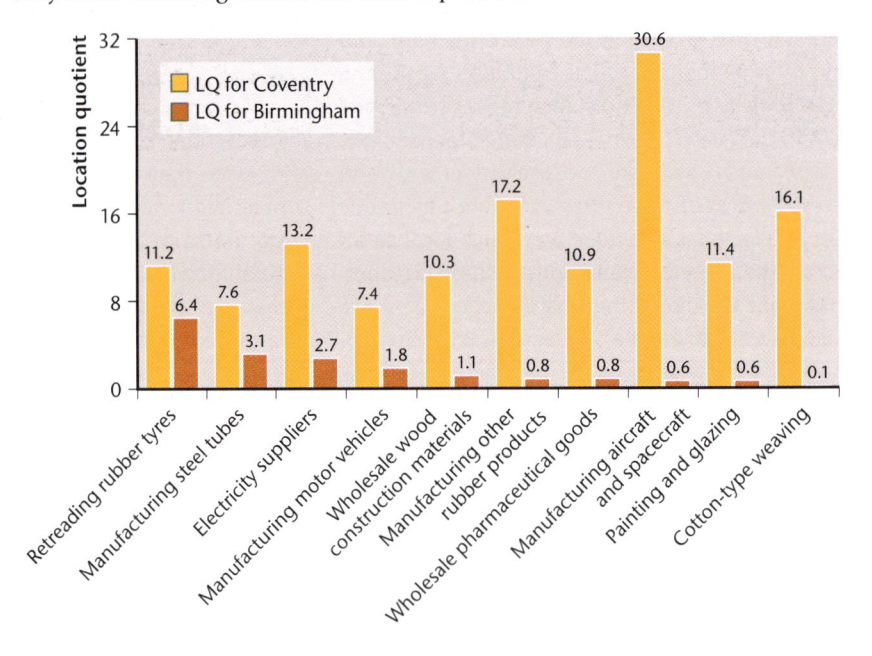

| Trade | LQ for Coventry | LQ for Birmingham |
|---|---|---|
| Retreading rubber tyres | 11.2 | 6.4 |
| Manufacturing steel tubes | 7.6 | 3.1 |
| Electricity suppliers | 13.2 | 2.7 |
| Manufacturing motor vehicles | 7.4 | 1.8 |
| Wholesale wood construction materials | 10.3 | 1.1 |
| Manufacturing other rubber products | 17.2 | 0.8 |
| Wholesale pharmaceutical goods | 10.9 | 0.8 |
| Manufacturing aircraft and spacecraft | 30.6 | 0.6 |
| Painting and glazing | 11.4 | 0.6 |
| Cotton-type weaving | 16.1 | 0.1 |

### Industrial cities

**Industrial cities** developed as manufacturing industries grew and drew people together as a labour force. The proto-industrial system declined and merchants either became factory owners or were slowly replaced by them. Mills or foundries were built close to power sources, such as rivers, and urban development took place as housing grew alongside theses factories. Nineteenth-century city growth in Europe and North America was largely based around the development of new industries, initially heavy (capital) industries and then consumer industries. Travel was difficult and expensive at this time, so housing was developed by factory

*Human systems, processes and patterns* **A2 Unit 5**

*Photograph 14.2 An example of poor-quality nineteenth-century housing, Market Court, Kensington High Street, London*

owners and other entrepreneurs close to the factory gates. The development of these cities was often rapid and without regulation or planning. Poor-quality housing and low standards of hygiene were dominant themes in the first half of the nineteenth century.

## Post-industrial cities

**Post-industrial cities** have developed more recently as the location of industry has shifted away from traditional centres and become less concentrated. Some cities have developed around certain types of economic activity, especially tertiary and quaternary services. More commonly, older cities have found non-manufacturing employment of growing significance and have lost the older city-forming activities, such as their traditional manufacturing core. These cities are characterised by low densities and, often, a high dependence on the car. The term post-industrial is rather misleading, because these cities do contain industry, but it is a different type of industry. In Table 14.5 the impact of tertiary and quaternary activity centred around the universities of Cambridge and Oxford shows up clearly in their relatively high location quotients for higher education employment.

| Rank | County | Location quotient |
|---|---|---|
| 1 | Cambridgeshire | 2.89 |
| 2 | Oxfordshire | 2.46 |
| 3 | Leicestershire | 2.09 |
| 4 | South Glamorgan | 1.73 |
| 5 | East Sussex | 1.67 |
| 6 | Gwynedd | 1.67 |
| 7 | Avon | 1.40 |
| 8 | Surrey | 1.36 |
| 9 | Central region | 1.35 |
| 10 | West Midlands | 1.31 |

Source: NOMIS (Employment Census, September 1993)

*Table 14.5 The top ten UK counties with higher education employment location quotients greater than 1 (1993)*

**Chapter 14** *Rural–urban interrelationships*

## Comparing cities

The terms 'pre-industrial', 'industrial' and 'post-industrial' are essentially chronological, in that they trace the evolution of cities. However, urbanisation without large-scale industrialisation centred around factories has been a significant theme in the last 50 years, especially in Asia and Africa. Thus many of the new 'million plus' cities are not copies of the cities that grew in nineteenth-century Europe and North America. It is also worth pointing out that many cities have not passed through the stages in a sequential manner.

- Reykjavik in Iceland is a largely twentieth-century creation, built around administration and other tertiary employment and a developing quaternary sector.
- Las Vegas is another unusual urban form; a city devoted almost entirely to the provision of leisure, supplied with power, water and all other resources from outside.

### Case study 14.3  A post-modern city: Las Vegas

- Mormons were the first Europeans to settle in the Las Vegas area in 1855. It had been avoided by the pre-European inhabitants because of the lack of water and very high temperatures.
- In 1905 the town of Las Vegas was established.
- In 1911 the settlement became a 'city'. At this time Las Vegas was just a stop along the railroad west and not a very significant one at that.
- The first period of rapid growth occurred in the 1930s. The construction of the Hoover Dam on the nearby Colorado River, funded by Washington DC, was begun in 1931.
- In 1931, gambling was legalised by the state of Nevada.

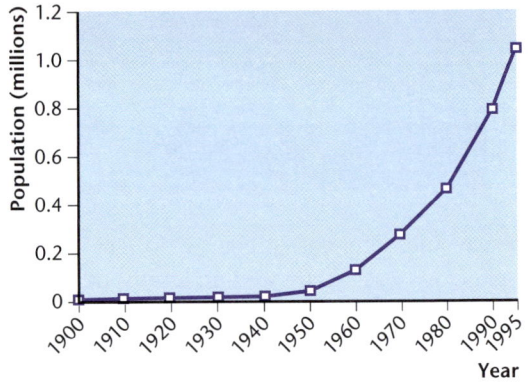

*Figure 14.5 Population graph for the Las Vegas valley, 1900–95*

The water supply offered by Lake Mead and the technical breakthrough in providing cheap domestic air-conditioning facilitated the growth of the city in such an unpromising desert environment. At the same time the growth of the southwest, itself heavily dependent on imported water and power, and the rapid development of an internal network of air transport gave Las Vegas the initial market for its particular brand of leisure.

The first hotels went up in the 1940s, both downtown and on what became known as the Strip. In this period the legalised gaming industry blossomed, putting Las Vegas on the tourist map with the development of large hotels.

In the 1960s and 1970s growth continued, fuelled by internal and international tourism, and the migration trend towards the sunbelt for both employment and retirement. Hotels continued to attract tourists.

Las Vegas became a major convention centre in the 1980s and 1990s. Entertainment was broadened to include amusement park rides and attractions, and many hotels became theme parks. Rivalled only by Orlando in Florida as a tourist destination, Las Vegas has 35 million visitors a year and the ten largest hotels in the world. Las Vegas has continued to grow. Today, the population of the area tops 1 million, but that doesn't include the tourist population which is estimated to average 582 000 per day. It is the fastest-growing metropolitan area in the country. One estimate is that the population will double by 2015.

*Human systems, processes and patterns* — **A2 Unit 5**

# Urbanisation today

## Global variations

Figure 14.6 shows the areas of the world where urban populations are growing rapidly. As you can see, the majority of the fastest-growing cities are in Africa and Asia.

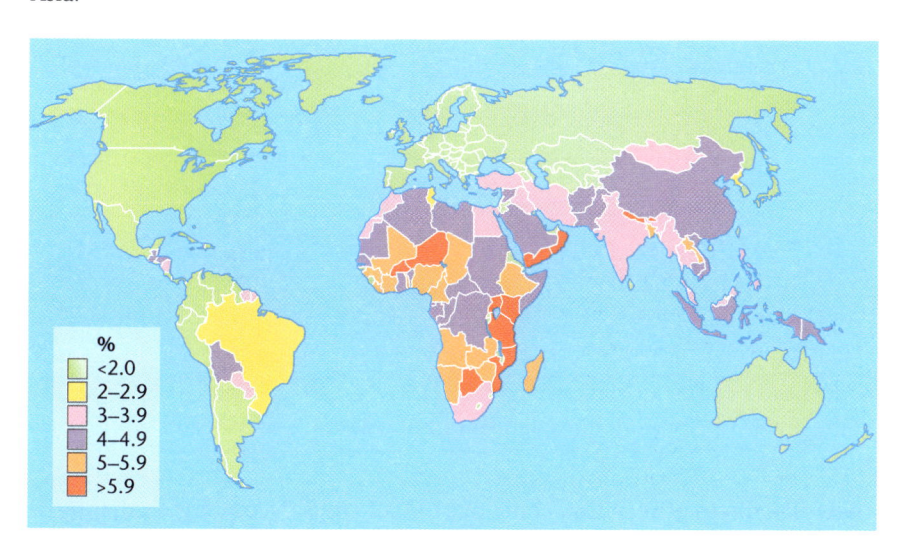

*Figure 14.6 Average annual rate of growth of urban population (%), 1995–2000*

Cities come and go. For example, Greater Zimbabwe is now in ruins and Sparta is just a dusty village. Catal Huyuk, which has a reasonable claim to be the first ever urban settlement, disappeared for reasons that remain obscure despite extensive archaeological research. However, some cities are remarkably resilient. Whether you consider London, Damascus or Beijing, it is obvious that some city-forming forces are powerful and long-lasting.

The forces that create and destroy cities today are generally global and dominated by the same economic forces as covered in Chapter 13. These forces have gathered pace in recent years. At some unidentified moment in 2004 the world reached a significant watershed — for the first time over 50% of the population lived in cities.

- In 1950 there were 86 million plus cities.
- By 2015 there will be at least 550 such cities.

Currently the number of people who live in cities is greater than the total world population in 1960. Ninety-five per cent of the growth of cities over the next few decades will be in LEDCs; examples such as Las Vegas are the exceptions that test the rule, not the norm. Table 14.6 shows the world's top ten cities for 1950 and 2000 and also contains the expected results for 2015.

The Canadian economist Jane Jacobs identifies five main forces by which cities affect both the regions in which they are set and those more distant from them:
- the power of city markets
- the power of city jobs

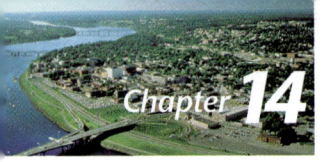

# Chapter 14  Rural–urban interrelationships

|  | 1950 |  | 2000 |  | 2015 |  |
|---|---|---|---|---|---|---|
| Rank | City | Population (millions) | City | Population (millions) | City | Population (millions) |
| 1 | New York, USA | 12.3 | Tokyo, Japan | 26.4 | Tokyo, Japan | 26.4 |
| 2 | London, UK | 8.7 | Mexico City, Mexico | 18.4 | Bombay, India | 26.1 |
| 3 | Tokyo, Japan | 6.9 | Bombay, India | 18.0 | Lagos, Nigeria | 23.2 |
| 4 | Paris, France | 5.4 | São Paulo, Brazil | 17.8 | Dhaka, Bangladesh | 21.1 |
| 5 | Moscow, Russia | 5.4 | New York, USA | 16.6 | São Paulo, Brazil | 20.4 |
| 6 | Shanghai, China | 5.3 | Lagos, Nigeria | 13.4 | Karachi, Pakistan | 19.2 |
| 7 | Essen, Germany | 5.3 | Los Angeles, USA | 13.1 | Mexico City, Mexico | 19.2 |
| 8 | Buenos Aires, Argentina | 5.0 | Calcutta, India | 12.9 | New York, USA | 17.4 |
| 9 | Chicago, USA | 4.9 | Shanghai, China | 12.9 | Jakarta, Indonesia | 17.3 |
| 10 | Calcutta, India | 4.4 | Buenos Aires, Argentina | 12.6 | Calcutta, India | 17.3 |

Source: United Nations, *World Urbanization Prospects*, 1999 Revision

*Table 14.6 The world's top ten cities*

- the power of city technology
- the power of city work transplanted out of cities
- the power of city capital

These five forces arise within some but not all cities. Jacobs argues that those cities that **import-replace** are the most successful economically. Import replacement takes place when goods previously imported into a country are gradually replaced by home-made products with significant design improvements and modifications. She uses the case of imported bikes, motorbikes and cars into Japan as some of her examples. One might equally well apply this theory to the disparities of Italian development. Here there are successful, import-replacing cities in northern Italy drawing labour from southern Italy; the north dominates the south with northern goods.

The American academic Saskia Sassen has developed the idea of the '**global city**'. In the present global economy, a world urban hierarchy exists in which cities have more or less specialised roles within specific regions, these regions varying in size. The precise function of each city is partly controlled by the policies of the nation state in which it is located.

At the top of the hierarchy, Sassen identifies New York, Tokyo and London as having a pivotal role in the world economy. These three cities, she argues, play the main role in the control and management of the global economy. While manufacturing industry and some office work might have dispersed to new locations, there has not been a parallel decentralisation of ownership. As a result there has been an emergence of a new class of highly paid professionals who work in the quaternary sector managing and developing the network of banking, insurance, legal and accountancy services that

> **Key terms**
>
> **Global cities** Cities that operate as the control centres of the world economy, characterised by a large number of headquarters of TNCs and a capital market that is growing fast and international.
>
> **Import-replacing cities** Cities that grow because of their capacity to innovate and adapt the design and construction of imported goods, producing ultimately more refined versions of these products.

*Human systems, processes and patterns* **A2 Unit 5**

oil the world economy. They develop new 'products' that, by and large, are intended to improve the efficiency by which capital is moved around the planet. There are other cities that have a subordinate role in this hierarchy, such as Miami which is an important centre for those **transnational corporations** that operate in Latin America and the Caribbean.

Many global cities have been remarkably adaptable and their success has been based on the speed of their adjustment to global changes. Thus London has been at various times, and in various combinations, a major manufacturing city, a major trading centre, a financial capital, a centre of government and a cultural capital of world significance. It has experienced population loss, particularly between 1951 and 1981 but, as with New York, the counterurbanisation process often spreads the influence of the city rather than being a sign of decline.

It is also obvious that cities are highly complex internally. Within a city such as London, one of Sassen's global cities, there are pockets of sweatshop production which seem more typical of the mega-cities of the LEDCs. Similarly, cities such as Accra or Mumbai have some characteristics that make the simple MEDC/LEDC division more or less useless.

Many LEDC cities have moved through four historical phases:

- pre-colonial
- colonial
- post-colonial 'national' phase
- global phase (the most recent)

This pathway of development is not evenly distributed and those cities that have some sort of gateway function as ports of entry have passed more rapidly into the global phase than those in more remote regions.

### Case study 14.4 An LEDC city: Mumbai

Greater Mumbai (formerly known as Bombay) occupies a peninsula surrounded by the Indian Ocean. It is about 500 km² in area. The current population of greater Mumbai is around 12 million people, averaging over 24 000 people km⁻². The Mumbai metropolitan region forms part of Western Maharashtra that is one of the most industrialised and urbanised regions of India.

Greater Mumbai is made up of two parts: the 'city' which is located south of Mahim Creek, and the suburbs, which have expanded northwards along the rail tracks onto the mainland. Limited space on the peninsula has meant that land prices have always been relatively high, and a steep gradient in land values exists from the south to the north. This was exacerbated in the 1980s as foreign companies began to arrive. There was a switch of Indian domestic policies at this time towards much greater integration with the global economy, often referred to as neo-liberalism.

*Photograph 14.3 Nariman Point, part of Mumbai's CBD*

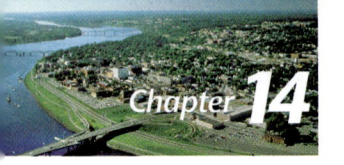

# Chapter 14  Rural–urban interrelationships

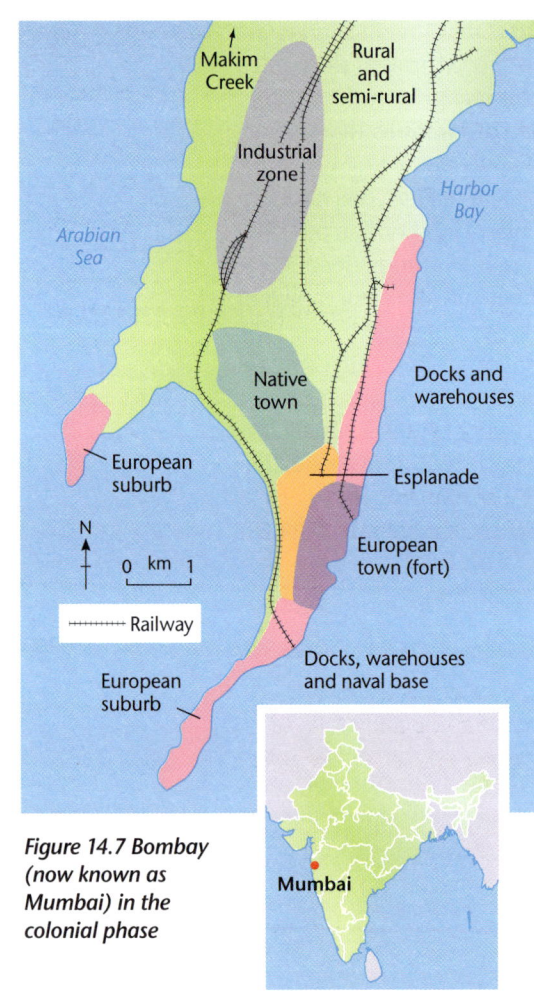

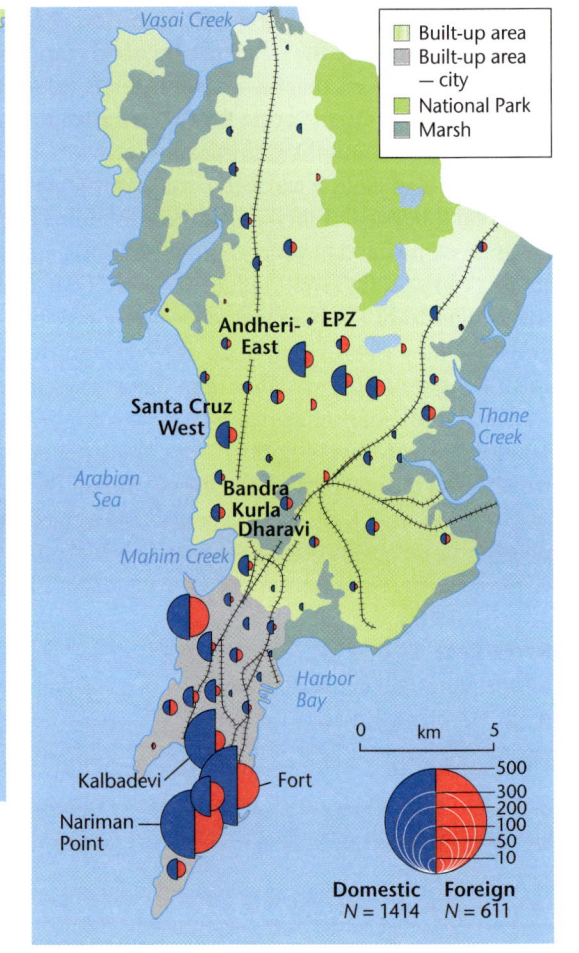

*Figure 14.7 Bombay (now known as Mumbai) in the colonial phase*

*Figure 14.8 Corporate activity in Mumbai, 1999*

Since that time Mumbai has changed its characteristics significantly, developing several CBDs, each with a distinctive character of its own. The present city has at least three distinct centres of business activity.

- Kalbadevi roughly corresponds to the old 'native town' area of colonial days and is dominated by bazaars but is also densely populated with over 100 000 people km$^{-2}$. The business district here contains a very large number of Indian businesses, mostly quite small, and relatively few foreign companies.
- By contrast Nariman Point, built on reclaimed land in the 1950s, is dominated by larger Indian companies and foreign companies, many of which specialise in financial services. These are housed in high-rise buildings built in the last 30 years which for all intents and purposes resemble a CBD in an MEDC.
- Alongside Nariman Point is the Fort area, a much larger area dominated by colonial-style architecture. This area also has many financial services, this time largely dominated by Indian firms. In neither the Fort area nor Nariman Point is there much housing.

These three CBDs are differently linked to the global economy. Nariman shows by far the greatest integration with the global economy, as measured by outgoing and incoming telephone calls, and Kalbadevi the least.

472

*Human systems, processes and patterns* **A2 Unit 5**

## Types of cities in the modern world

### Old industrial

In MEDCs, examples of old industrial cities are Sheffield (UK), Cleveland (USA) and Lille (France). They are characterised by:
- heavy loss of employment in the secondary or, occasionally, primary sector
- frequently high levels of social deprivation
- population loss, especially from the inner-city areas once dominated by heavy capital industries

### 'Born-again' industrial

In MEDCs, examples of born-again industrial cities are Boston (USA) and Milan (Italy). They are characterised by:
- successful import substitution, replacing old industries with new technologies
- some decline in the inner-city areas, reflecting loss of employment in old core industries, but growth in suburban and regional employment

### Post-industrial

In MEDCs, examples of post-industrial cities are Houston (USA), Grenoble (France), Las Vegas (USA) and Reykjavik (Iceland). They are characterised by:
- being based around new technologies and/or tertiary and quaternary industry
- rapid growth, although often in low-density sprawl into the surrounding region

### Global cities

In MEDCs, examples of global cities are London (UK), New York (USA) and Tokyo (Japan). They:
- are large cities that are the location of the control systems of the world economy
- have experienced recent growth, although their complex histories also feature some periods of decline

### NIC cities and city states

Examples of NIC cities are Seoul (South Korea), Taipei (Taiwan) and Singapore. They are characterised by:
- being major centres of administration, education and government as well as manufacturing cities
- rapid growth until recently; now growth has slowed down as fertility rates have fallen to MEDC levels and rural–urban migration has slowed

### Gateway cities in LEDCs

Examples are Mumbai (India) and Shanghai (China). They are:
- cities in the more recently industrialising countries (RICs) that exhibit clear signs of the growing strength of the global economy
- cities of rapid growth, although often controlled and not as fast today as 10–20 years ago
- experiencing considerable rural–urban migration

> **Key terms**
>
> **Shanty towns** Slum areas on the outskirts of major cities, frequently built on land to which there is no legal title.
>
> **Slums** Areas of poor housing, often characterised by minimal infrastructure (especially water supply, sewerage and power).

> **Examiner's tip**
>
> - Bear in mind that cities are complex and not always easy to categorise. Thus a city as large as London will exhibit features and functions that are drawn from several of the definitions and categories offered.

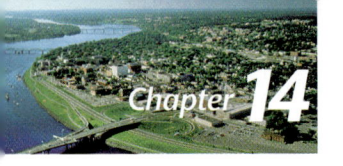

**Chapter 14** *Rural–urban interrelationships*

### Hyper-urbanising cities in the least developed countries

Examples are Khartoum (Sudan) and Banjul (The Gambia). These are characterised by:
- considerable levels of rural–urban migration driven by rural poverty and the commercialisation of agriculture even in the absence of broad-based economic growth — the landless poor become the urban poor
- rapid growth; Africa has the highest urban growth rate of all world regions, at over 4% per year

# The management of cities

## Economic and political processes

The management of cities has become a significant issue over the last 150 years, partly because of the rapid growth of urban populations.

During the rapid growth phase in the nineteenth century cities were not, by and large, managed at all. Birmingham and Manchester, for example, grew in an unrestrained and uncontrolled manner without planning controls. Manchester had a population of 160 000 in 1832 but no government other than a medieval manorial court.

Earlier, from the sixteenth and early seventeenth centuries, there had been a steady growth of urbanisation throughout Europe, which resulted in a breakdown of sanitation and hygiene. It was said of London that:

> The streets were filthy without, the houses filthy within. The rooms of the poor were more like pigsties than human habitations, unventilated, and strewn with rushes, which were seldom changed. There were no underground drains, and the soil of the town was soaked with the filth of centuries.
>
> Poore, G. V. (1889) *London (Ancient and Modern)*.

*Photograph 14.4 A Victorian slum in the City of Westminster, London (1872)*

*Human systems, processes and patterns*     **A2 Unit 5**

The problems in such cities were addressed initially by wealthy industrialists, who saw that the impact of disease and unsanitary conditions threatened the whole process of industrialisation and its profits. The workforce became less and less productive as ill-health and death rates rose owing to disease brought about by insanitary conditions. The wealthy industrialists also formed the dominant political group. In time, the responsibility for public health was shifted to the government; both local and national government funded public health improvements out of taxation.

The pivotal event in the UK was the Public Health Act of 1848, which laid down legislation controlling several aspects of public health. The government was reluctant to legislate, but probably had little choice given a severe cholera epidemic the previous year. There was also fear of Chartist radicalism gaining ground in the cities and threatening the whole edifice of industrial capitalism. The legislation followed the recommendations in Edwin Chadwick's *Sanitary Report* (1842) and included the following:

- the establishment of a Central Board of Health
- corporate boroughs were to assume responsibility for drainage, water supplies, removal of 'nuisances', paving, etc.
- finance was to be raised from the rates to pay for improvements
- Local Boards of Health *had to be set up* in places where the death rate was above 23 per 1000

### Case study 14.5 City management: Birmingham in the 1860s

In common with many other industrial cities in the mid-eighteenth century, death rates in Birmingham were rising despite the demographic trend being generally downwards. From 1851 to 1861 there were 34 517 infant deaths in a population of 290 000. Locally death rates reached as high as 27 per 1000 even as late as 1875, although the rate was only 13 per 1000 in leafier Victorian suburbs such as Edgbaston.

Smallpox, diphtheria and diarrhoea were the killing diseases, although among the adult population post-natal depression, alcoholism and epilepsy were often cited as causes of death. Sewers were rare and in this 'immense workshop, a huge forge, a vast shop' (de Toqueville 1835) most of the houses were back to back with multiple occupancy. As Charles Dickens expressed it, 'the din of the hammers, the rushing of team and the dead heavy clanking of engines was the harsh music which arose from every quarter' (*The Posthumous Papers of Mr Pickwick*). This 'harsh music' came from craft workers with a growing accumulation of skills using increasingly sophisticated machinery in a truly import-substituting city.

The initial reaction to this deteriorating urban environment was flight. The middle classes moved out to the first 'suburbs' and thus removed themselves from the noxious smells and risk of disease.

In 1842, Chadwick published his *Report on the Sanitary Condition of the Labouring Population of Great Britain*, in which he argued that it was unsanitary conditions and poverty, rather than moral defects (as previously argued by Malthus and others), that made the working class 'lazy and feckless'. For Chadwick, sanitary reform was cheaper than the Poor Law and more effective in creating a productive working class. It made perfect sense for the industrial families that dominated the town to begin the process of cleaning up the city. The cleaning process didn't really gather pace until the late 1860s, when the potential threat of civil disorder became all too apparent. This concern was subsequently reinforced by the violent uprising of the Paris commune in 1871.

## Chapter 14: Rural–urban interrelationships

Decades after the reform of housing and health conditions in most MEDC cities, many world cities are facing crises that are not dissimilar to the nineteenth-century crises of British cities. This is especially true of the most rapidly growing urban areas in Africa and Asia. The most compelling issues facing the managers of cities and their political masters are poverty and environmental issues.

- **Poverty:** up to 60% of the urban inhabitants of African cities live below the poverty line. Poverty is an issue in itself because of its demoralising impact on populations and the potential political impact of growing numbers of discontented citizens.
- **Environmental issues:** 40% of the global urban population have no access to safe drinking water. The results of this are obvious enough, with very high infant mortality a grim characteristic of large urban areas in much of the world. Air pollution, ground subsidence and water pollution are significant problems in many cities, as is the destruction of the natural environment partly caused by the need to manage waste disposal.

The stated aims of urban management set out by the United Nations Conference on Human Settlement (Istanbul, 1996) are to make cities healthy, safe, equitable and sustainable. The conference recognised three critical areas that need to be tackled.

- **Housing:** international law recognises basic housing as a human right. Up to 500 million urban inhabitants have no homes and the same number live in conditions that are insanitary and dangerous. Your own case studies should allow you to illustrate a range of management policies used to address this problem.
- **Clean water, sanitation and waste management:** the most important causes of death and disease are those related to a lack of clean water and the lack of a proper sewage disposal system.
- **Public transport:** in almost all the world's large cities, public transport systems are under considerable pressure. High levels of private vehicle usage have led to serious deterioration in air quality, and journey times have scarcely improved in many cities.

Not all management issues involve dealing with poverty and the related environmental problems. An increasing number of MEDC cities plan for growth to maintain competitiveness in the global economy, and are attempting to provide the infrastructure that is 'required' by international capital. In other words, how do you make a city attractive to foreign and domestic investors? Changes may include economic development such as transport facilities (Birmingham) or cultural facilities (Barcelona, Glasgow).

You should be familiar with the idea of **sustainable development**, which embraces not only the natural environment but also social and economic goals. For example, public housing provision has provided a base for economic development in Hong Kong through subsidies, while in Singapore the application of ethnic ratios has helped avoid high levels of racial and ethnic segregation, which are so common and so divisive in other cities.

*Human systems, processes and patterns* **A2 Unit 5**

## Case study 14.6 City management: Mumbai in the 2000s

The current conditions in many Indian cities are comparable to those of nineteenth-century cities in Britain. The descriptions of nineteenth-century Birmingham could be equally well applied to those in Calcutta, Delhi or Mumbai today. There are over 10 000 slum communities in these three cities, with a total population of more than 15 million.

Today in Mumbai slum dwellers make up 60% of the population, approximately 7 million people who are housed on about 5% of the land area. The slum dwellers have spread into areas of neighbouring Byculla, such as Mahim Creek and Matunga, where they find space, albeit by roadsides and on wasteland.

*Photograph 14.5 Slums in Mumbai, 2001*

In 1985, the local government tried to improve the problem by initiating the Slum Upgradation Project, which offered legal land ownership to slum households. By giving people an interest in their housing, and by ensuring home ownership, the authorities hoped that slums would slowly disappear. However, the programme was limited to only 10% of slum households and scarcely kept pace with the growth of the city. Again and again slum improvement schemes have failed to solve the most basic of problems.

Some observers have suggested that the reason for the consistent failure to address the problem is rooted in the absence of those very forces which led to the cleaning up of cities in MEDCs such as the UK.

- First, it is argued, there is an overabundance of labour streaming in from the countryside. This excess cannot be 'exported' to the New World; Europe exported some 45 million people during the 'long' nineteenth century (1805–1914). It ensures that businesses remain profitable without having to improve the working conditions of the poor.
- Second, the pressure from below is absent. Political power in India rests firmly in the hands of the middle classes, and they have ensured that the sanitation of their residential areas is paid for by the state. There is little political activism among the very poor, and they are divided and not unionised.
- Third, the governing classes are much less fearful of epidemics because of the Western medicines that can, to a large extent, immunise them against the risk.

Overall, the political need to improve the condition of the poor is largely absent — a situation that is rationalised by the very scale of the problem. If India were to provide basic sanitation for the 700 million Indians who currently lack this facility, the cost would run to about £100 billion.

Chapter 14 *Rural–urban interrelationships*

# Agriculture and rural areas

In most rural areas agriculture remains the dominant form of land use.

## Agricultural systems

### Subsistence farming

In a **subsistence farming** system, the aim is to grow enough food and fibre for your own needs, to collect your fuel and building materials from natural sources, and so hardly enter into the cash economy. About 30% of the world's population and about 60% of the world's farmers are living on small area subsistence farms. Many African countries and much of southern and eastern Asia contain a high proportion of intensive subsistence agricultural systems, of which the most important for human food supply is wet-rice agriculture (Figure 14.9).

*Figure 14.9 The distribution of subsistence agriculture*

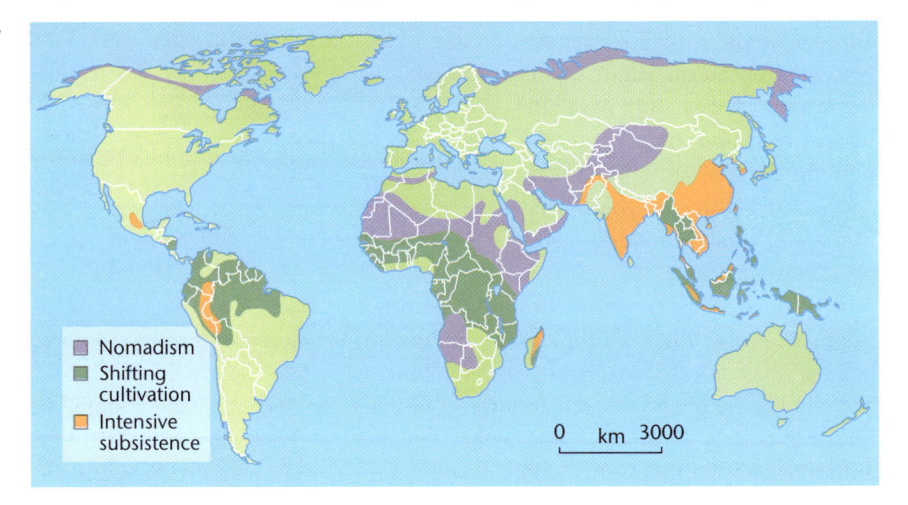

It used to be taken as a basic assumption that subsistence agriculture was inefficient and that the commercialisation of farming would bring instant returns in terms of increasing yield. Early development theorists suggested that a loss of labour from the countryside to the industries that developed in cities would not lead to any loss of output. This has not proved to be the case. Efficiency in agriculture, as practised in less-developed societies, almost always leads to more leisure time and thus more time for other activities.

### Commercial farming

**Commercial farming** systems produce agricultural commodities for sale, mostly food products but also fibres and oils. It does not matter to the producer where this food is sold and it may be to international markets. Capital is used, in varying degrees, to purchase items such as tractors and machinery, fertilisers, pesticides, improved plants, better breeds of animals and other technological innovations. Commercial farms tend to be larger than subsistence units, but very small farms may well be commercial. For example, some of the rice farms of Japan are less than 1 hectare. Not all commercial farms provide a living for their occupants, who may have other jobs to bridge the gap. This is not a common tradition in the UK, but is well-known in continental Europe and in south Asia. The large amount of capital required, including money, equipment, labour and management costs, tends to restrict ownership greatly and favours operators of large-scale units. Increasingly commercial farms are monocultural, producing only one crop, and the impact of agribusinesses is growing.

## Agribusiness

**Agribusiness** is a sub-set of commercial agriculture. It is a form of commercial agriculture in which ownership is generally divorced from management and where there are linkages with food retailers or traders. This type of farming system includes tropical and sub-tropical plantations such as LEDC banana and sugar cane growers and mid-latitude grain farming, such as in North America.

### Key terms

**Agribusiness** A type of commercial farming that is characterised by corporate ownership, monoculture and links with food retailers.

**Commercial farming** Farming in which the main motive is to grow crops for sale.

**Subsistence farming** Farming in which the main motivation is to provide food for the family unit.

### Examiner's tip

- It is the primary motive of farming that gives us the distinction between subsistence and commercial agriculture. Most subsistence farmers also produce a 'cash' crop for sale, but that is not their primary motive for agriculture — that is survival.

### Case study 14.7  Agribusiness: Cargill

Cargill is a US corporation based in Minnesota, but employing over 100 000 people in more than 50 countries. Operating with the motto 'Nourishing Ideas, Nourishing People', it is both the world's largest privately owned company and one of the world's largest agribusinesses.

Cargill began business as a wheat and corn trader, establishing silos across the USA and making a profit by buying and selling crops as it spread westwards in the nineteenth century. In recent years it has become much more involved with production, not so much of agricultural crops themselves but of derivatives from them. Examples are ethanol and fructose, from which it manufacturers a whole range of plastics and oils.

It is also one of the world's major players in the development of genetically modified crops. Cargill develops genetically modified organisms in its own laboratories, and last September gave $10 million to the University of Minnesota for a plant genetics research facility. Cargill has close ties with Monsanto, the biotechnology firm. It sold its international seed business to Monsanto in 1998 and has agreed to manufacture commercial livestock and poultry feeds using Monsanto's proprietary germplasm. Cargill aims to motivate its employees with the belief that the company's mission is to provide food to the world, and hence to tackle hunger. This mission statement is supported using two main arguments.

- Localised food production and distribution is vulnerable to instabilities in local conditions — from weather to pests to civil disorder. By contrast, a globalised system has a range of sources of supply, so there is no crisis if one fails; meanwhile the market allocates these fairly and efficiently.
- Greater agricultural production, achieved through economies of scale arising from specialisation in particular crops and through more efficient methods, will both increase farm incomes and reduce prices for consumers.

# Chapter 14  Rural–urban interrelationships

On 9 July 1999 Cargill won approval from the US government to acquire the grain-trading operations of its primary rival, Continental Grain. The approval came over the objections of attorney general's offices from farm states, the Farmers Union, consumer and green groups, which maintained that the takeover would create a near monopoly in the grain business. Combined, the two companies control 94% of the soybean and 53% of the corn market.

Cargill has come in for a considerable amount of criticism. The many critics of agribusiness giants claim that they benefit hugely from export subsidies from Northern governments, thus enabling them to undermine Southern farming systems.

'Already companies like Cargill have big chicken factories in many countries and they buy our cheap corn and soybeans and turn it into cheap manufactured feed that they can ship all over the world and feed their chickens,' explained George Nailer, President of the National Family Farm Coalition. 'Then (Cargill) floods the markets of various countries with cheap chickens and destroys their farmers,' he added.

### Collective farming

**Collective farming** systems are usually located in countries with a centrally planned economy, such as mainland China, North Korea, Vietnam and Cuba. In these societies agricultural production operates under a rigid system of collective and state farms. Land is collected into large units of operation under government supervision. The workers tend the land and receive a small salary but do not earn any profit for their produce. Some private ownership is allowed. In some countries, small private plots are provided for members of collective farms, state farm employees and other workers. These systems are generally not as productive as comparable privately owned operations.

## Agricultural efficiency

It is important to remember that different agricultural systems can co-exist in the same local and national areas. There is no doubt that commercial agriculture has grown at the expense of subsistence farming over recent years, but there has been much discussion about their relative efficiencies. It used to be believed that commercial agriculture was more efficient than subsistence farming but, in reality, a more complex picture emerges. It depends a great deal on the criteria used to measure efficiency.

The efficiency of any system means the ratio between the energy produced and the energy put into it. The more energy we get out per unit we put in, the more efficient the system. There is always some loss of energy, so no system is 100% efficient. The energy used in agricultural systems is of two types:
- solar energy, which varies according to latitude and day-length
- other sources of energy known as 'support' energy, which include labour, fossil fuels and the energy used in making equipment

To simplify the calculation, solar energy is omitted from the calculation. The second category is defined differently from textbook to textbook, from one piece of research to another. This reflects both the difficulty of the measurement and, sometimes, the outlook of the author. Those who defend modern commercial farming and agribusiness tend not to include the energy consumed in processing and distributing food. Some writers will also omit labour.

*Human systems, processes and patterns*  **A2 Unit 5**

Having made all these simplifications, we arrive at a rather different measure of efficiency: the **output/input ratio**. This figure can be larger than 1, suggesting we are getting more out of the system than we are putting in simply because some inputs are omitted. When growing plants, it is obvious that ignoring solar energy should lead to an output/input ratio of more than 1 because the plant is using solar energy to power its growth. A useful definition of **sustainable agriculture** is that the output/input ratio must be over 1. Table 14.7 gives output/input ratios for a range of agricultural systems.

| Farming system | Output/input ratio |
| --- | --- |
| UK agriculture in the 1850s | 40.3 |
| UK agriculture in 1971 | 2.1 |
| Shifting cultivation | 28 |
| Hunting and gathering | 4 |
| Maize, using oxen | 4 |
| Maize, using human labour only | 14 |
| Potato production, UK | 1.6 |
| Beef production, UK | 0.08 |

Table 14.7 suggests that agriculture hasn't necessarily become more efficient over time and that modern commercial farming is not inherently more efficient than subsistence agriculture.

Levels of intensity of agricultural systems also vary considerably.

- **Intensive farming** uses high levels of inputs (labour and/or capital) per unit area of land. It is thus characterised by high output per unit area.
- **Extensive farming** uses low levels of inputs per unit area of land and achieves low outputs per unit area.

*Table 14.7 Agricultural efficiency of selected crops and farming systems*

A common error is to assume that intensive agriculture is more profitable than extensive agriculture and hence, in some way, better. Of course it is more profitable per unit area, but agricultural systems that are extensive (cattle ranching) may involve large areas of land per farm and can be just as profitable as smaller, more intensive farms. Intensive farms can be intensive in their use of labour or intensive in their use of capital equipment. In the USA and several European countries there are more tractors than people involved in agriculture. Agricultural yields have risen in recent years, reflecting increasing intensity, especially in the use of fertilisers and improved crop varieties (Figure 14.10).

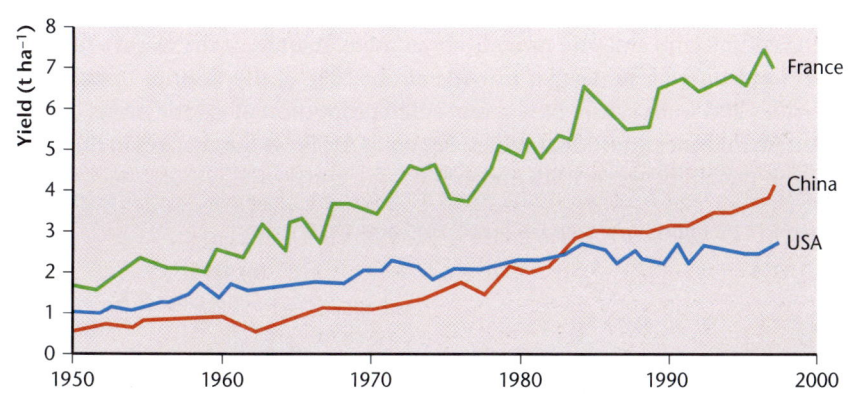

*Figure 14.10 Wheat yields in France, China and the USA, 1950–97*

### Examiner's tip

- Efficiency in agriculture is a complex topic. Historically, Western experts have tended to treat subsistence agriculture as though it is inherently inefficient. Recent work has led to a re-evaluation of that judgement.

# Chapter 14  Rural–urban interrelationships

A further distinction can be made between **pastoral** and **arable** systems. In almost all cases arable farming gives a higher output of food per unit area and is thus more efficient. Pastoral farming may often be a response to a constraining factor within the environment that prevents arable farming, usually a climatic constraint.

The global land area is roughly divided into thirds:
- 35% is used to produce agricultural products
- 31% is used to produce wood and timber
- 34% is in non-agricultural use

Of the area used for agricultural products, 24% is pasture and meadowland used for animal production, 10% is arable used for annual crops such as grains and cereals, and 1% is used for permanent crops such as fruit and nut trees. However, it needs to be added that many of the crops grown are not for human consumption. Animal feed is grown on 75% of EU and over 50% of US agricultural land.

A total of 3.1 billion hectares of land throughout the world is potentially arable. Of that, only about 1.5 billion hectares are being used for field and horticultural crops. Much of the unused land is in areas lacking essential transport or other infrastructure needed for the maintenance of commercial food production.

Virtually all human nutrients come from three sources: crops, animal products (meat, milk, eggs) and aquatic foods. About 3000 plant species have been used for food and more than 300 are widely grown. However, only a dozen different species of plants account for 90% of the world's food. Crops provide over 90% of the human population's calories and about two-thirds of its protein. The world currently produces 2 billion tonnes of cereals, which provide most of our current needs. This amount would feed 10 billion people following an Indian diet of mainly cereals, but it would only feed 2.5 billion Americans, who consume grains largely in the form of livestock and poultry.

Among crops, the cereals (wheat, rice, corn/maize and millet) occupy three-quarters of the crop area and provide about three-quarters of the world's calories. Root crops, oilseeds and sugar provide about 20% of the calories from crops. Vegetables and fruits satisfy only a very small proportion of caloric needs but are important for other nutritional values. Figure 14.11 shows the changes in the crops grown in the UK between 1945 and 1995.

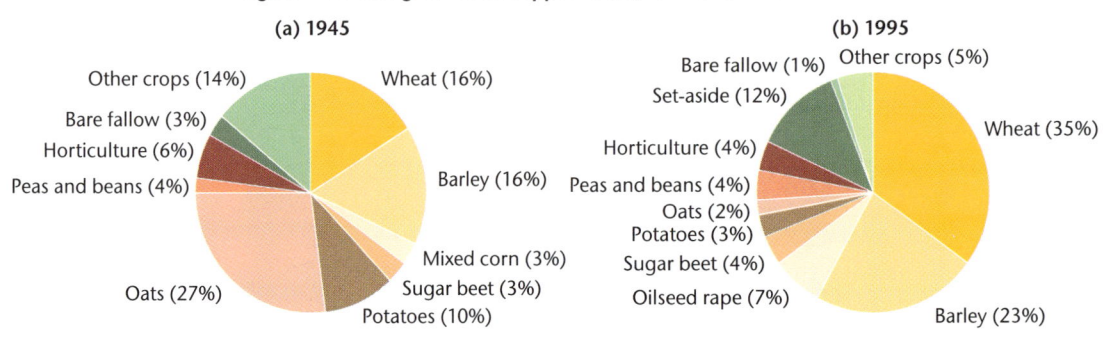

*Figure 14.11 UK agricultural cropped areas, 1945 and 1995*

(a) 1945
- Other crops (14%)
- Bare fallow (3%)
- Horticulture (6%)
- Peas and beans (4%)
- Oats (27%)
- Potatoes (10%)
- Sugar beet (3%)
- Mixed corn (3%)
- Barley (16%)
- Wheat (16%)

(b) 1995
- Bare fallow (1%)
- Other crops (5%)
- Set-aside (12%)
- Horticulture (4%)
- Peas and beans (4%)
- Oats (2%)
- Potatoes (3%)
- Sugar beet (4%)
- Oilseed rape (7%)
- Barley (23%)
- Wheat (35%)

*Human systems, processes and patterns*  **A2 Unit 5**

## The physical constraints of agricultural systems

The location of agriculture and the development of particular agricultural systems are inevitably influenced by both the physical and human environments. The significance of these factors depends on the type of agriculture and whether it is commercial or subsistence. The degree to which physical factors can be used to explain variation in location of different types of agriculture depends on the scale. On a global scale there is no question that climate is the dominant influence, especially precipitation and the length of the growing season, but on a more local scale soils and relief explain variation better.

### Physical location factors

This section will consider the following factors:
- soil
- climate
  - precipitation
  - length of thermal growing season
- relief
- aspect

On a global scale, the pattern shown in Figure 14.12 is largely explained by world patterns of precipitation and temperature. The world's climate zones are shown in Figure 14.13. However, the climatic zones do not allow for human intervention, especially in the form of irrigation systems or hosuing animals during colder seasons. Climate sets limits on natural vegetation, but agricultural practices can overcome these to an increasing extent. In addition plant breeding and the genetic manipulation of crops can introduce resistance to different climatic conditions.

At a more local scale, the variation of particular cropping patterns within a country can be explained by other physical factors. These days farmers can use sophisticated tools such as global imaging systems (GIS) to judge the possibilities and constraints of their environment.

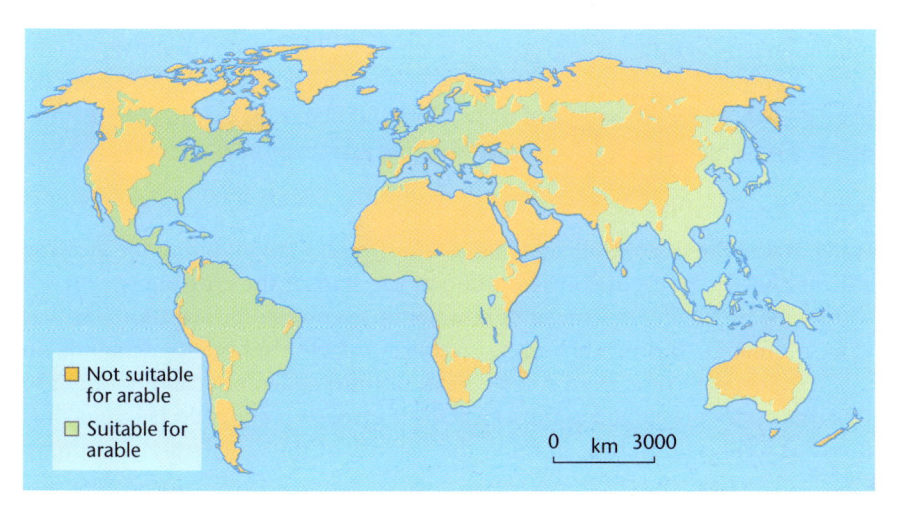

*Figure 14.12 Global distribution of available arable land*

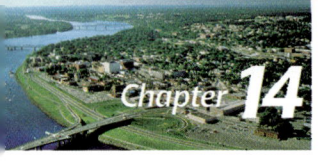

## Chapter 14  Rural–urban interrelationships

*Figure 14.13 World climate zones*

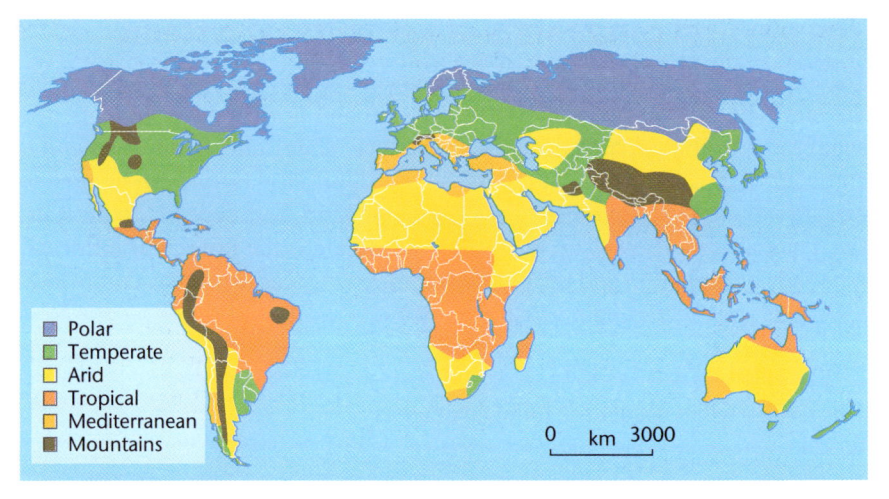

- Polar
- Temperate
- Arid
- Tropical
- Mediterranean
- Mountains

0 km 3000

The **agro-ecological zone** (AEZ) approach takes a hypothetical farmer who has the task of judging the suitability of a particular area for growing a range of crops. Details of the crop requirements are fed into the equation (or algorithm). The farmer would use a whole range of criteria to assess each unit of land, such as the local climate conditions, the quality of the soil, the slope and the possibilities of using different types of agricultural input (fertilisers, pesticides, machinery). The AEZ algorithm proceeds in the same way. Its use is described later in this section.

### Temperature

Plant development depends on temperature. Plants require a specific amount of heat to develop from one point in their life cycle to another. The critical temperature for the germination of the seed is the soil temperature, rather than the air temperature, and for many crops it is about 6°C. The number of days when temperatures exceed this is known as the **thermal growing season**. In reality many farmers use more sophisticated data than this by calculating **degree days**. This calculation uses the actual amount by which the daily temperature exceeds the minimum temperature for crop growth. It allows for the fact that a cold day might slow down plant development whereas a hotter than average July will speed up plant growth. In all this, you need to bear in mind that crops vary in their requirements. Day length is also critical in controlling germination and growth.

### Water availability

Water availability is controlled by the levels of precipitation during the growing season plus the levels of soil saturation at the start of the growing season. Of course, the possibility of using groundwater or river water for irrigation also has to be estimated. This is much more complex to factor in because demands on this resource will come from several areas, not just agriculture, and the right to take water from the ground or from local rivers involves political decision-making. It is also important to bear in mind the security of the water supply, as this is determined by the variability of the rainfall pattern from year to year and the intensity of the rainfall. The rainfall intensity is significant, in that intense convectional storms may lead to rapid runoff and little infiltration, and are therefore of less value than more gentle rain.

## Evapotranspiration

The amount of water that can potentially be evaporated and transpired is determined by the quantity of solar insolation received, which is controlled by temperature and day-length. The actual amount of water evaporated or transpired is determined by potential evapotranspiration (PET) and the water available. Where the PET is greater than the AET (actual evapotranspiration), plants will suffer stress unless irrigation can be provided. The AEZ algorithm uses a more detailed set of climate indicators, which include:
- monthly precipitation
- minimum and maximum temperature
- relative humidity
- sunshine hours per day
- wind speed

## Soils

Soils vary in their fertility and their water retentiveness. Some soils (such as chernozems) can store water much better than others (such as podzols). A given amount of rainfall might be sufficient to grow a particular crop on a soil with high water-holding capacity, while the same amount of rainfall might be insufficient if the soil lets the water infiltrate away into the groundwater.

Soil fertility can be judged using general classifications developed by the FAO (Food and Agriculture Organization) and adding other factors such as soil texture. This allows a soil rating which defines the suitability of each soil for each individual crop at defined levels of inputs and management. For instance, some soils may be very stony, which is a growing problem for root crops; others may have chemical problems such as shortage of key nutrients, which will reduce the potential crop production.

Soil texture combines with climatic factors to determine workability and wetness-related constraints. For example, the climate conditions might suggest that sowing is possible at a particular time. However, waterlogged ground might make it impossible to get onto the land without compacting the soil, and waterlogging might lead to the seed rotting in the ground.

## Slope and terrain

The slope and general terrain impose limits on crop production; slope angle is particularly important. For example, soils on steep slopes are much harder to cultivate than soils in floodplains. Steep slopes are also more susceptible to erosion and, consequently, fertility loss. The AEZ model takes into account a number of these constraints (specified by a terrain slope suitability classification) which further reduce the attainable grain production.

# Chapter 14 Rural–urban interrelationships

## Using the AEZ algorithm

After the factors discussed above have been mapped into the equation, the algorithm generates recommendations for the best crop or crops for a particular area. This involves more than one crop. For example, the best combination in the double-cropping region of eastern China might be rice or maize in the summer months with winter wheat for the remainder of the year. Each area can be classified against an ideal. For example, if the maximum wheat yield for a region is 10 tonnes per hectare, areas can be rated by how close they should be getting to that figure:

- very suitable = yields are equivalent to 80% or more of the overall maximum yield
- suitable = yields between 60% and 80%
- moderately suitable = yields between 40% and 60%
- marginally suitable = yields between 20% and 40%
- not suitable = yields between 0% and 20%

However sophisticated the algorithms of climatic constraints might be, the actual productivity of regions is influenced by many other factors. Human behaviour is seldom consistent, because of cultural and political constraints, and farmers are not always logical in their choices.

## Human characteristics

The impact of human behaviour on crop production is most easily studied by changing scale yet again and looking at variations in agricultural land-use within a local area. This is dominated by factors that are largely human and may include:

- systems of land tenure
- accessibility
- competition
- inheritance laws
- markets
- government

### Systems of land tenure

The various systems of land tenure range from large estates, often held for tax purposes (which are uncommon in modern Europe but are important in Latin America), through to share cropping and plantation systems. The great 'latifundia' estates were seldom farmed optimally because they generated rents sufficient to maintain the landowners in luxury without the need to innovate. In Europe, the key distinction is between the farmer as owner, the farmer as tenant and the farmer as manager. Agricultural land use is often determined by the differing access these various groups have to capital for investment.

### Inheritance laws

In general terms farm size has increased; very sharply so in some world regions such as Europe. This is often associated with increasing levels of commercialisation in agriculture. However, in some areas inheritance laws of both MEDCs and LEDCs and dowry customs mean that land becomes increasingly fragmented. These social factors have a negative effect on the development of commercial farming.

## Accessibility

Transport availability and cost, and distance from markets impact on price and on the possibility of preserving agricultural products. Even in a world in which transport costs have fallen sharply, this has a continued influence. The distribution and type of dairy production is an obvious example.

## Markets

The size and type of markets are very significant. Tastes in food and the impact of other cultures on these (e.g. McDonald's) clearly have a profound influence on what food is produced. A physiological intolerance of dairy products in some ethnic groups is an obvious explanation of the lack of large-scale milk production in, for example, much of China, even in areas where the climate is suitable. In contrast, the existence of a dairy industry in the sub-tropical southern USA is an indication of the popularity of milk products among Anglo-Saxons.

Increasing affluence has made a considerable impact on demand patterns in the UK and the rising popularity of organic food products represents a considerable shift in emphasis that is significant for producers.

In the high plains of the USA there has been a growth of feedlots where cattle are kept in artificial conditions and fed on maize rather than being allowed to graze. This reflects the preference in US kitchens for large tender steaks rather than the smaller, better flavoured product preferred by Europeans.

## Competition

Globalisation of food production and the removal of tariffs on imported foodstuffs has meant that farmers have had to face increasing competition from imported products. In the past, a country such as Uruguay specialised in the production of beef and beef products for export. However, Uruguay lost its market in the 1950s as other producers developed their own beef industries and the original market (mostly European) built itself a new agricultural policy in the period following the Second World War. The impact on Uruguay was catastrophic.

## Government

Government action is by far the most significant of the human variables that affect agriculture. Governments have interfered in food production for centuries, but during the twentieth century the extent and impact of government interference increased greatly. The best way to consider this is to look at the changes in EU policies towards agriculture over the last 50 years and the impact of these policies on food production and patterns of cropping. The important social changes that have taken place should also be addressed, including the decline of family farms and the rise of agribusinesses. However, the increasing role of the World Trade Organization (WTO) and the TNCs in developing global markets and global production patterns challenges some of the power of national governments in controlling their own agriculture.

## Four contrasting agricultural systems

The following four case studies illustrate many of these variables.

# Chapter 14 Rural–urban interrelationships

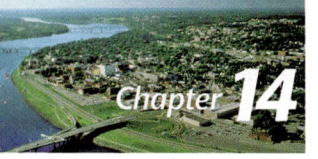

## Case study 14.8 Agricultural systems: shifting cultivation and deforestation

**Shifting cultivation** is a traditional method of subsistence agriculture in which the area cultivated rather than the planting of crops is rotated. Short cropping periods of a few years are followed by long fallow periods in which the plot is abandoned. Long-term fertility is maintained by allowing natural vegetation to regenerate on the fallow land. The clearance of new or previously cropped land is often accomplished by cutting and burning vegetation — hence the alternative name of 'slash and burn' agriculture.

Shifting cultivation systems encompass a wide range of land-use practices developed and modified by farmers in varied social, ecological, economic and political settings. They are usually used for subsistence, but often grow at least one cash-crop and have a complex relationship with the market economy. Shifting cultivation is the dominant agricultural system in tropical rainforests and thus has come under close scrutiny as an agent of deforestation.

An estimated 300–500 million people depend upon shifting cultivation. It used to be believed, especially by agricultural experts from MEDCs, that shifting cultivation was an outmoded system needing replacement by more modern and supposedly productive systems. However, Ester Boserup's classic work *The Conditions of Agricultural Growth* (1965) suggested that shifting cultivation was an efficient and sophisticated adaptation to conditions where labour, not land, is the limiting factor in agricultural production. It offered a sustainable 'solution' to the problems posed by the environmental conditions of the rainforest.

Forest farmers, in areas as diverse as Malaysia and Amazonia, are from cultures with a long history of shifting cultivation and have a profound understanding of their local environments. They practise sustainable agriculture, which has existed for many thousands of years, with no obvious damage to the ecosystems in which they live.

The negative view of shifting cultivation, with images of burning forest, come from a quite different type of forest farming, better known as 'shifted' cultivation. Shifted cultivation is usually a result of resettlement and transmigration schemes in which people are moved to new areas and have to survive as best they can.

A large part of the forest destruction in Amazonia can be explained by a land market which promotes forest clearance by migrants, often from rural southern Brazil, who remain on one plot of land until it is exhausted, and then sell it in an attempt to move on to better areas. The government might actually promote this sort of movement, in part to clear the way for large landowners, but also to relieve pressure on the cities where rapid in-migration is seen as undesirable for social, economic and political stability (see Case study 6.3 on Indonesian transmigration).

In reality, cultivation of primary as opposed to secondary forest is quite unusual. Of the 10 million hectares of tropical forests cut by shifting cultivators every year, only 17% is primary forest and in most of these cases the cutting is not permanent.

Most observers, including the World Bank, now agree that the causes of deforestation are much more complex than had previously been thought, and that it is wrong to blame the expansion of traditional shifting cultivation practices. Deforestation today is influenced by:

- the expansion of unfamiliar farming systems into forest areas, such as ranching or the growing of export crops
- government 'development' projects that use traditional subsistence areas for large-scale projects, pushing shifting cultivators into smaller and smaller areas of forest

*Human systems, processes and patterns*  **A2 Unit 5**

### Case study 14.9 Agricultural systems: wet-rice agriculture and the Green Revolution

*Photograph 14.6 Wet-rice agriculture in Vietnam*

Wet rice, so-called to contrast it from rice grown as any other cereal crop on 'dry' land, is the world's most important food crop. Most wet-rice farmers are subsistence farmers. Seeds are sown in small seedbeds and the seedlings are transplanted in prepared paddy-fields. While the plants are maturing, they must be kept irrigated, but as the rice ripens the fields are drained. The rice is then harvested and usually threshed by hand. Wet-rice agriculture is usually labour-intensive, although Japanese rice production has largely switched to the use of machinery.

Rice agricultural systems provide not only the food but also the main source of income and employment for about 1 billion people, principally in Asia, but also in Africa and Latin America. It is high yielding wherever it is grown, with the world average being about 4 tonnes per hectare, but the highest yields occur in regions where two or even three crops are planted a year. In these areas the water levels are controlled by irrigation, as opposed to 'rain-fed rice' which is dependent on the reliability of the rains.

There are about 50 million hectares under intensive rice-based cropping patterns, producing two to three crops per year, with each crop maturing in about 90 days from planting. Average yields are about 4–6 t ha$^{-1}$ per crop or about 10–15 t ha$^{-1}$ per year.

Wet-rice agriculture is associated with high population densities, sometimes more than 1000 people per km$^2$ (as in Java), because of the high demand for labour at harvest time when the 'new' transplanted seedlings need to be planted as well. This demand for labour led to traditionally high fertility rates in areas such as the Ganges valley, where a larger family was the obvious route to higher output of rice, acquisition of new land and family success.

In recent years the Green Revolution has transformed both the rice-growing regions and these

relationships. The Green Revolution was a response to the problem of feeding a growing world population. There are only two ways of increasing food output:
- increasing the area that we farm
- increasing the yield from the land that we farm

The Green Revolution focused on the second of these, by developing new hybrid varieties of rice (and other crops including wheat and maize). These new high-yielding varieties (HYVs) frequently require more rigorous growing conditions and higher fertiliser, pesticide and herbicide inputs than the older, indigenous varieties that they have replaced. They tend to be shorter stalked, sturdier and bear larger heads. They are also expensive to buy as seeds. The increases in yield have been significant, but critics have pointed out some problems.
- Some of the 'costs' are omitted when the yields are considered. For example, the World Health Organization estimates that half a million cases of acute pesticide poisoning occur every year in LEDCs. In the extreme case of bananas, over 30 kg of pesticide is applied to each hectare of land each year.
- The impact of the Green Revolution has been socially damaging, because only a small number of farmers have been able to afford the costly inputs.
- The Green Revolution allows governments to avoid the 'real' issue of unequal land distribution. Critics claim that what is needed is greater labour intensiveness, not the replacement of labour with machines. Small-scale projects like improved irrigation systems and growing 'green' manures would harness rather than discard rural labour.
- The main beneficiaries of the revolution are the small number of wealthy landowners who are largely preoccupied with export crops, and the MEDC companies who supply the HYV seeds and other inputs.
- The original research was carried out and paid for by companies with a vested interest in the Westernisation of traditional agricultural systems.

## Case study 14.10 Agricultural systems: cattle ranching on the high plains, USA

The word 'ranch' is derived from Spanish 'rancho', which means the land upon which stock is raised. In Texas, the word was first used to describe an enterprise engaged in livestock production on unimproved pastures, occasionally with additional arable crops to supplement the diet. Ranching included raising cattle, sheep and goats, and horses. Today ranching has a poor reputation with many environmentalists, yet it represents a logical adaptation to environments that do not support arable agriculture.

The ability of some animals to digest vegetation that is not digestible by man means that pastoral agriculture makes sense in areas where food crops cannot be grown. Beef cattle are able to digest plant materials that other animals cannot. They graze, so are not dependent on planting, harvesting or any other labour-intensive farm activity. Apart from a little branding and, of course, getting them to market, ranching is an extensive system in which, in the USA, cattle replaced bison as the dominant herbivores. However, in many MEDCs there has been a growing consumption of meat, encouraged by higher incomes and by marketing. This has had two impacts.
- The spread of pastoral agriculture to regions that may not be suited to such systems.
- The intensification of animal husbandry. An increasing proportion of land has been given over to the growth of crops to feed cattle, and feedlots have spread into areas that were once open grazing land.

The first of these impacts is not new. In the late nineteenth century the Aztec Land and Cattle Company expanded its Texas-based operation into Arizona as Texan ranches came under pressure from small farmers and land degradation. Arizona's arid climate provides land with a very low carrying capacity for cattle — about 1 animal for every 25 hectares — but

*Human systems, processes and patterns*   **A2 Unit 5**

Photograph 14.7 Cattle raised in the feedlot system

widespread overstocking led to a doubling or tripling of cattle numbers. Exacerbated by the droughts of the 1890s, this led to a disastrous transformation of the environment. From tall grass country it became a scrub landscape dominated by juniper bushes (juniper seeds only germinate after passing through an animal's gut) and bare ground where the topsoil had been stripped off by flash floods. Similarly, the grazing of cattle in tropical rainforests has caused widespread, and probably irreversible, destruction. It has been estimated that for every quarter-pounder burger made from rainforest cattle, an area of 13 km$^2$ of rainforest is destroyed, adding several kilograms of carbon dioxide to the atmosphere. If every American reduced their beef consumption by half, there would be no need to destroy the rainforest for cattle pasture.

The raising of 'corn-fed' beef in feedlots in the mid-west and high plains of the western USA amounts to a sort of 'battery' production. The American requirement for large steaks means that the cattle are housed in fenced areas and fed on locally grown maize and soybeans rather than letting them graze freely on the range. This leads to a high demand for water for irrigation, especially to grow the maize. The use of former grassland to grow crops for cattle feed has led to soil erosion and nutrient runoff which pollutes water supplies. The bottom line is profit, at least in the short term. Cows grazed on the plains are 4 or 5 years old before they are ready for slaughter, but a feedlot cow fed on corn and protein supplements is ready for slaughter at 14–16 months, growing from 30 kg to over 500 kg in its short life.

## Case study 14.11 Agriculture systems: wheat production in the UK

The UK has long been a producer of wheat. The long periods of daylight in the summer and the relatively moist climate combine to give the UK a comparative advantage in the production of this cereal crop.

Wheat remains the UK's most important arable crop. Around 15 million tonnes is harvested a year, of which 2.5 million tonnes is exported. Just over 40%, or 6 million tonnes of the crop, is used in animal feed

# Chapter 14  Rural–urban interrelationships

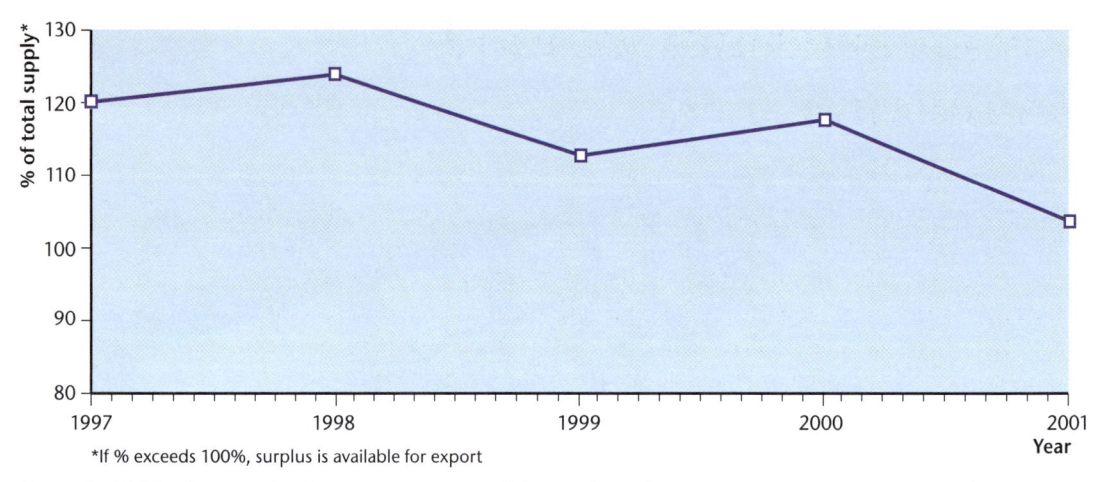

**Figure 14.14** UK wheat production as a percentage of the total supply

*If % exceeds 100%, surplus is available for export

for chickens, cows and pigs. The balance of the crop, 6.5 million tonnes, is used for human consumption, both directly in bread and related products and as an ingredient in a large number of other foodstuffs. Production remains concentrated in the drier eastern half of the country (Table 14.8), although with improved varieties there has been a gradual spread westwards.

- In 1982, wheat production covered 1.6 million hectares, yielding, on average, 6.2 tonnes per hectare at a price of £119 a tonne.
- In 2002, wheat production covered 1.9 million hectares yielding, on average, 8.05 tonnes per hectare at a price of £58 a tonne.

The fall in price is a consequence of changes in EU policy cutting price support, and the reduction in the level of import protection. Net farm income has reduced from £238 per hectare to £38 per hectare in the same period.

The key inputs into wheat production are:
- crop protection products
- fertilisers
- seeds

Crop protection with herbicides can cost anything up to £50 a hectare, depending on the previous crop in the ground. Wheat is so closely related to grasses that it is hard to protect against weeds such

| Region | Tonnes of wheat |
|---|---|
| Eastern | 4 480 000 |
| Southeast | 2 190 000 |
| Yorkshire and Humberside | 2 150 000 |
| Southwest | 1 520 000 |
| West Midlands | 1 250 000 |
| **Total UK** | **16 820 000** |

*Table 14.8 The top five UK wheat-producing regions*

as black grass. Fungicides can cost a further £50 a hectare. Pesticides might add a further £4 or £5 a hectare. Fertilisers are largely made up of nitrogen, phosphates and potassium, of which nitrogen is the most significant. Fertiliser costs have fallen slightly in recent years as farming has become less intensive, but they still amount to over £80 a hectare. Seed costs run at about £40 a hectare.

With profit margins being squeezed, there are several major issues facing UK agriculture, of which the most significant are:
- whether or not to embrace GM crops
- how to diversify rural employment
- whether or not to abandon those farm products for which there is no obvious comparative advantage
- what the relationship should be between agriculture and other uses of the countryside

*Human systems, processes and patterns* **A2 Unit 5**

# The globalisation of food production

The politics of global food production is not a new theme. During the seventeenth century the European North created a colonial system based on plantation agriculture in the South, with these colonies deliberately set up to grow agricultural products for export home. The colonies that lacked mineral wealth were settled by farmers who could extract wealth by exploiting the fertile soils and the tropical or subtropical species that could not be grown in Europe. After decolonisation, various policies were pursued, sometimes with the aim of increasing the independence of the former colonies but often in reality making them even more dependent on the richer North. For example, despite efforts to break out of this system (including a revolution), modern Cuba is still heavily reliant on the production of sugar. Ironically, sugar was first introduced to Cuba in the seventeenth century by Europeans using slaves for labour.

## Case study 14.12 Globalisation begins: the plantation system

In the seventeenth century Europeans began to establish settlements in the Americas. The division of the land into units under private ownership became known as the plantation system. Starting in Virginia, the system spread to the New England colonies. Crops grown on these plantations, such as tobacco, rice, sugar cane and cotton, were labour intensive and there was insufficient labour available.

European migrants had gone to America in order to own their own land and were reluctant to work for others. Convicts were sent over from Britain, but there were never enough to satisfy the tremendous demand for labour. Planters therefore began to purchase slaves. At first these came from the West Indies, where Africans had already been imported, but by the late eighteenth century they came directly from Africa. Busy slave markets were established in many southern cities, such as Richmond, Virginia.

Slaves were in the fields from sunrise to sunset and at harvest time they did an 18-hour day. Women worked the same hours as men and pregnant women were expected to continue until their child was born.

Many crops were significant in the early years of plantation agriculture, but none had quite the same impact as sugar. Paradoxically, sugar cane was first domesticated in New Guinea some 10 000 years ago.

It appears to have reached Egypt around 5000 years ago and ultimately came to Europe on the back of Arab expansion into Spain, where the Moors introduced not only the product but also the methods of refinement that they had developed. Sugar was taken from Spain to the New World in 1493 by Christopher Columbus. Sugar plantations began to develop in the Americas in the early 1500s. Production expanded rapidly, spurred on by an enormous appetite for the product in Europe.

The usual method for increasing output was to shift the production frontier further and further into the hinterland behind the original port of entry. Satisfying the European and North American sweet tooth involved the creation of a massive industry throughout the tropical and sub-tropical regions of the world over the centuries. When independence eventually came to many of these countries in the twentieth century, their position in the world economic system was already established and they found themselves locked into a relationship with the major sugar companies that was hard to escape.

In the world market, sugar-cane-producing countries are in direct competition with sugar-beet-producing countries in the world's temperate zones, such as Europe and North America. In these regions,

# Chapter 14  Rural–urban interrelationships

production is highly mechanised and farmers are heavily subsidised. As with many commodities, the terms of trade have moved and sugar prices have fallen sharply over the last 20 years. Meanwhile, the EU sugar policy, which is a notoriously complex system, produces a problem that is all too easy to recognise: too much sugar. Each year, Europe, which has no comparative advantage in sugar, produces an export surplus of approximately 5 million tonnes, which is 'dumped' on world markets, further depressing the price. The biggest beneficiaries are the large sugar processors. Some of these make the largest profits in the whole corporate sector. These companies benefit from export subsidies worth millions of euros each year, estimated by Oxfam to be worth a total of €819 million in 2003.

Globalisation of food production is obvious if you look around a supermarket. Vegetables from west Africa, fruit from New Zealand, wine from Argentina and tea from India compete for space on the shelves. The causes of such developments are:
- improvements in the technology of transporting perishable products
- reduction in the costs of transporting these products
- promotion of new products by large retailers and giant food companies
- standardisation of food production techniques
- reductions in tariffs and other barriers to trade

These processes need the consent of governments (even if reluctantly given) and hence the central role of politics in driving the globalisation of food production. At the heart of this debate lies a paradox.

Food is not in short supply; in truth, food products have never been so plentiful. Globally we produce enough to feed each human being 3500 calories a day. However, about 30 million people die each year of undernourishment or related diseases. Every year a further 800 million people suffer debilitating illnesses resulting from chronic malnutrition. And some of these people live in the very countries that are increasingly committed to exporting food products to stock the supermarket shelves in the richer MEDCs. Table 14.9 shows how the world's population divides into three groups.

*Table 14.9 The three socio-ecological classes*

|  | Overconsumers | Sustainers | Excluded |
|---|---|---|---|
| Number | 1.3 billion | 3.5 billion | 1.3 billion |
| Income per capita | >US$7500 | US$700–7500 | <US$700 |
| Consumption | Cars, meat, disposables | Living lightly | Deprivation |
| Travel | Car and air transport | Bicycle and public surface transport | Foot or donkey |
| Diet | High-fat, high-calorie, meat-based | Grains, vegetables and some meat | Nutritionally inadequate |
| Drink | Bottled water and soft drinks | Clean water, some tea and coffee | Contaminated water |
| Products and packaging | Throwaway products; discard substantial waste | Unpackaged goods; recycle waste | Local biomass; negligible waste |
| Housing | Spacious, climate controlled, single family | Modest, naturally ventilated, extended or multiple families | Rudimentary shelters or the open, lack secure tenure |
| Clothing | Image conscious | Functional | Second-hand or scraps |

*Human systems, processes and patterns*  **A2 Unit 5**

Famine and malnourishment occur because people can't afford to buy food, not because it is unavailable. This is a reminder of the population and resources debate that was begun by Malthus and still continues today (see Chapter 4).

The forces that drive globalisation are powerful and are certainly not restricted to the production of food. The effects of the globalisation of food production are debatable but include:

- increasing variety of products in MEDCs
- cheaper food for MEDCs
- increasing areas devoted to the growth of export crops in LEDCs
- increasing control of food production by large corporations
- increasing development of chains of production from agribusiness through to major retailers

Much more controversially, globalisation might be blamed for the following problems.

- Increasing dependence of LEDCs on MEDCs. The LEDCs are trapped in a system of dependence reinforced by trade rules to which the ruling groups in LEDCs have often signed up. In common with other primary producers, LEDCs are chasing falling prices by raising output, thus forcing world prices down even further.

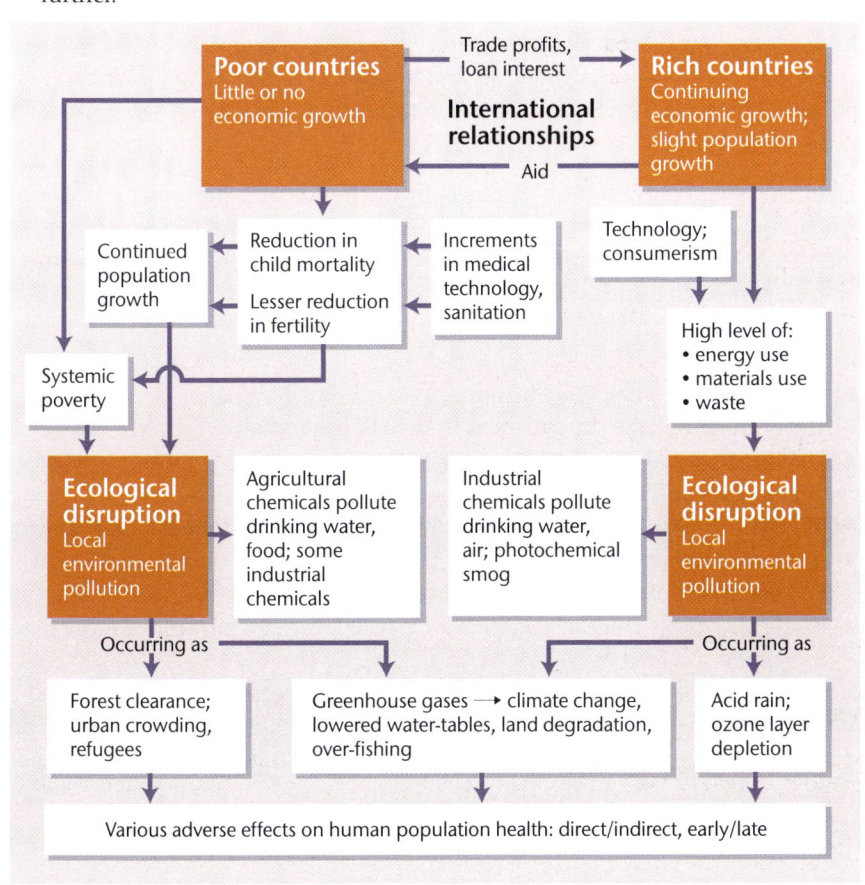

*Figure 14.15 Ecological disruption in MEDCs and LEDCs*

## Chapter 14: Rural–urban interrelationships

- Increasing rural poverty as poor peasant farmers lose their land. Poor farmers cannot afford sophisticated technology and often go into debt to a richer neighbour in an attempt to 'keep up'. When they cannot pay their debts they lose their land.
- Increasing urban malnourishment, as LEDCs switch land previously used for domestic food production to produce export crops.
- Changes in consumer tastes in LEDCs. The introduction of new products to replace traditional foods has meant a shift toward meat and wheat-based diets. Only rarely do these changes fit well into local climatic or cultural norms.

Figure 14.15, although complicated, attempts to summarise the context, causes and consequences of ecological disruption. As the diagram shows, this has effects beyond food production.

### Key terms

**Plantation agriculture** A commercial system growing one crop (monoculture) for export.

**Ranching** An extensive commercial agricultural system that replaces the 'natural' top herbivore with beef cattle.

**Shifting cultivation** A subsistent system of agriculture in which the area cultivated is rotated.

**Wet-rice agriculture** Either a commercial or subsistent system of agriculture that concentrates on the production of rice grown in shallow water.

## The rise of the supermarkets

The increasing concentration of power in the hands of a shrinking number of global corporations (TNCs) is of concern to many. This is especially evident in the food trade. The world's largest corporation is Wal-Mart (owners of Asda stores among many other businesses). Wal-Mart can take over a billion dollars on a decent day's trading. Some of the less controversial facts about modern supermarkets are as follows.

- In the UK eight small retail businesses close every day.
- The number of large supermarkets in the UK has tripled.
- The British Retail Planning Forum has discovered that 276 jobs are *lost* every time a new supermarket opens.
- In Britain, 75% of supermarket customers travel to and from the shop by car.
- A typical large supermarket in the UK can cost the local community over £25 000 a week in pollution and congestion externalities.
- In the UK, the average vegetable now travels over 600 km from producer to consumer.
- The UK imports 213 000 tonnes of pork a year and exports 272 000 tonnes.
- Fifty years ago, farmers in Europe and North America received between 45 and 60% of the money that consumers spent on food. Today, that proportion has dropped to just 7% in the UK and 3.5% in the USA, but remains at 18% in France.

*Human systems, processes and patterns* **A2 Unit 5**

- A typical US wheat farmer receives 6 cents of each dollar spent on bread, about the same as the cost of the wrapping.
- In the mid-west, 75% of sheep are slaughtered by ConAgra, Superior Packing, High Country and Denver Lamb.
- In the USA 80% of beef cattle are slaughtered by IBP, ConAgra, Cargill and Farmland Beef.
- In the USA 60% of pigs are slaughtered by Murphy Family Farms, Carroll's Foods, Continental Grain and Smithland Foods.
- Six firms process half of the chickens in the USA: Tyson Foods, Gold Kist, Perdue Farms, Pilgrim's Pride, ConAgra Poultry and Continental Grain.
- 95% of American broiler chickens are sold under contracts to fewer than 40 firms. Nationally, 76% of the grain (corn, wheat and soybeans) is sold to four companies: Cargill, Archer Daniels Midland, Continental Grain and Bunge.

Many people have serious worries about the following type of scenario in which the label 'organic' is attached to meat raised in a way that is difficult to see as logical, let alone sustainable.

> A male calf is born, castrated and, for two years, lives on a Texas range, feeding on grass that has been 'fertilized' with sewage sludge from New York City. The grown steer is then shipped to a feed lot and jammed in a pen. While there, he gets sick and is injected with a variety of antibiotics. Meanwhile, he fattens up on Purina cattle feed, made from Monsanto's genetically engineered grains and supplemented with animal protein that was rendered from the carcasses of diseased animals. He is eventually sent to the abattoir to be slaughtered. Then, the meat is irradiated, packaged and shipped to Chicago, where it is sold under the label, 'Meets USDA Organic Requirements'.
>
> Joel Bleifuss, *In These Times* magazine, February 1998
> quoted in http://www.thirdworldtraveler.com/
> Public_Relations/Organic_Corporations.html

### Case study 14.13 Anti-globalisation: the slow food movement

The 'slow food' movement began in Italy, arising out of concerns over the fast food sector's increasing homogenisation of taste and disregard of local distinctiveness in food styles. It was founded by journalist Carlo Petrini in the mid-1980s, and now has 70 000 members in 45 countries. They seek to protect local production from being driven into extinction by global brands. In 1999, the idea of slow food gave rise to the 'slow city' movement, which began in the four Italian cities of Orvieto, Greve, Bra and Positano. The concept of food which is slow to make and distinctive in flavour, resonant of place and people, has been taken up by local authorities. They have extended it with commitments to increase pedestrian zones, reduce traffic, encourage restaurants to offer local products, directly support local farmers, increase green spaces in cities and conserve local aesthetic traditions. Slow food and slow cities have given regional food systems and policies a focus that goes beyond but also incorporates farmers' markets and cooperative food production. The cities are known as 'Citta del Buon Vivere', all about creating a good life.

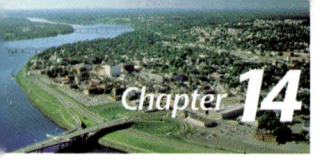

Chapter 14   *Rural–urban interrelationships*

A large number of food products are now dominated by corporate interests. One of the best known is the humble banana.

## Case study 14.14   Globalisation: banana wars

The banana trade is dominated by five TNCs that together account for over 80% of the world banana trade. These TNCs are Chiquita (USA), Dole (USA), Fyffes (Ireland), Del Monte (Grupo IAT, Chile) and Noboa (Ecuador). Most of the bananas that enter the world trade are grown as a monoculture crop, usually on plantations. In global terms the market has boomed, doubling in the 10 years between 1990 and 2000. The largest producers are Ecuador, Costa Rica, Colombia and the Philippines, all of which are dominated by large plantations. Meanwhile production in the Caribbean, largely carried out on family farms of not much more than 10 ha, has come under increasing pressure.

### Banana production

Europe, the largest export market, has tended to import its bananas from its former colonies in the Caribbean and, in the case of France, from its own territory in the Caribbean islands of Martinique and Guadeloupe. Given the mountainous nature of most Caribbean islands, the plantations cannot be expanded and economies of scale are unlikely. Plantations are small and family owned, requiring less machinery but more labour than those in Latin America. Export facilities in the Caribbean are not as good and costs are higher.

The plantations in Latin America, generally controlled or owned by TNCs, are frequently as large as 2000 ha and require a sophisticated infrastructure of power and communications. The plantations are vertically integrated and usually have packing plants. This makes them expensive in terms of inputs. However, low wages and limited workers' rights keep costs as low as possible.

There is a long and unsavoury political history behind bananas becoming a major food commodity in global markets. They were one of the first products

*Photograph 14.8 The dwarf Cavendish variety of banana*

for which there was a deliberate attempt to create a global market. Whole countries in Latin America were turned over to banana production, generally run by puppet regimes controlled by the USA in what became known as 'banana republics'.

The banana is a herbaceous plant rather than a tree and the fruits are produced all year around. They have been an important part of the diet in many tropical regions for centuries, although they are not native to the world's major producers. Like many other staple crops of the Americas, the banana was introduced from the Old World. Bananas exist in the wild in Asia, but were cultivated in north Africa over

*Human systems, processes and patterns* **A2 Unit 5**

2000 years ago and reached the Americas in the sixteenth century.

Most bananas are eaten locally and only some 14% enter world trade. That figure has risen sharply in recent years with the decreasing costs of air transport; bananas tend to ripen quickly and were difficult to transport by sea. Bananas are now the most popular fruit in many countries; they displaced the apple in the UK a few years ago. They are retailed through supermarkets, which dominate the trade. This includes an increasing number of organic bananas, which are difficult to grow given the banana's susceptibility to a wide range of pests and diseases.

## Conflicts and contoversies

The geography of supply has changed in recent years with an erosion of the traditional links to the Caribbean and the growing importance of the Latin American, so-called 'dollar' bananas. Fyffes (which also has a 50% share in the Geest brand) dominates the UK market and buys its bananas from Caribbean growers.

In recent years Caribbean banana production has been at the centre of a complex row between the EU and the USA. The USA accused the EU of unfairly restricting imports of cheaper 'dollar' bananas (so-called because of the significance of the USA-based TNCs). Meanwhile back in the Windward Islands and other parts of the Caribbean, producers have attempted to find an up-market niche for a product with a better taste than the ubiquitous Cavendish variety preferred on the large Latin American plantations.

The main issues attracting world media attention in the Latin American plantations are the very low wages, the restrictions on trade union membership and, above all, the poor working conditions with limited health and safety considerations despite the heavy doses of pesticides, fungicides and herbicides central to the management of these plantations.

On a broader level it is worth questioning the heavy dependence of a whole series of tropical economies on a fruit that has become progressively cheaper on world markets due to overproduction. To retain profit, the logical approach would be for the TNCs to squeeze out the small Caribbean farmers and to press down on their own costs. The attitude of national governments rather depends on who is in charge and what interest group they represent.

A topical and controversial issue in the politics of food production is the role and, more importantly, the control of technology in agriculture. The introduction of genetically modified (GM) seeds in the USA was hailed by some as an agricultural revolution. However, GM crops have run into some difficulties. European consumers have expressed concerns about the possible health implications of these new crops and the manner of their promotion by the large food corporations. Governments, which are responsible for their nation's health, have had to develop policies to deal with this issue and other health-related concerns.

At the other extreme of chemical intervention, there has been a marked rise in organic farming (Figure 14.16).

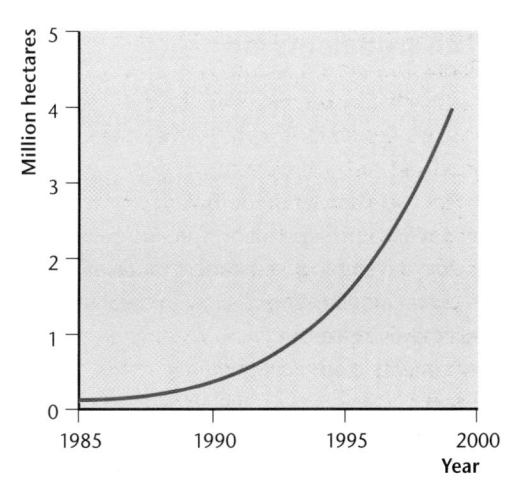

*Figure 14.16 The amount of agricultural land certified organic in the EU, 1985–99*

# Chapter 14  Rural–urban interrelationships

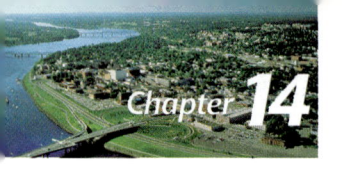

## Case study 14.15  Globalisation: genetically modified (GM) crops

The debate about GM crops goes to the heart of many of the issues surrounding the globalisation of food production.

GM is the latest in a series of technical 'fixes' to have a positive impact on yields and so increase the food supply, at least for those crops that benefit. In this respect it appears to be, like the Green Revolution before it, an unequivocally 'good' thing, yet the critics are loud and persistent in their attacks.

Since 1987, thousands of field trials of GM plants have been carried out in over 40 countries without apparent negative consequences for the environment. Most of these trials took place in the USA and Canada. However, many people have questioned the relevance of data obtained from field trials to the use of these crops on a much larger scale — millions of hectares of land.

More than 80 GM crops have been approved in the USA. This approval is both for large-scale sowing and for use in foods, and includes the important commercial food crops of maize, oil-seed rape and soybean.

### The arguments for

The arguments for developing GM crops are simple enough. The technology, it is claimed, will allow us to:

- increase output of food crops
- reduce our dependency on inorganic pesticides by developing resistance to disease through gene modification
- improve flavour
- simplify crop management

There is also the argument that if 'we' don't do it, someone else will and we will fall behind in the vital field of biotechnology, with serious economic impacts for all of us.

### The arguments against

The opponents of GM make a number of points.

- We cannot know what the long-term effects will be. The fact that there is, as yet, little evidence that GM foods cause damage does not mean that there will not be such evidence one day.
- Government assurances about scientific safety are not reliable given other food crises such as BSE. The government and its 'experts' did not acknowledge the problems their policies were creating until people were dying of BSE. This and other events, such as salmonella in eggs and the infection of haemophiliacs with HIV from contaminated blood, has led to a loss of public faith in government assurances.
- People should be able to make their own decisions about what they eat. It was this that drove the UK government to enforce labelling of GM products and forced both Tesco's and Monsanto into diplomatic silence over their labelling policy with GM soya.
- The main beneficiaries of GM are the large TNCs which control the global food industry — corporations such as Cargill and Monsanto. Opponents point out that such companies are only interested in profits and that morality, global politics and the fate of small farmers and rural life are hardly likely to figure in their decision-making.
- The other main beneficiaries are the large-scale producers of export crops. The millions of small and poor peasant farmers will only be further marginalised. The large-scale producers are either owned by large companies or are working to contract on their behalf.
- We do not need GM crops to feed the world; while the risks are not yet properly assessed, we would do better to concentrate on using established technology and science to improve distribution.

*Human systems, processes and patterns* **A2 Unit 5**

# The interdependence between urban and rural environments

## The changing resource base of the rural environment

Changes in rural areas have been profound in the past 200 years and those changes have taken place at an accelerating rate. It is obvious enough that rural areas are no longer quite so distinctive from cities as they once were.

In pre-capitalist societies rural could be defined as:
- areas dominated by the production of food and fibre crops both in terms of employment and area of land; these areas provide for a largely local rural and urban population
- areas dominated by settlement forms that were largely market towns, villages, hamlets or isolated farms with a mutual dependency on agriculture

The role of agriculture has declined in two senses.
- Less land is devoted to farming as other uses have developed. This is true even at farm level; significant amounts of land have been taken out of crop production in the past 20 years in both Europe and North America.
- Employment in agriculture has declined sharply, reducing the social, cultural and economic significance of this sector within local rural communities.

At the same time urban populations have grown rapidly in the past 200 years, representing something between 50 and 80% of the total population in most nation states. This means that populations have become more unevenly distributed. This process is still in full flow in sub-Saharan Africa.

Rural environments are changing and their resource base is changing. One such change is the growth of **amenity areas**.

The concept of amenity areas is ancient. For example, the New Forest in Hampshire was once a royal hunting area and its landscape of mixed forest and heathland still bears the imprint of this. The use of rural areas as an amenity increased rapidly in the twentieth century. The first National Parks were created in the USA and are certainly 'national' in that they are owned by the nation as a whole. The term 'park' is slightly more confusing.

The National Parks of the USA are wildernesses from which all permanent human activity is excluded. In some cases perfectly viable human communities that were already established in these areas were forcibly evicted to recreate the 'wilderness'. An example of this is the removal of the Cherokees from what became the Great Smoky Mountain National Park in the southern Appalachians.

In the UK the term National Park was used when creating protected areas, but these are not owned by the nation. They are made up of areas of largely privately owned land. The landscape of Britain's National Parks is not wilderness, but much

## Chapter 14  Rural–urban interrelationships

altered and affected by man. Human occupation is visible and permanent, but the area is regarded as exceptionally attractive. These areas are preserved by stricter planning controls than apply elsewhere. According to the 1949 Act of Parliament that established them, their function was:

- preserving and enhancing the natural beauty of the areas
- promoting their enjoyment by the public

In the UK the impact of National Parks on local inhabitants was considered, but it was not a statutory duty to do so at first. Nor was there any attempt to tackle the tricky issue of what is or what is not 'natural'. In reality, many of the landscapes in question were artificial but they were not 'industrial', at least not by the standards of the twentieth century, though many bore the scars of previous 'industrial' episodes. In a tradition of romanticising the countryside which began in the late eighteenth century, this concept of 'natural' landscapes has become deeply embedded in the British psyche. It includes attractive but 'unnatural' features such as dry-stone walls and the upland moors much modified by the introduction of grazing animals many millennia before. The old lead mines of the Yorkshire Dales have become a feature that visitors want to visit, but the development of modern quarries in the area is met with fierce resistance. Figure 14.17 shows the location of Britain's National Parks.

*Figure 14.17 Britain's National Parks*

### Ecological footprints

The most obvious cause of the changing resource base of rural areas is the growth of cities themselves. With 50% of the global population now living in cities, it is hardly surprising that the rural resource base is being used in more intensive and, sometimes, quite different ways. One way of measuring this impact is the concept of the **ecological footprint** of cities or countries, devised in 1992 by Dr William Rees and Mathis Wackernagel at the University of British Columbia (Table 14.10). They are firm believers that the notion of carrying capacity remains ecologically sound, although it clearly needs adapting to allow for the huge unevenness in the utilisation of the resource base.

A simple mental exercise serves to illustrate the ecological reality behind this approach. Imagine what would happen to any modern city or city region if it were enclosed in a glass or plastic hemisphere completely closed to material flows — nothing coming in and nothing going out. The city would cease to function and a large percentage of the population would be dead in a fairly short space of time.

That is why the figures for Hong Kong make such grim reading; they suggest that Hong Kong has a footprint that extends far beyond its jurisdiction because it is, to all intents and purposes, a city state. All of the resources which the people of Hong Kong use for their daily needs and activities come from somewhere else. Food, water, electricity and other basic amenities for human survival are all

## Human systems, processes and patterns — A2 Unit 5

|  | Population (1000s, 1997) | Footprint (ha per person) | Available capacity (ha per person) | Ecological deficit or surplus (ha per person) | Total footprint (km²) | Total available capacity (km²) |
|---|---|---|---|---|---|---|
| Australia | 18 550 | 9.0 | 14.0 | 5.0 | 1 669 500 | 2 597 000 |
| Bangladesh | 125 898 | 0.5 | 0.3 | −0.2 | 629 490 | 415 463 |
| Brazil | 167 046 | 3.1 | 6.7 | 3.6 | 5 178 426 | 11 192 082 |
| Canada | 30 101 | 7.7 | 9.6 | 1.9 | 2 317 777 | 2 889 696 |
| China | 1 247 315 | 1.2 | 0.8 | −0.4 | 14 967 780 | 9 978 520 |
| Ethiopia | 58 414 | 0.8 | 0.5 | −0.3 | 467 312 | 292 070 |
| France | 58 433 | 4.1 | 4.2 | 0.1 | 2 395 753 | 2 454 186 |
| Germany | 81 845 | 5.3 | 1.9 | −3.4 | 4 337 785 | 1 555 055 |
| Hong Kong | 5 913 | 6.1 | 0.0 | −6.1 | 360 693 | 0 |
| Iceland | 274 | 7.4 | 21.7 | 14.3 | 20 276 | 59 458 |
| India | 970 230 | 0.8 | 0.5 | −0.3 | 7 761 840 | 4 851 150 |
| Japan | 125 672 | 4.3 | 0.9 | −3.4 | 5 403 896 | 1 131 048 |
| Nigeria | 118 369 | 1.5 | 0.6 | −0.9 | 1 775 535 | 710 214 |
| Russian Federation | 146 381 | 6.0 | 3.7 | −2.3 | 8 782 860 | 5 416 097 |
| Singapore | 2 899 | 7.2 | 0.1 | −7.1 | 208 728 | 2 899 |
| Switzerland | 7 332 | 5.0 | 1.8 | −3.2 | 366 600 | 131 976 |
| UK | 58 587 | 5.2 | 1.7 | −3.5 | 3 046 524 | 995 979 |
| USA | 268 189 | 10.3 | 6.7 | −3.6 | 27 623 467 | 17 968 663 |
| **World** | **5 892 480** | **2.8** | **2.1** | **−0.7** | — | — |

produced using raw natural resources, however indirectly, but very few of these are mined, tapped or grown within the city itself.

Based on this relationship between humanity and the biosphere, an ecological footprint is a measurement of the land area necessary to sustain a population of any size. It measures the amount of arable land and water resources needed to sustain a population, based on its consumption levels at a given point in time. This incorporates water, energy use, uses of land for infrastructure and different forms of agriculture, as well as forests and all other forms of energy and material inputs that people require in their day-to-day lives. It also accounts for the land area required for waste disposal.

*Table 14.10 The ecological footprint of selected countries*

- The ecological footprint of London is around 125 times London's own surface.
- Most European cities have footprints approaching 3 ha per person.
- The more energy-consuming US cities in the southwest reach figures of 4 or even 5 ha per person.

This approach can be further formalised by adding some ideas.

- **Fair Earthshare**: the amount of ecologically productive land 'available' per capita on Earth, currently about 1.4 hectares (2000) — that is to say, how much we should have access to in a world of equal distribution.
- **Ecological deficit**: the difference between the ecological footprint of a given national, regional or city population and the geographic area it actually occupies.

# Chapter 14  Rural–urban interrelationships

- **Sustainability gap**: a measure of the decrease in consumption (or the increase in material and economic efficiency) required to eliminate the ecological deficit.

> **Key term**
>
> **Ecological footprint**
> The impact of urban areas on the surrounding area, measured as the area needed to support a city in fuel, food and other resources.

The ecological footprint (EF) approach is not without its critics. Some have pointed out that a distinction should be made between land use that is sustainable and land use that isn't. This would seem to be a primary requirement for an indicator of planetary sustainability. The ecological footprint approach also places a heavy emphasis on energy consumption and the amount of land necessary to soak up carbon dioxide emissions. Changes in the type of fuel used would achieve marked reductions in this area.

It has also been observed that the boundaries between places are essentially arbitrary. Obviously an area of high density has a high EF, so for a small country such as the Netherlands the EF is bound to be high. This is neither good nor bad; what matters is the nature of the relationship between the area with a high EF and the area that supports it. This would be the same for Hong Kong. It could be pointed out that as soon as urbanisation takes place, this unevenness is inevitable, and that international trade frees a country from the necessity to feed itself. Perhaps the nation state is an inappropriate tool of analysis in this context.

One could also ask whether putting a plastic dome over the rural hinterland that feeds and waters cities would enrich the rural area, or whether it would lose its economic function and the stimulus of city technology and city ideas so clearly described by another Canadian academic, Jane Jacobs (see Chapter 5).

## The impact of tourism on rural areas

The economic aim of tourist development is that it stimulates people to make greater use of an environment which, in turn, stimulates investment in more infrastructure — ranging from ski lifts to marinas. In other words, this is the multiplier effect. Tourism is also perceived by some as being beneficial in a sociocultural sense, because it may help 'modernise' a society. Problems brought by tourism include a negative impact on the environment; tourism involves many activities that can have adverse environmental effects.

- The construction of general infrastructure such as roads and airports, and of tourism facilities, including resorts, hotels, restaurants, shops, golf courses and marinas, may have a huge impact on the landscape and environment.
- Tourism development may destroy the environmental resources on which the tourism depends. There are easy contradictions to find in the advertising of some remote places. The Northumbria Tourist Board's description of the northeast's coastline and uplands as 'England's best-kept secret' is a good example of an area trading on this idea.

However, the dominant view is that tourism is an easy route to increasing the flow of capital, in particular foreign currency, into remote and often poor regions that have, historically, been losing residents. Tourism develops a specialist niche in the national and global division of labour for those areas whose natural resources

*Human systems, processes and patterns*  **A2 Unit 5**

| Main activity | All visits (%) | Town/city (%) | Countryside (%) | Seaside/coast (%) |
|---|---|---|---|---|
| Eat or drink out | 18 | 19 | 15 | 13 |
| Walk, hill-walk or ramble | 15 | 9 | 32 | 20 |
| Visit friends or relatives at their home | 14 | 16 | 10 | 9 |
| Shopping (not food and not regular) | 11 | 15 | 3 | 5 |
| Take part in sports, active pursuits — indoor, outdoor, field, water | 9 | 8 | 11 | 7 |
| Hobby or special interest | 8 | 8 | 8 | 6 |
| Entertainment (e.g. cinema, theatre, club) | 5 | 7 | 1 | 2 |
| Informal sports, games, relaxation and well-being | 4 | 4 | 3 | 2 |
| Visit leisure attraction, place of interest, special event/exhibition | 3 | 3 | 5 | 4 |
| Swimming | 3 | 3 | 1 | 3 |
| Visit park or garden | 3 | 3 | 3 | 1 |
| Watch live sport or attend a live event (not on television) | 2 | 3 | 2 | 1 |
| Drive, sightsee, picnic, pleasure boating | 2 | 1 | 2 | 6 |
| Cycle, mountain bike | 1 | 1 | 4 | 1 |
| Visit beach, sunbathe, paddle in sea | 1 | — | — | 21 |

*Table 14.11 Main activities by destination, UK, 2002–03*

such as mountains, deemed aesthetically pleasing in the modern world, are otherwise not being effectively linked to the rest of the economy.

The type of niche depends upon the rural resources available, but a number of newer categories have been identified to add to more traditional forms such as beach holidays and skiing.

- **Heritage** or **cultural tourism** refers to leisure travel for the purpose of visiting historic sites or works of art.
- **Nature-based tourism** may be either passive, in which visitors tend to be strictly observers of nature, or active (increasingly popular in recent years), where participants take part in outdoor recreation or adventure travel activities.
- **Agritourism** refers to visiting a working farm or any agricultural enterprise and includes a diverse range of activities from pick-your-own enterprises through to farm stays involving active participation in agricultural work.

Table 14.11 gives details of the main reasons why people visit tourist sites in the UK. Figure 14.18 gives the average duration of these visits and the average time spent at the destination. The difference between these two times gives an indication of the time spent travelling.

Average duration of visit (hours): Town/city 3.3, Countryside 3.1, Seaside/coast 3.9
Average time spent at destination (hours): Town/city 2.4, Countryside 2.3, Seaside/coast 2.6

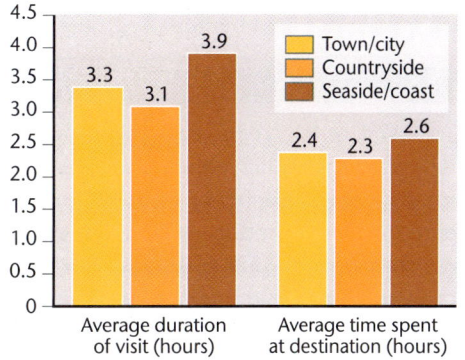

*Figure 14.18 Duration of visits and time spent at destination, UK, 2002/03*

# Chapter 14  Rural–urban interrelationships

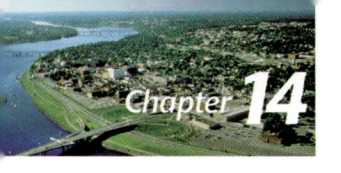

## Case study 14.16  Tourism in rural regions: the UK

Most of the remote rural regions of the UK have experienced a decline in their traditional economic base of agriculture. They are, for the most part, upland areas that have been severely hit by the decline in upland farming and loss of subsidy from the EU. These marginal areas, like so many other supply regions, have a long history of out-migration and many have seen the potential of tourism and leisure activities in arresting this long-term decline. The regions are, of course, in competition with one another in attempting to maintain their market share in a volatile market heavily dependent on factors such as general economic prosperity.

The model of rural development these areas pursue has been labelled the 'growth or death' approach. The emphasis is on continued development in order to remain attractive to visitors. The Cumbrian Tourist Board's *A Vision For Cumbria* report listed 12 objectives, of which nine were about promoting more tourism or gaining more income from it, even though the area received some 6 million tourist trips in 1995 and 13.9 million 'tourist nights' were spent there.

Tourism is also seen as a practical means of supplementing the incomes of local farmers and other rural inhabitants and providing some diversification in employment (e.g. by the conversion of surplus farm buildings to tourist usage). The Countryside Commission has been active in this field through its Recreation 2000 policy, in which such development is encouraged. The Countryside Commission paper *What Future For the Uplands?* stresses that more but better organised recreation would be good for both rural communities and the environment. A whole swathe of facilities has arisen in response to growing numbers of day-trippers, short-stay and long-stay visitors. These include:

- hotels and motels
- bed and breakfast accommodation
- camp sites
- cafés and restaurants
- souvenir shops, gift shops and antique shops

The growth in these amenities has been highly localised, and is often in marked contrast to the decline in more traditional rural services such as the village shop and the pub.

Some types of tourist and leisure activity are more demonstrably damaging to the environment than others. These 'damaging' activities raise more issues about sustainability and the complex relationship between the (usually) rural resource and the (usually) urban consumer of that resource. Some of the issues are well-illustrated by the largely twentieth-century phenomenon of skiing as a leisure activity.

## Case study 14.17  'Damaging' tourism: the Alps

Traditionally, the Alpine region of Europe relied on mountain agriculture, almost always pastoral, and forestry, supplemented by employment in mining and some local cottage industries. When these remote and inaccessible regions were first reached by modern communications systems and became linked with dynamic urban regions, they suffered rapid population loss through out-migration. Their local industrial base, stunted by small markets and difficulty of transportation, was overwhelmed by imported goods from the richer and more prosperous urban regions. They became supply regions relying on little more than the export of dairy products and other foodstuffs. Population loss continued until the middle of the nineteenth century. As in all remote regions, tourism was an obvious route by which to revive the economies. From limited beginnings in the nineteenth century, winter sports have become the most important source of income in many Alpine areas. By the late 1950s, mass tourism

*Human systems, processes and patterns* **A2 Unit 5**

Photograph 14.9
Arosa, Switzerland

had started as a result of increased leisure time, higher disposable income and an enormous growth in personal mobility through car ownership.

The original resorts were Alpine villages such as Chamonix but, as visitor numbers increased, these became insufficient to meet demand. This led to the development of purpose-built resorts closer to the snowfields. The most rapid period of resort building in the French Alps culminated in the building of purpose-built accommodation for the 1992 Olympic Games, spread through 13 Alpine villages in the Haute Savoie. The population of the region more or less doubles during the winter ski season and such have been the environmental problems that the French government has now banned further Alpine development.

Skiing holidays, now diversified to include snowboarding and other snow-related sports such as cross-country skiing or snow hiking, have become an affordable activity for the affluent middle classes in northern Europe. Some 100 million people visit the Alps every year, creating increasingly serious environmental problems.

- Large areas of Alpine pasture and woodland are now covered with concrete roads, hotels and other facilities. This increases runoff in spring and can increase erosion rates both from overland flow and in the rivers, which become 'peakier' in discharge.
- The construction of the 50 000 ski-runs in Alpine resorts has caused significant deforestation. This changes hydrological conditions and causes more frequent and more severe landslides and avalanches which strip off not just the snow but also significant amounts of vegetation. This risk has increased with the gradual shift of ski-runs into higher and more fragile areas.
- Soil erosion is further worsened by skiing on thin snow, which then becomes compacted to ice. The ice damages the vegetation below, leading to a longer recovery time in the spring and subjecting the slopes to much higher risks of erosion.
- Bird and mammal species have been badly hit, partly by habitat destruction but also by disturbance from passing skiers.
- Many small mountain villages have been overburdened by increased demand for water, sewage disposal and communications systems such as telephones, television and mail.
- The profitability of purpose-built resorts is dependent upon maximising bed occupancy for as long a period as possible. Prolonging the season by using artificial snow and a heavy use of herbicides and pesticides to clear the snow runs is obviously tempting.

The long-term future of Alpine tourism will be affected by climate change, for almost all Alpine

# Chapter 14  Rural–urban interrelationships

winter activities depend on snow. Mountain areas are particularly sensitive to climate change, and relatively small changes in temperature and precipitation can have a large impact. These areas are fragile: both in an environmental sense, with their steep slopes, varied and frequently extreme weather conditions, and in an economic sense, given their reliance on the income brought by tourists and the relatively few alternative sources of income.

The effects of climate change can be seen, for example, in less snow, receding glaciers, melting permafrost and more extreme events such as avalanches and landslides. In the European Alps, climate change in recent years has become a severe threat to snow-related sports such as skiing, snowboarding and cross-country skiing. In Austria it is forecast that the present snow line will rise 200–300 m higher in the next 30–50 years. The threat of reduced earnings in winter tourism will accentuate the economic disparities between urban areas and the rural Alpine regions. One response is to shift the resorts further up the mountains, developing areas that are even more fragile in order to 'guarantee the snow'.

## Other pressures on rural areas

There have been other pressures on rural areas, including the extraction of minerals, the development of reservoirs and the siphoning off of river water.

### Case study 14.18  Other conflicts: the Colorado River

In the middle of the twentieth century the Colorado delta was identified by many as North America's greatest oasis and one of its last lowland wildernesses. Covering an area of over 8000 km² just over the US border into Mexico, this immense wetland was home to 400 species of plants and animals, including the endangered Vacquita dolphin, and represented about 80% of the biodiversity of the whole Colorado River basin. It is also home to the Copacha Indians who, alongside the more recently arrived Mexican residents, made a decent living in the area from their fishing.

Today all that has changed. The Colorado River scarcely reaches the sea and the fierce competition between users of the water is an ongoing crisis in the southwest USA. Southern California is a desert. Water requirements of people living in the area mean that 95% of the Colorado's flow is retained in the USA before it crosses the border into Mexico for the final few kilometres to the sea. The Law of the River that divided up the river's water between the main states of the southwest was originally written to help out agricultural areas and impoverished farmers suffering from lack of water. However, the huge population increase in the

*Figure 14.19 The Colorado River basin*

region and its booming economy have given urban demands precedence. Agriculture too has changed, with the rapid growth of agribusiness.

More than half the river water is exported from the basin, much being used in urban centres such as Los Angeles, Denver and Salt Lake City. In some cases the in-basin recipients of water are selling their rights to out-of-basin users, although this is of questionable legality. The booming cities of Phoenix, Arizona, and Las Vegas, with their abundant golf courses, consume huge amounts of water. The per capita water use by the residents of California, Nevada and Arizona can be as much as 950 litres a day, more than three times as much as that in New York or London. The rest of the water is used in agriculture, mainly by agribusinesses growing alfalfa for cattle food, cotton and even potatoes in this arid desert climate. What is more, these businesses get their water at a subsidised rate. The result is that the Colorado delta is disappearing, with little more than a toxic and heavily saline trickle left to reach the sea. Its fragile ecosystem is under threat.

---

The other major impact on the rural resource base has been the further blurring of the distinction between urban and rural areas, as industry other than agriculture has developed in rural regions. This is discussed in the following section.

# The rural/urban fringe

The dominant theme when considering the patterns and processes operating on the rural/urban fringe has to be the intrusion of urban life into rural areas. This takes two forms:
- the physical growth of urban areas
- the sociocultural dominance of city life

The physical growth of urban areas means that they have increasingly sprawled out into the surrounding countryside. Such is the power of cities that they have generally overwhelmed the peri-urban fringes in countries as diverse as the USA and Indonesia. The problems of urbanisation in many sub-Saharan African countries, in Latin America and in parts of Asia are discussed in the section on 'Urban development in LEDCs' in Chapter 5 (page 189).

Cities tend to exert sociocultural dominance over the rural hinterland, with city 'ideas' and city images crowding out locally significant cultures.

There are at least three categories of urban growth, with contrasting processes, and they are explored in this section:
- growth of MEDC cities, especially those in the USA and Europe
- growth of cities in NICs, especially those of southeast Asia
- growth of cities in LEDCs

## MEDC cities

The physical growth of cities, the impact of commuting and the decline of inner-city populations have had a profound impact on the geography of the rural/urban fringe. There are distinctive differences between MEDCs. In Japan the density of city populations and the intensity of agricultural land use on the margin has made

# Chapter 14: Rural–urban interrelationships

urban growth much more restricted. At the other extreme, in the USA this boundary is now so blurred that two new urban forms have been identified: edge cities and post-suburban regions. At a smaller scale, there are also gated communities.

## Edge cities

According to most definitions, areas become **edge cities** when they have urbanised within the last 30 years, and have at least half a million m$^2$ of leased office space and 54 000 m$^2$ of retail space, along with a population that increases at 9 a.m. on workdays.

## Post-suburban regions

**Post-suburban regions** have either no specific centre or multiple centres. Such regions are not just multi-centred in their commercial activities. Their commerce, shopping, arts, residential life and religious activities are all conducted in different places on a network of interconnected travel paths linked primarily by private cars.

The following quote, taken from a resident of Fairfax County in Virginia, USA, is typical of the dependency on cars in many US households:

> My family has four cars. I drive only one, but the other three don't sit idle. My wife, my daughter in college, and my daughter in high school each use a car every day. There's not much chance I'll go back to two or even three cars, and one car is downright unthinkable. I don't believe anybody else is going to cut back either. There's a reason, and it's not just that Americans are car crazy (though they are). It's the freedom, convenience, and flexibility that comes from having a car at your disposal. The automobile is the most freeing instrument yet invented. It allows folks to take jobs far away from their homes. It enables them to live far from central cities and, if they're anti-social, far from other people. Two cars make the two-earner family possible. And the attachment has grown as cars have become an extension of home and office, with telephones, message pads, coffee cups, books (on tape), etc.

The European experience is not as extreme as that in the USA and nor is it so uniform, for there are more obvious contrasts within the continent.

An **urban growth boundary**, as used in France, has two main purposes: to rein in urban sprawl by promoting more compact development; and to protect wildernesses, agricultural land and other resource land from low-density, poorly regulated development.

**Green belts**, which attacked the same problem less comprehensively, worked by putting some restrictions on development within a belt around a city — hence the leapfrogging of the belts that took place in the UK.

The more simplistic idea of a 'belt' has been replaced in the USA with the general notion of corridors of open space and boundaries to development.

However, the processes that have led to these pressures are the same:
- fundamental changes in the location of economic activity
- increases in wealth and car ownership
- the development of urban highways and commuting transport systems

*Human systems, processes and patterns* **A2 Unit 5**

## Case study 14.19 Urban sprawl: Orange county, California, USA

*Photograph 14.10 Aerial view of suburban homes in Orange county, California*

Orange county is on the Pacific coast to the south of Los Angeles and to the north of San Diego. It is a large county, comprising 2066 km², with an average population density of 1392 per km². It is the second largest county in California and has experienced exceptionally rapid growth in the past 50 years, with the population rising from 216 000 in 1950 to nearly 3 million in the 2000 census. The fastest rate of growth was during the 1970s and 1980s. Recent growth is at a relatively modest annual rate of 1.6%, nowhere near the sorts of rates achieved in parts of Colorado and Nevada.

### Development

Orange county contains 34 'cities' but in reality much of the county is part of the new urban phenomenon of discontinuous urban development with no single focus — **post-suburbia**. Since the end of the Second World War, Orange county has evolved through the following stages.

- A largely rural county which befitted its name, with many orange groves and a largely rural environment.
- An industrial region and commuting zone to Los Angeles, driven by aeronautics and other defence-related industries, which experienced rapid population growth largely through in-migration from Los Angeles and other local areas.
- A complex region consisting of 34 separate municipalities with a substantially independent economy and cultural life. It has one of the highest rates of business creation in the USA and rivals Silicon Valley further north for the status of the world's leading area for information and communications related development. All this is set in a post-suburban landscape utterly reliant on private transport.

## Chapter 14 Rural–urban interrelationships

Beginning in the late 1960s, Orange county's information economy began to develop much faster than its older industrial economy, which in itself was a product of postwar growth of aeronautics and related hi-tech industries in California and other sunbelt states. The region's utopian promise of getting away from the city, reminiscent of the English garden city ideas of the early twentieth century, was a powerful incentive to the thousands of in-migrants and businesses who moved into the area. Orange county's immigrants included people who had previously lived and worked in nearby Los Angeles and who were seeking more open space and less expensive housing, as well as people from elsewhere in the nation who were drawn to the county's mixture of warm Mediterranean climate, open space and access to the coast.

By 1988, Orange county's growing hi-tech export business was so successful that an export licensing office was opened in Newport Beach, the first such branch outside Washington, DC. Orange county has an economy larger than that of many countries and the seventh highest 'dot.com' domain name incidence in the country (Figure 14.20).

### The post-suburban landscape

The social, economic and geographic descriptors of Orange County offer some insight into the future of urban development.

American suburbia from the 1920s through to the Second World War tended to be white, middle-class, family oriented and socially homogeneous. Population densities were lower than in inner-city areas, industrial land use was safely cordoned off or entirely absent and land use was primarily residential. Traditionally people travelled from these areas by train, bus or, increasingly, by private car into central city areas for work, although a working-class suburbanisation also developed to serve the more footloose industries that grew up in the cheaper land areas on city margins.

Post-suburban landscapes like Orange County differ markedly from this picture. This new type of urban form is a peripheral zone of many hundreds or perhaps thousands of square kilometres that has emerged as a viable social and economic unit. Spread out along its highway growth corridors are shopping malls and industrial parks interspersed with residential areas, all set in a 'greened' landscape that has no agricultural function. There are distinct zones within this landscape, specialising in one or other function. These specialised residential, commercial and industrial zones are almost impossible for pedestrians to navigate, but were designed to accommodate the car driver. In the southern sections of Orange county, city planners designed six-lane highways in anticipation of an increase in traffic flows, much of which has been realised. The average journey time to work is 26 minutes (Figure 14.21) and the vast majority commute alone.

*Figure 14.20 The distribution of US domain names, July 2000*

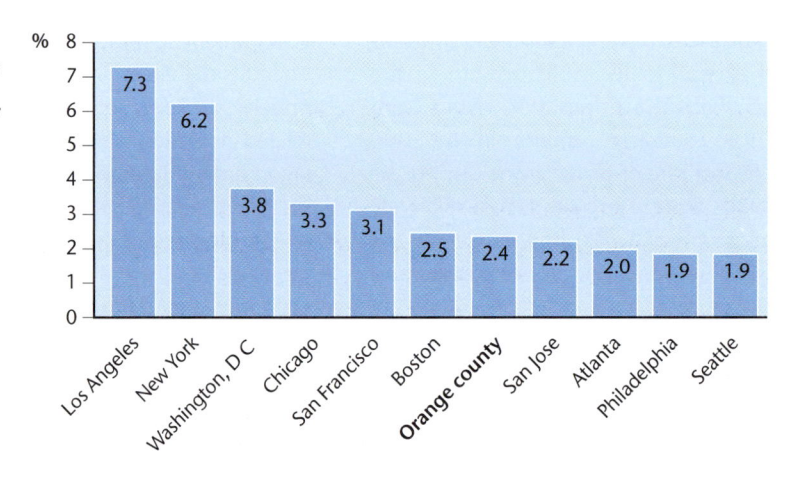

| City | % |
|---|---|
| Los Angeles | 7.3 |
| New York | 6.2 |
| Washington, DC | 3.8 |
| Chicago | 3.3 |
| San Francisco | 3.1 |
| Boston | 2.5 |
| Orange county | 2.4 |
| San Jose | 2.2 |
| Atlanta | 2.0 |
| Philadelphia | 1.9 |
| Seattle | 1.9 |

*Human systems, processes and patterns* **A2 Unit 5**

*Figure 14.21 Average commute to work times in the USA, 2001*

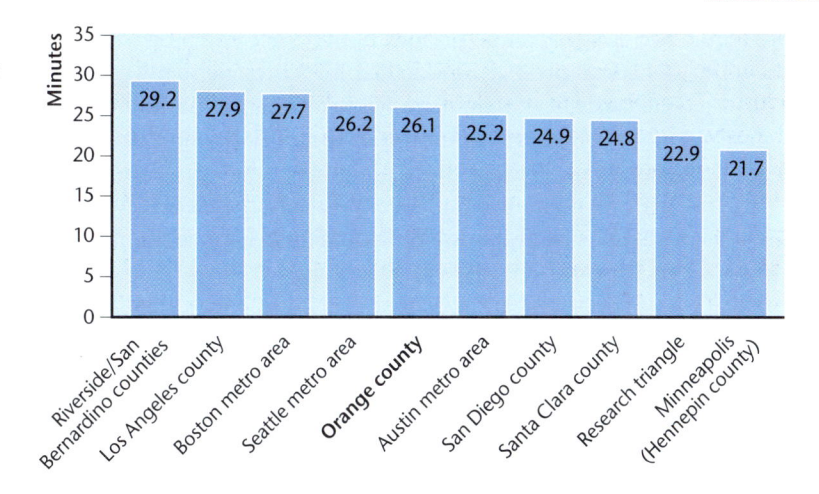

| Location | Minutes |
|---|---|
| Riverside/San Bernardino counties | 29.2 |
| Los Angeles county | 27.9 |
| Boston metro area | 27.7 |
| Seattle metro area | 26.2 |
| Orange county | 26.1 |
| Austin metro area | 25.2 |
| San Diego county | 24.9 |
| Santa Clara county | 24.8 |
| Research triangle | 22.9 |
| Minneapolis (Hennepin county) | 21.7 |

In most parts of Orange county, the pedestrian is likely to see a visually homogeneous landscape for block after block, with occasional changes to different, but also internally homogeneous, areas. The architecture can be quite pleasing, with large areas of attractive buildings of similar designs, but they have no variety.

By chance this recollects aspects of the UK urban landscape in New Towns where the 'neighbourhoods' were only identifiable by location or, bizarrely, the colour used for front doors, so similar were they in almost all other respects. This low-rise, low-density Orange county landscape spreads for many dozens of kilometres, as featureless in its own way as the desert. Despite the economic success of the region, serious questions arise about this sort of urban environment, with its heavy dependence on the car and its social exclusivity.

## Social exclusivity

Housing shortages are substantial; high land prices mean the median sale price in 2004 was $432 630. The annual salary for a nurse is $53 000. Buying an average home in Orange county with the usual 20% down payment is well beyond the budget of important city-serving employees such as teachers, fire fighters and nurses. This is frequently the case when the land market is left to operate with little state intervention. However, house prices are likely to continue to grow, whatever the economics and social logic might suggest. Suburbanites and neo-suburbanites comprise 50.5% of all Americans and, what is more, they are more inclined to vote. So the residents of Orange county and similar areas are likely to vote to protect their own interests, whether or not the public good is best served in this way.

## Gated communities

One notable urban form that has re-emerged in both MEDCs and LEDCs in recent years is the gated community. These are reminiscent of the heavily protected residential quarters of the Roman or Athenian aristocracies in the ancient world. They reappeared on the US urban scene 30 or 40 years ago and were examined in detail by Frances Fitzgerald in her popular account *Cities on a Hill*. Initially in the USA many of them were designed for a particular group, generally the elderly and wealthy white Americans who chose to retire southwards. Gated cities were also being built for those who wished to protect themselves from the risk, real or

# Chapter 14: Rural–urban interrelationships

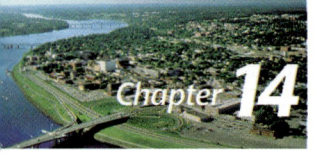

perceived, of crime (against both people and property), or for those who wished for social prestige or searched for a lifestyle that was not achievable in the city. Their rate of development has been exponential and by 1997 there were as many as 20 000 gated communities in the USA, amounting to more than 3 million properties. Perhaps 8 out of 10 new urban projects are for gated communities, many of them in suburban or post-suburban areas. By 1995 nearly one-half of the 120 housing projects in development in Orange county, California, were gated, an increase of over 300% in 10 years.

This privatisation of space, as it has been called, is not unique to the richer countries of the world. Nor is it uncontroversial.

- At one end of the spectrum lie writers such as Mike Davis who in *City of Quartz* describes the very high level of social segregation in Los Angeles, publicising the plight of those he sees being marginalised by the process.
- At the other end of the spectrum, gated communities have been defended as not just an expression of libertarian values in which people should be allowed to buy whatsoever they wish, but also an efficient allocation of resources.
- Somewhere in the middle is the detailed mapping by Blakely and Snyder who tried to identify the processes behind the movement and its root causes in *Fortress America*.

## Cities in NICs and RICs

Urban sprawl has also taken place in cities in NICs and RICs and the impact on the rural hinterland in these countries can be even more profound.

The hinterlands of most US cities have extremely low population densities and are farmed at relatively low levels of intensity. By contrast, the rural areas around many southeast Asian cities are densely populated and intensively farmed.

This difference means that land for road construction is relatively more expensive in Asia than in the USA. In Japan, which is also intensely cultivated and densely populated, $0.70 of every dollar for road investment is spent on land acquisition, compared to less than $0.25 in the USA. This difference in the relative value of land (due to the economic intensity of its use) makes the economic cost of a high level of private motor vehicle use much higher in Asia than in the USA.

The social and environmental costs of this sort of suburbanisation are also much greater in southeast Asia. The land consumed by new roads in suburban areas around Jakarta (Java) will displace more than 100 times as many people as it will in the USA. If all the roads which the World Bank claims are justified by economic analysis on the island of Java are actually built, some 800 000 people could be displaced. In Bangkok roughly 20 000 people a year face eviction from central areas and about half of these evictions are for road projects. This process of displacement is forcing an increasing percentage of the urban poor to relocate to distant suburban areas. Meanwhile, an estimated 250 km$^2$ of agricultural land, forest or wetlands are converted to urban uses every year on the island of Java, with enormous environmental consequences.

*Human systems, processes and patterns* **A2 Unit 5**

## Case study 14.20  Urban sprawl: Jakarta, Indonesia

Jakarta is one of the largest cities in southeast Asia and is the capital of the world's fourth most populous country, Indonesia. It is also one of the least densely populated cities, having sprawled outwards into the densely populated rural hinterland. The population of Greater Jakarta, or Jabotabek as it is known, is around 17–20 million, the imprecision of the figures being a consequence both of the rapidity of change and the difficulty of counting people in what are some of the worst slums in the world. Jakarta lacks a rapid transit system and is heavily dependent on an archaic road system. Land regulation is minimal and the provision of public utilities is patchy. Pollution levels are well above regional levels. According to almost all indicators other than growth rate, Jabotabek is in a mess, with problems similar to those of London in the nineteenth century.

In London, the urban problems were ultimately solved by private capital that saw the need to clean up cities in order to protect the profit margins of business. Later, improving conditions became a state enterprise carried out through public investment and a planning policy that put clear limits on urban growth through the use of green belts. In Indonesia, there is no benefit to the wealthy elite of such enterprise. Their need is to avoid rather than solve the urban problems of the mega-city, yet they still have to work in these cities.

So, 35 km down the road both west and southeast of Jakarta are two purpose-built cities with populations of about 30 000 each: Lippo Karawaci is to the west; Lippo Cikarang is to the southeast of Jakarta. The plan being executed by the Lippo Corporation is that each of these private cities will accommodate 1 million people by 2020. The core buildings in both sites are educational facilities designed to attract an upwardly mobile middle class to the area. The incomers can choose from well over 100 house designs that mirror the sort of developments seen in Orange county.

In essence these are 'company towns' that hark back to the model villages built by enlightened industrialists for their workers in the nineteenth century (e.g. Port Sunlight). This ended when resentment against paternalist employers and the inevitable ups and downs of industrial capitalism made them unworkable. The Lippo Corporation is pursuing profit through land development, but it is also trying to build enclave cities. With their golf courses, universities and a short hop to the international airport, these will become 'informational' cities which have more links with the global economy than with their national roots.

## Cities in LEDCs

The processes in LEDCs are quite different. For the most part, the dominant driving force remains rural–urban migration. The results of this are well known. On the margins of many cities in LEDCs there are recognisable areas of:

- squatter settlements
- spontaneous settlements

The distinction between these terms needs clarifying.
- The residents of **squatter** settlements have no legal right to the land on which they build.
- **Spontaneous** settlement is unplanned, usually without any infrastructure of water, sewerage, electricity or transport. The residents may or may not have legal title.

### Key terms

**'Gated' communities**
Areas of expensive housing that have entry restricted by gates.

**Urban sprawl**
The process by which cities grow into their surrounding regions driven by cheaper land prices and the ownership of cars.

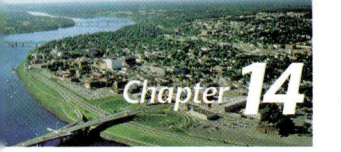

## Chapter 14  Rural–urban interrelationships

Rio de Janeiro, like most LEDC cities, is experiencing a dramatic increase in population. This increase has come mostly in the form of the rural poor migrating to the city. Because of the high land values and the enormous demand for space, these poor are forced into squatter settlements known as favelas. (The term comes from the location of the first such settlement, the hill Morro da Favela.) These settlements usually occur in two areas of Rio:
- along the steep hillsides otherwise avoided by housing
- along the outer fringes of urban expansion

*Photograph 14.11 A favela in Rio de Janeiro*

LEDC cities can also be explored through the case studies about Cairo (Case study 5.11, page 192) or Mumbai (Case study 14.4, page 471)

# The influence of urban economies on the socioeconomic characteristics of rural areas

It is obvious that urban influence extends beyond the limits of the urban area. These influences are both social and economic. The most obvious economic impacts are those of urban demand on rural regions. They may stretch well beyond any conventional rural/urban boundary and even beyond national frontiers.

Within MEDCs, the influences include:
- impact on the social character of rural settlements through counterurbanisation and changing values and norms
- changes in provision of services and commerce

*Human systems, processes and patterns* **A2 Unit 5**

- growth of second-home ownership and seasonality of populations
- increased numbers of commuters and impact on local transport systems
- changing employment structure within rural areas meeting distant demand from cities

The distance of the rural area under consideration from the major urban centres, and the relative importance of these centres in the urban hierarchy, will make a difference, especially in the social impact of cities on rural settlements. Thus the villages close to a major world city, such as those close to London or Paris, are much more profoundly influenced than those that are an equivalent distance from lesser cities. On the other hand, the huge economic demands imposed by the heavy ecological footprints of cities and city regions stretch to the remotest corners of national territories.

### Case study 14.21  Ecological footprint: quarries and superquarries

The rapid growth of the infrastructure in the UK has led to a heavy demand on the production of stone and stone products for road building and construction. Booming regions such as the southeast of England cannot produce their own supplies in anything like sufficient quantities, so the large companies that dominate the industry look elsewhere, often in quite remote regions where the opposition to quarrying may be expected to be less.

In early 2004 one of the longest running sagas in planning history came to an end when the controversial plans by Lafarge to develop a £70 million superquarry on one of the Western Isles in Scotland finally ended. The story began when the local authority gave planning permission for a quarry at Lingerbay on the island of Harris. That quarry was not developed, but in 1991 Redland Aggregates, one of the largest such companies in the UK, applied for planning permission to remove 600 million tonnes of stone from the site over a 60-year period. That application was approved by Western Isles Council.

Redland Aggregates was then taken over in 1997 by the French company Lafarge. With its headquarters in Paris, Lafarge is the world's biggest producer of building materials. It employs 77 000 people in 75 countries, its sales in 2004 totalled more than £10 billion and it has 674 rock quarries around the world.

The plans for the site were considerable: their quarry would be 50 times larger than conventional quarries and would cover an area of 459 ha. Overall, a Scottish mountain would disappear, leaving a sea loch and a white scar six times the height of the white cliffs of Dover. The plans, if they had been approved, would have brought 33 jobs to the area with some multiplier impact beyond. The rationale for the planned quarry was the rising demand for aggregates within Europe. This was put at 400 million tonnes per

*Photograph 14.12 A photomontage of the proposed superquarry development at Lingerbay, Isle of Harris by Redland Aggregates*

# Chapter 14  Rural–urban interrelationships

annum initially, but the figure was scaled down by the industry itself to 220 million tonnes per annum.

The company, which claimed that its plans were 'sustainable', invoked the European Court of Human Rights when appealing against the rejection of its plans. The planning battle and its aftermath lasted for 9 years, but 4 years later, in 2004, Lafarge was still fighting the rejection of its plans by the Environment Minister. The company finally announced that it had abandoned such plans in April 2004.

> **Examiner's tip**
>
> - It is our perception of rural and urban areas that drives our behaviour and consumption patterns and informs our views of what is or is not 'natural' in a landscape.

Urbanisation has also had profound impacts on rural areas in LEDCs:

- rural–urban migration has drained rural areas of some of the most enterprising members of the population
- rural areas have lost population, particularly young people
- increasing demand for food from rapidly growing cities has led to a switch to cash crops, sometimes non-food crops, to satisfy distant demand
- an informal manufacturing sector with linkages to rural settlements has grown up in urban economies

Some of these themes can be explored by looking at the case study on Indonesian transmigration (Case study 6.3, page 219).

## Examination questions

1. (a) Distinguish between arable and pastoral farming. (5)
   (b) Examine the view that the physical environment controls the type of farming systems found in a region. (20)

   *Total = 25 marks*

2. (a) Outline the ways in which urban growth can impact on rural communities. (5)
   (b) Assess the view that urban growth has a negative impact on rural communities. (20)

   *Total = 25 marks*

# Synoptic link
# Rapid change has created pressures on rural–urban interdependence

- The management of waste in cities — the problems of water and air quality control, and waste disposal.
- The growth of the leisure and tourist industry, and its social, cultural and economic impact on rural environments.

## China's air quality

Official statistics have shown that only eight of China's 47 major cities were up to the state environmental requirements for both air and groundwater in 1999.

*Human systems, processes and patterns*  **A2 Unit 5**

According to the state Environmental Protection Administration, out of the 46 state-designated major cities, 28 met the environmental protection requirements in groundwater quality but failed for air quality. Fifteen cities, including Beijing, failed to meet discharge requirements for total suspended particles (TSP), sulphur dioxide and nitrogen oxide. All of these cities have experienced rapid recent population growth and industrialisation as China modernises.

Lanzhou was listed as the most TSP-polluted city in China, ranking first in terms of excessive TSP discharge among 29 cities that failed to meet the state TSP discharge requirement. The TSP level in Lanzhou was 2.47 times more than the state requirement.

Lanzhou is toying with a radical solution: demolishing a mountain to let in some fresh air. The scheme is based on the theory that the surrounding mountains block the movement of air into the city, causing a build-up of pollution in its natural depression.

The proposed solution is proving controversial. It is popular with developers, but not so popular with the local people — who have expressed doubts about whether it would work. Water was used to erode the mountain, pumping 45 000 litres a day from the Yellow River (Huang He) in pipes to the summit, then allowing it to flow down a zigzag path as workers shovelled soil into the flow. This would be by far the cheapest way to shift so much material. There were further protests from local people whose ancestors were buried in tombs on the hill.

## China's water quality

Every year, large amounts of pollutants are dumped into China's water bodies from agricultural, domestic and industrial sources.

### Agriculture

China is the world's largest consumer of synthetic nitrogen fertilisers, largely to increase agricultural yield to feed its huge and still growing population. As a result, pollution is widespread in China's rivers and lakes. This has worsened in recent years, with the pollution adjacent to industrially developed cities and towns being particularly severe.

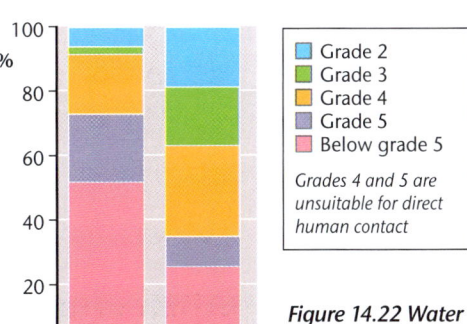

*Figure 14.22 Water quality at 135 monitored urban river sections, 1995*

Grade 2
Grade 3
Grade 4
Grade 5
Below grade 5

*Grades 4 and 5 are unsuitable for direct human contact*

### Industrial wastewater

Only 77% of industrial wastewater received any treatment at all in 1995, and nearly 50% of the industrial wastewater discharged failed to meet government standards. Industrial discharges contain a range of toxic pollutants including petroleum, cyanide, arsenic and heavy metals. This type of pollution is difficult to control, because it is scattered widely over a very large country. But more importantly, local authorities are reluctant to tighten control over pollution when pursuit of economic benefits is their first goal.

## Rural–urban interrelationships

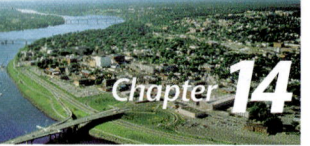

### Domestic waste

Treatment of domestic sewage is even less advanced. In 1995, China had only 100 modern wastewater treatment plants. Beijing had only one secondary sewage treatment plant, with a capacity of 500 000 tonnes. This plant cannot keep pace with the increasing amounts of sewage in the city.

### What can be done?

As is often the case, the push for rapid development has come at the expense of a clean environment. China's low production costs make it competitive in world markets. If pollution controls were in place and rigorously applied, production in China would become more expensive and less competitive. However, both the short-term and long-term external costs of this type of development are very high.

## Fuel-cell vehicles

Six of the world's cities most affected by smog are to benefit from the introduction of fuel-cell powered buses. The 5-year, $60 million programme announced by the Global Environment Facility will provide 46 buses powered by fuel cells for Mexico City, São Paulo, Cairo, New Delhi, Shanghai and Beijing.

Smog forms when pollutants released from petrol and diesel-powered vehicles react with heat and sunlight. Smog can inflame breathing passages, decreasing the working capacity of the lungs, and is especially harmful for elderly people, children, asthmatics and people with heart and lung conditions.

A fuel cell powers vehicles using electricity which it creates by combining hydrogen with oxygen. The only emission from these vehicles is water vapour, so they are significantly cleaner than existing petrol and diesel vehicles. Recent studies have linked the tiny particles emitted by diesel engines directly to several health problems, including heart attacks and childhood asthma. By emitting only water vapour, fuel-cell technology can alleviate those health risks and reduce greenhouse gas emissions that contribute to global warming. (On average, a car pumps out more than 3 tonnes of carbon dioxide, the foremost greenhouse gas, every year.) However, energy is required to make the hydrogen used in fuel cells. Unless this energy is derived from renewable sources such as HEP, use of fuel cells only reduces pollution locally, not globally.

## Criteria for sustainable tourism

**Sustainable tourism** seeks to minimise ecological and sociocultural impacts while providing economic benefits to local communities and host countries. In any scheme, the criteria used to define sustainable tourism should address minimum standards in the following (as appropriate).

### Overall

- Undertaking environmental planning and impact assessment which considers social, cultural, ecological and economic impacts (including cumulative impacts and strategies to limit impact).
- A commitment to environmental management by tourist businesses.

*Human systems, processes and patterns* **A2 Unit 5**

- Staff training, education, responsibility, knowledge and awareness in environmental, social and cultural management.
- Mechanisms for monitoring and reporting environmental performance.
- Accurate, responsible marketing, leading to realistic expectations.
- Consumer feedback.

## Social and cultural aspects

- Impacts upon social structures, culture and economy (on both local and national levels).
- Appropriateness of land acquisition/access processes and land tenure.
- Measures to protect the integrity of the local community's social structure.
- Mechanisms to ensure rights and aspirations of local and/or indigenous people are recognised.

## Ecological

- Appropriateness of location and sense of place.
- Biodiversity conservation and integrity of ecosystem processes.
- Site disturbance, landscaping and rehabilitation.
- Drainage, soils and storm water management.
- Sustainability of energy supply and minimisation of use.
- Sustainability of water supply and minimisation of use.
- Sustainability of wastewater treatment and disposal.
- Noise and air quality (including greenhouse emissions).
- Waste minimisation and sustainability of disposal.
- Visual impacts and light.
- Sustainability of materials and supplies (recyclable and recycled materials, locally produced, certified timber products, etc.).
- Minimal environmental impacts of activities.

## Economic

- Requirements for ethical business practice.
- Mechanisms to ensure labour arrangements and industrial relations procedures are not exploitative, but conform to local laws and international labour standards (whichever are higher).
- Mechanisms to ensure negative economic impacts on local communities are minimised and that there are preferably substantial economic benefits to local communities.

See also Case Study 14.16 (Tourism in rural regions: the UK, page 506) and Case Study 14.17 ('Damaging' tourism: the Alps, page 506).

# Chapter 15

# Development processes

## Global variations in development

Development is a complex term and there is no one view of what it means. According to Raymond Williams in his book *Keywords*, it came into English in the seventeenth century following an earlier word, *disvelop*, which had its origins in the old French *desveloper*. The original meaning of 'unwrapping' or 'unrolling' something came to be associated with evolution in biology. Since development is a controversial subject, it is hardly surprising that the terminology itself is disputed. The following list of popular terms gives a flavour of the problem:

- developed and developing
- North and South
- First World, Second World and Third World
- core and periphery

The terms **developed** and **developing country** have fallen into disuse. In the case of 'developed' this is because it makes it sound as though the processes involved in development have stopped. A developed project is something that is complete. It seems odd to apply this to a country or region that is still changing, growing and developing. To call Britain a developed country would be to imply that whatever processes have been taking place, they have now stopped. It also suggests that there is a predictable end-point in the process, a view associated with modernisation theorists such as Rostow.

By extension, the term 'developing' is applicable to all countries except, one might think, those that are getting poorer. For a large number of Latin American or sub-Saharan countries the term 'developing' would appear to be a bad joke, given the reduction in their national wealth that has taken place in recent years.

The term **underdeveloped** is rarely used, although it does still appear. It is generally avoided because it has negative connotations that are not far removed from 'backward' or even 'savage' in meaning.

Thus the terms **more economically developed country** (MEDC) and **less economically developed country** (LEDC) have become common, with the occasional use of **least economically developed** to identify those countries that are a major cause for concern in global politics. It is significant that 'economically' has retained its place, despite the fact that an increasing number of development scientists are inclined to view simple economic measures as potentially distorting. However, in order to make sense of these terms we need some way of recognising the differences. It is important to realise that there is a continuum here from 'more to less', in much the same way as there is from tall to short or from fat to thin. In

*Human systems, processes and patterns*  **A2 Unit 5**

addition, the terms depend on the context. Thus a fat friend doesn't look fat when surrounded by Sumo wrestlers; similarly an MEDC such as Portugal may have more similarity to an LEDC such as Algeria than it does to the USA or Canada. These are, in other words, relative concepts. One of the least well-developed regions of Europe is very wealthy when compared with the richer areas of Afghanistan.

Other terminology has been used. The North/South division of the planet needs a bit of artistic licence with the position of the equator, but the idea is one of economic division (Figure 15.1). The terms were popularised by the Brandt Report published in 1980, which was one of the first recognitions made by the 'establishment' that there was a problem of disparity in wealth.

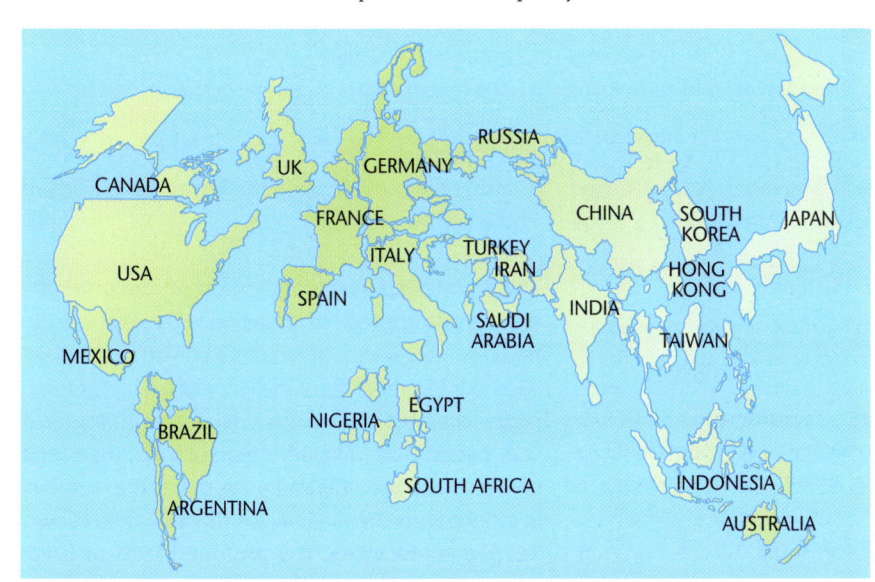

*Figure 15.1 Map showing countries according to the size of their GDPs (1997).*

*Note:* Relative size of country indicates proportional size of gross domestic product measured by purchasing power parities. Not all countries are identified and many are not represented.

The Brandt Report sought a balance in development policies and demanded that the countries of the South be integrated into the global economic system. This would, the authors argued, bring about necessary improvements in economic and social conditions in disadvantaged countries. The rich industrial countries of the North were called upon to share their means and power with the countries of the South. The report contained a number of proposals for the reform and transformation of the world economic system. It concluded that these reforms would be an important contribution to the survival of humanity.

The use of the term **Third World** is also common. Its origins were in a division of the world into political groupings first identified in French sociology:
- the First World, which was capitalist, and in which the USA was the dominant military power
- the Second World, which was socialist or communist, and in which the USSR was the dominant power
- the Third World, made up of non-aligned countries over which the first two worlds competed for influence

# Chapter 15 Development processes

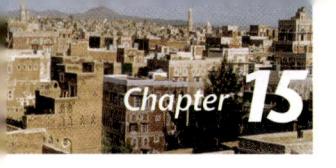

## Case study 15.1 The Brandt Report

One of the first attempts by the development establishment to clarify a view of the future of the world was the so-called Brandt Report. In 1977, Robert McNamara, then president of the World Bank, asked Willy Brandt, who had been chancellor of West Germany, to assume the chairmanship of the Independent Commission for International Developmental Issues, or the North–South Commission as it became known.

Through his personal contacts with politicians from every continent, Brandt was able to involve well-known statesmen and experts from both developing and industrial countries in the work of the North–South Commission. The report was published after 2 years of submissions and visits to a large number of countries.

Willy Brandt wrote in the foreword:

> This Report is based on the most simple of common interests — humanity wants to survive and, one can add, it has the moral duty to survive. This raises not only classical questions of war and peace, but also the questions of how can one defeat hunger in the world, overcome mass misery, and meet the challenge of the inequality in living conditions between rich and poor. To express it in a few words: this report is about peace.

The key conclusions were the following.

- The South had to be integrated into the global economic system and not left outside it — this was a plea for more and fairer trade.
- The institutions of the North, especially the United Nations (UN), the International Monetary Fund (IMF) and the World Bank, had to be reformed in order to allow the smaller nations more input and more control.
- Aid from countries in the North should be increased to about 1% of their GDP (in 2003 the figure was 0.21%).
- The free market would not solve all problems and governments would need to take a leading role.
- A clear relationship exists between the interests of armaments industries in the North and poverty in the South.

The Brandt Report was an important document in that it attempted to redefine the relationship between rich and poor nations that was, and still is, hard to interpret without the colonial legacy looming large. By one of those occasional ironies of history, it was swept away, like all other such visions of the future, by the neo-liberal regimes that came to power in 1979 (under Thatcher in the UK) and 1980 (under Reagan in the USA). The profound faith of these administrations in the free market meant that they rejected the blend of social democracy and cultural relativism in the Brandt Report.

### Key terms

**Core** That part of the world economy that is the focus of wealth acquisition and power.

**Periphery** That part of the world economy that supplies the core with raw materials.

**Semi-periphery** That part of the world that supplies the core with some manufactured goods.

According to this, the Third World was, essentially, the rest. Despite the fact that the term has been overtaken by the end of the totalitarian regimes of the Soviet bloc, and thus the disappearance of the Second World, it has entered the general vocabulary.

The use of the terms **core**, **periphery** and **semi-periphery** is common in some writings. It refers, of course, to a controlling, powerful and usually rich core and a poorer and frequently dependent periphery. The processes which interconnect these two are complex and highly controversial. One of the best known developments of this idea is the **world systems approach**. This abandons the nation state as being an outmoded and inefficient tool of analysis given that there is a wealthy elite in many LEDCs and many 'peripheral' workers in MEDCs.

*Human systems, processes and patterns*  **A2 Unit 5**

## Case study 15.2  Development divisions: poverty in the USA

According to the US Census Bureau, 35.9 million Americans were living in poverty in 2003. This is an increase of 1.3 million over the previous year, but the increase is not the real issue. What is important is that 12.2% of all Americans live in poverty (using a US population of 294.1 million as estimated by the US Census Bureau). The bureau defined poverty in 2003 as a family of two adults and two children with an income of less than $18 660 a year.

The top 1% richest Americans are now estimated to own 40–50% of the nation's wealth, more than the combined wealth of the bottom 95%. Over the past 30 years the disparities have become more rather than less marked. Real wages are lower in the USA today than they were in the 1960s and family incomes have only maintained their 1960s levels because of longer working hours and more members of the family being at work. In 1998, 30% of US citizens worked for wage rates of under $8 an hour.

The journalist Barbara Ehrenreich spent a year undercover in the sorts of jobs that many Americans have to take in an economy that is increasingly polarised between the wealthy inhabitants of Orange county or Silicon Valley earning $200 000 a year (see Case study 14.19) and the multitude who serve these more fortunate, more talented or simply luckier neighbours. Ehrenreich worked in Wal-Mart, as a waitress and as a house cleaner. She found enormous reserves of human compassion and much to admire in the attitudes of the people she worked with, but she did not find security and she did not find much evidence that a booming economy benefits all. The 'trickle down' thesis suggests that wealth creation in one sector eventually reaches all sectors of the economy, but Ehrenreich saw little evidence of this. The people she worked with:

- were not home-owners
- had no health insurance
- could not afford to save
- often held several jobs
- had to put up with authoritarian and dictatorial management
- were discouraged from union membership
- had no job protection
- had no faith in the political system and generally didn't trouble to register to vote

One of most fascinating aspects of Ehrenreich's work is that the worst performing areas of the USA in terms of poverty are those associated with the boom of the

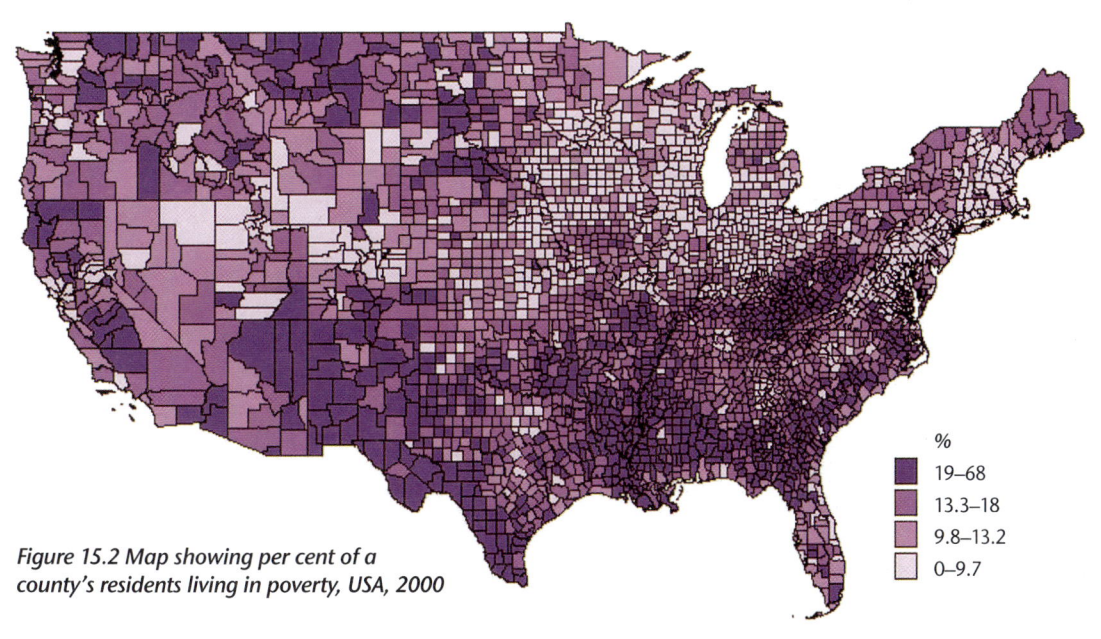

*Figure 15.2 Map showing per cent of a county's residents living in poverty, USA, 2000*

%
- 19–68
- 13.3–18
- 9.8–13.2
- 0–9.7

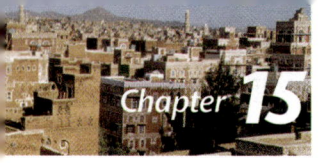

# Chapter 15 Development processes

US economy in recent years, the south and the west, the so-called sunbelt (Figure 15.2).

Ehrenreich offers no solutions, but she does point out that a situation in which 30% of the population are effectively left out of the good times is not healthy, either morally or politically. One of the few legitimate lessons of history is that you cannot afford to ignore the disadvantaged.

Two new terms have emerged to describe nation states whose development fits uncomfortably into any of the dominant categories. These terms are **newly industrialised countries** (NICs) and **recently industrialised countries** (RICs). In both cases there is no irony intended, but it seems strange to refer to China and India as RICs given their long history of industrial dominance before European development.

A further complication with all of these categories is that there is a natural tendency to ignore those variables or factors that cannot be measured easily. We might agree that human happiness is a fair indicator of development or degree of freedom, but it is not easily measured. This important qualification needs to be borne in mind when reading the next section.

### Key terms

**LEDC** Less economically developed country.
**MEDC** More economically developed country.
**NIC** Newly industrialised country.
**RIC** Recently industrialised country.

### Examiner's tip

- When defining these terms, remember to add that this is a vigorously debated topic that lies at the heart of political debate. Why the poor are poor is the core issue.

## Case study 15.3 Measuring development: happiness

Richard Layard has recently offered up a novel approach to looking at development and progress. Layard is co-director of the Centre for Economic Performance at the London School of Economics. There has long been a view that we might be over-obsessed with wealth and income as measures of development. Indeed, the human development index (HDI) takes this into account when it adds a social variable to GDP. However, the core idea that a more developed society is a richer society *in economic terms* has only been challenged periodically, often from the margins. Layard points out that we are now much better equipped to evaluate and quantify 'happiness', given recent advances in neuroscience. He argues that we have to face the accumulating body of evidence that in several respects 'money doesn't bring happiness'.

- It would appear that, within a country, richer people are, unsurprisingly, happier than the poor. Forty-six per cent of those in the top 25% of income report themselves as being very happy compared with only 26% of those in the bottom 25%.
- However, these proportions have not changed over time in any country, despite huge increases in personal wealth and material possessions in almost all societies.
- On the contrary, there is some evidence that each generation is successively less happy than their parents, who are in turn less happy than their grandparents. Clinical depression, suicide and addiction are increasingly common in the young.
- Comparisons between countries are quite startling. The relationship between average income and happiness is positive up to about $15 000 a

*Human systems, processes and patterns* — **A2 Unit 5**

year, but in countries above that level of average income the relationship disappears.

- It would appear that most of us adjust our responses to what we want in terms of what other people have, at least as far as income is concerned. It is our *relative* income that seems to matter, so it is hardly surprising that greater disparities of income in almost all MEDCs have led to decreasing contentment for many, even though they are better off in absolute terms.

A whole range of issues is raised by Layard's work, which is still in its infancy. One example is the utilitarian tradition whereby, if you take a pound from someone who will hardly miss it (say Bill Gates, but one need not be so extreme) and give it to someone who needs it, society as a whole is better off. This offers a reminder that regressive tax systems might be good for society, which is hardly the neo-liberal consensus. Perhaps it helps explain why the Icelanders consistently win the international 'happiness' league, living in a society that is among the least polarised in the world, and why the Bangladeshis (75% happy) do so much better than the Russians (37% happy) despite being poorer. The spirit of the last 30 years has been one of individuals and individualism and not, by and large, society as a whole. Famously, 'society' was not a meaningful concept for Margaret Thatcher, the Conservative prime minister through the 1980s.

# Measurements of development

Until recently development was equated with economic growth, partly because it was easier to measure but also because almost all other measures involve a normative judgment that runs into trouble quite quickly. There will be very different attitudes to 'freedom of speech' or the 'status of women' depending upon one's cultural and religious background. Most commentators accept the need to incorporate economic, social, cultural and political aspects of a society in any sensible evaluation of development. However, it is unlikely that the status of women will be evaluated in the same way by female British academics as it is by male Iranian clerics.

Neither the cultural nor the political variables are easily measured (quantified) and they tend to be ignored — except as footnotes. The usual measures are dominated by social and economic variables. The human development index (HDI) is the most commonly used, although there are other initiatives. New surveys have been devised to measure different aspects of human development, including the following:

- the living standard measurement study (LSMS), supported by the World Bank
- the core welfare indicators questionnaire (CWIQ), also supported by the World Bank
- the demographic and health survey (DHS), mostly financed by US-AID
- the multiple indicator cluster surveys, sponsored by UNICEF and WHO

There is no doubt that this multiplicity of statistics is helpful, although it is worth remembering that the data chosen may very well reflect something of the beliefs and ideology of the person quoting them.

The data used for such measurements are not perfect. Here are some of the obvious problems.

# Chapter 15 Development processes

- The comparability of economic data from country to country. For example, different systems are used to measure occupational status in different countries.
- How reliable are the initial data? Data are frequently quite old by the time they get into the public arena.
- The difficulty of measuring economic activity that is not official or legal — the 'black economy'. Estimating this important area of economic activity is obviously problematic, but in many countries it represents a substantial fraction of all economic activity — perhaps between 20% and 40%.
- The variations within a country can make some averages very unreliable. Per capita income is an obvious example; the mean can rise, but more people can be in poverty — both absolutely and relatively.
- Most of the data assume the dominance of a market economy whereby goods and services are offered for sale. Poverty cannot easily be assessed in these terms in subsistence economies where much is provided through family labour without money changing hands. From the small allotment in MEDCs through to the family farm in many LEDCs, this production is simply not counted.

Following the lead of other nations, a few years ago the USA replaced gross national product (GNP) with gross domestic product (GDP) as the measure of the value of the nation's output of goods and services. The difference between them is that, while GDP measures all the output produced domestically (within the borders of the USA), GNP only measures the output of firms owned by Americans. However, GNP includes the output of US-owned firms located abroad, while GDP does not. In recent years domestic savings have been inadequate to finance the desired level of domestic investment. GDP and GNP may therefore diverge even more in the future because of a rise in foreign ownership of firms located in the USA. GDP is the best-known measure of national income. It is a summary measure used to quantify national well-being. Many economists today complain that the government spends too little on collecting these and related data and, as a result, the figures are not adequate or accurate enough even in a country as rich as the USA.

Tables such as Table 15.1 can be used to display the usual component parts of the HDI and to highlight variations between the different indicators. A correlation test such as Spearman rank correlation coefficient can be done to test the relationship between these indicators, especially the economic measures and the social indicators. The last column, which shows the difference between ranking by GDP per capita and by HDI, is particularly instructive.

- Sweden's relatively high position reflects its welfare state and tax policy, which reduce income levels but help eradicate social and economic inequalities.
- The USA performs less well in the social indicators despite its very high GDP per capita. This reflects its national decision to lower taxes and remove the safety net of state help for the disadvantaged.

*Human systems, processes and patterns* **A2 Unit 5**

| HDI rank | Life expectancy (years) | Combined 1st, 2nd, 3rd grade school enrolment ratio | Real GDP per capita (US$) | Adult literacy | HDI | Real GDP per capita rank minus HDI rank |
|---|---|---|---|---|---|---|
| 1 Norway | 78.7 | 98 | 29 620 | 100 | 0.944 | 4 |
| 2 Iceland | 79.6 | 91 | 29 990 | 100 | 0.942 | 2 |
| 3 Sweden | 79.9 | 113 | 24 180 | 100 | 0.941 | 15 |
| 4 Australia | 79.0 | 114 | 25 370 | 100 | 0.939 | 8 |
| 5 Netherlands | 78.2 | 99 | 27 190 | 100 | 0.938 | 3 |
| 6 Belgium | 78.5 | 107 | 25 520 | 100 | 0.937 | 5 |
| 7 USA | 76.9 | 94 | 34 320 | 100 | 0.937 | −5 |
| 8 Canada | 79.2 | 94 | 27 130 | 100 | 0.937 | 1 |
| 9 Japan | 81.3 | 83 | 25 130 | 100 | 0.932 | 5 |
| 10 Switzerland | 79.0 | 88 | 28 100 | 100 | 0.932 | −3 |

*Table 15.1 A comparison of GDP and HDI data*

## Case study 15.4 *Measuring development: the HDI in practice*

In 2002 the United Nations (UN) published an Arab Human Development Report (AHDR) looking for a 'holistic development' approach to the application of the HDI. The methodology is based on an expanded version of the HDI designed by Nobel prize-winning economist Amartya Sen. During the 1990s, Sen's index revolutionised debate about 'development', which had previously been measured by economic indicators alone. The new version included measures of life expectancy, literacy, schooling and per capita real GDP in an attempt to achieve a more complete picture of poverty, growth and inequality. The AHDR authors added more to Sen's index, including data on life-long knowledge acquisition, especially regarding information technology, women's access to societal power and 'human freedom'. In a phrase that became well known, the AHDR found that the Arab world is 'richer than it is developed'.

Arab countries have greater resources than some developing countries that rank above them in various indices of human development, particularly if those indices measure women's status. For example, the maternal mortality rate in Arab countries is double that in Latin America. In spite of clear improvements in the post-independence era, women's literacy in Arab countries still stands at only 50%. And Arab countries score lower than any other region on the index of freedoms which include political and civil rights, independent press and television, and the accountability of rulers to the ruled.

However, the data also show that the Middle East and north Africa have the most equal income distribution and lowest level of absolute poverty in the world, a feat made possible both by remittances from Arabs working abroad and by a strong system of family and social responsibility within Arab society. Yet these remittances are unreliable and social cohesion is strained by a system that forces migration on some and isolation on others, usually the women who are left at home. The AHDR urges:

- employment creation and poverty-reducing growth
- increased attention to healthcare, education and knowledge acquisition related to information technology
- a massive effort to give women social and economic equality with men

## Chapter 15 — Development processes

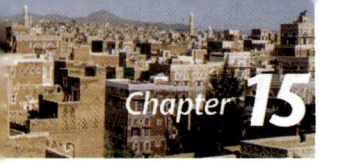

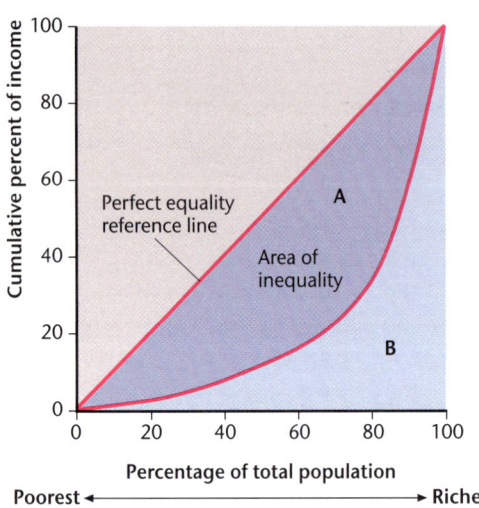

*Figure 15.3 A Lorenz curve for distribution of income*

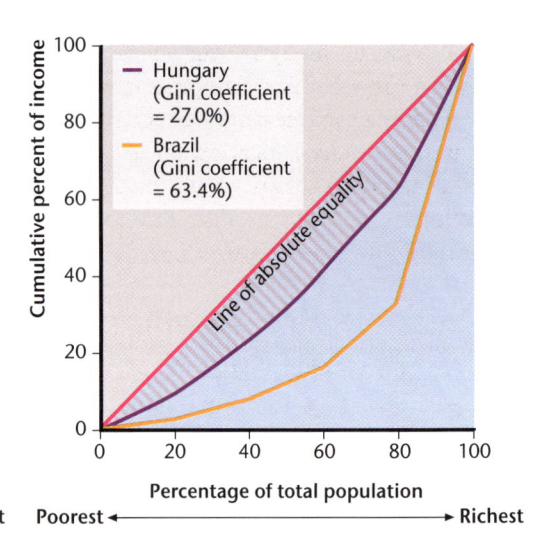

*Figure 15.4 Lorenz curves for income in Hungary and Brazil*

## Measuring inequalities

A tool often used to measure unequal distributions is the Lorenz curve. This can be applied to distribution of income: if 50% of the population earn 50% of the income, then the 'curve' will be a straight line and this will be a society where all citizens have the same income. From these curves a Gini coefficient can be derived, which expresses Area A in terms of Area B (Figure 15.3). Thus the higher the Gini coefficient, the greater is the inequality within the country (Figure 15.4). There is more about Lorenz curves in Chapter 8.

Income disparities vary from country to country and from time to time, reflecting different economic management. Figure 15.5 shows changes in Gini coefficients with time for a range of countries and Table 15.2 gives figures for world regions. Remember that an increase in the Gini coefficient means that the distribution has become less equal.

|  | Gini coefficient, 1993 (%) | Gini coefficient, 1998 (%) |
| --- | --- | --- |
| Africa | 47.2 | 42.7 |
| Asia | 61.8 | 55.9 |
| Latin America and the Caribbean | 55.6 | 57.1 |
| Eastern Europe and the former USSR | 46.4 | 25.6 |
| Western Europe, North America and Oceania | 36.6 | 37.1 |

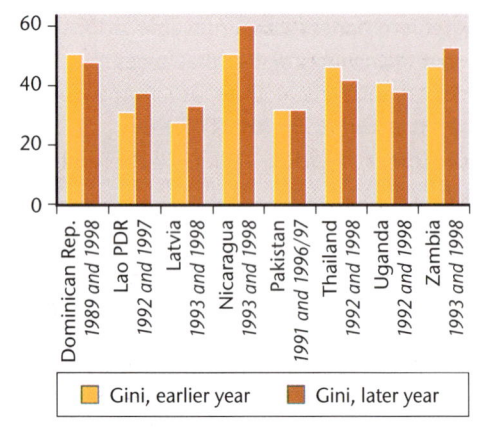

*Figure 15.5 Changes in Gini coefficients calculated for world development indicators, 1990s*

*Table 15.2 Estimated Gini coefficients for household-level distribution of income and consumption*

*Human systems, processes and patterns* **A2 Unit 5**

This type of analysis can also be applied to individual variables. For example, the World Bank has devised an index for measuring educational attainment from basic primary education up to degree level. In India a large percentage of the population has no schooling at all and the distribution of education remains very unequal (Figure 15.6).

## Sustainable development

The term **sustainable development** has become so widely used that its meaning is sometimes lost. The phrase became current in the early 1980s and appeared in the World Conservation Strategy conference in 1984, and in a book of collective authorship entitled *Building a Sustainable Society* the following year. Its adoption as a key catchphrase at the 1992 UN Conference on Environment and Development in Rio marked its coming of age as the mantra for development policies. In its initial form it called for a halt to economic development altogether — the zero-growth response. However, the usual definition goes well beyond this, suggesting that sustainability is a relationship between people and their environment. A common definition is:

> Meeting the needs of the present without compromising the ability of future generations to meet their own needs.
> World Commission on Environment and Development

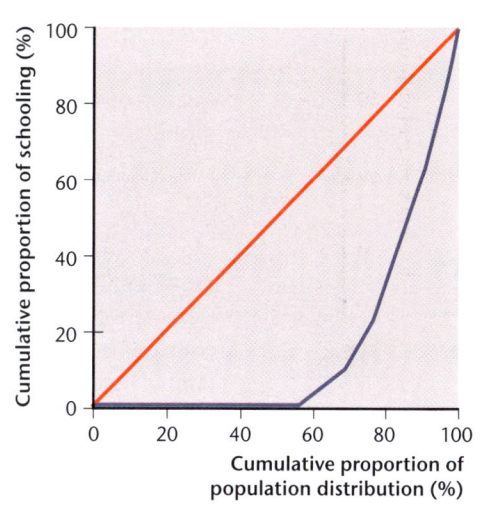

*Figure 15.6 Lorenz curve for education in India, 1990*

This reinforces the link between current behaviour and future consumption, but the view that economic development can be halted has largely been abandoned. It is now conventional wisdom that there exists a vicious circle whereby poverty leads to environmental degradation which in turn leads to greater poverty. Environmental challenges arise both from the unintended effects of some forms of economic growth, such as displacing people into unsuitable environments where their farming practices lead to environmental destruction, and from the lack of development. The zero-development concept gave way to a dynamic concept of sustainable development, a form of progress that tries to ensure human development without destroying the environment for future generations.

This can be interpreted in a number of ways. Here are some more quotes.

> Sustainable production and consumption is the use of goods and services that respond to basic needs and bring a better qualify of life, while minimising the use of natural resources, toxic materials and emissions of waste and pollutants over the life cycle, so as not to jeopardise the needs of future generations.
> *Symposium: Sustainable Consumption*, Oslo, 1994

The emphasis of sustainable production is on the supply side of the equation, which concentrates on improving environmental performance in key economic sectors, such as agriculture, energy, industry, tourism and

### Key term

**Sustainable development**
Development that accepts the need to compromise present demands in order to allow future generations to meet their own needs.

# Chapter 15  Development processes

transport. Sustainable consumption addresses the demand side, looking at how the goods and services required to meet basic needs and improve quality of life — such as food and health, shelter, clothing, leisure and mobility — can be delivered in ways that reduce the burden on the Earth's carrying capacity.

Robins, N. and Roberts, S., Changing Consumption and Production Patterns: Unlocking Trade Opportunities, 1997

Sustainable consumption implies that the consumption of current generations as well as future generations improves in quality. Such a concept of consumption requires the optimisation of consumption subject to maintaining services and quality of resources and the environment over time.

Salim, E., The challenge of sustainable consumption as seen from the South, Symposium: Sustainable Consumption, Oslo 1994

All the above approaches subdivide people and nature in an interesting way that separates the two quite clearly.

The website of 'sustain-ed' (**http//:www.uyseg.org/sustain-ed/**), one of the many bodies that promote sustainable development, features Photograph 15.1, which shows flooding in York. The text with the photograph reads as follows:

> Emissions of carbon dioxide and other gases cause the Greenhouse Effect. This has caused the temperature of the planet to rise. Scientists estimate that the planet has not been this warm for over 1000 years. Since records began, the seven warmest years have occurred since 1990.
>
> Global warming causes climate change. Droughts kill off crops in developing countries and cause famines, while at the same time there are floods causing damage in others.

The implication is that the floods in York are a direct consequence of global warming. While this may be the case, it is possible for a rational human to argue that the flooding of York is not primarily the result of global warming. This neither diminishes the potential significance of global warming nor suggests that other global events cannot be attributed to it.

*Photograph 15.1 Flooding in York*

# Human systems, processes and patterns — A2 Unit 5

# People and nature

## Four views

The relationship between humankind and the natural world is seen in a number of ways. Four of these are discussed below.

### Robust

One view is that nature is extremely robust. The system is forgiving of human impacts and is virtually inexhaustible as a resource base. In its purest form, this view also regards changes to the global environment as a positive challenge, with new opportunities for human ingenuity. It assumes that green technology will prevent, correct or even restore any unanticipated damage to the environment.

This is essentially a conservative view of the world, in which the invisible hands of markets are seen to be the only necessary regulatory mechanism for the system. It suggests that the sustainable development ideology is a thinly veiled disguise for opposition to capitalism in general — a conspiracy of the political left.

### Fragile

An opposite view is to see nature as extremely fragile — vulnerable to irreversible collapse due to ecological degradation or natural resource exploitation. This, in systems terms, is a ridge position in which a very small change will induce positive feedback and global catastrophe of the sort popularised by the film *The Day after Tomorrow*. Those who hold this view believe that we have to change the very basis of economic, political and social interactions to create harmony between humans and the environment or be doomed to ultimate disaster. For them, capitalism is at odds with planetary equilibrium and has to be modified.

### A middle view

As one might imagine there is a middle view, which is embraced by most of the mainstream. According to this, nature is robust within limits that must not be surpassed. This view assumes that global catastrophe can be avoided through scientific understanding of ecological limits and the establishment of sensible management procedures. It is popular with governments, which envisage a type of global bureaucracy to manage environmental change. Economic growth can, according to this, be maintained through rational management.

The strategy proposed in the Brundtland Report, published in 1987 by a UN Commission set up to consider sustainable development, provides a good example of this. This report was confident that global environmental change could be managed by means of an institutional framework made up of numerous advisers, government committees and large institutions such as the UN. Managing institutions, therefore, should be composed of experts employing 'top down' management strategies. Critics suggest that this is just a thinly veiled mechanism for ensuring the continuation of the global *status quo*.

### Chaos

A fourth and rather disconcerting view sees nature as essentially random. The system is both chaotic and unpredictable. Needless to say, there is little point in management strategies, other than prayer, if natural change is random.

# Chapter 15    Development processes

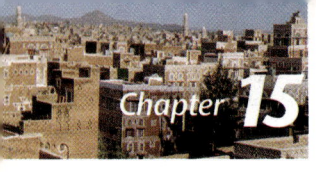

### More about the middle view

The third view, which is the dominant one in the Western, capitalist world, embraces a whole series of terms, each of which carries a deeper set of meanings.

**Carrying capacity** is the maximum number of individuals of a defined species that a given environment can support over the long term at a reasonable quality of life. The idea of limits is fundamental to the concept of carrying capacity. For human beings, calculating carrying capacity is problematic since we alter our environment and our expectations of a reasonable quality of life. Some argue that the concept is meaningless as free market conditions and technological innovation can extend limits indefinitely, although this is obviously more arguable at a global level.

**Ecological footprint** is the area of land and water required to support a defined economy or population at a specified standard of living. Industrialised economies require far more land than they have, so through trade they have an impact on resources in other countries. This is also known as 'appropriated carrying capacity'. The ecological footprint of the richest MEDCs is substantially above that of most LEDCs (see Table 14.10, page 503).

**Natural resources** are those environmental 'goods', including the so-called 'free' goods such as air and water, which are available in nature. It refers to a stock (e.g. a forest), which produces a flow of goods (e.g. new trees). Natural resources can be divided into renewable and non-renewable; the level of flow of non-renewable resources (e.g. fossil fuels) is determined economically and politically. Natural resources are culturally defined — we can change our minds about what to use as well as changing our technical ability to exploit the environment.

**Environmental debt** is the cost of restoring previous environmental damage. Unless measures are taken to prevent environmental degradation, environmental debt is bound to rise and the debt is transferred to future generations. However, some environmental damage, such as species extinction, is not restorable and therefore cannot be included in the environmental debt even if it were readily measurable. Of course, given that the only sure future for any species is extinction, it becomes quite difficult to tease apart what might be termed 'normal extinction' from accelerated 'human driven' extinctions.

**Inter-generational equity** is the principle of fairness between people alive today and future generations. The implication is that unsustainable production and consumption now will degrade and destroy the ecological, social and economic basis for our children. Sustainability involves ensuring that future generations will have the means to achieve a quality of life equal to or better than the one we enjoy today. It is impossible to evaluate quite what technical changes future generations will make to release them from a dependence on resources that we now consider at risk from over-consumption.

**Intra-generational equity** is the principle of equity between different groups of people alive today. Like inter-generational equity, it implies that consumption and production in one community should not undermine the ecological, social and economic basis for other communities to maintain or improve their quality of life. This

---

**Key terms**

**Capital (material) resources** The equipment, buildings and infrastructure of a society.

**Human resources** The ability and potential of the population, frequently expressed in terms of health, literacy and educational attainment.

**Natural resources** Those materials found in nature of use to us and which we choose to exploit.

*Human systems, processes and patterns*   **A2 Unit 5**

is a far from universally accepted goal, because it suggests that competition between different groups, whether it be classes or countries, is to be avoided. In a global economy based on competition and with the increasing gap between rich and poor, it sits uneasily.

## The relationship between natural resources and development

Most commentators recognise three types of resources:
- natural resources
- material or capital resources
- human resources

**Natural resources** are things found in the natural world that humans have the ability and the desire to use. These change over time. Flints were a resource once, but are not so any longer. Uranium is a resource while we decide to use it for nuclear power and weaponry, but we might decide otherwise. As Figure 15.7 shows, few parts of the world owe much of their wealth to natural resources.

**Material resources** are stocks of capital held in the form of machinery, equipment and the built environment. These tend to increase over time as labour, or human resources, are replaced by stocks of machines and equipment that are the product of past labour. Societies with labour shortages tend to be technically innovative, whereas those with abundant labour tend not to be. The importance of material resources is about equal the world over (Figure 15.7).

**Human resources** are the abilities and potential of the human population in terms of their educational levels, their skills and their capacity to innovate and invent. In most countries, human resources are the most important (Figure 15.7). These have often been neglected in studies of development. Early campaigns to send tractors to India overlooked the question of what would become of the labour that such machinery replaced.

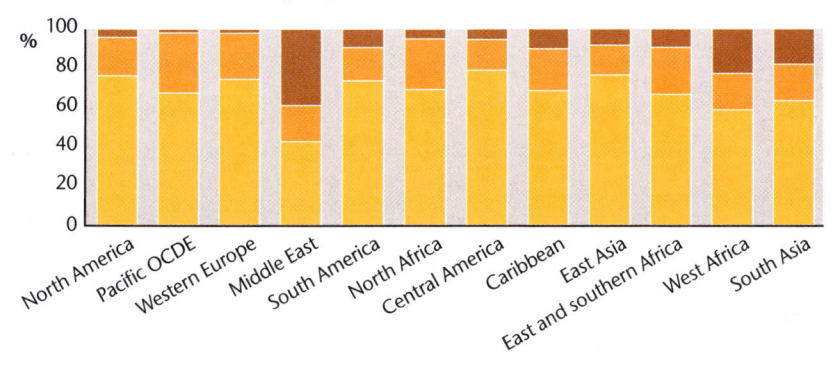

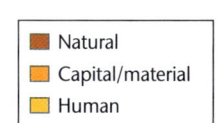

*Figure 15.7 The composition of the wealth of a selection of global regions, 1999*

There are three possible relationships between natural resources and development.
- Natural resources are an essential element in the development of a region or state.
- Natural resources are a useful but not essential element in economic development.

# Chapter 15 Development processes

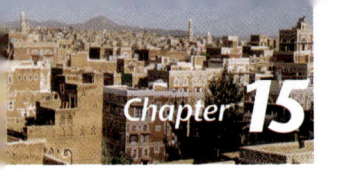

- Natural resources are an obstacle to the economic development of a region or state.

At an intuitive level it seems obvious that the more resources there are in a country, the better off it will be, but on a global scale that is clearly not supported. In reality the correlation is very weak. Modern nation states with abundant natural resources are not, by and large, among the wealthiest. Zambia, Botswana and Chile are obvious examples. By contrast, listed among the world's richest nation states are Japan and the Netherlands, which are poorly resourced. In Japan's case this includes a lack of land available for agriculture. It is argued that states which lack natural resources are obliged to exploit their human resources as a means of improving the national wealth, so education becomes a significant feature in their national policies. This has been the case in Singapore and Taiwan, neither of which are well resourced in terms of natural wealth.

## Resource curse theory

The most challenging position is sometimes described as **resource curse theory**. This is the counter-intuitive view that natural resource endowment can actually obstruct balanced economic development because it leads to too great a dependence on one or more natural resources. This then discourages industrialisation and the development of the other two main categories of resources because the raw materials are sufficient to generate high incomes, at least for some, while they endure.

According to analysis by Jeffrey Sachs and fellow economists at the Harvard Institute for International Development, the more a developing country's economy depended on resource exports, the less it grew in GDP per capita between 1965 and 1990. Statistically, at least, it seemed that resource-rich countries were performing less well than resource-poor countries.

Table 15.3 shows some countries where a single product is responsible for a large proportion of export earnings.

*Table 15.3 Export earnings due to a single product*

| Product | More than 80% | 60–79% | 40–59% |
|---|---|---|---|
| Agriculture and fishing | Chad (cotton) | Comoros (spices)<br>Iceland (seafood)<br>Malawi (tobacco)<br>Rwanda (coffee)<br>Somalia (seafood)<br>Uganda (coffee) | Benin (cotton)<br>Burma (timber, opium)<br>Ethiopia (coffee)<br>Fiji (sugar)<br>Guadeloupe (bananas)<br>Pakistan (cotton) |
| Crude oil and petroleum products | Algeria<br>Brunei<br>Iran<br>Iraq<br>Nigeria<br>Saudi Arabia<br>Syria | Egypt<br>Georgia<br>Venezuela | Cameroon<br>Democratic Republic of Congo<br>Norway<br>Russia<br>Trinidad and Tobago |
| Metals and minerals | | Botswana (diamonds)<br>The Gambia (diamonds)<br>Surinam (aluminium) | Liberia (diamonds)<br>Mauritania (iron ore)<br>Togo (phosphates) |

Prices of these commodities are highly variable, driving governments into lavish programmes of expenditure in the good times, which are then hard to maintain as prices fall. Resource extraction apparently does little to impart skills to the managers and labourers it employs, who may well be foreign personnel employed by TNCs. The workforce employed in extractive industries does not need to be educated. On the contrary, too high a level of education might lead to them migrating in search of more rewarding and better paid jobs.

Extractive economies, geared to exports, do not generate cities — and cities are where innovation and knowledge transfers take place. The import-substituting cities are seen by some as central to sustained economic growth. Modern extractive industries, in contrast, usually fail to enrich the local economic fabric. The skills do not transfer well to other industries, and extractive industries do not stimulate local growth in related industries such as mining machinery and food processing. Most mineral extraction or agricultural production is handled by joint ventures of local capital, often with an elite group taking an active part, and foreign capital, often in the form of TNCs turning these regions into economic enclaves.

Recent World Bank studies have found that concentration of land ownership, which can be translated into concentrated control of natural resources, correlates with slower economic growth. There is no incentive for land reform, whether in the Mezzogiorno or Brazil, if a wealthy elite class is comfortably provided for by export-led extractive industries. Thus currency is fixed in order to maximise the value of exports and minimise the costs of imports, and this acts as a further disincentive to the establishment of a broader industrial base in the country. Any such industry would find it impossible to compete with cheap imports.

A variety of resource curse is the so-called 'Dutch disease' in which the value of a currency is driven upwards because foreigners need to buy the currency in order to purchase an essential export such as oil and gas. The high value of the currency makes all other exports, such as agricultural products and manufactured goods, expensive and thus hard to sell abroad, which in turn leads to unemployment and dependency upon state aid within the country. Conversely it makes imported manufactured goods cheap.

The resource-controlling elite class is usually the dominant government class, which tends to under-invest in education, healthcare and agricultural extension for the poor. None of these improvements are of any direct benefit to the elite, but instead represent a threat to their interests. For example, control of logging rights in the Philippines has turned a few hundred well-connected families into the nation's wealthy elite. They are, unsurprisingly, also the dominant political class since they alone have the resources to run for political office.

The American economist William Easterly has developed this idea further by adding that the nature of the original colonialisation is an important key to the rate and nature of development. If a country or region was simply a resource provider, then it developed a distorted land ownership system with a ruling class supplied by the 'mother' country, as in Brazil. However, where countries did not have abundant resources that were immediately obvious, as in the early history of the USA, they were settled by colonists who developed a market system and domestic industries to serve their needs.

# Chapter 15  Development processes

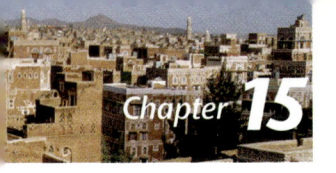

There is, of course, the wider problem that an over-reliance on natural resources in a world where the terms of trade have moved against primary producers is unlikely to bring broad-based economic growth in a country. The fall in world prices of a whole swathe of commodities, from cocoa to copper, leads to attempts to increase production to maintain income or to cut costs. In the first case this is likely to lead in a further fall in prices and in the second to a reduction in the living standards of the population as a whole.

## Case study 15.5  The rise and fall of a supply economy: Uruguay

The history of supply economies such as Uruguay illustrates the resource curse theory. Some statistics are shown in Table 15.4.

Uruguay is a Latin American country squashed between Brazil and Argentina, two large and better-known neighbours. Migrant populations arrived mainly from Italy and Germany at the end of the nineteenth century and the country developed, rather unusually for Latin America, as a country of small farmers with relatively few large estates (*haciendas*).

Labour was in short supply so wages were high, and family farms tended to dominate. A society developed with a relatively even income distribution and no great distortions of wealth. The only resource was the land, which was plentiful and reasonably fertile. A pastoral economy developed, specialising in production of cattle and sheep for export. Beef, leather and wool were exported in large quantities, mainly to Europe which was still experiencing rapid population growth and rapid industrialisation.

Uruguay flourished in the early twentieth century and was frequently held up as model of Latin American development, much as South Korea is used today as a model of NIC development. However, Uruguay was no NIC. It developed no industries of its own but fulfilled the same function in the world economy as the southern states of the USA had done before and Saudi Arabia does today — the supply of raw materials to a distant market. The wealth created was re-invested in a highly advanced welfare system, health service and infrastructure. This led contemporary observers to describe the country as the Switzerland of South America ( in other words, small and rich!). It was, according to some estimates, the eleventh wealthiest nation in the 1930s.

| | |
|---|---|
| Growth rate | 0.77% |
| Birth rate | 17.42 per 1000 |
| Death rate | 9.06 per 1000 |
| Infant mortality | 14.14 per 1000 live births |
| Fertility rate | 2.37 |
| Literacy | 97.3% |

*Table 15.4 Data for Uruguay*

Montevideo was the capital and only urban centre with a primacy of ×20. Its functions were administrative, educational and as a port for trade on which the country depended. Following the Second World War the Uruguayan 'miracle' began to unravel.

- The markets disappeared as postwar Europe established a protected trade area (the EEC — the forerunner of the EU) that encouraged increases in farm output within the area but put tariffs on imported foodstuffs.
- Competition from other distant suppliers increased, specifically Australia, South Africa and New Zealand.
- The emergence of substitute products, especially artificial fibres, further reduced demand.

As the markets disappeared in the 1950s, Uruguay embarked upon a crash industrialisation programme, but it had:
- no industrial 'tradition' and an education system that was geared to producing administrators and not engineers
- a city without any industrial structure — no workshops, no 'culture' of industrialisation
- high costs of starting up from scratch, which increases the import bill as machine tools all have to be imported

*Human systems, processes and patterns* **A2 Unit 5**

Thus the industrialisation programme incurred very high costs for few returns, so the following sequence occurred.

- The government was obliged to cut back on welfare and education and to increase taxation.
- More money was printed, which led to hyper-inflation.
- Strikes and civil unrest led to the rise of the Tupamoras guerrilla movement.
- 1 million people fled into neighbouring countries such as Brazil, as civil war broke out.
- In 1973 the military took over.
- Uruguay reverted to an export-oriented economy specialising in cheap semi-processed products, e.g. leather goods, for a market largely in Brazil and Argentina.
- Uruguay slid down the 'league' table of wealth and development.

# Development theory

## Economic, social and cultural factors

Studies of economic development, as it is now understood, really started in the 1930s. This was when studies in the 'colonies' showed that many people did not appear to live in an advanced capitalist economic system. Early ideas of development were dominated by colonial and, ultimately, post-colonial views of the world. These tended to concentrate on possible obstructions within 'under-developed' countries.

Some of these were unapologetically racist in tone, a view carried in textbooks up to the 1950s. These attitudes still show through, for example in the 'blame-the-victim' type of analysis when *The Economist* magazine describes Africa as 'the hopeless continent' or when throw-away clichés such as 'basket case' are applied to whole nation states, many of which were damaged by their colonial histories. The 'backward' nature of these countries was frequently equated with human development models in which Africans in particular were 'like children' and couldn't be trusted.

Economic development theory equated development with growth and industrialisation from an exclusively economic point of view. As a result, Latin American, Asian and African countries were seen as primitive, backward or infant versions of European nations that could, with time (and help), develop the institutions and standards of living of Europe and North America. The analogy with human development from embryo to adult was often made.

## Development and underdevelopment: the theoretical battleground

### Why are some countries richer than others?

Development theory is a relatively new field that emerged in post-colonial times. Insofar as a theory existed until then, it was largely based on the inherent superiority of one race over another — a view that was periodically 'supported' by scientific research into skull size, brain weight and ultimately IQ. Early theories of

# Chapter 15 Development processes

development tended to emphasise the problems within countries that might be obstructive to growth, including the qualities of the population. According to Abbe Prevost, 'They [the Africans] are naturally sly and violent. They cannot live in peace with each other.' Just in case the point was not clearly understood, the eighteenth-century French traveller Bougainville added, 'Blacks were much more savage than the Indians, whose colour approached that of whites.'

In the period after the Second World War, a rather more enlightened set of ideas began to emerge that suggested that poor countries were trapped in a vicious circle of poverty. Low incomes led to low savings, which in turn led to low investment and low incomes. It was suggested that the best route out of this was to improve agricultural production by adopting more modern production methods, thereby increasing output from the land and increasing income. The 'surplus labour' could then take up the factory jobs that would start to appear as the rate of saving and thus the rate of investment increased. Development was seen as a process of removing internal obstacles to industrialisation within societies. Savings could be encouraged by governments, making industrialisation feasible and laying the foundations for further development. Thus, government involvement, whether by planning, socioeconomic engineering or effective demand management, was regarded as a critical tool of economic development.

However, a new (or not so new) idea became more widespread at the same time. The structuralist thesis, in brief, called attention to the *distinct* structural problems of LEDCs. It argued that LEDCs were not merely primitive versions of developed countries; rather they had distinctive aspects of their own and thus required a different set of policies for development.

One of these aspects was that, unlike MEDC industrialisation, LEDC industrialisation was expected to occur while these countries existed alongside already industrialised MEDCs and were tied to them by trade. It was argued that this could give rise to distinct structural problems for development.

Although capital formation and investment remained the central part of these theories, an increasing amount of attention was paid to human resources. This led to an emphasis on education and training as prerequisites of growth, and the identification of the problem of the brain drain from LEDCs to MEDCs.

Social development as a whole, notably to do with education, health and fertility, would establish the foundations for growth, by improving human capital. In this view, industrialisation, if it came at the cost of social development, could never be sustainable.

By the 1960s development was taken to include the elimination of poverty, unemployment and inequality, as well as the presence of economic growth. Thus, structural issues such as dualism, population growth, inequality, urbanisation, agricultural changes, education, health and unemployment began to be treated as a central part of development.

An important debate on the very desirability of growth was stimulated by Schumacher in his book, *Small is Beautiful* (1973). He argued against the desirability of industrialisation and extolled the merits of handicraft economies. As the world environmental crisis became clearer in the 1980s, this debate took a new

*Human systems, processes and patterns* **A2 Unit 5**

twist. The sustainability and the desirability of economic development began to be questioned.

In a complex field it is possible to identify two dominant schools of thought.
- **Modernisation theory**, associated with the work of **Rostow**.
- **Dependency theory**, associated with the work of **Gunder Frank**.

As with much of this discussion of development, there is a problem with definitions.

According to modernisation theory, most LEDCs are economically *un*developed in that they are developing towards an end stage much as human beings develop as they grow up into adulthood. This means that they are at the same *stage* of 'undevelopment' that present-day industrialised countries experienced before the process of industrialisation got under way. In this sense LEDCs are, basically, 'unmodern' compared with metropolitan countries (USA, Japan, Britain).

According to dependency theory, most LEDCs are economically *under*developed, using the word developed as a verb rather than an adjective. This is because the relationship they have with the metropolitan centre or core is one of exploitation. As long as this relationship lasts, LEDCs will never become developed. So the problem they face is not one of economic 'undevelopment'. On the contrary, they need to find a means of uncoupling or disengaging their economies from an international pattern of trade and commerce which contributes to their continuing underdevelopment.

## W. W. Rostow: modernisation theory

Walter Whitman Rostow was an American economic historian who, rather ironically given his conservative reputation, was named after one of America's great radical poets and thinkers, Walt Whitman. His best known book is *The Stages of Economic Growth: A Non-Communist Manifesto* (1960). As the subtitle of this text suggests, Rostow's work had a clear political purpose. He explicitly saw his theory of stages of economic growth as an alternative to Marx's own stage model, although the Marxist model did have a theory behind it, albeit one that was deeply challenging for many members of the ruling class. Rostow was both an academic and a political figure working inside the US administrations of the 1960s to develop strategies for dealing with the perceived threat of communism, especially in southeast Asia.

Based on his work on the growth of industrialisation in the UK and the USA, Rostow put together a model that incorporated five stages through which all developing societies must pass. These are discussed below.

### Stage 1 Traditional

Rostow was acutely aware that he conflated a huge variety of societies covering most of human history and prehistory, ranging from stone age cultures to, say, France on the eve of the revolution. However, he believed that for the purposes of his model of development their common features mattered more than their obvious differences. These features include:
- pre-Newtonian science and technology

> **Key terms**
>
> **Dependency theories**
> The theories that build upon the idea that the rate of development of a country is largely a consequence of its relationships with other countries.
>
> **Modernisation theories**
> Theories that build on the idea that the rate of development of a country is largely to do with the internal structures, governance and culture of that country.

# Chapter 15  Development processes

- a subsistence agricultural economy
- a hierarchical social structure, usually based on birth rather than ability
- a fatalistic belief system that has no concept of progress

### Stage 2  Preconditions for take-off
Changes begin across a whole range of institutions, often as a result of some impulse from outside (e.g. invasion). In the economy, some examples are the following:
- agriculture outputs increase through improvements in technology
- trade between regions starts to develop as improved communications facilitate the growth of the economy
- some industries develop, especially extractive industries such as coal and iron ore
- socially, these processes are related to the emergence of an elite group, able and willing to reinvest their wealth rather than spending it on luxuries or display goods
- rational scientific ideas begin to emerge with the foundation of scientific societies
- the idea that progress is possible gains ground — leading to rationality rather than fatalism

### Stage 3  Take-off
Rostow described this critical stage in both quantitative and qualitative terms:
- investment as a proportion of national income rises to at least 10% of GDP
- one or more manufacturing sectors (but not the whole range) emerge, possibly including iron and steel and textiles
- the infrastructure is rapidly improved
- political and social institutions are reshaped in order to permit the pursuit of growth to take root

Rostow suggested that this takes place over about 20 years. In Britain it happened between 1783 and 1803, in Japan between 1878 and 1900, in Russia between 1890 and 1914.

### Stage 4  The drive to maturity
This is a period of consolidation, of making sure that the development made possible in the take-off period is sustained:
- modern science and technology are extended to most, if not all, branches of the economy and mechanisation develops
- the rate of investment remains high, at 10–20% of national income
- political reform continues, with the extension of voting rights and the development of democratic institutions
- international trade begins to become more significant

### Stage 5  The age of high mass consumption
This involves yet further consolidation and advance:
- the growth of consumer industries which become economically significant
- rapid gains in levels of personal ownership of property and material wealth

- stable political and social systems
- high levels of education in the population
- increasingly significant international trade

Rostow suggested that there were broadly three possible variants available at this last stage.
- Wealth can be concentrated in individual consumption, as in the USA.
- Wealth can be channelled into a welfare state, as in western Europe.
- Wealth can be used to build up global power and influence, which is how Rostow described the USSR at the time.

So Rostow presented a highly descriptive model of development in which much was left unsaid about how one stage leads to another. However, he did commit himself to a number of testable hypotheses.
- Rostow's theory is **evolutionist**. It sees social and economic change as unfolding through a fixed set of stages.
- It is **unilinear**. All countries *must* pass by the same route, in the same order. There can be no 'leaps', short cuts, choices or alternative routes to development.
- It is **internalist**. The factors that determine whether or not a country can develop are internally generated within any one society. So if countries do not develop in the manner described, it is because of problems with themselves and, of course, nothing to do with the world system.

## Andre Gunder Frank: dependency theory

Andre Gunder Frank is also an American economist. After studying economics at Chicago University he went to Latin America in the early 1960s and dramatically revised his largely conventional economic views under the impact of both the Cuban revolution and the emerging 'dependency' school of economic thought. His most famous book is *Capitalism and Underdevelopment in Latin America* (1969). Frank's position is an explicit critique of Rostow's. In particular, his key term 'the development of underdevelopment' may be seen as the radical counterpart of Rostow's 'take-off' stage.

The major points of Frank's approach are as follows.
- Rather than taking a particular society as his unit of analysis (i.e. unlike Rostow), Frank sees national economies as structural elements in the global capitalist system. It is this system, not individual societies, which is the necessary unit of analysis, according to Frank. In order to understand a particular country's state of development, we must look at its place in the international capitalist economy.
- This system is set up in favour of a small number of metropolitan or core countries. Dominated by the USA, Europe and Japan, there exists a whole series of intermediate places which are simultaneously both core and periphery. For example, São Paulo contains intermediate figures who represent and belong to the core, living in gated suburbs and enjoying luxurious lifestyles. These individuals exploit the landless rural labourer who has nothing and no one to exploit. Thus Frank is not terribly interested in nation states, seeing them as incidental to the relationships in the system.

# Chapter 15  Development processes

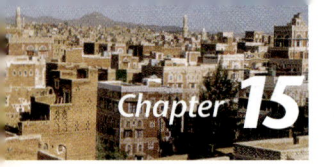

- In Frank's language, which is borrowed from Marx, 'surplus' is continuously moved upwards and outwards, at all levels, from bottom to top. This occurs because each metropolis has monopoly economic power in its part of the system, rather than a free market. The system has been like this, says Frank, since the sixteenth century in Latin America, and remains so. Given this, any real development will require a revolutionary break from the system.

On the basis of this model, Frank outlines a number of more specific hypotheses. First, the development of satellites is limited simply because they are satellites. Development is not possible for satellites, given their subordinate position in the international system.

Dependency theory also states that satellites can develop only when their ties with the metropolis are relatively weakened. Frank offers two different sorts of example of this.

- Development can occur where geographical and/or economic isolation allows some measure of development free of 'satellisation' — Frank cites Paraguay and Japan as examples here.
- Once a country has become a satellite, its only chance is to seize brief opportunities when the grip of the metropolis temporarily weakens, whether because of war or recession. An example of this, for Frank, was the industrialisation a few Latin American countries achieved in the twentieth century, made possible by the two world wars and the 1930s depression.

Frank's third hypothesis is perhaps the boldest of all. The regions most 'ultra-undeveloped' today are, he says, those which had the closest ties to the metropolis in the past.

In summary:

- Frank's version of dependency theory is externalist. All changes, at least for the LEDCs, come from and are imposed by outside forces, sometimes forcibly.
- The theory also suggests that the wealth of the MEDCs is dependent upon the poverty of the LEDCs. Because we are pumping out wealth from poorer countries, the world system redistributes from poor to rich rather than vice-versa.

The dependency theory suggests that a core–periphery relationship has developed among nations, with LEDCs forced into becoming the producers of raw materials for MEDCs. LEDCs are thus condemned to a peripheral and dependent role in the world economy. The logic of this position is to argue for some degree of protectionism in trade, which would be necessary if these countries were to enter a self-sustaining development path. Import substitution, enabled by protection and government policy, rather than trade and export-valorisation, has been the preferred strategy. Historical examples of government-led industrialisation, such as Japan and Soviet Russia, have been held up as proof that there was more than one path to development, contrary to that implied by stage theories such as Rostow's.

Many LEDC governments adopted the language and policies of the structuralists and/or the Marxists who supported this broad structuralist view in the 1960s

*Human systems, processes and patterns* **A2 Unit 5**

and 1970s. 'Neo-colonialism', 'core–periphery' and 'dependency' were the key words of this period.

The neo-classicists (or neo-liberal or neo-conservative) thesis was built upon Rostow's, and remains simple: government intervention, it argued, not only impeded development, in fact it more or less stopped it altogether. The emergence of huge bureaucracies and state regulations, they argued, suffocated private investment and distorted prices, making developing economies extraordinarily inefficient. In their view, the ills of unbalanced growth and dependency could all be ascribed to too much government interference, not too little. Similar arguments have been voiced in the defence of globalisation by Johan Norberg (*In Defence of Global Capitalism*, 2003) in which he argues that too much regulation is still obstructing the full benefits of free trade.

In recent years, the neo-classical thesis has gained greater support, particularly from governments in Latin America. However, the evidence is still ambiguous and disputed and the argument is by no means over. Both structuralists and neo-classicists use rapid east Asian development and the subsequent economic crisis, as well as the disastrous African experience, as proofs of their directly opposing theses.

> **Examiner's tip**
>
> ■ When writing about theories of development it is best to gather them together into 'schools of thought' rather than attributing all ideas to one or two individuals. Thus 'the modernisers, including Rostow, argue that....'

## Case study 15.6  *Import substitution in Latin America*

Up to the 1930s most Latin American economies had been characterised by free trade, with a stress on exporting raw materials and cash crops mostly to the USA and Europe. During the great depression, Latin American governments were obliged to rethink radically their philosophies of political economy. Latin American exports declined from an average of about US$5000 million in 1928–29 to US$1500 million in 1933. The reactions to this were:

■ to raise tariffs in order to protect domestic manufacturing industry
■ to impose import quotas on foreign goods entering the country
■ to limit the flow of capital overseas, thus imposing restrictions on the use of foreign exchange

Since the 1930s Latin America has changed from a group of free-trading economies to one of highly protected economies. Latin American businesses, taking advantage of the scarcity of goods and the level of protection, began to produce or increase the domestic production of goods previously imported (manufactured goods). Their new political and economic policy was based upon the following reasoning:

■ all major industrial countries (especially the USA and Japan) had industrialised behind high protective tariffs
■ a country needed to develop a mature industrial structure before it could become involved in the free trading of manufactured goods
■ protective policies should promote a wide rather than a specialised range of industries
■ protective policies create more opportunities for employment at a time of rapid growth in both national populations and labour markets

By the 1950s this set of policies became known as **import substitution industrialisation** (ISI for short) and was developed by the influential ECLA (Economic Commission for Latin America). The basic assumption was that colonisation had turned the economies of former colonies into 'structures that specialised in producing raw materials, cash crops and foodstuffs at low prices to meet the needs of the coloniser's economies'. This created economically divided societies in which a modern, industrial sector was being held back by international trade, with which it could not compete, and a traditional sector based on agriculture or mining blocked any process

545

# Chapter 15  Development processes

of modernisation by exercising political power. The land-owning class with political power and a strong control of the military proved resistant to change that might threaten its power and income. This type of elite class often had strong connections to the USA or other MEDCs. These structures were creating a dynamic that was impoverishing former colonies, or at least large sections of their populations, instead of promoting modernisation.

It was also true that the prices of manufactured goods imported by peripheral countries were rising faster than the prices of raw materials, cash crops and foodstuff they exported. In other words, the terms of trade were shifting against raw material suppliers. They were forced to increase output to maintain income. This might be logical for an individual producer, but collectively it makes a difficult situation worse as any increase in global supply forces prices down even further. ECLA foresaw two potential barriers to ISI:

- international trade forcing their economies to concentrate on producing more raw materials and cash crops/foodstuffs, which would contribute to further deterioration in terms of trade
- the absence of a strong capitalist class in Latin America and the other former colonies to shift policy away from landed interests (who were of course keen to preserve their control)

ECLA suggested a four-stage route to industrialisation:

**Stage 1** Concentration on the production of cheaply produced consumer goods, such as textiles, foodstuffs and pharmaceuticals.

**Stage 2** Specialisation on more complex products, known as consumer durables, including white goods such as gas/electrical cookers, radios and television sets, and motor vehicles using but adapting imported technologies and parts.

**Stage 3** Promotion of heavy industries such as steel, petrochemicals and aluminium, to enable the production of a wide range of the parts and components needed by the consumer goods industries.

**Stage 4** The development of domestic technology through a growing base of capital goods that produce the machine tools needed by manufacturing industry. This process would require foreign capital, private domestic capital and state capital.

By the 1960s this very recognisable type of 'dependent' industrialisation was in place in Latin America, as an outcome of an alliance between:

- state enterprises, which invested in capital goods because there were insufficient savings in their economies to generate industrialisation without state intervention
- national private enterprise, often in the form of large conglomerates with multiple interests in a whole range of secondary and tertiary activities
- TNCs, whose involvement both in the extractive industries and the early stages of the development of large industrial enterprises was vital

There were many critics of this model, but the two most obvious came from the political right (the neo-liberals) and the political left (the Marxists).

- The neo-liberals saw this type of ISI as the antithesis of the free market. Why produce cars in Brazil or televisions in Chile if it is cheaper to buy Japanese or American goods, which might also be of better quality? Do what you are best at.
- The Marxists argued that this type of development remained wholly within the capitalist mode of production, and the dependency on foreign capital and TNCs made it just as unacceptable as the colonial dependency, which it was supposed to replace. This critique led to the development of 'dependency theory'.

---

As is frequently the case, the development history of a country can be read quite differently depending upon the political perspective adopted. The next case study contains two versions of Zambian history.

*Human systems, processes and patterns* **A2 Unit 5**

## Case study 15.7 Development theories: two versions of Zambian history

| Modernisation theory | Dependency theory |
| --- | --- |
| **Pre-colonial**<br>Subsistence societies living simply, relatively peacefully and in balance with the environment but without modern resources and in great poverty. | Centuries of conflict are accentuated by European struggle for control of long-distance trade (especially for slaves and ivory). |
| **Colonial (1900–64)**<br>European capitalists help to develop the resources of the region, mainly copper, laying the basis for future development.<br><br>Growth in demand for African labour leads to phased migration, largely rural–urban, and eventually to urbanisation.<br><br>Urbanisation leads to profound sociocultural changes, from tribal to class and national identities, bringing modernism and a more educated workforce. | Industrial methods permit mass plunder of natural resources, largely for the benefit of Europeans and a few local allies.<br><br>Migration takes place because of changes in rural areas imposed by colonial systems of export agriculture. Ties with rural areas are not lost and many return home after unsuccessful experiences.<br><br>Traditional values and the relationship between tribal peoples and their land are systematically destroyed by exposure to other cultural norms. |
| **Post-colonial (1964–)**<br>Failure to convert copper wealth into sustained economic growth can be blamed on internal factors, local corruption and mismanagement by politicians. Better use of international agencies would have helped. | Failure can be blamed on unfavourable external factors: adverse terms of trade, regional conflicts (including the fight against apartheid), dependence on aid and the growing debt burden which is imposed by the IMF and other agencies. |
| **The future**<br>International agencies and foreign investment can bring about a resumption of economic growth with the proper use of policies geared to the global economy, specifically through the implementation of structural adjustment programmes.<br><br>The shift to a modern or Western identity and standard of living can be completed. | Integration with global capital will continue to be on unfavourable terms for the majority of Zambians, especially as the terms of trade move against their primary exports.<br><br>Modernity is a myth that helps to maintain popular acquiescence to the existing economic and political order. |

## The challenge of the NICs
### The newly industrialised countries

There is no commonly agreed list of newly industrialised countries, and indeed the terminology itself is subject to subtle shifts; in some texts it is newly industri-*alising* countries. In recent years a new group of recently industrialised countries (RICs) has been added to this growing list of acronyms. It is not hard, however, to provide an outline of what has happened (Table 15.5).

# Chapter 15 Development processes

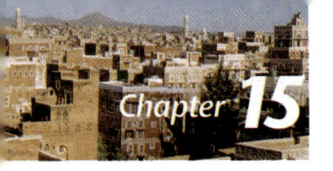

Table 15.5 World GDP growth, 1981–95

|  | Annual average growth of real GDP (%) | | | |
|---|---|---|---|---|
|  | 1981–90 | 1991–93 | 1994 | 1995 |
| OECD countries | 3.1 | 1.2 | 2.9 | 2.4 |
| Eastern and central Europe | 2.1 | –9.0 | –7.5 | –0.7 |
| Former Soviet Union | 1.0 | –15.5 | –12.6 | –4.0 |
| Developing countries | 3.3 | 4.6 | 4.6 | 4.9 |
| East Asia | 7.6 | 8.7 | 9.3 | 9.2 |
| China | 9.9 | 12.3 | 12.2 | 10.2 |
| South Asia | 5.7 | 3.2 | 4.7 | 5.5 |
| Sub-Saharan Africa | 1.7 | 0.6 | 2.2 | 3.8 |
| Latin America and the Caribbean | 1.7 | 3.2 | 3.9 | 0.9 |
| Middle East and North Africa | 0.2 | 3.4 | 0.3 | 2.5 |
| **World total** | 3.2 | 1.2 | 2.9 | 2.8 |

## The history

Forty or so years ago, anybody predicting the social and economic progress that has been achieved in east Asia would have been regarded as eccentric at best, insane at worst. Average per capita incomes in South Korea, for example, were lower than those in Zaire or Afghanistan. The level of poverty and deprivation in Indonesia and Malaysia was similar to that in much of sub-Saharan Africa, with both countries dependent on exports of primary commodities such as rubber and copper. Indonesia was a particularly bad case — it had huge debts, received massive aid and was dependent on imported food. The parallels with sub-Saharan Africa today are difficult to avoid. Many commentators saw these countries as beyond hope of foreseeable development and a major threat to world stability as this region became more and more backward. Public statements to that effect were common, including those of the neo-Malthusians who saw the future of India and China in bleak terms. What has happened since is an astonishing story of economic and social development.

The **Asian Tigers** (Hong Kong, Singapore, South Korea and Taiwan) have grown at unprecedented rates, sustaining growth in GNP at 10% per annum over many years (Table 15.6). These countries have been followed by other **tiger-cubs** in the region, including Malaysia, Indonesia and Thailand.

These growth rates represent a major achievement and a major challenge to those who argued that development was unlikely, if not impossible, both in this region and, just as importantly, in theory. In several of the Asian Tigers, such as South Korea, Singapore and Taiwan, wage levels and labour costs grew by almost 100% in the 1980s and are approaching Western levels. There are strong indications that social and labour standards have improved in these countries, although there is evidence of much non-compliance in protecting labour in the export processing zones that have been central to success.

*Human systems, processes and patterns* **A2 Unit 5**

|  | Annual average growth of real GDP (%) |  |  |  |
|---|---|---|---|---|
|  | 1980–1990 | 1990–94 | 1995 | 1996 |
| Bangladesh | 4.3 | 4.2 | 4.4 | 4.7 |
| India | 5.8 | 3.8 | 7.1 | 6.8 |
| Nepal | 4.6 | 4.9 | 2.9 | 6.1 |
| Pakistan | 6.3 | 4.6 | 4.4 | 6.1 |
| Sri Lanka | 4.2 | 5.4 | 5.6 | 3.8 |
| China | 10.2 | 12.9 | 10.2 | 9.7 |
| Indonesia | 6.1 | 7.6 | 8.2 | 7.8 |
| Malaysia | 5.2 | 8.4 | 10.1 | 8.8 |
| Philippines | 1.0 | 1.6 | 4.8 | 5.5 |
| Thailand | 7.6 | 8.2 | 8.7 | 6.7 |
| Vietnam | n.a. | 8.0 | 9.5 | 9.5 |
| Hong Kong | 6.9 | 5.7 | 4.7 | 4.7 |
| South Korea | 9.4 | 6.6 | 9.7 | 7.0 |
| Singapore | 6.4 | 8.3 | 8.8 | 7.0 |
| Papua New Guinea | 1.9 | 11.5 | –2.9 | 2.3 |

*Table 15.6 Rates of economic growth in selected countries*

## What was the formula for success?

The World Bank and the IMF frequently cite the region's record over the past 40 years as proof of the success of free-market policies of the type associated with export-led growth or **export valorisation**. These are in marked contrast with the policies carried out by many countries up until the 1970s. For example, Japan followed the import substitution investment (ISI) programme with astonishing results; the goal is to produce goods for the domestic (home) market, slowly replacing a dependence on imports.

As one might guess, no single east Asian model exists. The countries of the region have followed a wide variety of policies, reflecting their individual historical, political and economic circumstances. The contrasts between the 'city states' of Hong Kong and Singapore and the much larger countries of South Korea and Taiwan are obvious. Equally obvious is the heavy degree of Western, especially American and European, interest in the region. Both South Korea and Taiwan were important 'windows' of capitalism to flaunt at near neighbours China and North Korea. At the height of the cold war the success of these countries was almost obligatory. The USA invested heavily in all four, but especially in South Korea and Taiwan. With varying degrees of success, most of these countries have combined growth with equity and poverty reduction. But different countries have followed different routes — and thus they offer different lessons.

The major generalisations that can be made are as follows.
- All the NICs have a high rate of domestic saving, for reasons that may be to do with culture, the development of institutions and economic management.

# Chapter 15  Development processes

- They have been successful in attracting another source of money for the development of industry — foreign direct investment (FDI).
- They have been successful in export markets. Whatever the balance between domestic markets and exports, it is exports that have dominated in every case.
- They have invested in development of skills through education and training, often with huge government support.
- Land reform and agricultural reform have been important. The attention given to land reform programmes is frequently overlooked.

A common myth attached to the east Asian 'model' is that governments in the region have followed free-market policies. In fact nearly the opposite is true. A key feature of economic policy in most of the east Asian countries is a rejection of the simple free-market models of the neo-classicists. Indeed, many of the policies associated with structural adjustment are inconsistent with the policies that have achieved rapid growth and poverty reduction in east Asia. It is also true that almost all the east Asian NICs have pursued policies that combined import substitution with export-led growth.

Nor did the east Asian countries copy one another. Taiwan did not follow the policies of South Korea, Indonesia did not follow the path of Taiwan, and China and Vietnam have not adopted the policies of Malaysia. Each country has developed its own policies and plans depending on its own circumstances. There are lessons from each country, but they vary through the region and they vary over time.

### Key term

**Structural adjustment programmes (SAPs)**
A controversial set of policies imposed by lending agencies such as the IMF on debtor countries. They generally require a re-orientation towards export growth to earn foreign currency and a reduction in welfare and education spending, which give no immediate benefits.

## Case study 15.8 Development of an NIC: South Korea

### Background information

Korea is an ancient nation with a homogeneous population (i.e. same ethnic group). Historically China was dominant in the region.

The Japanese occupation of Korea in 1910–45 choked development. Koreans were not allowed to hold key jobs and levels of education were poor. However, an infrastructure was built.

After the Second World War Korea split into North and South Korea. North Korea was communist and South Korea was capitalist. Most industry at the time was in North Korea and valuable Japanese markets were lost after the war. Despite enormous US aid, South Korea remained a struggling economy.

The Korean War broke out in 1950 when North Korea invaded South Korea. The USA and much of the Western world sent troops in support of the South, while China supported North Korea. Over 5 million people were killed and the ceasefire established a

## Human systems, processes and patterns — A2 Unit 5

truce between the two countries that remains today, although the border is still closed.

The military coup of 1961 brought a powerful military government into South Korea which started the push towards industrialisation.

## 1960s: an export-led growth strategy

The military government took control of the economy and intervened in markets.

- The currency was devalued, so South Korean exports became cheaper for other countries to buy.
- Export firms were given access to cheap capital (loans to buy machinery, discounted prices, etc.).
- Foreign investment into the country was encouraged, especially to the 'chaebols'. These are large family-run enterprises with multiple areas of production. Daewoo and Hyundai are two of the best known.
- Relations with Japan were normalised and it became an export market.
- Tariffs (taxes) were reduced on imports of raw materials. Therefore companies could reduce their costs and produce more profitably. Prices of goods became lower and if prices are lower, demand is generally higher.
- There was a focus on manufacturing for export rather than raw materials (greater added value).
- Cheap loans, tax benefits and subsidies for export firms were provided.
- Cheap electricity and water were provided for export firms. Development of transport and communications networks was supported by government funding.

## 1970s

The focus on export growth in the 1960s resulted in inequalities between rural and urban areas. The government committed money to invest in rural areas. This substantially reduced inequality and removed some of the obstacles for rural development.

In the late 1970s there was a focus on social equity. A new tax system was set up to redistribute income.

## 1980s

The economy became more open in the 1980s. Tariffs on foreign imports were lowered and government intervention reduced. At the same time there was a focus on training workers, developing the education system and investing in foreign technology — becoming more hi-tech.

## 1990s

Rapid economic growth halted as the Asian economies started to fail in 1997. Confidence in the Asian markets fell and investors withdrew money. A Japanese recession meant that the export markets shrank. Observers began to talk of the 'tired tigers'. The government had intervened too heavily in the banking sector and, with limited knowledge of world markets and lack of world confidence in them, some banks collapsed. Those banks that survived restructured, became more commercial and were less reliant on foreign debt.

## 2000s

The South Korean economy continues to grow in a more open market.

See also Case study 13.11, page 436.

---

### Examiner's tip

- Theories of development have been very influential in guiding governments. Broadly speaking the modernisers can be linked to free-trade arguments and export-led growth models, whereas the dependency school can be linked to import substitution policies and greater levels of protectionism.

# Chapter 15 Development processes

# Variations in economic growth and development within countries

## Regional variations in economic, social and political development

The various theories of development concentrate on the nation state. That is to say, comparisons are drawn between South Korea and Taiwan, or between the UK and the USA. It is obvious that these comparisons are flawed in one vital respect. A 'country' is generally far from uniform. There are variations of wealth within countries that are profound and deeply entrenched. Some of those variations have a geographical dimension, in that the poor tend to live in one area while the wealthy live somewhere else. In other words, the geographical segregation mirrors the economic segregation. This can operate on a regional level: usually the dominant city and its region are considerably wealthier than remoter regions on the periphery.

Tables 15.7 and 15.8 give details of regional disparities between countries in the EU and between their least prosperous regions.

Table 15.7 Regional disparities in the EU, 1998

|  | GDP per capita (EU average = 100) | GDP of most prosperous region (EU average = 100) | GDP of least prosperous region (EU average = 100) |
|---|---|---|---|
| Austria | 112 | 127 | 90 |
| Belgium | 112 | 173 | 89 |
| Denmark | 119 | – | – |
| Finland | 97 | 119 | 97 |
| France | 104 | 160 | 85 |
| Germany | 108 | 192 | 61 |
| Greece | 68 | 77 | 65 |
| Ireland | 97 | – | – |
| Italy | 103 | 133 | 66 |
| Luxembourg | 169 | – | – |
| Netherlands | 107 | 115 | 93 |
| Portugal | 70 | 71 | 50 |
| Spain | 79 | 101 | 59 |
| Sweden | 101 | – | – |
| UK | 100 | 140 | 81 |

According to classical economic theory, regional differences should not persist because regions with a high demand for labour will ultimately become high-cost regions, whereas regions that may start poorer will, it is argued, become cheaper

*Human systems, processes and patterns* **A2 Unit 5**

| Region | GDP per capita (EU average = 100) | % GDP from agriculture | % GDP from industry | % GDP from services |
|---|---|---|---|---|
| Azores (Portugal) | 50 | 11.8 | 34.5 | 61.4 |
| Madeira (Portugal) | 54 | 3.8 | 18.2 | 78.0 |
| Kentriki Ellada (Greece) | 58 | 24.6 | 27.2 | 48.2 |
| South (Spain) | 59 | 8.6 | 22.8 | 68.6 |
| Mecklemburg-Vorpommern (Germany) | 61 | 2.9 | 29.8 | 67.3 |
| Sachsen-Anhalt (Germany) | 61 | 2.1 | 37.0 | 60.9 |
| Thuringen (Germany) | 61 | 1.9 | 36.1 | 62.0 |
| Sachsen (Germany) | 64 | 1.2 | 36.9 | 61.9 |
| Vareia Ellada (Greece) | 65 | 22.1 | 28.8 | 49.2 |
| Sicily (Italy) | 66 | 6.2 | 19.1 | 74.7 |
| Campina (Italy) | 66 | 3.8 | 20.1 | 76.1 |
| South (Italy) | 67 | 7.4 | 20.2 | 72.4 |
| Northeast (Spain) | 67 | 7.1 | 30.4 | 62.5 |

*Table 15.8 The least prosperous areas in the EU, 1998*

areas for industry as the demand falls. This is the famous 'hidden hand' of the market that should even out differences. For some theorists this has been reason enough to disregard regional policy. However, regional differences have persisted in many countries and it was Gunnar Myrdal who provided the first coherent explanation of this in his model of **cumulative causation** (Figure 15.8).

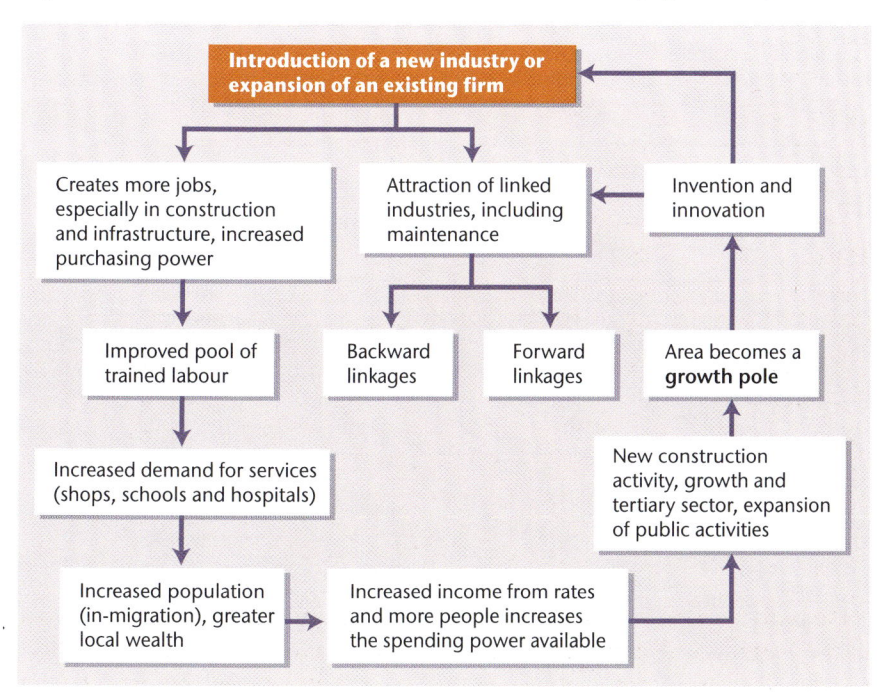

*Figure 15.8 Myrdal's model of cumulative causation*

# Chapter 15 Development processes

For Myrdal, the reasons for the persistence of regional disparities are that initial investments lead to a positive feedback sequence that makes a region even more attractive to new investment, which amplifies the original investment. This works on both the private level and the state level.

- At the private level, the creation of a pool of skilled labour attracts other companies (a well-known phenomenon — for example, the clustering of manufacturers in Silicon Valley), or the arrival of a large company (for example, Honda in Swindon) leads to the growth of other companies that are component suppliers.
- At a state level, increased receipts from local taxation allow better investment in infrastructure, improved schools with smaller class sizes, better street lighting and leisure facilities, which again stimulate further inward investment. This is a powerful argument for why, once a region has gained an initial advantage, it tends to retain it.

Myrdal also identified an important set of forces that make it difficult for peripheral, poorer regions to break out of what becomes an unequal relationship.

## Backwash effects

Suppose there are two regions, one (A) more developed and populated than the other (B) — perhaps because of natural resource advantages or preferable location.

Firms in A have larger markets, so they are able to realise **economies of scale** not available to firms in B. Thus firms in A may have a cumulative and growing advantage over firms in B. Higher incomes in A mean there is an incentive in that region for a shift into more advanced production of more sophisticated products, further increasing the income gap. This higher demand may also induce technical change, generating further divergence between the two regions. In other words,

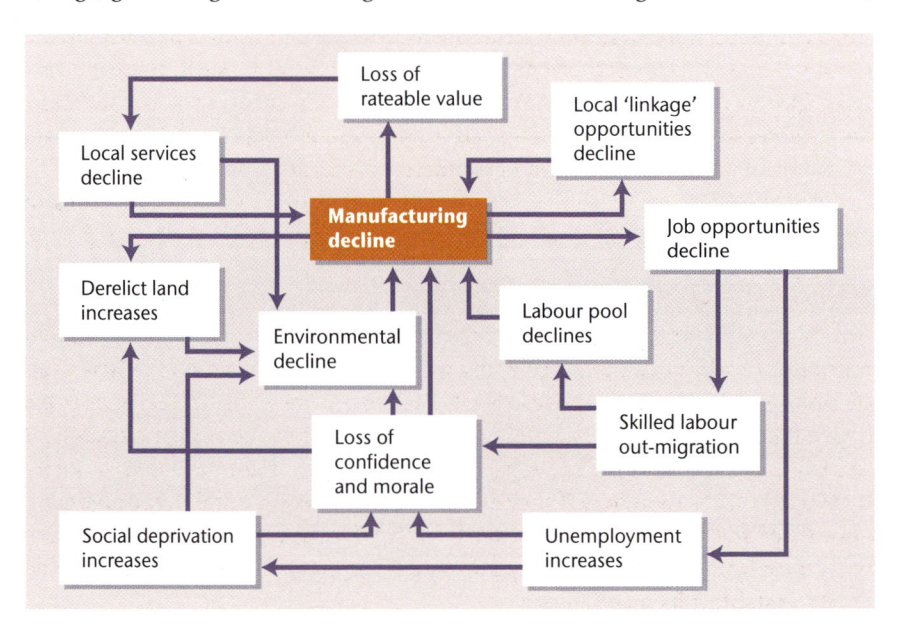

*Figure 15.9 How two regions diverge*

*Human systems, processes and patterns*  **A2 Unit 5**

the products will be better because of the higher research and development investment and they will be cheaper because of larger markets and greater economies of scale. Thus region B's manufacturing industry is unable to compete and begins to disappear through loss of market share as the products from rich region A flood into the shops in B. This does not happen, of course, if they are protected by barriers to trade. Decline is inevitable without such protection. Figure 15.9 develops this sequence.

> **Key terms**
>
> **Backwash effects** The negative impacts on peripheral regions caused by the growth of core regions.
>
> **Spread effects** The positive impacts on peripheral regions caused by growth of core regions.

Region B may be increasingly limited to producing cheap, low value-added goods or raw materials for which the income elasticity of demand is low (as income rises the demand rises more slowly, as with agricultural goods and primary products). Obvious examples would be coal, rubber or copper. Demand for products from B may also be dampened by the development of substitutes in A. Better investment opportunities in A may attract the more efficient elements of capital and labour away from region B, further limiting its production possibilities.

### Spread or trickle-down effects

On the other hand, region A may have a growing demand for the resources of B and the ability and wealth to exploit and control these resources. This is a stimulus for the development of B, but the effect may not spread far if the profits of resource extraction largely go to region A. A quarry company with its head office in London, extracting road stone in the Pennines, is an obvious example. The profits may also go to a client elite group in region B, who may well be absentee landlords, who live in Region A.

There may be some consolation for region B (the poorer region) in that it gains an improved transport system so the resource(s) can be exported at lower costs. This may also allow B to export its own products at lower costs. At the same time, competition from imports from the more efficient region A may lead to the closing down of the least efficient enterprises in region B. This will cause productivity in B to rise and allow the employed to earn higher incomes. In the longer term, increasing congestion and pollution in region A, coupled with rising factor costs, may cause some economic activities to relocate from A to B. In reality, however, the real effort will be directed towards relieving these pressures in region A by building new infrastructure rather than supporting moves to region B.

### A global comment

Although Gunnar Myrdal developed his theory in the context of regional economies within a nation, they apply well at an international level. There is an obvious similarity to the ideas developed by Andre Gunder Frank and his dependency theory or Jane Jacobs' 'supply regions'. Thus the rich core regions of the USA, western Europe and Japan can out-compete manufacturing industry developed on the periphery if they are allowed to. This is central to the argument about free trade. The removal of tariff barriers would, of course, remove any protection from region A's better quality and cheaper goods, making it impossible for B to establish its own industry.

# Chapter 15 Development processes

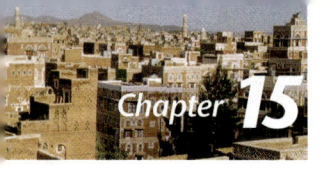

## Case study 15.9 Regional differences at a global level: two car companies

Hindustan Motors of India and Toyota of Japan began production in 1937. Since then their histories have diverged. While Toyota has competed on the global market, Hindustan Motors was sheltering from competition behind substantial Indian tariffs.

- At the Hindustan Motors plant near Calcutta, over 11 000 workers make 18 000 Ambassadors a year.
- At the Toyota City Lexus factory, 66 workers use robots to produce over 100 000 cars a year.

As a result of India's protected economy, imported cars became extremely expensive. Thus, in reality, Indian consumers who needed a car were offered a choice of two domestically produced models: the Padmini, which was a copy of a 1960s Fiat, or an Ambassador, a copy of a 1950s Morris Oxford. Consumers could wait as long as 7 years for delivery of a car they ordered, and then high taxes would double the price.

|  | **Hindustan Motors** | **Toyota** |
| --- | --- | --- |
| Year production began | 1937 | 1937 |
| Number of models in 2002 | 2 | 18 |
| Maximum horsepower | 55 hp at 4500 rpm (Ambassador) | 240 hp at 4800 rpm (Sequoia) |
| Design changes since inception | Very few | Too many to list |
| Total cars made in 2003 | 95 000 | 5.98 million |
| Financial results for 2003 | Losses of over $25 million | Profits of $3.9 billion |

Table 15.9 A comparison of Hindustan Motors and Toyota

The reaction to regional imbalances on the part of governments has been variable, but it is generally politically dangerous to do nothing about them. The risk, in a democracy, is that these regions will become bastions of opposition to the government and vote accordingly when they get the chance. Worse than this, under more or less any form of regime, democratic or otherwise, they may become areas of serious unrest. The most dramatic form of government intervention has been the development of **growth poles**.

Growth poles are locations developed explicitly with the intention of stimulating remote regions in an attempt to even out regional disparities and avoid some of the imbalances outlined and explained by Myrdal. They have usually involved major investment from governments, which often develop the infrastructure, with private capital being encouraged to join in. The core of these poles has usually been a large-scale industrial development, for example a hydro-electric project or the exploitation of a previously neglected natural resource supply. The theory is that, once established, these investments lead to the **multiplier effect** and **cumulative causation**, attracting other industries until self-sustained growth is achieved. This often involves a major regional development plan. There is a long and rather variable history of these projects, but many have been modelled on the first of them all, the Tennessee Valley Authority (TVA) in the USA.

## Regional policies

Regional policies involve the improven[t of] standards in specific regions or areas o[f...] scales from the local (e.g. Cardiff Ba[y...] (Highlands of Scotland, Carajas in Braz[il...] schemes was the TVA.

### Case study 15.10 Regional p[olicies]

*Photograph 15.2 The Tennessee River near Chattanooga in 1933*

In 1933 US President Franklin Roosevelt signed the legislation that created the Tennessee Valley Authority (TVA). As the most ambitious part of his New Deal, it was the first and one of the most successful of all regional projects. At that time fewer than 3% of the households in the Tennessee Valley had electricity. Malaria afflicted up to 30% of the population in some areas, and the average expenditure per child for education was about one-third of that of the USA as a whole. The average income in the valley was $639, about a third of the national average. This was a deprived region, which was battered periodically by the forces of nature, making a difficult situation worse. The periodic flooding of the Tennessee River prevented the development of cities along the river's banks, leaving small and isolated towns. Unchecked fires burned 10% of the woodlands every year, and because upwards of 3 million hectares were under threat of soil depletion, the agricultural base of the region was threatened.

The TVA changed all that. The idea was elegant and simple:
- provide power for agriculture and industry
- provide water for agriculture and industry
- improve navigation
- install flood control

…VA, the … went from … sion of power … r. It also allowed … e Second World War, … m to build bombs and … nium plants require large … ity. To provide power for such … stries, TVA engaged in one of the … power construction programmes ever … en in the USA. In its first 20 years, the TVA … 20 dams to generate HEP. Nearly 200 000 men … d women were employed by the TVA during this two-decade period of dam construction. At that time, the TVA was the largest construction project in the world and did much to bring the USA out of the depression.

In order to build the intricate systems of dams and reservoirs to tame the Tennessee River, 15 000 families had to be moved from the areas that were to be flooded. Entire towns and villages were relocated, or physically reorganised, to make way for the lakes that were created behind the dams.

The instigators of the TVA knew that electricity would be the most important factor in improving the standard of living of the people in the valley. With electricity came the possibility of eliminating much of the labour-intensive work in farming, which, before the TVA, had been conducted with nineteenth century technology. Once electricity and fertiliser factories were available, agricultural productivity in the valley tripled.

As increased productivity created an excess of manpower on the farms, the cheap electricity made it possible for new and modern factories to spring up. Between 1933 and 1950, nearly half a million jobs in industry were created in the Tennessee Valley.

The TVA also:
- established its own health and safety department, to rid the valley of its endemic diseases
- established libraries at every dam construction site, which connected otherwise isolated communities to the rest of the world
- established 'model farms', where farmers were taught modern farming methods
- supplied free fertiliser from the TVA for those who offered to pass on techniques to neighbours

It was this plan to transform the population as well as the landscape that made the TVA the model for development in nations around the world.

There have been many copies of the TVA, but few have managed the trick of combining economic and social development with such success. The dam building continued in the USA, not least in the southwest where the Boulder Dam on the Colorado River (now known as the Hoover Dam) provided much of the water that facilitated the growth of the economy in that arid region. The TVA has also posed problems for those who believe in the free market and regard the USA as the template for neo-liberal policies. However, it is worth noting that the annual income taxes now paid into the federal treasury from the TVA are almost six times the government's yearly investment in the TVA.

Elsewhere, the history of large-scale development strategies has been mixed. One of the best known is the long history of attempts to create economic growth in southern Italy, the Mezzogiorno.

The policies pursued in the Mezzogiorno have changed over time. Regional imbalances are generally seen as damaging in that they divide countries and make them harder to govern. Thus, whatever their political perspective, most democratic governments try to deliver convergence whereby the gap closes between rich regions and poor regions. These policies have a variable history.

Human systems, processes and patterns  **A2 Unit 5**

## Case study 15.11  Regional development: the Mezzogiorno

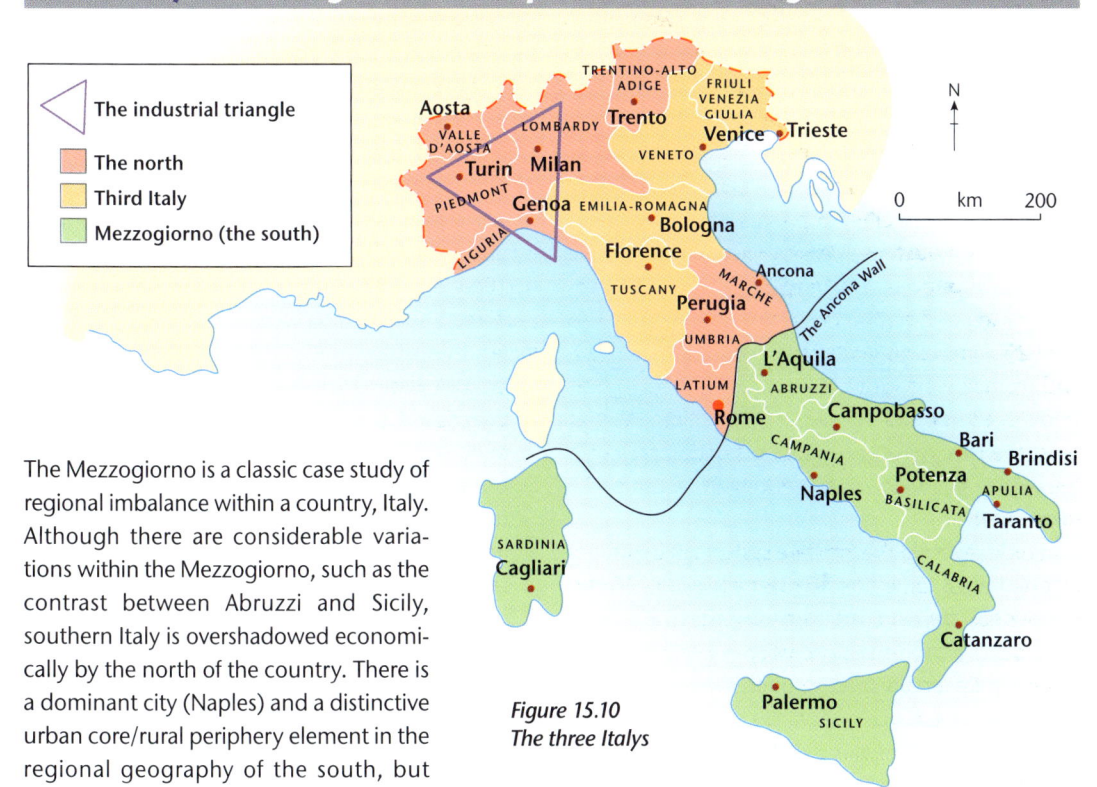

**Figure 15.10** The three Italys

Legend:
- The industrial triangle
- The north
- Third Italy
- Mezzogiorno (the south)

The Mezzogiorno is a classic case study of regional imbalance within a country, Italy. Although there are considerable variations within the Mezzogiorno, such as the contrast between Abruzzi and Sicily, southern Italy is overshadowed economically by the north of the country. There is a dominant city (Naples) and a distinctive urban core/rural periphery element in the regional geography of the south, but whatever indicators are taken, the south remains poor by Italian standards. It is true that regional convergence has not happened despite enormous expenditure by central government, but two things should be remembered.

First, poverty and economic backwardness are relative concepts. In most statistical measures the poverty line is taken to be 50% of mean earnings. So, as mean earnings rise, so does the poverty line. Southern Italy in the second half of the twentieth century experienced significant growth. Real GDP per capita rose by nearly 3% per annum and the volume of per capita consumption, thanks in part to generous transfers from Rome, grew at an annual rate of nearly 3.5%, or increased nearly five-fold over 50 years. The gap between the north and the south may not have decreased (Figure 15.11), but absolute poverty fell substantially and living standards today bear no comparison to what they were soon after the Second World War.

Second, it is impossible to evaluate what state the Mezzogiorno would be in without the regional policies of the past 40 years. The success or failure of the policies is therefore hard to measure. Perhaps the policies applied have worked in that they have prevented the Mezzogiorno being even more behind the rest of Italy.

The reasons for the Mezzogiorno's relative poverty have been ascribed to two sets of causes: physical constraints and the human response.

### The physical constraints

- The Mezzogiorno is 80% mountainous (the Apennines).
- Only 20% is lowland plain, often poorly drained.
- Soils are thin, alkaline and frequently infertile.
- The upper slopes have experienced erosion and overgrazing since Roman times.

559

# Chapter 15 Development processes

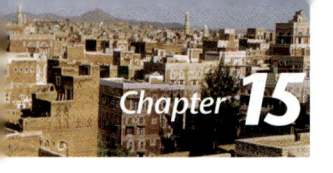

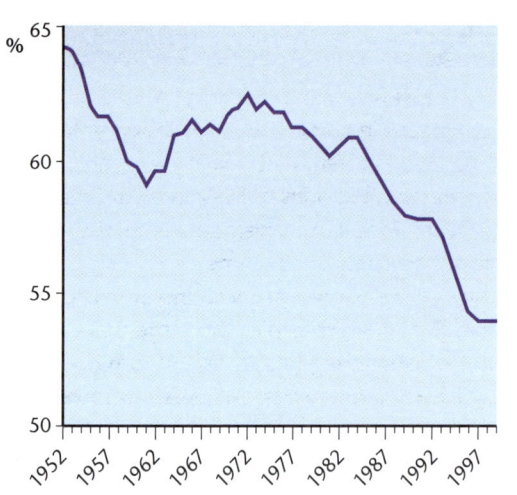

*Figure 15.11 The GDP of southern Italy as a percentage of the GDP of northern Italy, 1952–99*

- The climate is warm Mediterranean (summer average 23°C, winter 10°C), with winter rainfall concentrated on the western coast. Summer aridity is a problem.
- The landscape is poorly suited to intensive agriculture.
- Much of the area is tectonically active, with frequent hazards from volcanic and earthquake activity: Etna and Vesuvius are the best-known volcanoes.

Although these conditions may appear difficult, they are not deterministic. The obvious contrast here is with southern California, which has very similar conditions and yet has become one of the richest corners of one of the wealthiest countries in the world. Whatever else physical conditions do, they clearly don't prevent economic development.

## The human responses and limitations

- Settlement on the plains was avoided until early modern times because of:
  - the threat of piracy
  - malaria
- The area has been long dominated by large estates (*latifundia*) owned by absentee landlords. This social system pushed tenant farmers onto poorer parcels of land.
- The cash crops have been olives and wine since Roman times.
- Southern Italy was ruled by Norman-Swabian princes who retained strong control of an essentially feudal society until well into the eighteenth century. In contrast the north has a long tradition of cities and industrial innovation dating back to the Renaissance and before.
- Industries and modernisation were discouraged by southern landowners who feared a loss of their income as peasants deserted the land.
- The phylloxera crisis of the 1870s, which wiped out almost all the vines in the region and followed a series of bad harvests, stimulated huge out-migration, some to other parts of Italy, some overseas.
- This initiated a long period of out-migration, particularly by the younger and more able sections of the population.
- The unification of Italy in the 1870s brought the south into closer contact with the north along with a modern infrastructure which facilitated out-migration.
- In a typical backwash effect, southern industries, such as textiles, were badly affected by cheaper and better products imported from the north.

Large-scale projects like the TVA and the Cassa per il Mezzogiorno are typical of the 'top-down' approach led by national governments or economic unions (the European Union), which usually involve major investments in infrastructure and a concentration of resources on a growth pole or growth corridor.

The 'bottom-up' approach usually takes a small-scale people-centred approach, drawing funds from charities, and is often highly local. These projects aim to be self-sufficient but still need to be integrated into wider (and more expensive) national policies.

*Human systems, processes and patterns*  **A2 Unit 5**

## Case study 15.12 Regional policy: Cassa per il Mezzogiorno

The Cassa per il Mezzogiorno was a regional development agency established by the Italian government. The first stage of the Cassa's work in the 1950s was to attempt to improve the condition of the rural poor in the south. Policies were directed at the rural infrastructure in particular. Roads were improved, drainage provided on the plains and irrigation systems developed. After many years of effort it became clear that, despite some localised improvements in agricultural output, there had been limited reductions in unemployment, only small increases in income and no convergence — in other words, there was little closing of the economic gap between the south and the rest of Italy (see Figure 15.11).

### Industrial development

Experts had insisted all along that only rapid industrial development would create the jobs and prosperity needed to break the vicious circle of unemployment and poverty. Put simply, industry was to be moved to the area of surplus labour rather than labour moving to the area of industrial activity. This would slow down south–north migration.

Since there was no southern money for such industrialisation, state investment funds were moved from agriculture to industry. Growth poles were designated in the Naples/Salerno area and giant basic industries were to be established in the Taranto/Bari/Brindisi triangle (see Figure 15.10). It was believed that these would in time attract further, secondary industries and create a broad and diversified industrialisation through the operation of the multiplier. The south would then be on its way to autonomous and sustainable economic development.

The fund's spending on infrastructure was therefore cut from 42% in 1957 to 13% in 1965 while assistance to industrial enterprises was stepped up from 48% to 80%. To encourage private investment the government offered generous grants and tax incentives. It also required the state holding companies, IRI and ENI, to place increasing amounts of capital in the south. IRI, the major investor, injected over $20 billion into the Mezzogiorno between 1963 and 1978. In this way a steel industry (at one time the most modern in Europe) was developed at Taranto and an Alfa Romeo plant near Naples, as well as petrochemical factories in Sicily and Sardinia. The volume of industrial development in the south increased seven-fold in this time.

Even though some of these investments were productive and profitable, the large-scale heavy industrialisation programme did not turn out to be the solution to the south's problems, any more than it was in Uruguay and in many other cases. The new industries were not labour intensive and created relatively few jobs. The managerial and technical personnel were mostly recruited from the north and eventually returned there, leaving the south without a local managerial and technical class, much as Myrdal had argued would happen. And the industries themselves were subsidiaries or branch plants of northern companies, which never developed links to the local economy.

### The 1970s

Laws were therefore enacted in 1971 and 1976 requiring the development fund to concentrate on a new range of projects. These included the further development of industrial infrastructure, the exploitation of natural resources, social projects in metropolitan areas, water-reclamation projects, cleaning the Gulf of Naples, extending the road network into remote mountainous areas, reforestation and promotion of citrus fruit production.

Many of these programmes had to be scaled down or scrapped as a result of the economic crisis following the oil price increases after 1973. Some of them were gigantic mistakes. Over $800 million was spent on cleaning the Gulf of Naples without producing a litre of unpolluted water. By the end of the decade the era of big infrastructure projects was over.

### The 1980s

Emphasis then shifted to the selective encouragement of small- and medium-sized firms, which in the early 1980s accounted for over 90% of the projects

# Chapter 15 Development processes

*Figure 15.12 GDP in Italy, 1995*

*Figure 15.13 Growth rates in GDP, 1991–95*

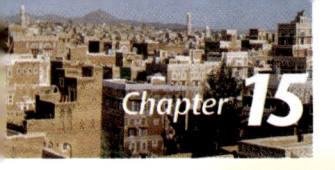

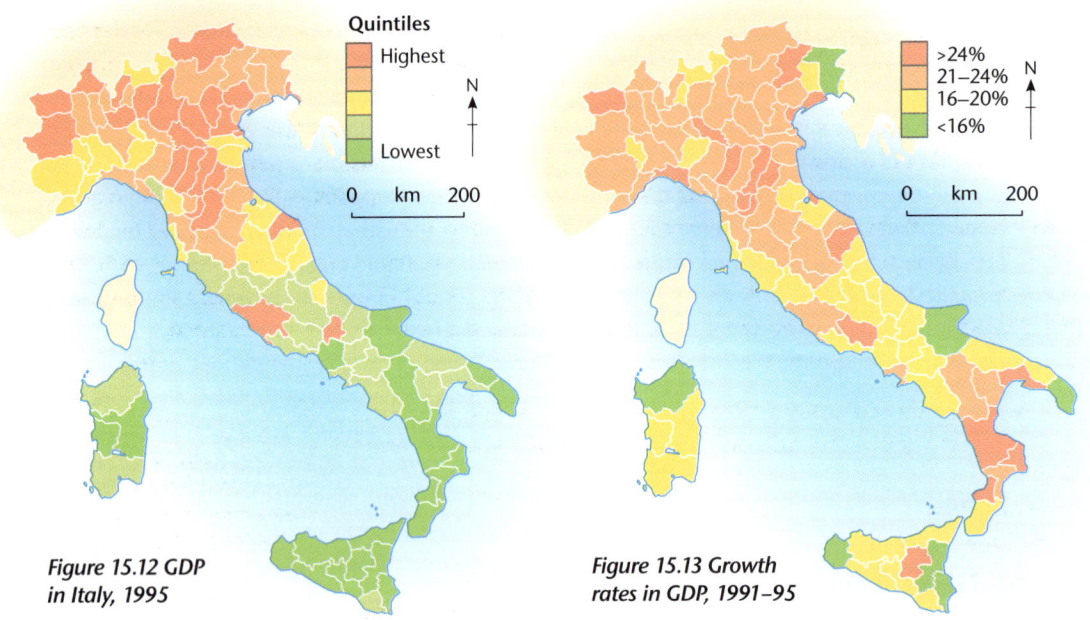

and 85% of the investment of the development fund. For many Italians the most obvious impact has been the creation of a large bureaucracy in the south, with far more government officials and local government officers than elsewhere in the country.

The fund, originally established for 15 years, was extended year after year until 1984, when it was finally terminated. During the 30 years between 1950 and 1980, it spent about $20 billion of the total $30 billion allocated to it. The European Community's regional fund added over $2 billion of its own, making the Mezzogiorno its single main beneficiary.

Many Italians are doubtful about the benefits of these policies. The south remains a poor region relative to the rest of Italy and the difference between north and south is still large (Figure 15.12). There are clear improvements in some areas, but they are not spread evenly in the region (Figure 15.13).

## Case study 15.13 'Bottom-up' development in The Gambia

The Gambia is the smallest country in continental Africa, with an area of just over 10 000 km² (480 km from east to west and on average 48 km from north to south). It is surrounded by Senegal. In 2000 it had a population of 1.4 million.

'Concern Universal' is one of many NGOs (non-government organisations) operating in The Gambia, using charitable donations to support local and largely small-scale projects. In The Gambia, as in most of Africa, the bulk of the farming is carried out by women.

### The Gambian is good! project

Gambian is good! is a joint initiative between Haygrove (the UK's largest soft fruit producer and marketer), Concern Universal, Gambia Tourism Concern and the National Women Farmers' Association. The project's aim is to establish a horticultural marketing company in The Gambia that will provide economic and social benefits to poor Gambian communities. The primary objectives are:

- to create a Gambian fresh produce market that develops local livelihoods, inspires entrepreneurship and reduces the environmental and social cost of imported produce
- to develop sustainable rural livelihoods among small gardeners and community groups in The Gambia. This initiative has been funded by Business Linkage Challenge Fund (UK government Department for International Development)

*Human systems, processes and patterns*  **A2 Unit 5**

## Smallholder irrigation for livelihood enhancement (SMILE)

The Smallholder irrigation for livelihood enhancement scheme (SMILE) is designed to help women vegetable farmers increase their output, thus improving their livelihoods. It works by introducing intermediate irrigation technologies such as treadle pumps that have been used successfully by women farmers elsewhere in Africa. To ensure that as many women farmers as possible can benefit, local manufacturers are trained to make good-quality irrigation equipment. The women farmers can buy the pumps for themselves without the need for continued support from outside agencies, thus creating sustainable development within the region. A group of ten local partners is being developed to deliver the SMILE initiative across the country, including farmers' associations, rural development organisations and a national credit organisation.

Iceland is an unusual country. It is peripheral to Europe in almost every sense, but is also a useful reminder of why stereotypes are risky. Iceland:

- is very wealthy
- scores highly in the HDI
- has the smallest per capita prison population in the world
- relies on primary products for its wealth, fish being the dominant export
- urbanised late and is still urbanising at a rapid rate
- has the happiest population on the planet

Table 15.10 provides some key economic data for Iceland.

### Examiner's tip
- Try to get some complexity into your discussion of regional differences. Countries such as Iceland illustrate that poor regions can become transformed into very wealthy regions as challenges are overcome and turned into opportunities.

|  | 1999 | 2000 | 2001 | 2002 | 2003 |
|---|---|---|---|---|---|
| Population | 277 184 | 281 154 | 284 600 | 286 877 | 289 172 |
| Population growth (%) | 1.2 | 1.4 | 1.2 | 0.8 | 1.0 |
| GDP (billion US$) | 8.487 | 8.462 | 7.676 | 8.410 | 8.828 |
| GDP per capita (US$, current exchange rate) | 30 617 | 30 099 | 26 972 | 29 316 | 30 528 |
| GDP growth (% change from previous year) | 3.6 | 5.6 | 3.0 | −0.8 | 2.4 |
| GDP by sector (%) |  |  |  |  |  |
| Agriculture | 1.7 | 1.6 | 1.6 | 1.5 | – |
| Fisheries and fish processing | 12.6 | 11.5 | 11.3 | 12.1 | – |
| Manufacturing, construction and utilities | 19.4 | 18.6 | 20.0 | 20.8 | – |
| Private services | 43.9 | 46.0 | 44.5 | 43.0 | – |
| Public services | 22.4 | 22.3 | 22.7 | 22.6 | – |
| Main export categories (%) |  |  |  |  |  |
| Marine products | 45.9 | 40.8 | 40.1 | – | – |
| Energy intensive products | 12.1 | 13.6 | 14.6 | – | – |
| Tourist revenues | 12.9 | 13.1 | 12.4 | – | – |
| Other | 29.1 | 32.5 | 32.9 | – | – |

*Table 15.10 Key economic indicators for Iceland*

# Chapter 15  Development processes

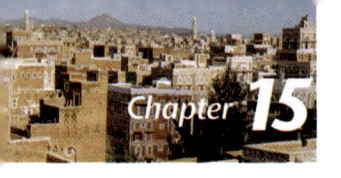

## Case study 15.14  Regional development: Iceland

There are two scales at which Iceland can be used as a regional case study.

### A European region

Iceland can be treated as a region of Europe, albeit one that is some distance removed from the European mainland as it sits astride the tectonic divide between Europe and North America. In that context, Iceland is clearly peripheral in a geographical sense. For much of its history as a Danish colony, it was peripheral in an economic sense as well, being one of the least-developed economic regions in Europe. However, its rapid growth in the twentieth century and its high incomes suggest that it is no longer economically peripheral. Iceland can be contrasted with the Mezzogiorno at the other end of Europe, a region that has not had a history of spectacular development in the twentieth century (see Case studies 15.11 and 15.12).

### Regions within Iceland

Iceland has a serious internal regional problem. It is a small country of 100 000 km$^2$, dominated by its capital and the surrounding region to an extent rare in other modern states. Iceland urbanised late and has never really industrialised, skipping 'Rostowian' stages to become one of the wealthiest countries in the world with a sophisticated population. Its modern economy is based on primary exports, in itself unusual for a wealthy MEDC, and a growing quaternary sector. Remote regions such as the Vestfjords have experienced rapid out-migration. For example, about 20% of the population moved out in the period 1987–97. In the last decade, migration to the capital area has increased again, to become one of Europe's highest rates of rural–urban migration. Meanwhile, employment opportunities have been reduced, especially in agriculture and the fishing industry, stimulating further out-migration, especially of the young.

The areas of loss are characterised by:
- physical remoteness and a poorer quality of life
- dependency on agriculture and the fishing industry

Regional policy in Iceland has concentrated on the following.
- Strengthening innovation and economic and residential development in the provincial areas.
- Offering advice to firms through economic and residential analyses, and providing training courses and education in cooperation with research and technical institutions and universities.
- The Institution of Regional Development supplies regions with grants and assistance with economic development. These approaches include simple measures such as heating-cost allowances, subsidising education costs and improving the highway system.
- Strengthening the cooperation between distant regions and with national institutions in the capital area.
- Attempting to divert some of the high investment in the knowledge sector back to the provincial areas, thereby increasing their economic diversity.
- Introducing a teleconference system called the Regional Bridge to solve the problem of distance from the capital. This is used for passing on information and for distance education, which offers residents in fringe locations the same educational opportunities as people in the larger towns.
- Assisting with international and domestic cooperation projects, and leading some working groups that deal with tourism and residential factors such as environmental matters, communications, education, culture, retail and services.
- Increased emphasis on self-help in the provincial areas. Financial firms located there are increasingly used to provide loans to businesses and buy shares.

Replicating the TVA has not proved to be easy. The formula, which revolves around big dam projects, has been tried in many countries with what are best described as patchy results. Some of the most notable catastrophes have occurred in Amazonia.

*Human systems, processes and patterns*  **A2 Unit 5**

## Case study 15.15  Regional development: Amazonia

In the recent past there have been a number of spectacular failures in the attempts to develop the Amazon. This vast region, much of which is in Brazil, is one of the last resource frontiers in the world. Dreams of riches have driven development onto sometimes bizarre paths...

### Fordlândia

In 1927 Henry Ford purchased 1 million hectares of land in the eastern Amazonian state of Pará, most of it rainforest. He was encouraged by the governments at both ends of the process and rather unimaginatively he named the area Fordlândia. He established a monoculture of rubber trees on the plantation model, with housing, a school and other facilities shipped in from the USA. The aim was to provide rubber for tyres. The poor soils were unsuitable for the project, labour was scarce, and the American corporate culture did not endear the company to the local population. In 1945, Ford pulled out, losing nearly $10 million.

### Ludwig

Failing to learn from this example, the eccentric billionaire Ludwig lost an even greater fortune trying to make money from the region. Seeing that this 'green Eldorado' was full of trees and noting the world shortage of wood pulp for paper, he set about buying more than 1.5 million hectares on the Rio Jari. He had it planted as a monoculture of *Gmelina*, a fast-growing pulpwood species. Like Ford, he had not taken into account the poverty of the region's soils. His trees did not produce as expected, the soil was depleted, and Ludwig lost an estimated US $1.5 billion in the process.

### Grande Carajas Programme

The Grande Carajas Programme is one of the largest integrated development schemes ever undertaken in tropical rainforests. It covers an area of 880 000 km$^2$, almost 11% of the country of Brazil. Major projects include iron-ore mines, two aluminium plants, a hydroelectric power plant, and the extension of roads and railways. One of the most controversial aspects of this project is the environmental degradation it has caused in the Amazon basin. The deforestation of Amazonia leapt from 2.4% in 1978 to 10% in 1988 and, at the heart of the area, the figure is estimated at over 30%. Another important criticism is about the social impact on rural peasants and indigenous people. The project involved huge population displacement and accelerated problems of food supply.

On the other hand, the project generates and processes 45 million tonnes of iron ore annually, as well as 1.5 millions tonnes of manganese and 10 000 kg of gold. It is preparing to produce copper, bauxite and other metals. All these activities are monitored by an environmental policy that, officially at least, strives to conserve and regenerate delicate Amazon ecosystems.

The plan was approved in 1980, and the World Bank, Japan and the EEC agreed to lend money. Japanese advisers played a significant role in persuading the Brazilian government and produced several influential reports. They started aid even though none of the 12 smelters approved by 1987 had carried out the environmental impact assessment required by the World Bank. Like most large-scale projects, there is a dam at the centre of the operation, in this case the Tucuruí dam.

### The Tucuruí dam

The Tucuruí dam was built to serve the growing city of Belem, the state of Pará, the massive Grande Carajas mining project and the national electricity grid. Its construction led to rapid urban growth in the immediate vicinity of the dam, both in a planned company town and in the village of Tucuruí itself. The building of the dam and the development of the lake displaced 30 000 people from 17 upstream villages. The company contracted to clear vegetation in advance of the flooding used large quantities of the defoliant dioxin (Agent Orange), exposing thousands of rural residents to toxins. The creation of a large body of standing water in the reservoir has increased the incidence of malaria, and increases in schistosomiasis are expected as a result of the

# Chapter 15  Development processes

construction. Finally, water quality and constraints on migration have resulted in the disappearance of several species of fish from rivers in the vicinity of Tucuruí.

## Do such schemes work?

There is considerable debate about the effectiveness of the Amazonian schemes. Very few agencies have assessed the costs and benefits involved and the problems they have faced. The externalities of these schemes are not always positive and one could suggest that many of them have only worked for a limited group of investors and TNCs without bringing much long-term benefit to the region in terms of balanced and sustainable growth. On the other hand, the TVA model suggests that development is possible, although most of the Amazonian projects have judiciously ignored the local population rather than involving them in determining their own destiny.

# Trade and aid

The idea that countries can develop economically and socially without outside links is hard to sustain. **Autarkic development**, as this is known, did take place in the UK (England, to be more precise), as this country was the first to experience an Industrial Revolution. However, even here the links with a growing overseas empire were highly important. The USA can also be cited: given its huge land area and high level of natural resources, it could be largely independent of the rest of the world should it wish to pursue this path.

For the rest of the world, some degree of linkage is inevitable. These links involve:

- trade in goods
- international aid
- international loans
- flows of capital (foreign direct investment, FDI)

Table 15.11 gives levels of FDI received by developing and transitional countries. Figure 15.14 shows changes in FDI between the 1970s and 1997.

*Table 15.11 The top 20 recipients of FDI in 2002*

| Country | FDI ($US millions) |
| --- | --- |
| China | 45 300 |
| Brazil | 16 330 |
| Mexico | 12 101 |
| Singapore | 10 000 |
| Argentina | 6 327 |
| Russian Federation | 6 241 |
| Chile | 5 417 |
| Indonesia | 5 350 |
| Poland | 5 000 |
| Venezuela | 4 893 |
| Malaysia | 3 754 |
| Thailand | 3 000 |
| India | 3 264 |
| Hong Kong | 2 600 |
| Colombia | 2 447 |
| South Korea | 2 341 |
| Taiwan | 2 248 |
| Hungary | 2 085 |
| Peru | 2 000 |
| Kazakhstan | 1 320 |

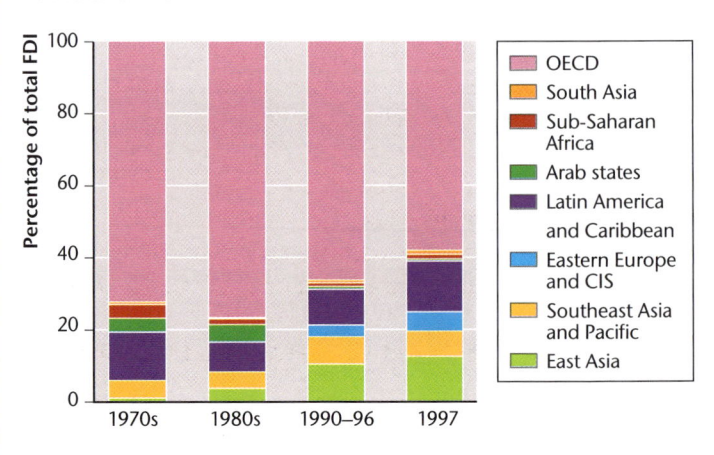

*Figure 15.14 Changes in FDI, 1970s–1997*

*Human systems, processes and patterns*  **A2 Unit 5**

Trade is a deceptively simple term. Humans have traded over long distances since early times. Obsidian (volcanic glass) is found in southern England in the form of axe heads and knife blades, although the nearest source of this material is southern Turkey. Thus 'international' trade took place long before 'nations' or international borders existed. The development of trade has gone through distinct stages:

- **Stage 1** Trading in local areas dominant, with a few exotics appearing from outside local areas but at great expense and restricted to a social elite. International trade restricted to long-distance land routes (the Silk Road).
- **Stage 2** The inter-linking of national regions through the development of an infrastructure allowing greater regional specialisation and much more internal trade. At the same the rapid development of international trading routes across oceans and the decline of the land routes.
- **Stage 3** The development of empires and widespread growth of international trade in these regions behind tariff walls. Internal trading gathers pace with the development of consumer industries. Backwash effects have an increasing impact upon local producers.
- **Stage 4** Globalisation of markets and increasing specialisation on a global level with an international division of labour emerging. Rapid development of international trade. Increasing pressure for the deregulation of trade and removal of barriers. Free trade seen either as a panacea or a fatal global disease.

The development of trade can be justified by a number of arguments which amount, in total, to the view that the more trade there is, the better off everyone will be. It follows that anything that obstructs trade, or slows down the flow of goods and services, is bad for national economies and bad for the population as a whole. Thus international trade, so it is argued, should be free trade.

# Theories of trade

## Factor endowments

Countries have different types and amounts of resources that determine what they can or cannot produce. Natural resources are unevenly distributed and both material and human resources are developed to different extents. The combination of these resources is referred to as a country's **factor endowment**.

Factor endowments are determined by:
- geographical features such as climatic conditions and natural resources
- historical development and political stability
- social and demographic issues
- political systems and development policies
- economic development, size and quality of the workforce and access to capital
- entrepreneurial skills and the freedom to pursue entrepreneurial activities

The UK, for example, has a large supply of natural resources such as coal, iron ore, oil and fertile land. Japan has limited natural resources, but it

> **Key terms**
>
> **Fair trade** Trade that allows for differences in the power and influence that small producers have in affecting prices or markets.
>
> **Free trade** Trade without restrictions in the flow of goods.

# Chapter 15 Development processes

has developed a highly skilled workforce that uses advanced technology to produce cars and electrical equipment. More mundanely, Costa Rica has a climate in which bananas can be grown outside and the UK does not. As a result, the UK buys bananas from Costa Rica rather than grow them here, at great expense, in greenhouses.

At a very basic level trade is likely to develop because 'we' have something that they want and 'they' have something that we want. There are two other simple bits of economics that help develop this idea: specialisation and economies of scale.

### Specialisation

By specialising in one form of production rather than another, a country can develop skills in its workforce. Costs of production, as measured in the amount of time and resources used to produce the goods or services, should decline. This allows the price of the finished products to be lower than those produced without this degree of specialisation.

### Economies of scale

There are two sets of costs involved in production. The so-called **fixed costs** cover heating, lighting, rent of buildings and some of the salaries paid to management. These costs do not vary as output changes. The **variable costs**, which do vary with output, include the costs of the raw materials or parts used in production, and the wages of the workforce. As production is increased, the burden of the fixed costs is spread more thinly over each unit produced. As a result, the cost of production slowly declines as output increases. This makes large-scale production cheaper than small-scale production.

## The theory of absolute advantage

The Scottish economist Adam Smith first outlined the theory of absolute advantage in 1776. He showed that a country has an absolute advantage in the production of a good when it can produce more of that good with a given amount of resources than another country. The reasons for this advantage are determined by the varying factor endowments of the country.

A simple economic model can be used to illustrate the principle of absolute advantage. This model is based on the following assumptions:

- there are only two countries, England and France
- these two countries each produce only wheat and cloth
- each country has the same amount of resources (land, labour and capital), although the quality differs
- resources can be transferred between the production of wheat and cloth
- production costs for each country are fixed
- there are no trade barriers, such as tariffs, between the two countries

Table 15.12 Absolute advantage: production before specialisation

|  | Wheat (units) | Cloth (units) |
| --- | --- | --- |
| England | 20 | 30 |
| France | 25 | 5 |
| Total output | 45 | 35 |

Table 15.12 shows the production for each country before specialisation.

With a given amount of resources England can produce 30 units of cloth and 20 units of wheat while France can produce 5 units of cloth and 25 units of wheat. England then has an absolute advantage in the production of cloth and France an absolute advantage in the production of wheat.

Table 15.13 shows the production gains after specialisation.

|  | Wheat (units) | Cloth (units) |
|---|---|---|
| England | 0 (−20) | 60 (+30) |
| France | 50 (+25) | 0 (−5) |
| Total output | 50 (+5), net gain | 60 (+25), net gain |

Table 15.13

When each country specialises in the production of the goods they have an advantage in, greater production of both goods can occur, so, in principle, both countries are better off.

This is illustrated in Table 15.13, where the production of wheat has increased by 5 units and production of cloth by 25 units.

There is no difficulty in accepting that some countries have an absolute advantage in the production of some goods.

### The theory of comparative advantage

The Nobel prize-winning economist Paul Samuelson was once challenged by mathematician Stanislas Ulam to name a single proposition to do with social science that was both true and non-trivial. After some time he came up with Ricardo's law of comparative advantage. This law complicates Smith's original proposition by examining the situation that arises when one country has an absolute advantage in the production of all goods and services. Does trade still benefit them?

In 1817, David Ricardo, a classical economist, developed the law of comparative advantage to explain this situation. The idea is based on the relative efficiencies of production in a situation where each country has a comparative advantage in producing the commodity in which it has the lower opportunity cost. The opportunity costs are what must be given up in order to consume or produce another good. For example, going on a holiday to Australia may involve giving up the purchase of a new car.

What Ricardo tried to show was that even when a country was more efficient than any other country in the production of two products, it should still concentrate on the one it produced most efficiently of the two. It should use the resources it had once devoted to the other product to make the first product. It should then export the surplus produced and use that money to buy the other product from the other country.

Of course the moral of these theoretical approaches is not the same for all commentators. Many people see comparative advantage and economies of scale as the twin supports for free trade — trade without obstacles and impediments such as quotas, tariffs or any other artificial barriers to the free flow of goods and services across borders.

### Key terms

**Comparative advantage** The theory that all parties will benefit when countries specialise in the production of the goods and services that they produce relatively efficiently.

**Economies of scale** The advantages enjoyed by large companies because high output allows them to spread some costs, thus reducing average costs of production.

# Chapter 15 Development processes

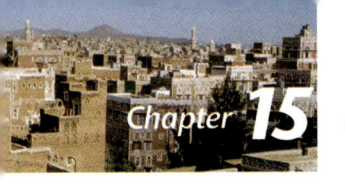

**Examiner's tip**

- Don't muddle up arguments about free trade with arguments about international trade in general. One can argue for the latter without being convinced about the merits of the former.

However, the critics point out that a large number of Ricardo's assumptions are no longer applicable, especially his insistence on the immobility of capital and labour. Since capital is now much more mobile it is quite difficult to establish the advantages apparent in one country over another. What gives countries a comparative advantage these days sometimes boils down to not much more than the labour costs involved in production. There are two other difficulties with the move towards a world of specialisation in those products that one produces most efficiently.

- Who owns the companies that produce the raw materials in which resource-rich countries have an obvious comparative advantage? If these companies are TNCs, then the benefits of pursuing the comparative advantage that, for example, Costa Rica has in bananas may only end up increasing the profitability of companies which have headquarters in New York. In this example, there is not much benefit for Costa Ricans in general. The nation states used by Ricardo are no longer especially significant in the international division of labour.
- The second problem is even more serious. In the past 70 years the **terms of trade** have moved against primary producers. The terms of trade are the comparative values of primary products such as bananas, cotton, copper and coal relate to the cost of manufactured goods such as cars, microwave ovens and mobile phones. An example is shown in Figure 15.15. Figure 15.16 shows how commodity prices are related to the total level of international debt in the countries of the South.

The terms of trade have continued to move against primary producers as demand declines when substitutes are found (e.g. fibre optics replace copper in wiring) or new competitors and over-production force down prices. For example, in 1975 the sale of 8 tonnes of Kenyan coffee could buy a tractor; it now takes 40 tonnes to buy an equivalent tractor.

This arithmetic forces many of the poorest LEDCs into a spiral of environmental exploitation that is almost certainly not sustainable. In response to falling prices, an individual producer can do one of two things:

- reduce costs by cutting wages or taking short-cuts with health and safety
- increase production and thus maintain income

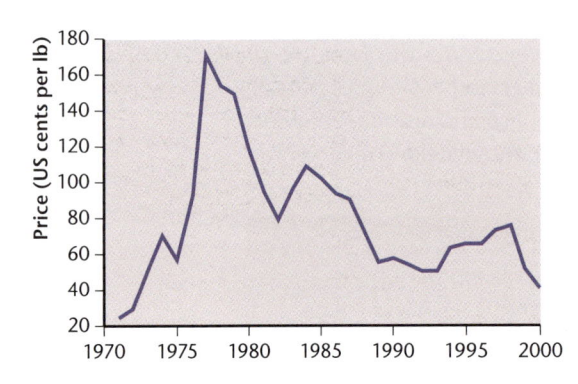

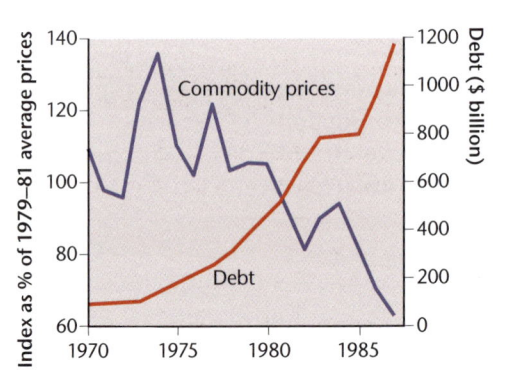

*Figure 15.15 The annual average price of cocoa, 1971–2000*

*Figure 15.16 Commodity prices and total debt, 1970–87*

In the first case the impact is to reduce rather than increase the wealth of the population as a whole. Through the multiplier effect, operating here in a negative manner, this response reduces the income and wealth of the mass of the population. In the second case the increase in supply, if widely pursued by farmers (which is likely), will further reduce the price of the product. Thus the 'logical' response of the individual becomes wholly illogical when followed by other producers. This 'spiralling down to the bottom' is a risk faced by many sub-Saharan African countries.

## The role of aid

The boundary between **aid** and loans that require repayment is somewhat blurred. A common means of defining foreign aid is:

> Assistance in the form of grants and loans given by one government or multi-lateral organisation to an LEDC, which in the case of a loan must have at least 25% in the form of grant (or gift). This also includes the costs of providing, for example, expert training or providing advice on economic reforms.

In other words, it is a flow of capital (money) at a concessional rate. The degree of concession will vary greatly. At one extreme is emergency relief aid, when private individuals and governments, through official charities (like Oxfam or Christian Aid), give food and medicine or the money to buy it for no charge. At the other extreme are cases in which the 20% grant element of the aid is almost invisible behind a screen of conditions and restrictions.

Aid can also be classified in terms of the recipient and the donor. **Bilateral aid** is government to government assistance, whereas **multilateral aid** has many donors and many recipients.

The impact of aid is no clearer than the impact of any other inward investment in LEDCs, whether it be a normal loan or a gift. Some countries, such as Chile, have managed respectable rates of economic growth in recent years despite very low levels of aid. Bilateral aid seldom comes without some political motive. The USA has by its own statistics spent over $1 trillion on foreign aid since 1945. It freely admits that the motives for this aid are:

- protecting its political and strategic interests
- promoting US exports
- rebuilding war-damaged economies
- providing relief during humanitarian crises

Only the last of these is entirely altruistic, that is to say without any self-serving element. The other three bring possible benefits to the USA.

### Examiner's tip

- Remember that aid is a great deal more than the charitable private donations to disaster relief that many of us have made. Much that is called aid comes with considerable conditions attached.

## The alternative view

Among neo-liberals the most vociferous opponent of foreign aid in recent decades has been Peter Bauer. Bauer was, and is, a fierce opponent of the conventional view that foreign aid helps less-developed societies to develop. He took the view that protectionism would destroy entrepreneurial spirit. His own research in

# Chapter 15  Development processes

Malaysia convinced him that the role of government was to ensure that property rights and contractual obligations were enforced, so that all people are treated equally under the law. Apart from this, governments should stay well away from the free-market forces that ensure an efficient distribution of resources.

Most controversially he argued that there would be no concept of the Third World if it were not for the invention of foreign aid. Aid politicised economies, directing money into the hands of governments rather than towards profitable businesses. Interest groups then fought to control this money rather than engage in productive activity. Indeed, aid had proved 'an excellent method for transferring money from poor people in rich countries (through taxation) to rich people in poor countries'.

Bauer argued that aid is simply a 'guilt trip' driven by a mistaken view that colonialism was a bad thing. He argued that former colonies are better off now than they were before colonialism. The most developed of the poorer countries are those that have the most interaction with rich countries, through trade and the exchange of ideas. To support this he pointed to the rise of the NICs.

Bauer is still regarded by many as a maverick. He opposes policies aimed at reducing income inequality because he regards them as likely to reduce the rate of economic development. He is also in favour of population growth, rather than population control, seeing it as far more likely to stimulate economic growth than impede it.

Foreign aid is supposed to turn vicious circles of poverty into virtuous circles by allowing countries to break out of the cycle:

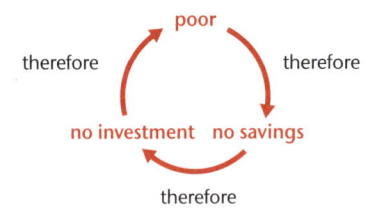

```
           poor
therefore  ↑  ↓  therefore
    no investment ← no savings
           therefore
```

The World Bank is generally content with the recent record on aid.

> Foreign aid has been highly successful in reducing poverty in countries with sound economic management and robust government institutions. The list of countries that now meet the criteria for using aid well has increased dramatically in the 1990s. Yet aid has fallen to its lowest point in more than 50 years, at a time when it could be helping hundreds of millions of people escape a hand-to-mouth existence.
>
> World Bank, *Assessing Aid, Policy Research Report 7*

However, William Easterly (himself a former researcher at the World Bank) has analysed data from 88 countries and found there to be no relationship at all between levels of foreign aid and rates of investment. He found only six countries where the relationship was positive, including two (China and Hong Kong) which had received only tiny amounts of aid. The other four were Tunisia, Malta, Morocco and Sri Lanka.

There are sceptics from the political left too:

> Aid policies, unaccountable to African producers and pastoralists, have generally bypassed their needs in favour of expensive, large-scale projects. Africa has historically received less aid for agriculture than any other continent, and only a fraction of it has reached rainfed agriculture, on which the bulk of grain production depends. Most of the aid has backed irrigated, export-oriented, elite-controlled production.
>
> Lappé and Collins (1978), *Food First: Beyond the Myth of Scarcity*

## The institutional framework

The key players are the following.
- The International Monetary Fund (IMF) is a lending institution, set up in 1944 to help stabilise exchange rates and give credit to countries with a balance of payments problem. Its role today is to promote trade liberalisation.
- The World Bank was set up at the same time to provide loans for the reconstruction of the war-damaged European economies which were labelled, at the time, developing economies. Since that time its brief has widened considerably.
- The World Trade Organization (WTO) replaced GATT. GATT itself replaced the International Trade Organization. The WTO implements the agreements made in 1994 by 117 governments in the Uruguay Round of negotiations. The thrust of this agreement is a legally binding commitment to free trade.

The funding for the IMF and the World Bank comes from MEDCs. The policies are formulated in similar fashion, with voting rights given in proportion to levels of donation. Hence Europe, Japan and the USA dominate both institutions. The WTO settles disputes through panels of experts and lawyers who make decisions that can be overturned by unanimous vote. This unanimity is very hard to achieve, so the dominance of the MEDCs is still all too apparent.

The main policy instrument in the governance of international links is the **structural adjustment programme** (SAP). Sets of policies are imposed upon LEDCs as a condition of the loans offered to help them repay their debts. The key features include:
- privatisation of state enterprises to try to end inefficiency and to attract foreign investment
- opening of the economy to foreign direct investment and imports by removing any regulations and subsidies which protect local industries and agriculture from foreign competition
- raising hard currency to help repay loans by promoting exports and encouraging tourism
- a reduction in government spending by cutting services, introducing charges for health and education, and cutting back on government employees
- devaluation of the local currency to make exports more attractive by reducing their price on the world market, and increasing the price of imported goods
- increasing interest rates to encourage savings and reduce local spending

## Chapter 15 Development processes

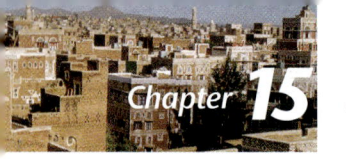

Debtor LEDCs have no choice but to accept these SAP conditions because loans will only be granted to countries that have the official IMF/World Bank seal of approval. The impact of such policies on many of these countries is highly controversial. One view is that the poor have suffered from the reductions in budget expenditures on health and education, especially in countries such as Mozambique and Tanzania where interest payments exceed the combined health and education budgets. On the other hand, the lending institutions rigorously defend the polices by pointing out that, where properly applied, the policies lead to economic growth which soon generates wealth.

The UN has appointed independent commissioners to investigate the impact of SAPs, but unsurprisingly they could not agree on the findings. Thirty were in favour, fifteen opposed and eight abstained on the following statements which:

- acknowledged that the serious problem of the foreign debt burden remains one of the most critical factors adversely affecting economic, social, scientific and technical development and living standards in many developing countries
- stressed that the economic globalisation process creates new challenges, risks and uncertainties for the implementation and consolidation of development strategies
- expressed concern that, despite repeated rescheduling of debt, developing countries continue to pay out more each year than the amount they receive in official development assistance
- noted the relationship between the heavy foreign debt burden and the considerable increase in poverty at the world level and especially in Africa
- stated that foreign debt constitutes one of the main obstacles preventing developing countries from fully enjoying their right to development

# The consequences of development

The positive impact of development is self-evident — people become better off, fewer people live in poverty and all the measurable indicators of human progress become positive. It is important to evaluate how these benefits are distributed within a society. An elite group might benefit hugely from some types of development, but at the expense of the continuing poverty of the majority of the population. This elite group may very well be content to promote policies that are clearly in their interests but not necessarily in the interest of other groups.

## Neo-colonialism

At the end of the nineteenth century much of the world was divided into empires that consisted of the home or mother country and a number of colonies. The function of these colonies was to produce raw materials for the industries of the mother country.

*Human systems, processes and patterns*  **A2 Unit 5**

## Case study 15.16 *The colonial influence: India*

The British prime minister Disraeli called India a 'jewel in the crown of England'. In the 1880s India took 20% of British exports and overseas investment. In 1850 all tea imported to Britain came from China. By 1900 most of it came from India. India also provided a cheap source of labour following the abolition of slavery, with over 50 000 workers being sent overseas in the peak year of 1858/59 alone. The possession of India, as Churchill remarked, made all the difference between Britain being a first-rate and a third-rate world power. The role of India in the creation of British power was and is incalculable.

As tariff walls were erected and protectionist policies were imposed in western Europe and the USA during the nineteenth century, British business was obliged to look for other markets for its goods. Britain still needed to import agricultural products, but was finding it harder to establish markets for its goods. India solved the problem. British business exported its own finished products to India. In particular, India was a captive market for the Lancashire textile industry.

During the 1880s about two-thirds of British exports to India were cotton goods. In the words of Karl Marx: 'The homeland of cotton was inundated with cotton'. This market was created by imposing tariffs on the movement of Indian 'home'-produced textile goods from state to state within the sub-continent while allowing British goods to move freely. In addition, India's export surplus with countries other than Britain, through the sale of agricultural products and raw materials, was used to offset British deficits elsewhere. As the English Historian Eric Hobsbawm put it:

India was the 'brightest jewel in the imperial crown' and the core of British global strategic thinking precisely because of her very real importance to the British economy...anything up to 60% of British cotton exports went to India and the far east, to which India was key — 40–45% went to India alone — and the international balance of payments of Britain hinged on the payments surplus which India provided.

This colonial system provided the structure for the modern world system. The colonies exported one or two **primary commodities** (raw materials) with very low **value-added**. Examples are tea (Kenya) and copper (Zambia). The colonies then imported the higher value-added manufactured goods made from these raw materials, such as electrical wire made from the exported copper.

The mother country often prevented industrialisation from taking place in colonies or, even more extreme, broke up existing industry if it posed a threat to the colonial power. The Indian textile industry is a frequently quoted example of the latter.

An alternative view is that, whatever may have been the shortcomings of colonial rule, the overall effect was positive. It reduced the economic gap between Europe, the USA and the rest of the world. It brought, so some argue, enlightenment where there was ignorance. It suppressed slavery and other barbaric practices such as pagan worship and cannibalism. Formal education and modern medicine were brought to people who had limited understanding or control of their physical environment. The introduction of modern communications, exportable agricultural crops and some new industries provided a foundation for economic development. Africans, for example, received new and more efficient forms of political and economic organisation.

# Chapter 15  Development processes

In the postwar world (since 1945) almost all of the former colonies were given their independence — at least in the usual sense of being able to govern themselves. The critics argue, however, that 'neo-colonialism' adopted new methods of extracting the wealth of the former colonies using TNCs as the agents, and, most commonly, FDI as the weapon. Today, for example, Africa owes $227 billion to Western creditors and this crushing debt burden is keeping the continent impoverished. These countries are tied into a system that more or less compels them to take on a dependent role in the world economy, concentrating on the production of raw materials and cheap commodities much as they did in colonial times — hence the term **neo-colonialism**.

## Case study 15.17  Global inequalities

The world is characterised by enormous inequalities in wealth, income and resources. That is the fundamental conclusion of the annual *Human Development Report*, issued by the United Nations Development Programme.

The world's 225 richest individuals have wealth that is equal to the annual income of the poorest 47% of the entire global population. Spending just 4% of that wealth, or about $40 billion a year, is all it would take to maintain universal access to basic education, healthcare, reproductive healthcare, adequate food, clean water and safe sewers for everyone in the world. Taking care of the basic needs of the poor could prevent a global depression.

- The three richest people have assets that exceed the combined GDP of the 48 least-developed countries.

|  | Annual expenditure |
| --- | --- |
| Basic education for all | $6 billion |
| Cosmetics in the USA | $8 billion |
| Water and sanitation for all | $9 billion |
| Ice cream in Europe | $11 billion |
| Reproductive health for all women | $12 billion |
| Perfumes in Europe and the USA | $12 billion |
| Basic health and nutrition | $13 billion |
| Pet foods in Europe and the USA | $17 billion |
| Business entertainment in Japan | $35 billion |
| Cigarettes in Europe | $50 billion |
| Alcoholic drinks in Europe | $105 billion |
| Military spending in the world | $780 billion |

*Table 15.14 The world's spending priorities*

- The 15 richest individuals have assets that exceed the entire GDP of sub-Saharan Africa.
- The wealth of the 32 richest individuals exceeds the total GDP of south Asia.

In terms of consumption, the report finds that the richest 20% consume 16 times more than the poorest 20%.

> Globally, the 20% of the world's people in the highest-income countries account for 86% of total private consumption expenditures — the poorest 20% a minuscule 1.3%.
> 
> *Human Development Report*, UN

The report also highlights disparities in consumption of particular goods.

- The richest 20% use 17 times as much energy as the poorest 20%.
- The richest 20% of the world's population consume 11 times as much meat as the poorest 20%.
- The richest 20% eat seven times as much fish as the poorest 20%.
- The richest 20% consume more than 45% of all meat and fish, the poorest 20% a mere 5%.
- With 74% of all telephone lines, the richest 20% have 49 times as many telephone lines as the poorest 20%.
- The richest 20% use 84% of all paper, 77 times as much as the poorest 20%.
- The richest 20% own 145 times more cars than the poorest 20%.

## Dependency theory

Dependency theory arose in the 1960s and argues that the growth and development of the MEDCs is, ultimately, dependent on pumping out the wealth of the LEDCs. The LEDCs are dependent upon the MEDCs and our wealth is dependent upon the process of transferring wealth from them to us.

Despite the rise of the NICs, which challenges this view, the evidence for dependency theory is quite strong.

The gap between the MEDCs and LEDCs is more pronounced than it has ever been. The UN *Human Development Report* for 1997 shows that the share of world trade for the 48 least-developed nations, representing 10% of the world's population, has halved in the past two decades to just 0.3%, with over 50% of all developing countries not receiving any FDI. Two-thirds of the FDI went to just eight developing countries.

In fact, around 100 LEDCs are experiencing slow economic growth, stagnation or outright decline, and the incomes of more than a billion people no longer reach levels attained 10 or even 30 years ago.

# The debt crisis

The end of the last wave of economic growth is generally dated at the end of the 1960s. The world economy was facing a difficult period with too much capacity (too many factories) and too little demand for goods. Increasing ownership of cars and household electrical equipment had fuelled a long period of growth, but now most who could afford these things had them, so demand declined. Factories geared up to produce a certain number of cars, refrigerators or whatever per year found themselves facing a smaller demand as market saturation was reached and people only purchased goods to replace existing goods. New consumers were few and far between.

Production was cut back and a long period of rationalisation began which, of course, led to unemployment and a further reduction in demand. This set in motion the following chain of events.

- Partly in response to the falling demand and partly as a consequence of the Arab/Israeli war, the price of oil was raised on world markets by 500% in 1973.
- Huge quantities of so-called petro-dollars (profits from oil sales) were accumulated by individuals and governments in the oil-producing countries.
- Since they couldn't spend them fast enough, the oil-producing countries put much of the money into Western banks. Effectively money was transferred from MEDCs, largely from the individuals and companies who had to pay more for fuel, into the bank accounts of a much smaller number of individuals and companies which couldn't possibly spend the money. This further reduced demand for goods and services in MEDCs.
- Banks have to lend money to make money; that is their function and their reason for existence — if they don't lend, they cannot survive. Thus the banks

## Chapter 15 Development processes

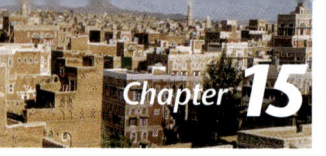

faced a crisis because the depression in the MEDCs meant that neither individuals nor industries were much inclined to borrow money. There was an economic recession, people feared unemployment and industries were experiencing a reduction in profits as demand for their products fell. This negative multiplier effect led to varying but increasingly severe levels of economic recession in most MEDCs.

- The banks hunted around for 'new' customers and found them in LEDCs. They offered loans at relatively low rates of interest to recycle their large reserves of petro-dollars. Much of this money was poured into large-scale economic development schemes, including major dams, transport systems and crash industrialisation programmes. The aim was that these countries produce enough goods for sale to pay back their loans from the profits made from exports. Other loans were used to buy armaments or spent on extravagant prestige projects (international airports were a favourite). Corruption was common and the management of the loans was often very poor.

- The arrival of neo-liberal (neo-conservative) administrations in the USA and the UK in the early 1980s began a period of monetary policy which, among other things, involved large increases in interest rates in order to 'defeat inflation'. In the view of political leaders like Reagan and Thatcher, inflation was the main economic 'enemy'. The initial impact of high interest rates was to further cut back demand in the UK and USA, deepening the crisis of consumption.

- For many LEDCs which had borrowed money, this was a catastrophic combination of circumstances. They were forced to pay higher interest rates on loans. But they had little chance of repaying the loans because the markets for their goods were now in economic recession and demand was stagnant or even falling. This was forcing down prices, especially of raw materials. The terms of trade moved sharply against the primary producers. Figure 15.17 shows how the total foreign debt rose between 1970 and 1997.

*Figure 15.17 The increasing debt mountain, 1970–97*

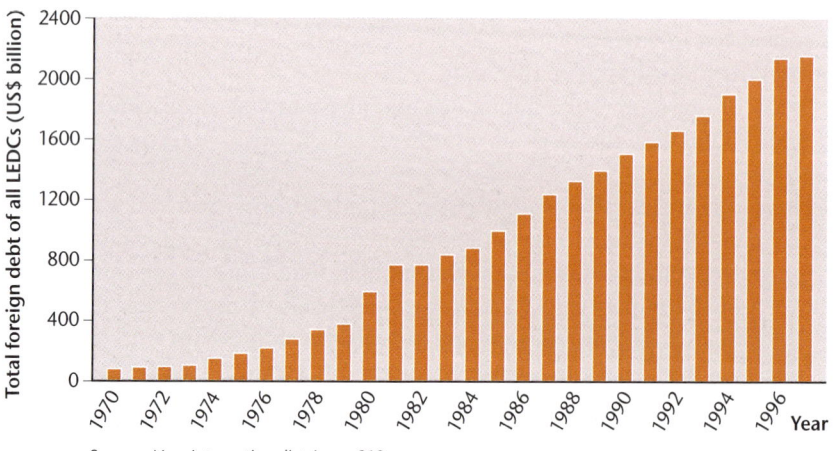

Source: *New Internationalist*, Issue 312

- The most alarming events took place in the summer of 1982. Mexico, which owed $80 billion but had no cash even to pay the interest, let alone repay any part of the loan itself, threatened to default. Nine large US banks had no less than 44% of their entire capital tied up in Mexican loans. If Mexico defaulted (refused to pay the interest), these banks would have become bankrupt, with untold consequences for the world economy. The crisis was averted with a last minute (almost literally) rescue package of new loans of $8 billion from the IMF. The first structural adjustment programme (SAP) was born.

The debt crisis has not gone away and recent events in east Asia (1997) and Russia (1998) suggest that recurring crises might become an increasingly trouble-some feature of the world economy. The management of that debt has profound implications for all of us.

## Countries as experiments in economics

There is probably less agreement about the issues raised in this chapter than any other section of your studies. The profound political divisions make balanced discussion improbable and the interpretation of the same 'facts' problematic. There are few better examples of this than the South American state of Chile (Case study 15.18).

### Case study 15.18 Economic experiments: Chile

#### Vital statistics

- Chile has the world's driest desert and the world's longest mountain chain.
- Chile's population is 16 million with a growth rate of 1% per year.
- Nearly half the population is under 25 years of age and 72% is under 40. Life expectancy is the highest in South America at 75 years.
- Literacy levels are high at 95%.
- 80% of the population lives in urban areas.
- 3% of the land area is arable.
- Chile is a major global supplier of copper. Reserves are estimated at 193.8 million tonnes; 26.5% of world reserves. Output averages 4.5 million tonnes a year.
- Chileans are a mix of Spanish, Irish, English, German and Scottish immigrants and native Amerindians.
- 'Chile' means 'where the world ends' in the language of the indigenous peoples.

#### The Chicago Boys

Chile has become a battleground of development ideology over the past 40 years. Like most Latin American countries, it pursued policies of import substitution. These were driven forward rapidly under the democratically elected President Allende who embarked upon 'the Chilean road to socialism' in 1970. His regime was overthrown by a foreign-backed military coup in 1973 and Allende himself was assassinated. Sixteen years of military rule under General Pinochet followed, with a return to democracy in the 1990s.

As far as the development theorists were concerned, Pinochet's dictatorship was also a period in which a series of ideas that had come to be known as the Chicago School was implemented in a living laboratory. The Chicago Boys were a group of postgraduate Chilean economists who had trained under Milton Friedman and others in Chicago. Heavily influenced by Frederich Hayek's

## Chapter 15 Development processes

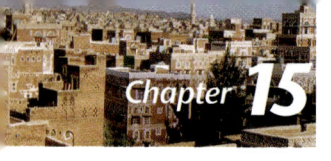

*Road to Serfdom*, published in 1944, the argument was the familiar Adam Smith view that the market knew best. Hayek took the view that socialism, even the very watered down version of social democracy that was popular in Europe in the 1940s, represented a step along the path to the kind of tyranny that held sway in Nazi Germany and the Soviet Union.

The Chicago School believed in free markets with minimal government intervention and had a golden opportunity to trial their ideas in Chile. While Pinochet dealt with issues of law and order and popular resistance, the Chicago Boys were given a relatively free rein in implementing major changes to Chile's economy. These can be boiled down to ten policies.

### Ten economic policies

1. **Removal of quotas and tariffs.** These had been implemented to protect Chile's manufacturing base from foreign competition. They created what the Chicago School regarded as 'market distortions' whereby some goods and services cost more than they should.
2. **Tax reform.** Company tax and income tax were reduced, especially for higher earners. This was seen as a way of removing barriers to investment and saving among the group of people most likely to invest.
3. **Privatisation.** State-run industries were sold off to the highest bidder. They were seen as inefficient.
4. **Monetary policy.** The Central Bank was given increased powers to control the money supply through the use of interest rates; this supply-side economics was thought to be the best way of tackling inflation.
5. **Deregulation of banking.** Banks and pension systems were privatised.
6. **Encouraging foreign investment.** Chile withdrew from the 'Andean pact', which had restricted the amount of foreign capital flowing into the region.
7. **Wages and prices.** Regulation of prices and wages was ended. Collective bargaining with trade unions ceased and the private sector was allowed to determine wages.
8. **Property rights.** These were reinforced and the rule of law was imposed.
9. **Employment legislation.** There was substantial deregulation of the labour market. It became easier to hire and fire labour.
10. **The black market.** Chile's black economy was firmly dealt with, making it one of the smallest in Latin America at 19.8% of all economic activity.

These classic steps contain the core economic ideas that became known as neo-liberalism to some and neo-conservatism to others. They provided the template for economic reform that was implemented widely in the late 1970s in MEDCs such as the USA under Reagan and the UK under Thatcher. Chile provided the first major global experiment in this free market ideology.

The economic, social, political and environmental consequences of the Chilean experiment remain controversial, both in Chile and well beyond. It is a good example of how the selective use of evidence can support two conflicting opinions. Table 15.15 gives both sides of the argument.

There is clearly no consensus here, but some facts are undisputed:
- the economic miracle of Chile was introduced under a military dictatorship which reduced personal freedoms and used torture and murder as political weapons
- Chile is now an export-led economy that concentrates on raw materials, such as copper, fruit, timber or fish, or manufactured goods that are processed from these, such as woodchip and paper
- most, but not all, Chileans are better off now than they were in 1974

*Human systems, processes and patterns* **A2 Unit 5**

| Miracle | Disaster |
|---|---|
| **Economics**<br>■ Chile has grown faster than any other Latin American economy<br>■ Poverty levels have fallen sharply since 1990.<br>■ Chile is a major export economy and attracts large volumes of foreign investment, especially from the USA<br>■ In the HDI Chile is ranked 43rd | **Economics**<br>■ Disparities in wealth between rich and poor have risen sharply and Chile's inequalities are increasing, not reducing<br>■ Unemployment is high at 15%<br>■ Real wage rates have fallen<br>■ Much of the growth is based on new export industries that are ultimately unsustainable and environmentally damaging, specifically forestry, fruit and fishing |
| **Social and political**<br>■ Chile is a stable democracy with a freely elected government that contains a broad spectrum of Chilean society<br>■ Trade unions are legal and the press is relatively free<br>■ Most of Chile's economic progress has come since the return of democracy in 1990 | **Social and political**<br>■ The Pinochet regime is accused of having murdered up to 3000 political opponents and used torture as a political weapon |
| **Environmental**<br>■ Central Santiago is one of the most modern cities in Latin America<br>■ Progress has been made in recent years cleaning up the environment<br>■ The huge forestry industry largely uses plantation timber and not natural forests | **Environmental**<br>■ Copper mining involves huge environmental damage, with 98% of material mined being waste.<br>■ Copper smelting produces sulphur dioxide<br>■ The expanding fruit industry depends heavily upon 4 million tonnes of pesticides, mostly highly toxic<br>■ The natural forests are under pressure because this is the fastest growing sector, producing hardwood woodchip products<br>■ Depletion of fish stocks is becoming alarming. Not only is the average size of the fish falling, but the fleets need to stay out longer to keep fish catches constant. Both suggest over-fishing |

*Table 15.15 Chile: miracle or disaster?*

# Examination questions

**1** (a) Outline the problems of measuring development. (5)
   (b) Examine the view that it is not possible for all countries to enjoy high standards of living. (20)

*Total = 25 marks*

**2** (a) Outline the differences between types of aid. (5)
   (b) Criticise the view that international aid does more harm than good. (20)

*Total = 25 marks*

# Chapter 15
*Development processes*

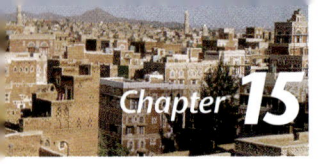

# Synoptic link
## The world is increasingly interdependent

- The role of values and attitudes in determining the type and rate of development.
- The concept of sustainable and unsustainable development at national and global scales.

## Attitudes to development

Development is a multi-dimensional process that involves, social, economic, political and cultural elements. The ultimate goal is to improve living standards, and perhaps even human happiness — although this tends to be forgotten in the rush to equate wealth with contentment.

Almost all attempts to define development include the following objectives:
- to increase the availability and widen the distribution of basic life-sustaining goods
- to raise the standard of living of all, including, in addition to higher incomes, the provision of shelter, food, education and employment
- to expand the range of economic and social choices available to individuals and nations

All of this is uncontroversial, although different priorities are often given to these three objectives and there is considerable disagreement about the detail. For example, the nature of 'social choice' may be disputed between the protagonists in the debate about abortion.

The real disagreements begin when the three main ideologies that dominate this subject are reviewed.
- **Liberalism** assumes that politics and economics exist in separate spheres.
- **Nationalism** assumes and advocates the primacy of politics over economics.
- **Marxism** holds that economics drives politics.

### The liberals

According to the **liberal perspective,** development should be committed to free markets and the price mechanism as the most efficient ways to allocate resources and maximise individual welfare. It advocates minimal state intervention, and then only to fix a market failure or to provide a public good, like open space. The path to growth, according to this view, is by way of savings. The more savings in a society, the more capacity to invest, and thus the society generates more income and economic growth. When economic growth is achieved, individuals can save again and the circle is repeated. The basic promise of this ideology lies in the assumption that individuals behave rationally and attempt to maximise or satisfy certain values at the lowest possible cost to themselves.

### The nationalists

The **nationalist perspective** believes in political intervention in defence of the state's interests. Economic nationalism develops from the tendency of markets to concentrate wealth in a few hands and to establish dependency or power relations between strong and weak economies. Nationalists frequently emphasise national self-sufficiency in contrast to economic interdependence.

## The Marxists

The Marxist starting point is class struggle and the proletariat revolution. It is composed of four essential elements:
- a dialectic approach to knowledge and society
- a materialist approach to history
- a general view of capitalist development
- a normative commitment to socialism

Marxist theory focuses on the problems of production, information and distribution, with the dependency and world system schools as its most recent approaches. The ultimate development goal is the socialist society, which is a utopian society.

## Applying the ideologies

These views are not readily compatible and one cannot easily take a little from each school of thought. Their significance is that the preferred ideology clearly influences the chosen route to development. Thus for all practical purposes the policies adopted within a country by the government are determined by ideology.

Few people dispute the increasing disparity of world income shown in Table 15.16. However, the interpretation of this disparity varies according to the ideological position held.

| World income inequality | 1960 | 1970 | 1980 | 1989 | 1998 |
|---|---|---|---|---|---|
| 20% 'low' | 2.3 | 2.3 | 1.7 | 1.4 | 1.2 |
| 60% 'middle' | 27.5 | 23.6 | 22.0 | 15.9 | 9.8 |
| 20% 'high' | 70.2 | 73.9 | 76.3 | 82.7 | 89.0 |
| Gini coefficient | 54 | 57 | 60 | 65 | 70 |

Table 15.16 Changes in share of world income, 1960–98 (%)

For the liberals:
- the world is a much richer place, so *absolute* poverty has decreased
- wealth has to be created by the few before it can trickle down to the poor
- the speed of this trickle down has not been helped by foolish and misguided opponents of globalisation
- there is plenty of evidence of a successful narrowing of the 'gap': Taiwan and South Korea are obvious examples

For the nationalists:
- there is some validity in these data because countries have been sucked into a world system, which has impoverished them
- states need to be able to develop their own industrial base or they risk being trapped by declining terms of trade and increasing dependency on the manufactured goods from richer countries, hence the widening gap

# Chapter 15 Development processes

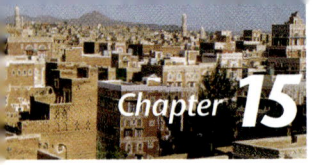

For the Marxists:
- there are poor people in every country and they are getting poorer — look at average wage rates in the USA
- the nation state is not a useful tool of analysis — there is no 'United Kingdom' because there are different classes that are fundamentally opposed to one another
- the disparity of income is an inevitable part of advanced capitalism and will, in time, lead to its downfall

## Who benefits from development?

In many developing countries, the social system is characterised by a stratification based not only on gender but also on class. Rural Bangladesh has an especially pronounced social hierarchy.
- The economic and political power is rooted primarily in the ownership of land.
- The uppermost class, possessing 2 ha or more of land, accounts for only about 8% of all rural households in the country, yet it controls almost 48% of all the cultivable land.
- Lowest in the scale are the 50% of marginal landowners and landless people, who hold only about 3.7% of the total land area. Lacking formal education and any kind of resources, they enjoy scant social prestige. Most of them earn their living as daily workers and represent an enormous oversupply of labour, so that they are locked into dependence on the rural elite.
- Given such social structures, it is not surprising that the socially and economically privileged wield political influence in addition to economic power, and also control whatever development aid comes to the rural areas.
- In bilateral projects between foreign donors and non-governmental organisations (NGOs), those who have more influence than the poor may siphon off funds.
- To keep the money flowing, when applying for funds organisations are quick to adopt the latest development jargon, from 'sustainable development' to 'empowerment of women' and 'participation of target groups'.

## Sustainability

The concept of sustainable development is not new, although the term itself is much more recent. The apparent tension between population growth and the natural environment was known to the Greek historian Herodotus (c. 484–420 BCE). The term entered common usage following the publication of the Brundtland Report (1987). This commission also generated the best-known definition of the concept:

> To meet the needs of the present without compromising the ability of future generations to meet their own needs.

But what are these 'needs'?
- **economic needs** including access to an adequate livelihood or productive assets; and economic security when unemployed, ill or otherwise unable to secure a livelihood

## Human systems, processes and patterns — A2 Unit 5

- **social, cultural and health needs** including shelter which is healthy, safe, affordable and secure; and access to basic healthcare, education, transport and protection from environmental hazards
- **political needs** including the freedom to participate in national and local politics in a broad framework that ensures respect for civil and political rights and the implementation of environmental legislation

It is worth noting that generally accepted definitions do not exist for many basic concepts used by society, from 'fairness' to 'freedom'. This is especially true of those terms which concern our well-being. The idea of sustainable development, without a commonly accepted definition, therefore appeals to virtually all groups who choose to participate in the environmental debate because they can simply 'feed in' their preferred version of 'adequate livelihood' or 'economic security'. Under such conditions, being 'pro' sustainable development entails no risk or commitment to a specific set of goals or conditions since none are agreed upon. It is much like being against 'sin', which is an easy position to adopt just so long as 'sin' is not defined.

Sustainable development has become an article of faith that is often used but seldom explained. In broad terms it encompasses:

- help for the very poor because they are left with no option other than to destroy their environment
- self-reliant development, within natural resource constraints
- cost-effective development using differing economic criteria to the traditional approach; that is to say, development should not degrade environmental quality, nor should it reduce productivity in the long run
- the great issues of health control, appropriate technologies, food self-reliance, clean water and shelter for all
- the notion that people-centred initiatives are needed; human beings, in other words, are the resources in the concept

A further complication occurs because the phrase contains two important words: 'sustainable' and 'development'.

- **Sustainable** suggests the relationship between people and the carrying capacity of the planet.
- **Development** can be interpreted as compatible with economic growth with little or no reference to the natural resource base.

The following quote makes clear something of the problem:

> Development is a whole; it is an integral, value-loaded, cultural process; it encompasses the natural environment, social relations, education, production, consumption, and well-being.
>
> Dag Hammarskjold Foundation

There are sceptics who reject the whole concept of sustainable development. One of the best known is Wilfred Beckerman. He argues that:

- sustainability is an unnecessary concept because the welfare of future generations is already accounted for — natural capital (the environment) and

## Chapter 15 Development processes

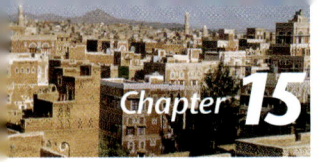

human capital are infinitely substitutable and the free market will 'protect' natural capital
- economic growth works better than sustainable development programmes in improving levels of world poverty
- sustainability outlines a programme of impractical, selfish and fuzzy 'requirements to preserve intact the environment as we find it today in all its forms'
- sustainability is ethically unacceptable because it requires us to choose between directing funds to save a species of beetles or to provide sanitary water for the poor

# A2 Unit 6

## Synoptic: People and their environments

# Chapter 16

## Section A

Unit 6 is the synoptic unit. The purpose of this unit is to assess students' ability to draw on their understanding of the connections between different aspects of the subject represented in the specification.

- The synoptic paper consists of five questions.
- The examination lasts for 2 hours.
- Question 1 (Section A) is a compulsory question divided into four sub-sections. The total mark for this question is 50.
- Questions 2, 3, 4 and 5 (Section B — see Chapter 17) are essay questions from which the candidate chooses one. The total mark for this essay is 25 marks.
- Each essay question corresponds to a synoptic theme.

## Synoptic links

### Unit 1 Physical environments

#### 1.1 There are relationships between people and tectonic environments

- Positive and negative impacts of earthquakes and volcanoes on human activity, both short term and long term.
- The importance of risk assessment, prediction and monitoring, and their limitations.

#### 1.2 The hydrological cycle is used and managed

- The reasons for and methods of groundwater and river management in countries at different stages of development.
- Decision-making issues related to the management of the hydrological cycle.

#### 1.3 Coastal processes can be managed

- The need for coastal management schemes.
- Issues of management, including methods and strategies used, and their possible impact.

### Unit 2 Human environments

#### 2.1 Governments have a direct and indirect influence on population change

- Data collection (e.g. census) and its role in planning and providing services for a changing population.

# Synoptic: People and their environments — A2 Unit 6

- Government policies to increase and reduce birth rates — reasons, effects and relative success.

## 2.2 Government policies influence settlement characteristics and patterns
- Policies for managing changing rural and urban settlement.
- Issues, rationales and outcomes of policies.

## 2.3 Government policies influence migration patterns
- Reasons for, effects of and issues associated with government policies to influence migration into, out of and within a country.
- The causes and consequences of recent migrations and refugee movements.

# Unit 4  Physical systems, processes and patterns

## 4.1 Weather and human activity are interdependent
- Possible impacts of humans on weather and climate, including pollution, ozone destruction, global warming, cloud seeding.
- Weather hazards and their impact on human activity, to include hurricanes, tornadoes and drought, with an emphasis on recent events.

## 4.2 Glaciated and periglacial environments offer opportunities and challenges for human activity
- Opportunities and challenges exist in upland areas (either glaciated in the past and/or currently active) for tourism, energy production, quarrying, transport, agriculture, settlement, etc.
- Periglacial and permafrost environments present their own challenges and opportunities.

## 4.3 Ecosystems offer opportunities and challenges for human activity
- The management opportunities and challenges associated with grassland and forest ecosystems.
- The causes and management of soil erosion.

# Unit 5  Human systems, processes and patterns

## 5.1 Industrialisation and de-industrialisation have an impact on the physical and human environment
- The interrelationships between the physical environment and industrial location and the management of environmental impact.
- The role of government in the control and modification of the relationship between industry and the environment, and the response to changing values and attitudes within society as a whole.

# Chapter 16 Section A

### 5.2 Rapid change has created pressure on rural–urban interdependence
- The management of waste in cities — the problems of water and air quality control, and waste disposal.
- The growth of the leisure and tourist industry, and its social, cultural and economic impact on rural environments.

### 5.3 The world is increasingly interdependent
- The role of values and attitudes in determining the type and rate of development.
- The concept of sustainable and unsustainable development at national and global scales.

# The synoptic assessment

## Question 1

Question 1 involves responding to 'unseen' resources which usually include a map, text, graphs, photographs and diagrams. It tests your ability to interpret these resources and to apply your geographical understanding and skills to the material.

Typical questions might be:

**a** With reference to…assess the impact of the physical environment on settlement and population in….
**b** Using the map and Item…examine the impact of human activity on the physical environment in….
**c** With reference to…evaluate the success of the management of the….
**d** In what ways can economic development be regarded as sustainable in…?

These questions are drawn from the four synoptic themes but do not presume detailed knowledge of those themes. They do assume a broad understanding of the key ideas and terminology.

- Question 1 consists of four sections (a)–(d): all are compulsory. It is worth 50 marks out of 75, so should occupy about two-thirds of the allotted time — 80 minutes out of 120.
- This one question carries 80 uniform marks towards the final A-level result, almost as much as the 90 uniform marks carried by the whole of either Unit 4 or Unit 5.
- There is usually a map of the region or area under consideration, not necessarily an OS map.
- A resource booklet is provided containing information on the area. This booklet is *not* seen before the examination itself. It will probably include:
  - photographs
  - text: usually several items, including an introduction
  - tables of data
  - graphs
  - diagrams

## Synoptic: People and their environments — A2 Unit 6

It is suggested that before answering the question, candidates should:
- Read the question to get a general idea of the tasks required.
- Ring or highlight key command words. If necessary deconstruct the question. Watch out for any obvious restrictions.
- Look at the mark allocation in order to recognise the important sections. Note down the time that should be spent on the answer. This works out at approximately 1.5 minutes per mark, so a section worth 16 marks should take about 24 minutes to complete, including time to read and absorb the relevant resources.
- Read the text items and highlight and/or note comments about them. Pay particular attention to the introduction, which often contains useful general information. Make comments that extend from the text.
- Go back to the first part of the question and begin your answer.

### Skills to practise for Question 1

#### Annotation

As you read through the resource booklet, make comments and highlight text that might be useful. Keep the wording of the question in mind so that you are able to determine which pieces of information are relevant and important.

> A *World Bank* funded development programme in *Vietnam's Mekong Delta* aims to shift the delta's water management regime from the current system, which is relatively *open and adapted to the natural flows* of the Mekong to one which relies on *large-scale infrastructure* for the *control of salinity intrusion and flood protection*. The World Bank's plans are partly intended as a response to the *vulnerability of the delta* to the impacts of economic development in the river basin. They also aim to support the national objective of *increasing rice production* in the delta.

Annotate photographs by identifying the key features and by trying to offer explanation or comments about their significance. The wording of the question again needs to be borne in mind to ensure relevancy.

*Figure 16.1 Annotating a photograph*

Annotations on photograph:
- Cloud cover — orographic
- Limited vegetation cover — too steep?
- V-shaped valley — vertical erosion
- Steep, bare slopes — soil erosion?

#### Description of trends in data

Try to spot general trends (up/down/variable) and then pick out anomalies or obvious inconsistencies. Rather than quoting figures from the data, try to quantify change or difference, possibly expressing it as a fraction or percentage. When dealing with changes over time, it is best to avoid starting at the beginning of the time period and describing the change at each time interval step-by-step.

## Chapter 16  Section A

*Table 16.1  Retail price of rice (in rupees) in Bangladesh in 1974*

| J | F | M | A | M | J | J | A | S | O | N | D |
|---|---|---|---|---|---|---|---|---|---|---|---|
| 114 | 122 | 145 | 169 | 168 | 172 | 175 | 212 | 263 | 311 | 264 | 234 |

Overall trend: an increase in price during the year from 114 rupees to 234 rupees, but at a variable rate of change.

Detail of trend: increased steadily from Jan to June then more rapidly from July to a peak in Oct of 311 rupees.

Anomalies: very small decrease from April to May of 1 rupee; significant drop from Oct to Dec of 75 rupees, a fall of about 25%.

### Description of patterns/distributions on maps
Try to spot general patterns (even/uneven) and then pick out details, anomalies or obvious inconsistencies.

*Figure 16.2  Map showing population density of the UK, 1997*

Persons km$^{-2}$
- 1500+
- 1000–1499
- 750–999
- 450–749
- 250–449
- 150–249
- 75–149
- under 75

General pattern: the highest population densities are in the south and east, with London having over 1500 km$^{-2}$, the lowest in the north and west, with most of northern Scotland having less than 75 km$^{-2}$.

Detail: coastal locations generally have higher densities than adjacent inland areas. The coast of south Wales around Cardiff and Newport has densities as high as 75–150 km$^{-2}$ whereas the area immediately inland to the north has under 75 km$^{-2}$.

Anomalies: the coasts of northern Scotland have densities that are the same as the adjacent inland areas; less than 75 km$^{-2}$. The English midlands have very high densities, as much as 1500 km$^{-2}$ or more around Birmingham.

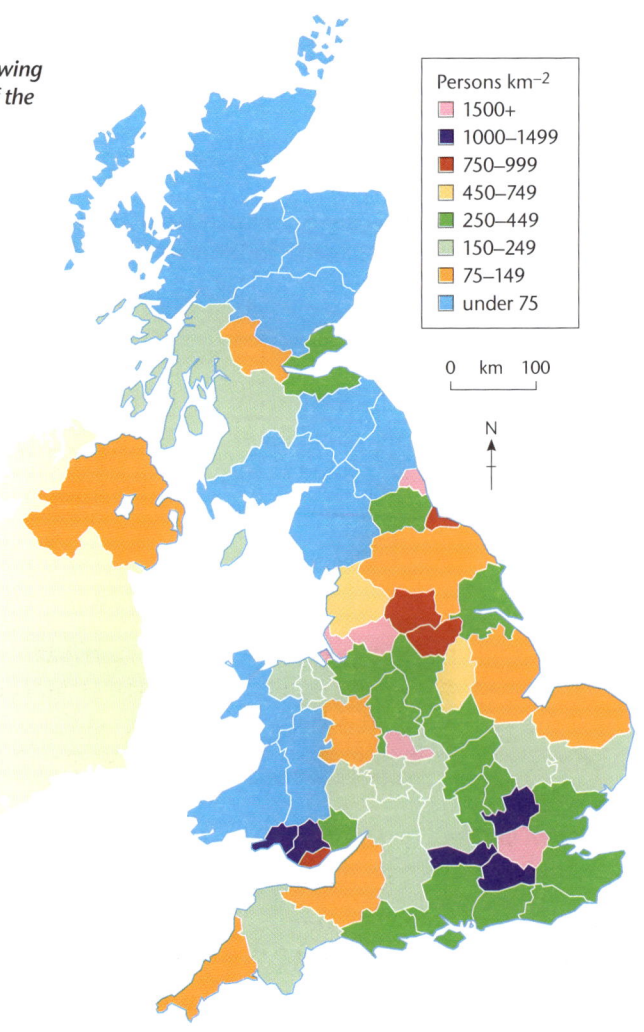

## Synoptic: People and their environments  A2 Unit 6

### Drawing sketch maps
Sketch maps are unlikely to be required by the question, but a well-annotated sketch map is often an economical way of answering a question. If one is drawn, then it should be referred to in the answer. This may be a more appropriate technique for the essay question in Section B, but it could be used in Section A. Very little credit will be gained for simply copying a map that has been given as a resource item. However, a sketch map which selects relevant elements of the map provided in order to answer the question will gain credit.

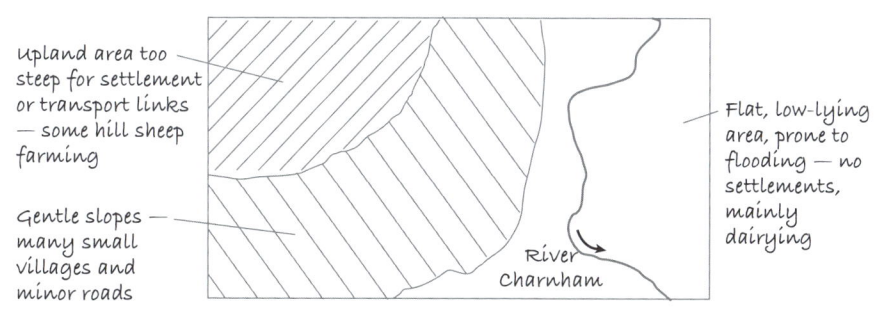

*Figure 16.3* Sketch map showing the relationship between settlement type and physical geography

- upland area too steep for settlement or transport links — some hill sheep farming
- Gentle slopes — many small villages and minor roads
- Flat, low-lying area, prone to flooding — no settlements, mainly dairying
- River Charnham

## Producing quality responses to Question 1
- Use all the material — it is all there for a reason, so check at the end that each item has been used in some way.
- Try to use personal words — don't lift great chunks of text. For every piece of text used, make a comment about how it relates to the question asked: for example, 'This passage shows…' or 'Item 2 suggests that…'. In other words, be explicit.
- Obey the command words and phrases absolutely. Never do more than they require and try not to do less. The hardest thing to do is to resist the temptation to *explain* when only asked to *describe*. You will not lose marks for this but you will waste time. The command words most likely to be encountered are:
  - **describe**: what it is like
  - **explain**: give reasons for it being like it is
  - **examine**: describe, explain and comment
  - **assess**: how significant/important it is
  - **identify**: what it is
  - **comment on**: give some personal observations/views
  - **critically evaluate** (or **examine**): discuss the arguments in favour of and against an idea using evidence; deliver a judgement
  - **to what extent**: in what ways is something true and in what ways is it not; overall, is it more true than not or vice versa?

## Accessing the top bands in the mark scheme
In order to reach higher bands in the mark scheme try to:
- Make links between different resources and explicitly cross-reference when doing so. Thus 'as Item 3 suggests, the pattern of land use shown in Item 1 might be explained by…'.

## Section A

- Apply understanding of processes from other parts of the specification. Again be explicit when doing this. For example, 'the river delta, formed by deposition of sediment as the speed of the flow is checked,…'.
- Apply knowledge of other areas that have been studied. It is perfectly legitimate to draw contrasts with other regions or areas that have been covered, but be careful not to stray from the question asked. Thus 'Examine the possible impact of the Three Gorges project on …' is obviously about this specific project, but comparisons could be offered with other large dam schemes such as the Hoover dam. An understanding of the hydrological changes that dams cause can also be applied.
- Use words like 'therefore' and 'however' to qualify comments. Use words like 'partial' and 'tentative' to suggest that the conclusions drawn are by no means certain.
- Look at the mark allocation and allocate time carefully. The long last section, often worth 14 or 16 marks, should not be neglected or rushed because of shortage of time.

### Typical features of a top-level answer

- Wide-ranging and accurate knowledge.
- Thorough explanation and balanced discussion of all dimensions of the question.
- An understanding of concepts which are used to support argument.
- Detailed examples, together with a range of diagrams, maps and data used appropriately.
- Comparisons of scale, in space and time, support discussion.
- English is fluent, expressing complex ideas clearly and with few, if any, errors in grammar, punctuation and spelling.

## Examination question
# The disappearing Aral Sea

### Item 1 Introduction

The Aral Sea is located in central Asia in the lowlands of Turan. Administratively the water body is divided between the republics of Kazakhstan in the north and Karakalpakistan in the south. The latter is an autonomous republic within the republic of Uzbekistan. The republics themselves are divided into a large number of smaller districts, so called 'oblasts' and 'rayons'. The two rivers Amu-Dar'ja and Syr-Dar'ja are heavily used for irrigation by these republics. The area around the lake and the islands was used for military testing until the 1980s (especially testing of chemical weapons), mainly in Kazakhstan and on the Ust-Jurt plateau west of the lake.

Synoptic: People and their environments  **A2 Unit 6**

## Item 2  Location

*Boats left stranded by the receding Aral Sea*

## Item 3  Changes to the Aral Sea since 1960

| Year | Area (km$^2$) | Volume (km$^3$) | Sea level (m) |
| --- | --- | --- | --- |
| 1960 | ≈68 000 | ≈1040 | 53 |
| 1985 | 45 713 | 468 | 41.5 |
| 1986 | 43 630 | 380 | 40.5 |
| 1987 | 42 650 | 354 | 40 |
| 1988 | 41 134 | 339 | 39.5 |
| 1989 | 40 680 | 320 | 39 |
| 1990 | 38 817 | 282 | 38.5 |
| 1991 | 37 159 | 248 | 38 |
| 1992 | 36 087 | 231 | 37.5 |
| 1993 | 35 654 | 248 | 37 |
| 1994 | 35 215 | 248 | 37 |
| 1995 | 35 374 | 248 | 37 |
| 1996 | 31 516 | 212 | 36 |
| 1997 | 29 632 | 190 | 35 |
| 1998 | 28 687 | 181 | 34.8 |
| 2010 (projected) | 21 058 | ≈124 | 32.4 |

## Section A

### Item 4  The disappearing sea

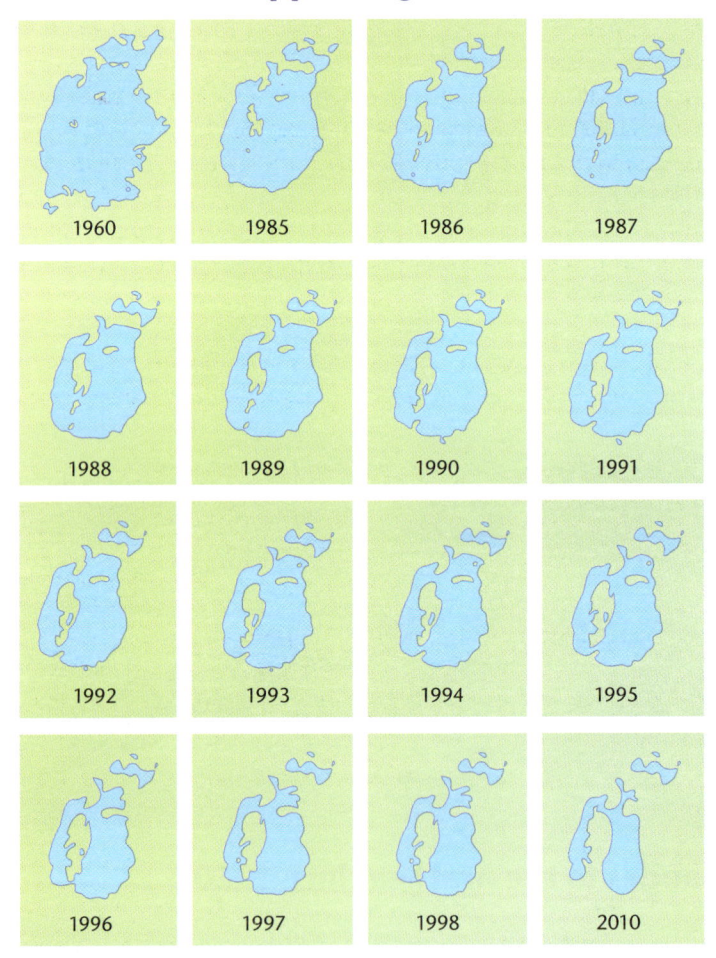

| 1960 | 1985 | 1986 | 1987 |
| --- | --- | --- | --- |
| 1988 | 1989 | 1990 | 1991 |
| 1992 | 1993 | 1994 | 1995 |
| 1996 | 1997 | 1998 | 2010 |

### Item 5  Irrigation

The two rivers Amu-Dar'ja and Syr-Dar'ja are heavily used for irrigation by the republics in which the Aral Sea lies. In the last 35 years irrigation has been greatly intensified. During this period the desiccation of the Aral Sea has been dramatic, with the result that the former fourth-largest lake of the world is now the world's eighth-largest lake. One of the large irrigation areas is the Amu-Dar'ja delta to the south of the Aral Sea. The irrigated area extends over approximately 28 000 km$^2$, and is mainly used for the cash crops cotton and rice. In the twentieth century the irrigated area in the Aral Sea region increased from about 3 million hectares to over 8 million hectares. This increased demand for water meant that the discharge to the Aral Sea by the two rivers dropped from 60 km$^3$ per year to almost 0 km$^3$ per year.

In addition there are high water losses due to the age of the irrigation network. About 80% of all irrigation channels lack any kind of base sealing, which results in high water losses from evaporation and infiltration.

Synoptic: People and their environments  A2 Unit 6

## Item 6  Consequences

Along the former shoreline, salt has accumulated due to evaporation, and solonchaks (saline soils) have developed. As a result of the strong northeasterly winds in this area, the salt is picked up and transported by aeolian processes. Through deflation it lands on the irrigated fields in the south. The north of the Amu-Dar'ja delta used to be an important ecosystem with a large variety of flora and fauna. The increasing salinity and the water shortage have led to vast degradation of this area. At the same time the productivity of the agricultural fields has dropped significantly due to secondary salinisation as a result of capillary uprising of soil water. The local people are frequently inundated with clouds of pesticide-laden dust.

The Aral Sea itself has experienced significant increases in salinity, rising from approximately $10\,g\,l^{-1}$ in 1960 to $30\,g\,l^{-1}$ in 1989. It is predicted that it will reach $70\,g\,l^{-1}$ by 2010.

| Climatic consequences | Ecological/economic consequences | Health consequences |
|---|---|---|
| Mesoclimatic changes (increase of continentality) | Degeneration of the delta ecosystems | Increase of serious diseases (e.g. cholera, typhus, gastritis, leukaemia) |
| Increase of salt and dust storms | Total collapse of the fishing industry (originally 44 000 tonnes $yr^{-1}$) | Increase of respiratory system diseases (asthma, bronchitis) |
| Shortening of the vegetation period | Decrease of productivity of agricultural fields | Birth defects and high infant mortality |

Source of Items 1–6: The Aral Sea Homepage http://www.dfd.dlr.de/app/land/aralsee/ back_info.html

## Item 7  Recovery?

Since 1987, the contraction of the Aral Sea has created two lakes. The larger one grows drier by the year and appears to be beyond saving. But the smaller, northern lake may be salvageable, according to local and international experts.

In 1997, the local government in the town of Aralsk took matters into its own hands. It deployed residents and earth-moving equipment to scoop sand from the seabed and build a dike 19 km long and 25 m wide between the two lakes.

Protected from the larger, contaminated body, the smaller lake's shoreline began to stretch again toward the ships' cemetery. Birds reappeared, including gulls, swans and pheasants. Danish scientists analysed fresh sole from the lake's waters and were amazed to find them clean enough to eat.

'We had assumed that the lake was completely poisoned by the chemicals and pesticides. This means it wasn't,' says Altynbek Meldebekov, deputy executive director of the International Aral Sea Rehabilitation Fund, an Almaty-based group linked to the Kazakh government.

The water has gone from a depth of 35 m to 38 m — and is still rising. So far local authorities have spent about $535 000 on the dike. They want to build a proper drainage system in the surrounding delta.

The project has attracted the attention of international experts, who are

## Chapter 16  Section A

mustering more support. The World Bank is considering funding to make the dike a permanent fixture. The United Nations and European donors have granted more than $1 million to help clean the lakeside area and revive traditional livestock and fishing.

One country that isn't forthcoming is Russia. As the biggest remaining part of the former Soviet Union, Kazakhstan says Russia should take responsibility. 'We asked for compensation, but Russia refused in 1996 on the grounds of financial difficulties,' says Mr. Meldebekov sadly. 'With the current economic crisis there, we can forget about any help from them today.'

Source: The Christian Science Monitor
http://csmonitor.com/cgi-bin/durableRedirect.pl?/durable/1999/02/05/fp8s2-csm.shtml

## Question

a  Using only Items 3 and 4, describe the changes to the Aral Sea since 1960. (12)
b  Explain why these changes have occurred. (8)
c  Assess the economic consequences of the changes to the Aral Sea. (16)
d  Critically examine the view that the changes to the Aral Sea have been an ecological catastrophe. (14)

*Total = 50 marks*

# Chapter 17

# Section B

There are four synoptic essays to choose from on the examination paper, each set on one synoptic theme. You have to choose *one* question. Preparation for this essay takes the form of acquiring knowledge and understanding from the synoptic links relevant to each theme. The specification is designed in such a way that each of the Units 1, 2, 4 and 5 have six synoptic links, two for each section; so Unit 1, Section 1, Earth Systems, has two 'links'. These are covered in the synoptic section at the end of each chapter of this book.

## Preparation

It is most unlikely that you will have been prepared for all four essay titles in the examination. Most centres teach with a maximum of *two* titles in mind, so that you have a choice on the day. You will therefore have concentrated much more on some synoptic links than others. You need to know which links are really important to you and how they relate to the synoptic theme. The best way of doing this is to look at the following information.

# Synoptic themes for essays

There are four synoptic themes, each containing six synoptic links from Units 1, 2, 4 and 5.

### Theme 1   Physical environments influence the human activity

- Positive and negative impacts of earthquakes and volcanoes on human activity, both short-term and long-term.
- The reasons for and methods of groundwater and river management in countries at different stages of development.
- The need for coastal management schemes.
- Weather hazards and their impact on human activity, to include hurricanes, tornadoes and drought, with an emphasis on recent events.
- Opportunities and challenges exist in upland areas (either glaciated in the past and/or currently active) for tourism, energy production, quarrying, transport, agriculture, settlement, etc.
- Periglacial and permafrost environments present their own challenges and opportunities.

# Chapter 17 Section B

## Theme 2 Human activities modify physical environments

- The reasons for and methods of groundwater and river management in countries at different stages of development.
- The causes and consequences of recent migrations and refugee movements.
- Possible impacts of humans on weather and climate, including pollution, ozone destruction, global warming, cloud seeding.
- The causes and management of soil erosion.
- The interrelationships between the physical environment and industrial location, and the management of environmental impact.
- The growth of the leisure and tourist industry, and its social, cultural and economic impact on rural environments.

## Theme 3 Physical and human resources may be exploited, managed and protected

- The importance of risk assessment, prediction and monitoring, and their limitations.
- Decision-making issues related to the management of the hydrological cycle.
- Issues of management, including methods and strategies used, and their possible impact.
- The management opportunities and challenges associated with grassland and forest ecosystems.
- The causes and management of soil erosion.
- The management of waste in cities — the problems of water and air quality control, and waste disposal.
- The concept of sustainable and unsustainable development at national and global scales.

## Theme 4 Communities and their governance influence geographical interrelationships at a range of scales

- Data collection (e.g. census) and its role in planning and providing services for a changing population.
- Government policies to increase and reduce birth rates — reasons, effects and relative success.
- Policies for managing changing rural and urban settlement.
- Issues, rationales and outcomes of policies.
- Reasons for, effects of and issues associated with government policies to influence migration into, out of and within a country.
- The role of government in the control and modification of the relationship between industry and the environment, and the response to changing values and attitudes within society as a whole.
- The role of values and attitudes in determining the type and rate of development.

# Synoptic: People and their environments — A2 Unit 6

## Analysing the question

Once you have identified the relevant sections for detailed learning of the material and understanding of the processes, it is worth paying attention to the nature of the questions asked.

The essay titles are designed so you can draw on knowledge gathered from several links. They will not ask for any particular process or any particular environment. They will not be huge and unmanageable titles such as:

> 'Assess the impact of the physical environment on man.'

Neither will they be narrowly drawn and too limited, such as:

> 'Assess the impact of hurricanes on the economy and environment of LEDCs.'

They will be somewhere in between, such as:

> 'Compare the short-term and long-term impact of hazards on human activity.'

In constructing the titles for all four topics, a series of critical words and phrases will be used to define the *focus* of the essay. The topic of an essay is what the essay is about, in broad terms; the focus gives you a more precise idea of what is wanted. The following words and phrases are vital for you and need to be applied to whichever themes you have studied in depth.

## Focus 1  Short term and long term

It may be that you are asked to compare the short-term and long-term effects of a particular set of events, actions or policies. Obvious examples would be the 'short-term and long-term impact of hazards …'. It is worth making the point that these terms are highly dependent on the context. Short term on a human scale may be a matter of hours, days, weeks or months, whereas on the level of cultural changes or changes in values it may be measured in generations. For geologists, short term is measured in many thousands or even hundreds of thousands of years.

## Focus 2  Direct and indirect

Some impacts are direct. Farming has a direct impact on the physical environment through processes such as ploughing, planting and otherwise modifying the 'natural' environment. There are indirect impacts as well, which are a consequence of the direct impacts, and which may be geographically more dispersed. These might include greater sedimentation in rivers or changing relationships between animals on the food web. An indirect effect of building dams and regulating river flow is to increase channel erosion downstream of the dam because the rivers are, initially at least, without load and thus have more energy. A volcanic eruption has the direct impact of damaging the environment and perhaps killing people in the immediate area but may also have indirect impacts. For example, in the Laki eruption, it was the indirect or secondary impact that did most damage when the toxic ash fall destroyed the natural vegetation and undermined the pastoral economy of Iceland, leading to starvation for many people.

## Section B

### Focus 3  Intentional and unintentional

The question of intent is closely related to direct and indirect impacts. There are many intentional modifications of the environment and many policies and management decisions that are made with these intentions in mind. However, there are also unpredictable and unintentional consequences of almost all such modifications. The management of rivers often reduces sediment discharge to the sea; this in turn reduces the amount of sediment on the shoreline, which inevitably leads to greater erosion. Eutrophication of waterways is another obvious example of an unintentional impact. Similarly river management by one country may impact, unintentionally, on a country further downstream; this can create serious international problems. Although unintentional and 'indirect' are closely related, there are cases where an indirect result can be quite intentional. An example would be the building of wing dykes (direct), which effectively narrows a river, thus increasing the velocity, thus increasing the erosion (indirect), so deepening the channel (intentional).

### Focus 4  MEDC and LEDC

These abbreviations are very familiar, maybe too familiar. It is worth reminding ourselves that there is no clear demarcation between 'more …' and 'less…'. There are bound to be uncertainties as to the appropriate category, but more importantly the categories themselves are based on huge generalisations about the 'countries'. It is worth pointing out that there is enormous variation *within* almost all countries. Thus the CBD of Mumbai is not very different from CBDs in any large city, with the same array of services and a similar pattern of high buildings. However, a few kilometres down the road the slums and shanty towns are distinctively LEDC. At the same time, within a few kilometres of London's global and extraordinarily expensive CBD are Bangladeshi immigrant communities and sweatshops that are reminiscent of LEDCs as well. Any contrasts at this level of generalisation are subject to qualification.

### Focus 5  Positive and negative

Few actions or events are purely 'good' or 'bad'. It is well known that the impact of war on an economy, if the country is not devastated by aerial bombardment or invading armies, can be positive. The US economy emerged from the Second World War with a GNP that had doubled in the space of 4 years. Disasters of all varieties provide niches for some activity, some creatures or some economic entrepreneurs. What is positive for one set of individuals may be negative for another. Flood control in Country A by building higher levèes is positive for most (although not all) enterprises within that country. However, it may increase the risk of flooding downstream, thus having a negative impact for some people in Country B.

### Focus 6  Challenges and opportunities

All environments pose both challenges to be overcome and opportunities to be exploited for personal or social gain. Note that a gain can be spiritual or moral as

*Synoptic: People and their environments*  **A2 Unit 6**

well as economic. These challenges and opportunities are not fixed in either time or space. For example, Iceland's highly unpredictable pro-glacial rivers are subject to violent changes in discharge, are wide, braided and liable to sudden shifts in course, and were hard to bridge for centuries. These challenges meant that remote communities were isolated and civic society was fractured. This was a significant obstacle to development.

In the twentieth century these same rivers were exploited for their hydroelectric potential, bridged effectively (although not without problems) and the power created contributed to Iceland's rapid rate of development — an impressive list of opportunities. However, another aspect of this has been the rapid loss of population from the outlying regions.

### Focus 7  Rural and urban

Rural–urban contrasts are frequently made, but the distinction between the two categories is growing smaller all the time, especially in MEDCs. What remains true is that there are distinctive urban cultures that dominate in many countries and distinctive rural ways of life that are threatened. This cultural division may be valuable in essays on management themes or when considering values and attitudes for Question 5. Rural backwardness and urban modernism have a long tradition in European culture, while recent dramatic social and economic experiments, such as the Cultural Revolution in China, saw urban not just as modern but also as Western and decadent and so threatening to the true values of Chinese society.

## A central problem for 'synoptic' geography

A synopsis is a summary in which various strands of an argument are synthesised and brought together. In geography this is often achieved by considering the physical environment and the human environment together. Unfortunately, this division between physical geography and human geography is not quite as simple as it first appears. Consider the following.

- The 'physical environment' can be broken down into:
  - the lithosphere: land surface, bedrock, soils etc.
  - the hydrosphere: water, principally the oceans
  - the atmosphere: the air
  - the biosphere: living organisms
- We (human beings) belong to the last category and as such we are a part of the physical environment as well as being part of the 'human' environment.
- It can therefore be argued that anything we do is as natural as the behaviour of any other species. In this sense, slaughtering bison in their millions or destroying rainforests becomes 'natural'. Opponents of this view would say that for most of our history we have not greatly modified the environment(s) in which we have lived, but we are now held responsible for hunting to extinction a variety of creatures from mammoths to moas, some of them many millennia ago, so this may not be a convincing distinction.

**Section B**

# Developed ideas for synoptic essay answers

## Uniformitarianism and catastrophism

There is no universally agreed theory about the nature of the processes that operate on the planet, including the processes that are frequently addressed in physical geography.

- **Uniformitarianism** or **gradualism**: This theory says that the natural processes that change the Earth in the present have operated in the past at the same gradual rate, and that geological formations and structures can be interpreted by observing present-day actions. This has often been shortened to the phrase: 'the present is the key to the past'.
- **Catastrophism**: This theory says that the main features of the Earth were formed by a series of sudden, violent catastrophes rather than a gradual evolutionary process.

### The debate

Since the 1830s conventional geological theory has revolved around the concept of uniformitarianism (or gradualism), assuming that the processes of the Earth have always been the same as those we can observe today. This can be summarised as the view that the 'present is the key to the past'. By understanding and measuring current processes we can infer what happened in the past. The originator of these ideas was the Scottish geologist James Hutton (1726–97), although it took the efforts of Charles Lyell (1797–1875) and his *Principles of Geology* (1830) for the theory to become widespread. This gradualist viewpoint, involving time-spans of millions of years, gave rise to the modern ideas of continental drift and the ice ages, and considerably influenced Charles Darwin's theory of evolution.

Any visit to a beach or valley shows that the processes of erosion, transport and weathering gradually change the face of the planet. Waves gently move grains of sand up and down the beach, and rivers slowly carve their valleys into the landscape. Obviously we have to allow for the small-scale catastrophes that we witness, such as volcanoes and hurricanes. However, the majority of Earth-forming processes, such as continental drift, the formation of mountains, and ice ages, are non-catastrophic, and *must* happen gradually over millions of years. This translates into the view that the landscape we observe and measure is dominated by slowly acting processes.

Uniformitarianism defeated the original version of catastrophism, which disallowed gradual change because it was at odds with Christian teaching about the creation — a sudden and thus catastrophic event. Uniformitarianism dominated scientific thought in the fields of biology, geology and the associated fields of geomorphology and palaeontology for much of the last century. In a nice example of how a theory can be so dominant that it influences 'facts', the meteor crater of Arizona, which has a width of 1.2 km, was only recognised to be the result of a meteor in the last century. In the 1800s it was believed to be a volcanic crater, because uniformitarianism did not take into account sudden extraterrestrial causes.

## A middle way? Punctuated evolution

The idea of **punctuated evolution** was developed by the great Harvard geologist and palaeontologist, Stephen Jay Gould (one of the very few practising scientists to be so well known as to rate an appearance on *The Simpsons*). He argued that, instead of a slow, continuous movement, evolution tends to be characterised by long periods of virtual standstill (equilibrium or statis), punctuated by episodes of very fast development of new forms stimulated by external events, probably climatic change. Although the theory originated in biology, it has been applied in other fields, not least in physical geography where the idea that landscape change might be episodic is not new. Punctuated evolution is widely challenged but provides a useful way of looking at the world. The theory is easy to explain using some general insights from the systems approach.

Imagine a system, say the atmosphere, as a typical landscape in which there are valleys separated by ridges (Figure 17.1). If the evolving system has reached the bottom of a deep valley (A or C), there will be almost no change, since variation will fail to pull the system out of that valley. This is a negative feedback regime, in which fluctuations are counteracted, pulling the system back to its equilibrium position at the bottom of the valley. In other words, the system is resilient to change.

*Figure 17.1*

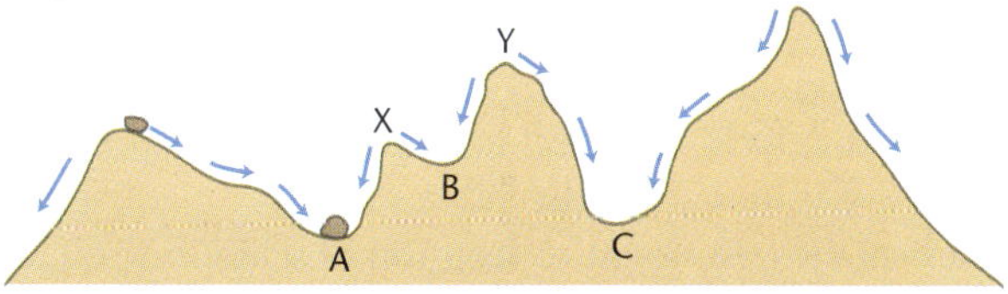

On the other hand, if there is only a small ridge separating a valley (B) from a neighbouring, deeper valley (A), then a chance event may be enough to push the system over the edge so that it enters the other valley. Once over the ridge, the descent into the new valley will go very fast; in other words, the system will change very quickly. This is a positive feedback regime in which deviations from the previous position are amplified. This means that the system will evolve very quickly to a new configuration. If the system is delicately balanced at the top of a ridge (X or Y), then very small changes can lead to a substantial systemic change, with positive feedback accelerating that process almost immediately. The status of the system thus becomes vital and it is true that we now take much more notice of catastrophic events in our explanation of landforms and landscapes.

## Paradigm shifts

In Thomas Kuhn's book, *The Structure of Scientific Revolutions*, he argued that scientific research and thought are defined by 'paradigms', or conceptual world-views, that consist of theories, supportive experiments and trusted methods.

Scientists typically accept a prevailing paradigm. In other words, they accept the conventions of the time rather than challenging the consensus view of the academic establishment. They need to do this in order to gain access to the academic posts that they seek. Rather than sit at their first interview for a university teaching job pouring scorn on the dominant theories in a particular field, they nod wisely and fall into line. This is sometimes referred to as conventionalism.

It is suggested that academics work *within* these dominant conventional theories, trying to extend their scope by refining them, explaining puzzling data and establishing more precise measures of standards and phenomena. 'Facts' that seem to contradict these mainstream theories are frequently ignored or explained away unconvincingly and opponents are ridiculed. Eventually, however, these opponents may generate insoluble theoretical problems or experimental anomalies that expose a paradigm's central weaknesses. This happens, so Kuhn argued, when those with doubts about the central core of beliefs eventually rise to the top of their respective fields and are then able to dust off their youthful criticisms of central theories without risking their careers. At this moment a **paradigm shift** or switch occurs, and in the space of a very short time the 'establishment' view changes completely.

The rise of plate tectonic theory is an excellent case study of these processes. The geological community clung on to a 'fixed Earth' view long after the weight of evidence against it was overwhelming. The campaign against Wegener, who devised the plate tectonic theory, was vitriolic and personal, like the campaigns against Darwin and Einstein. It is worth remembering that not all dissenting voices from the outside turn out to be right — but those that do remind us that what is taught as 'the truth' is subject to sudden changes over time according to the dominant set of values and beliefs in a society.

This contradicts the traditional conception of scientific progress as a gradual, cumulative acquisition of knowledge based on rationally chosen experimental frameworks. Instead, Kuhn argued that the paradigm determines the kinds of experiments scientists perform, the types of questions they ask, and the problems they consider important. In other words, science spends a good deal of its time shoring up old theories rather than ruthlessly exposing them as inadequate. And, more worryingly, scholars believe what it is 'conventional' to believe because challenging it won't help their career prospects.

## Externalities

Externalities are the effects, intentional or otherwise, that the acts of one group of people, including governments, have on other groups of people. They can be divided into external costs (which are largely negative) and external benefits (which are largely positive). These effects may operate at an international scale; thus the emission of toxic fumes in one country can have a negative impact on the population of another country. Externalities range from pollution and technological inventions to the change in the range of options available to consumers. They are the side-effects borne by third parties. In each case the firms or the individuals bear some form of cost known as the **external cost**. Externalities can be positive or negative.

- **Producer on producer externalities:** e.g. the use of insecticides and herbicides on cotton crops in Uzbekistan to increase yields reduces fish stocks in the Aral Sea. The gains made by one group of producers should be assessed in terms of the losses incurred by another group.
- **Producer on consumer externalities:** e.g. the savings in fixed costs and in wages for ancillary staff such as receptionists made by doctors by operating in groups in large centralised health clinics rather than in lots of single-doctor surgeries distributed more widely. There may be a **positive externality** in the form of an improved healthcare service, and thus fewer days off work for patients, but there is also a **negative externality** in the time taken, the fuel used and the lost income caused by longer journeys of individual patients to see the doctor.
- **Consumer on consumer externalities:** e.g. car drivers releasing fumes that cause asthma-related ailments in non-drivers.
- **Consumers on producer externalities:** e.g. drivers on the 'school run' causing congestion and slowing business traffic.

Externalities are by no means always negative and they do not always result in external costs. There are many positive externalities — benefits accruing to non-participants in the market place arising from the consumption and production of goods and services — but it is the external costs that are contentious and create social and political tension. For example, a cement works might ask permission to use old tyres as a fuel in order to cut its fuel costs and increase the profit that is shared among shareholders and executives. However, it may well be that the increased smoke emissions cause ill-health in the local community that costs more in time off work, medicines and loss of wealth creation, and that these losses add up to a larger figure. What is good for private companies is not always good for the community as a whole. This example is complicated, as are many of these, by the further problem that the benefits are relatively easy to measure whereas the externalities may be *long term* and difficult to quantify. You need to be able to answer questions such as 'How much asthma would occur normally?', 'How many years does it take for lung diseases to show as a significant cluster in the area downwind of the cement works?' and so on.

## Short term and long term; costs and benefits

The issue of externalities is made more complex when effects over *time* are considered. For example, cutting wages may bring a short-term economic benefit by increasing profits for a company, but lead to long-term costs if the workforce becomes disenchanted, sick and less efficient.

This idea might be developed by suggesting that the vigorous campaign against Nike employing cheap LEDC labour, which would have cut costs, lost them sales worldwide in the longer term. Conversely, the short-term cost of improving these working practices helped restore Nike's global reputation and therefore generated a long-term benefit.

Another example is the installation of basic public health systems such as sewers, which were paid for by industrialists in cities such as Birmingham in the nineteenth century. This was a short-term cost to the industrialists, but it led to improvements in the health of their workforce which ultimately generated improved output and thus improved profits. What subsequently happened was that the factory owners, influential men in the local community and often senior politicians nationally, made sure that governments took charge of public health. The financial burden was shifted from themselves as individuals to the taxpayer at large.

In the 1920s, the introduction of new production line techniques by companies such as Ford hugely increased productivity per worker and thus increased profits. The shift from skilled workshop-type production to assembly lines involved a *reduction* in the skills needed in the workforce. Normally this would lead to a reduction in wage rates, but this was not the case. The boredom endured by workers doing repetitive tasks led to some resistance among them to the introduction of these new techniques. As a consequence workers were paid *more*, out of the increased profit. These higher wages led to increased demand for the products that they made and brought major long-term benefits as 'Fordism' spread.

### Inter-generational equity and intra-generational equity

A central proposition of 'sustainability' is that future generations have a right to an inheritance (including natural resources, capital and human skills) sufficient to allow them to generate a level of well-being no less than that of the current generation. The set of resources inherited will not be exactly the same as those of today — for example, species may die out and different technologies may arise. Each generation has the right to inherit the same diversity in natural and cultural resources as that enjoyed by previous generations and to have equitable access to the use and benefits of these resources. At the same time, the present generation is a custodian of the planet for future generations. In this way, inter-generational equity extends the scope of social justice into the future.

The concept of inter-generational equity is compromised by our attitudes to *intra-generational* equity. It is a good deal easier to sign up to inter-generational equity if it seems that one is handing on a privileged and comfortable lifestyle to future generations. This is what 1.5 billion wealthy inhabitants of the planet can do, secure in the knowledge that their children will have many opportunities to enjoy the planet, given their privileged status.

Inter-generational equity is less of a mantra in the slums of Mumbai or the favelas of Rio, for quite obvious reasons. The concerns of the 1.5 billion disinherited people who live in poverty have less to do with inter-generational equity. They are more worried about present-day inequalities — intra-generational equity. They argue that equity in the world today is more of a priority. Why are we so concerned about future generations when we cannot be bothered to do much about our fellow global citizens?

An example of this kind of conflict is the emphasis an individual places on protecting an endangered species. If you are being brought up in Bengal on a small subsistence farm where your ox is a potential supper for the local tiger, your attitude towards the Bengal tiger will be different from that of someone living in a leafy suburb in south London with a poster of a Bengal tiger on the kitchen wall. The fact that south London was cleared of dangerous animals centuries ago as Britain developed is worth remembering.

## Environmentalism and capitalism

Issues to do with environmentalism and capitalism are at the core of many of the most critical debates going on in the world today.

### Criticisms of capitalism

Some people believe capitalism is an economic and political system that *inevitably* leads to destruction of the environment.

Capitalism is a system whereby there is private ownership of the means of production by a small number of people. This is the basic reason why it creates environmental problems: while the majority of the world's people have a material interest in maintaining a healthy planet, the small capitalist ruling class is not accountable to this majority. The pursuit of profit is essentially short term, especially when driven by shareholders who are only interested in the dividend they receive.

A second reason why capitalism creates environmental problems is that, although it means the world's resources are controlled by a relatively small number of people, planning is not centralised. Instead, production is uncontrolled; it is centred on making profits, not around meeting basic human needs. In other words, what is good for a company or a shareholder is not necessarily good for all. It can be argued that much of what is produced by the capitalist system is unnecessary and wasteful, and the system is not capable of incorporating long-term human survival as a need.

An obvious example would be the Nestlé baby milk scandal when this large corporation tried to promote formula milk in Africa. Advertised as 'modern' and 'hygienic', formula milk appealed to many women, who abandoned breast-feeding in favour of bottle-feeding. However, a lack of clean water meant that bottle-fed babies risked infection — the water used to make up their feed from powdered milk often contained pathogens. Many children died.

A final point is that the capitalist system does not distribute resources equitably. Under capitalism, many people have inadequate resources for survival and this leads to environmental problems. The poor, too politically enfeebled to do much else, are pushed into unsuitable forests or infertile deserts by the spread of commercial agriculture, and they are then blamed for the deforestation or the desertification that results.

### Criticisms of environmentalism

In the view of some of its opponents, environmentalism is not a benevolent movement seeking to improve human life by cleaning up the air and water. It is, they argue, a doctrine which attacks the ideals of Western civilisation. Opposed to scientific, technological and economic development, environmentalism, it is argued, holds that the non-human has value but the human does not. Environmentalism, according to this argument, has become the gravest threat to human survival. Under the guise of advocating clean air and water, and being protective of animals, environmentalists aim to retard and then dismantle our industrial society.

Environmentalists believe that economic growth as a world system is unsustainable. According to their opponents, they use false scientific claims, from global warming to desertification, to frighten the unwary. With a doomsday mentality, environmentalists try to block new inventions and economic development, from the local to the international level. If environmentalism is successful in its assault on Western values, life on Earth will become increasingly difficult, as wealth and freedom are eroded. Environmentalists are, it is claimed, latter-day Luddites, seeking to hold back technological advances. Capitalism, it is argued, is productive and very innovative, just so long as the free market is allowed to operate without state interference or constraint.

### Other views

In between these polarised views, there is a whole spectrum of opinions. Considering this spectrum takes us neatly on to the next pair of ideas.

## Consensus and conflict

**Consensus theorists**, who believe that compromise is both necessary and possible, point out that societies cannot survive unless their members share at least some common sets of perceptions, attitudes and values. They assume that societies are built on consensus, with a broad agreement on important values and beliefs. Of course, they recognise that in reality there will be considerable variance and that it is much more difficult to achieve consensus in advanced industrial states than in smaller and supposedly less complex societies. Consensus theorists point out, however, that in any society, modern or ancient, large or small, there is a shared set of abstract and complex assumptions about the world without which social life would be impossible. This shared set of values and beliefs becomes the foundation of society and is transmitted through socialisation. In modern societies, the function of socialisation is primarily performed by the family and the schools. They also argue that consensus around certain values, beliefs and norms is in the best interests of all the members of a society.

**Conflict theory** stresses that some people are getting richer at the expense of the majority who are getting poorer. In all societies, one dominant group prevails over the others. In advanced industrial societies, power may be more equally distributed among a complex set of elite groups than it was in the past, but many groups are significantly underrepresented. Stemming from Marxist theory, conflict theory emphasises that in societies where there are many competing groups, a dominant group takes control through economic rather than political power. Becoming dominant, it is argued, inevitably involves conflict. Once in power the dominant group works hard to develop consensus to justify and legitimise the new social order. Marx called this false-consciousness.

*Synoptic: People and their environments*  **A2 Unit 6**

From the conflict perspective, ideas are important, but not as important as human competition for the ownership of property and material wealth. Values and ideas are primarily used to justify an existing social structure, so the less privileged in society will accept the unequal distribution of the good things of life, including power, status and wealth. In this sense, values and ideas are an extension of the power struggle between social groups. Believing something to be 'wrong' is therefore a function of the economic rules that dominate. Thus slavery only became perceived as 'wrong' when it was no longer economically profitable. Similarly, women got the vote not because it was suddenly realised that this was a good idea but because to remain competitive industrial states had to use all their human resources, including women, obliging them to become more democratic.

## Eurocentrism and the problem with maps

Eurocentrism means looking at the world from a European perspective. It is seldom deliberate, but it does lead to a distorted view of the world. For example, some development material contains the assumption that 'we (i.e. Europeans) discovered America' and that the history of world development begins with the expansion overseas of European empires. This is a view that is clearly partial at best and racist at worst.

We cannot help our own background, nationality and education, nor the way that these socialisation processes inform our view of the world, but we should be aware that there are other perspectives and that they may be quite different. The problem with maps is a good example of this. Figure 17.2 shows the Mercator projection, which is one solution to the problem of showing the surface area of a globe in two dimensions. Look at the area of Africa compared to the area of Greenland. In reality Greenland is 2.2 million km² in area and the African mainland is 30.1 million km².

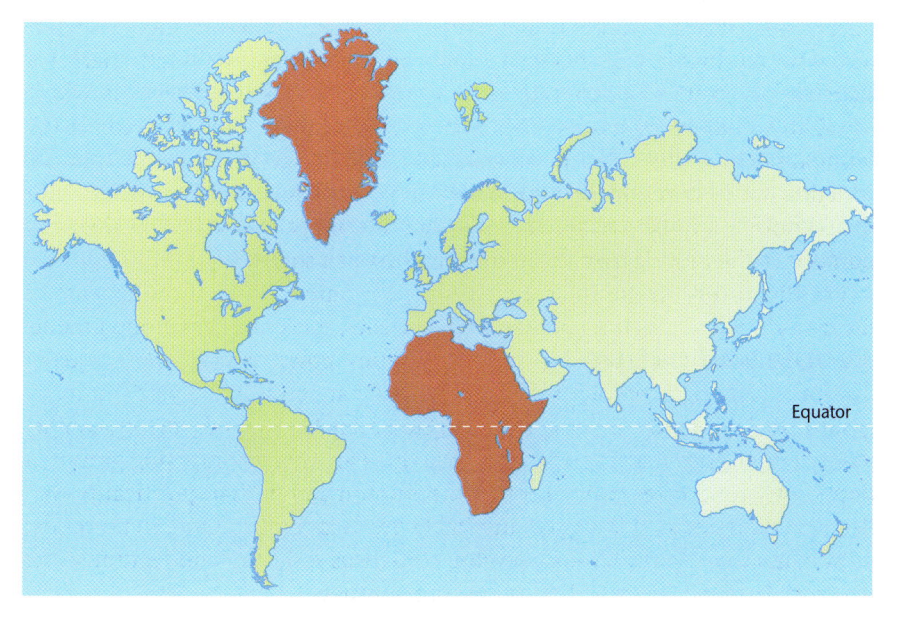

*Figure 17.2 Comparing Greenland and Africa*

**Section B**

The Mercator projection creates increasing distortions of size as it moves away from the equator. Close to the poles the distortion becomes severe. Because the Mercator distorts size so much at the poles, it is common practice to leave Antarctica off the map. This results in the northern hemisphere appearing much larger than it really is. The equator tends to appear about 60% of the way down the map, diminishing the size and importance of the countries in the southern hemisphere. This technique also happens to place the UK more or less neatly in the middle of the world, an image that many children in the UK have grown up with.

As Gunder Frank has pointed out, the idea that Europe is a continent seems a fine exercise in delusion. In reality, Europe is little more than a peninsula of Asia. 'European subcontinent' would perhaps be more appropriate, but there is something vaguely demeaning about the term, although we use it freely enough to refer to the 'Indian subcontinent'. Once again it seems that we want to put ourselves (i.e. Europeans) at the centre of the world.

## Models and theories

**Models** come in many forms but they are, essentially, descriptions of reality.
- The Rostow model describes the changes that take place in nation states as they develop.
- The demographic transition model describes the changes in birth and death rates as that process of development takes place.
- The tri-cellular model describes atmospheric circulation.

There is no claim that these models *explain* what they are describing.

On the other hand, **theories** do make such a claim. Not only do they explain the reasons for what they describe, they are also:
- **predictive**, in that they tell you what will happen next — which models do not
- **testable** — because they are predictive, one can test their validity by experiment

There are problems with both of these claims because predictive reliability depends greatly on what we agree on as acceptable tests and acceptable limits to reliability. Plate tectonics is clearly a theory rather than a model. It predicts a whole series of physical phenomena and can be tested — we can set up experiments to show that the mid-Atlantic ridge is spreading. But this is not quite as obvious as it might seem. How do we respond to evidence that does not confirm our theory? Do we throw out the theory or do we hold on to it until a better one comes along?

## People and nature

There is a valley in the US state of Colorado known as Paradox Valley because the river crosses the valley at right-angles to its axis. Paradox Valley provides a fascinating metaphor for the contradictory attitudes that we hold about man and nature. For two centuries the US west was the frontier where man took on nature and tamed it, slaughtering buffalo along the way, building highways and railroads and, ultimately, air-conditioned hotels and housing complexes that are serviced by water brought in, often from many hundreds of kilometres away.

# Synoptic: People and their environments — A2 Unit 6

The cultural icons of the west remain the cowboy and ranchers, and we romanticise a past in which the landscape was untouched by human activity. This is the 'man against the wilderness' idea, which is central to many of our views about the natural world. This is really a false opposition, for *Homo sapiens* is a species, and is a rather successful member of that part of the natural world that we call the biosphere. It seems perverse to put us in a different category.

Recognising the risk to their 'natural' heritage, the Americans created National Parks from which, in most cases, they evicted the human inhabitants to re-create a wilderness: a 'natural' landscape. When Europeans first arrived in the region, the Great Smokey Mountain National Park was populated by Cherokee Indians, an advanced culture that had well-defined settlements and roads throughout the region. When the park was formed they were evicted, along with the more recently arrived European settlers. Even so, these parks are not 'wilderness' — the impact of visitors by the thousands alters their ecosystems, as it does the most remote landscapes on the planet such as Antarctica. This may not be so dramatic as the deliberate destruction of aspects of the wilderness that displeased us, such as the native wolves and bears in medieval Europe, but it is intrusion nonetheless. Quite apart from our conscious intrusion on the natural environment, we also introduce new species in a largely unconscious way. For example, the hemlock forests of the Appalachians are fast disappearing as European species succeed.

Taking this idea further, most of us do not extend our enthusiasm to preserve 'nature' to include rats, let alone bed bugs, head-lice or malarial mosquitoes. So which 'natural' world do we have in mind when we set up natural as an opposite to man-made? Do we wish to return the world to the state that it was in before we first swung down from the trees?

If we reject the opposition of man and nature, and so include man as part of nature, does this exonerate us from the impact that our very destructive species has had on the planet? The view taken depends on how we judge the planetary ecosystem.

## Sample questions

*Question 2*
Critically examine the view that the varying impact of hazards is determined more by human behaviour than it is by the strength of the hazard event.

*Question 3*
Assess the view that most human modification of the environment is indirect and unintentional.

*Question 4*
Examine the concept of **sustainable** development as a strategy for the exploitation and management of the natural world.

*Question 5*
Describe and explain the impact of new forms of communications and employment on rural and urban communities.

# Index

## A

abandoned (relict) cliffs  97
ablation  324, 325, 327, 337
ablation till  338
abrasion  58, 63, 80, 329, 330, 331, 332, 336
absolute advantage  568
accumulation  324, 325, 327
active layer  348, 349, 350, 354, 361
actual fertility rate  134
adiabatic  307
adult literacy rate  140
advance the line  100
advection fog  298, 306
aeolian  336, 354
African plate  7
aftershock  41
ageing population  144, 146
ageing world  145
agglomeration  408, 410
agribusiness  479
agritourism  505
aid  566, 571
AIDS  127
air mass  305
Alaska  355
albedo  287, 288, 312
Aleutian islands  11, 14
Aleutian trench  13
allogenic factor  377, 379
Alps  12
Alyeska pipeline  355
amenity area  501
anaerobic  385
Andes  8, 12, 13, 358
anemometer  309
annual budget  325
anticyclone  296, 297, 298, 299, 305, 309, 310
anti-natal policies  155
aquifer  73, 75
arable agriculture  482
Aral Sea  594
arch  86
arctic maritime  306
arête  333
ash  36
ashfall  37
Asian Tigers  431, 432, 436
aspect  114, 288, 382
assembly-line production  420
asthenosphere  5, 6, 8, 10, 14
asylum seekers  211, 245
Atlantic-Indian ridge  14
Atlantic slave trade  221

Atlas mountains  12
atmospheric circulation  289
attrition  58, 80
autarkic development  435, 566
autogenic factor  377
autotroph  365, 378
avalanche  324, 336
azonal soil  382, 384

## B

backwash  80, 81, 82, 83
backwash effect  554, 555
bar chart  258, 265
barometer  309
basal sliding  328
basalt  4, 12
base flow  45, 51, 52
basic industry  399
batholith  18
bay  83, 84, 85
beach  83, 88, 89
beach nourishment  110
beach replenishment  85
bed deformation  328
behaviouralism  412, 414
Benioff zone  9
berm  88
best-fit line  263, 265, 278
bid-rent curve  182
bid-rent theory  181
bilateral aid  571
bioconstruction  101
biological weathering  24
biomass  366, 367, 368, 371, 372, 375, 378
biome  365, 367, 369, 373
birth rate  133
black earth  374
black economy  528
blockfield  26, 27, 354
blocking anticyclone  298
bluff  62, 66
Boscastle  53
Boserup, Ester  149
boss  18
bottom-set bed  69
braided stream  347, 354
braiding  67
breakpoint analysis  269
British car industry  428
brown earth  383
building regulations  42
Burren  24

## C

Cairo  192
calcification  374, 385
California  11
call centres  430
Campbell Stokes recorder  309
Canada  392
capacity  59
capillary action  374, 385
capital industry  399
capitalism  609
capital (material) resources  534
carbonation  21, 23, 27, 28
carbonation weathering  27
Cargill  479
Carlsberg ridge  14
carnivore  365
carrying capacity  502, 534
catastrophism  604
catchment area  169
cave  86
cavitation  58
CBD  122, 187
central place  122, 169
centrifugal force  119
centripetal force  119, 176, 291, 292, 464
chalk  50, 57, 85, 88
changes in the CBD  196
changing land use  192
chelation  22
cheluviation  384
chemical weathering  23, 24
chernozem  374, 375, 383
Chesil Beach  91
Chicago  183, 184, 185
Chichester  98, 99
china clay  29
chi-squared test  273, 276
choropleth map  263, 265
Christaller  169
Christaller's hierarchy  171
Christaller's theory  173, 174
Churchill  387
circulation  207
cirque (corrie)  323, 332, 333
city centre redevelopment  195
city-forming  465
city-forming functions  165
city functions  191
city-serving  465
city-serving functions  165
classical theory  412
clay  57, 85
cliff  87

614

# Index

climate   22
climatic dome   310, 311
climax community   100, 376, 377, 378, 379
clints   27
clitter   26, 27
closed-system pingo   350
clustering   408, 410, 430
coalescence   295
Cocos plate   32
cold front   299, 300, 315
cold (polar) glacier   326, 327
collective farming   480
collision margin   4, 9, 10, 13, 17
co-location (agglomeration)   407, 411
colonialism   167, 445, 448
Colorado River   49, 79, 508
commercial farming   478, 479
comparative advantage   443, 569
competence   59
competition   183
compressing flow   328
concrete mattress revetments   71
condensation   44, 45, 47, 296, 298, 302, 307, 309
    level   307, 314
    nuclei   314
conditional instability   308, 309
conduction   286
conflict theory   610
consensus theorists   610
conservation   391, 393
conservative margin   4, 5, 9, 11, 17
constructive margin   4, 5, 6, 7, 9, 14, 15
constructive wave   80, 83
consumer industry   120, 399, 419
continental crust   5, 8, 13
continental drift   2
continents   3
continuity of settlement   165
contraception   141
contract workers   211
convection cell   6, 8, 14
convection currents   4, 5, 6
convectional uplift   295
convergence   8, 9, 13, 558
core   6, 524
Coriolis force   291, 296, 302
correlation   270, 271, 273
cost minimisation   407, 408
Cotopaxi   8, 15
counterurbanisation   119, 121, 194, 234, 463
creep   328
crevasse   329
critical value   271, 272, 274
cross profile   62
cross-sectional area   56, 57
crust   4, 5

crustal plate   4
cumulative causation   553
cumulative line graph   257, 265
cusp   88
cuspate foreland   92
cyclone   303
*Cynagnathus*   2

## D

Dalmatian coast   97
DALR   308
dam   49, 76, 77
Dartmoor   19, 26
death rate   133, 136
debris flows   336
debt crisis   423
Deccan plateau   17
decomposer   370, 378
decomposition   388
deforestation   24, 52, 371, 488
de-industrialisation   455
delta   69
delta kames   346
demographic transition model   141
dependency   144
dependency ratio   144
dependency theory   433, 541, 543, 577
deposition   58, 59, 60, 63, 69, 337
depression   296, 299, 301, 305, 309, 310, 315
desertification   220
desire lines   261
desired fertility rate   134
destructive margin   4, 5, 8, 9, 13, 15, 16
destructive wave   80, 83
detritivore   365, 366
developed country   522
development   522
development theory   539, 547
dew-point temperature   307
diagenesis   324
dilatation   20, 330
direct costs   38
discharge   49, 50, 51, 55, 57, 272, 273
dispersed settlements   158
dispersion   266
diurnal   312
divergence   6
divided bar chart   260, 265
doldrums   292
doline   27, 28
doubling time   134
drag   6
drainage basin   47, 54
drainage basin cycle   46
drift   337
drift south   121, 209
drought   316, 321

drumlin   337, 340, 341, 348
dry adiabatic lapse rate   307
dry-point site   162
Dungeness   93
Dutch disease   537
dyke   18, 20
Dyson   413

## E

earthquake   6, 7, 8, 9, 11, 31, 32, 33, 34, 37, 39, 41, 319
Earth's heat budget   288
east African rift valley   6, 7
east Pacific ridge   14
Easter Island   150
easterlies   292
ecological footprint   502, 503, 504, 517, 534
economies of scale   179, 442, 443, 554, 568, 569
ecosystem   364, 365, 390
edge city   510
efficiency   55, 56
El Niño   293, 320, 321
ELR   308
eluviation   385, 389
emergence   93, 97
emergency service provision   43
emigrants   134
employment structure   400, 423
englacial   337
environmental debt   534
environmental lapse rate   307
environmentalism   609
equilibrium   325
    line   324
erosion   58, 60, 63, 330
erratic   342
esker   346, 347
ethnicity   124, 131
Etna   17
eurasian plate   7, 10
Eurocentrism   611
European migration   223
eustatic   93, 94
evacuation drill   43
evaporation   44, 45, 52, 286, 317, 382
evapotranspiration   44, 45, 50, 311, 374, 375, 385, 386, 389
export-led growth   434
export valorisation   434, 549
extending flow   328
extensive farming   481
externalities   178, 606
extrusive   19
extrusive igneous landforms   15, 16

## F

factor endowment   567
fair trade   567

615

falling limb  51, 52
fault  6
feldspar  21, 23
felsenmeer  354
female employment  401
Ferrel cell  289, 291
fertility  135, 139
fertility rate  117, 133, 137, 140
fetch  82
fieldwork  279, 280, 281, 283
fire  34
firn  324
  line  324, 325
fissility  23
fissure  6, 15, 17
fjard  97
fjord  96
flocculation  69
flood  53, 76, 77, 319, 320, 321
floodplain  54, 57, 62, 63, 69, 77
flow lines  261, 265
fluvial deposition  58
fluvial erosion  58
fluvial transportation  58
fluvioglacial deposition  337
fluvioglaciation  343, 337, 343
focus  6, 9
fog  300, 306
fold mountains  8, 9, 10, 12
food chain  365, 366
food web  365, 366
footloose industry  404
forced migration  221
forces of attraction  186
forces of repulsion  186
Ford Motor Company  420
Fordism  419, 420
foreign direct investment  429, 435
fore-set bed  69
foreshock  39
form  122
fossil  2, 10
free trade  443, 567
freeze–thaw  20, 22, 23, 24, 25, 27, 28, 330, 348, 354
freezing  45
friction  291, 292
frontal uplift  295
frost heave  348, 351
frost pull  351
frost push  351
Fujita scale  322
function  122, 158, 161

## G

gabion  110
gated community  510, 513, 515
GDP  528
general circulation  289
genetically modified (GM) crop  500

gentrification  177
geology  22, 23, 45, 84
geothermal  19
geothermal energy  36
Geysir  36
glacial budget  324, 326
glacial deposition  337
glacial transportation  337
glacial trough  334
glacier  323, 359
glacier system  324
gley  383, 385, 388
global city  210, 470
global hydrological cycle  44
global region  423
global warming  24, 286, 287, 316, 318, 319, 363
globalisation  425, 448, 452, 453, 454
glocalisation  453
*Glossopteris*  2
GNP  528
Gondwanaland  3
Goredale  28
gorge  63
government policies  238
granite  8, 19, 20, 21, 23, 26, 29, 52, 84, 85
gravity model  269, 214
greenhouse effect  24, 286, 316
Griggs, David  23
gross primary productivity (GPP)  367
ground contraction  352
ground displacement  31
groundwater  73, 75
groundwater flow  44, 45, 55
growth pole  556
groyne  80, 81, 85, 110
grykes  27
guyots  11

## H

Hadley cell  289, 296
Haeimaey  15, 16
halosere  101, 105
hanging valley  334, 335
Hawaiian islands  11, 15, 16
Hayward fault  11
headland  83, 84, 85
heat budget  287
heather  380, 381
Hekla  15, 36, 37
helical flow  65
herbivore  365
heritage  505
Hess, Harry  3
heterotroph  365
high-order goods  170
Himalayas  10, 12, 13
histogram  258, 265
hi-tech industry  399, 439, 441

HIV/AIDS  127
Hjulström curve  60
hold the line  100, 107
home-working  431
Honda  416
horizon  386, 387
hot spot  10, 11, 15, 16
Hoyt's sector model  186
human resources  534, 535
humidity  300
humification  384
humus  384, 386, 388
hurricane  296, 302, 303, 316, 319, 320, 321
Hurricane Mitch  303, 304, 320
hydration  21
hydraulic action  58, 63, 80
hydraulic radius  55, 56, 57, 272, 273
hydroelectric power  77
hydrograph  53
hydrological cycle  44, 46, 73
hydrolysis  21, 22, 23, 26, 29
hydrosere  376, 379
hygrometer  309

## I

ice cap  323
ice-contact drift  344
Iceland  7, 16, 19, 36, 37, 68, 359, 395
ice sheet  323
ice shelf  323
ice-wedge polygons  352, 353, 354
igneous activity  15, 19
igneous intrusion  26
illegal immigrants  211, 245
illuviation  385
immigrants  134
immigration  239, 240, 242
impact of migration  219
impact of tourism  504
impermeable  52, 54
import replacement  470
import substitution  434, 473, 544, 545
import substitution industrialisation  545
India  34
indirect cost  38
indirect investment  435
Indo-Australian plate  10, 12, 33
Indonesia  33, 34, 35
Indonesian transmigration  219
industrial city  465, 466
industrial land use  187
industrial location  402, 403, 404, 412
industrialisation  404, 455, 465
industry cluster  411
inertia  119, 121
infant mortality  133, 139, 140

# Index

infiltration 44, 45, 46, 52, 54
infiltration capacity 46
inheritance law 486
inner core 4
insolation 21, 286, 288, 311
instability 308, 309
integrated steelworks 404
intensive farming 481
interception 47, 52
inter-generational equity 534, 608
intergranular flow 328
internal deformation 328
internal migration 207, 208
international division of labour 445, 446
international migration 207, 210
International Monetary Fund (IMF) 573
inter-quartile range 267
inter-tropical convergence zone (ITCZ) 289, 292
intervening opportunities 215
intra-generational equity 534, 608
intrazonal soil 382, 383, 387, 388
intrusive 19
intrusive igneous landforms 16, 18, 19
intrusive igneous rocks 8
invasion 183, 185
iron pan 385, 387, 388
irrigation 75, 79
island arc 9, 13
Isle of Arran 19, 378
Isle of Purbeck 388
isopleth map 264, 265
isostatic 93, 94

## J

Japan 14, 31, 35
Japan trench 13
Java 35
Java trench 13
jet stream 290
'Jim Crow' laws 225
Jökulhlaup 37, 68

## K

kame terrace 347, 348
Kansas 322
kaolin (china clay) 21
kaolinite 26, 29
karren 27
Kerala 140
Kerala model 140
kettle hole 345, 348
kettle lake 379
Keyhaven 104
Keyhaven Marshes 103
'key-settlement' policy 161
Kobe 31, 38

Kondratieff cycle 420, 421
Kuril islands 14
Kuril trench 13

## L

La Niña 294, 320
lacustrine delta 69
lag time 51, 52
lahars 36, 37
laminar flow 326, 328
land ownership 177
land tenure 486
landslide 41, 54
land-use patterns 187
land-use zoning 42
latent heat 286, 307
latosol 370, 383
Laurasia 3
lava 10, 15, 16, 17, 36, 39
lava flow 37
lava plateau 7, 17
law of comparative advantage 442
leaching 368, 370, 371, 372, 374, 375, 378, 384, 385, 386
LEDC 522, 526
Lee 213
Leeward islands 14
less economically developed country 522
levée 62, 68, 76, 77
life expectancy 133, 140
limestone 18, 21, 23, 24, 27, 50, 73, 84, 85
limestone pavement 27, 28
line graph 256, 265
liquefaction 32
lithology 84
lithosere 376, 378
lithosphere 5, 6, 8
litter 370, 372, 375, 378, 386, 387, 388
location 413
  quotient 466
  theory 406
lodgement till 337
logarithmic graph 257, 265
long profile 61
longshore drift 80, 81
low-order goods 170
*Lystrosaurus* 2

## M

MacKenzie and Palmer 5
magma 6, 9, 10, 16, 17, 18, 36, 37, 40
Malham 28, 29
malnourishment 136
Malthus, Thomas 147
managed retreat 100, 107
managing urban decay 204
mantle 4, 5, 9, 37

manufacturing 399
Maoke Range 12
Maquiladoras 444
marble 18
Mariana Islands 9, 13, 14
Mariana trench 13
marram grass 101
Marxism 414
mass balance 324
material index 407
Mauna Loa 15, 16
Meadowhall 199
mean 266, 267
meander 64
meander scar 66
mechanisation 424
MEDC 522, 526
median 266, 267
mega-city 462, 515
Mekong delta 79
Mekong River 78
melting 45
meltwater channel 344
*Mesosaurus* 2
metamorphic rock 18
Mexican immigration to the USA 230
Mexico 32, 41
Mezzogiorno 559
Micron 418
mid-Atlantic margin 7
mid-Atlantic ridge 4, 6, 14, 19, 36
middle-order goods 170
mid-Indian ridge 14
mid-oceanic ridge 6
migration 206, 207, 391
migration patterns 243
migration trends in Romania 233
mineral–nutrient cycle 367, 369, 371, 375
minerals 35
Mississippi River 70, 71, 76
mist 298, 300
mode 266, 267
modernisation theory 541
mono-climax theory 376
Montreal 313
Montserrat 218
mor 385
moraine 340, 342
  ground 342
  lateral 341
  medial 341
  push 342
  recessional 341, 348
  terminal 341
more economically developed country 522
mor humus 387, 388
mortality 135
Mt St Helens 31, 32, 38

617

muirburn 381
mull 388
multilateral aid 571
multiple nuclei 186
multiplier 178, 179, 504
multiplier effect 20, 77, 556
Mumbai 471, 477
Myrdal's model 553

## N

NAFTA 443
natural environment 403
natural increase 134
natural resource 79
natural resources 534
nature-based tourism 505
Nazca plate 8
nearest neighbour analysis 268
negative externalities 183
negative multiplier 422
neighbourhood principle 178
neo-colonialism 448, 574
net migration 134
net primary productivity (NPP) 367
neutral stability 309
Nevado del Ruiz 15
neve 324
newly industrialised country 421, 526, 547
new towns 203
NIC 431, 526, 547
nivation 331, 354
normal 266
North American plate 7, 11, 32
North Yorkshire Moors 380
nucleated settlements 158
nutrient cycle 368

## O

occluded front 299, 300, 315
oceanic crust 4, 6, 8, 9, 13
oceanic plate 14
ocean ridge 14
ocean trench 9, 13
offshore bar 92
Ogallala Aquifer 74
ognip 351
omnivore 365, 366
one-child policy 156
onshore bar 91
open-system pingo 350
open V-shape 63
optimum location 410
optimum population 149
Orford Ness 90
orographic effect 314
orographic (relief) rainfall 114
orographic uplift 295
orthogonal 83, 85

outer core 4
out-of-town shopping 172, 198, 199
output/input ratio 481
outwash 337, 343, 344, 345
outwash plain 348
overland flow 45
overpopulation 149
overspill channel 344
oxbow lake 66
oxidation 21
ozone 316, 317, 318

## P

Pacific-Antarctic ridge 14
Pacific Ocean 14
Pacific plate 8, 9, 11
palaeomagnetism 4
Pangaea 3, 10, 14
Papua New Guinea 320
paradigm shifts 605
parasitic cone 17
pastoral agriculture 482
patterned ground 351
pattern of migration 221
peat 383, 388
percolation 44, 46, 52
periglacial 349, 357
periphery 524
peri-urban fringes 234
permafrost 348, 349, 354, 357, 361, 362, 363
permanent migration 207, 208
permeable 50
personal enquiry 248
Peru 8
Peru–Chile trench 8, 13
Philippines 14, 35
Philippines plate 9
Philippines trench 13
photosynthesis 378
physiological drought 374
piedmont 323
pie graph 259, 265
pingo 350, 351, 354
pioneer species 377
plagioclimax 376, 380
planning in the UK 202
plantation agriculture 496
plantation system 493
plant community 102, 376, 378, 381
plant succession 100, 101, 103, 105, 376
plate 3, 5, 6, 9, 13
plate margin 5, 6, 10, 15, 16
plate tectonic theory 3, 5
plucking 329, 330, 336
plume 10
podzol 383, 387
podzolisation 385
point bar 65

Polar cell 289
polar continental 306
polar maritime 301, 306
political causes and consequences 217
pollution 109
pollution dome 311
poly-climax theory 376
Popocatépetl 17
population
   change 133
   debate 147
   density 112, 123
   distribution 112
   global distribution 117
   movement 207
   movements 206
   pyramids 124, 128, 129, 131
   structure 124
Port Kembla steelworks 404
positive deindustrialisation 424
post-industrial 473
post-industrial city 465, 467
post-suburban landscape 512
post-suburban region 510
pounding 81
precipitation 44, 45
prediction 39
pre-industrial city 464, 465
pressure gradient force 291
pressure melting point 326, 327, 328
pressure release 20
primacy 167
primary activity 398
primary industry 400
primary succession 377
primate city 158
primate city: Dublin 168
professionals 211
profile 387, 388
profit crisis 425
profit maximisation 408
pro-natal policies 155
proportional symbols 260, 265
proto-industrialisation 465
psammosere 100, 105
pull factors 215
punctuated evolution 605
push factors 215
pyramidal peak 333, 334
Pyrenees 12
pyroclastic flow 36, 37

## Q

quaternary activity 399
quaternary industry 400
questionnaire 282, 283

## R

radiation 286, 287, 311, 317
radiation fog 296

# Index

radiation window   286
radon   39
rainfall   47
  condensation   294, 295
  convectional   48, 294, 295
  frontal   47, 294, 295
  orographic   47, 294, 295
rain shadow   48
rain shadow effect   295
raised beach   97
rampart   351
ranching   496
range   169, 266, 267
rank–size rule   166, 167, 171
rapids   64
Rates of migration   232
rationalisation   425
Ravenstein   212
raw material   404
receiving area   207
recently industrialised country   526
recession limb   51
recreational land use   182
redevelopment   195
reforestation   24, 52
refugees   211, 245
regelation   328
regime   49, 50
region   423
regional development   559
regional policy   417, 557
regolith   25
Reilly's law of retail gravitation   174
rejuvenation   61
relief   24, 49, 50
rendzina   383, 388
reserve army   224
resource curse theory   536
returning polar maritime   306
re-urbanisation   194
ria   95
ribbon lake   343
RIC   431, 526
riffles   65
rift valley   6
ripples   89
rising limb   51, 52
risk assessment   39
River Avon   66
river cliff   65
River Hooke   57
River Thames   50
river velocity   280
River Wye   62
roche moutonnée   335, 336
rock basin   334
rockfall   336, 354
Rockies   12
rock step   334
rofabard   396

Rossby wave   291, 300
Rostow, Walter Whitman   541
Ruhr valley   403
runoff   44, 45, 50, 52, 54, 368
rural areas   122
rural settlements   158
rural/urban fringe   509
rural–urban migration   235, 463
rust-belt   415

## S

salinisation   98, 385
SALR   308
salt crystallisation   21, 22
salt marsh   101, 102, 103, 104, 107
saltation   59
sample   255, 281, 282, 283, 284
  random   255, 256, 281
  stratified   255, 256, 281
  systematic   255, 256, 281
San Andreas fault   11
sand dune   100, 102
sandstone   18, 19, 21, 26, 73
sandur   345
San Francisco   34, 38, 41
saturated adiabatic lapse rate   307
scale   322
scattergraph   263, 265, 278
schist   18, 19
scree   25, 28, 29, 354
scree slope   354
sea-floor spreading   3, 4
sea level   93, 94, 95, 97, 98, 100, 107, 109
seamount   11
secondary activity   399
secondary industry   400
secondary succession   377
segregation   176
Selsey   98, 99
semi-periphery   524
sere   376
Serengeti   390
service sector   399
sesquioxide   385, 387, 388
settlement patterns   158, 161
settlers   210
shake hole   27, 28
shanty town   473
shield   16
shifting cultivation   488, 496
sichelwannen   344
significance   277
silica   16, 36
sill   18, 20
Singapore   157, 439
site   158, 161, 403
situation   158, 161, 403
Six's thermometer   309
skewed   266

Slapton Sands   91
slip-off slope   65
slippage   328
slip plane   328
slum   473
slums   189
smog   314
snow   298, 315, 317, 324
soil   26, 35, 54, 368
  erosion   390, 393, 395, 396
  horizon   382
  profile   382, 386
solar constant   287
solifluction   353, 354, 363
solifluction lobe   353
solution   21, 58, 59, 61, 81
solution hollow   27
Sony   414
source area   207
source region   305
South American plate   8
Southern Alps   12
Spearman rank correlation coefficient   270, 271, 280, 282
specialisation   568
spit   89, 101
spread effect   555
spring-line settlement   162
spur   62
Sri Lanka   33, 34
stack   86
statistical significance   271, 273, 274
stemflow   47
Stephenson screen   310
stone garland   352
stone polygon   351, 354
stone stripe   352
storm beach   88
storm hydrograph   50, 51, 54
Stouffer   215
stratified   282
striation   335
structural adjustment programme (SAP)   550, 573
structuralism   413, 414
Studland   102, 105
stump   86
sub-climax community   376
subduction   8, 9, 17
subduction zone   4, 6, 9, 13
subglacial   337
sublimation   45, 295, 324, 337
submergence   93
sub-optimal location   410
subsistence farming   478, 479
suburban industry   122
suburbanisation   194
succession   184, 185, 376, 378, 379, 380
sunbelt   226, 227

619

sunbelt states 229
Sunda islands 14
supermarkets 198
superstores 198
supply economy 538
supply region 506
supraglacial 337
surface runoff 44, 45, 50, 52, 54
surge 329
Surtsey 16
suspension 59
sustainability 584, 608
sustainable agriculture 481
sustainable development 476, 531
sustainable management 393
sustainable tourism 520
swash 80, 81, 82, 83

## T

talik 349, 350
tarn 333
tectonic plate 4
teleworking 430
temperate grassland 373
temperature inversion 296, 311
temporary migration 207
Tennessee Valley Authority (TVA) 77, 557
terms of trade 570
terrestrial radiation 289
tertiary activity 399
tertiary industry 400
Tethys Sea 10
Thailand 33, 34
thalweg 65
Thatcher years 426
theories of migration 212
theory of comparative advantage 434, 569
thermal expansion 21, 23
thermal growing season 113, 119
thermometer 309
Third World 523
Three Gorges dam 78
threshold 169
threshold population 170
throughfall 47
throughflow 45
thunderstorm 298, 299, 300, 306, 315
till 337, 338, 343
till sheet 340
tombolo 91
Tonga trench 13
top-set bed 69
tor 26, 354
tornado 316, 319, 320, 321, 322
tourism 20, 29, 36, 107, 109, 359, 391, 392, 506

tourist 28, 99, 104
Town and Country Planning Act 121, 178
Toyotaism 429
track 305
traction 59
trade 566, 567
trade wind 289, 292, 302
trading bloc 441
transform margin 9
transitional zone 197
transnational corporation 448, 471
transpiration 44, 45
transportation 58, 59, 60, 81, 337
triangular graph 262, 265
tri-cellular model 289, 290
trickle-down effect 555
trophic level 365
trophic pyramid 365, 366
tropical continental 306
tropical maritime 301, 306
tropical rainforest 369, 371, 372
tropopause 289, 302
truncated spur 334, 335
tsunami 33, 34, 37, 38, 41
tunnel-valleys 344
typhoon 303

## U

UK census 154
undernourishment 136
underpopulation 149, 152
uniformitarianism 604
Urals 13
urban areas 122
urban boundary layer 311
urban canopy 310, 311
urban development in LEDCs 189
urban heat island 310, 311
urban hierarchy 167, 175
urbanisation 461, 462, 463, 469
urbanisation of poverty 189
urban land-use 183
urban microclimate 310
urban planning 202
urban settlements 165
urban sprawl 201, 511, 515
U-shaped valley 334

## V

varve 345, 346
velocity 55, 57, 59, 60, 61, 69
vent 17
Vesuvius 17, 36, 38
vicious circle of poverty 540
viscous 16
visibility 300, 306, 314

volcanic activity 9, 15, 16
volcanic cone 15
  acid 16
  basic 16
  caldera 16
  composite 16
  fissure 16
  shield 15
volcanic eruption 8, 10, 40
volcanic islands 7, 9, 10, 11, 13
volcano 6, 7, 11, 15, 16, 31, 35, 37, 39, 40
volume of migration 221
voluntary migration 221, 223, 225
V-shaped valley 63

## W

Walker circulation 293
warm front 299, 300, 301, 315
warm (temperate) glacier 326, 327
waterfall 63
watershed 47
waterspout 322
water table 45, 52, 54, 73
wave 82
  energy 82
  height 82
  length 82
  period 82
  power 82
  refraction 83, 85, 86
  steepness 82
wave-cut notch 87
wave-cut platform 87
weathering 20, 22, 26, 27, 29, 86, 273, 274, 330, 370, 384
  biological 20
  chemical 20
  physical 20
Weber, Alfred 406
Wegener, Alfred 2, 3
westerlies 292
wet-point site 162
wet-rice agriculture 115, 489, 496
wetted perimeter 56
willy-willies 303
wing dyke 71, 76
World Bank 572
world systems approach 524

## Y

Yorkshire Dales 25, 28, 29
youthful populations 144, 147

## Z

Zelinsky 214
zonal concept 383
zonal soil 382, 383, 387
zone in transition 187, 197